Anti-Inflammatory Drug Discovery

RSC Drug Discovery Series

Editor-in-Chief:
Professor David Thurston, *London School of Pharmacy, UK*

Series Editors:
Dr David Fox, *Pfizer Global Research and Development, Sandwich, UK*
Professor Salvatore Guccione, *University of Catania, Italy*
Professor Ana Martinez, *Instituto de Quimica Medica-CSIC, Spain*
Professor David Rotella, *Montclair State University, USA*

Advisor to the Board:
Professor Robin Ganellin, *University College London, UK*

Titles in the Series:

1: Metabolism, Pharmacokinetics and Toxicity of Functional Groups
2: Emerging Drugs and Targets for Alzheimer's Disease; Volume 1
3: Emerging Drugs and Targets for Alzheimer's Disease; Volume 2
4: Accounts in Drug Discovery
5: New Frontiers in Chemical Biology
6: Animal Models for Neurodegenerative Disease
7: Neurodegeneration
8: G Protein-Coupled Receptors
9: Pharmaceutical Process Development
10: Extracellular and Intracellular Signaling
11: New Synthetic Technologies in Medicinal Chemistry
12: New Horizons in Predictive Toxicology
13: Drug Design Strategies: Quantitative Approaches

14: Neglected Diseases and Drug Discovery
15: Biomedical Imaging
16: Pharmaceutical Salts and Cocrystals
17: Polyamine Drug Discovery
18: Proteinases as Drug Targets
19: Kinase Drug Discovery
20: Drug Design Strategies: Computational Techniques and Applications
21: Designing Multi-Target Drugs
22: Nanostructured Biomaterials for Overcoming Biological Barriers
23: Physico-Chemical and Computational Approaches to Drug Discovery
24: Biomarkers for Traumatic Brain Injury
25: Drug Discovery from Natural Products
26: Anti-Inflammatory Drug Discovery

How to obtain future titles on publication:
A standing order plan is available for this series. A standing order will bring delivery of each new volume immediately on publication.

For further information please contact:
Book Sales Department, Royal Society of Chemistry, Thomas Graham House, Science Park, Milton Road, Cambridge, CB4 0WF, UK
Telephone: +44 (0)1223 420066, Fax: +44 (0)1223 420247,
Email: booksales@rsc.org
Visit our website at http://www.rsc.org/Shop/Books/

Anti-Inflammatory Drug Discovery

Edited by

Jeremy I Levin
Boehringer Ingelheim Pharmaceuticals, Inc., Ridgefield, CT, USA

Stefan Laufer
Institute of Pharmacy, Tuebingen, Germany

RSC Publishing

RSC Drug Discovery Series No. 26

ISBN: 978-1-84973-413-4
ISSN: 2041-3203

A catalogue record for this book is available from the British Library

© The Royal Society of Chemistry 2012

Published by The Royal Society of Chemistry,
Thomas Graham House, Science Park, Milton Road,
Cambridge CB4 0WF, UK

Registered Charity Number 207890

For further information see our web site at www.rsc.org

Printed in the United Kingdom by Henry Ling Limited, at the Dorset Press, Dorchester, DT1 1HD

Preface

Driven by the need for new orally active anti-inflammatory medicines, intensive research is ongoing in both industry and academia on a diverse set of biological pathways and targets with the goal of delivering therapeutics to improve the lives of patients with conditions such as asthma, rheumatoid arthritis, atherosclerosis, inflammatory bowel disease, psoriasis, allergic diseases, and pain. It was the goal of this volume to bring together respected scientists who have been directly involved in these drug discovery efforts to present the rationale for prosecuting specific targets, a current review of the medicinal chemistry strategies that have been used, and the outcome of those efforts. For some of these targets drugs with exciting potential have been delivered to clinical trials, while others have encountered unexpected roadblocks, and both of those outcomes provide important lessons for today's drug hunters. It is our hope that the stories presented herein will serve as a valuable resource for anyone interested in the discovery of new anti-inflammatory drugs.

In addition, we would like to express our great appreciation to all of the chapter authors for delivering clear, comprehensive reviews that will inform and enlighten readers on the state of the art in their respective areas of research. We would also like to express our appreciation to Professor David Rotella (Montclair State University), and Gwen Jones and Amaya Camara at RSC Publishing for their help, support and patience in the course of putting together this book.

Jeremy I. Levin, Ph.D.
Stefan Laufer, Ph.D.

RSC Drug Discovery Series No. 26
Anti-Inflammatory Drug Discovery
Edited by Jeremy I Levin and Stefan Laufer
© The Royal Society of Chemistry 2012
Published by the Royal Society of Chemistry, www.rsc.org

Contents

RSC Drug Discovery Series No. 26
Anti-Inflammatory Drug Discovery
Edited by Jeremy I Levin and Stefan Laufer
© The Royal Society of Chemistry 2012
Published by the Royal Society of Chemistry, www.rsc.org

Section 4: Sphingolipids

Section 5: Steroid Hormone Receptors

Introduction

The discovery of new and novel anti-inflammatory drugs is an area of intense interest within both the pharmaceutical industry and academic laboratories. Significant advances have been made in the treatment of inflammatory diseases such as rheumatoid arthritis (RA) and multiple sclerosis, but most dramatically with new biologic agents. Perhaps due in part to the mixed experiences with COX-2 inhibitors, very few small molecule anti-inflammatory drugs with novel modes of action have made it to the market in the last decade. Therefore, there remains an enormous unmet medical need for new, effective and safe small molecule disease-modifying therapies to expand treatment options for these and other indications, including asthma and chronic obstructive pulmonary disease (COPD), allergic diseases, atherosclerosis, psoriasis, inflammatory bowel disease and pain.

Inflammatory diseases are characterized by numerous symptoms and, on a mechanistic basis, multiple pathways are involved. This multi-factorial nature of inflammation is both an opportunity and a challenge for drug discovery. The aim of this book is to review recent noteworthy medicinal chemistry approaches to a variety of important therapeutic targets and so provide a key reference for those interested in the prosecution of modern drug-discovery programs directed at anti-inflammatory mechanisms of action. The biology, pharmacology and medicinal chemistry literature of a collection of topics ranging from components of the arachidonic acid cascade, to kinases, GPCRs, sphingolipids and steroid hormone receptors have been summarized by highly respected scientists from academia and industry offering new insights on major advances and issues related to bringing new anti-inflammatory therapies to market.

In the first section of this text, drug targets in the arachidonic acid (AA) cascade are described. This pathway, although it is the molecular target of very old drugs like aspirin, is still not fully understood. For example, COX-1b ("COX-3") might be the molecular target for paracetamol (acetaminophen) in

RSC Drug Discovery Series No. 26
Anti-Inflammatory Drug Discovery
Edited by Jeremy I Levin and Stefan Laufer
© The Royal Society of Chemistry 2012
Published by the Royal Society of Chemistry, www.rsc.org

dogs, but not in man. Taking into account the lessons of COX-2 drug development, an important downstream target is the microsomal prostaglandin E2 synthase-1 (mPGES1), and inhibition of this mechanism has the potential to yield a third generation of non-steroidal anti-inflammatory drugs (NSAIDs).

The opposite approach, going upstream, leads to cytosolic phospholipase A_2 alpha (cPLA$_2\alpha$) as a target for therapeutic intervention. The rate-limiting step in the generation of prostaglandins, leukotrienes and platelet activating factor (PAF), all highly active substances with diverse biological actions in inflammation, is the cleavage of the *sn*-2-ester of membrane phospholipids by a phospholipase A_2 (PLA$_2$). Among the superfamily of PLA$_2$ enzymes, cPLA$_2\alpha$ is thought to play the primary role in this biochemical reaction. Therefore, the inhibition of cPLA$_2\alpha$ activity is an attractive approach for the control of inflammatory disorders. Furthermore, on the leukotriene branch of the arachidonic acid cascade is the cytosolic enzyme leukotriene A_4 hydrolase (LTA$_4$H), which catalyzes the formation of the pro-inflammatory mediator LTB$_4$. The inhibition of LTA$_4$H offers an alternative to blocking the targets 5-lipoxygenase (5-LO) and 5-lipoxygenase activating protein (FLAP) for modulating LTB$_4$ levels and has potential utility for treating a variety of inflammatory disorders including coronary artery disease, though this awaits clinical validation.

The AA cascade section concludes with CRTH2, the chemoattractant receptor homologous molecule expressed on T helper type 2 cells, though this target could have also been placed in Section 3 since it is a G-protein coupled receptor (GPCR). Small-molecule antagonists of CRTH2, the receptor for prostaglandin D_2 (PGD$_2$), have been the subject of extensive research in the pharmaceutical industry and more than ten of these have been advanced to clinical trials for indications ranging from seasonal allergic rhinitis to asthma and atopic dermatitis, making it a very promising therapeutic target.

The second section of this volume focuses on protein kinase inhibitors as anti-inflammatory agents, although this field still suffers from important unanswered questions. From 518 protein kinases in our genome, about 250 are disease related. However, the fundamental role of kinases in signal transduction raises questions about how safe and free of side-effects kinase inhibitors can be. Another important question concerns selectivity within the human kinome. Thus, how selective is selective enough, and is the dogma "one mechanism, one target, one disease" still valid? The inhibition of multiple kinases might be beneficial in cancer, but it remains to be demonstrated that similarly promiscuous inhibitors provide the necessary safety margin for utility in treating inflammation. Despite excellent pre-clinical evidence, many kinase targets still suffer from a lack of clinically proven target validation.

Nevertheless, there has been an explosion of academic and industrial research in this field, fuelled by the groundbreaking success of anti-TNF biologicals for the treatment of rheumatoid arthritis. In the search for small-molecule drugs that impact the biosynthesis and/or release of pro-inflammatory cytokines, prominent targets include the kinases p38, MK2, Syk, JAK, IKKβ and Btk. Many p38 MAP kinase inhibitors have entered clinical trials but, despite excellent pre-clinical validation, these have failed clinical validation in RA, although they have recently shown very promising results in clinical studies for chronic obstructive pulmonary disease (COPD). On the other hand, several inhibitors of the tyrosine kinase JAK, including tofacitinib (CP-690,550) and INCB28050, have demonstrated exciting clinical activity in Phase II and III trials for RA, psoriasis and inflammatory bowel disease, and may be poised to become orally dosed alternatives to the anti-TNF biologics.

Similarly, inhibitors of spleen tyrosine kinase (Syk), an important target in B-cell receptor signalling, have advanced to RA clinical trials. Fostamatinib has demonstrated encouraging activity in Phase II studies, and another Syk inhibitor with a different kinase selectivity profile, PRT062607, is scheduled to begin Phase II studies in 2012. A second kinase target affecting B-cell receptor signalling that has generated considerable excitement in the pharmaceutical industry for its potential to treat inflammatory and autoimmune disease as well as haematopoietic cancers is Btk, or Bruton's tyrosine kinase. Two distinctly different approaches have been taken for the discovery of small-molecule Btk inhibitors, one relying on the reversible inhibition of an inactive conformation of the kinase and another focusing on irreversible, covalent binding to the ATP binding site of the protein, and it will be fascinating to see whether either or both of these can offer a new modality for the treatment of RA and related diseases.

Two additional kinases, MK2 and IKKβ, that have long held promise as targets that might yield novel therapeutic options for treating inflammatory disease are discussed in Section 2. However, these targets have struggled to achieve clinical validation for inflammatory disease indications and it is instructive to consider the impediments that have delayed the transformation of potent small-molecule inhibitors of these targets into drugs.

The third section of this volume is dedicated to the review of three GPCR targets. The first two of these are CCR1 and CCR2, members of the widely studied CC chemokine receptor family that have been awaiting clinical validation. The recent disclosure of data from a Phase II study of the ChemoCentryx CCR1 antagonist CCX-354 in patients with moderate to severe RA indicates that this target may yet fulfill its promise as a novel treatment for autoimmune disease. Also, while CCR2 antagonists have had some recent setbacks in trials for RA and multiple sclerosis, there have been promising results in studies for type 2 diabetes and atherosclerosis that offer continued hope for this target. The third GPCR drug target reviewed is the cannabinoid receptor CB2. Agonists of this receptor are under clinical investigation as treatments for pain and inflammation, and it has been a considerable challenge to identify and develop candidates that are highly selective for CB2 over CB1 in order to avoid dose limiting CNS side-effects and enable target validation in clinical trials.

Sphingolipids are the subject of the fourth section of this volume and this is the only area represented herein in which a significant new drug has recently made it to market. The approval of fingolimod, a prodrug of an S1P (sphingosine-1 phosphate) receptor agonist, by the FDA in 2010 as the first orally bioavailable, disease-modifying agent for the treatment of relapsing-remitting multiple sclerosis provides an enormous boost for the entire field of sphingo-lipid drug discovery research. The development of new S1P agonists remains a very competitive area as work continues in order to ascertain the ideal selectivity profile within the S1P family, combined with the challenge of finding molecules with good physicochemical properties and receptor potency, to provide the safest and most effective drug. Other important targets in this field are three enzymes that modulate S1P levels, S1P lyase (S1PL) and the two isoforms of sphingosine kinase, SK1 and SK2. In particular, an inhibitor of S1PL should soon advance to Phase II RA clinical trials.

The final chapter of this volume covers the search for clinically useful non-steroidal glucocorticoid receptor (GR) agonists with functional selectivity that will allow them to achieve the clinical anti-inflammatory efficacy of steroids but with significantly reduced side-effects. Here the progression of systemically available glucocorticoid receptor agonists into the clinic will enable the testing of hypotheses that could lead to a new class of anti-inflammatory drugs.

In summary, these reviews cover a variety of the most active and promising areas of research ongoing today in small-molecule anti-inflammatory drug discovery, and it is our hope that they can provide useful perspective and insight into the challenges encountered and surmounted in the progression from the identification of a target biological mechanism of action, to the identification of a lead molecule and its optimization to afford a marketed drug that can impact patients' lives.

Acknowledgement

The editors would like to thank John Proudfoot, Steven Taylor, Domnic Martyres, Derek Cogan, Matthias Hoffman, Steve Brunette, Rob Hughes, Todd Bosanac, Dan Kuzmich, Christian Harcken, Dave Rotella, Renee Zindell, Doris Riether, Josh Horan, Rene Lemieux, Louise Modis, John Huber, John Ginn and Jennifer Swantek for their prompt, thorough and thoughtful review of chapters in this volume.

Section 1
Arachidonic Acid Cascade

CHAPTER 1

Microsomal Prostaglandin E₂ Synthase-1

ANDREAS KOEBERLE AND OLIVER WERZ*

Chair of Pharmaceutical/Medicinal Chemistry, Institute of Pharmacy,
University Jena, Philosophenweg 14, D-07743 Jena, Germany
*Email: oliver.werz@uni-jena.de

Prostaglandins (PGs) are pivotal bioactive lipid mediators that mediate inflammatory reactions but also contribute to various homeostatic biological processes.[1] The cyclooxygenase (COX) isoenzymes namely COX-1 (constitutively expressed in numerous cell types and thought to provide PGs mainly for physiological functions) and COX-2 (an inducible isoform in inflammatory cells, primarily producing PGs relevant for inflammation, fever and pain[2]) initiate PG biosynthesis from arachidonic acid. The non-steroidal anti-inflammatory drugs (NSAIDs) or COX-2 selective drugs (coxibs) inhibit COX activities and thus exert potent anti-inflammatory and analgesic effects but may also cause severe side-effects such as gastrointestinal[3] and renal complications.[4] The latter effects have been ascribed to the general suppression of constitutively formed prostanoids such as COX-1-derived cytoprotective PGE₂ and prostacyclin (PGI₂) in gastroduodenal epithelium.[5] Even though coxibs have an improved gastrointestinal tolerance, recent clinical studies revealed small but significantly increased risks for cardiovascular events such as myocardial infarction, stroke, systemic and pulmonary hypertension, congestive heart failure and sudden cardiac death.[6] Apparently, the imbalance between the anti-thrombotic and vasodilatory PGI₂ on one hand

RSC Drug Discovery Series No. 26
Anti-Inflammatory Drug Discovery
Edited by Jeremy I Levin and Stefan Laufer
© The Royal Society of Chemistry 2012
Published by the Royal Society of Chemistry, www.rsc.org

and the pro-thrombotic TxA_2 on the other are responsible for those side-effects. Since side-effects are particularly problematic in the therapy of chronic pathologies, such as rheumatoid arthritis, requiring long-term drug application, there is a strong need for novel potent and safe anti-inflammatory drugs lacking such toxicity.

1.1 Function of PGE_2 as Bioactive Mediator

Among the PGs, PGE_2 is the most prominent mediator in inflammation, fever and pain but also has physiological functions in the gastrointestinal tract, the kidney and in the immune and the central nervous system.[7,8] PGE_2 mediates its bioactivities essentially by four G-protein coupled PGE_2 receptor subtypes (EP1–EP4) in diverse tissues, supported by experiments with knockout mice deficient (-/-) in EP receptor subtypes and selective EP receptor antagonists.[8] For example, deletion of respective EP receptor subtypes significantly reduced exudate formation in carrageenan-induced mouse pleurisy (EP2 and EP3),[9] diminished arachidonic acid-induced cutaneous inflammation (EP3)[10] and inflammation and joint destruction in collagen (EP4),[11] as well as collagen-antibody-induced arthritis (EP4 and EP2).[12] In lipopolysaccharide-challenged mice, activation of the hypothalamo-pituitary-adrenal axis[13] and systemic illness including the febrile response[14] in response to PGE_2 was mainly ascribed to the EP3 receptor (although other subtypes are also involved). Regarding pain sensation, all four EP receptors seemingly contribute, but the EP-receptor subtype involved depends on the nociceptive stimulus and/or the pre-treatment of the animals. For example, the EP1 receptor mediates the acute pain response in the acetic acid-induced writhing test in mice,[15] whereas EP3 is the critical receptor for LPS-induced hyperalgesia.[16] However, PGE_2 also exerts immunosuppressive effects contributing to the resolution of inflammation,[17–21] mediates protection of gastrointestinal mucosa[22–26] and regulation of natriuresis and regulates renal blood flow and blood pressure.[15,27–30]

1.2 PGE_2 Biosynthesis by mPGES-1

PGE_2 synthases (PGES) perform the terminal step in the biosynthesis of PGE_2, that is, the isomerization of the COX-derived peroxide PGH_2 to PGE_2.[31] Three terminal isoforms of PGE_2 synthases have been cloned and characterized. Co-transfection of COX-1 and -2 with PGES isoenzymes in mammalian cells suggests that molecular interactions may cause preferential functional coupling.[32] While the constitutively expressed cytosolic PGE_2 synthase (cPGES) was found to be coupled to COX-1,[33] mPGES-1 is predominantly involved in COX-2-mediated PGE_2 production.[34] mPGES-2 is also constitutively expressed, uses PGH_2 supplied by COX-1 and COX-2[35] and contributes to basal PGE_2 synthesis but its functional role in physiology and patho-physiology is still elusive.

1.3 Structure and Biochemical Properties of mPGES-1

Human mPGES-1 (16 kDa) is a member of the membrane-associated proteins involved in eicosanoid and glutathione metabolism (MAPEG) family[36] and is characterized by a high turnover number for PGH_2 ($k_{cat} = 50 s^{-1}$). The enzyme is membrane-bound and localized to the microsomal fraction after subcellular fractionation.[36,37] K_m values of 14–160 µM and 710–750 µM were reported for PGH_2 and glutathione, respectively.[37,38]

The homotrimeric structure of mPGES-1 was recently determined at low resolution by electron crystallography.[39] The monomers consist of four transmembrane helices (TM1-4) and enclose an inner core with a funnel-shaped opening towards the cytoplasm. The essential cofactor glutathione is bound in a U-shaped conformation at the interface between the subunits.[39] Glutathione was proposed to attack the peroxide of PGH_2 in the active site *via* its thiol-group during the catalytic cycle.[39–41] Mutation studies suggest Arg126, which is located near the thiol of glutathione, as catalytic residue[41] and Thr-131, Leu-135 and Ala-138 (all are located within the transmembrane-helix 4) as gate keepers for the active site.[42] The smaller size of these gate keepers in the human enzyme compared to those in the rat enzyme may explain the failure of several mPGES-1 inhibitors to inhibit rat mPGES-1. For the related MAPEG enzyme LTC_4 synthase, conformational changes from a closed to an open conformation are required for substrate entry into the active site, and a comparable mechanism might also account for mPGES-1.[39,43,44] Molecular dynamic simulations strengthened by site-directed mutagenesis and subsequent hybridization suggest only one PGH_2 substrate pocket of the trimer being occupied at the same time during the reaction cycle.[45] While the low-resolution crystal structure of mPGES-1 seems to represent the closed confirmation, Hamza *et al.* proposed an mPGES-1 model in the open confirmation based on the crystal structures of MGST1 and ba3-cytochrome c oxidase.[46]

1.4 Regulation of mPGES-1 Expression

Pro-inflammatory stimuli such as interleukin-1β, tumour necrosis factor-α or lipopolysaccharide strongly induce expression of both mPGES-1 and COX-2 in a variety of tissues and cell-types (including human lung carcinoma A549 cells,[36] macrophages,[47] endothelial cells[48] and others[31]), whereas glucocorticoids reverse their up-regulation.[49] However, differences occur in the kinetics of mPGES-1 and COX-2 expression.[50–56]

Thus, mPGES-1 is constitutively expressed in diverse tissues at a low level, for example in seminal vesicles,[36] ovary,[57] kidney,[57,58] male reproductive organs,[59] placenta,[60] lung, spleen and gastric mucosa[61] of mice and rats, but not in other tissues such as heart, liver and brain.[35,62] The expression of mPGES-1 is markedly induced in inflamed rodent tissue and brain under diverse pathological conditions including inflammation, fever, pain and seizure.[56,62–69] In humans, mPGES-1 was found to be up-regulated among others in arthritic synovial tissue,[35,70] in the cartilage and chondrocytes of osteoarthritic patients,[71–73] in inflamed intestinal mucosa from patients with inflammatory bowel disease,[74] in atherosclerotic carotid plaques,[75–77] in Alzheimer's disease tissues,[78] in heart tissue after acute myocardial infarction[35] and in liver tissue from patients with hepatitis.[35] Thus, mPGES-1 seems to play a putative role in various diseases related to inflammation.

Increasing experimental evidence implies that PGE_2 also regulates critical steps in tumourigenesis by stimulating cell proliferation and angiogenesis, preventing apoptosis and inducing cancer cell migration.[79] In fact, elevated expression of mPGES-1 and increased PGE_2 levels are characteristic for many human tumours and cancer cell lines derived from colon,[80] intestine,[81] stomach,[82–84] oesophagus,[85,86] larynx,[87] lung,[36,88] liver,[89,90] pancreas,[91] breast,[92] ovary,[93] squamous epithelium[94,95] and brain.[96–98]

1.5   Redirection of the mPGES-1 Substrate PGH_2 Due to the Action of mPGES-1 Inhibitors

Selective inhibition of PGE_2 formation by pharmacological interference with mPGES-1 as therapeutic strategy has been questioned because of a redirection of the substrate PGH_2 to other prostanoid synthases leading to an increased formation of PGI_2,[99–102] $PGF_{2\alpha}$,[102,103] TxB_2[103] or PGD_2.[104] The redirection pattern strongly depends on the cell type and assay conditions. In mPGES-1-deficient mice for example, the above-mentioned prostanoids (PGI_2,[99–102] $PGF_{2\alpha}$,[102,103] TxB_2[103] or PGD_2[104]) are increased in the stomach but not in other tissues.[61] An increased biosynthesis of thromboxanes is associated with cardiovascular diseases such as myocardial ischemia[105] or heart failure.[106] However, the redirection of prostanoids might also be beneficial in some cases because of their inflammation-resolving and cytoprotective effects.[107] Taken together, the clinical safety of mPGES-1 inhibitors is not readily known so far.

1.6   Determination of mPGES-1 Activity

In general, two different types of assays for analysis of mPGES-1 activity can be distinguished: cell-free assays and cell-based test systems. Cell-free mPGES-1 activity can be assessed in microsomal preparations of cells highly expressing mPGES-1 such as interleukin-1β-stimulated human lung carcinoma A549[36] or human cervix carcinoma HeLa cells,[133] but also cell lines transfected with recombinant rodent or human mPGES-1 were used as source of mPGES-1.[34,62] PGH_2 (1–20 μM) is added to a reaction mix containing glutathione (2.5 mM) and microsomal preparations of mPGES-1-expressing cells. After 0.5–5 min on ice or at room temperature, the reaction is stopped by converting the remaining PGH$_2$ to PGF$_{2\alpha}$ using mild reducing agents such as SnCl$_2$[62] or FeCl$_2$.[48] Formed PGE$_2$ is quantified by either EIA[47] or RP-HPLC combined with UV or radiometric detection (when radioactively labelled PGH$_2$ is supplied).[36]

Determination of mPGES-1 activity in cell-based assays is less convenient. Test systems are based on isolated cells or whole blood stimulated with lipopolysaccharide (LPS) for 5 to 24 h. PGE$_2$ is quantified either directly by EIA or after separation of PGE$_2$ by RP-HPLC prior to immunological detection.[108,109] Incubation times exceeding 24 h increase the contribution of mPGES-1 to total PGE$_2$ synthesis, which is efficiently inhibited by selective mPGES-1 inhibitors (*e.g.* MF-63, 70–80% inhibition).[110] Incubation times <5 h may minimize effects of the test compounds on gene expression, which results, however, in a higher portion of mPGES-1-independent PGE$_2$ synthesis. Diverse mPGES-1 inhibitors such as MK-886 (30 µM)[108,111] and many others maximally inhibited PGE$_2$ formation by 40–60% under these experimental conditions. The remaining PGE$_2$ formation was suggested to derive from constitutively expressed PGE$_2$ synthases such as cPGES and mPGES-2.[111]

1.7 mPGES-1 in Disease and Homeostasis – Results from KO Studies

The knockout of mPGES-1 in mice does not result in phenotypic, behavioural or histological differences.[112] Several excellent reviews have summarized the role of mPGES-1 in inflammation, neurologic diseases, cardiovascular disease and tumourigenesis.[31,113–115,116]

1.7.1 Inflammation, Fever and Pain

Knockout studies in mice have revealed a role of mPGES-1 for the induction and progression of arthritis[99,112,117] and other inflammatory diseases,[99,112] hyperalgesia[99,112,117–119] and the febrile response.[120,121] The role of mPGES-1 in nociception is still under discussion due to conflicting observations. Thus, mPGES-1-deficient mice have reduced pain sensation after intra-peritoneal injection of acetic acid,[99,112,118] during collagen-induced arthritis[117] and in a neuropathic pain model.[119] However, knockout of mPGES-1 was without effect on the nociceptive behaviour in the formalin test and during zymosan-evoked mechanical hyperalgesia.[118] Since COX-inhibitors were effective under the same experimental conditions and the mPGES-1 knockout reduced PGE$_2$ levels in the spinal cord, the authors explained the differences by a redirection of PGH$_2$ to other prostanoids (PGD$_2$, PGF$_{2\alpha}$ and 6-keto PGF$_{1\alpha}$).

1.7.2 Neurological Diseases

mPGES-1-deficient mice show a reduced severity of infarction, edema and apoptotic cell death in the cortex after transient focal ischemia,[55] of symptoms during experimental autoimmune encephalomyelitis (a mouse model for multiple sclerosis)[122] and of neuronal loss in response to kainaic acid.[123] However, priming of mPGES-1-deficient mice by intra-thecal injection of LPS suggests a role of mPGES-1 also for resolving neuroinflammation.[124]

1.7.3   Cancer

mPGES-1 is overexpressed in various tumours.[36,81–92] Knockout of mPGES-1 reduced cell proliferation, migration and invasion *ex vivo* and in mouse tumour xenocraft models.[125,126] Deletion of mPGES-1 counteracts azoxymethane-induced preneoplastic aberrant crypt foci formation[127] and colorectal carcinogenesis[128] as well as intestinal cancer growth in APC-mutant mice.[127] However, another study using APC-mutant mice found an increased tumour formation in mPGES-1-deficent mice which is not readily understood.[129] Moreover, the knockdown of mPGES-1 prevented apoptosis and thus induced tumourigenesis in nude mice injected with human glioblastoma cells.[97] The growth and angiogenesis of endometrial implants is also increased in mPGES-1-deficient mice.[130] Taken together, mPGES-1 seems to have both anti- and pro-tumoural effects depending on the experimental model.

1.7.4   Renal and Cardiovascular System

The knockout of mPGES-1 in mice increases neither thrombogenesis nor blood pressure at standard diet in contrast to a knockout of COX-2.[101] However, the constitutive expression of mPGES-1 along the nephron and collecting duct[58,131,132] suggests a physiological role for water and salt resorption. In fact, mPGES-1-deficient mice were afflicted with hypertension, impaired natriuretic responses and worsened cardiac parameters when challenged with a high salt-diet,[133] angiotensin II,[133–135] aldosterone[136] or DOCA-salt.[137] In contrast, the blood pressure was not increased at high salt diet in a study published by Cheng *et al.*[101] Moreover, knockdown of mPGES-1 resulted in a detrimental left ventricular remodelling after myocardial infarction but was not associated with an increased pulmonary edema or mortality.[138] Deletion of mPGES-1 reduced plaque formation and increased plaque stability in a mouse atherosclerosis model,[102] reduced cisplatin-induced nephrotoxicity[139] and retarded disease progression in a mouse model of renal mass reduction.[140] Taken together, adverse effects of mPGES-1 inhibition might occur when risk factors are present such as chronic salt loading or cardial ischemic injury.

1.8   Direct Inhibitors of mPGES-1

The first mPGES-1 inhibitors described were rather unselective. Among them were several COX-2-selective drugs, such as celecoxib ($IC_{50} = 22\,\mu M$) and lumiracoxib (IC$_{50}$ = 33 μM), and endogenous fatty acids and eicosanoids, such as arachidonic acid (IC$_{50}$ = 0.3 μM)[141] and 15-deoxy-Δ^{12,14}-PGJ$_2$ (IC$_{50}$ = 0.3 μM).[141] Inhibition of COX-2 by celecoxib could be abolished by methylation yielding dimethylcelecoxib.[142] Interestingly, dimethylcelecoxib still moderately inhibited mPGES-1 activity (IC$_{50}$ = 16 μM) but also selectively prevented mPGES-1 protein up-regulation in interleukin-1β- and tumour necrosis factor-α-treated HeLa cells (IC$_{50}$ = 10–20 μM).[143] The latter effect was related to a reduced expression of the transcription factor EGR1.[144,145]

celecoxib

lumiracoxib

arachidonic acid

15-deoxy-Δ^{12,14}-PGJ$_2$

dimethylcelecoxib

Structural Formula 1.1

1.8.1 MK886 Derivatives

As a potent inhibitor of FLAP,[146] MK-886 suppresses leukotriene formation in intact leukocytes (IC$_{50}$ = 2.5 nM).[147] mPGES-1 is inhibited at low micromolar concentrations (IC$_{50}$ = 2.4 μM).[56] MK-886 was used as lead structure for the design of a series of potent and selective indole-2-propanionic acid-based mPGES-1 inhibitors.[109] Introduction of substituted biphenyl moieties in the 5-position yielded the most potent inhibitors of this series, which are compounds **1** and **2** (IC$_{50}$ = 7 and 3 nM, respectively). Direct binding of these inhibitors to FLAP was investigated and found to be negligible. Inhibition of mPGES-1 is selective over mPGES-2 (IC$_{50}$ > 1 μM) and TxA$_2$ synthase (IC$_{50}$ = 1 μM). Both compounds **1** and **2** remarkably lost potency in cell-based assays (in the presence of 2% FCS) and were inactive in human whole blood (IC$_{50}$ > 40 μM), probably due to their high plasma protein binding. Accordingly, increasing the percentage of FCS in the cellular test systems strongly shifted the IC$_{50}$ value of compounds **1** and **2**. Hence, this series did not enter pre-clinical trials.

MK-886

1

2

Structural Formula 1.2

3D-quantitative structure-activity relationship (3D-QSAR)[148,149] and docking studies[149] using a computational 3D structural model of mPGES-1[150] were performed to characterize the potential binding mode of indole-type inhibitors. The authors suggested that among others an anionic substituent (*e.g.* the carboxylic group) has to interact with Arg110 and that the bulky hydrophobic substituents in the 5-position of the indole have to reach a hydrophobic pocket formed by Val37, Val128 and Thr131.

1.8.2   Phenanthreneimidazoles and Related Compounds

Phenanthreneimidazoles and heterocyclic derivatives constitute a rather promising class of potent and specific mPGES-1 inhibitors that tolerate plasma protein. The lead compound was the JAK kinase inhibitor azaphenanthrenone[151] (IC$_{50}$ for mPGES-1 = 0.14 µM) with an azaphenanthrenone core.[152] Replacement of azaphenanthrenone by phenanthrenone, introduction of a 2,6-di-cyano-phenyl residue in the 2-position and replacement of the 6-fluoro substituent by chlorine potentiated the inhibition of mPGES-1 in cell-free and cellular assays (MF-63: IC$_{50}$ = 1 nM and 0.05–0.42 µM, respectively).[152] MF-63 interferes with the assembly of mPGES-1 complexes as shown by photo-crosslinking.[153] The authors suggested that MF-63 binds mPGES-1 near the interface of the subunits of the mPGES-1 homotrimer. Backbone amide H/D exchange mass spectrometry was also applied to map the binding sites of mPGES-1 inhibitors (including MK-886 and MF-63). The binding sites differ between the inhibitors in detail but seem to be predominantly located in the hydrophobic cleft of the suggested PGH$_2$ binding pocket.[154] He *et al.* found a similar binding mode for MF-63 (and also for other mPGES-1 inhibitors like MK-886 and pirinixic acid derivatives) by molecular docking.[155] The binding site of MF-63 apparently overlaps with the glutathione and PGH$_2$ pocket. This binding model is strengthened by kinetic studies showing a competition of the inhibitors with both glutathione and PGH$_2$.[155]

azaphenanthrenone   MF-63   3

4   5

Structural Formula 1.3

MF-63 inhibited PGE$_2$ formation in interleukin-1β-stimulated A549 cells and in isolated lipopolysaccharide-stimulated peritoneal macrophages from human mPGES-1 knockin mice with IC$_{50}$ values of 46 nM and 10 to 100 nM, respectively.[156] The potency of MF-63 is slightly plasma protein-shifted. In the presence of plasma protein (50% FCS) and in lipopolysaccharide-stimulated whole blood, the IC$_{50}$ values were determined as 0.8–1.3 μM.[152,156] MF-63 is highly selective for human mPGES-1 over other recombinant prostanoid synthases (mPGES-2, PGI$_2$ synthase, PGD2 synthase, TxA$_2$ synthase)[152,156] and isolated JAK isoenzymes (JAK2, JAK3) – the main target of the lead compound. Accordingly, MF-63 selectively inhibits PGE$_2$ formation in cellular models but not COX-2-derived PGF$_{2\alpha}$ formation in interleukin-1β-stimulated A549 cells and COX-1-derived TxB$_2$ formation in lipopolysaccharide-stimulated whole blood.[156] Levels of PGF$_{2\alpha}$ were rather increased, probably due to redirection of PGH$_2$.

MF-63 inhibits neither mouse nor rat mPGES-1.[152] This is a major disadvantage for its assessment in pre-clinical studies. A mouse model expressing human mPGES-1 instead of mouse mPGES-1 was therefore generated. Alternatively, guinea pigs could be used as test organism. Guinea pig mPGES-1 is equipotently inhibited by MF-63 *versus* human mPGES-1.[152] Using these animal models, the authors could demonstrate that intra-peritoneal injection of MF-63 suppresses lipopolysaccharide-induced thermal hyperalgesia (EC$_{50}$ = 10 mg/kg), iodoacetate-induced osteoarthritic pain (EC$_{50}$ < 10 mg/ml) and lipopolysaccharide-induced pyresis ($EC_{50} \approx 15$–$50\,mg/kg$).[156] MF-63 selectively inhibited PGE_2 formation (but not $PGF_{2\alpha}$, 6-keto $PGF_{1\alpha}$ and TxB_2 formation) in the inflamed tissues. Peak plasma and brain levels of 4.1 and $20\,\mu M$ were observed for MF-63 after 3 h when orally administered at $30\,mg/kg$.[156] The anti-inflammatory and analgesic potency of MF-63 was comparable to established COX-inhibitors. However, in contrast to NSAIDs, gastrointestinal toxicity was not observed for MF-63 up to $100\,mg/kg$ daily, although PGE_2 levels were reduced in the stomachs of mice.[156] The ductus arteriosus contracted after a single dose of MF63 ($10\,mg/kg$), but repeated dosing resulted in dilatation of the ductus arteriosus and delayed postnatal closure.[157]

Further structural optimization yielded the disubstituted phenanthrene imidazoles **3** and **4** ($IC_{50} = 1\,nM$, each), which are comparably selective for mPGES-1 as MF-63 but more potent in whole blood ($IC_{50} = 0.20$ and $0.14\,\mu M$, respectively).[158] Surprisingly, anti-analgesic effects were observed at $14\,mg/kg$ oral administration in an LPS-induced guinea pig hyperalgesia model. The half-life of **4** (2.3 h) is considerably lower than that of compound **3**. Shiro *et al.* reported a novel class of mPGES-1 inhibitors based on a substituted imidaquinoline core, which resembles the structure of the phenanthrene imidazole series in the substitution pattern.[159] The most potent imidaquinoline **5** inhibited mPGES-1 with an IC_{50} of $9.1\,nM$ and showed a more than 1000-fold selectivity over COX-1 and COX-2.

1.8.3   Others

Several mPGES-1 inhibitors were recently identified by high-throughput and virtual screening of compound libraries. Computational screening identified, for example, the mPGES-1 inhibitor **6** ($IC_{50} = 3.2\,\mu M$),[160] compound **7** ($IC_{50} = 0.5\,\mu M$)[161] and compound **8** ($IC_{50} = 3.5\,\mu M$).[162] As a non-acidic compound, **7** failed to inhibit COX isoenzymes. According to the 3D-structural model of mPGES-1, compound **8** binds at the interface of the homotrimeric mPGES-1 subunits and forms a salt bridge to Arg110 and a hydrogen bond to Thr129.[162] Biaryl-imidazoles[163] and the benzothiazine PF-9184[164,165] were identified as mPGES-1 inhibitors from high-throughput screening. The biaryl-imidazole **9** inhibited cell-free mPGES-1 with an IC_{50} of 1 nM, and PGE_2 formation of activated A549 with IC_{50} values of 13 and 160 nM in the presence of 2 and 50% BSA, respectively.[163] PGE_2 formation in LPS-stimulated human whole blood was inhibited with an IC_{50} of $1.6\,\mu M$. PF-9814 inhibited cell-free human mPGES-1 ($IC_{50} = 16.5\,nM$) without inhibiting COX-1 and COX-2.[164] PGE_2 formation was suppressed in intact cells (IL-1β-stimulated human fibroblasts from patients with rheumatoid arthritis) with an IC_{50} of approx. $0.8\,\mu M$ and in LPS-stimulated whole blood with an IC_{50} of approximately $4\,\mu M$. PGF_1, PGF_2 and TxB_2 formation were not suppressed. As observed for phenanthrene imidazoles, inhibition of rat mPGES-1 requires considerably higher concentrations ($IC_{50} = 1.1\,\mu M$) than inhibition of the human enzyme.

Structural Formula 1.4

1.9 Dual Inhibitors of mPGES-1 and 5-LO

5-Lipoxygenase (5-LO) is the key enzyme in the biosynthesis of leukotrienes.[166] Arachidonic acid is released by cPLA$_2$ and transferred to 5-LO *via* FLAP. 5-LO converts arachidonic acid into the epoxid LTA$_4$, which is further metabolized to LTB$_4$ and cysteinyl-leukotrienes. Interestingly, both FLAP and mPGES-1 belong to the MAPEG family of proteins.[31] Leukotrienes are important lipid mediators in inflammatory and allergic diseases,[167] and increasing evidence also suggests a role of leukotrienes in cancer[168–170] and cardiovascular diseases.[171,172] Dual inhibitors of prostanoid and leukotriene formation, and in particular dual mPGES-1 and 5-LO inhibitors, were suggested superior over COX-1 and COX-2 inhibitors.[115,173] They might be not only more efficient but also less afflicted with gastrointestinal and cardiovascular side-effects. Clinical trials have not been performed so far but will be necessary as proof of concept.

1.9.1 MK-886 and Derivatives

MK-886 inhibits cellular leukotriene formation by interference with FLAP (IC$_{50}$ = 2.5 nM).[147] At higher concentrations, MK-886 also inhibited isolated LTC$_4$ synthase (IC$_{50}$ = 3.1 μM)[174] and isolated ovine COX-1 (IC$_{50}$ = 8 μM).[175]

Cell-free mPGES-1 is inhibited with an IC_{50} of 1.6–2.4 μM.[56,109,176] The potency of MK-886 to interfere with PGE_2 formation is diminished in cellular studies[71,176] and even more so in LPS-stimulated human whole blood ($IC_{50} = 30$ μM).[176] Interestingly, MK-886 was also described to selectively down-regulate mPGES-1 expression in interleukin-1β- and tumour necrosis factor-α-stimulated human gingival fibroblasts[177] and in HL-60 cells.[178]

The three-dimensional structure of licofelone resembles MK-886, as shown by flexible alignment.[179] Licofelone recently passed Phase III trials for osteoarthritis. Like MK-886, licofelone inhibits COX-1,[180] mPGES-1[176] and FLAP.[179] COX-2 and 5-LO are not targeted by licofelone, although previously stated.[181,182] Cell-free and cellular mPGES-1 activity is inhibited by licofelone with an IC_{50} of 5–6 μM.[176] A loss in potency was not observed in whole blood. Licofelone has proven anti-inflammatory effectiveness in a number of clinical and pre-clinical studies.[173,183] Whether mPGES-1, 5-LO or COX-1 is the most relevant target remains unclear. Notably, licofelone lacks gastrointestinal toxicity, which is usually observed for COX-1 inhibitors. Structural optimization of licofelone led to the tolyl sulfonimide derivative **10**, which inhibited cell-free mPGES-1 with an IC_{50} of 2.1 μM.[184]

licofelone   **10**   **11**

Structural Formula 1.5

SAR studies of a series of benzo[g]indole-3-carboxylates yielded the *m*-chloro-benzyl-substituted compound **11** as most potent mPGES-1 inhibitor ($IC_{50} = 0.6$ μM).[185] Compound **11** suppressed PGE_2 synthesis in IL-1β-treated A549 cells ($IC_{50} = 2$ μM) as well as in rat carrageenan-induced pleurisy and mouse carrageenan-induced paw edema after intra-peritoneal application. COX-2-derived formation of 6-keto PGF_{1a} was not affected under the same conditions in A549 cells. Semi-purified 5-LO was inhibited with an IC_{50} of 0.09 μM, and 5-LO product formation in neutrophils was suppressed with an IC_{50} of 0.23 μM.[186]

1.9.2 Pirinixic Acid Derivatives

The PPARα agonist pirinixic acid (WY-14,643), used as lead structure, inhibited neither mPGES-1 nor 5-LO.[176] However, introduction of long and bulky lipophilic substituents (hexyl-, octyl- or naphthyl-residues) yielded potent dual inhibitors of mPGES-1 and 5-LO. The α-hexyl-substituted compound YS121 inhibited cell-free mPGES-1 activity (IC$_{50}$ = 3.4 μM) and directly bound to mPGES-1 protein (K$_D$ = 10–14 μM) as determined by plasmon resonance spectroscopy.[187] PGE$_2$ formation in LPS-stimulated whole blood was inhibited with an IC$_{50}$ of 2 μM, whereas TxB$_2$ and 6-keto PGF$_{1a}$ formation were not significantly reduced. Moreover, YS121 suppressed exudate formation, leukocyte infiltration and eicosanoid formation (PGE$_2$, LTB4, 6-keto PGF$_{1a}$) during carrageenan-induced rat pleurisy (1.5 mg/kg i.p.). An optimized synthetic route of YS121 for large-scale preparation was recently published.[188]

pirinixic acid   YS121

12   **13**

14

Structural Formula 1.6

The α-naphthyl pirinixic acid **12** showed an improved *in-vitro* pharmacological profile compared to the α-alkyl series.[189] mPGES-1 is inhibited with an IC$_{50}$ of 1 μM and 5-LO with an IC$_{50}$ of 0.1 μM in cell-free assays. Modification of the pirinixic acid backbone yielded dual mPGES-1/5-LO inhibitors based on the structure of diphenethoxypyrimidine.[190] The most favourable pharmacological profile was observed for compound **13** (5-LO: IC$_{50}$ = 0.8 μM; mPGES-1: IC$_{50}$ = 1.1 μM).

Also the 2-mercaptohexanoeic acid **14** has no longer a pirinixic acid core, though it shares the medium-chain 2-mercapto-carboxylic acid moiety of the pirinixic acid series described above.[191] Compound **14** inhibited cell-free mPGES-1 and 5-LO with IC$_{50}$ values of 2.2 and 3.5 μM, respectively. COX-1, COX-2, 12-LO, 15-LO and cPLA$_2$ were not inhibited at 10 μM.[191]

1.9.3   Acylphloroglucinols

The natural products myrtucommulone (from myrtle), hyperforin (from St. John's wort), garcinol (camboginol; from the rind of Guttiferae) and arzanol (from *Helichrysumitalicum*) were identified as dual inhibitors of mPGES-1 and 5-LO.[192–195] The four compounds share an acylphloroglucinol core but have no further obvious structural similarities. mPGES-1 activity was inhibited with IC_{50} values of 0.3–1.3 µM and cellular leukotriene formation with IC_{50} values of 1–3 µM.[194–197] All four compounds directly interfere with 5-LO. At higher concentration, the acylphloroglucinols also inhibited isolated ovine COX-1 ($IC_{50} \geq 12$ µM). PGE_2 formation in LPS-stimulated human whole blood is significantly inhibited at low micromolar concentrations (1–30 µM). In contrast, the concomitant formation of TxB_2 or 6-keto PGF_{1a} was not affected. A number of reports describe anti-inflammatory and anti-carcinogenic effects of these acylphloroglucinols in animal models.[193,195,198,199] Among others, myrtucommulone, hyperforin and arzanol impaired exudate volume and leukocyte numbers in carrageenan-induced rat pleurisy (at 3–5 mg/kg, given i.p.).[193,195,199] However, only hyperforin and arzanol reduced the PGE_2 levels in the exudates.

hyperforin   myrtucommulone

garcinol   arzanol

Structural Formula 1.7

1.9.4   Others

Besides acylphloroglucinols, a small number of other natural products have been described as dual mPGES-1/5-LO inhibitors. Among them are the β-diketon curcuminfrom turmeric[200] and boswellic acids from frankincense.[201] Cell-free mPGES-1 was inhibited with IC_{50} values between 0.3 and 10 µM[200,201] and 5-LO with IC_{50} values between 0.7 and 24 µM by curcumin and boswellic acids.[202–204] Both curcumin and boswellic acids (*e.g.* β-boswellic acid) are rather selective for mPGES-1 within prostanoid biosynthesis.[200,201] Inhibition of cellular COX-1 and COX-2-derived prostaglandin formation required higher concentrations ($> 10\,\mu M$). Due to the variety of molecular targets described for curcumin and boswellic acids, it remains unclear to what extent the dual inhibition of mPGES-1 and 5-LO contributes to the anti-inflammatory potential observed in animal and human studies.[205–207]

curcumin   β-boswellic acid

15

16   17

Structural Formula 1.8

Virtual screening approaches were also successfully adapted to identify dual inhibitors of mPGES-1 and 5-LO. The triazol **15** partially inhibited cell-free mPGES-1 starting at $3\,\mu M$ and semi-purified 5-LO with an $IC_{50} = 0.8\,\mu M$.[160] Compounds **16** and **17** were found to inhibit cell-free mPGES-1 with an IC_{50} of 0.4 and $0.5\,\mu M$, respectively.[208] Isolated 5-LO and 5-LO product formation in human neutrophils were also inhibited by compounds **16** and **17** with IC_{50} values of $2–3\,\mu M$.

1.10 Selective Suppressors of mPGES-1 Expression

The expression of mPGES-1 and COX-2 is often co-induced,[31] which makes sense because both enzymes are required for an efficient conversion of arachidonic acid to PGE$_2$. However, different downstream signalling pathways and transcription factors regulate the induction of mPGES-1 and

COX-2.[52,74,209] Exploiting these differences led to selective inhibitors of mPGES-1 expression. Among a series of γ-hydroxy-butenolide derivatives with aromatic and heteroaromatic substituents in the β-position, the benzo[b]thiophen-2-yl-substituted compound **18** suppressed PGE_2 formation and mPGES-1 induction ($IC_{50} = 1.8 \mu M$ and $< 1 \mu M$, respectively) in LPS-stimulated RAW264.7 macrophages.[210,211] mPGES-1 expression and the inflammatory response were also repressed by compound **18** (0.05–0.5 μmol/pouch) in azymosan-induced mouse air pouch model and during collagen-induced arthritis in mice (5 mg/kg i.p.).[212] Moreover, the compound showed an analgesic effect during acetic acid-induced hyperalgesia in LPS-sensitized mice ($ED_{50} = 6.2$ mg/kg i.p.). COX-2 expression was not affected. Further structural optimization of the benzothiophene γ-hydroxybutenolide **18** by removal of the 3-bromo-substituent and esterification of the 5-hydroxy-group yielded more potent suppressors of mPGES-1 induction (**19**: $IC_{50} = 0.85 \mu M$; **20**: $IC_{50} = 0.79 \mu M$).[213]

18   **19**   **20**

minocyclin   **21**   **22**

resveratrol

Structural Formula 1.9

Bastos *et al.* describe that the tetracyclic antibiotic minocycline significantly impairs mPGES-1 expression but not that of COX-2 in LPS-activated primary rat microglial at concentrations between 3 and 100 μM.[214] The furo-naphthoimidazole derivate **21** – which is structurally related to the phenanthreneimidazoles described above – inhibited LPS-induced PGE_2 synthesis with an $IC_{50} = 47$ nM by selective suppression of mPGES-1 expression.[215] Sepsis-induced PGE_2 formation in rat serum was strongly reduced after oral administration of compound **21** at 10 mg/kg. The polyphenol

2-(3,4-dihydroxyphenyl)-ethanol (**22**) from olive oil reduced tumour growth, vessel lumina and blood perfusion to tumour in a murine HT-29 xenograft model.[216] The IL-1β-induced increase of mPGES-1 (but not of COX-2, cPGES and mPGES-2) expression was reduced in HT-29 cells at 100 μM.

Controversial results were obtained for the polyphenolic antioxidant resveratrol and the endogenous lipid PPARγ ligand 15-deoxy-$\Delta^{12,14}$-PGJ$_2$. Both compounds selectively blocked the induction of mPGES-1 (but not of COX-2) in lipopolysaccharide-stimulated rat microglial cells (IC$_{50}$ = 10–25 μM)[217] and in MC615 cartilage cells (at 3 μM),[218] respectively. However, in other experimental models, mPGES-1 expression was induced and COX-2 repressed.[219–226]

1.11 Conclusion

Massive PGE$_2$ formation during inflammation predominantly depends on the concomitant induction of COX-2 and mPGES-1. Genetic and pharmacological inhibition of mPGES-1 suppresses the inducible PGE$_2$ formation and counteracts inflammation, fever, pain and neurodegeneration. The role of mPGES-1 in the cardiovascular and renal system and for tumour progression is not fully understood. A detailed understanding of these implications is, however, essential to estimate the safety of mPGES-1 inhibitors. The favourable pharmacological profile of mPGES-1 resulted in intensive research on the identification and characterization of selective mPGES-1 inhibitors within the last years. Several potent and selective inhibitors of mPGES-1 activity and suppressors of mPGES-1 expression were identified but only a few of them entered pre-clinical studies. Most advanced are the phenanthrene imidazole, pirinixic acid and γ-hydroxy-butenolide series. The mPGES-1 inhibitors are comparably potent to NSAIDs but seem to lack gastrointestinal side-effects, which might be a major advantage of this class of compounds as anti-inflammatory drugs.

References

1. C. D. Funk, *Science*, 2001, **294**, 1871.
2. K. D. Rainsford, *Subcell Biochem.*, 2007, **42**, 3.
3. F. Buttgereit, G. R. Burmester and L. S. Simon, *Am. J. Med.*, 2001, **110**(3A), 13S.
4. H. F. Cheng and R. C. Harris, *Curr. Pharm. Des.*, 2005, **11**, 1795.
5. A. Burke, E. Smyth and G. A. FitzGerald, *Goodman & Gilman's: Analgesic-anti-pyretic and Anti-inflammatory Agents and Drugs Employed in the Treatment of Gout*, McGraw-Hill, New York, 2005.
6. P. McGettigan and D. Henry, *JAMA*, 2006, **296**, 1633.
7. W. L. Smith, *Biochem. J.*, 1989, **259**, 315.
8. Y. Sugimoto and S. Narumiya, *J. Biol. Chem.*, 2007, **282**, 11613.

9. K. Yuhki, A. Ueno, H. Naraba, F. Kojima, F. Ushikubi, S. Narumiya and S. Oh-ishi, *J. Pharmacol. Exp. Ther.*, 2004, **311**, 1218.

10. J. L. Goulet, A. J. Pace, M. L. Key, R. S. Byrum, M. Nguyen, S. L. Tilley, S. G. Morham, R. Langenbach, J. L. Stock, J. D. McNeish, O. Smithies, T. M. Coffman and B. H. Koller, *J. Immunol.*, 2004, **173**, 1321.

11. J. M. McCoy, J. R. Wicks and L. P. Audoly, *J. Clin. Invest.*, 2002, **110**, 651.

12. T. Honda, E. Segi-Nishida, Y. Miyachi and S. Narumiya, *J. Exp. Med.*, 2006, **203**, 325.

13. Y. Matsuoka, T. Furuyashiki, H. Bito, F. Ushikubi, Y. Tanaka, T. Kobayashi, S. Muro, N. Satoh, T. Kayahara, M. Higashi, A. Mizoguchi, H. Shichi, Y. Fukuda, K. Nakao and S. Narumiya, *Proc. Natl Acad. Sci. USA*, 2003, **100**, 4132.

14. F. Ushikubi, E. Segi, Y. Sugimoto, T. Murata, T. Matsuoka, T. Kobayashi, H. Hizaki, K. Tuboi, M. Katsuyama, A. Ichikawa, T. Tanaka, N. Yoshida and S. Narumiya, *Nature*, 1998, **395**, 281.

15. J. L. Stock, K. Shinjo, J. Burkhardt, M. Roach, K. Taniguchi, T. Ishikawa, H. S. Kim, P. J. Flannery, T. M. Coffman, J. D. McNeish and L. P. Audoly, *J. Clin. Invest.*, 2001, **107**, 325.

16. A. Ueno, H. Matsumoto, H. Naraba, Y. Ikeda, F. Ushikubi, T. Matsuoka, S. Narumiya, Y. Sugimoto, A. Ichikawa and S. Oh-ishi, *Biochem. Pharmacol.*, 2001, **62**, 157.

17. L. Yang, N. Yamagata, R. Yadav, S. Brandon, R. L. Courtney, J. D. Morrow, Y. Shyr, M. Boothby, S. Joyce, D. P. Carbone and R. M. Breyer, *J. Clin. Invest.*, 2003, **111**, 727.

18. D. M. Aronoff, C. Canetti and M. Peters-Golden, *J. Immunol.*, 2004, **173**, 559.

19. T. Kunikata, H. Yamane, E. Segi, T. Matsuoka, Y. Sugimoto, S. Tanaka, H. Tanaka, H. Nagai, A. Ichikawa and S. Narumiya, *Nat. Immunol.*, 2005, **6**, 524.

20. H. Yamane, Y. Sugimoto, S. Tanaka and A. Ichikawa, *Biochem. Biophys. Res. Commun.*, 2000, **278**, 224.

21. A. Fennekohl, Y. Sugimoto, E. Segi, T. Maruyama, A. Ichikawa and G. P. Puschel, *J Hepatol*, 2002, **36**, 328.

22. K. Takeuchi, Y. Ogawa, S. Kagawa and H. Ukawa, *Aliment. Pharmacol. Ther.*, 2002, **16**(2), 74.

23. K. Takeuchi, S. Kato, M. Takeeda, Y. Ogawa, M. Nakashima and M. Matsumoto, *J. Pharmacol. Exp. Ther.*, 2003, **304**, 1055.

24. T. Kunikata, H. Araki, M. Takeeda, S. Kato and K. Takeuchi, *J. Physiol. Paris*, 2001, **95**, 157.

25. Y. Komoike, M. Nakashima, A. Nakagiri and K. Takeuchi, *Digestion*, 2003, **67**, 186.

26. H. Araki, H. Ukawa, Y. Sugawa, K. Yagi, K. Suzuki and K. Takeuchi, *Aliment Pharmacol. Ther.*, 2000, **14**(1), 116.

27. C. R. Kennedy, Y. Zhang, S. Brandon, Y. Guan, K. Coffee, C. D. Funk, M. A. Magnuson, J. A. Oates, M. D. Breyer and R. M. Breyer, *Nat. Med.*, 1999, **5**, 217.

28. S. L. Tilley, L. P. Audoly, E. H. Hicks, H. S. Kim, P. J. Flannery, T. M. Coffman and B. H. Koller, *J. Clin. Invest.*, 1999, **103**, 1539.

29. Y. Guan, Y. Zhang, J. Wu, Z. Qi, G. Yang, D. Dou, Y. Gao, L. Chen, X. Zhang, L. S. Davis, M. Wei, X. Fan, M. Carmosino, C. Hao, J. D. Imig, R. M. Breyer and M. D. Breyer, *J. Clin. Invest.*, 2007, **117**, 2496.

30. J. Chen, M. Zhao, W. He, G. L. Milne, J. R. Howard, J. Morrow, R. L. Hebert, R. M. Breyer and C. M. Hao, *Am. J. Physiol. Renal Physiol.*, 2008, **295**, F818.

31. B. Samuelsson, R. Morgenstern and P. J. Jakobsson, *Pharmacol. Rev.*, 2007, **59**, 207.

32. I. Kudo and M. Murakami, *J. Biochem. Mol. Biol.*, 2005, **38**, 633.

33. T. Tanioka, Y. Nakatani, N. Semmyo, M. Murakami and I. Kudo, *J. Biol. Chem.*, 2000, **275**, 32775.

34. M. Murakami, H. Naraba, T. Tanioka, N. Semmyo, Y. Nakatani, F. Kojima, T. Ikeda, M. Fueki, A. Ueno, S. Oh and I. Kudo, *J. Biol. Chem.*, 2000, **275**, 32783.

35. M. Murakami, K. Nakashima, D. Kamei, S. Masuda, Y. Ishikawa, T. Ishii, Y. Ohmiya, K. Watanabe and I. Kudo, *J. Biol. Chem.*, 2003, **278**, 37937.

36. P. J. Jakobsson, S. Thoren, R. Morgenstern and B. Samuelsson, *Proc. Natl Acad. Sci. USA*, 1999, **96**, 7220.

37. M. Ouellet, J. P. Falgueyret, P. H. Ear, A. Pen, J. A. Mancini, D. Riendeau and M. D. Percival, *Protein Expr. Purif.*, 2002, **26**, 489.

38. S. Thoren, R. Weinander, S. Saha, C. Jegerschold, P. L. Pettersson, B. Samuelsson, H. Hebert, M. Hamberg, R. Morgenstern and P. J. Jakobsson, *J. Biol. Chem.*, 2003, **278**, 22199.

39. C. Jegerschold, S. C. Pawelzik, P. Purhonen, P. Bhakat, K. R. Gheorghe, N. Gyobu, K. Mitsuoka, R. Morgenstern, P. J. Jakobsson and H. Hebert, *Proc. Natl Acad. Sci. USA*, 2008, **105**, 11110.

40. T. Yamada, J. Komoto, K. Watanabe, Y. Ohmiya and F. Takusagawa, *J. Mol. Biol.*, 2005, **348**, 1163.

41. T. Hammarberg, M. Hamberg, A. Wetterholm, H. Hansson, B. Samuelsson and J. Z. Haeggstrom, *J. Biol. Chem.*, 2008, **284**, 301.

42. S. C. Pawelzik, N. R. Uda, L. Spahiu, C. Jegerschold, P. Stenberg, H. Hebert, R. Morgenstern and P. J. Jakobsson, *J. Biol. Chem.*, 2010, **285**, 29254.

43. D. Martinez Molina, A. Wetterholm, A. Kohl, A. A. McCarthy, D. Niegowski, E. Ohlson, T. Hammarberg, S. Eshaghi, J. Z. Haeggstrom and P. Nordlund, *Nature*, 2007, **448**, 613.

44. H. Ago, Y. Kanaoka, D. Irikura, B. K. Lam, T. Shimamura, K. F. Austen and M. Miyano, *Nature*, 2007, **448**, 609.

45. S. He, Y. Wu, D. Yu and L. Lai, *Biochem. J.*, 2011, **440**, 13.

46. A. Hamza, M. Tong, M. D. M. AbdulHameed, J. J. Liu, A. C. Goren, H. H. Tai and C. G. Zhan, *J. Phys. Chem. B*, 2010, **114**, 5605.

47. M. Murakami, H. Naraba, T. Tanioka, N. Semmyo, Y. Nakatani, F. Kojima, T. Ikeda, M. Fueki, A. Ueno, S. Oh-ishi and I. Kudo, *J. Biol. Chem.*, 2000, **275**, 32783.

48. W. Uracz, D. Uracz, R. Olszanecki and R. J. Gryglewski, *J. Physiol. Pharmacol.*, 2002, **53**, 643.
49. S. Thoren and P. J. Jakobsson, *Eur. J. Biochem.*, 2000, **267**, 6428.
50. D. O. Stichtenoth, S. Thoren, H. Bian, M. Peters-Golden, P. J. Jakobsson and L. J. Crofford, *J. Immunol.*, 2001, **167**, 469.
51. M. Ek, D. Engblom, S. Saha, A. Blomqvist, P. J. Jakobsson and A. Ericsson-Dahlstrand, *Nature*, 2001, **410**, 430.
52. F. Kojima, H. Naraba, Y. Sasaki, R. Okamoto, T. Koshino and S. Kawai, *J. Rheumatol.*, 2002, **29**, 1836.
53. A. I. Ivanov, R. S. Pero, A. C. Scheck and A. A. Romanovsky, *Am. J. Physiol. Regul. Integr. Comp. Physiol.*, 2002, **283**, R1104.
54. Y. Shinji, T. Tsukui, A. Tatsuguchi, K. Shinoki, M. Kusunoki, K. Suzuki, T. Hiratsuka, K. Wada, S. Futagami, K. Miyake, K. Gudis and C. Sakamoto, *Am. J. Physiol. Gastrointest. Liver Physiol.*, 2005, **288**, G308.
55. Y. Ikeda-Matsuo, A. Ota, T. Fukada, S. Uematsu, S. Akira and Y. Sasaki, *Proc. Natl Acad. Sci. USA*, 2006, **103**, 11790.
56. D. Claveau, M. Sirinyan, J. Guay, R. Gordon, C. C. Chan, Y. Bureau, D. Riendeau and J. A. Mancini, *J. Immunol.*, 2003, **170**, 4738.
57. Y. Guan, Y. Zhang, A. Schneider, D. Riendeau, J. A. Mancini, L. Davis, M. Komhoff, R. M. Breyer and M. D. Breyer, *Am. J. Physiol. Renal Physiol.*, 2001, **281**, F1173.
58. A. L. Fuson, P. Komlosi, T. M. Unlap, P. D. Bell and J. Peti-Peterdi, *Am. J. Physiol. Renal Physiol.*, 2003, **285**, F558.
59. M. Lazarus, C. J. Munday, N. Eguchi, S. Matsumoto, G. J. Killian, B. K. Kubata and Y. Urade, *Endocrinology*, 2002, **143**, 2410.
60. N. Alfaidy, M. Sun, J. R. Challis and W. Gibb, *Endocrine*, 2003, **20**, 219.
61. L. Boulet, M. Ouellet, K. P. Bateman, D. Ethier, M. D. Percival, D. Riendeau, J. A. Mancini and N. Methot, *J. Biol. Chem.*, 2004, **279**, 23229.
62. J. A. Mancini, K. Blood, J. Guay, R. Gordon, D. Claveau, C. C. Chan and D. Riendeau, *J. Biol. Chem.*, 2001, **276**, 4469.
63. K. Yamagata, K. Matsumura, W. Inoue, T. Shiraki, K. Suzuki, S. Yasuda, H. Sugiura, C. Cao, Y. Watanabe and S. Kobayashi, *J. Neurosci.*, 2001, **21**, 2669.
64. D. Engblom, M. Ek, I. M. Andersson, S. Saha, M. Dahlstrom, P. J. Jakobsson, A. Ericsson-Dahlstrand and A. Blomqvist, *J. Comp. Neurol.*, 2002, **452**, 205.
65. J. Guay, K. Bateman, R. Gordon, J. Mancini and D. Riendeau, *J. Biol. Chem.*, 2004, **279**, 24866.
66. A. H. Moore, J. A. Olschowka and M. K. O'Banion, *J. Neuroimmunol.*, 2004, **148**, 32.
67. Y. Ozaki-Okayama, K. Matsumura, T. Ibuki, M. Ueda, Y. Yamazaki, Y. Tanaka and S. Kobayashi, *Crit. Care Med.*, 2004, **32**, 795.
68. R. Schuligoi, R. Ulcar, B. A. Peskar and R. Amann, *Neuroscience*, 2003, **116**, 1043.
69. N. P. Turrin and S. Rivest, *Neurobiol. Dis.*, 2004, **16**, 321.

70. M. Westman, M. Korotkova, E. af Klint, A. Stark, L. P. Audoly, L. Klareskog, A. K. Ulfgren and P. J. Jakobsson, *Arthritis Rheum.*, 2004, **50**, 1774.
71. F. Kojima, H. Naraba, S. Miyamoto, M. Beppu, H. Aoki and S. Kawai, *Arthritis Res. Ther.*, 2004, **6**, R355.
72. K. Masuko-Hongo, F. Berenbaum, L. Humbert, C. Salvat, M. B. Goldring and S. Thirion, *Arthritis Rheum.*, 2004, **50**, 2829.
73. X. Li, H. Afif, S. Cheng, J. Martel-Pelletier, J. P. Pelletier, P. Ranger and H. Fahmi, *J. Rheumatol.*, 2005, **32**, 887.
74. K. Subbaramaiah, K. Yoshimatsu, E. Scherl, K. M. Das, K. D. Glazier, D. Golijanin, R. A. Soslow, T. Tanabe, H. Naraba and A. J. Dannenberg, *J. Biol. Chem.*, 2004, **279**, 12647.
75. F. Cipollone, C. Prontera, B. Pini, M. Marini, M. Fazia, D. De Cesare, A. Iezzi, S. Ucchino, G. Boccoli, V. Saba, F. Chiarelli, F. Cuccurullo and A. Mezzetti, *Circulation*, 2001, **104**, 921.
76. A. Gomez-Hernandez, J. L. Martin-Ventura, E. Sanchez-Galan, C. Vidal, M. Ortego, L. M. Blanco-Colio, L. Ortega, J. Tunon and J. Egido, *Atherosclerosis*, 2006, **187**, 139.
77. F. Cipollone, A. Iezzi, M. Fazia, M. Zucchelli, B. Pini, C. Cuccurullo, D. De Cesare, G. De Blasis, R. Muraro, R. Bei, F. Chiarelli, A. M. Schmidt, F. Cuccurullo and A. Mezzetti, *Circulation*, 2003, **108**, 1070.
78. U. A. Chaudhry, H. Zhuang, B. J. Crain and S. Dore, *Alzheimers Dement.*, 2008, **4**, 6.
79. D. Wang and R. N. Dubois, *Gut*, 2006, **55**, 115.
80. K. Yoshimatsu, D. Golijanin, P. B. Paty, R. A. Soslow, P. J. Jakobsson, R. A. DeLellis, K. Subbaramaiah and A. J. Dannenberg, *Clin. Cancer Res.*, 2001, **7**, 3971.
81. B. H. von Rahden, B. L. Brucher, C. Langner, J. R. Siewert, H. J. Stein and M. Sarbia, *Br. J. Surg.*, 2006, **93**, 1424.
82. B. P. van Rees, A. Sivula, S. Thoren, H. Yokozaki, P. J. Jakobsson, G. J. Offerhaus and A. Ristimaki, *Int. J. Cancer*, 2003, **107**, 551.
83. T. J. Jang, *Virchows Arch.*, 2004, **445**, 564.
84. K. Gudis, A. Tatsuguchi, K. Wada, T. Hiratsuka, S. Futagami, Y. Fukuda, T. Kiyama, T. Tajiri, K. Miyake and C. Sakamoto, *Hum. Pathol.*, 2007, **38**, 1826.
85. T. Soma, Y. Shimada, A. Kawabe, J. Kaganoi, K. Kondo, M. Imamura and S. Uemoto, *Dis. Esophagus*, 2007, **20**, 123.
86. B. H. von Rahden, H. J. Stein, S. A. Hartl, J. Theisen, B. Stigler, J. R. Siewert and M. Sarbia, *Dis. Esophagus*, 2008, **21**, 304.
87. R. Kawata, S. Hyo, T. Maeda, Y. Urade and H. Takenaka, *Acta Otolaryngol.*, 2006, **126**, 627.
88. K. Yoshimatsu, N. K. Altorki, D. Golijanin, F. Zhang, P. J. Jakobsson, A. J. Dannenberg and K. Subbaramaiah, *Clin. Cancer Res.*, 2001, **7**, 2669.
89. Y. Takii, S. Abiru, H. Fujioka, M. Nakamura, A. Komori, M. Ito, K. Taniguchi, M. Daikoku, Y. Meda, K. Ohata, K. Yano, S. Shimoda, H. Yatsuhashi, H. Ishibashi and K. Migita, *Liver Int.*, 2007, **27**, 989.

90. M. Breinig, R. Rieker, E. Eiteneuer, T. Wertenbruch, A. M. Haugg, B. M. Helmke, P. Schirmacher and M. A. Kern, *Int. J. Cancer*, 2008, **122**, 547.

91. S. Hasan, M. Satake, D. W. Dawson, H. Funahashi, E. Angst, V. L. Go, H. A. Reber, O. J. Hines and G. Eibl, *Pancreas*, 2008, **37**, 121.

92. S. Mehrotra, A. Morimiya, B. Agarwal, R. Konger and S. Badve, *J. Pathol.*, 2006, **208**, 356.

93. K. Rask, Y. Zhu, W. Wang, L. Hedin and K. Sundfeldt, *Mol. Cancer*, 2006, **5**, 62.

94. E. G. Cohen, T. Almahmeed, B. Du, D. Golijanin, J. O. Boyle, R. A. Soslow, K. Subbaramaiah and A. J. Dannenberg, *Clin. Cancer Res.*, 2003, **9**, 3425.

95. D. Golijanin, J. Y. Tan, A. Kazior, E. G. Cohen, P. Russo, G. Dalbagni, K. J. Auborn, K. Subbaramaiah and A. J. Dannenberg, *Clin. Cancer Res.*, 2004, **10**, 1024.

96. T. Payner, H. A. Leaver, B. Knapp, I. R. Whittle, O. C. Trifan, S. Miller and M. T. Rizzo, *Mol. Cancer Ther.*, 2006, **5**, 1817.

97. L. Lalier, P. F. Cartron, F. Pedelaborde, C. Olivier, D. Loussouarn, S. A. Martin, K. Meflah, J. Menanteau and F. M. Vallette, *Oncogene*, 2007, **26**, 4999.

98. N. Baryawno, B. Sveinbjornsson, S. Eksborg, A. Orrego, L. Segerstrom, C. O. Oqvist, S. Holm, B. Gustavsson, B. Kagedal, P. Kogner and J. I. Johnsen, *Neuro. Oncol.*, 2008, **10**, 661.

99. D. Kamei, K. Yamakawa, Y. Takegoshi, M. Mikami-Nakanishi, Y. Nakatani, S. Oh-Ishi, H. Yasui, Y. Azuma, N. Hirasawa, K. Ohuchi, H. Kawaguchi, Y. Ishikawa, T. Ishii, S. Uematsu, S. Akira, M. Murakami and I. Kudo, *J. Biol. Chem.*, 2004, **279**, 33684.

100. M. Kapoor, F. Kojima, M. Qian, L. Yang and L. J. Crofford, *Faseb J.*, 2006, **20**, 2387.

101. Y. Cheng, M. Wang, M. Yu, J. Lawson, C. D. Funk and G. A. Fitzgerald, *J. Clin. Invest.*, 2006, **116**, 1391.

102. M. Wang, A. M. Zukas, Y. Hui, E. Ricciotti, E. Pure and G. A. FitzGerald, *Proc. Natl Acad. Sci. USA*, 2006, **103**, 14507.

103. C. E. Trebino, J. D. Eskra, T. S. Wachtmann, J. R. Perez, T. J. Carty and L. P. Audoly, *J. Biol. Chem.*, 2005, **280**, 16579.

104. S. U. Monrad, F. Kojima, M. Kapoor, E. L. Kuan, S. Sarkar, G. J. Randolph and L. J. Crofford, *Prostaglandins Leukot. Essent. Fatty Acids*, 2011, **84**, 113.

105. J. A. Oates, G. A. FitzGerald, R. A. Branch, E. K. Jackson, H. R. Knapp and L. J. Roberts, 2nd, *N. Engl. J. Med.*, 1988, **319**, 761.

106. S. Castellani, B. Paladini, R. Paniccia, C. Di Serio, B. Vallotti, A. Ungar, S. Fumagalli, C. Cantini, L. Poggesi and G. G. Neri Serneri, *Am. Heart J.*, 1997, **133**, 94.

107. A. N. Hata and R. M. Breyer, *Pharmacol. Ther.*, 2004, **103**, 147.

108. A. Koeberle, U. Siemoneit, U. Buhring, H. Northoff, S. Laufer, W. Albrecht and O. Werz, *J. Pharmacol. Exp. Ther.*, 2008, **326**, 975.

109. D. Riendeau, R. Aspiotis, D. Ethier, Y. Gareau, E. L. Grimm, J. Guay, S. Guiral, H. Juteau, J. A. Mancini, N. Methot, J. Rubin and R. W. Friesen, *Bioorg. Med. Chem. Lett.*, 2005, **15**, 3352.
110. D. Xu, S. E. Rowland, P. Clark, A. Giroux, B. Cote, S. Guiral, M. Salem, Y. Ducharme, R. W. Friesen, N. Methot, J. Mancini, L. Audoly and D. Riendeau, *J. Pharmacol. Exp. Ther.*, 2008, **326**, 754.
111. A. Koeberle, F. Pollastro, H. Northoff and O. Werz, *Br. J. Pharmacol.*, 2009, in press.
112. C. E. Trebino, J. L. Stock, C. P. Gibbons, B. M. Naiman, T. S. Wachtmann, J. P. Umland, K. Pandher, J. M. Lapointe, S. Saha, M. L. Roach, D. Carter, N. A. Thomas, B. A. Durtschi, J. D. McNeish, J. E. Hambor, P. J. Jakobsson, T. J. Carty, J. R. Perez and L. P. Audoly, *Proc. Natl Acad. Sci. USA*, 2003, **100**, 9044.
113. S. Hara, D. Kamei, Y. Sasaki, A. Tanemoto, Y. Nakatani and M. Murakami, *Biochimie*, 2010, **92**, 651.
114. M. K. O'Banion, *Prostaglandins Other Lipid Mediat.*, 2010, **91**, 113.
115. O. Radmark and B. Samuelsson, *J. Intern. Med.*, 2010, **268**, 5.
116. M. Nakanishi, V. Gokhale, E. J. Meuillet and D. W. Rosenberg, *Biochimie*, 2010, **92**, 660.
117. F. Kojima, M. Kapoor, L. Yang, E. L. Fleishaker, M. R. Ward, S. U. Monrad, P. C. Kottangada, C. Q. Pace, J. A. Clark, J. G. Woodward and L. J. Crofford, *J. Immunol.*, 2008, **180**, 8361.
118. C. Brenneis, O. Coste, R. Schmidt, C. Angioni, L. Popp, R. M. Nusing, W. Becker, K. Scholich and G. Geisslinger, *J. Cell Mol. Med.*, 2008, **12**, 639.
119. T. Mabuchi, H. Kojima, T. Abe, K. Takagi, M. Sakurai, Y. Ohmiya, S. Uematsu, S. Akira, K. Watanabe and S. Ito, *Neuroreport*, 2004, **15**, 1395.
120. D. Engblom, S. Saha, L. Engstrom, M. Westman, L. P. Audoly, P. J. Jakobsson and A. Blomqvist, *Nat. Neurosci.*, 2003, **6**, 1137.
121. S. Saha, L. Engstrom, L. Mackerlova, P. J. Jakobsson and A. Blomqvist, *Am. J. Physiol. Regul. Integr. Comp. Physiol.*, 2005, **288**, R1100.
122. Y. Kihara, T. Matsushita, Y. Kita, S. Uematsu, S. Akira, J. Kira, S. Ishii and T. Shimizu, *Proc. Natl Acad. Sci. USA*, 2009, **106**, 21807.
123. T. Takemiya, K. Matsumura, H. Sugiura, M. Maehara, S. Yasuda, S. Uematsu, S. Akira and K. Yamagata, *J. Neurosci. Res.*, 2010, **88**, 381.
124. C. Brenneis, O. Coste, K. Altenrath, C. Angioni, H. Schmidt, C. D. Schuh, D. D. Zhang, M. Henke, A. Weigert, B. Brune, B. Rubin, R. Nusing, K. Scholich and G. Geisslinger, *J. Biol. Chem.*, 2011, **286**, 2331.
125. D. Lu, C. Han and T. Wu, *Oncogene*, 2011, **31**, 842.
126. D. Lu, C. Han and T. Wu, *Gastroenterology*, 2011, **140**, 2084.
127. M. Nakanishi, D. C. Montrose, P. Clark, P. R. Nambiar, G. S. Belinsky, K. P. Claffey, D. Xu and D. W. Rosenberg, *Cancer Res.*, 2008, **68**, 3251.
128. Y. Sasaki, D. Kamei, Y. Ishikawa, T. Ishii, S. Uematsu, S. Akira, M. Murakami and S. Hara, *Oncogene*, 2011, in press.
129. N. Elander, J. Ungerback, H. Olsson, S. Uematsu, S. Akira and P. Soderkvist, *Biochem. Biophys. Res. Commun.*, 2008, **372**, 249.

130. A. Numao, K. Hosono, T. Suzuki, I. Hayashi, S. Uematsu, S. Akira, Y. Ogino, H. Kawauchi, N. Unno and M. Majima, *Biomed. Pharmacother.*, 2011, **65**, 77.

131. V. Campean, F. Theilig, A. Paliege, M. Breyer and S. Bachmann, *Am. J. Physiol. Renal Physiol.*, 2003, **285**, F19.

132. H. Vitzthum, I. Abt, S. Einhellig and A. Kurtz, *Kidney Int.*, 2002, **62**, 1570.

133. Z. Jia, A. Zhang, H. Zhang, Z. Dong and T. Yang, *Circ. Res.*, 2006, **99**, 1243.

134. Z. Jia, X. Guo, H. Zhang, M. H. Wang, Z. Dong and T. Yang, *Hypertension*, 2008, **52**, 952.

135. D. J. Zhang, L. H. Chen, Y. H. Zhang, G. R. Yang, D. Dou, Y. S. Gao, X. Y. Zhang, X. M. Kong, P. Zhao, D. Pu, M. F. Wei, M. D. Breyer and Y. F. Guan, *Acta Pharmacol. Sin.*, 2010, **31**, 1284.

136. Z. Jia, T. Aoyagi, D. E. Kohan and T. Yang, *Am. J. Physiol. Renal Physiol.*, 2010, **299**, F155.

137. Z. Jia, T. Aoyagi and T. Yang, *Hypertension*, 2010, **55**, 539.

138. N. Degousee, S. Fazel, D. Angoulvant, E. Stefanski, S. C. Pawelzik, M. Korotkova, S. Arab, P. Liu, T. F. Lindsay, S. Zhuo, J. Butany, R. K. Li, L. Audoly, R. Schmidt, C. Angioni, G. Geisslinger, P. J. Jakobsson and B. B. Rubin, *Circulation*, 2008, **117**, 1701.

139. Z. Jia, N. Wang, T. Aoyagi, H. Wang, H. Liu and T. Yang, *Kidney Int.*, 2011, **79**, 77.

140. Z. Jia, H. Wang and T. Yang, *Hypertension*, 2012, **59**, 122.

141. O. Quraishi, J. A. Mancini and D. Riendeau, *Biochem. Pharmacol.*, 2002, **63**, 1183.

142. X. Song, H. P. Lin, A. J. Johnson, P. H. Tseng, Y. T. Yang, S. K. Kulp and C. S. Chen, *J. Natl Cancer Inst.*, 2002, **94**, 585.

143. I. Wobst, S. Schiffmann, K. Birod, T. J. Maier, R. Schmidt, C. Angioni, G. Geisslinger and S. Grosch, *Biochem. Pharmacol.*, 2008, **76**, 62.

144. K. Deckmann, F. Rorsch, R. Steri, M. Schubert-Zsilavecz, G. Geisslinger and S. Grosch, *Biochem. Pharmacol.*, 2010, **80**, 1365.

145. K. Deckmann, F. Rorsch, G. Geisslinger and S. Grosch, *Cell Signal.*, 2012, **24**, 460.

146. J. A. Mancini, M. Abramovitz, M. E. Cox, E. Wong, S. Charleson, H. Perrier, Z. Wang, P. Prasit and P. J. Vickers, *FEBS Lett.*, 1993, **318**, 277.

147. J. Gillard, A. W. Ford-Hutchinson, C. Chan, S. Charleson, D. Denis, A. Foster, R. Fortin, S. Leger, C. S. McFarlane, H. Morton, H. Piechuta, D. Riendeau, C. A. Rouzer, J. Rokach, R. Young, D. E. MacIntyre, L. Peterson, T. Bach, G. Eiermann, S. Hopple, J. Humes, L. Hupe, S. Luell, J. Metzger, R. Meurer, D. K. Miller, E. Opas and S. Pacholok, *Can. J. Physiol. Pharmacol.*, 1989, **67**, 456.

148. A. A. San Juan and S. J. Cho, *J. Mol. Model.*, 2007, **13**, 601.

149. M. D. AbdulHameed, A. Hamza, J. Liu, X. Huang and C. G. Zhan, *J. Chem. Inf. Model.*, 2008, **48**, 179.

150. X. Huang, W. Yan, D. Gao, M. Tong, H. H. Tai and C. G. Zhan, *Bioorg. Med. Chem.*, 2006, **14**, 3553.

151. J. E. Thompson, R. M. Cubbon, R. T. Cummings, L. S. Wicker, R. Frankshun, B. R. Cunningham, P. M. Cameron, P. T. Meinke, N. Liverton, Y. Weng and J. A. DeMartino, *Bioorg. Med. Chem. Lett.*, 2002, **12**, 1219.

152. B. Cote, L. Boulet, C. Brideau, D. Claveau, D. Ethier, R. Frenette, M. Gagnon, A. Giroux, J. Guay, S. Guiral, J. Mancini, E. Martins, F. Masse, N. Methot, D. Riendeau, J. Rubin, D. Xu, H. Yu, Y. Ducharme and R. W. Friesen, *Bioorg. Med. Chem. Lett.*, 2007, **17**, 6816.

153. P. O. Hetu, M. Ouellet, J. P. Falgueyret, C. Ramachandran, J. Robichaud, R. Zamboni and D. Riendeau, *Arch. Biochem. Biophys.*, 2008, **477**, 155.

154. E. B. Prage, S. C. Pawelzik, L. S. Busenlehner, K. Kim, R. Morgenstern, P. J. Jakobsson and R. N. Armstrong, *Biochemistry*, 2011, **50**, 7684.

155. S. He and L. Lai, *J. Chem. Inf. Model.*, 2011, **51**, 3254.

156. D. Xu, S. E. Rowland, P. Clark, A. Giroux, B. Cote, S. Guiral, M. Salem, Y. Ducharme, R. W. Friesen, N. Methot, J. Mancini, L. Audoly and D. Riendeau, *J. Pharmacol. Exp. Ther.*, 2008, **326**, 754.

157. B. Baragatti, F. Scebba, D. Sodini, E. Pagni, E. Ciofini, D. Xu and F. Coceani, *Neonatology*, 2011, **100**, 139.

158. A. Giroux, L. Boulet, C. Brideau, A. Chau, D. Claveau, B. Cote, D. Ethier, R. Frenette, M. Gagnon, J. Guay, S. Guiral, J. Mancini, E. Martins, F. Masse, N. Methot, D. Riendeau, J. Rubin, D. Xu, H. Yu, Y. Ducharme and R. W. Friesen, *Bioorg. Med. Chem. Lett.*, 2009, **19**, 5837.

159. T. Shiro, H. Takahashi, K. Kakiguchi, Y. Inoue, K. Masuda, H. Nagata and M. Tobe, *Bioorg. Med. Chem. Lett.*, 2012, **22**, 285.

160. R. De Simone, M. G. Chini, I. Bruno, R. Riccio, D. Mueller, O. Werz and G. Bifulco, *J. Med. Chem.*, 2011, **54**, 1565.

161. F. Rorsch, I. Wobst, H. Zettl, M. Schubert-Zsilavecz, S. Grosch, G. Geisslinger, G. Schneider and E. Proschak, *J. Med. Chem.*, 2010, **53**, 911.

162. A. Hamza, X. Zhao, M. Tong, H. H. Tai and C. G. Zhan, *Bioorg. Med. Chem.*, 2011, **19**, 6077.

163. T. Y. Wu, H. Juteau, Y. Ducharme, R. W. Friesen, S. Guiral, L. Dufresne, H. Poirier, M. Salem, D. Riendeau, J. Mancini and C. Brideau, *Bioorg. Med. Chem. Lett.*, 2010, **20**, 6978.

164. G. Mbalaviele, A. M. Pauley, A. F. Shaffer, B. S. Zweifel, S. Mathialagan, S. J. Mnich, O. V. Nemirovskiy, J. Carter, J. K. Gierse, J. L. Wang, M. L. Vazquez, W. M. Moore and J. L. Masferrer, *Biochem. Pharmacol.*, 2010, **79**, 1445.

165. J. Wang, D. Limburg, J. Carter, G. Mbalaviele, J. Gierse and M. Vazquez, *Bioorg. Med. Chem. Lett.*, 2010, **20**, 1604.

166. T. Shimizu, *Annu. Rev. Pharmacol. Toxicol.*, 2009, **49**, 123.

167. O. Werz and D. Steinhilber, *Pharmacol. Ther.*, 2006, **112**, 701.

168. I. Avis, S. H. Hong, A. Martinez, T. Moody, Y. H. Choi, J. Trepel, R. Das, M. Jett and J. L. Mulshine, *FASEB J.*, 2001, **15**, 2007.

169. M. Romano, A. Catalano, M. Nutini, E. D'Urbano, C. Crescenzi, J. Claria, R. Libner, G. Davi and A. Procopio, *FASEB J.*, 2001, **15**, 2326.

170. J. Ghosh and C. E. Myers, *Proc. Natl Acad. Sci. USA*, 1998, **95**, 13182.

171. M. Mehrabian and H. Allayee, *Curr. Opin. Lipidol.*, 2003, **14**, 447.

172. C. D. Funk, *Nat. Rev. Drug Discov.*, 2005, **4**, 664.

173. F. Celotti and S. Laufer, *Pharmacol. Res.*, 2001, **43**, 429.

174. B. K. Lam, J. F. Penrose, J. Rokach, K. Xu, M. H. Baldasaro and K. F. Austen, *Eur. J. Biochem.*, 1996, **238**, 606.

175. A. Koeberle, U. Siemoneit, H. Northoff, B. Hofmann, G. Schneider and O. Werz, *Eur. J. Pharmacol.*, 2009, **608**, 84.

176. A. Koeberle, U. Siemoneit, U. Buehring, H. Northoff, S. Laufer, W. Albrecht and O. Werz, *J. Pharmacol. Exp. Ther.*, 2008, **326**, 975.

177. T. Bage, T. Modeer, T. Kawakami, H. C. Quezada and T. Yucel-Lindberg, *Biochim. Biophys. Acta*, 2007, **1773**, 1589.

178. Y. Li, S. Yin, D. Nie, S. Xie, L. Ma, X. Wang, Y. Wu and J. Xiao, *Int. J. Hematol.*, 2011, **94**, 472.

179. L. Fischer, M. Hornig, C. Pergola, N. Meindl, L. Franke, Y. Tanrikulu, G. Dodt, G. Schneider, D. Steinhilber and O. Werz, *Br. J. Pharmacol.*, 2007, **152**, 471.

180. S. Tries, W. Neupert and S. Laufer, *Inflamm. Res.*, 2002, **51**, 135.

181. S. A. Laufer, J. Augustin, G. Dannhardt and W. Kiefer, *J. Med. Chem.*, 1994, **37**, 1894.

182. S. Laufer, S. Tries, J. Augustin and G. Dannhardt, *Arzneimittelforschung*, 1994, **44**, 629.

183. S. K. Kulkarni and V. P. Singh, *Curr. Top. Med. Chem.*, 2007, **7**, 251.

184. A. J. Liedtke, P. R. Keck, F. Lehmann, A. Koeberle, O. Werz and S. A. Laufer, *J. Med. Chem.*, 2009, **52**, 4968.

185. A. Koeberle, E. M. Haberl, A. Rossi, C. Pergola, F. Dehm, H. Northoff, R. Troschuetz, L. Sautebin and O. Werz, *Bioorg. Med. Chem.*, 2009, **17**, 7924.

186. E. M. Karg, S. Luderer, C. Pergola, U. Buhring, A. Rossi, H. Northoff, L. Sautebin, R. Troschuetz and O. Werz, *J. Med. Chem.*, 2009, in press.

187. A. Koeberle, A. Rossi, H. Zettl, C. Pergola, F. Dehm, J. Bauer, C. Greiner, S. Reckel, C. Hoernig, H. Northoff, F. Bernhard, V. Dotsch, L. Sautebin, M. Schubert-Zsilavecz and O. Werz, *J. Pharmacol. Exp. Ther.*, 2010, **332**, 840.

188. M. Gabler and M. Schubert-Zsilavecz, *Molecules*, 2011, **16**, 10013.

189. M. Hieke, C. Greiner, T. M. Thieme, M. Schubert-Zsilavecz, O. Werz and H. Zettl, *Bioorg. Med. Chem. Lett.*, 2011, **21**, 1329.

190. M. Hieke, C. Greiner, M. Dittrich, F. Reisen, G. Schneider, M. Schubert-Zsilavecz and O. Werz, *J. Med. Chem.*, 2011, **54**, 4490.

191. C. Greiner, H. Zettl, A. Koeberle, C. Pergola, H. Northoff, M. Schubert-Zsilavecz and O. Werz, *Bioorg. Med. Chem.*, 2011, **19**, 3394.

192. A. Koeberle, F. Pollastro, H. Northoff and O. Werz, *Br. J. Pharmacol.*, 2009, **156**, 952.

193. A. Koeberle, A. Rossi, J. Bauer, F. Dehm, L. Verotta, H. Northoff, L. Sautebin and O. Werz, *Front. Pharmacol.*, 2011, **2**, 7.

194. A. Koeberle, H. Northoff and O. Werz, *Biochem. Pharmacol.*, 2009, **77**, 1513.
195. J. Bauer, A. Koeberle, F. Dehm, F. Pollastro, G. Appendino, H. Northoff, A. Rossi, L. Sautebin and O. Werz, *Biochem. Pharmacol.*, 2011, **81**, 259.
196. C. Feisst, L. Franke, G. Appendino and O. Werz, *J. Pharmacol. Exp. Ther.*, 2005, **315**, 389.
197. D. Albert, I. Zundorf, T. Dingermann, W. E. Muller, D. Steinhilber and O. Werz, *Biochem. Pharmacol.*, 2002, **64**, 1767.
198. S. Sosa, R. Pace, A. Bornancin, P. Morazzoni, A. Riva, A. Tubaro and R. Della Loggia, *J Pharm. Pharmacol.*, 2007, **59**, 703.
199. A. Rossi, R. Di Paola, E. Mazzon, T. Genovese, R. Caminiti, P. Bramanti, C. Pergola, A. Koeberle, O. Werz, L. Sautebin and S. Cuzzocrea, *J. Pharmacol. Exp. Ther.*, 2008, in press.
200. A. Koeberle, H. Northoff and O. Werz, *Mol. Cancer Ther.*, 2009, **8**, 2348.
201. U. Siemoneit, A. Koeberle, A. Rossi, F. Dehm, M. Verhoff, S. Reckel, T. J. Maier, J. Jauch, H. Northoff, F. Bernhard, V. Doetsch, L. Sautebin and O. Werz, *Br. J. Pharmacol.*, 2011, **162**, 147.
202. J. Hong, M. Bose, J. Ju, J. H. Ryu, X. Chen, S. Sang, M. J. Lee and C. S. Yang, *Carcinogenesis*, 2004, **25**, 1671.
203. H. Safayhi, T. Mack, J. Sabieraj, M. I. Anazodo, L. R. Subramanian and H. P. Ammon, *J. Pharmacol. Exp. Ther.*, 1992, **261**, 1143.
204. U. Siemoneit, C. Pergola, B. Jazzar, H. Northoff, C. Skarke, J. Jauch and O. Werz, *Eur. J. Pharmacol.*, 2009, **606**, 246.
205. B. B. Aggarwal, A. B. Kunnumakkara, K. B. Harikumar, S. T. Tharakan, B. Sung and P. Anand, *Planta Med.*, 2008, **74**, 1560.
206. H. Hatcher, R. Planalp, J. Cho, F. M. Torti and S. V. Torti, *Cell Mol. Life Sci.*, 2008, **65**, 1631.
207. D. Poeckel and O. Werz, *Curr. Med. Chem.*, 2006, **13**, 3359.
208. B. Waltenberger, K. Wiechmann, J. Bauer, P. Markt, S. M. Noha, G. Wolber, J. M. Rollinger, O. Werz, D. Schuster and H. Stuppner, *J. Med. Chem.*, 2011, **54**, 3163.
209. A. C. de Oliveira, E. Candelario-Jalil, H. S. Bhatia, K. Lieb, M. Hull and B. L. Fiebich, *Glia*, 2008, **56**, 844.
210. M. D. Guerrero, M. Aquino, I. Bruno, M. C. Terencio, M. Paya, R. Riccio and L. Gomez-Paloma, *J. Med. Chem.*, 2007, **50**, 2176.
211. M. Aquino, M. D. Guerrero, I. Bruno, M. C. Terencio, M. Paya and R. Riccio, *Bioorg. Med. Chem.*, 2008, **16**, 9056.
212. M. D. Guerrero, M. Aquino, I. Bruno, R. Riccio, M. C. Terencio and M. Paya, *Eur. J. Pharmacol.*, 2009, **620**, 112.
213. R. De Simone, R. M. Andres, M. Aquino, I. Bruno, M. D. Guerrero, M. C. Terencio, M. Paya and R. Riccio, *Chem. Biol. Drug Des.*, 2010, **76**, 17.
214. L. F. Silva Bastos, A. C. Pinheiro de Oliveira, J. C. Magnus Schlachetzki and B. L. Fiebich, *Immunopharmacol. Immunotoxicol.*, 2011, **33**, 576.

215. C. H. Tseng, C. C. Tzeng, P. K. Shih, C. N. Yang, Y. C. Chuang, S. I. Peng, C. S. Lin, J. P. Wang, C. M. Cheng and Y. L. Chen, *Mol. Divers.*, 2011, in press.

216. E. Terzuoli, S. Donnini, A. Giachetti, M. A. Iniguez, M. Fresno, G. Melillo and M. Ziche, *Clin. Cancer Res.*, 2010, **16**, 4207.

217. E. Candelario-Jalil, A. C. de Oliveira, S. Graf, H. S. Bhatia, M. Hull, E. Munoz and B. L. Fiebich, *J. Neuroinflammation*, 2007, **4**, 25.

218. V. Ulivi, R. Cancedda and F. D. Cancedda, *J. Cell Physiol.*, 2008, **217**, 433.

219. A. Bernardo, M. A. Ajmone-Cat, G. Levi and L. Minghetti, *J. Neurochem.*, 2003, **87**, 742.

220. J. J. Heynekamp, W. M. Weber, L. A. Hunsaker, A. M. Gonzales, R. A. Orlando, L. M. Deck and D. L. Jagt, *J. Med. Chem.*, 2006, **49**, 7182.

221. S. Das and D. K. Das, *Inflamm. Allergy Drug Targets*, 2007, **6**, 168.

222. K. Subbaramaiah, D. T. Lin, J. C. Hart and A. J. Dannenberg, *J. Biol. Chem.*, 2001, **276**, 12440.

223. H. Inoue, T. Tanabe and K. Umesono, *J. Biol. Chem.*, 2000, **275**, 28028.

224. H. Fahmi, J. P. Pelletier, F. Mineau and J. Martel-Pelletier, *Osteoarthritis Cartilage*, 2002, **10**, 845.

225. K. Farrajota, S. Cheng, J. Martel-Pelletier, H. Afif, J. P. Pelletier, X. Li, P. Ranger and H. Fahmi, *Arthritis Rheum.*, 2005, **52**, 94.

226. E. B. Berry, J. A. Keelan, R. J. Helliwell, R. S. Gilmour and M. D. Mitchell, *Mol. Pharmacol.*, 2005, **68**, 169.

CHAPTER 2

Inhibitors of Cytosolic Phospholipase A₂α as Anti-inflammatory Drugs

MATTHIAS LEHR

Institute of Pharmaceutical and Medicinal Chemistry,
University of Münster, Hittorfstrasse 58–62, 48149 Münster, Germany
Email: lehrm@uni-muenster.de

2.1   Classification of the Phospholipase A₂ Enzymes

Phospholipase A_2 (PLA$_2$) constitutes a large and diverse family of enzymes that catalyze the hydrolysis of membrane phospholipids at the *sn*-2 position to generate free fatty acids and lysophospholipids.[1–7] When the liberated fatty acid is arachidonic acid, subsequent metabolism through the cyclooxygenase (COX) and the lipoxygenase (LOX) pathways leads to the formation of eicosanoids, including prostaglandins and leukotrienes. A subset of lyso-phospholipids (1-*O*-alkyl-substituted choline glycerophospholipids) can be acetylated to the platelet-activating factor (PAF). Prostaglandins, leukotrienes, lysophospholipids and the PAF are potent mediators of inflammation.[8–11] Thus, inhibition of PLA$_2$ is considered as an interesting target for the design of new anti-inflammatory drugs.[12–20] The special attraction of this approach is based on the evidence that, unlike cyclooxygenase inhibitors, inhibitors of PLA$_2$ not only reduce the formation of prostaglandins, but also suppress the generation of leukotrienes, lysophospholipids and the PAF. Therefore, it was expected that inhibitors of PLA$_2$ have improved therapeutic activities in

RSC Drug Discovery Series No. 26
Anti-Inflammatory Drug Discovery
Edited by Jeremy I Levin and Stefan Laufer
© The Royal Society of Chemistry 2012
Published by the Royal Society of Chemistry, www.rsc.org

comparison to the cyclooxygenase inhibitors therapeutically applied today, such as aspirin, indomethacin or celecoxib.

One problem associated with the development of anti-inflammatory PLA_2 inhibitors is the fact that a plurality of different PLA_2 enzymes are present in the organism.[1-7] A current classification scheme divides the PLA_2s into 12 groups (I–XII) with different subgroups based on their structures, enzymatic characteristics and subcellular distribution. Alternatively, the PLA_2s can be ordered into such enzymes that utilize a catalytic histidine and in enzymes that have a catalytic serine. The small secretory PLA_2s ($sPLA_2$), which require millimolar Ca^{2+} for the phospholipid cleavage, belong to the first category. In humans meanwhile 11 distinct $sPLA_2$s have been identified (groups IB, IIA, IIC, IID, IIE, IIF, III, V, X, XIIA and XIIB).[6] The serine branch of the PLA_2 family composes larger enzymes that do not contain a Ca^{2+} in the active site. Members of this are the cytosolic PLA_2s ($cPLA_2$), the calcium-independent PLA_2s ($iPLA_2$) and the PAF-acetylhydrolases. The $cPLA_2$s comprise six enzymes, group IVA – also referred as $cPLA_2\alpha$ – characterized first, and groups IVB, IVC, IVD, IVE and IVF ($cPLA_2\beta$, γ, δ, ε, ζ) described more recently. The PAF-acetylhydrolase family consists of four isoenzymes (VIIA, VIIB, VIIIA and VIIIB), while the $iPLA_2$ enzyme class has nine members,[6] from which group VIA $iPLA_2$ is the best investigated.

The development of PLA_2 inhibitors as anti-inflammatory drugs started in the 1980s. For the first years the focus was directed on the low-molecular-weight $sPLA_2$, especially on group IIA $sPLA_2$s isolated from synovial fluid of patients with rheumatoid arthritis, and later on group V $sPLA_2$. However, the role of these enzymes in eicosanoid and PAF formation remained unclear and potent inhibitors of the enzymes developed did not show clear effects on arachidonic acid metabolism during inflammatory processes. The presence of soluble and membrane $sPLA_2$ binding proteins in the organism suggests that mammalian $sPLA_2$s not only have PLA_2 activity but also act as ligands for specific proteins.[21] A very likely physiological function of some $sPLA_2$s, in particular $sPLA_2$ IIA, is the participation in host defence against pathogen bacteria.[22,23]

With the discovery of $cPLA_2\alpha$ in 1990 a competing target for treatment of inflammation arose.[24-28] In contrast to $sPLA_2$s, $cPLA_2\alpha$ selectively cleaves phospholipids with arachidonoyl-residues at the *sn*-2 position. The central role of $cPLA_2\alpha$ in the arachidonic acid cascade and during inflammatory response was supported by experiments with cells over-expressing the enzyme[29] and by investigations performed with $cPLA_2\alpha$ knockout animals.[30-37] The physiological impact of the other isoforms of $cPLA_2$ is still unclear.

The $iPLA_2$s are supposed to play a house-keeping role in phospholipid remodelling. It has also been widely accepted that arachidonic acid and lysophospholipids generated by $iPLA_2$ act as signalling molecules in cellular functions such as glucose-induced insulin secretion and Fas-induced apoptosis.[38,39] Furthermore, $iPLA_2$ VIA seems to be involved in spermatogenesis and in bone formation as shown by analysis of the phenotype of $iPLA_2$-knockout mice.[40,41]

An important member of the PAF-acetylhydrolase family is the group VIIA enzyme, which is also called lipoprotein-associated PLA₂ (LP-PLA₂).[42] It is found in blood plasma bound to low-density lipoprotein (LDL) and high-density lipoprotein (HDL). Several clinical studies have shown that LP-PLA₂ is a definitive marker of coronary heart disease.[43] Darapladib (SB480848), a specific inhibitor of the enzyme, is actually in Phase III of clinical trials for treatment of coronary atherosclerosis.[44]

In the following, the focus will be directed towards cPLA₂α, which plays a key function in the development of inflammatory processes.

2.2 Properties and Function of Cytosolic Phospholipase A₂α

The cPLA₂α enzyme is a ubiquitous 85 kDa protein that is regulated mainly by an increase in intra-cellular Ca^{2+} concentrations and by phosphorylation on multiple serine residues.[1–7,25–27] On agonist-induced stimulation it translocates from the cytosol to cellular membranes, where it is bound *via* its calcium-phospholipid binding domain. During the inflammatory response, cPLA₂α is activated especially by tumour necrosis factor-α (TNF-α) and interleukin-1β (IL-1β) (Figure 2.1).[45–47] Furthermore, its expression is increased by these cytokines.

The catalytic mechanism of cPLA₂α is thought to be much like that of serine proteases: the enzyme forms an acyl enzyme intermediate between the *sn*-2 fatty

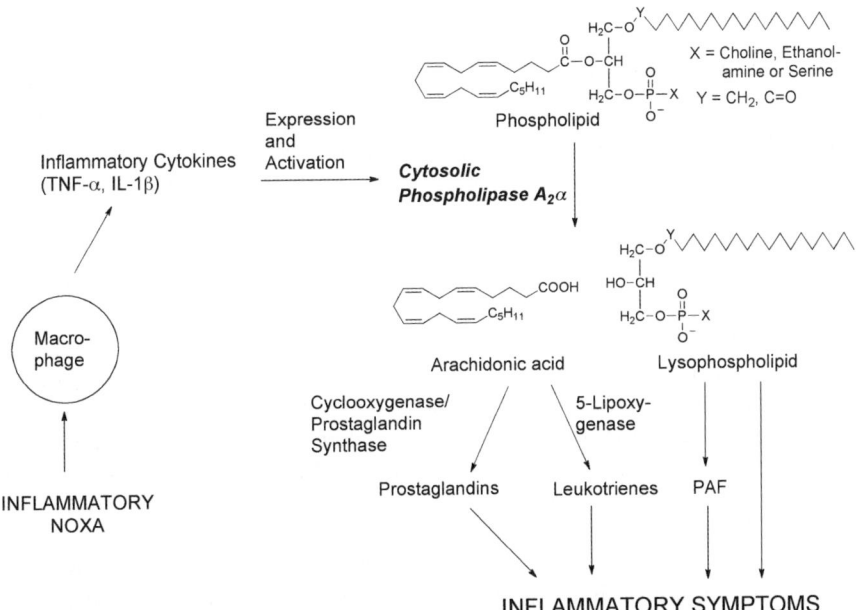

Figure 2.1 Role of cytosolic phospholipase A₂α in the inflammatory cascade.

acid of the phospholipid substrate and an active site serine residue.[48] cPLA$_2\alpha$ is unique among the PLA$_2$ enzymes in regard of its high selectivity for phospholipids containing arachidonic acid in the *sn*-2 position.

Despite the presence of several phospholipases A$_2$ in the mammalian organism, the pre-eminence of cPLA$_2\alpha$ for lipid mediator generation was demonstrated especially by studies with cPLA$_2\alpha$-deficient mice. These animals, which display a reduced eicosanoid production, are resistant to disease in a variety of models of inflammation, including collagen-induced arthritis.[30–36] Therefore, cPLA$_2\alpha$ is considered as an interesting target for the treatment of inflammatory diseases.[37] Importantly, cPLA$_2\alpha$-deficient mice are generally healthy. Possible side-effects of drugs inhibiting this enzyme can be deduced by the properties of the knockout mice too. Such animals lacking the enzyme were normally healthy, with the exception that labour induction was derogated and fertility was reduced, as in COX-1 and COX-2 deficient mice. Furthermore, they had small bowel ulcers. In a human male patient possessing a genetic functional deficiency of cPLA$_2\alpha$ such ulcers also occurred, but they were much more pronounced in ileum and jejunum. Interestingly, no gastroduodenal ulcers as caused by non-selective COX inhibition could be detected in this man.[49,50] Taken together, as in the therapy with inhibitors of COX, intestinal ulcers can be expected to be complications of cPLA$_2\alpha$ inhibitors administered systemically over a longer period. However, since such compounds not only inhibit the formation of prostaglandins, but also suppress the generation of leukotrienes, lysophospholipids and PAF, it can be supposed that they possess improved therapeutic activities in comparison to the cyclooxygenase inhibitors clinically used today.

2.3   Assessment of cPLA$_2\alpha$ Inhibitors

For evaluation of cPLA$_2\alpha$ inhibitors many different *in-vitro* assays have been applied.[16] They can be divided into assays using the isolated enzyme and assays measuring substrate cleavage in intact cells. Both types of test systems are associated with some problems.

The activity of cPLA$_2\alpha$ against monomerically dispersed substrate molecules is slow, while aggregated phospholipid molecules are cleaved much faster. Therefore, in assays applying isolated cPLA$_2\alpha$, aggregated substrates are widely used. However, when substrate concentrations are close to the concentrations of the inhibitors tested, there is the danger that the test compounds lead to a reduced substrate cleavage, not by binding to the active site of cPLA$_2\alpha$ but merely by altering the quality of the membrane/water interface.[12,13,17] To devoid such an undesired non-specific inhibition, the substrate should be presented at high concentrations or embedded in a large concentration of other lipids or detergents.[51] Under these conditions true cPLA$_2\alpha$ inhibitors show inhibition of the enzyme even at concentrations that are a small fraction of the total amount of lipid and detergent present in the assay solution.

Because for many cell types it has been proven that cPLA$_2\alpha$ is the enzyme responsible for agonist-induced arachidonic acid release, also cell-based assays have been developed for inhibitor screening. However, when using such cellular assays it has to be kept in mind that certain compounds including lipophilic acids and bases can readily cause cell lysis or loss of viability.[52] This may result in a reduced substrate cleavage and thus an erroneously indicated enzyme inhibition.[53,54] Furthermore, an observed inhibition of cellular arachidonic acid release can be due to an interaction of the test compound with the agonist-induced biochemical processes, which trigger the activation of cPLA₂α. Therefore, off-target effects have to be considered when cellular systems are applied.

2.4 Inhibitors of cPLA₂α

2.4.1 General

In the last years several reviews have been published that highlighted the field of cPLA₂α inhibitors.[7,12–20] In the following only the most interesting of these compounds will be discussed. These comprise substances with a pronounced *in-vitro* activity and compounds that have been evaluated *in vivo* in animal models. Beforehand, one of the major obstacles in the development of perorally active cPLA₂α inhibitors shall be addressed. The substrate molecules of cPLA₂α are phospholipids, which contain an arachidonoyl residue in the *sn-2* position and a second long lipophilic acyl or alkyl residue in position 3. Thus, these substrates possess a high molecular mass (600–700) and a substantial lipophilicity. Compounds that tightly bind to the active site of the enzyme therefore *a priori* can be expected to have similar properties. Furthermore, cPLA₂α in its active status works at the phospholipid membrane/water interface and its inhibitors have to partition into the lipophilic membrane before they can compete with the phospholipid substrate for enzyme binding. For enrichment in the phospholipid bilayer, the inhibitors therefore must possess some lipophilicity too. However, such properties lead to substances that are not very drug like and suffer from bioavailability problems.

2.4.2 Trifluoromethyl Ketones and Methyl Fluorophosphonates

Trifluoromethyl ketones are a well-established class of serine protease inhibitors.[55,56] These activated ketones, which form hydrates in aqueous solution, can inhibit such enzymes through the generation of covalent hemiketal adducts with the active site serine, and as such belong to the broad class of serine traps. The trifluoromethyl ketone analogue of arachidonic acid **1** (Figure 2.2) – also known as AACOCF₃ – was found to be a reversible, slow and tight binding inhibitor of cPLA₂α.[57,58] It showed spectroscopic characteristics with the enzyme that are consistent with the formation of a hemiketal with a serine residue in the active site of cPLA₂α.

AACOCF₃ (**1**) blocked arachidonic acid release from platelets stimulated with calcium ionophore A23187 with an IC₅₀ of 2 μM[59] and 3.3 μM,[54]

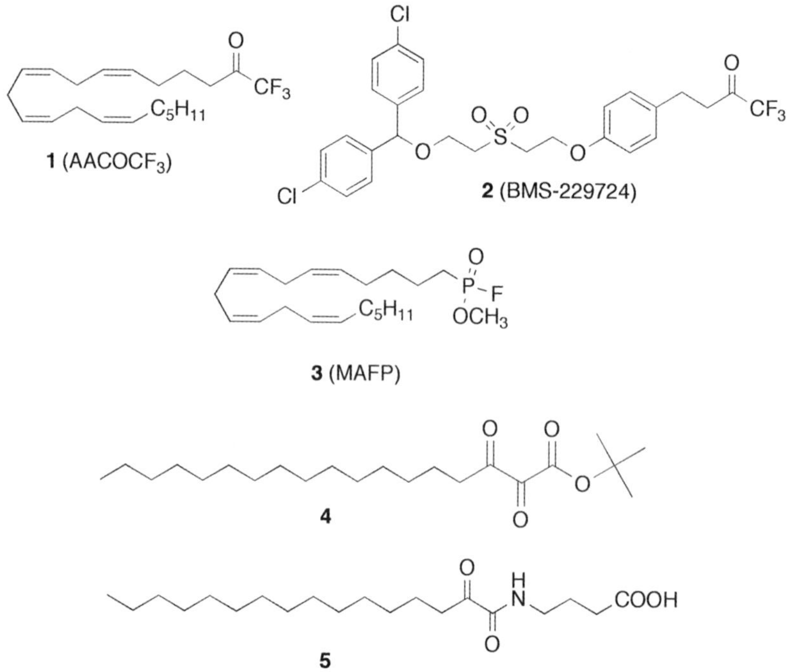

Figure 2.2 Inhibitors of cPLA$_2\alpha$ with trifluoromethyl ketone, methyl fluorophosphonate, α,β-dioxoester and α-ketoamide structure.

respectively. Since the compound is commercially available, it has been widely used to determine the role of cPLA$_2\alpha$ in the mediation of arachidonic acid release in various cell types. However, a deficiency of many of these studies is that attention was not focused on the cytotoxic properties of this amphiphilic substance. It was shown that AACOCF$_3$ lyses platelets[53,54] and monocytes[60] at concentrations not far higher than its IC$_{50}$ value. Such destruction of the cells may misleadingly indicate enzyme inhibition. Furthermore, AACOCF$_3$ is known to inhibit the activity of other serine containing enzymes, such as iPLA$_2$,[51,61] fatty acid amide hydrolyse (FAAH)[62,63] and monoacylglycerol lipase (MAGL).[64] The two latter enzymes are part of the cannabinoid system, which is also involved in inflammatory processes. Therefore, cellular and *in-vivo* effects observed for AACOCF$_3$ do not undoubtedly result from a cPLA$_2\alpha$ inhibition.

Besides, a series of analogues of AACOCF$_3$ were synthesized and tested for cPLA$_2\alpha$ inhibition.[51,60,65] Two of these compounds, a linolenic acid and a linoleic acid trifluoromethyl ketone derivative, inhibited cPLA$_2\alpha$ in a cellular assay to about the same extent as AACOCF$_3$ concomitantly showing less cytotoxic properties.[60] After intra-peritoneal administration, both substances were effective in the rat carrageenan-induced hind paw edema and the rat adjuvant arthritis model.[65] However, these AACOCF$_3$ derivatives also inhibit other serine hydrolases such as iPLA$_2$ and FAAH.[51,62]

Bristol-Myers Squibb developed the trifluoromethyl ketone cPLA$_2$α inhibitor BMS-229724 (2) (Figure 2.2).[66–68] Enzyme inhibition was measured by an assay applying human recombinant cPLA$_2$α and [^{3}H]-arachidonate labelled U937 membranes as substrate. The IC$_{50}$-value of BMS-229724 in this assay was 2.8 μM. The compound was also active in inhibiting arachidonate and eicosanoid production in U937 cells, neutrophils, platelets, monocytes and mast cells (IC$_{50}$: 3–15 μM). Furthermore, BMS-229724 was found to possess anti-inflammatory activity when applied topically in the phorbolester-induced chronic ear inflammation model in the mouse, reducing edema and inflammatory cell infiltration along with inhibition of prostaglandin and leukotriene biosynthesis in the skin. The systemic activity against skin inflammation was evaluated in hairless guinea pigs, in which skin erythema was induced by ultraviolet B (UVB) irradiation. When dosed orally (10 mg/kg), BMS-229724 reduced the erythema. The response was similar to the positive control ibuprofen (10 mg/kg). The pharmacokinetic profile of BMS-229724 showed an oral bioavailability of approximately 10%. Since an intra-portal-administered dose showed > 90% bioavailability, the low oral bioavailability was assumed to be an absorption problem rather than a first-pass hepatic metabolism effect.

Methyl arachidonyl fluorophosphonate (MAFP) (3) (Figure 2.2) has been reported to inactivate cPLA$_2$α irreversibly, possibly by phosphorylation of a serine residue of the enzyme.[69,70] In cell assays MAFP blocked cPLA$_2$α-mediated arachidonic acid formation triggered by A23187 with an IC$_{50}$ of 0.6 μM[69] and 2.9 μM,[54] respectively. MAFP possesses cytotoxic properties at concentrations not far above its IC$_{50}$.[53,54] *In-vivo* data for methyl fluorophosphonates are not accessible. Like the activated ketone AACOCF$_3$ (1), MAFP is not a selective cPLA$_2$α inhibitor but also inhibits iPLA$_2$, FAAH and MAGL.[71–75]

2.4.3   α,β-Dioxoesters and α-Ketoamides

Besides trifluoromethyl ketones, α,β-dioxoesters and α-ketoamides are other known ketones with enhanced electrophilic reactivity, which can act as serine traps. Therefore, these structural elements were used in the design of cPLA$_2$α inhibitors too.

The α,β-dioxoester 4 and the α-ketoamide 5 (Figure 2.2) were found to inhibit cPLA$_2$α with approximately the same potency as that of the trifluoromethyl ketone inhibitor 1.[19,76,77] Compound 5 also blocked the lipopolysaccharide (LPS)-stimulated arachidonic acid liberation in murine P388D1 macrophage-like cell line with an IC$_{50}$-value of 8.6 μM. In the rat paw carrageenan-induced edema assay 5 gave an ED$_{50}$ value of 3.6 mg/kg after intra-peritoneal administration.[78] Furthermore, derivatives of 5 were reported to have therapeutic effects in animal models of pain and inflammation after intra-peritoneal or intra-thecal administration.[79,80]

2.4.4   Benzhydrylindoles

Some of the pioneers in the development of inhibitors of cPLA$_2$α had been scientists of Wyeth (now Pfizer, formerly American Home Products).

Their compound WY-48,489 (**6**) (Figure 2.3), which had been claimed as inhibitor of secretory PLA$_2$ from platelets in 1988,[81] was later found also to inhibit cPLA$_2\alpha$.[82] Beside several other substance classes, in the early 2000s Wyeth developed a series of 1-benzhydryl-substitued indoles like **7**.[83]

These compounds proved to block the cleavage of a soluble monomeric coumarin substrate by cPLA$_2\alpha$. However, when a more lipophilic analogue of this substrate was used in the presence of larger amounts of the micelle-forming detergent Triton X-100, inhibitory potency of **7** dropped drastically as shown by an increase of the IC$_{50}$ from 0.8 to 160 μM.[84] This behaviour again reflected the problems in cPLA$_2\alpha$ inhibitor screening. In their strategy to increase the potency of **7**, the scientists of Wyeth planned to incorporate an electrophilic ketone moiety between the indole and the benzene scaffolds. This structural element had been found to be a very effective pharmacophoric element in a class of potent ketone based cPLA$_2\alpha$ inhibitors first published by AstraZeneca in 1998.[85–87] Since the alcohol precursor **8** of drafted ketone (Figure 2.3) already showed weak but reproducible activity in the Triton X-100 micelle assay (IC$_{50}$: 40 μM) as well as in a whole blood assay, this compound was used as a starting point for further inhibitor development. Systematic variation of **8** led to a series of considerably active inhibitors such as **9** (IC$_{50}$ in the micelle assay: 0.5 μM). These compounds were suggested to have a two-part

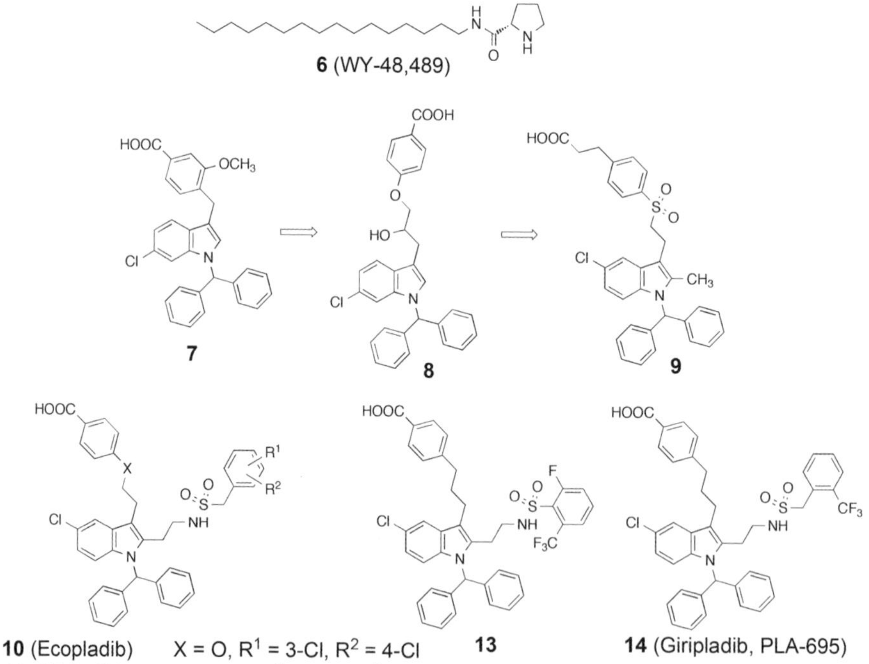

6 (WY-48,489)

7

8

9

10 (Ecopladib)   X = O, R^1 = 3-Cl, R^2 = 4-Cl
11 (Efipladib)   X = CH$_2$, R^1 = 3-Cl, R^2 = 4-Cl   13
12 (WAY-196025) X = CH$_2$, R^1 = 2-CH$_3$, R^2 = 6-CH$_3$

14 (Giripladib, PLA-695)

Figure 2.3   Benzhydrylindol inhibitors of cPLA$_2\alpha$.

pharmacophore for cPLA$_2\alpha$ inhibition in the benzhydryl group and the benzoic acid functionality. Subsequently, Wyeth started efforts to synthesize more potent cPLA$_2\alpha$ inhibitors by increasing the size of the substituent in position 2 of the indole.[88] The new residue should allow the molecules to create a third interaction point with the enzyme. Introduction of a benzyl sulfonamide residue produced the desired increase of activity. Compound **10** (ecopladib) was found to inhibit the enzyme in the micelle assay and a whole blood assay with submicromolar potency (IC$_{50}$: 0.15 μM and 0.11 μM, respectively).

To examine selectivity of ecopladib, its inhibition of the β- and γ-isoforms of cPLA$_2\alpha$ and of group II sPLA$_2$ was evaluated.[88] At a reasonable concentration, it was unable to inhibit these isoenzymes of cPLA$_2\alpha$. In pharmacokinetic studies with rats, ecopladib showed a bioavailability of 29% at an oral dose of 5 mg/kg in a standard methylcellulose (0.5%)/Tween (2%) formulation. Since the oral bioavailability decreased by approximately four-fold to 8% when dosed perorally at 20 mg/kg in methylcellulose/Tween, a scalable formulation was looked for in the evaluation of the oral efficacy of ecopladib in animal models. A suitable alternative found was a lipid-based vehicle consisting of Phosal 53 medium chain triglyceride, Tween 80, Labrosol and propylene carbonate. With this formulation an oral bioavailability of ecopladib ranging from 6% at a dose of 5 mg/kg to 19% at a dose of 150 mg/kg was measured, and *in-vivo* activity after oral application could be observed in two animal models of inflammation. In the rat carrageenan air pouch model it displayed an ED$_{50}$ of 8 mg/kg, while in the rat carrageenan paw edema model an ED$_{50}$ of 40 mg/kg was determined.

Further optimization of the structure of ecopladib led to the two structurally related compounds efipladib (**11**) and WAY-196025 (**12**) (Figure 2.3), which demonstrated increased potency over ecopladib, reduced clearance *in vivo* and efficacy in several *in-vivo* models of inflammation and asthma.[89] Thus, the inhibitory potency of efipladib and WAY-196025 in the micelle assay with the isolated enzyme was about four- and fifteen-fold, respectively, higher than that of ecopladib (IC$_{50}$ efipladib: 0.04 μM; IC$_{50}$ WAY-196025: 0.01 μM). In the rat whole blood assay, in which thromboxane B$_2$ formation after stimulation with calcium ionophore A23187 was monitored, the increase of activity was two- and five-fold, respectively. More detailed studies with human whole blood showed that efipladib and WAY-196025 inhibited the formation of several arachidonic acid metabolites (thromboxane B$_2$, leukotriene B$_4$, prostaglandins E$_2$ and F$_{2\alpha}$) and of PAF, as it could be expected for cPLA$_2\alpha$ inhibitors. Despite a low oral bioavailability of efipladib and WAY-196025 in rats, the plasma levels of reasonable peroral doses exceeded the rat whole blood IC$_{50}$ values and therefore *in-vivo* efficacy of these compounds was examined. The two compounds showed improved activity over ecopladib in the rat carrageenan air pouch model and in the rat carrageenan paw edema model, which are both short-term models of inflammation. In the latter model, the near maximum effect of WAY-196025 and the COX-2 inhibitor celecoxib were also determined and it was found that both compounds inhibit edema formation to the same maximal extent (34% *vs.* 37% at 25 mg/kg). The *in-vivo* efficacy was further assessed in two long-term models of inflammation. Both efipladib and WAY-196025 tested at a dose level of 100 mg/kg twice daily gave a significant reduction in the clinical disease severity relative to the vehicle-treated group in the mouse collagen-induced arthritis model (CIA). In the adjuvant induced arthritis (AIA) in rats at doses as low as 5 mg/kg, Celecoxib, efipladib and WAY-196025 rapidly reduced the mean clinical joint scores. Furthermore, after peroral application WAY-196025 prevented experimental autoimmune encephalomyelitis development in mice, an animal model of multiple sclerosis.[90] In all *in-vivo* experiments with efipladib and WAY-196025, a non-conventional lipid-based Phosal formulation was applied.

Because the scientists of Wyeth had concerns that the use of lipid-based formulations may change the plasma lipoprotein profile, they searched for derivatives of efipladib and WAY-196025 that show good peroral activity even when dosed in a conventional methylcellulose/Tween (MC/TW) formulation. With compound **13** (Figure 2.3) a substance was found that showed a much higher bioavailability in rats and dogs than WAY-195025 in such a vehicle (13% and 16% *vs.* 1.0 and 0.08%, respectively).[91] Dosed perorally in MC/TW, **13** led to an efficacious decrease of paw volume in the rat carrageenan paw edema model (44% at 25 mg/kg and 34% at 10 mg/kg). With a dose of 25 mg/kg of **13**, almost the same reduction of paw swelling could be achieved as with 10 mg/kg Naproxen in Phosal and 25 mg/kg Celecoxib in MC/TW. However, these data also show that in the rat carrageenan paw edema model a superior activity for the $cPLA_2\alpha$ inhibitor **13** over the COX inhibitors Naproxen and Celecoxib could not be registered. Notably, with glucocorticoids in this animal model a significantly higher inhibition of edema formation can be achieved than with COX inhibiting compounds.[92]

With giripladib (PLA-695) (**14**) (Figure 2.3) a further structurally closely related compound was investigated for *in-vivo* activity.[93] Oral treatment with 30 and 100 mg/kg twice a day in the murine collagen-induced arthritis model a significant reduction of clinical and microscopic diseases scores was measured. Furthermore, this compound was also active in the K/BxN mouse model of inflammation[94] after oral treatment of 3 or 10 mg/kg twice a day and 10 or 20 mg/kg once daily. Besides, several efforts were undertaken to reduce the undesirable high lipophilicity of this substance class.[95,96]

Additionally, the structurally related benzhydrylindole PF-5212372 was found to be active in several pre-clinical asthma models.[97]

Taken together, the tremendous work of the scientists of Wyeth has led to a series of $cPLA_2\alpha$ inhibitors that show efficacy in a wide range of *in-vitro* and *in-vivo* models of inflammation and asthma. For ecopladib a Phase I trial had been under the way in the USA. In 2007 Wyeth conveyed that the development of this compound has been discontinued.[97] Efipladib was reported to be in Phase II of clinical investigation with the indications rheumatoid arthritis, osteoarthritis, pain and asthma in 2005.[98] Giripladib had launched a Phase II clinical trial,[99] which was terminated in 2007 due to "gastrointestinal and lipase events".[93] Actually, no clinical trials for these and related compound are announced.[100]

2.4.5 Thiazolidinediones

Shionogi has developed a series of highly active cPLA$_2$α inhibitors consisting of a 1,2,4-trisubstituted pyrrolidine scaffold and a thiazolidienedione moiety.[101–105] One of the most active compounds found was RSC-3388 (**15**) (Figure 2.4). It inhibited isolated cPLA$_2$α with an IC$_{50}$ of 1.8 nM and arachidonic acid release in human monocytic cells (THP-1) stimulated with A23187 with an IC$_{50}$ of 22 nM.[101] In comparison with AACOCF$_3$ (**1**), RSC-3388 was about 230-fold more active in the isolated enzyme assay and about 3900-fold more active in the cellular assay.

The closely related pyrrolidine derivative **16** potently inhibited the arachidonic acid liberation in calcium ionophore A23187-stimulated CHO cells stably transfected with cPLA$_2$α, in zymosan- and okadaic acid-stimulated mouse peritoneal macrophages, and in ATP- and A23187-stimulated MDCK cells.[102] It was much less potent on iPLA$_2$ and groups IIA, V and X sPLA$_2$. Another highly active pyrrolidine inhibitor published is **17**, meanwhile designed as pyrrophenone.[103,104] This compound strongly inhibited arachidonic acid release in A23187-stimulated THP-1 cells with an IC$_{50}$ value of 24 nM, and it also showed inhibition of interleukin-1-induced prostaglandin E$_2$ synthesis in human renal mesangial cells (IC$_{50}$: 8.1 nM). Furthermore, pyrrophenone was also active in human whole blood stimulated with A23187, showing an about three-fold more potent inhibition of arachidonic acid release (IC$_{50}$: 0.19 µM) than RSC-3388 in this assay.[104]

RSC-3388 was also tested topically in two models of skin inflammation in mice.[106] In the 2,4,6-trinitro-1-chlorobenzene (TNCB)-induced ear inflammation model, RSC-3388 significantly inhibited ear swelling after application as acetone solution. At the end-point of the test after 21 days, the maximum

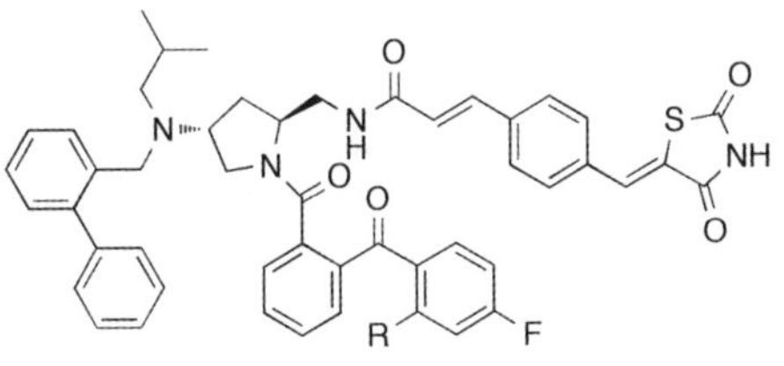

15 R = F (RSC-3388)
16 R = H

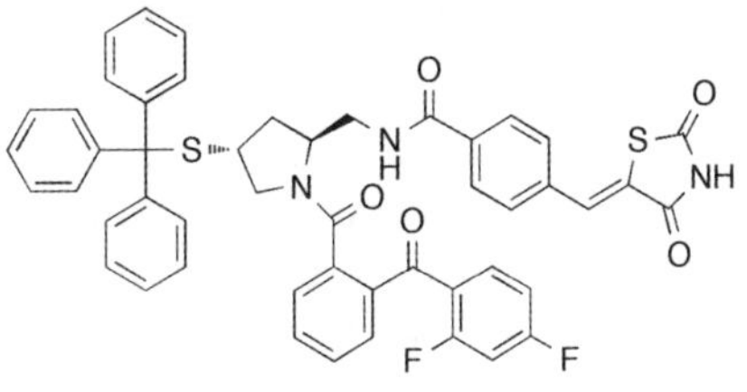

17 (Pyrrophenone)

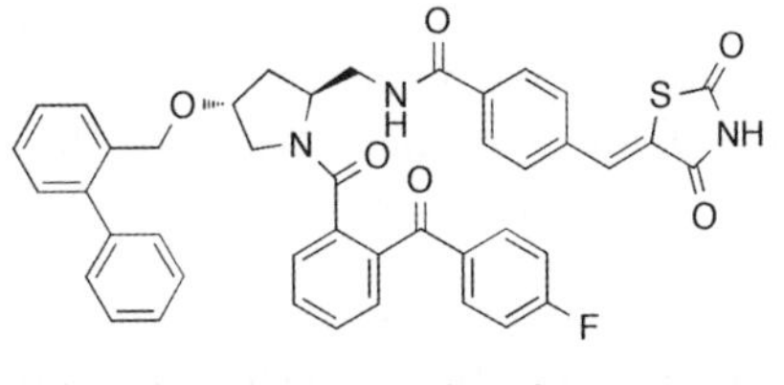

18 (Pyrroxyphene)

Figure 2.4 Thiazolidinedione inhibitors of cPLA$_2$α.

reduction of ear swelling was 60% at a dose of 0.3 mg per ear twice daily. In the histological examination, RSC-3388 inhibited epidermal and dermal thickness and inflammatory cell infiltration. Moreover, it reduced formation of the arachidonic acid metabolite prostaglandin E_2 as well as expression and protein levels of the inflammatory cytokine interleukin-1β. A two-fold application of a 3% ointment of RSC-3388 per day was equipotent to a 0.1% ointment of the calcineurin antagonist tacrolimus applied once daily both in the TNCB-induced ear inflammation model and in the mite antigen-induced dermatitis model. The test duration here was 21 and 13 days, respectively.

Additionally, the effects of another thiazolidinedione derivative, called pyrroxyphene (**18**) (Figure 2.4), was evaluated *in vivo* in the collagen-induced arthritis model in mice after peroral administration.[107] Although pyrroxyphene was less active than pyrrophenone against the isolated enzyme (IC_{50}: 78 nM *vs.* 1.8 nM) and in cellular situation (IC_{50} of inhibition of arachidonic acid release in A23187-stimulated THP-1 cells: 320 nM *vs.* 22 nM), pyrroxyphene was chosen for these experiments because of a better absorption/distribution/ metabolism/excretion (ADME) profile. Arthritis was induced by intra-cutaneous injection with collagen on day 0 and 21 and accelerated by subcutaneous injection of interleukin-1β. Pyrroxyphene was perorally administered twice a day from days 21 to 41 in a formulation containing the solvent dimethylformamide and the medium-chain triglyceride Miglyol. It significantly inhibited the arthritis index from day 26, and the inhibitory effect lasted throughout the experiment with 83% and 70% inhibition of paw swelling on day 41 at doses of 30 and 100 mg/kg, respectively. Moreover, pyrroxyphene was found to inhibit both the increase in levels of $cPLA_2\alpha$ and the eicosanoids prostaglandin E_2 and leukotriene B_4 as well as the mRNA-expression of COX-2 and several matrix metalloproteinases.

2.4.6 (Aryloxy)propan-2-ones

With the 1,3-disubstituted propan-2-one AR-C73346XX (**19**) (Figure 2.5), described by AstraZeneca in 1998,[85,86] the first highly potent $cPLA_2\alpha$ inhibitor arose. The structurally related compound AR-C70484XX (**20**) published several years later was found to be more than 20-fold more active against the isolated enzyme (IC_{50}: 0.03 μM *vs.* 0.80 μM) than the $cPLA_2\alpha$ inhibitor AACOCF$_3$ (**1**), and also greater than 10-fold more active than AACOCF$_3$ against the cellular production of arachidonic acid by HL60 cells (IC_{50}: 2.8 μM *vs.* 29 μM).[87] A three-point model for the binding of this kind of inhibitors to the $cPLA_2\alpha$ enzyme was proposed: the carboxylic acid group might interact with the binding site of the phosphate group of the phospholipid substrate molecules, the electrophilic propan-2-one could serve as a trap for the serine of the $cPLA_2\alpha$ active site and the lipophilic alkoxyphenyl chain should be accommodated in the hydrophobic site, where the *sn*-2 arachidonoyl residue of the natural substrate binds. In efforts to moderate the undesirable very high lipophilicity caused by the extended alkoxyphenyl chain, AstraZeneca synthesized a series of less lipophilic, more "drug-like" inhibitors such as **21**.[108]

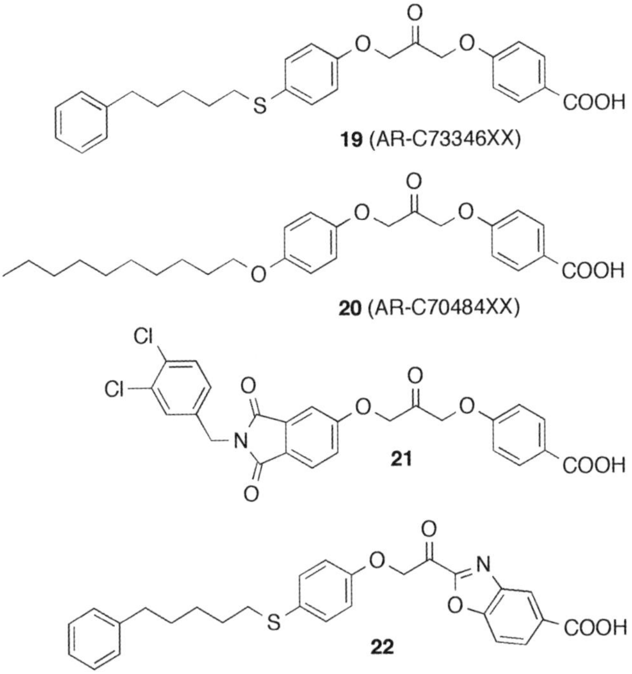

Figure 2.5   Inhibitors of cPLA$_2$α containing (aryloxy)propan-2-one and α-keto heterocycle backbones.

This compound displayed improved whole cell (HL60) potencies against
cPLA$_2$α (IC$_{50}$: 1.0 µM) along with a significant lower lipophilicity than
AR-C70484XX (logP: 5.5 *vs.* 7.8). *In-vivo* data of these propan-2-one inhibitors
of cPLA$_2$α have not been reported.

Besides, AstraZeneca has developed another series of cPLA$_2$α inhibitors
bearing an α-keto-heterocycle (**22**) (Figure 2.5) instead of the 1,3-disubstituted
propan-2-one as serine trap.[109,110] Bristol-Myers Squibb later claimed compounds structurally related to AR-C70484XX.[111]

In our group we started cPLA$_2$α inhibitor screening with a cellular test
system measuring arachidonic acid release in intact platelets after stimulation
of cPLA$_2$α with calcium ionophore A23187. A series of acylated indole and
pyrrole-carboxylic acid derivatives like ML3176 (**23**) (Figure 2.6) showed
activity in this assay.[112] ML3176 displayed an IC$_{50}$ of 1.6 µM, without having
cell lytic properties at this concentration. Furthermore, the compound was
found to give high plasma levels up to 100 µg/mL after peroral application of
100 mg/kg in a methyl cellulose formulation.[113] In the rat carrageenan edema
model 100 mg/kg ML3176 reduced edema formation by about 20% when
applied perorally and by about 68% after intra-peritoneal administration.[113,114] However, in an assay with isolated cPLA$_2$α ML3176 and related
compounds proved to be inactive at the highest test concentrations (10 µM).[115]
Thus, the reduced liberation of arachidonic acid in the cellular assay and the

in-vivo effects were obviously not produced by a direct inhibition of cPLA$_2\alpha$ but by an off-target effect of the compound.

In an effort to gain true inhibitors of cPLA$_2\alpha$, we synthesized indole derivatives with a 3-aryloxy-2-oxopropyl-residue as it is present in the structure of the potent cPLA$_2\alpha$ inhibitor AR-C70484XX (**20**) from AstraZeneca (Figure 2.5). This approach led us to the discovery of compound **24** (Figure 2.6), which showed an IC$_{50}$ of 0.035 µM in an isolated enzyme assay. Under the

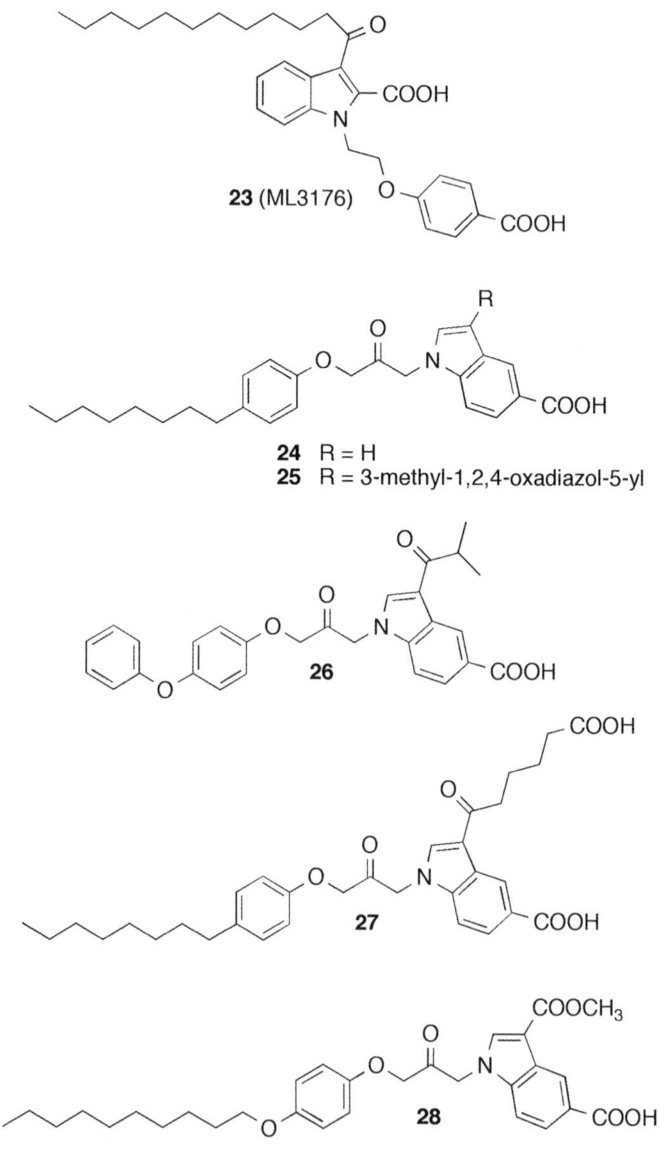

Figure 2.6   Indolylpropan-2-one inhibitors of cPLA$_2\alpha$.

same conditions AR-C70484XX and AACOCF$_3$ exhibited IC$_{50}$ values of 0.011 μM and 2.3 μM, respectively.[116]

With an IC$_{50}$ of 0.41 μM **24** inhibited cPLA$_2$α mediated arachidonic acid release stimulated by calcium ionophore A23187 in intact platelets nearly as potent as AR-70484XX (IC$_{50}$: 0.25 μM). Using the phorbol ester 12-*O*-tetradecanoylphorbol-13-acetate (TPA) to elicit cPLA$_2$α activity in platelets, significantly lower IC$_{50}$s were obtained for **24** and AR-C704784XX (0.018 μM and 0.0047 μM, respectively). These values now lay in the same order of magnitude as the IC$_{50}$s obtained in the enzymatic assay with isolated cPLA$_2$α. Obviously, an increased incubation time (60 min. with TPA like in the isolated enzyme assay *vs.* 1 min. with A23187) was the reason for this observed increase of activity in cellular situation. The time dependence of inhibitory potency of these compounds reflects that the proposed covalent bonds between the propan-2-one inhibitors and the enzyme are formed rather slowly.

With the introduction of certain electron-withdrawing substituents in indole 3 position of **24**, more potent inhibitors could be obtained. The derivative with the highest *in-vitro* activity of this series was the oxadiazole substituted indole **25** (IC$_{50}$ against isolated cPLA$_2$α: 0.0021 μM; IC$_{50}$ in TPA-stimulated platelets: 0.0006 μM) (Figure 2.6).[117]

Like other known potent cPLA$_2$α inhibitors, the above-described indoles possess a high lipophilicity leading to a low aqueous solubility, which is detrimental for a good bioavailability. A second shortcoming of these compounds is the electrophilic ketone moiety, because this functionality is susceptible for metabolic reduction to an alcohol and such a reduction results in a loss of cPLA$_2$α inhibitory activity.[118] Therefore, efforts were made to synthesize derivatives with optimized properties. With compound **26** (Figure 2.6) a substance was found that possesses a high cPLA$_2$α inhibitory activity (IC$_{50}$: 0.012 μM in an isolated enzyme assay), an adequate lipophilicity and water solubility comparable to that of indomethacin, combined with a moderate metabolic stability *in vitro* against rat liver microsomes.[119] Besides, selectivity of cPLA$_2$α inhibition over inhibition of group IIA PLA$_2$ and rat brain iPLA$_2$ by **26** was established. Despite passable physicochemical properties, in pharmacokinetic studies in mice perorally dosed with 100 mg/kg of **26** in a methylcellulose formulation, only very low plasma levels of this compound and its alcohol metabolite were measured (< 1 μg/mL after 2 h) while the reference indomethacin gave much higher values even at a peroral dose of only 10 mg/kg (34 μg/mL after 2 h). The low peroral bioavailability obviously was not due to an impaired absorption but rather a result of an excessive glucuronidation of **26** and its alcohol metabolite followed by a very rapid biliar excretion.[117,119]

Compound **26** was also evaluated in a murine model of contact dermatitis after topical application.[119] In this model ear swelling induced by benzalkonium chloride was measured over 24 h. While substance doses of 0.1 and 1.0 mg/ear produced a nearly total inhibition of edema formation, ear swelling was less reduced by the lowest dose of 0.01 mg/ear, reflecting the dose dependency of the effect. In a second independent experiment the anti-inflammatory properties of **26** in the acute irritant contact dermatitis model at a dose of 0.1 mg/ear was compared with that of the potent glucocorticoid clobetasol-17-propionate at a dose of 0.005 mg/ear. At the peak of edema formation, which appeared 7 h after application of the irritant benzalkonium chloride, **26** produced similar effects to the glucocorticoid. Inhibition of ear swelling by **26** was 84%, while clobetasol-17-propionate reduced edema formation by 90%. The structurally related derivative **27** gave similar data in pharmacokinetic experiments and in the benzalkonium chloride contact dermatitis model in mice.[117,120]

The indolepropanone derivative **28** (CAY10502) (Figure 2.6) has been tested for inhibition of neovascularization in the rat model of oxygen-induced retinopathy (OIR), since it was known that prostaglandins, formed from the cPLA$_2$α product arachidonic acid can exert a proangiogenic influence by inducing vascular endothelial growth factor (VEGF) production and by stimulating angiogenic behaviour in vascular endothelial cells. Intra-vitreal injection of **28** in rat eyes using a formulation containing DMSO decreased oxygen-induced retinal neovascularization by about 50%.[121]

2.4.7 Aryl Thiazolidinones

Inhibitors of cPLA$_2$α with oxazolidinone scaffold were described by Nippon Soda.[122] One of the most active compounds of this series was **29** (Figure 2.7). It inhibited the enzyme activity in a cellular assay by 86% at a concentration of 0.01 µM, while the reference inhibitor AACOCF$_3$ here only showed 65% inhibition at 3 µM. In the TPA-induced ear edema model of the mouse **29** reduced ear swelling 6 h after induction of the inflammation by 84% when 0.3 mg of the compound were applied twice (30 min. before and 15 min. after TPA). Under the same conditions, the reference dexamethasone acetate inhibited edema formation after a two-fold application of 1 mg and 0.3 mg by 80% and 49%, respectively. For some compounds data obtained in the picryl chloride induced delayed contact dermatitis model in mice, which was monitored over 48 h, were published. The most active compound presented was **30**. It inhibited ear swelling by 32% in this model when applied twice in an amount of 1.5 mg/ear, while dexamethasone here was significantly more potent reducing edema by 80% at a two-fold dose of 0.02 mg/ear.

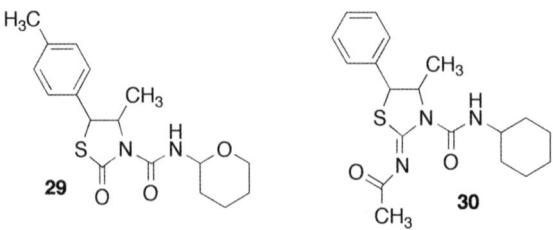

Figure 2.7   Aryl thiazolidinone inhibitors of cPLA$_2$α.

2.5 Conclusions

Due to the eminent role cPLA$_2$α is obviously playing in the patho-physiology of inflammation, it was expected that major improvements in the therapy of inflammatory diseases could be achieved with inhibitors of this enzyme. Intense efforts have been made to find clinical active drug candidates. However, actually no clinical trials are performed with cPLA$_2$α inhibitors.[100] Furthermore, looking at the actual publications and patent applications one could get the impression that the pharmaceutical companies involved in cPLA$_2$α inhibitor development have terminated their programs. Only some last Mohicans at academia are still trying to find new and better compounds. The main reason for the missing success seems to lie in the pharmacokinetic properties of the inhibitors found. Regarding the structure-activity relationship studies performed with these compounds, it is obvious that a substantial lipophilicity is necessary to achieve a good inhibitory potency. However, this lipophilicity, probably essential to bind to the enzyme or to partition sufficiently into the phospholipid substrate bilayers (= cell membranes), mostly results in low aqueous solubility and, as a consequence of this, in low oral absorption. One of the few cPLA$_2$α inhibitors with good activity in combination with a moderate lipophilicity is the indole derivative **26**. Unfortunately, this substance also possesses only a very poor peroral bioavailability due to an excessive metabolism and biliar excretion.

The described bioavailability problems do not arise when the compounds are applied topically for treatment of dermal diseases. Since several cPLA$_2$α inhibitors have shown good effects in topical models of inflammation,[66,67,106,119,120] further development of such agents for external application could be promising. At present, for the therapy of inflammatory skin diseases mainly glucocorticoids are used. However, these compounds lead to skin atrophy when applied over a longer period. Administration in the face is not recommended. The calcineurin antagonists pimecrolimus and tacrolimus used as an alternative for glucorticoids in the last years also have severe side-effects. Thus, there is an urgent need for new topical active antiphlogistics. Inhibitors of cPLA$_2$α could be such drugs.

In the last years it has become more and more evident that cPLA$_2$α does not solely play an important role in the pathogenesis of inflammation but is also involved in tumour progression.[123–129] In several cancers, such as non-small-cell lung cancer, cholangiosarcomas, oesophageal cancers and cancers of the colon and small intestine, an over-expression of cPLA$_2$α can be found. In many cases this is accompanied by the increased production of other cancer-associated stimuli such as epidermal growth factor receptor, transforming growth factor, COX-2 and microsomal prostaglandin E synthase-1 (mPGES-1). In concurrence with cPLA$_2$α, the latter two enzymes lead to an increased formation of prostaglandin E$_2$ (PGE$_2$) in tumours, which is often associated with a poor tumour prognosis. PGE$_2$ stimulates the formation of pro-angiogenic factors such as VEGF. A second oncologically important lipid mediator produced by cPLA$_2$α is lysophosphatidylcholine, formed beside arachidonic acid during cleavage of choline phospholipids. Recent studies suggest that this mediator also significantly contributes to the formation of tumour vasculature. Thus, cPLA$_2\alpha$ seems to be an important player in tumour development and progression. Possibly, these findings will stimulate the search for new and clinical active cPLA$_2\alpha$ inhibitors again.

References

1. D. A. Six and E. A. Dennis, *Biochim. Biophys. Acta*, 2000, **1488**, 1.
2. I. Kudo and M. Murakami, *Prostaglandins Other Lipid Mediat.*, 2002, **68–69**, 3.
3. R. H. Schaloske and E. A. Dennis, *Biochim. Biophys. Acta*, 2006, **1761**, 1246.
4. J. E. Burke and E. A. Dennis, *J. Lipid Res.*, 2009, **50**, S237.
5. J. E. Burke and E. A. Dennis, *Cardiovasc. Drugs Ther.*, 2009, **23**, 49.
6. M. Murakami, Y. Taketomi, Y. Miki, H. Sato, T. Hirabayashi and K. Yamamoto, *Prog. Lipid Res.*, 2011, **50**, 152.
7. E. A. Dennis, J. Cao, Y. H. Hsu, V. Magrioti and G. Kokotos, *Chem. Rev.*, 2011, **111**, 6130.
8. D. D. Funk, *Science*, 2001, **294**, 1871.
9. M. P. Wymann and R. Schneiter, *Nat. Rev. Mol. Cell Biol.*, 2008, **9**, 162.
10. C. C. Yost, A. S. Weyrich and G. A. Zimmerman, *Biochimie*, 2010, **92**, 692.
11. J. R. Sundaram, E. S. Chan, C. P. Poore, T. K. Pareek, W. F. Cheong, G. Shui, N. Tang, C. M. Low, M. R. Wenk and S. Kesavapany, *J. Neurosci.*, 2012, **32**, 1020.
12. S. Connolly and D. H. Robinson, *Curr. Opin. Ther. Patents*, 1993, **3**, 1141.
13. S. Connolly and D. H. Robinson, *Expert Opin. Ther. Patents*, 1995, **5**, 673.
14. U. Tibes and W. G. Friebe, *Expert Opin. Investig. Drugs*, 1997, **6**, 279.
15. R. J. Mayer and L. A. Marshall, *Emerging Drugs*, 1998, **3**, 333.
16. M. Lehr, *Expert Opin. Ther. Patents*, 2001, **11**, 1123.
17. J. D. Clark and S. Tam, *Expert Opin. Ther. Patents*, 2004, **14**, 937.
18. M. C. Meyer, P. Rastogi, C. S. Beckett and J. McHowat, *Curr. Pharm. Design*, 2005, **11**, 1301.
19. M. Lehr, *Anti-Inflammatory Anti-Allergy Agents Med. Chem.*, 2006, **5**, 149.
20. V. Magrioti and G. Kokotos, *Expert Opin. Ther. Patents*, 2010, **20**, 1.
21. M. Rouault, J. G. Bollinger, M. Lazdunski, M. H. Gelb and G. Lambeau, *Biochemistry*, 2003, **42**, 11494.
22. V. J. Laine, D. S. Grass and T. J. Nevalainen, *Infect. Immun.*, 2000, **68**, 87.
23. J. O. Gronroos, V. J. Laine, M. J. Janssen, M. R. Egmond and T. J. Nevalainen, *J. Immunol.*, 2001, **166**, 4029.
24. S. Connolly and D. H. Robinson, *Drug News Perspect.*, 1993, **6**, 584.
25. J. D. Clark, A. R. Schievella, E. A. Nalefski and L. L. Lin, *J. Lipid Mediat. Cell Signal.*, 1995, **12**, 83.

26. R. M. Kramer and J. D. Sharp, *FEBS Lett.*, 1997, **410**, 49.
27. Y. Kita, T. Ohto, N. Uozumi and T. Shimizu, *Biochim. Biophys. Acta*, 2006, **1761**, 1317.
28. M. Ghosh, D. E. Tucker, S. A. Burchett and C. C. Leslie, *Prog. Lipid Res.*, 2006, **45**, 487.
29. L. L. Lin, A. Y. Lin and J. L. Knopf, *Proc. Natl Acad. Sci. USA*, 1992, **89**, 6147.
30. J. V. Bonventre, Z. Huang, M. R. Taheri, E. O'Leary, E. Li, M. A. Moskowitz and A. Sapirstein, *Nature*, 1997, **390**, 622.
31. N. Uozumi, K. Kume, T. Nagase, N. Nakatani, S. Ishii, F. Tashiro, Y. Komagata, K. Maki, K. Ikuta, Y. Ouchi, J. Miyazaki and T. Shimizu, *Nature*, 1997, **390**, 618.
32. T. Nagase, N. Uozumi, S. Ishii, K. Kume, T. Izumi, Y. Ouchi and T. Shimizu, *Nat. Immunol.*, 2000, **1**, 42.
33. A. Sapirstein and J. V. Bonventre, *Biochim. Biophys. Acta*, 2000, **1488**, 139.
34. M. Hegen, L. Sun, N. Uozumi, K. Kume, M. E. Goad, C. L. Nickerson-Nutter, T. Shimizu and J. D. Clark, *J. Exp. Med.*, 2003, **197**, 1297.
35. C. Miyaura, M. Inada, C. Matsumoto, T. Ohshiba, N. Uozumi, T. Shimizu and A. Ito, *J. Exp. Med.*, 2003, **197**, 1303.
36. S. Marusic, M. W. Leach, J. W. Pelker, M. L. Azoitei, N. Uozumi, J. Cui, M. W. Shen, C. M. DeClercq, J. S. Miyashiro, B. A. Carito, P. Thakker, D. L. Simmons, J. P. Leonard, T. Shimizu and J. D. Clark, *J. Exp. Med.*, 2005, **202**, 841.
37. J. Bonventre, *Trends Immunol.*, 2004, **25**, 116.
38. S. Akiba and T. Sato, *Biol. Pharm. Bull.*, 2004, **27**, 1174.
39. J. Balsinde and M. A. Balboa, *Cell Signal.*, 2005, **17**, 1052.
40. S. Bao, D. J. Miller, Z. Ma, M. Wohltmann, G. Eng, S. Ramanadham, K. Moley and J. Turk, *J. Biol. Chem.*, 2004, **279**, 38194.
41. S. Ramanadham, K. E. Yarasheski, M. J. Silva, M. Wohltmann, D. V. Novack, B. Christiansen, X. Tu, S. Zhang, X. Lei and J. Turk, *Am. J. Pathol.*, 2008, **172**, 868.
42. D. M. Stafforini, *Cardiovasc. Drugs Ther.*, 2009, **23**, 73.
43. A. Lerman and J. P. McConnell, *Am. J. Cardiol.*, 2008, **101**, 11F.
44. M. A. Corson, *Ther. Adv. Cardiovasc. Dis.*, 2010, **4**, 241.
45. M. Krönke and S. Adam-Klages, *FEBS Lett.*, 2002, **531**, 18.
46. O. J. Jupp, P. Vandenabeele and D. J. MacEwan, *Biochem. J.*, 2003, **374**, 453.
47. C. W. Lee, I. T. Lee, C. C. Lin, H. C. Lee, W. N. Lin and C. M. Yang, *J. Cell. Biochem.*, 2010, **109**, 1045.
48. A. Dessen, J. Tang, H. Schmidt, M. Stahl, J. D. Clark, J. Seehra and W. S. Somers, *Cell*, 1999, **97**, 349.
49. D. H. Adler, J. D. Cogan, J. A. Phillips, N. Schnetz-Boutaud, G. L. Milne, T. Iverson, J. A. Stein, D. A. Brenner, J. D. Morrow, O. Boutaud and J. A. Oates, *J. Clin. Invest.*, 2008, **118**, 2121.
50. D. H. Adler, J. A. Phillips, J. D. Cogan, T. M. Iverson, N. Schnetz-Boutaud, J. A. Stein, D. A. Brenner, G. L. Milne, J. D. Morrow, O. Boutaud and J. A. Oates, *J. Gastroenterol.*, 2009, **44**, 1.

51. F. Ghomashchi, R. Loo, J. Balsinde, F. Bartoli, R. Apitz-Castro, J. D. Clark, E. A. Dennis and M. H. Gelb, *Biochim. Biophys. Acta*, 1999, **1420**, 45.

52. M. Aranzazu Partearroyo, H. Ostolaza, F. M. Goni and E. Barbera-Guillem, *Biochem. Pharmacol.*, 1990, **40**, 1323.

53. M. Lehr and K. Griessbach, *Pharm. Pharmacol. Commun.*, 1999, **5**, 389.

54. D. Risse, A. Schulze Elfringhoff and M. Lehr, *J. Chromatogr. B*, 2002, **769**, 185.

55. P. D. Edwards and P. R. Bernstein, *Med. Res. Rev.*, 1994, **14**, 127.

56. P. E. J. Sanderson, *Med. Res. Rev.*, 1999, **19**, 179.

57. L. A. Trimble, I. P. Street, H. Perrier, N. M. Tremblay, P. K. Weech and M. A. Bernstein, *Biochemistry*, 1993, **32**, 12560.

58. I. P. Street, H. K. Lin, F. Laliberte, F. Ghomashchi, Z. Wang, H. Perrier, N. M. Tremblay, Z. Huang, P. K. Weech and M. H. Gelb, *Biochemistry*, 1993, **32**, 57.

59. D. Riendeau, J. Guay, P. K. Weech, F. Laliberté, L. Yergey, C. Li, S. Desmarais, H. Perrier, S. Liu, D. Nicoll-Griffith and I. P. Street, *J. Biol. Chem.*, 1994, **269**, 15619.

60. E. Amandi-Burgermeister, U. Tibes, B. M. Kaiser, W. G. Friebe and W. V. Scheuer, *Eur. J. Pharmacol.*, 1997, **326**, 237.

61. E. Ackermann, K. Conde-Frieboes and E. A. Dennis, *J. Biol. Chem.*, 1995, **270**, 445.

62. B. Koutek, G. D. Prestwich, A. C. Howlett, S. A. Chin, D. Salehani, N. Akhavan and D. G. Deutsch, *J. Biol. Chem.*, 1994, **269**, 22937.

63. S. Maurelli, T. Bisogno, L. De Petrocellis, A. Di Luccia, G. Marino and V. Di Marzo, *FEBS Lett.*, 1995, **377**, 82.

64. G. G. Muccioli, G. Labar and D. M. Lambert, *ChemBioChem*, 2008, **9**, 2704.

65. U. Tibes, A. Vondran, E. Rodewald, W. G. Friebe, W. Schäfer and W. Scheuer, *Int. Arch. Allergy Immunol.*, 1995, **107**, 432.

66. J. R. Burke, L. B. Davern, P. L. Stanley, K. R. Gregor, J. Banville, R. Remillard, J. W. Russell, P. J. Brassil, M. R. Witmer, G. Johnson, J. A. Tredup and K. M. Tramposch, *J. Pharmacol. Exp. Ther.*, 2001, **298**, 376.

67. J. R. Burke, *Current Opin. Invest. Drugs*, 2001, **2**, 1549.

68. J. Banville, G. Johnson, F. Zusi and J. Burke, PCT WO 9915129, 1999.

69. B. Kennedy, P. Payette, D. Riendeau, P. Weech and M. Gresser, *Mediat. Inflamm.*, 1994, **3**, 307.

70. Z. Huang, P. Payette, K. Abdullah, W. A. Cromlish and B. P. Kennedy, *Biochemistry*, 1996, **35**, 3712.

71. Y. C. Lio, L. J. Reynolds, J. Balsinde and E. A. Dennis, *Biochim. Biophys. Acta*, 1996, **1302**, 55.

72. D. G. Deutsch, R. Omeir, G. Arreaza, D. Salehani, G. D. Prestwich, Z. Huang and A. Howlett, *Biochem. Pharmacol.*, 1997, **53**, 255.

73. L. De Petrocellis, D. Melck, N. Ueda, S. Maurelli, Y. Kurahashi, S. Yamamoto, G. Marino and V. Di Marzo, *Biochem. Biophys. Res. Commun.*, 1997, **231**, 82.

74. B. R. Martin, I. Beletskaya, G. Patrick, R. Jefferson, R. Winckler, D. G. Deutsch, V. Di Marzo, O. Dasse, A. Mahadevan and R. K. Razdan, *J. Pharmacol. Exp. Ther.*, 2000, **294**, 1209.

75. T. P. Dinh, D. Carpenter, F. M. Leslie, T. F. Freund, I. Katona, S. L. Sensi, S. Kathuria and D. Piomelli, *Proc. Natl Acad. Sci. USA*, 2002, **99**, 10819.

76. K. Conde-Frieboes, L. J. Reynolds, Y. C. Lio, M. R. Hale, H. H. Wasserman and E. A. Dennis, *J. Am. Chem. Soc.*, 1996, **118**, 5519.

77. G. Kokotos, S. Kotsovolou, D. A. Six, V. Constantinou-Kokotou, C. C. Beltzner and E. A. Dennis, *J. Med. Chem.*, 2002, **45**, 2891.

78. G. Kokotos, D. A. Six, V. Loukas, T. Smith, V. Constantinou-Kokotou, D. Hadjipavlou-Litina, S. Kotsovolou, A. Chiou, C. C. Beltzner and E. A. Dennis, *J. Med. Chem.*, 2004, **47**, 3615.

79. T. L. Yaksh, G. Kokotos, C. I. Svensson, D. Stephens, C. G. Kokotos, B. Fitzsimmons, D. Hadjipavlou-Litina, X. Y. Hua and E. A. Dennis, *J. Pharmacol. Exp. Ther.*, 2006, **316**, 466.

80. D. A. Six, E. Barbayianni, V. Loukas, V. Constantinou-Kokotou, D. Hadjipavlou-Litina, D. Stephens, A. C. Wong, V. Magrioti, P. Moutevelis-Minakakis, S. F. Baker, E. A. Dennis and G. Kokotos, *J. Med. Chem.*, 2007, **50**, 4222.

81. W. H. McGregor and J. Y. Chang, PCT WO 8806885, 1988.

82. F. Märki, W. Breitenstein, E. Beriger, R. Bernasconi, G. Caravatti, J. E. Francis, R. Paioni, H. U. Wehrli and R. Wiederkehr, *Agents Actions*, 1993, **38**, 202.

83. C. McKew, F. Lovering, J. D. Clark, J. Bemis, Y. Xiang, M. Shen, W. Zhang, J. C. Alvarez and D. Joseph-McCarthy, *Bioorg. Med. Chem. Lett.*, 2003, **13**, 4501.

84. J. C. McKew, M. A. Foley, P. Thakker, M. L. Behnke, F. E. Lovering, F. W. Sum, S. Tam, K. Wu, M. W. Shen, W. Zhang, M. Gonzalez, S. Liu, A. Mahadevan, H. Sard, S. P. Khor and J. D. Clark, *J. Med. Chem.*, 2006, **49**, 135.

85. A. Mete, R. Austin, C. Bennion, M. Bernstein, S. Connolly, C. G. Jackson, S. King, L. Lawrence, R. Lewis, D. H. Robinson, L. Stein, I. Walters and W. J. Withnall, Poster presented at XVth International Symposium on Medicinal Chemistry, Edinburgh, UK, 1998.

86. P. Norman, *IDrugs*, 1998, **1**, 49.

87. S. Connolly, C. Bennion, S. Botterell, P. J. Croshaw, C. Hallam, K. Hardy, P. Hartopp, C. G. Jackson, S. J. King, L. Lawrence, A. Mete, D. Murray, D. H. Robinson, G. M. Smith, L. Stein, I. Walters, E. Wells and W. J. Withnall, *J. Med. Chem.*, 2002, **45**, 1348.

88. K. L. Lee, M. A. Foley, L. Chen, M. L. Behnke, F. E. Lovering, S. J. Kirincich, W. Wang, J. Shim, S. Tam, M. W. Shen, S. Khor, X. Xu, D. G. Goodwin, M. K. Ramarao, C. Nickerson-Nutter, F. Donahue, M. S. Ku, J. D. Clark and J. C. McKew, *J. Med. Chem.*, 2007, **50**, 1380.

89. J. C. McKew, K. L. Lee, M. W. Shen, P. Thakker, M. A. Foley, M. L. Behnke, B. Hu, F. W. Sum, S. Tam, Y. Hu, L. Chen, S. J. Kirincich,

R. Michalak, J. Thomason, M. Ipek, K. Wu, L. Wooder, M. K. Ramarao, E. A. Murphy, D. G. Goodwin, L. Albert, X. Xu, F. Donahue, M. S. Ku, J. Keith, C. L. Nickerson-Nutter, W. M. Abraham, C. Williams, M. Hegen and J. D. Clark, *J. Med. Chem.*, 2008, **51**, 3388.

90. S. Marusic, P. Thakker, J. W. Pelker, N. L. Stedman, K. L. Lee, J. C. McKew, L. Han, X. Xu, S. F. Wolf, A. J. Borey, J. Cui, M. W. Shen, F. Donahue, M. Hassan-Zahraee, M. W. Leach, T. Shimizu and J. D. Clark, *J. Neuroimmunol.*, 2008, **204**, 29.

91. K. L. Lee, M. L. Behnke, M. A. Foley, L. Chen, W. Wang, R. Vargas, J. Nunez, S. Tam, N. Mollova, X. Xu, M. W. Shen, M. K. Ramarao, D. G. Goodwin, C. L. Nickerson-Nutter, W. M. Abraham, C. Williams, J. D. Clark and J. C. McKew, *Bioorg. Med. Chem.*, 2008, **16**, 1345.

92. B. B. Newbould, *Br. J. Pharmacol. Chemother.*, 1963, **21**, 127.

93. *Drug Data Report*, 2008, **30**, 611.

94. H. J. Ditzel, *Trends Mol. Med.*, 2004, **10**, 40.

95. L. Chen, W. Wang, K. L. Lee, M. W. Shen, E. A. Murphy, W. Zhang, X. Xu, S. Tam, C. Nickerson-Nutter, D. G. Goodwin, J. D. Clark and J. C. McKew, *J. Med. Chem.*, 2009, **52**, 1156.

96. S. J. Kirincich, J. Xiang, N. Green, S. Tam, H. Y. Yang, J. Shim, M. W. Shen, J. D. Clark and J. C. McKew, *Bioorg. Med. Chem.*, 2009, **17**, 4383.

97. C. A. Hewson, S. Patel, L. Calzetta, H. Campwala, S. Havard, E. Luscombe, P. A. Clarke, P. T. Peachell, M. G. Matera, M. Cazzola, C. Page, W. M. Abraham, C. M. Williams, J. D. Clark, W. L. Liu, N. P. Clarke and M. Yeadon, *J. Pharmacol. Exp. Ther.*, 2012, **340**, 656.

98. *R & D Focus Drug News*, January 15, 2007.

99. V. Khurdayan and M. Cullell-Young, *Drug News Perspectives*, 2005, **18**, 277.

100. http://www.clinicaltrials.gov

101. K. Seno, T. Okuno, K. Nishi, Y. Murakami, F. Watanabe, T. Matsuura, M. Wada, Y. Fujii, M. Yamada, T. Ogawa, T. Okada, H. Hashizume, M. Kii, S. Hara, S. Hagishita, S. Nakamoto, K. Yamada, Y. Chikazawa, M. Ueno, I. Teshirogi, T. Ono and M. Ohtani, *J. Med. Chem.*, 2000, **43**, 1041.

102. F. Ghomashchi, A. Stewart, Y. Hefner, S. Ramanadham, J. Turk, C. C. Leslie and M. H. Gelb, *Biochim. Biophys. Acta*, 2001, **1513**, 160.

103. K. Seno, T. Okuno, K. Nishi, Y. Murakami, K. Yamada, S. Nakamoto and T. Ono, *Bioorg. Med. Chem. Lett.*, 2001, **11**, 587.

104. T. Ono, K. Yamada, Y. Chikazawa, M. Ueno, S. Nakamoto, T. Okuno and K. Seno, *Biochem. J.*, 2002, **363**, 727.

105. N. Flamand, S. Picard, L. Lemieux, M. Pouliot, S. G. Bourgoin and P. Borgeat, *Br. J. Pharmacol.*, 2006, **149**, 385.

106. M. Yamamoto, T. Haruna, K. Imura, I. Hikita, Y. Furue, K. Higashino, Y. Gahara, M. Deguchi, K. Yasui and A. Arimura, *Pharmacology*, 2008, **81**, 301.

107. N. Tai, K. Kuwabara, M. Kobayashi, K. Yamada, T. Ono, K. Seno, Y. Gahara, J. Ishizaki and Y. Hori, *Inflamm. Res.*, 2010, **59**, 53.

108. I. Walters, C. Bennion, S. Connolly, P. J. Croshaw, K. Hardy, P. Hartopp, C. G. Jackson, S. J. King, L. Lawrence, A. Mete, D. Murray, D. H. Robinson, L. Stein, E. Wells and W. J. Withnall, *Bioorg. Med. Chem. Lett.*, 2004, **14**, 3645.
109. S. Connolly and A. Mete, PCT WO 0034254, 2000.
110. A. Mete, G. Andrews, M. Bernstein, S. Connolly, P. Hartopp, C. G. Jackson, R. Lewis, I. Martin, D. Murray, R. Riley, D. H. Robinson, G. M. Smith, E. Wells and W. J. Withnall, *Bioorg. Med. Chem. Lett.*, 2011, **21**, 3128.
111. J. Banville, R. Remillard, N. Balasubramanian, G. Bouthillier and A. Martel, US 2002037875, 2002.
112. M. Lehr, *J. Med. Chem.*, 1997, **40**, 2694.
113. S. Tries, W. Neupert, W. Albrecht, M. Lehr and S. Laufer, Abstract of Papers, 10th National Meeting of the Inflammation Research Association, Hot Springs, USA, 2000.
114. S. Tries, K. Neher, S. Laufer, W. M. Abraham and M. Lehr, *Mediat. Inflamm.*, 1999, **8**, S123.
115. A. Ghasemi, A. Schulze Elfringhoff and M. Lehr, *J. Enzyme Inhib. Med. Chem.*, 2005, **20**, 429.
116. J. Ludwig, S. Bovens, C. Brauch, A. Schulze Elfringhoff and M. Lehr, *J. Med. Chem.*, 2006, **49**, 2611.
117. S. Bovens, A. Schulze Elfringhoff, M. Kaptur, D. Reinhardt, M. Schäfers and M. Lehr, *J. Med. Chem.*, 2010, **53**, 8298.
118. J. Fabian and M. Lehr, *J. Pharm. Biomed. Anal.*, 2007, **43**, 601.
119. A. Drews, S. Bovens, K. Roebrock, C. Sunderkötter, D. Reinhardt, M. Schäfers, A. van der Velde, A. Schulze Elfringhoff, J. Fabian and M. Lehr, *J. Med. Chem.*, 2010, **53**, 5165.
120. K. Roebrock, M. Wolf, S. Bovens, M. Lehr and C. Sunderkötter, *Br. J. Dermatol.*, 2012, **166**, 306.
121. J. M. Barnett, G. W. McCollum and J. S. Penn, *Invest. Ophthalmol. Vis. Sci.*, 2010, **51**, 1136.
122. M. Takagi, K. Ishimitsu and T. Nishibe, EP1277743, 2001.
123. A. Linkous and E. Yazlovitskaya, *Cell. Microbiol.*, 2010, **12**, 1369.
124. W. C. Jeong, K. J. Kim, H. W. Ju, H. K. Back, H. K. Kim, S. Y. Im and H. K. Lee, *Anticancer Res.*, 2010, **30**, 3421.
125. F. Caiazza, N. S. McCarthy, L. Young, A. D. Hill, B. J. Harvey and W. Thomas, *Br. J. Cancer*, 2011, **104**, 338.
126. F. Caiazza, B. J. Harvey and W. Thomas, *Mol. Endocrinol.*, 2010, **24**, 953.
127. S. C. Lim, H. Cho, T. B. Lee, C. H. Choi, Y. D. Min, S. S. Kim and K. J. Kim, *Yonsei Med. J.*, 2010, **51**, 692.
128. A. G. Linkous, E. M. Yazlovitskaya and D. E. Hallahan, *J. Natl Cancer Inst.*, 2010, **102**, 1398.
129. G. Tosato, M. Segarra and O. Salvucci, *J. Natl Cancer Inst.*, 2010, **102**, 1377.

CHAPTER 3

Leukotriene A₄ Hydrolase: Biology, Inhibitors and Clinical Applications

CHERYL A. GRICE,* ANNE M. FOURIE AND
ALICE LEE-DUTRA

Johnson & Johnson Pharmaceutical Research & Development, 3210
Merryfield Row, San Diego, California 92121, USA
*Email: cheri@abidetx.com

3.1 Background: Eicosanoids and Leukotrienes

Arachidonic acid (AA), an essential fatty acid, is the metabolic precursor for several classes of bioactive molecules, including leukotrienes, lipoxins, prostaglandins and thromboxanes (Figure 3.1). These molecules, members of the eicosanoid family, are synthesized from membrane phospholipids *via* the action of cytosolic phospholipase A_2 alpha (cPLA$_2$α) and are implicated in many inflammatory disorders.[1–3]

The prostaglandins, thromboxanes and prostacyclins, collectively known as the prostanoids, signal through GPCRs and play an active role in the inflammatory process. Significant work has focused on inhibition of the cyclooxygenase 1 and 2 enzymes (COX-1 and COX-2) that lead to their production and the receptors that are involved in the signalling pathways. The role of the prostanoids in the resolution phase of inflammation and other homeostatic

RSC Drug Discovery Series No. 26
Anti-Inflammatory Drug Discovery
Edited by Jeremy I Levin and Stefan Laufer
© The Royal Society of Chemistry 2012
Published by the Royal Society of Chemistry, www.rsc.org

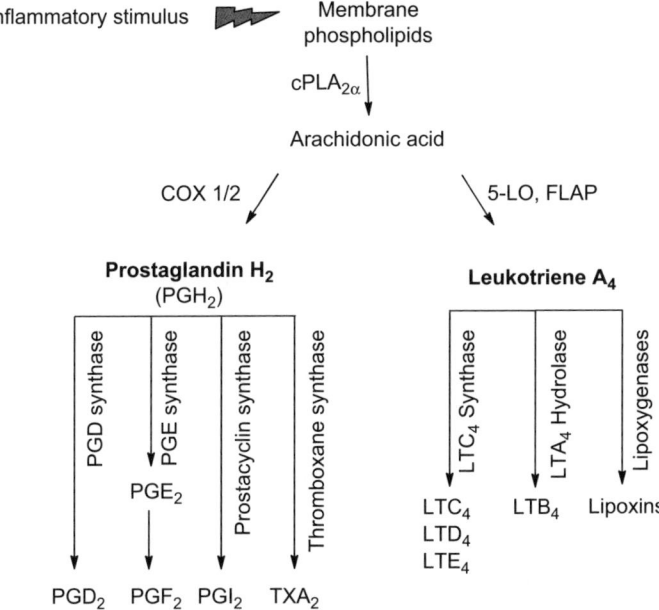

Figure 3.1   Arachidonic acid cascade.

processes is a topic of intense investigation with particular focus on elucidating the optimal point for intervention within this part of the arachidonic pathway.[4]

Similar to the prostanoids, the leukotriene component of the arachidonic acid cascade also contains numerous soluble mediators that signal through GPCRs and contribute to the inflammatory process.[5] The name leukotriene is derived from the original isolation source, leukocytes, and the common structural feature, a conjugated triene.[6] The pro-inflammatory mediator LTB₄ is a potent chemoattractant and activator of neutrophils and a chemoattractant of eosinophils, macrophages, mast cells, T-cells, dendritic cells (DC), smooth muscle cells and keratinocytes.[7–15] Increased levels of LTB₄ are implicated in disorders such as asthma,[3,16–24] chronic obstructive pulmonary disease (COPD),[25–27] inflammatory bowel disease (IBD),[28–30] rheumatoid arthritis (RA),[31,32] cardiovascular disease[33–42] and cancer.[43–50] The leukotrienes LTC₄, LTD₄ and LTE₄, also known as cysteinyl leukotrienes because of the presence of a cysteine in their structure, were referred to as the slow-reacting substance of anaphylaxis (SRS-A) by Brocklehurst.[51] A review of the work describing that SRS-A is comprised of the cysteinyl leukotrienes was published in 1981.[52] Several more recent reviews have appeared with a strong focus on the cysteinyl leukotriene receptors and marketed antagonists.[53–55] The lipoxins, LXA₄ and LXB₄, are endogenous anti-inflammatory agents that are thought to participate in the inflammation resolution phase.[56]

The biosynthetic pathway detailing the formation of the leukotrienes is shown in Figure 3.2. Starting at the top of the cascade, cPLA₂α plays an

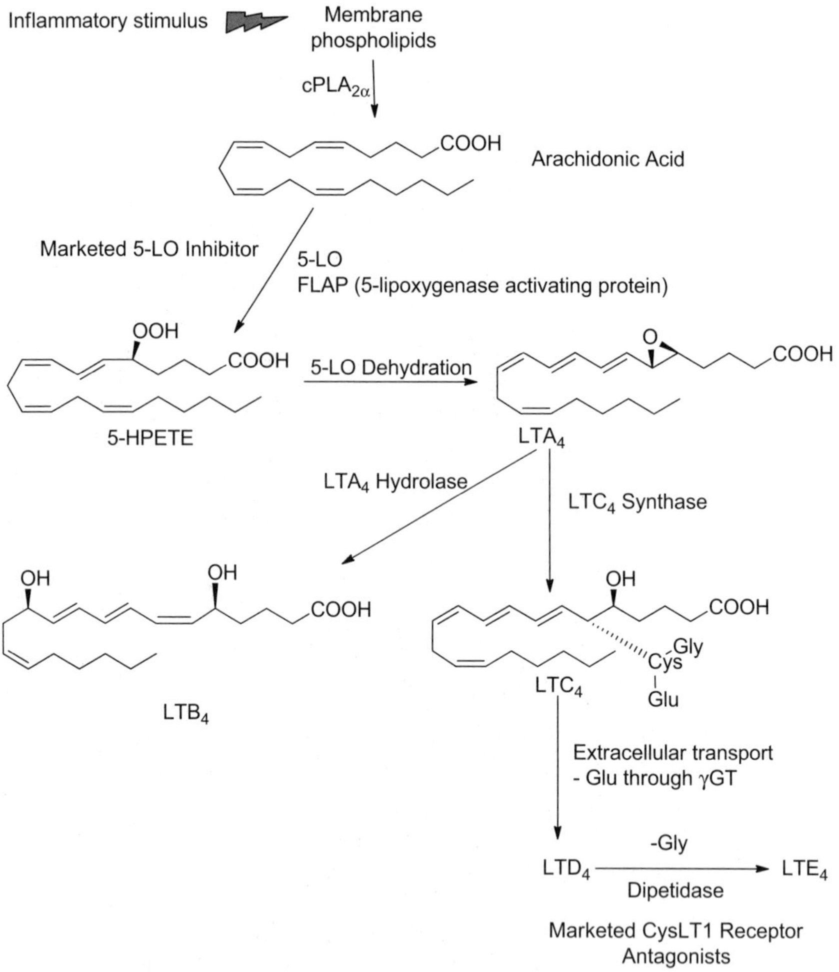

Figure 3.2   Biosynthetic pathway of the leukotrienes.

essential role in the initiation of AA metabolism and its activation is tightly regulated by Ca^{2+} and kinase mediated phosphorylation. Knockout mice ($cPLA_2\alpha^{-/-}$) have been generated, and mast cells and macrophages from those animals show reduced production of prostaglandins and leukotrienes.[57,58] Several groups have targeted $cPLA_2\alpha$ for the treatment of inflammation, RA, asthma and osteoarthritis (OA), given its central role in the arachidonic acid cascade.[59] However, a human genetic deficiency[60] and clinical trial results[61] have revealed gastrointestinal side-effects associated with decreased $cPLA_2\alpha$ activity, thus limiting the potential of pharmacological intervention at this point in the AA pathway.

   Many pharmaceutical companies have evaluated inhibitors of 5-lipoxygenase (5-LO) and 5-lipoxygenase activating protein (FLAP).[62] These inhibitors prevent or reduce the formation of both LTB_4 and the cysteinyl leukotrienes by either inhibiting the 5-LO enzyme or, in the case of FLAP inhibitors, preventing the association of cytosolic 5-LO with arachidonic acid.[63,64] Merck & Co., Inc. (Merck) and Bayer AG (Bayer) progressed three FLAP inhibitors, MK-886, MK-591 and BAY X-1005, into clinical trials in the 1990s.[65–67] Numerous clinical trials have been conducted on inhibitors of 5-LO, and several reviews highlight the history of this area.[62,68–70] A few additional publications have appeared disclosing selective inhibitors of 5-LO[71–73] with the bulk of the published work focusing on dual inhibition of 5-LO and other enzymes within the arachidonic acid pathway. There has been a recent resurgence of activity within the 5-LO and FLAP arena,[74,75] which may be due in part to findings by scientists at deCODE Genetics (deCODE) demonstrating a genetic link between a haplotype in the ALOX5AP gene (the gene encoding FLAP) with myocardial infarction and stroke.[37]

Antagonists of the binding of LTB_4 and the cysteinyl leukotrienes to their receptors have also received significant attention over the past several decades. The pre-clinical and clinical results of these efforts have been reviewed.[54,55,76–79] Much of the early reported work in this area, and many of the compounds that entered clinical trials, pre-dated the publications detailing the cloning and characterization of the GPCRs responsible for downstream signalling.[80–86] The LTB_4 receptors are termed BLT1 (high affinity, cloned in 1997, expressed on neutrophils and T-cells)[9,15,85,87] and BLT2 (low affinity, expressed in the liver, spleen and peripheral leukocytes).[84] Recent reports also established the presence of BLT1 and BLT2 receptors on mast cells[13] and dendritic cells.[7] Likewise, the publications detailing the cloning and characterization of the cysteinyl leukotriene receptors, CysLT1 (expressed on spleen, smooth muscle, lung, intestine and leukocytes including eosinphils, monocytes and B lymphocytes)[5,81,87] and CysLT2 (expressed on spleen, adrenal medulla, lung, heart, brain and leukocytes including eosinophils)[5,79] appeared during the same timeframe.

In stark contrast, LTC_4 synthase, the enzyme responsible for converting LTA_4 to LTC_4 has received comparatively little attention until more recently. Biolipox AB has disclosed inhibitors of LTC_4 synthase as recently as 2010.[88–92] The potential advantage of a synthase inhibitor *versus* montelukast, which selectively antagonizes the high-affinity LTD_4 receptor CysLT1, is that it prevents the formation of all of the cysteinyl leukotrienes and consequently the associated biological signalling.

3.2 Leukotriene A₄ Hydrolase

LTA_4 is also a substrate of the enzyme leukotriene A_4 hydrolase (LTA_4H), which is a ubiquitously distributed 69 kDa bifunctional, zinc-containing, cytosolic enzyme. As a hydrolase, its more widely known function, LTA_4H stereospecifically catalyzes the transformation of the unstable epoxide LTA_4 to the diol leukotriene B_4 (LTB_4) as shown in Figure 3.2. In addition, LTA_4H functions as an anion-dependent aminopeptidase, processing arginyl di- and tri-peptides with high efficiency. Recently, a role for this aminopeptidase activity in the degradation and inactivation of the tripeptide proline-glycine-proline (PGP) was described.[93]

It is important to note that LTA_4H-deficient mice ($LTA_4H^{-/-}$) have been generated and that these mice develop normally and are healthy.[94] Studies of LTA_4H-deficient mice in several models of inflammation clearly establish the contribution of LTB_4 to the inflammatory process and strongly support the pursuit of LTA_4H inhibitors as potential treatments for inflammatory diseases.[31,94] Additionally in 2006, scientists at deCODE showed that a haplotype spanning LTA_4H (HapK) also conferred modest risk of myocardial infarction and stroke,[95] and this observation, in conjunction with their other findings regarding FLAP, may have contributed to an increased interest in targets within the arachidonic acid pathway.

Inhibitors of LTA_4H may have advantages over 5-LO or FLAP inhibitors, both of which prevent the formation of LTA_4, a source of the anti-inflammatory lipoxins.[56] In contrast, the inhibition of LTA_4H would not hinder the production of the lipoxins, as it acts downstream of this branch point in the cascade. In fact, researchers at Johnson & Johnson Pharmaceutical Research & Development, L.L.C. (J&J PRD, L.L.C.) have shown that an inhibitor of LTA_4H, when tested in murine whole blood and when dosed orally at 30 mg/kg in a zymosan-induced peritonitis model, selectively inhibited LTB_4 production without affecting cysteinyl leukotriene production and maintained the production of the anti-inflammatory mediator LXA_4.[96] Averting the formation of LTB_4 eliminates the potential challenges associated with antagonizing two signalling receptors, BLT1 and BLT2, which may be needed in order to achieve full efficacy *in vivo*.[97]

3.3 LTA_4H – Pre-clinical and Human Genetic Rationale

3.3.1   Respiratory Disease

Leukotrienes are well-established contributors to inflammation and allergic broncho-constriction in respiratory diseases, and a number of leukotriene-modifying drugs are approved for use in the treatment of asthma and/or allergic rhinitis. These are the 5-lipoxygenase inhibitor, zileuton (now available as extended release form ZyfloCR®) and the cysteinyl LT receptor antagonists, montelukast (Singulair®), pranlukast (Onon®/Ultair®) and zafirlukast (Accolate®). While cysteinyl leukotriene antagonists are effective in a subset of mild asthmatics, inhibition of LTB_4 synthesis may be beneficial in severe asthma and viral exacerbations, as well as COPD, where neutrophilic inflammation is more prominent.[98–100] However, recent data[93] have brought into question the potential of LTA_4H as a target for neutrophilic asthma. LTA_4H has been identified in mice as having a role in the peptidase degradation of PGP.[93] Similar to LTB_4, PGP and the more potent acetylated tripeptide

AcPGP (not susceptible to LTA_4H degradation) are neutrophil chemoat-tractants. This suggests a disparate role of LTA_4H in the pathogenesis of inflammatory diseases, both promoting neutrophilia through LTB_4 production, as well as decreasing neutrophilic inflammation through the degradation of PGP. It has been reported, however, that neutrophilia in the bronchoalveolar lavage (BAL) fluid in a mouse model of allergic airway inflammation is reversed with an inhibitor of LTA_4H.[23] Additionally, neutrophilia was inhibited in rat colitis and mouse peritonitis models when animals were dosed with LTA_4H inhibitors.[96,101] Additional work is needed to further elucidate the specific relevance of LTB_4 and PGP/AcPGP in acute neutrophil recruitment and chronic inflammation. Recently identified compounds that inhibit only the peptidase, but not the hydrolase activity, of LTA_4H, may be useful in such studies.[102] It will be particularly important to evaluate the role of PGP in different human respiratory diseases and utilize highly selective LTA_4H inhibitors, which do not inhibit other enzymes, to evaluate the potential role of other proteases in PGP degradation.

Several pre-clinical findings suggest that the inhibition of LTB_4 synthesis by LTA_4H inhibitors may have utility in respiratory diseases.[17,103] A number of different studies in BLT1-deficient or LTA_4H-deficient mice, as well as in antisense-mediated down-regulation of BLT2, have demonstrated a role for LTB_4 in the development of airway inflammation, hyper-responsiveness and cytokine production in mice in response to allergen sensitization and challenge.[24,104–109] Pharmacological studies using selective inhibitors of LTA_4H have added additional findings to further elaborate this potential. In a chronic model of airway inflammation, induced by intra-nasal sensitization with house dust mite extract, a relevant allergen for human asthma, pharmacological LTA_4H inhibition reduced airway inflammation, hyper-responsiveness, mucin secretion and airway LTB_4, cytokines and chemokines.[110] In a mast cell dependent model of allergic airway inflammation, Rao *et al.*[23] from J&J PRD, L.L.C. showed that treatment with a selective inhibitor of LTA_4H improved multiple parameters in airway hyper-responsiveness (AHR) and lung function, as well as decreased airway inflammation, cytokines and mucin, but increased levels of the anti-inflammatory lipoxin A_4. Inhibiting LTA_4H also decreased antigen-specific IgE and IgG1 in serum, as well as recruitment of both CD4 and CD8 cells to the lung tissue, and migration of dendritic cells from the lung to lymph nodes. A recent study by Waseda *et al.*[111] in established disease in mice showed that pharmacological blocking of BLT1 with CP-105,696 ((+)-1-(3S,4R)-[3-(4-phenylbenzyl)-4-hydroxychroman-7-yl]cyclopentane carboxylic acid) had no effect on the early asthmatic response (EAR), but significantly suppressed the late asthmatic response (LAR), including AHR and inflammation. The same BLT1 antagonist had also been shown to reduce AHR induced by multiple antigen challenges in a primate model.[112]

While the pre-clinical data on the role of LTA_4H and LTB_4 in respiratory disease are compelling, it will be important to test this hypothesis in carefully designed human clinical trials. Further support for a role of LTA_4H in human respiratory disease is provided by a number of human genetic associations.

A particular SNP as well as a particular haplotype for the LTA$_4$H gene have been linked with an increased risk for the development of asthma in 341 Caucasian families in the United Kingdom, and this risk is further increased in individuals with "at-risk" SNPs for both LTA$_4$H and ALOX5AP genes.[113] In an additional study, variation in the LTA$_4$H gene in 61 US Caucasian subjects was shown to be associated with a risk of exacerbations while on treatment with the CysLT1 antagonist, montelukast.[114] Similarly, a recent study in 52 Japanese patients showed significant associations between SNP genotypes in the LTA$_4$H gene and changes in peak expiratory flow (PEF) and forced expiratory volume in 1 second (FEV1) in response to montelukast treatment.[115]

Two recent publications have examined the role of ALOX5AP and LTA$_4$H polymorphisms in the risk for asthma, and in augmentation of bronchodilator responsiveness to albuterol in Latinos.[116,117] SNPs in the ALOX5AP and LTA$_4$H genes were analyzed for association with asthma in 687 Mexican and Puerto Rican families. Five new polymorphisms in LTA$_4$H were identified, two of which were protective for asthma. In Mexican patients, LTA$_4$H polymorphisms were associated with baseline lung function and IgE levels, while in the Puerto Rican families a minor allele for ALOX5AP was protective for asthma. A gene-gene interaction was observed between LTA$_4$H and ALOX5AP SNPs[117] similar to observations in the Caucasian UK families.[113] In Puerto Rican, but not Mexican, asthmatics, LTA$_4$H polymorphisms modified the augmentation of bronchodilator response by leukotriene modulators, and this effect was increased by ALOX5AP polymorphisms. In these studies, the use of cysteinyl leukotriene antagonists or the 5-LO inhibitor zileuton were not distinguished, despite their mechanisms of action and effects on LTB$_4$ being clearly distinct. Future studies to separate these two mechanisms and their relative influence on bronchodilator responses may be useful. Clinical trials with leukotriene modulators in pulmonary diseases are summarized in the last section of this chapter.

A recent study in zebrafish identified associations between mutants in the LTA$_4$H gene and susceptibility to Mycobacterium marinum.[118] The investigators found that heterozygosity at the human LTA$_4$H locus correlated with protection from tuberculosis in a Vietnamese cohort (692 TB patients and 759 controls), and from severe leprosy in a Nepalese cohort, suggesting roles for LTA$_4$H in resistance to mycobacterial infection. In contrast, a subsequent study in 3703 pulmonary tuberculosis patients and 5412 healthy controls in the UK found no role for polymorphisms in the LTA$_4$H gene in susceptibility to clinical pulmonary tuberculosis.[119] Further studies are needed to determine whether the findings in the smaller study were specific to the particular ethnic groups studied, and whether other factors may have influenced the outcomes.

3.3.2 Gastrointestinal Disease

Inflammatory bowel disease is characterized by elevated inflammatory mediators, including LTB$_4$, in colon tissue, and increased LTA$_4$H-positive cells have been observed in colonic biopsies from ulcerative colitis and Crohn's patients.[28]

However, the precise role of LTB_4 in the mechanisms underlying Crohn's disease or ulcerative colitis is uncertain. While an LTA_4H inhibitor, SC-57461A (compound **12**, Figure 3.6), was reported to reduce spontaneous colitis in cotton-top tamarins, reducing LTB_4 levels through the inhibition of FLAP or 5-LO did not yield positive results in human clinical trials.[120] This may be related to inhibition of anti-inflammatory lipoxin synthesis by these upstream inhibitors, while LTA_4H inhibitors have been shown to maintain or increase lipoxin synthesis.[23,96] Two different orally available and selective LTA_4H inhibitors have been tested by J&J PRD, L.L.C. in a 2,4,6-trinitrobenzene sulfonic acid (TNBS)-induced colitis model in rats. Compounds **29**[101] and **30**[121] (Figure 3.9) dose-dependently reduced inflammatory damage to the colon, an effect that correlated with decreased levels of colonic LTB_4, TNFα, IL-6 and neutrophil myeloperoxidase. Compound **29** achieved superior efficacy relative to the 5-LO inhibitor zileuton in reducing LTB_4 and IL-6 levels in colon tissue.[101] These data support a role for LTB_4 in driving colonic inflammation.

Interestingly, a recent study comparing mild dextran sulfate sodium (DSS)-induced colitis in wild-type, BLT1-deficient and BLT2-deficient mice showed that BLT1 knockout mice did not develop colitis, similarly to wild-type mice, while BLT2 knockouts showed aggravated DSS colitis, with severe weight loss and inflammation.[122] BLT2 is a low-affinity receptor for LTB_4 and a high-affinity receptor for 12-(S)-hydroxyheptadeca-5Z,8E,10E-trienoic acid (12-HHT), a COX-1-derived fatty acid that is abundant in intestine.[123] The aggravated DSS colitis suggests that BLT2 plays a protective role, possibly through 12-HHT-mediated enhancement of colon epithelial barrier function. Thus, it is possible that LTB_4 drives colonic inflammation through BLT1-mediated recruitment and activation of leukocytes, while 12-HHT plays a protective role through BLT2. These data are not inconsistent with a potential application of LTA_4H inhibitors as therapeutics in IBD, as LTA_4H is required for LTB_4 synthesis, but not 12-HHT.

3.3.3 Inflammatory Arthritis

LTB_4 is elevated in the synovial fluid and serum of patients with rheumatoid arthritis[124,125] and BLT1 and BLT2 are expressed in synovial leukocytes and synovial tissue, respectively, of RA, but not OA patients.[126] In pre-clinical models of inflammatory arthritis, BLT1-deficient mice, as well as mice lacking both BLT1 and BLT2, are protected from development of collagen-induced arthritis.[127] Similarly BLT1$^{-/-}$ mice, or wild-type mice dosed with the BLT1 antagonist CP-105,696, are protected in the K/BxN serum transfer model.[128] In a recent serum transfer model study, BLT2-deficient mice showed a significant reduction in disease incidence and severity, bone and cartilage loss, all of which could be reconstituted by a bone marrow transplant of BLT2-expressing cells.[97] Considering the previous data showing that BLT1-deficient mice do not develop arthritis, and that BLT1 expression on neutrophils is critical for delivery of IL-1 into the joint to initiate arthritis,[129] the overall conclusion is that BLT1 and BLT2 have non-redundant roles in the development of murine inflammatory arthritis. While BLT2 has lower affinity for LTB_4 than for the COX-1 derived ligand 12-HHT, LTB_4 appears to be the major chemotactic factor for BLT2.[97] Interestingly, both BLT1 and BLT2 are expressed in human synovial tissue and leukocytes of patients with RA.[126] This may have important implications for therapeutic applications of antagonists which are selective for one of the two receptors.

The non-redundant roles of BLT1 and BLT2, and apparent dominant role of LTB_4 *versus* 12-HHT for chemotaxis through BLT2, suggest that decreasing LTB_4 through LTA_4H inhibition may have therapeutic potential in human RA. Consistent with this hypothesis, $LTA_4H^{-/-}$ and $5\text{-}LO^{-/-}$ mice are protected from disease development in the K/BxN serum transfer model of arthritis.[31] In murine collagen-induced arthritis, researchers at J&J PRD, L.L.C. have found that oral dosing of a selective inhibitor of LTA_4H at 30 mg/kg q.d., attenuated clinical score and decreased inflammation, pannus formation and damage to bone and cartilage.[130] In contrast, a study by Anderson *et al.*[131] showed little impact of the LTA_4H inhibitor SC-57461A when dosed alone in a collagen-induced arthritis model, but did demonstrate a synergistic therapeutic effect from the combination of SC-57461A and the COX-2 inhibitor SC-58236. Clinical trials with potent and selective LTA_4H inhibitors will be needed to fully evaluate the contribution of LTB_4 to human RA.

3.3.4   Cardiovascular Disease

Considerable human and animal experimental data exist linking LTB_4-driven mechanisms and atherosclerosis.[11,42,132] For example, an increased level of LTB_4 and increased expression of its receptors in human atherogenic plaques have been observed.[133,134] This is consistent with the increased expression of components of the 5-lipoxygenase pathway including 5-LO, FLAP and LTA_4H in these lesions.[39,134] More strikingly, the LTB_4 receptor antagonist CP-105696 markedly reduced lesion size in atherosclerosis prone mice.[132] In addition, in an atherosclerosis model in $BLT1^{-/-}$ mice crossed with $ApoE^{-/-}$ mice, reduced lesions were observed in the early stages of disease progress.[11] Furthermore, Back *et al.*[135] showed that a FLAP inhibitor significantly reduced atherosclerotic lesion size and LTB_4 production in apoE-deficient mice expressing dominant negative TGFβRII in T-cells. While these findings establish a role of LTB_4 in atherogenesis, little is known about the molecular mechanisms involved. Aiello *et al.*[132] showed that LTB_4 influenced monocyte recruitment into atherosclerotic lesions by up-regulating CD11b expression, thereby mediating monocyte adhesion, and that monocyte chemotactic protein-1 (MCP-1) played a role in the LTB_4-mediated atherogenic effects. In support of these data, Huang *et al.* showed that LTB_4 induced MCP-1 production in primary human monocytes,[136] and Subbarao *et al.*[42] observed LTB_4-induced up-regulation of MCP-1 gene expression in rat basophilic leukaemia cells transfected with human BLT1.

The leukotriene pathway has also been linked through multiple human genetic associations to cardiovascular disease. Thus, promoter variants of 5-LO showed associations with atherosclerosis that were modulated by diet.[137] Variants of the ALOX5AP gene (HapA and HapB) were shown to be associated with the risk of myocardial infarction and stroke in Icelandic and Scottish subjects, respectively,[37,138] and with in-stent restenosis in a group of 92 subjects.[139] A specific haplotype in LTA$_4$H (Hap K) was shown to confer risk for myocardial infarction, particularly in African-Americans.[95] Recently Crosslin *et al.*[140] performed genetic association studies of early onset coronary artery disease (EOCAD) in familial and non-familial datasets. These authors found that ALOX5AP HapA and LTA$_4$H HapK showed significant associations with EOCAD in the non-familial dataset, but no significant associations with ALOX5AP HapB were detected. The study also identified an association of SNPs in 5-LO with cardiovascular disease. Hartiala *et al.*[141] recently evaluated the genetic contribution of leukotriene pathway genes to coronary artery disease (CAD) in 4512 Caucasian and African-American subjects. Their studies showed an association of ALOX5 promoter polymorphisms with the risk of CAD in African-Americans, and association of the HapK haplotype and a particular SNP in LTA$_4$H with increased risk in Caucasians. Monocytes from subjects with at-risk variants produced higher LTB$_4$ levels upon stimulation. While similar data were found for granulocytes from Icelandic HapK individuals,[95] clear functional correlates for the genetic associations in the leukotriene pathway, and for LTA$_4$H in particular, remain to be clarified.

3.3.5 Cancer

There is considerable evidence that inflammation underlies malignancy, although the exact mechanisms are unclear.[142] In particular, arachidonic acid metabolites have been implicated in carcinogenesis.[143] It has been demonstrated that inhibition of the cyclooxygenase pathway of AA metabolism, particularly inhibition of COX-2, decreases cancer incidence and inhibits tumour growth, particularly in the case of gastrointestinal epithelial tumours.[143] In addition to a role for prostaglandins, it has been shown that 5-lipoxygenase-derived leukotrienes, including LTB$_4$, play a role in tumourigenesis.[144] 5-LO is over-expressed in tumours, and leukotrienes activate tumour cell proliferation and inhibit apoptotic pathways and the inhibition of 5-LO or leukotriene receptor signalling inhibits tumour growth.[144]

In addition, BLT1 and BLT2 are up-regulated in pancreatic cancer, and BLT2 has been shown to play a role in the invasion, metastasis and survival of bladder cancer cells.[145,146] Furthermore, LY293111, a BLT1 antagonist, 5-LO inhibitor and PPARγ agonist was shown to inhibit pancreatic tumour growth *in vitro* and *in vivo*.[142] Unfortunately, when LY293111 was tested in combination with gemcitabine in pancreatic and non-small-cell lung cancer trials, no enhancement of gemcitabine efficacy was observed.[147-149]

The enzyme LTA$_4$H has also been connected to cancer cell growth; LTA$_4$H is over-expressed in several cancer cell lines, including colorectal cancers. Knockdown of LTA$_4$H, or inhibition of its enzymatic activity by [6]-gingerol, has been shown to inhibit anchorage-independent growth of colon cancer cells.[46] Recently, resveratrol, an active compound found in red wine which slows tumour growth, has also been shown to bind and inhibit LTA$_4$H.[48] In this study, knockdown of LTA$_4$H, or treatment with resveratrol, suppressed anchorage-independent growth of pancreatic cancer cells and LTB$_4$ production. In cells where LTA$_4$H had been knocked down by shRNA, the inhibition by resveratrol on anchorage-independent growth was reduced, suggesting that LTA$_4$H may be the target for the anti-cancer effects of resveratrol.

3.4   Small-molecule Approaches Targeting LTA$_4$H

Early inhibitors of LTA$_4$H focused on the synthesis and characterization of LTA$_4$ structural mimics, as well as known zinc-containing metalloproteinase inhibitors and related analogues, as illustrated by bestatin,[150] kelatorphan,[151] SA6541[152] and **1** (Figure 3.3).[153] This work has been previously reviewed by Penning *et al.*,[120] and given the lack of cellular activity achieved by these compounds and the unlikely event that they would advance to approval, the focus of this section will be on inhibitors of LTA$_4$H that were disclosed after 1996.

Most enzymatic data discussed below represent inhibition of the hydrolase activity of LTA$_4$H using human recombinant enzyme pre-incubated with test compound prior to addition of LTA$_4$ unless otherwise noted. The production of LTB$_4$ was quantified by enzyme linked immunosorbent assay (ELISA; G. D. Searle & Co., Janssen Pharmaceutica NV and select data from Schering AG) or mass spectrometry (deCODE Genetics). This measurement is abbreviated as "hLTA$_4$H IC$_{50}$" and is likely comparable across sections. In the section reviewing work published by Schering AG, enzymatic data represent inhibition of the peptidase activity, quantitated using a colorimetric readout measuring the conversion of L-alanine-*p*-nitroanilide to alanine and highly coloured *p*-nitroaniline. The measurement for this work is abbreviated "peptidase IC$_{50}$". In the section reviewing work published by Santen Pharmaceutical Co., Ltd, the source of the LTA$_4$H enzyme is guinea-pig lung and the quantification of LTB$_4$ production was conducted by mass spectrometry. This measurement is abbreviated as "LTA$_4$H IC$_{50}$" and the comparability across sections is unknown.

The whole blood data discussed below represent inhibition of LTB$_4$ production after stimulation with calcium ionophore, A23187 in either human (1:4 dilution with culture medium, G. D. Searle and Schering AG) or mouse whole blood (1:15 dilution with culture medium, Janssen Pharmaceutica NV). This is denoted as either "hWB IC$_{50}$" or "mWB IC$_{50}$", respectively. Whole blood data generated within the same species are likely comparable while correlation of data across species is unknown. The specific experimental details for the

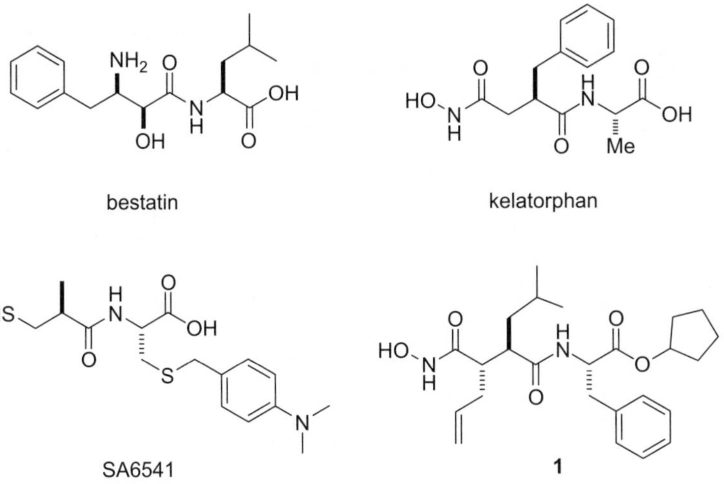

bestatin

kelatorphan

SA6541

1

Figure 3.3 Early LTA$_4$H inhibitors containing potential zinc-chelating groups.

enzymatic and whole blood assays are available within the referenced publications, *vide infra*.

3.4.1 X-ray Crystallography

The high-resolution (1.95 Å) crystal structure of human LTA$_4$H complexed with bestatin, a zinc-chelating aminopeptidase inhibitor, was published in 2001.[154] The structure revealed that the protein is folded into three domains (N-terminal, catalytic and C-terminal) and, although the domains are packed closely together, they create a deep cleft or binding pocket allowing access to the catalytic Zn^{2+} site. From this crystal structure, key observations were made with regard to the nature of the binding pocket. The Zn^{2+} is coordinated to three amino acids, His295, His299 and Glu318. The N-terminal domain defines the entry to the cavity, and a large portion of the surface appears to be solvent exposed. More importantly, a narrow hydrophobic cavity in the C-terminal domain stretches beyond the catalytic site. The shape and curvature of the cavity provides important insight into how LTA$_4$ binds to the enzyme, as well as potentially useful information for the design of small molecule inhibitors. A later publication by Thunnissen *et al.* describes the crystal structures of other inhibitors complexed with the enzyme, including captopril (*vide infra*), offering additional insight into potential binding interactions.[155] Figure 3.4 shows a cartoon schematic of bestatin and compound **26** bound to LTA$_4$H; key residues have been labelled. The importance of these residues in driving the SAR of various programs will be presented below.

Notably, since the publication of the crystal structure of LTA$_4$H, the X-ray crystal structures of several other proteins within the pathway have been

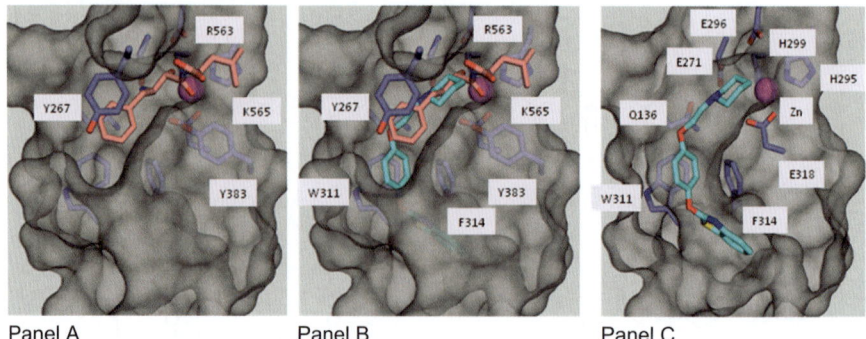

Panel A   Panel B   Panel C

Figure 3.4   Docking representations of LTA$_4$H using pdb1HS6[153] with key residues
noted. Panel A: Bestatin complexed with LTA$_4$H; Panel B: Bestatin with
compound **26** docked in LTA$_4$H; Panel C: Compound **26** docked in
LTA$_4$H. Graphics were prepared courtesy of Scott D. Bembenek using
Pymol v.1.4.1, Delano Scientific L.L.C.

disclosed, further increasing interest in the pathway: FLAP in 2007,[156] LTC$_4$S
also in 2007[157,158] and 5-LO in 2011.[159]

3.4.2   G. D. Searle & Co.

G. D. Searle & Co described non-chelating inhibitors of LTA$_4$H in the late
1990s in a series of patent applications[160–165] that appeared prior to the publication of the LTA$_4$H crystal structure, and which were further detailed in
subsequent manuscripts. The original lead structure, SC-22716, shown in
Figure 3.5, was identified through screening and showed inhibition of recombinant human LTA$_4$H hydrolase and peptidase (*not shown*) activity and inhibition of cellular LTB$_4$ production in human whole blood.[166] This compound,
however, lacked activity in an *ex vivo* mouse pharmacodynamic model (PK/
PD), prompting further investigation to identify orally active inhibitors.[166]
Optimization of this initial lead, summarized below, resulted in a series of
related analogues, which demonstrated good *in-vitro* activity as well as oral
activity in the *ex vivo* mouse PK/PD model.[166]

Modification of the biphenyl moiety of SC-22716 with various linkers
between the two aromatic rings identified a selection of one- and two-atom
linkers that maintained or enhanced potency. In the cases of CH$_2$ (**3**), O (**4**) and
NHCH$_2$ (not shown) linkers, increased activity in the *ex vivo* mouse PK/PD
model was noted. Based on these results and for ease of synthesis, the
remaining SAR were determined using either the CH$_2$ or the O linker.[166]

Investigation of the amine moiety of these inhibitors revealed that neither the
enzymatic nor the whole blood assay was significantly affected by sterics and
that both cyclic and acyclic amines were tolerated. Compound **5** demonstrated
notably improved enzymatic and cellular potency. Investigation of the linker

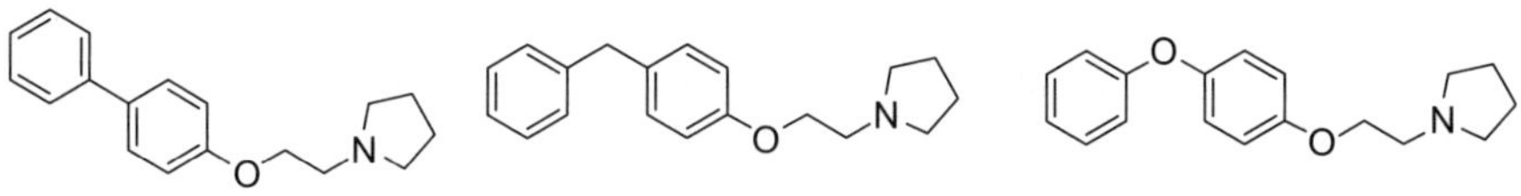

2, SC-22716
hLTA$_4$H IC$_{50}$ = 200 nM
hWB IC$_{50}$ = 790 nM

3
hLTA$_4$H IC$_{50}$ = 26 nM
hWB IC$_{50}$ = 120 nM

4
hLTA$_4$H IC$_{50}$ = 30 nM
hWB IC$_{50}$ = 79 nM

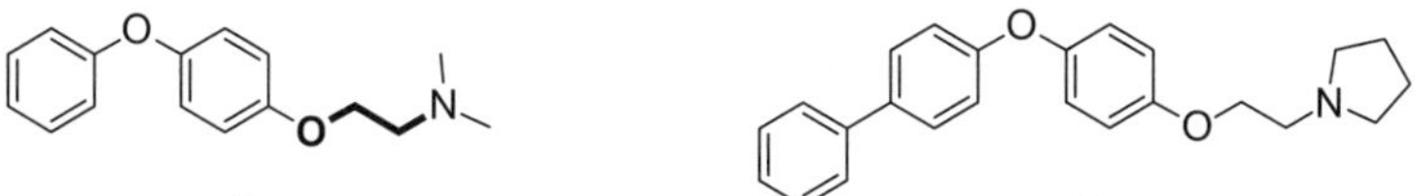

5
hLTA$_4$H IC$_{50}$ = 6 nM
hWB IC$_{50}$ = 73 nM

6
hLTA$_4$H IC$_{50}$ = 7 nM
hWB IC$_{50}$ = 140 nM

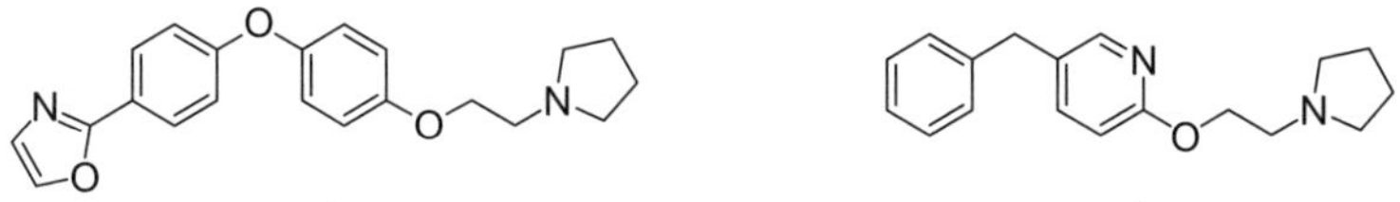

7
hLTA$_4$H IC$_{50}$ = 43 nM
hWB IC$_{50}$ = 55 nM

8
hLTA$_4$H IC$_{50}$ = 26 nM
hWB IC$_{50}$ = 120 nM

9
hLTA$_4$H IC$_{50}$ = 36 nM
hWB IC$_{50}$ = 530 nM

10
hLTA$_4$H IC$_{50}$ = 2.9 μM
hWB IC$_{50}$ = 4.3 μM

Figure 3.5   First-generation non-chelating LTA$_4$H inhibitors reported by G. D. Searle.

connecting the central aromatic ring to the amine (oxyethyl, highlighted in bold in compound **5**), including modifications to the heteroatom and the linker length were unsuccessful in increasing activity. Investigation of the terminal aromatic ring was more fruitful. Several alternatives to the phenyl ring were identified, most aromatic or heteroaromatic in nature. However, a cyclohexyl group was also tolerated and demonstrated an IC$_{50}$ value in the enzymatic assay of 40 nM. The biphenyl analogue, **6**, revealed that additional steric bulk in the *para*-position was well tolerated. Compound **7** showed very little shift in activity between the enzymatic and whole blood assay and superior efficacy when dosed at 10 mg/kg p.o. in the *ex vivo* mouse PK/PD model with 93% inhibition of LTB$_4$ production at one hour post-dose.[166]

Modification of the central aromatic ring was also explored. Heterocyclic 5- and 6-membered rings were tested, as was the relative orientation of the connection points to the terminal aromatic group and the amine. Generally 6-membered rings, exemplified by pyridine (as shown in compound **8**) or phenyl, were preferred, and further substitution with fluorine was tolerated.

Interestingly, the thiophene analogue **9** was similar in potency to **3**, supporting the idea that thiophene and phenyl are isosteric in nature. However, thiazole analogue **10** shows substantially diminished potency.[166]

Using the encouraging data for compounds **3**, **5** and **7**, the Searle group sought to further expand the scope of the SAR.[167] To this end the linkers denoted by x and y in generic compound **11** (Figure 3.6) were investigated, and it was determined that $R^1 = CH_3$, x = 3 and y = 2, as in SC-57461A (**12**), offered the best combination of enzymatic and cellular potency. SC-57461A showed efficacy in an 8-week model of spontaneous colitis in cotton-top tamarins when dosed orally at 10 mg/kg b.i.d. but failed to advance to Phase I clinical trials due to toxicity in a pre-clinical safety study.[167]

Using this result, further optimization to determine the optimal left-hand side moiety was undertaken. The phenoxy (**13**) and ether-linked biphenyl (**14**) and phenyloxazole (**15**) analogues of **12** resulted in reduced potency and efficacy in the human whole blood assay and *ex vivo* mouse PK/PD model. Replacements of the acid moiety were also investigated, and data were provided for various methyl and ethyl esters (**11**, R^2 = Me or Et); presumably these would be converted to the corresponding acids *in vivo*. The most potent and efficacious acid replacement was the acyl sulfonamide **16**.

Subsequent publications from Searle further explored replacement of the amine of SC-57461A. Specifically, the group reported on substituted

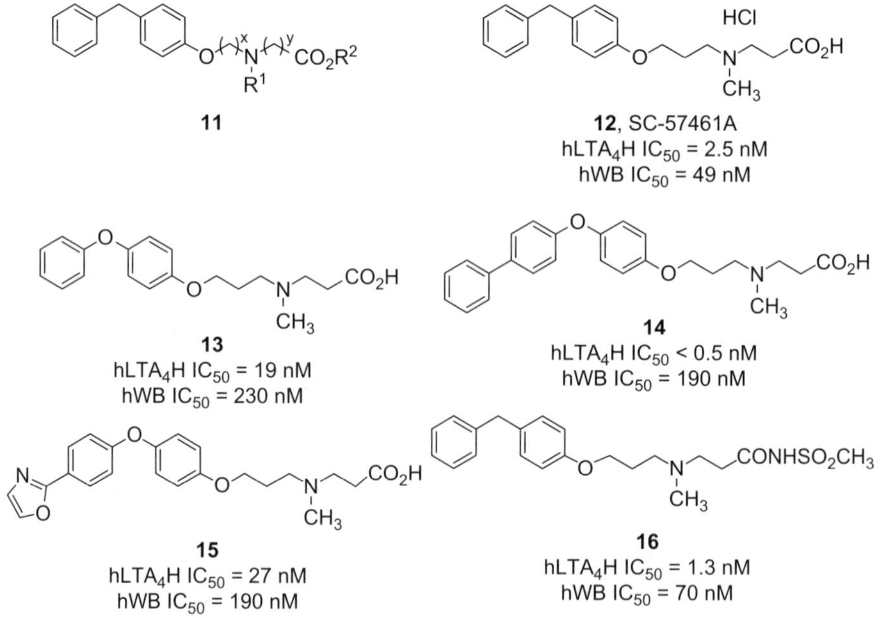
O
N
CO2R2
x
y
R1
11
HCl
O
N
CO2H
CH3
12, SC-57461A
hLTA4H IC50 = 2.5 nM
hWB IC50 = 49 nM
O
O
N
CO2H
CH3
13
hLTA4H IC50 = 19 nM
hWB IC50 = 230 nM
O
O
N
CO2H
CH3
14
hLTA4H IC50 < 0.5 nM
hWB IC50 = 190 nM
N
O
O
O
N
CO2H
CH3
15
hLTA4H IC50 = 27 nM
hWB IC50 = 190 nM
O
N
CONHSO2CH3
CH3
16
hLTA4H IC50 = 1.3 nM
hWB IC50 = 70 nM

Figure 3.6 Second-generation non-chelating LTA_4H inhibitors reported by G. D. Searle. Definitions of x, y, R^1 and R^2 for compound **11** as defined in the publication reference 167.

pyrrolidine and piperidine analogues, reverting to the oxyethyl linker found in SC-22716 (**2**).[168] None of the compounds in Figure 3.7 demonstrated the *in-vitro* potency seen with SC-57461A but several showed inhibition of LTB$_4$ production when measured one hour post dose in the *ex vivo* mouse PK/PD model with an oral dose of 10 mg/kg. The data for compounds **17** (75% inhibition), **18** (97% inhibition) and **19** (95% inhibition) compared favourably to SC-57461A (% inhibition > 98%). Notably, the acid analogue of **18**, compound **20** (80% inhibition), was more potent in the *in-vitro* assays but less efficacious in the same *ex vivo* PK/PD model.[168] SC-56938 (**18**), was selected for further development but no reports on progression to clinical trials have appeared in the literature.

The last publication to appear from this group regarding their work on LTA$_4$H disclosed analogues such as **21** (Figure 3.8), which contained bicyclic heterocycles as amine replacements.[169] Further functionalization was tolerated, as exemplified by **22**. In several cases these heterocyclic analogues maintained good enzymatic and whole blood potency but did not offer an advantage over the disclosed biological profiles of either SC-57461A or SC-56938.

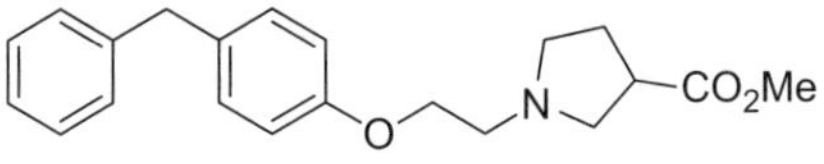

17
hLTA$_4$H IC$_{50}$ = 4.1 µM
hWB IC$_{50}$ = 230 nM

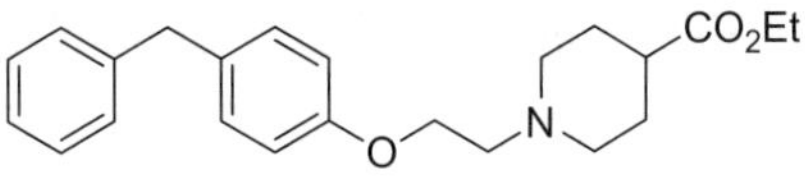

18, SC-56938
hLTA$_4$H IC$_{50}$ = 1.8 µM
hWB IC$_{50}$ = 820 nM

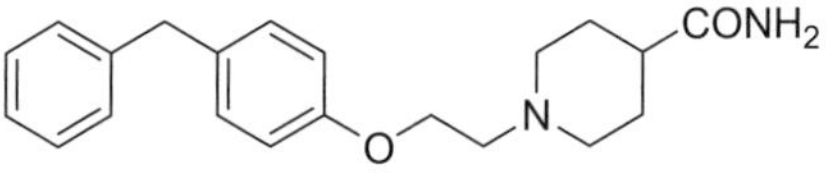

19
hLTA$_4$H IC$_{50}$ = 130 nM
hWB IC$_{50}$ = 640 nM

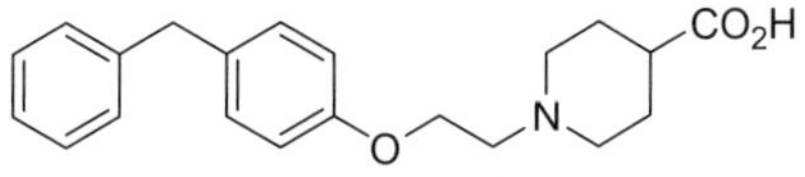

20
hLTA$_4$H IC$_{50}$ = 280 nM
hWB IC$_{50}$ = 290 nM

Figure 3.7 Amine modifications of SC-57641.

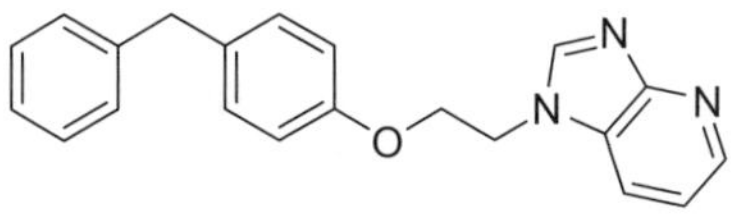

21
hLTA$_4$H IC$_{50}$ = 8 nM
hWB IC$_{50}$ = 330 nM

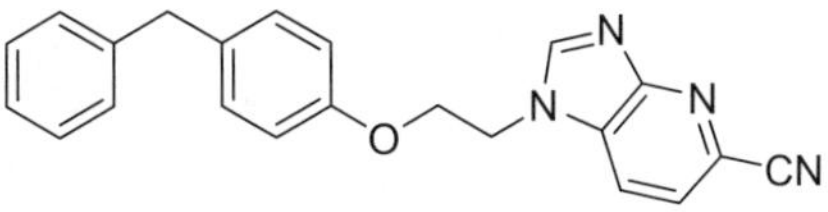

22
hLTA$_4$H IC$_{50}$ < 1 nM
hWB IC$_{50}$ = 150 nM

Figure 3.8 Heterocyclic amine replacements for SC-57641 and SC-56398.

3.4.3 Janssen Pharmaceutica NV

Two patent applications from Janssen Pharmaceutica NV (Janssen) published in 2005[170,171] pertaining to molecules containing a bicyclic aromatic component defined as benzothiazole, benzimidazole and benzoxazole as LTA$_4$H inhibitors. Both applications contain the same exemplified compounds but the definitions differ with respect to the linker between the central aromatic core and the basic amine moiety. This work was also described in a manuscript published in 2008, which details the synthesis and biological activity of this class of LTA$_4$H inhibitors.[121] Examples of these compounds are shown in Figure 3.9.

Compounds **23**, **25** and **27** were identified using an *in-silico* screening method, which emphasized the potential importance of the Glu271 and Glu296 residues to form key interactions with the basic amine of known LTA$_4$H inhibitors.[121] Compounds **24**, **26** and **28** were prepared in an effort to minimize undisclosed off-target cross-reactivity seen with compounds **23**, **25** and **27**. Further work was confined to the benzothiazole series, given the noted drop in activity observed with the benzimidazoles and the lack of low-pH stability of the benzoxazoles.[121] Settling on the benzothiazoles, analogues with both CH$_2$CH$_2$ and CH$_2$ linkers were prepared. These changes were well tolerated and minimized inhibition of the hERG channel, as measured by astemizole binding. Further functionalization of the amine moiety led to the identification of

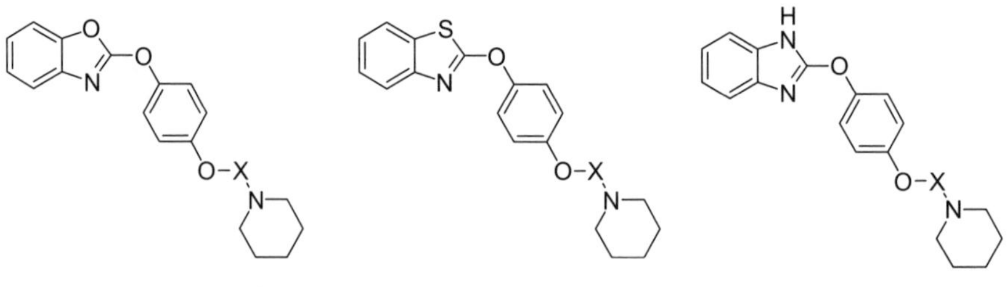

23, X = CH$_2$CH$_2$CH$_2$
hLTA$_4$H IC$_{50}$ = 27 nM
24, X = CH$_2$CH$_2$
hLTA$_4$H IC$_{50}$ = 11 nM

25, X = CH$_2$CH$_2$CH$_2$
hLTA$_4$H IC$_{50}$ = 60 nM
26, X = CH$_2$CH$_2$
hLTA$_4$H IC$_{50}$ = 54 nM

27, X = CH$_2$CH$_2$CH$_2$
hLTA$_4$H IC$_{50}$ = 17 nM
28, X = CH$_2$CH$_2$
hLTA$_4$H IC$_{50}$ = 110 nM

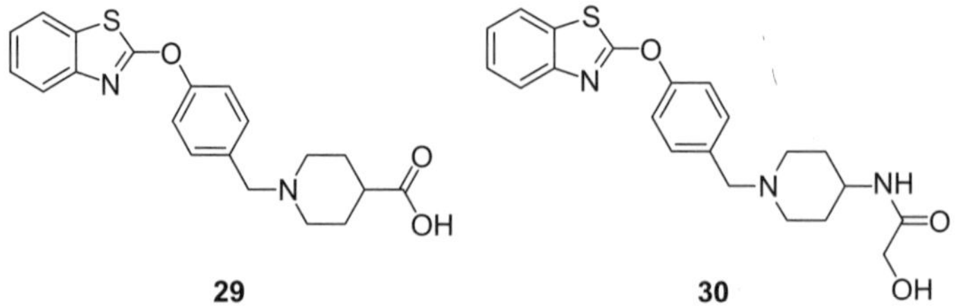

29
hLTA$_4$H IC$_{50}$ = 11 nM
mWB LTA$_4$H IC$_{50}$ = 88 nM

30
hLTA$_4$H IC$_{50}$ = 8 nM
mWB LTA$_4$H IC$_{50}$ = 104 nM

Figure 3.9   Inhibitors from Janssen Pharmaceutica, NV.

compounds **29** and **30**, which were profiled as potential pre-clinical candidates.[23,96,101,110,121] Briefly, compounds (**29** and **30**) were evaluated in models of IBD (TNBS-induces colitis model in rat) and acute and chronic allergic asthma (OVA (ovalbumin)-induced lung inflammation and house dust mite (HDM) in mice respectively). In the TNBS-induced colitis model, dosing compounds **29** and **30** at 30 mg/kg orally b.i.d. (24 hours prior to treatment with TNBS and on day 1 and day 2 post challenge), significantly reduced levels of LTB₄, IL-6 and TNF-α as well as reduced areas of necrosis and weight loss, were noted in LTA₄H inhibitor treated animals compared to control.[101,121] In the OVA model, compound **29** at doses of 25 or 50 mg/kg b.i.d. during the challenge phase, resulted in modulation of airway inflammation and airway hyper-responsiveness (AHR) as measured by BAL inflammation and invasive pulmonary function. Additionally, inhibition of LTB₄ formation, T-cell and DC chemotaxis and reduced mucous secretion were observed. Notably, in this model an increase in the anti-inflammatory lipoxin LXA₄ was also noted.[23] Treatment with another LTA₄H inhibitor (structure not disclosed) also showed decreases in AHR and BAL inflammation in the chronic HDM model, with a reduction of both Th2 cytokines and IL-17A.[110]

A third patent application, published in 2006,[172] disclosed phenyl- and pyridyl-based compounds. In this instance, analogues containing aryl or alkyl ether, amide, carbamate, urea, carbonate and ester moieties linked to the central phenyl or pyridyl ring were prepared. Two examples, **31** and **32** and their associated data, are shown in Figure 3.10. Other disclosed examples contained phenethyl or phenpropyl ethers and carbamates. Compounds disclosed in this and subsequent patent applications contain biological data, including IC₅₀ values for recombinant human LTA₄ hydrolase and calcium ionophore stimulated LTB₄ production in murine whole blood. Additionally, the *in-vivo* activity of compounds is illustrated using the murine arachidonic acid induced inflammation model, and the data reported include percent inhibition of LTB₄ production measured *ex vivo* and myeloperoxidase activity (a measure of neutrophil influx) when dosed orally at 30 mg/kg one hour before challenge, with PD measurements taken four hours post-dose.

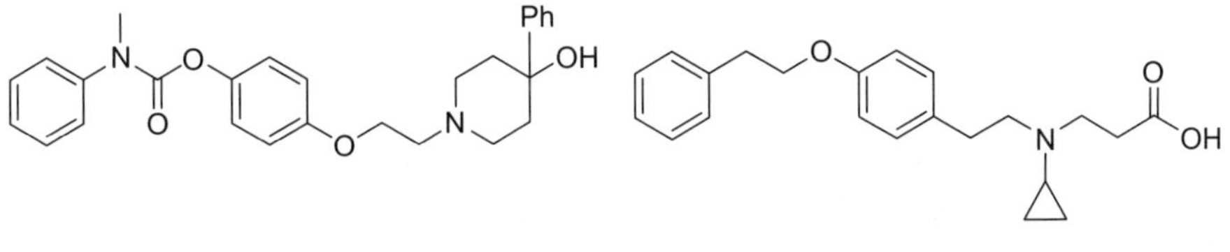

31
hLTA₄H IC₅₀ = 73 nM
mWB LTA₄H IC₅₀ = 150 nM
% inh LTB₄ = 79%
% inh MPO = 75%

32
hLTA₄H IC₅₀ = 0.5 nM
mWB LTA₄H IC₅₀ = ND
% inh LTB₄ = 84%
% inh MPO = 92%

Figure 3.10 Compounds disclosed in Janssen patent application WO2006/105304.

In 2009 two more patent applications were published from Janssen.[173,174] One of these applications describes benzothiazole- and benzoxazole-containing bridged bicyclic and fused amines and included analogues with varying linkers and substituents on the terminal and central aromatic ring.[173] From the data included in the application it is evident that the addition of the bridge is beneficial (compare **33** with **34**, Figure 3.11).[121] Addition of fluorine to the benzoxazole 4-position (**35**) is tolerated and beneficial in some cases, but methyl (**36**) is deleterious. Modifications to the linker length and the nature of the amine are also well tolerated, as evidenced by compounds **37** and **38**. Modification of the atom between the two aromatic rings from oxygen to carbon, or addition of fluorine to the central aromatic ring, results in a modest loss of potency, as illustrated by compounds **39** and **40** (compare to the des-fluoro compound **41**). It is worth noting that the potency of the analogues contained within the application varies depending on the combination of changes specific to each molecule. This is illustrated by compound **42**, which shows that the

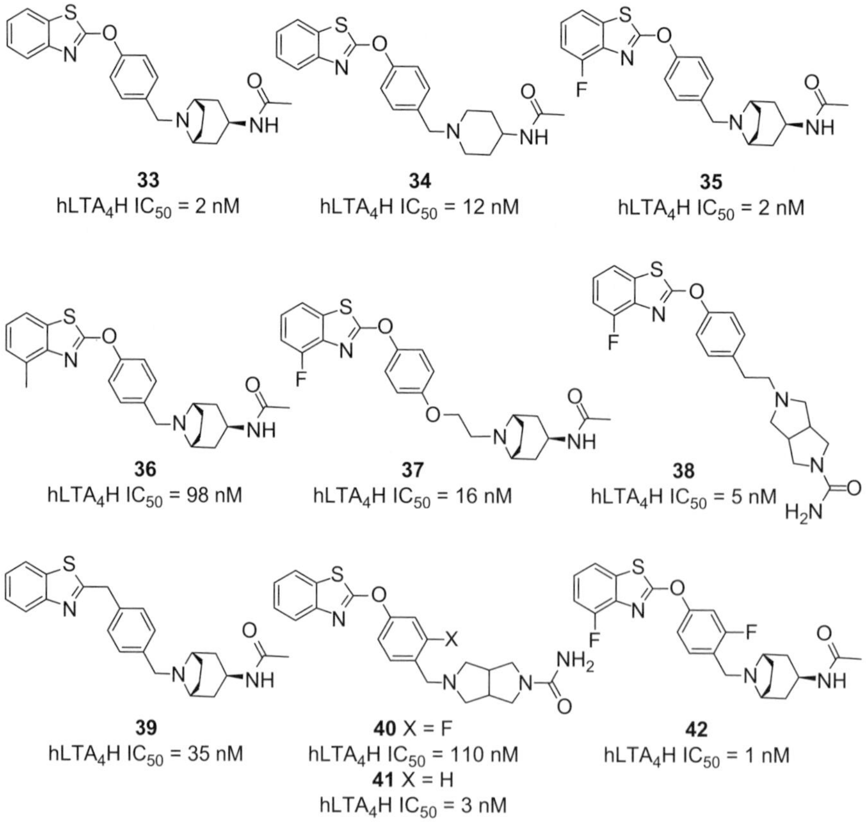

33
hLTA$_4$H IC$_{50}$ = 2 nM

34
hLTA$_4$H IC$_{50}$ = 12 nM

35
hLTA$_4$H IC$_{50}$ = 2 nM

36
hLTA$_4$H IC$_{50}$ = 98 nM

37
hLTA$_4$H IC$_{50}$ = 16 nM

38
hLTA$_4$H IC$_{50}$ = 5 nM

39
hLTA$_4$H IC$_{50}$ = 35 nM

40 X = F
hLTA$_4$H IC$_{50}$ = 110 nM
41 X = H
hLTA$_4$H IC$_{50}$ = 3 nM

42
hLTA$_4$H IC$_{50}$ = 1 nM

Figure 3.11 Inhibitors from Janssen Pharmaceutica containing fused and bridged amines.

addition of fluorine to the central aromatic ring has little impact when combined with a judicious change in the amine.

The other Janssen 2009 patent application describes molecules that contain one or more nitrogens in the phenyl portion of the benzothiazole ring system.[174] In addition to fused and bridged bicyclic amines that were disclosed in the previously described 2009 application, simpler cyclic and acyclic amines are also included. The linkers investigated included OCH_2CH_2, CH_2CH_2 and CH_2. Comparison of compounds **43**, **44** and **45** in Figure 3.12 illustrates a tolerance for changes in linker length. Review of the biological data in the application with respect to the position of the nitrogen within the pyridyl ring of the

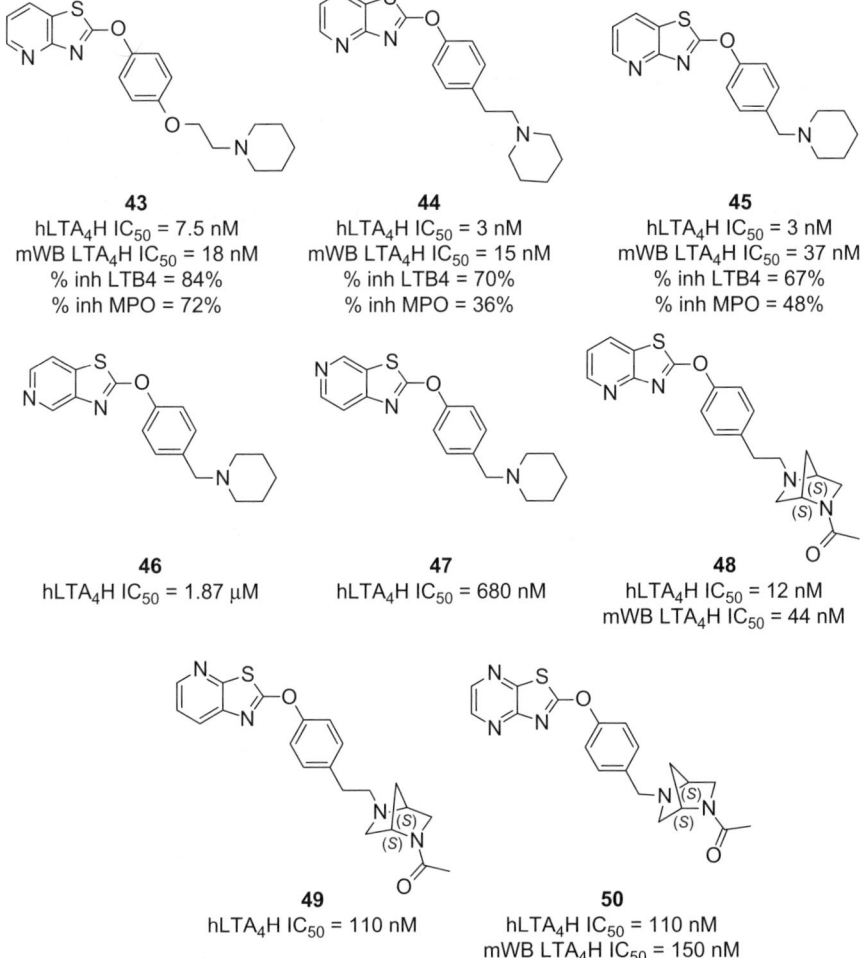

Figure 3.12 Inhibitors from Janssen Pharmaceutica containing thiazolopyridines and thiazolopyrazines.

azabenzothiazole, as illustrated by comparing compound **45** to **46** and **47** and comparing compound **48** to **49**, suggests that the 4-position is optimal. However, addition of a second nitrogen atom (compound **50**) in the 4-position of compound **49** does not impart the potency seen in compounds **45** and **48**. The stereochemistry of the diamine does not influence the inhibitory activity as the *R,R* enantiomer of compound **48** (structure not shown) demonstrates an hLTA$_4$H IC$_{50}$ = 6 nM and mWB IC$_{50}$ = 20 nM.

Notably, the addition of substituents to the benzothiazole, such as methyl, chloro and fluoro, was tolerated in some cases. Compounds **52–54** in Figure 3.13 show the effect of placing substituents in the 6-position of compound **51**. Although there is a drop in potency, these modifications, in most cases, are acceptable. In contrast, placing a substituent in the 5-position, as illustrated by compound **55**, has a negative impact but substitution at the 7-position is tolerated, as shown by compound **56**.

The data shown in Figure 3.14 illustrate that the enzyme does not exhibit a significant preference for the type of amine. Monocyclic, fused bicyclic and bridged bicyclic amines are all well tolerated, as exemplified by **57–59**.

The remaining patent application from Janssen discloses compounds that contain a fused bicyclic heterocycle as the central aromatic component.[175] Specifically, the application allows for the incorporation of indoles,

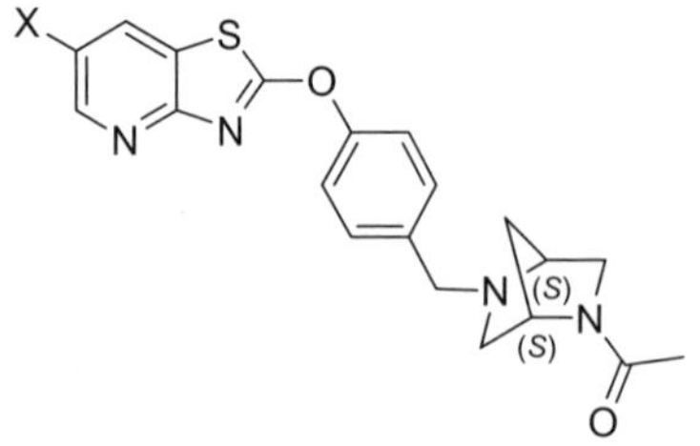

51, X = H
hLTA$_4$H IC$_{50}$ = 5 nM
mWB LTA$_4$H IC$_{50}$ = 7.5 nM

52, X = CH$_3$
hLTA$_4$H IC$_{50}$ = 38 nM
mWB LTA$_4$H IC$_{50}$ = 120 nM

53, X = F
hLTA$_4$H IC$_{50}$ = 23 nM
mWB LTA$_4$H IC$_{50}$ = 5 nM

54, X = Cl
hLTA$_4$H IC$_{50}$ = 48 nM
mWB LTA$_4$H IC$_{50}$ = 270 nM

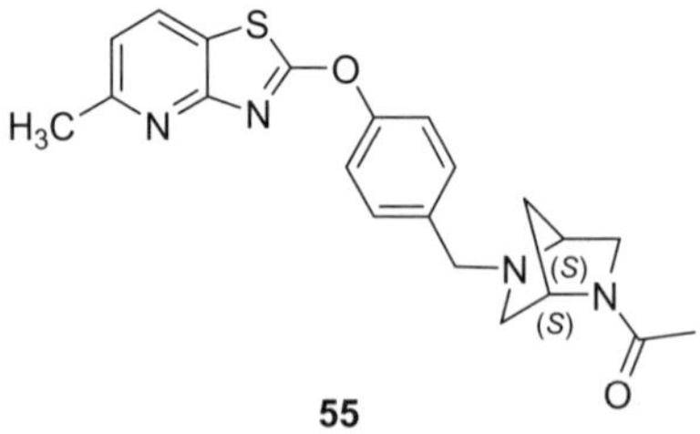

55
hLTA$_4$H IC$_{50}$ = 920 nM

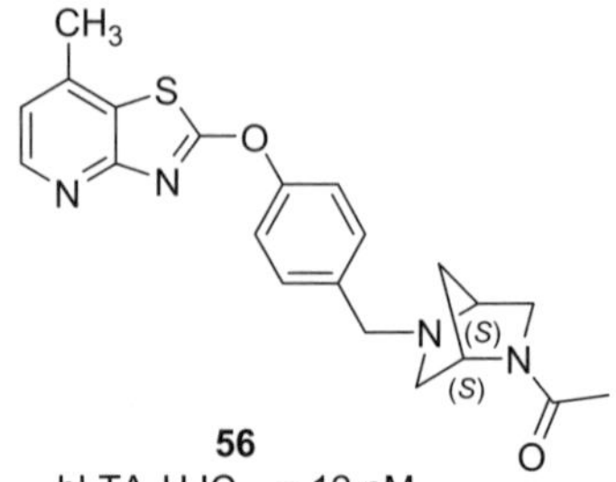

56
hLTA$_4$H IC$_{50}$ = 12 nM

Figure 3.13   Effect of adding substituents to the thiazolopyridine ring.

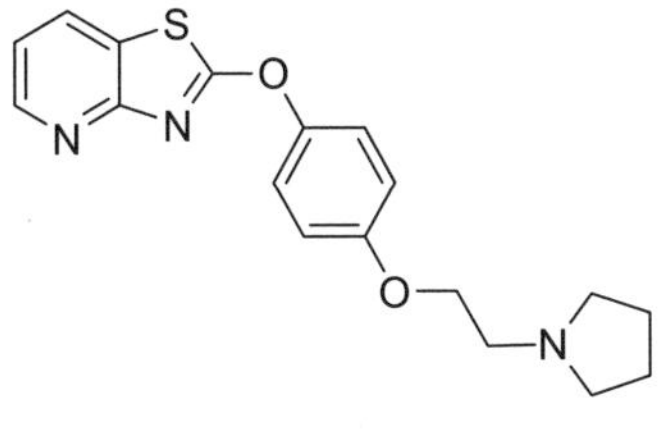

57
hLTA$_4$H IC$_{50}$ = 6 nM
mWB LTA$_4$H IC$_{50}$ = 17 nM

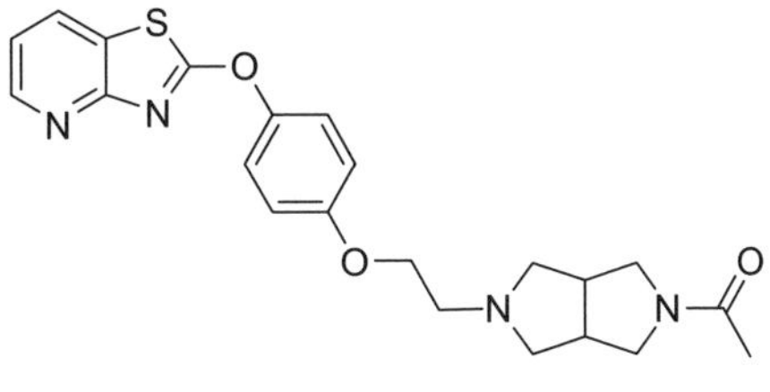

58
hLTA$_4$H IC$_{50}$ = 5 nM
mWB LTA$_4$H IC$_{50}$ = 32 nM

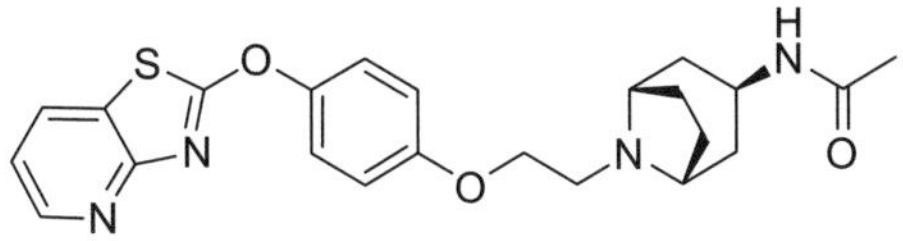

59
hLTA$_4$H IC$_{50}$ = 3 nM
mWB LTA$_4$H IC$_{50}$ = 26 nM

Figure 3.14   Inhibitors from Janssen Pharmaceutica with *in vivo* data.

benzofurans and benzthiophenes. Within the indole series, substitution in the 3,6-orientation is preferred over the 3,5-orientation, as illustrated by compounds **60** and **61** (Figure 3.15). Incorporation of the 4-azabenzothiazole enhances potency, whereas incorporation of the 4-azaindole results in a modest loss (compounds **62** and **63**, respectively). Alkylation of the indole nitrogen, as represented by compound **65**, is deleterious (compare with **64**). Further change in the orientation of substituents to the 2,5-position (**66**) results in a loss of potency, but the 2,6-orientation (**67**) is better tolerated. Incorporation of a CH$_2$CH$_2$ linker is not beneficial to potency as evidenced by compound **68**. Substitution of fluorine at the 4-position of the benzothiazole or inclusion of bridged amines is again beneficial (not shown).

The benzofuran analogues demonstrated several of the same trends seen with the indoles. Bridged, fused and monocylic amines are tolerated, as detailed in Figure 3.16. Incorporation of the 4-azabenzothiazole, as shown by the comparison of **71** and **72**, is also beneficial. Substitution on the benzothiazole, as illustrated by the 4-fluoro analogue **76**, is tolerated as compared to the desfluoro analogue (IC$_{50}$ = 70 nM, compound not shown). Similar to the indoles, there is also a preference for the 3,6-orientation, as can be seen by comparing compounds **71** and **75** with **73** and **77**, respectively.

The benzthiophene analogues disclosed contain only the 2,6-orientation of substituents. Generally, these compounds are less potent than their indole and benzofuran counterparts (Figure 3.17). This is evident in comparing analogues **78** and **79** in Figure 3.17 with **67**, **71** and **72** in Figures 3.15 and 3.16.

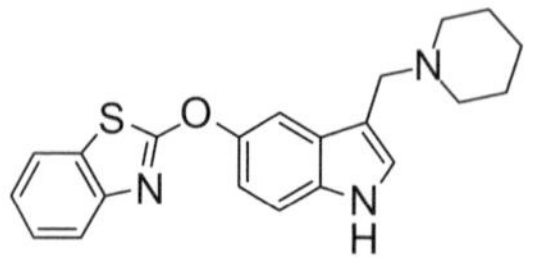

60
hLTA$_4$H IC$_{50}$ = 415 nM

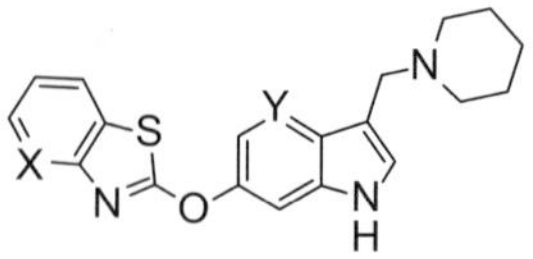

61, X = CH, Y = CH
hLTA$_4$H IC$_{50}$ = 19 nM
62, X = N, Y = CH
hLTA$_4$H IC$_{50}$ = 3 nM
63, X = CH, Y = N
hLTA$_4$H IC$_{50}$ = 55 nM

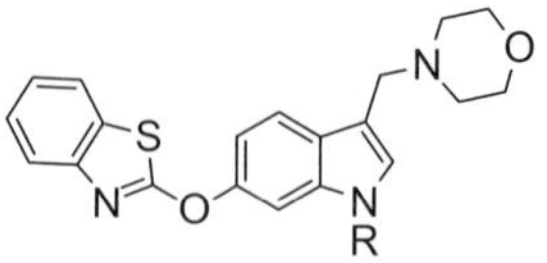

64, R = H
hLTA$_4$H IC$_{50}$ = 16 nM
65, R = CH$_3$
hLTA$_4$H IC$_{50}$ = 610 nM

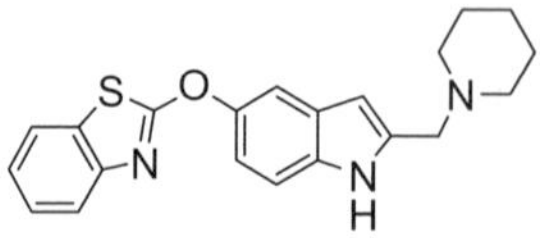

66
hLTA$_4$H IC$_{50}$ = 100 nM

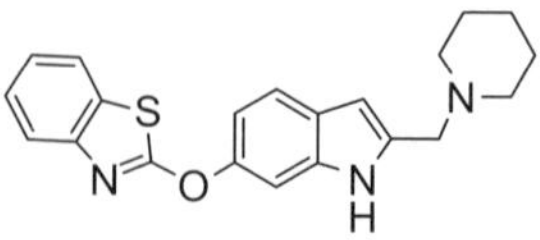

67
hLTA$_4$H IC$_{50}$ = 34 nM

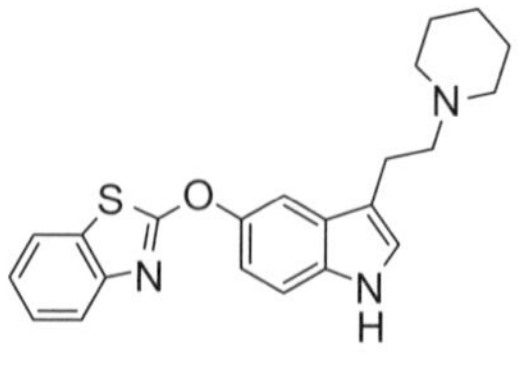

68
hLTA$_4$H IC$_{50}$ = 2 μM

Figure 3.15   Inhibitors from Janssen Pharmaceutica with indole moieties.

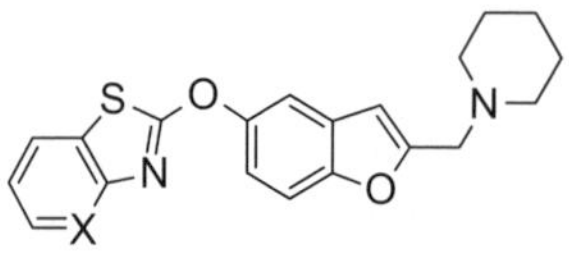

69, X = CH
hLTA$_4$H IC$_{50}$ = 190 nM
70, X = N
hLTA$_4$H IC$_{50}$ = 68 nM

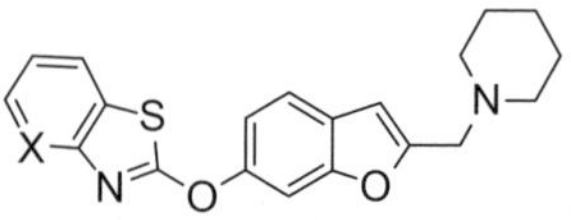

71, X = CH
hLTA$_4$H IC$_{50}$ = 492 nM
72, X = N
hLTA$_4$H IC$_{50}$ = 43 nM

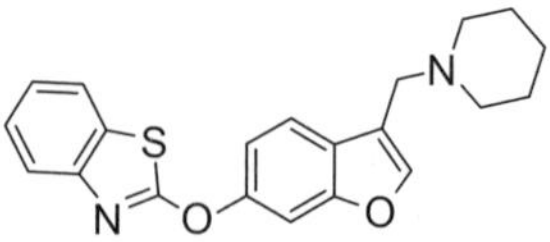

73
hLTA$_4$H IC$_{50}$ = 77 nM

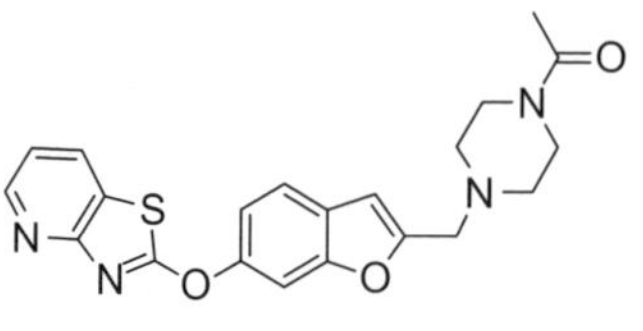

74
hLTA$_4$H IC$_{50}$ = 59 nM

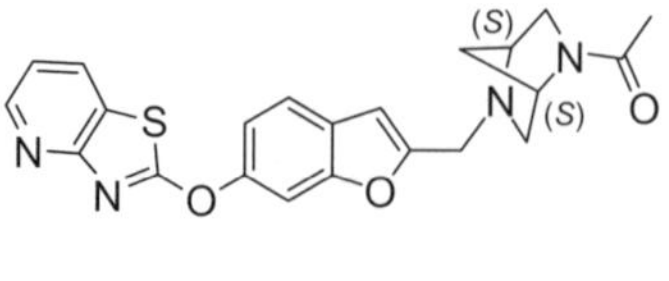

75
hLTA$_4$H IC$_{50}$ = 36 nM

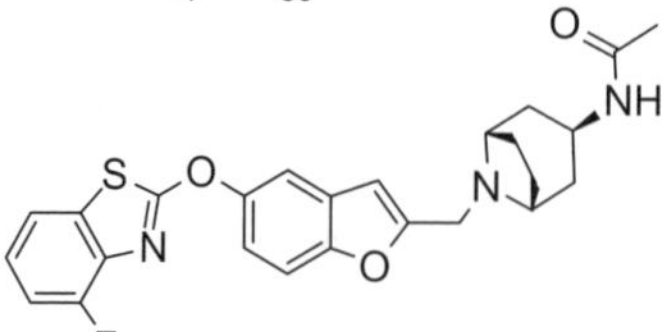

76
hLTA$_4$H IC$_{50}$ = 17 nM

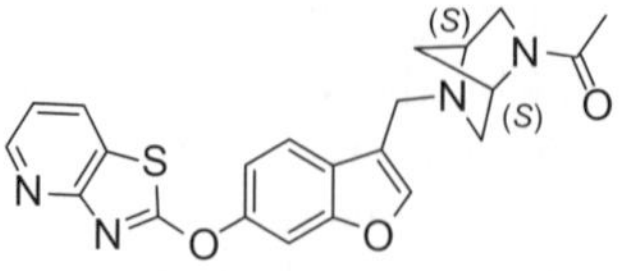

77
hLTA$_4$H IC$_{50}$ = 2 nM

Figure 3.16   Inhibitors from Janssen Pharmaceutica with benzofuran moieties.

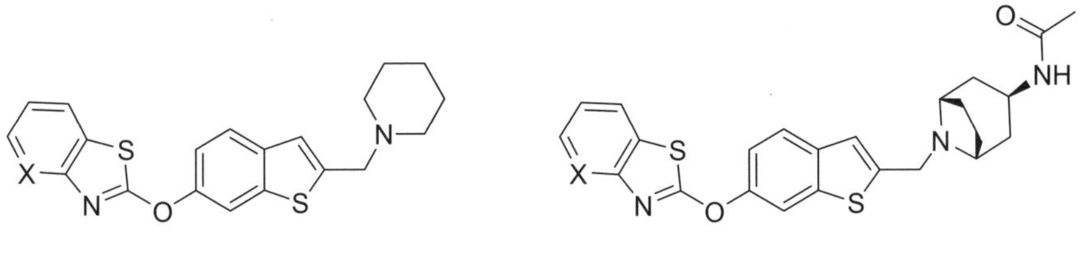

78, X = CH
hLTA$_4$H IC$_{50}$ = 850 nM
79, X = N
hLTA$_4$H IC$_{50}$ = 120 nM

80, X = CH
hLTA$_4$H IC$_{50}$ = 43 nM
81, X = N
hLTA$_4$H IC$_{50}$ = 3 nM
mWB LTA$_4$H IC$_{50}$ = 109 nM

Figure 3.17   Inhibitors from Janssen Pharmaceutica with benzothiophene moieties.

An exception to this trend is the bicyclic analogue **81**, which shows very promising enzymatic and mWB activity.

3.4.4   deCODE Genetics

In 2007 and 2008, patent applications from deCODE published disclosing LTA$_4$H inhibitors.[176–181] Evaluation of the applications reveals that the central aromatic portion of the generic structures in each application overlap and this group is typically a phenyl ring. However, two of the applications allow for the incorporation of a heteroaryl ring, with select examples containing pyridine.[179,180] These patent applications contain activity ranges for enzymatic and cellular assays for the majority of exemplified compounds, with the most potent range being "IC$_{50}$ <5 µM". The work has been subsequently published in a series of manuscripts, and the discussion below will be limited to those publications.[55,182–184]

The first publication from deCODE describes a fragment based drug design (FBDD) approach using an internal library termed "fragments of life" (FOL) and X-ray crystallography.[182] The FOL library consisted of 1329 fragments, which generally obeyed the "rule of three" criteria set forth by Congreve *et al.*[185] Fifteen percent of this library was randomly selected and pooled prior to soaking with the LTA$_4$H protein to identify crystals with unaccounted for electron density within the binding pocket (molecules that bound outside of this region were deprioritized).[182] The paper discloses the crystal structures for 19 compounds, of which 13 were from the FOL library and 6 resulted from further modification to optimize potency. Binding was noted throughout the active site pocket, divulging a diverse set of binding modes.

Different methods of elaboration were evaluated using information from all of the fragment crystal structures, specifically "fragment plus fragment" linking or fragment evolution.[182] Compounds **82**, **83** and **84**, which were crystallized in the presence of bestatin (Figure 3.18), were chosen for further optimization using the fragment evolution process. Compounds **85**, **86** and **87** are shown

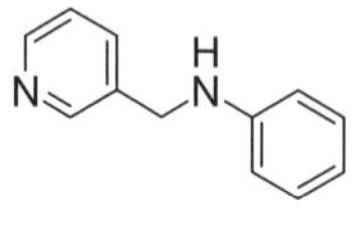

82
hLTA$_4$H IC$_{50}$ = 1667 μM

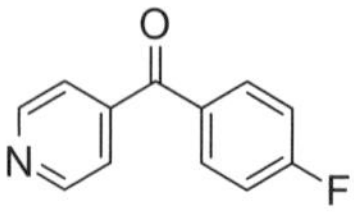

83
hLTA$_4$H IC$_{50}$ = 5308 μM

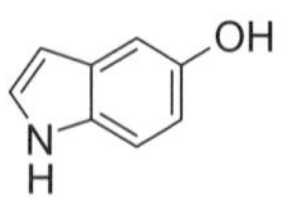

84
hLTA$_4$H IC$_{50}$ > 2000 μM

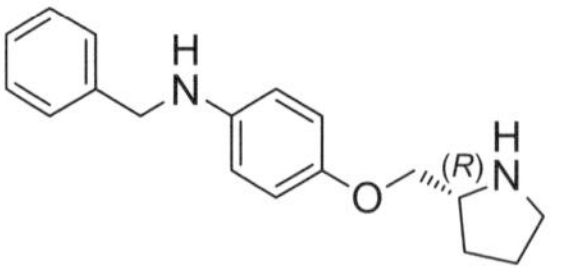

85
hLTA$_4$H IC$_{50}$ = 157 nM

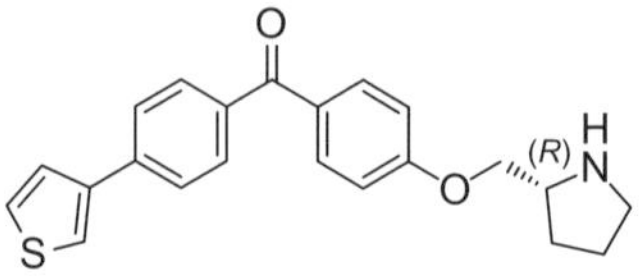

86
hLTA$_4$H IC$_{50}$ = 189 nM

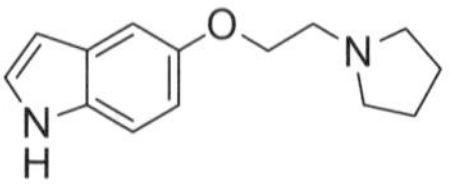

87
hLTA$_4$H IC$_{50}$ = 234 μM

Figure 3.18   FOL library compounds that bind LTA$_4$H and optimized inhibitors.

below their respective fragment and represent the most potent compound from each fragment discussed in the publication. Compounds **85** and **86** have enzymatic and whole blood activity that approaches those of other published inhibitors.

Compound **92** (DG-051, Figure 3.19) was developed through a combination of fragment evolution and a fragment plus fragment approach starting from **82** and **83**.[186] The evolution of these fragments was introduced by Davies *et al.*[182] and advanced further by Sandanayaka *et al.*[184] The fragment plus fragment approach was augmented by the observation that acetate ions in the crystallization solution bound to the catalytic zinc in the resolved crystal structures.[184] The goal was to increase potency by optimizing interactions in the hydrophobic pocket with residues Tyr267, Trp311 and Phe314, and by evaluating linking strategies to combine the fragments with acetate as mentioned above.[184] Specifically, the progression of the series involved the addition of the *N*-hydroxyethyl pyrrolidine or 4-hydroxyethyl pyridine moiety to fragment **83** and removal of the carbonyl to give either **88** or **89**. Combining components of each of these molecules to avoid disruptive interactions (between the carbonyl of **89** with Trp311 and the pyridyl nitrogen of **89** and a water molecule) and further modifying the orientation of the nitrogen led to **90**, an analogue very similar to compound **3** from Searle (Figure 3.5). As can be seen in Figure 3.19, each modification led to an enhancement of potency. Finally, incorporation of chlorine on the terminal aromatic ring led to **91**, and the subsequent addition of a carboxylate through a tether that appropriately bridges the gap to the active site Zn^{2+} led to **92**. Importantly, the (*S*)-prolinol component of **91** was changed to the corresponding (*R*)-prolinol in compound **92**. The *R*- and *S*-enantiomers of **92** are equipotent in the hWB assay (IC$_{50}$ = 37 nM *vs.* 33 nM) and given the high cost of D-proline needed to prepare the *R*-enantiomer, the *S*-enantiomer was chosen for further advancement.[184]

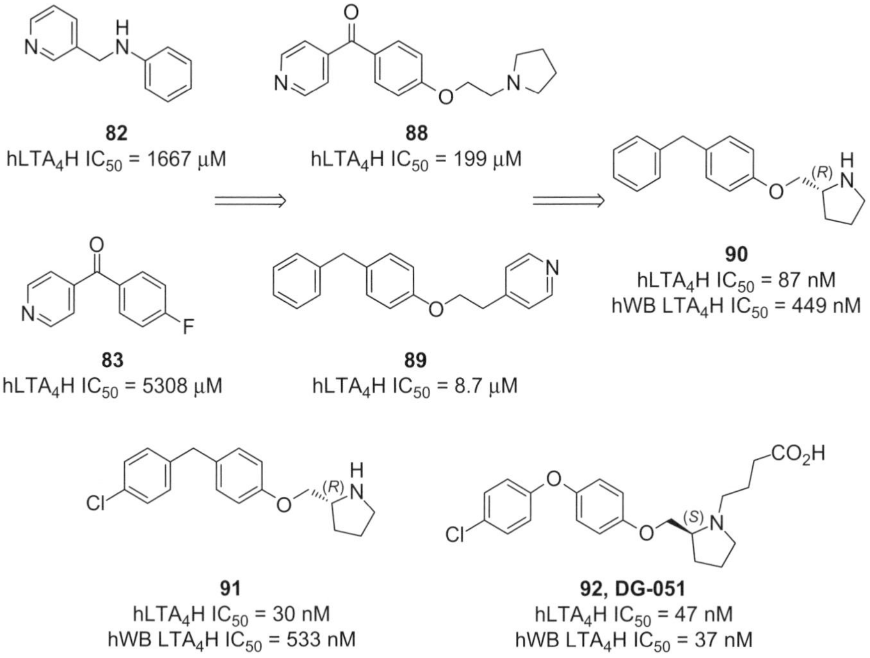

82
hLTA₄H IC$_{50}$ = 1667 µM

88
hLTA₄H IC$_{50}$ = 199 µM

90
hLTA₄H IC$_{50}$ = 87 nM
hWB LTA₄H IC$_{50}$ = 449 nM

83
hLTA₄H IC$_{50}$ = 5308 µM

89
hLTA₄H IC$_{50}$ = 8.7 µM

91
hLTA₄H IC$_{50}$ = 30 nM
hWB LTA₄H IC$_{50}$ = 533 nM

92, DG-051
hLTA₄H IC$_{50}$ = 47 nM
hWB LTA₄H IC$_{50}$ = 37 nM

Figure 3.19　Evolution of fragments to yield DG-051.

DG-051 possesses good pharmacokinetic properties across multiple species (mouse, rat, dog and monkey oral $t_{1/2} > 2.5$ h; %F > 85) with a profile that is consistent with once daily oral dosing.[184] The compound is selective when evaluated against a panel of > 50 targets including ion channels, GPCRs, nuclear receptors and peptidases, and does not demonstrate greater than 50% inhibition of any of the major P$_{450}$ isozymes at 10 µM.[184] DG-051 was also reported to have oral activity in a rat *ex vivo* assay with 90% inhibition of LTB$_4$ production when dosed at 10 mg/kg.[187] In May 2007, deCODE reported positive top line results from their Phase I trial, which evaluated DG-051 at doses up to 320 mg/day for a period of 7 days and showed a dose-dependent reduction of LTB$_4$ levels.[188] A Phase IIa trial in patients with a history of coronary artery disease also showed a dose-dependent decrease in LTB$_4$.[189] In both trials the compound showed favourable pharmacokinetics (oral $t_{1/2} = 9$ h; $t_{max} = 2.5$ h) and no major adverse events.[188,189]

Using the information gleaned from the X-ray crystallography studies and iterative medicinal chemistry, the deCODE group later disclosed work aimed at further optimizing DG-051.[183] To this end, their goals were to exploit the hydrophobic region of the binding pocket with the specific aims of displacing conserved water molecules and engaging the active site Zn^{2+} through accommodation of the bend in the active site separating the hydrophobic and hydrophilic regions of the binding pocket.[183] Compound **93** (Figure 3.20)

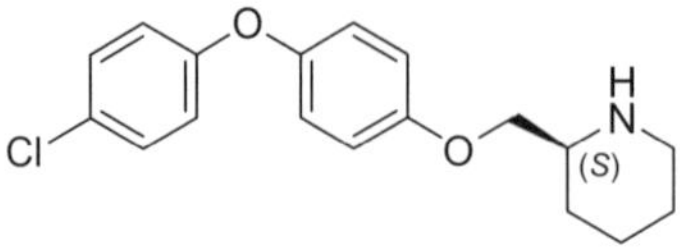

93
hLTA$_4$H IC$_{50}$ = 110 nM
hWB LTA$_4$H IC$_{50}$ = 40 nM

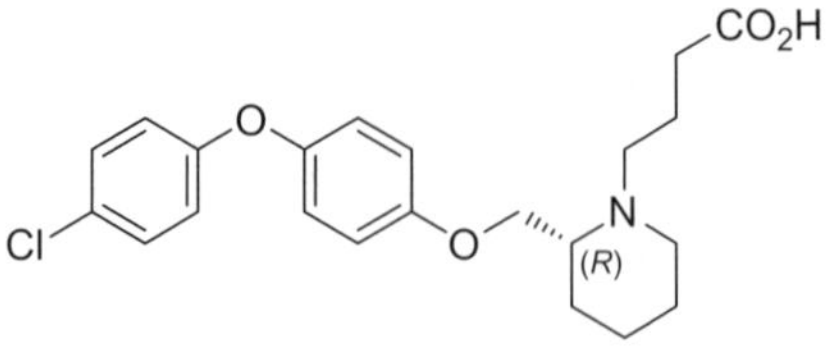

94
hLTA$_4$H IC$_{50}$ = 53 nM
hWB LTA$_4$H IC$_{50}$ = 6 nM

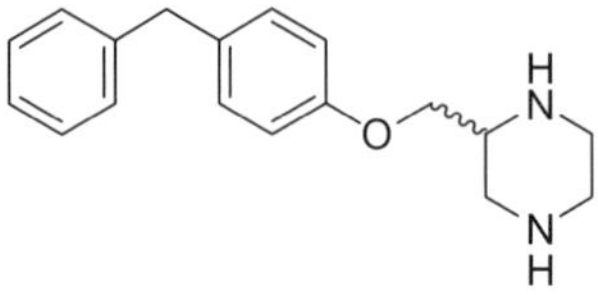

95
hLTA$_4$H IC$_{50}$ = 342 nM
hWB LTA$_4$H IC$_{50}$ = ND

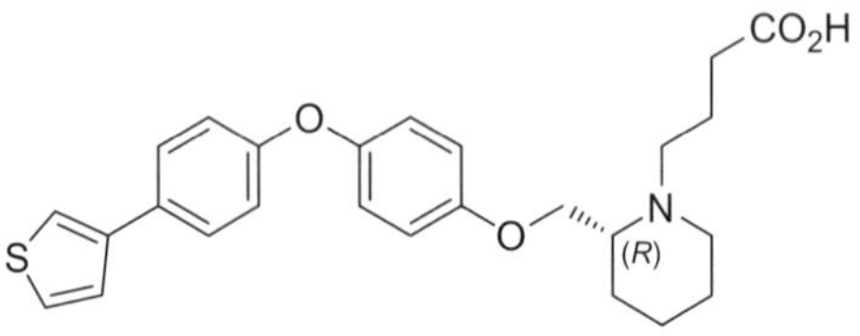

96
hLTA$_4$H IC$_{50}$ = 49 nM
hWB LTA$_4$H IC$_{50}$ = 6 nM

Figure 3.20 Piperidine and piperazine analogues of DG-051.

served as a template for this work (analogous to **91**). The incorporation of various linkers led to the identification of the butyric acid derivative **94** as optimal. Continued modification of the terminal phenyl ring to more fully occupy the hydrophobic region led to compound **96**. Compound **95**, the piperazine analogue of **93**, was less potent in the enzymatic assay, and further optimization was not pursued.[183]

3.4.5 Schering AG

Patent applications published in 2007 from Schering AG, the parent company of Berlex Biosciences, describe the use of diamine and amide derivatives as LTA$_4$H inhibitors for inflammatory disorders.[190,191] These applications include the descriptions for *in-vitro* (LTA$_4$H inhibition, peptidase and whole blood assays) and *in-vivo* (rat and mouse experimental autoimmune encephalomyelitis) assays but contain no biological data. Subsequent publications appeared in 2008 and the discussion below is limited to these disclosures.

A high-throughput screening campaign targeting the peptidase activity of LTA$_4$H led to the identification of compound **97** (Figure 3.21).[192] This compound also possessed moderate activity in the human whole blood assay. Efforts to optimize this hit focused on amino acid replacements of the glycine moiety highlighted in bold.[192] Incorporation of amino acids with hydrophobic side chains, such as isoleucine and phenylalanine, or polar groups, such as glutamine or histidine, were ineffective at increasing activity. X-ray crystallography of aspartic acid derivative **98** revealed that the amine did not form any hydrogen bonds with the protein, but the carboxylic acid group was nearing the range to interact with the active site Zn^{2+}. Elongation of the linker through the use of glutamic acid with concomitant reversal of the attachment point, through the acid side chain, led to the (*S*)-amino acid derivative **99**, which

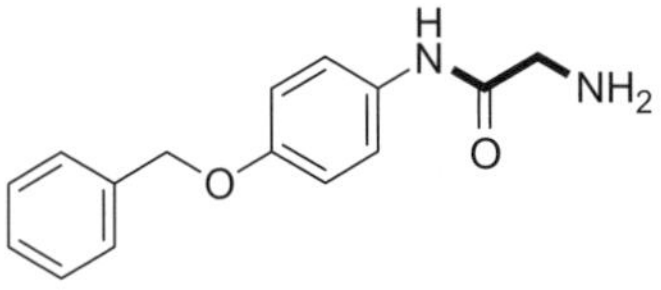

97
peptidase IC$_{50}$ = 280 nM
hWB LTA$_4$H IC$_{50}$ = 7 μM

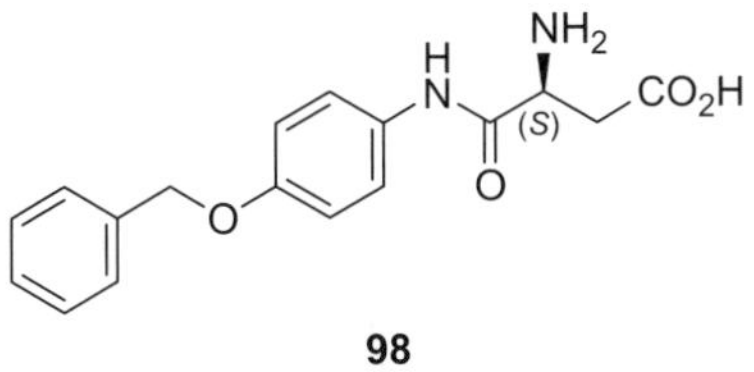

98
peptidase IC$_{50}$ = 5.4 μM
hWB LTA$_4$H IC$_{50}$ = ND

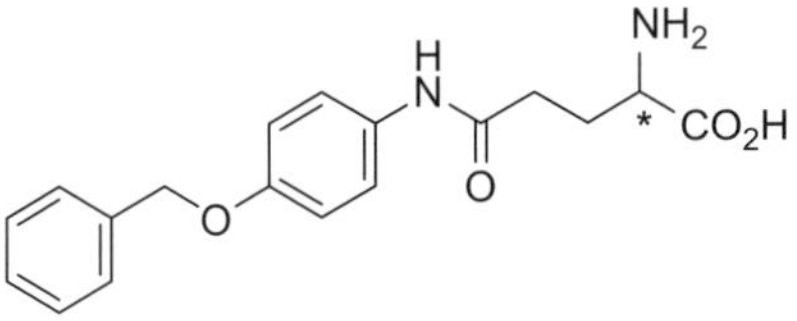

99 (S)
peptidase IC$_{50}$ = 20 nM
hWB LTA$_4$H IC$_{50}$ = 8 μM
100 (R)
peptidase IC$_{50}$ = 1.6 μM
hWB LTA$_4$H IC$_{50}$ = ND

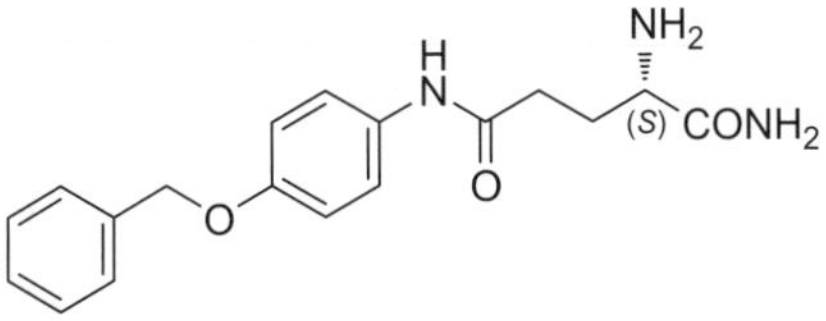

101
peptidase IC$_{50}$ = 23 nM
hWB LTA$_4$H IC$_{50}$ = 85 nM

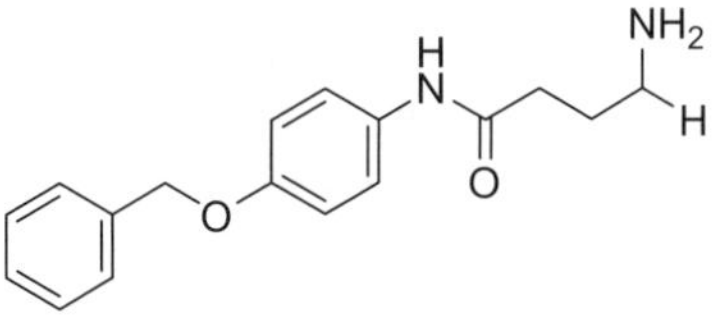

102
peptidase IC$_{50}$ = 61 nM
hWB LTA$_4$H IC$_{50}$ = 380 nM

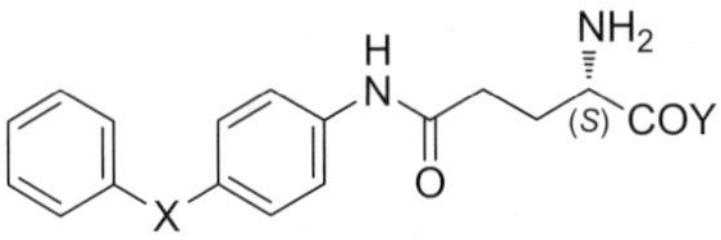

103, X = CH$_2$, Y = NH$_2$
peptidase IC$_{50}$ = 21 nM
hWB LTA$_4$H IC$_{50}$ = 110 nM
104, X = O, Y = OH
peptidase IC$_{50}$ = 19 nM
hWB LTA$_4$H IC$_{50}$ = 3.3 μM
105, X = NCH$_3$, Y = OCH$_3$
peptidase IC$_{50}$ = 8 μM
hWB LTA$_4$H IC$_{50}$ = ND

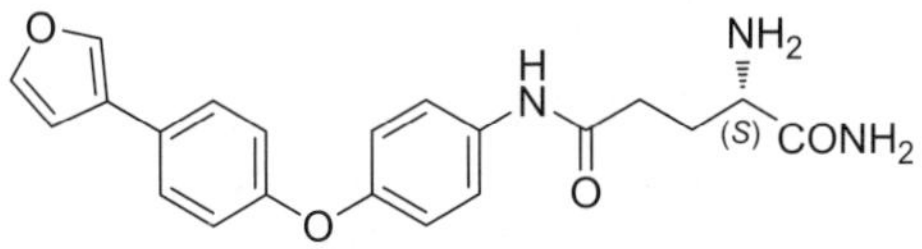

106, X = CH$_2$, Y = NH$_2$
peptidase IC$_{50}$ = 29 nM
hWB LTA$_4$H IC$_{50}$ = 92 nM

Figure 3.21   Examples of amino acid derived inhibitors of LTA$_4$H.

possessed significantly improved peptidase inhibitory activity. Subsequent X-ray crystallography showed hydrogen bonds between the amine of **99** and key residues in the active site as well as interaction of the carboxylic acid with the active site Zn^{2+}.[192] Thus, the amine forms hydrogen bonds with Glu271 and Gln136 and the carboxylate forms hydrogen bonds with Glu296 and Tyr383. Comparison of **99** to the (*R*)-enantiomer **100** revealed that the peptidase activity is sensitive to the stereochemistry at this position. The conversion of acid **99** to amide **101** led to a 100-fold improvement in whole blood activity. Review of compound **102** reveals only a modest loss in enzymatic potency relative to **99** associated with removal of the carboxylate, which questions the contribution of the Zn^{2+} interaction. The authors proposed that this minimal shift in enzymatic activity between **99** and **102** may be explained by the Circe effect,[193] which suggests that an enzyme has reduced affinity for products (*i.e.* the carboxylate product of peptide cleavage) *versus* substrates (peptides) given the aminopeptidase activity associated with LTA_4H.

Modification of the linker connecting the aromatic groups as shown in compounds **103** and **104** is tolerated, in contrast to incorporation of the tertiary amine linker represented by **105**.[192] Further substitution at the *para*-position of the terminal aromatic ring is tolerated as illustrated by compound **106**. Additionally, the acid functionality is generally not tolerated with respect to whole blood activity, regardless of other changes.[192]

In another publication, the Berlex group described additional studies derived from hit **97**.[194] In this work an early lead, compound **107** (Figure 3.22), was optimized using iterative medicinal chemistry and X-ray crystallography in an effort to access interactions with the Arg563 and Lys565 binding pocket residues located near the active site Zn^{2+}.[194] To that end, the side chain of compound **107** was extended with various groups containing carboxylic acids as exemplified by compound **108**.[194] Additional variations in the hydrophobic region, as depicted in compound **109**, are tolerated but did not offer any significant improvement in activity. Attempts to constrain the diamine portion of the molecule were met with mixed results. The piperizine analogue **110** is only weakly active whereas the aminomethyl pyrrolidine analogue **111** showed improved activity. Interestingly, the enzyme is sensitive to the stereochemistry associated with this change, with the (*R*)-enantiomer **111** more than 10-fold more potent than (*S*)-enantiomer **112**.[194] The authors note that compounds **108** and **111** also offer the advantage of greater microsomal stability (from <33% to >90% remaining after one hour in rat microsomes) and bioavailability (from negligible to ~45% F) over compounds **97** and **107**.[194]

In a final publication, the Berlex team looked to optimize activity by removing the amino amide linker between the benzoic acid and the central aromatic ring, maintaining only the diamine component.[195] The initial optimization started with the 2-phenyl oxazole derivative **113** (Figure 3.23). Evaluation of a variety of acyclic, cyclic and bridged diamines led to the identification of compound **114** that had increased activity in both the molecular assay and the whole blood assay. Further investigation determined that the oxazolylphenyl group was optimal within the scope of this study.

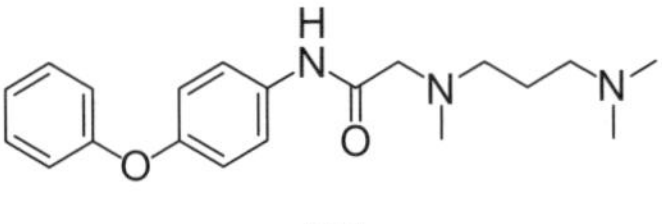

107
hLTA$_4$H IC$_{50}$ = 160 nM
hWB LTA$_4$H IC$_{50}$ = 135 nM

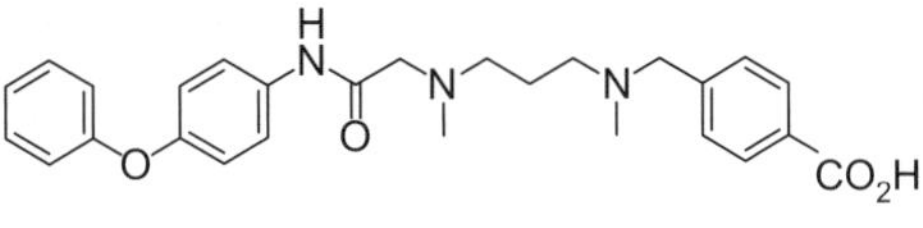

108
hLTA$_4$H IC$_{50}$ = 6 nM
hWB LTA$_4$H IC$_{50}$ = 120 nM

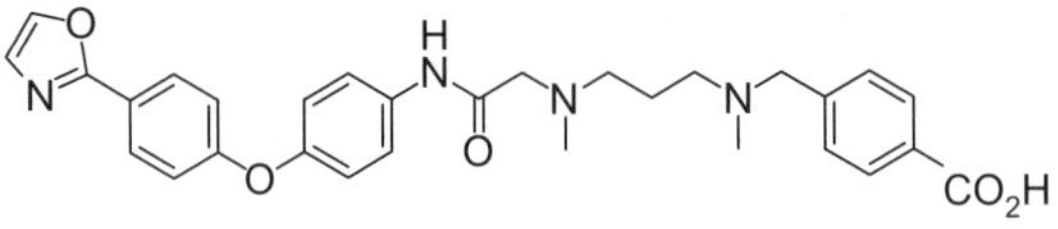

109
hLTA$_4$H IC$_{50}$ = 5 nM
hWB LTA$_4$H IC$_{50}$ = 60 nM

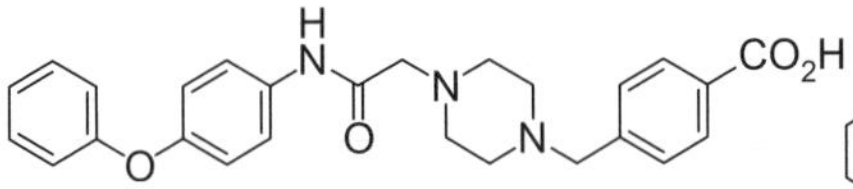

110
hLTA$_4$H IC$_{50}$ = 1 μM

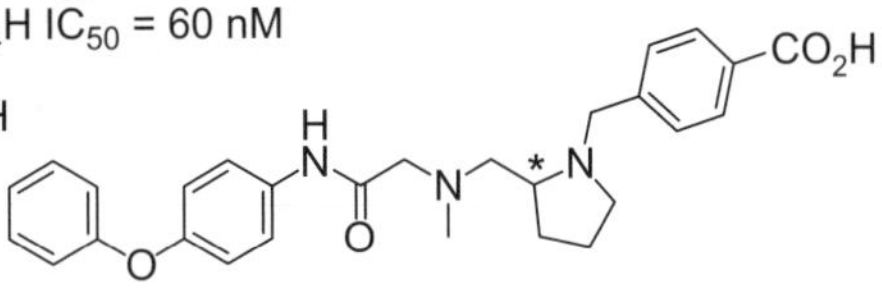

111 (*R*)
hLTA$_4$H IC$_{50}$ = 7 nM
hWBLTA$_4$HIC$_{50}$ = 30 nM
112 (*S*)
hLTA$_4$H IC$_{50}$ = 100 nM
hWBLTA$_4$H IC$_{50}$ = 200 nM

Figure 3.22   Diamine inhibitors of LTA$_4$H from Berlex.

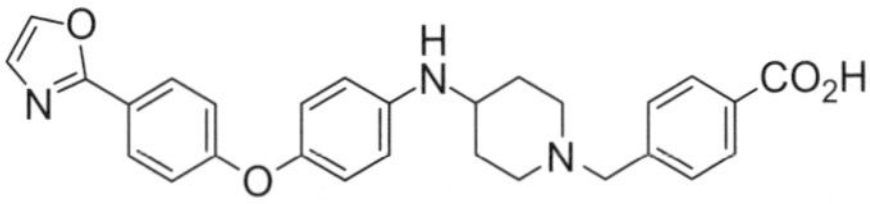

113
hLTA$_4$H IC$_{50}$ = 30 nM
hWB LTA$_4$H IC$_{50}$ = 80 nM

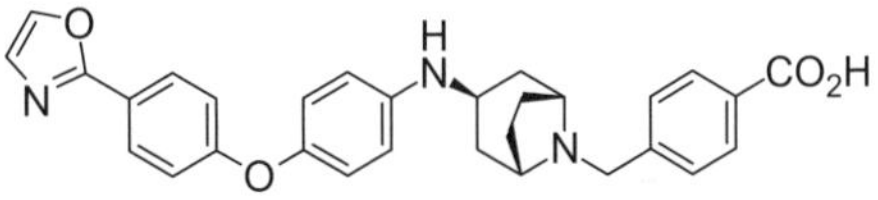

114
hLTA$_4$H IC$_{50}$ = 5 nM
hWB LTA$_4$H IC$_{50}$ = 40 nM

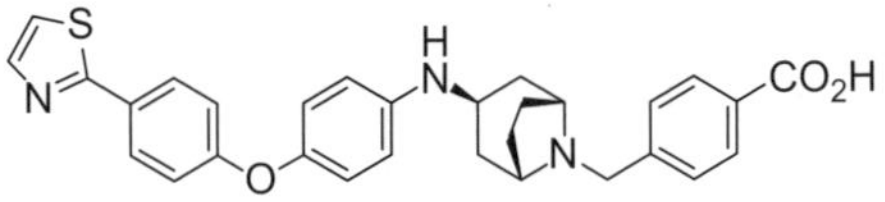

115
hLTA$_4$H IC$_{50}$ = 8 nM
hWB LTA$_4$H IC$_{50}$ = 19 nM

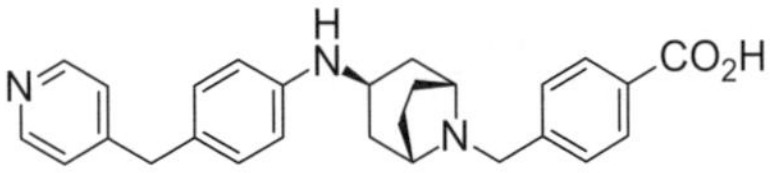

116
hLTA$_4$H IC$_{50}$ = 75 nM
hWB LTA$_4$H IC$_{50}$ = 440 nM

Figure 3.23   LTA$_4$H inhibitors from Berlex containing cyclic diamine linkers.

As examples, the thiazolylphenyl **115** was equipotent and the pyridyl analogue **116** was less active when compared to **114**. The authors also indicate that compound **114** has a superior PK profile relative to the acyclic diamine analogues tested, though data were not reported.[195]

3.4.6   Santen Pharmaceutical Co., Ltd

Santen Pharmaceutical Co., Ltd (Santen) published the effect of SA6541 (Figure 3.24) in a carrageenan-induced murine dermatitis model in 1998.[152] More recently, they have published on the medicinal chemistry efforts towards optimization of the LTA$_4$H inhibitory activity of captopril (**117**) and SA6541.[196,197] Initially, modification of the proline ring in captopril and the thiazolidine ring in **118** was evaluated.[196] Modifications to the R^1 side chain in either ring system, or addition of substituents at R^2 in the thiazolidine series, failed to produce molecules with inhibitory activity superior to captopril. Substitution at the 4-position of the proline ring with various *para*-substituted (R^3) thiobenzyl groups did provide analogues with substantially increased potency against LTA$_4$H, as illustrated by compounds **119–121**, while minimizing activity against angiotensin converting enzyme (ACE), the original target of captopril.[196] This is illustrated in Figure 3.24, which denotes

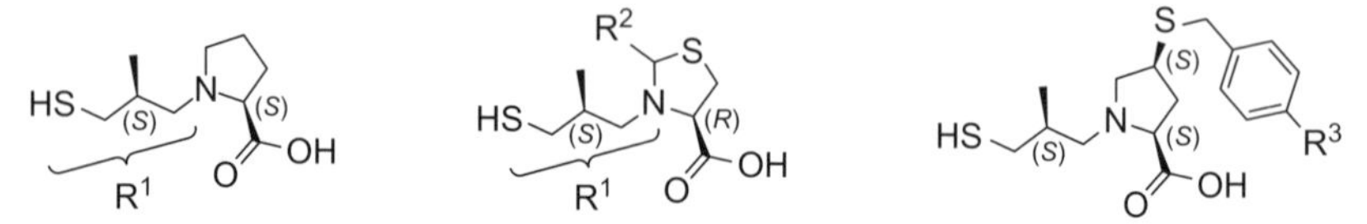

117, Captopril
LTA$_4$H = 59% inh at 1 mM

118, R^2 = H
LTA$_4$H = 58% inh at 1 mM

119, R^3 = *t*-Bu
LTA$_4$H IC$_{50}$ = 31 nM
13-fold selective over ACE
120, R^3 = cyclohexyl
LTA$_4$H IC$_{50}$ = 34 nM
8-fold selective over ACE
121, R^3 = *i*-Pr
LTA$_4$H IC$_{50}$ = 52 nM
<1-fold selective over ACE

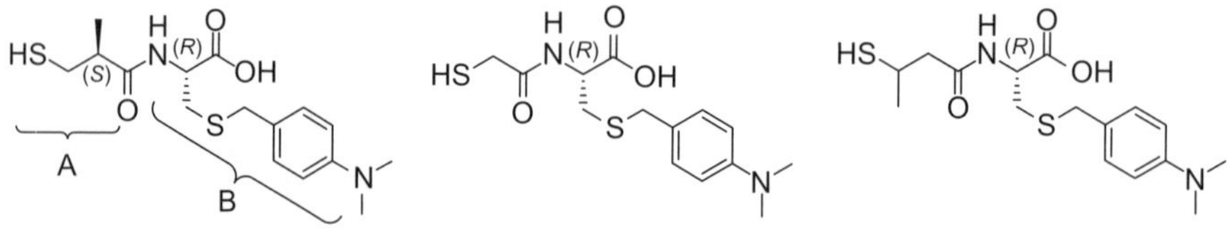

122, SA6541
LTA$_4$H IC$_{50}$ = 270 nM

123
LTA$_4$H IC$_{50}$ = 470 nM

124
LTA$_4$H IC$_{50}$ = 7.4 µM

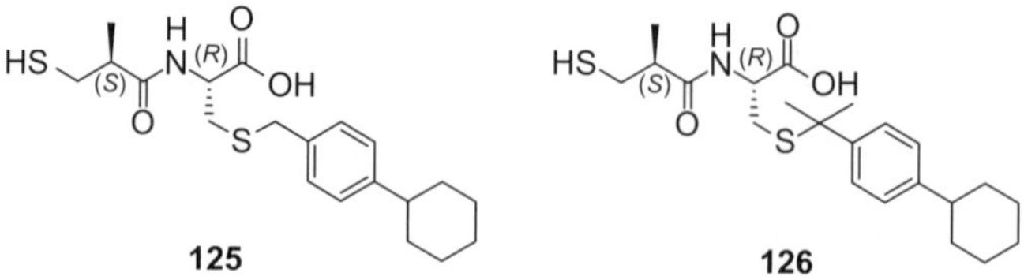

125
LTA$_4$H IC$_{50}$ = 79 nM
50-fold selective over ACE

126
LTA$_4$H IC$_{50}$ = 55 nM
50-fold selective over ACE

Figure 3.24   Inhibitors reported by Santen.

increasing selectivity for LTA_4H over ACE for compounds **119** and **120** with increasing bulkiness of R^3.

Starting with region A of SA6541, removal of the methyl group and concomitant shortening of the distance between the amide and the thiol led to compound **122**, which showed a minimal decrease in LTA_4H inhibitory activity.[197] Additional modifications to the methyl group including inversion, movement along the chain or elongation are deleterious to activity, as illustrated by compound **123**.

Moving the dimethyl amino group from the benzyl substituent in region B of SA6541 to the *meta*-position or removal altogether is detrimental to activity, as was inversion of the stereochemical centre (denoted in region B) or elongation of the thioether chain (data not shown). However, replacement of the dimethyl amino group with other substituents is beneficial and, in select cases, led to improved selectivity over ACE. Compound **125**, which contains a cyclohexyl group in the 4-position of the benzyl ring, shows 50-fold selectivity over ACE. Addition of geminal methyl groups at the benzylic position of the thioether, illustrated by **126**, yields a similarly 55-fold selective compound.[197]

The authors developed a model using a combination of X-ray crystal structures and docking to understand the increased selectivity of compounds **119**, **120** and **124–126** between ACE and LTA_4H. This model relies on the premise that the R^1/A substituents (and the 5-membered ring in captopril and compounds **119** and **120**) in all analogues are overlapping and occupied a similar conformation in both ACE and LTA_4H. They identified an active conformation for the molecules in which the hydrophobic group or region B extended into the C-terminal domain towards Phe340 in LTA_4H. Similar conformations within the ACE active site are not as accessible due to its restricted size and altered electronic characteristics.

3.5 Clinical Evaluation of LTB₄ Modulators

In spite of the extensive medicinal chemistry effort devoted to the identification of potent and selective inhibitors of LTA_4H, there has been only one public disclosure of an LTA_4H inhibitor (DG-051 from deCODE) progressing through Phase II clinical development, and no further information has been released on this compound.[198] Select compounds that inhibit LTB_4 production or signalling at other points in the AA cascade (5-LO, FLAP, BLT1/BLT2) and have reached or advanced beyond Phase II clinical development are listed in Table 3.1. Of these compounds, only zileuton has received marketing approval as a drug. This 5-LO inhibitor is indicated for the prophylaxis and chronic treatment of asthma in adults and in children 12 years of age and older.[199] The recommended dosage of zileuton for the treatment of patients with asthma is two 600 mg extended-release tablets twice daily for a total daily dose of 2400 mg. Because zileuton can cause elevation of liver enzymes in a subset of patients, liver enzyme test monitoring is necessary while taking this drug.

Table 3.1 Compounds against LTB_4-related targets that reached Phase II/III clinical trials.

Drug name	Company	Target	Disease indications in clinical studies
DG-051	deCODE	LTA_4H	Myocardial infarction[189]
Zileuton	Cornerstone Therapeutics/ Abbott	5-LO	Asthma,[200,201] other immune disorders*[98]
MK-0633/setileuton	Merck	5-LO	COPD,[202,203] asthma,[204] atherosclerosis[205]
VIA-2291/ABT-761/ atreleuton	VIA Pharmaceuticals/ Abbott	5-LO	Cardiovascular inflammation,[73] asthma[206,207]
ZD-2138	Zeneca	5-LO	Asthma[208]
MK-886	Merck	FLAP	Asthma,[67] psoriasis[209]
MK-591	Merck	FLAP	Asthma,[62,66] ulcerative colitis[210]
DG-031/BAY X-1005/veliflapon	decode/Bayer	FLAP	Asthma,[62,65] myocardial infarction[211]
AM-803/fiboflapon	Amira/GSK	FLAP	Asthma[212]
BIIL-284/amelubant	Boehringer Ingelheim	BLT1/BLT2	RA,[213,214] cystic fibrosis[215]
LY-293111/etalocib	Eli Lilly/Vanguard	BLT1/BLT2	Atopic asthma,[216] psoriasis,[217] pancreatic cancer,[142,149] non-small-cell lung carcinoma,[147] ulcerative colitis[218]
CP-195543	Pfizer	BLT1/BLT2	RA[219]
LTB-019/ CGS-25019C	Novartis	BLT1/BLT2	COPD,[220] asthma[218]
Biomed-101	Searle	BLT1/BLT2	Reduction of side-effects of IL-2 treatment for renal cell carcinoma[221]
ONO-4057	Ono Pharmaceutical	BLT1/BLT2	Ulcerative colitis[218]

*Includes ulcerative colitis, allergic rhinitis, chronic nasal polyposis, sinusitis, atopic dermatitis, urticaria and COPD.

3.6 Summary

Despite over two decades of research into the leukotriene pathway, only a handful of commercially available drugs have emerged. CysLT1 antagonists block LTD_4 signalling but do not affect the parallel LTB_4 pathway. Zileuton, the only marketed 5-LO inhibitor, is efficacious in treating moderate to severe asthma, but the toxicological profile of this compound limits its use, thus highlighting the need for additional therapeutics. Furthermore, though there exists much pre-clinical data linking LTA_4H with inflammatory disorders, LTA_4H has had the least clinical development publicly reported of the LTB_4-related targets, with just one disclosure of a Phase II trial in 2010,[222] and no compound yet launched to market.

   A number of companies have conducted research into small-molecule LTA_4H inhibition during the past decade, work that appears to have been influenced by two key events: Searle's patent publications in the 1990s and the X-ray crystal structure of LTA₄H complexed with bestatin that appeared in 2001. The various series that have been disclosed by J&J PRD, L.L.C., deCODE and Berlex share a common motif with the compounds reported by Searle: an aromatic core flanked by another aromatic region on one side with a functionalized amine on the other.[223] All three companies appear to have used knowledge gleaned from the LTA₄H crystal structure for lead generation and optimization. On the other hand, Santen's compounds stemmed from captopril, though this company also relied heavily on the LTA₄H crystal structure for lead optimization. While clinical validation for LTA₄H in human disease remains elusive, human genetic associations and a significant body of pre-clinical data, support its potential involvement in multiple disease areas.

Acknowledgements

The authors would like to thank J&J PRD, L.L.C. colleagues and management. Support from Jake Wiener and Scott D. Bembenek with manuscript preparation and valuable discussions are gratefully acknowledged.

References

1. N. Kim and A. D. Luster, *Scientific World Journal*, 2007, **7**, 1307.
2. M. Peters-Golden and W. R. Henderson, Jr., *N. Engl. J. Med.*, 2007, **357**, 1841.
3. P. Rubin and K. W. Mollison, *Prostag. Oth. Lipid M.*, 2007, **83**, 188.
4. E. Ricciotti and G. A. FitzGerald, *Arterioscler. Thromb. Vasc. Biol.*, 2011, **31**, 986.
5. M. Nakamura and T. Shimizu, *Chem. Rev.*, 2011, **110**, 6231.
6. P. Borgeat, M. Hamberg and B. Samuelsson, *J. Biol. Chem.*, 1976, **251**, 7816.
7. A. Del Prete, W.-H. Shao, S. Mitola, G. Santoro, S. Sozzani and B. Haribabu, *Blood*, 2007, **109**, 626.
8. E. J. Goetzl, D. W. Goldman, P. H. Naccache, R. I. Sha'afi and W. C. Pickett, *Adv. Prostaglandin Thromboxane Leukot. Res.*, 1982, **9**, 273.
9. K. Goodarzi, M. Goodarzi, A. M. Tager, A. D. Luster and U. H. von Andrian, *Nat. Immunol.*, 2003, **4**, 965.
10. J. Z. Haeggstrom, F. Kull, P. C. Rudberg, F. Tholander and M. M. Thunnissen, *Prostag. Oth. Lipid M.*, 2002, **68–69**, 495.
11. E. A. Heller, E. Liu, A. M. Tager, S. Sinha, J. D. Roberts, S. L. Koehn, P. Libby, E. R. Aikawa, J. Q. Chen, P. Huang, M. W. Freeman, K. J. Moore, A. D. Luster and R. E. Gerszten, *Circulation*, 2005, **112**, 578.
12. Y. Iizuka, T. Yokomizo, K. Terawaki, M. Komine, K. Tamaki and T. Shimizu, *J. Biol. Chem.*, 2005, **280**, 24816.
13. K. A. Lundeen, B. Sun, L. Karlsson and A. M. Fourie, *J. Immunol.*, 2006, **177**, 3439.

14. N. M. Munoz, I. Douglas, D. Mayer, A. Herrnreiter, X. Zhu and A. R. Leff, *Am. J. Respir. Crit. Care. Med.*, 1997, **155**, 1398.
15. V. L. Ott, J. C. Cambier, J. Kappler, P. Marrack and B. J. Swanson, *Nat. Immunol.*, 2003, **4**, 974.
16. N. P. Duroudier, A. S. Tulah and I. Sayers, *Allergy*, 2009, **64**, 823.
17. A. M. Fourie, *Curr. Opin. Investig. Drugs*, 2009, **10**, 1173.
18. C. Lemiere, S. Pelissier, C. Tremblay, S. Chaboillez, M. Thivierge, J. Stankova and M. Rola-Pleszczynski, *Clin. Exp. Allergy*, 2004, **34**, 1684.
19. A. D. Luster and A. M. Tager, *Nat. Rev. Immunol.*, 2004, **4**, 711.
20. P. Montuschi, *Pharmaceuticals*, 2010, **3**, 1792.
21. P. Montuschi and M. L. Peters-Golden, *Clin. Exp. Allergy*, 2010, **40**, 1732.
22. P. Montuschi, A. Sala, S. E. Dahlen and G. Folco, *Drug Discov. Today*, 2007, **12**, 404.
23. N. L. Rao, J. P. Riley, H. Banie, X. Xue, B. Sun, S. Crawford, K. A. Lundeen, F. Yu, L. Karlsson, A. M. Fourie and P. J. Dunford, *Am. J. Respir. Crit. Care Med.*, 2010, **181**, 899.
24. K. Terawaki, T. Yokomizo, T. Nagase, A. Toda, M. Taniguchi, K. Hashizume, T. Yagi and T. Shimizu, *J. Immunol.*, 2005, **175**, 4217.
25. P. J. Barnes, *Nat. Rev. Drug Discov.*, 2002, **1**, 437.
26. S. Gompertz and R. A. Stockley, *Chest*, 2002, **122**, 289.
27. E. Marian, S. Baraldo, A. Visentin, A. Papi, M. Saetta, L. M. Fabbri and P. Maestrelli, *Chest*, 2006, **129**, 1523.
28. J. Jupp, K. Hillier, D. H. Elliott, D. R. Fine, A. C. Bateman, P. A. Johnson, A. M. Cazaly, J. F. Penrose and A. P. Sampson, *Inflamm. Bowel Dis.*, 2007, **13**, 537.
29. J. Rask-Madsen, *Drugs Today (Barc.)*, 1998, **34**, 45.
30. P. Sharon and W. F. Stenson, *Gastroenterology*, 1984, **86**, 453.
31. M. Chen, B. K. Lam, Y. Kanaoka, P. A. Nigrovic, L. P. Audoly, K. F. Austen and D. M. Lee, *J. Exp. Med.*, 2006, **203**, 837.
32. F. Tsuji, K. Oki, K. Fujisawa, A. Okahara, M. Horiuchi and S. Mita, *Life Sci.*, 1999, **64**, PL51–56.
33. M. Back, D. X. Bu, R. Branstrom, Y. Sheikine, Z. Q. Yan and G. K. Hansson, *Proc. Natl Acad. Sci. USA*, 2005, **102**, 17501.
34. M. Back and G. K. Hansson, *Ann. Med.*, 2006, **38**, 493.
35. E. B. Friedrich, A. M. Tager, E. Liu, A. Pettersson, C. Owman, L. Munn, A. D. Luster and R. E. Gerszten, *Arterioscler. Thromb. Vasc. Biol.*, 2003, **23**, 1761.
36. C. D. Funk, *Nat. Rev. Drug Discov.*, 2005, **4**, 664.
37. A. Helgadottir, A. Manolescu, G. Thorleifsson, S. Gretarsdottir, H. Jonsdottir, U. Thorsteinsdottir, N. J. Samani, G. Gudmundsson, S. F. Grant, G. Thorgeirsson, S. Sveinbjornsdottir, E. M. Valdimarsson, S. E. Matthiasson, H. Johannsson, O. Gudmundsdottir, M. E. Gurney, J. Sainz, M. Thorhallsdottir, M. Andresdottir, M. L. Frigge, E. J. Topol, A. Kong, V. Gudnason, H. Hakonarson, J. R. Gulcher and K. Stefansson, *Nat. Genet.*, 2004, **36**, 233.
38. V. R. Jala and B. Haribabu, *Trends Immunol.*, 2004, **25**, 315.

39. H. Qiu, A. Gabrielsen, H. E. Agardh, M. Wan, A. Wetterholm, C.-H. Wong, U. Hedin, J. Swedenborg, G. K. Hansson, B. Samuelsson, G. Paulsson-Berne and J. Z. Haeggstrom, *Proc. Natl Acad. Sci. USA*, 2006, **103**, 8161.
40. G. Riccioni, M. Back and V. Capra, *Curr. Drug Targets*, 2010, **11**, 882.
41. E. Sanchez-Galan, A. Gomez-Hernandez, C. Vidal, J. L. Martin-Ventura, L. M. Blanco-Colio, B. Munoz-Garcia, L. Ortega, J. Egido and J. Tunon, *Cardiovasc. Res.*, 2009, **81**, 216.
42. K. Subbarao, V. R. Jala, S. Mathis, J. Suttles, W. Zacharias, J. Ahamed, H. Ali, M. T. Tseng and B. Haribabu, *Arterioscler. Thromb. Vasc. Biol.*, 2004, **24**, 369.
43. X. Chen, S. Wang, N. Wu and C. S. Yang, *Curr. Cancer Drug Targets*, 2004, **4**, 267.
44. P. Gao, L. Guan and J. Zheng, *Biochem. Biophys. Res. Commun.*, 2010, **402**, 308.
45. A. Ihara, K. Wada, M. Yoneda, N. Fujisawa, H. Takahashi and A. Nakajima, *J. Pharmacol. Sci.*, 2007, **103**, 24.
46. C. H. Jeong, A. M. Bode, A. Pugliese, Y. Y. Cho, H. G. Kim, J. H. Shim, Y. J. Jeon, H. Li, H. Jiang and Z. Dong, *Cancer Res.*, 2009, **69**, 5584.
47. S. Larre, N. Tran, C. Fan, H. Hamadeh, J. Champigneulles, R. Azzouzi, O. Cussenot, P. Mangin and J. L. Olivier, *Prostag. Oth. Lipid M.*, 2008, **87**, 14.
48. N. Oi, C. H. Jeong, J. Nadas, Y. Y. Cho, A. Pugliese, A. M. Bode and Z. Dong, *Cancer Res.*, 2010, **70**, 9755.
49. W. G. Tong, X. Z. Ding, M. S. Talamonti, R. H. Bell and T. E. Adrian, *Biochem. Biophys. Res. Commun.*, 2005, **335**, 949.
50. F. H. Tsuji, M. Enomoto and H. Aono, *Curr. Top. Pharm.*, 2005, **9**, 71.
51. W. E. Brocklehurst, *Prog. Allergy*, 1962, **6**, 539.
52. P. Borgeat and P. Sirois, *J. Med. Chem.*, 1981, **24**, 121.
53. V. Capra, M. Ambrosio, G. Riccioni and G. E. Rovati, *Curr. Med. Chem.*, 2006, **13**, 3213.
54. J. F. Evans, *Prostag. Oth. Lipid M.*, 2002, **68–69**, 587.
55. D. R. Davies, B. Mamat, O. T. Magnusson, J. Christensen, M. H. Haraldsson, R. Mishra, B. Pease, E. Hansen, J. Singh, D. Zembower, H. Kim, A. S. Kiselyov, A. B. Burgin, M. E. Gurney and L. J. Stewart, *J. Med. Chem.*, 2010, **53**, 2330.
56. C. N. Serhan, M. Hamberg and B. Samuelsson, *Proc. Natl Acad. Sci. USA*, 1984, **81**, 5335.
57. J. V. Bonventre, Z. Huang, M. R. Taheri, E. O'Leary, E. Li, M. A. Moskowitz and A. Sapirstein, *Nature*, 1997, **390**, 622.
58. N. Uozumi, K. Kume, T. Nagase, N. Nakatani, S. Ishii, F. Tashiro, Y. Komagata, K. Maki, K. Ikuta, Y. Ouchi, J. Miyazaki and T. Shimizu, *Nature*, 1997, **390**, 618.
59. M. Lehr, *Anti-Inflamm. Anti-Allergy Agents Med. Chem.*, 2006, **5**, 149.
60. D. H. Adler, J. D. Cogan, J. A. Phillips, 3rd, N. Schnetz-Boutaud, G. L. Milne, T. Iverson, J. A. Stein, D. A. Brenner, J. D. Morrow, O. Boutaud and J. A. Oates, *J. Clin. Invest.*, 2008, **118**, 2121.

61. NCT00396955 from www.clinicaltrials.gov
62. C. D. Brooks and J. B. Summers, *J. Med. Chem.*, 1996, **39**, 2629.
63. M. Peters-Golden and T. G. Brock, *Prostaglandins Leukot. Essent. Fatty Acids*, 2003, **69**, 99.
64. P. J. Vickers, *J. Lipid Mediat. Cell Signal*, 1995, **12**, 185.
65. B. Dahlen, M. Kumlin, E. Ihre, O. Zetterstrom and S. E. Dahlen, *Thorax*, 1997, **52**, 342.
66. Z. Diamant, M. C. Timmers, H. van der Veen, B. S. Friedman, M. De Smet, M. Depre, D. Hilliard, E. H. Bel and P. J. Sterk, *J. Allergy Clin. Immunol.*, 1995, **95**, 42.
67. B. S. Friedman, E. H. Bel, A. Buntinx, W. Tanaka, Y. H. Han, S. Shingo, R. Spector and P. Sterk, *Am. Rev. Respir. Dis.*, 1993, **147**, 839.
68. J. L. Masferrer, C. Wegner and M. Graneto, ed., 5-Lipoxygenase Inhibition, S. Karger, Basel, 2010, p. 201.
69. C. Pergola and O. Werz, *Expert Opin. Ther. Pat.*, 2010, **20**, 355.
70. O. Werz and D. Steinhilber, *Pharmcol. Therpeut.*, 2006, **112**, 701.
71. Y. Ducharme, M. Blouin, C. Brideau, A. Chateauneuf, Y. Gareau, E. L. Grimm, H. Juteau, S. Laliberte, B. MacKay, F. Masse, M. Ouellet, M. Salem, A. Styhler and R. W. Friesen, *ACS Med. Chem. Lett.*, 2010, **1**, 170.
72. J. L. Masferrer, B. S. Zweifel, M. Hardy, G. D. Anderson, D. Dufield, L. Cortes-Burgos, R. A. Pufahl and M. Graneto, *J. Pharmacol. Exp. Ther.*, 2010, **334**, 294.
73. J. C. Tardif, P. L. L'Allier, R. Ibrahim, J. C. Gregoire, A. Nozza, M. Cossette, S. Kouz, M. A. Lavoie, J. Paquin, T. M. Brotz, R. Taub and J. Pressacco, *Circ. Cardiovasc. Imaging*, 2010, **3**, 298.
74. J. F. Evans, A. D. Ferguson, R. T. Mosley and J. H. Hutchinson, *Trends Pharmacol. Sci.*, 2008, **29**, 72.
75. A. P. Sampson, *Curr. Opin. Investig. Drugs*, 2009, **10**, 1163.
76. A. Hicks, S. P. Monkarsh, A. F. Hoffman and R. Goodnow, Jr., *Expert Opin. Investig. Drugs*, 2007, **16**, 1909.
77. J. H. Hutchinson, *Expert Opin. Ther. Pat.*, 2010, **20**, 707.
78. T. R. Jones, K. Metters and J. Evans, *Clin. Exp. Allergy Rev.*, 2001, **1**, 205.
79. H. Shirasaki, *Expert Opin. Ther. Targets*, 2008, **12**, 415.
80. C. E. Heise, B. F. O'Dowd, D. J. Figueroa, N. Sawyer, T. Nguyen, D. S. Im, R. Stocco, J. N. Bellefeuille, M. Abramovitz, R. Cheng, D. L. Williams, Jr., Z. Zeng, Q. Liu, L. Ma, M. K. Clements, N. Coulombe, Y. Liu, C. P. Austin, S. R. George, G. P. O'Neill, K. M. Metters, K. R. Lynch and J. F. Evans, *J. Biol. Chem.*, 2000, **275**, 30531.
81. M. Kamohara, J. Takasaki, M. Matsumoto, T. Saito, T. Ohishi, H. Ishii and K. Furuichi, *J. Biol. Chem.*, 2000, **275**, 27000.
82. K. R. Lynch, G. P. O'Neill, Q. Liu, D. S. Im, N. Sawyer, K. M. Metters, N. Coulombe, M. Abramovitz, D. J. Figueroa, Z. Zeng, B. M. Connolly, C. Bai, C. P. Austin, A. Chateauneuf, R. Stocco, G. M. Greig, S. Kargman, S. B. Hooks, E. Hosfield, D. L. Williams, Jr., A. W. Ford-Hutchinson, C. T. Caskey and J. F. Evans, *Nature*, 1999, **399**, 789.

83. Y. Tryselius, N. E. Nilsson, K. Kotarsky, B. Olde and C. Owman, *Biochem. Biophys. Res. Commun.*, 2000, **274**, 377.

84. S. Wang, E. Gustafson, L. Pang, X. Qiao, J. Behan, M. Maguire, M. Bayne and T. Laz, *J. Biol. Chem.*, 2000, **275**, 40686.

85. T. Yokomizo, T. Izumi, K. Chang, Y. Takuwa and T. Shimizu, *Nature*, 1997, **387**, 620.

86. T. Yokomizo, K. Kato, K. Terawaki, T. Izumi and T. Shimizu, *J. Exp. Med.*, 2000, **192**, 421.

87. A. M. Tager, S. K. Bromley, B. D. Medoff, S. A. Islam, S. D. Bercury, E. B. Friedrich, A. D. Carafone, R. E. Gerszten and A. D. Luster, *Nat. Immunol.*, 2003, **4**, 982.

88. P. Nilsson, Biolipox AB, *PCT Pat. Appl.*, WO 2010/07566, 2010.

89. P. Nilsson, B. Pelcman and M. Katkevics, Biolipox AB, *PCT Pat. Appl.*, WO 2010/103279, 2010.

90. P. Nilsson, B. Pelcman and M. Katkevics, Biolipox AB, *PCT Pat. Appl.*, WO 2010/03297, 2010.

91. P. Nilsson, B. Pelcman and M. Katkevics, Biolipox AB, *PCT Pat. Appl.*, WO 2010/103278, 2010.

92. P. Nilsson, B. Pelcman, M. Katkevics, E. Suna and I. Popovs, Biolipox AB, *PCT Pat. Appl.*, WO 2010/103283, 2010.

93. R. J. Snelgrove, P. L. Jackson, M. T. Hardison, B. D. Noerager, A. Kinloch, A. Gaggar, S. Shastry, S. M. Rowe, Y. M. Shim, T. Hussell and J. E. Blalock, *Science*, 2010, **330**, 90.

94. R. S. Byrum, J. L. Goulet, J. N. Snouwaert, R. J. Griffiths and B. H. Koller, *J. Immunol.*, 1999, **163**, 6810.

95. A. Helgadottir, A. Manolescu, A. Helgason, G. Thorleifsson, U. Thorsteinsdottir, D. F. Gudbjartsson, S. Gretarsdottir, K. P. Magnusson, G. Gudmundsson, A. Hicks, T. Jonsson, S. F. Grant, J. Sainz, S. J. O'Brien, S. Sveinbjornsdottir, E. M. Valdimarsson, S. E. Matthiasson, A. I. Levey, J. L. Abramson, M. P. Reilly, V. Vaccarino, M. L. Wolfe, V. Gudnason, A. A. Quyyumi, E. J. Topol, D. J. Rader, G. Thorgeirsson, J. R. Gulcher, H. Hakonarson, A. Kong and K. Stefansson, *Nat. Genet.*, 2006, **38**, 68.

96. N. L. Rao, P. J. Dunford, X. Xue, X. Jiang, K. A. Lundeen, F. Coles, J. P. Riley, K. N. Williams, C. A. Grice, J. P. Edwards, L. Karlsson and A. M. Fourie, *J. Pharmacol. Exp. Ther.*, 2007, **321**, 1154.

97. S. P. Mathis, V. R. Jala, D. M. Lee and B. Haribabu, *J. Immunol.*, 2010, **185**, 3049.

98. W. Berger, M. T. De Chandt and C. B. Cairns, *Int. J. Clin. Pract.*, 2007, **61**, 663.

99. P. A. Wark, S. L. Johnston, I. Moric, J. L. Simpson, M. J. Hensley and P. G. Gibson, *Eur. Respir. J.*, 2002, **19**, 68.

100. R. A. Peleman, P. H. Rytila, J. C. Kips, G. F. Joos and R. A. Pauwels, *Eur. Respir. J.*, 1999, **13**, 839.

101. B. J. R. Whittle, C. Varga, A. Berko, K. Horvath, A. Posa, J. P. Riley, K. A. Lundeen, A. M. Fourie and P. J. Dunford, *Br. J. Pharmacol.*, 2008, **153**, 983.

102. X. Jiang, L. Zhou, D. Wei, H. Meng, Y. Liu and L. Lai, *Bioorg. Med. Chem. Lett.*, 2008, **18**, 6549.
103. H. Ohnishi, N. Miyahara and E. W. Gelfand, *Allergol. Int.*, 2008, **57**, 291.
104. K. J. Cho, J. M. Seo, Y. Shin, M. H. Yoo, C. S. Park, S. H. Lee, Y. S. Chang, S. H. Cho and J. H. Kim, *Am. J. Respir. Cell Mol. Biol.*, 2009, **42**, 294.
105. N. Miyahara, S. Miyahara, K. Takeda and E. W. Gelfand, *Allergol. Int.*, 2006, **55**, 91.
106. N. Miyahara, H. Ohnishi, H. Matsuda, S. Miyahara, K. Takeda, T. Koya, S. Matsubara, M. Okamoto, A. Dakhama, B. Haribabu and E. W. Gelfand, *J. Immunol.*, 2008, **181**, 1170.
107. N. Miyahara, H. Ohnishi, S. Miyahara, K. Takeda, S. Matsubara, H. Matsuda, M. Okamoto, J. E. Loader, A. Joetham, M. Tanimoto, A. Dakhama and E. W. Gelfand, *Am. J. Respir. Cell Mol. Biol.*, 2009, **40**, 672.
108. N. Miyahara, K. Takeda, S. Miyahara, S. Matsubara, T. Koya, A. Joetham, E. Krishnan, A. Dakhama, B. Haribabu and E. W. Gelfand, *Am. J. Respir. Crit. Care Med.*, 2005, **172**, 161.
109. N. Miyahara, K. Takeda, S. Miyahara, C. Taube, A. Joetham, T. Koya, S. Matsubara, A. Dakhama, A. M. Tager, A. D. Luster and E. W. Gelfand, *J. Immunol.*, 2005, **174**, 4979.
110. H. Banie, J. Riley, C. Jaramillo, A. P. de Leon, A. Fourie, L. Karlsson and N. L. Rao, American Thoracic Society, San Diego, CA, 2009, A4245.
111. K. Waseda, N. Miyahara, A. Kanehiro, G. Ikeda, H. Koga, Y. Fuchimoto, E. Kurimoto, Y. Tanimoto, M. Kataoka, M. Tanimoto and E. W. Gelfand, *Am. J. Respir. Cell Mol. Biol.*, 2011, **45**, 851.
112. C. R. Turner, R. Breslow, M. J. Conklyn, C. J. Andresen, D. K. Patterson, A. Lopez-Anaya, B. Owens, P. Lee, J. W. Watson and H. J. Showell, *J. Clin. Invest.*, 1996, **97**, 381.
113. J. W. Holloway, S. J. Barton, S. T. Holgate, M. J. Rose-Zerilli and I. Sayers, *Allergy*, 2008, **63**, 1046.
114. J. J. Lima, S. Zhang, A. Grant, L. Shao, K. G. Tantisira, H. Allayee, J. Wang, J. Sylvester, J. Holbrook, R. Wise, S. T. Weiss and K. Barnes, *Am. J. Respir. Crit. Care Med.*, 2006, **173**, 379.
115. H. Kotani, R. Kishi, A. Mouri, T. Sashio, J. Shindo, A. Shiraki, T. Hiramatsu, S. Iwata, H. Taniguchi, O. Nishiyama, M. Iwata, R. Suzuki, H. Gonda, T. Niwa, M. Kondo, Y. Hasegawa, H. Kume and Y. Noda, *J. Clin. Pharm. Ther.*, 2011, in press.
116. H. Tcheurekdjian, M. Via, A. De Giacomo, H. Corvol, C. Eng, S. Thyne, R. Chapela, W. Rodriguez-Cintron, J. R. Rodriguez-Santana, P. C. Avila and E. G. Burchard, *J. Allergy Clin. Immunol.*, 2010, **126**, 853.
117. M. Via, A. De Giacomo, H. Corvol, C. Eng, M. A. Seibold, C. Gillett, J. Galanter, S. Sen, H. Tcheurekdjian, R. Chapela, J. R. Rodriguez-Santana, W. Rodriguez-Cintron, S. Thyne, P. C. Avila, S. Choudhry and E. Gonzalez Burchard, *Clin. Exp. Allergy*, 2010, **40**, 582.
118. D. M. Tobin, J. C. Vary, Jr., J. P. Ray, G. S. Walsh, S. J. Dunstan, N. D. Bang, D. A. Hagge, S. Khadge, M. C. King, T. R. Hawn, C. B. Moens and L. Ramakrishnan, *Cell*, 2010, **140**, 717.

119. J. Curtis, L. Kopanitsa, E. Stebbings, A. Speirs, O. Ignatyeva, Y. Balabanova, V. Nikolayevskyy, S. Hoffner, R. Horstmann, F. Drobniewski and S. Nejentsev, *Tuberculosis (Edinb.)*, 2010, **91**, 22.

120. T. D. Penning, *Curr. Pharm. Des.*, 2001, **7**, 163.

121. C. A. Grice, K. L. Tays, B. M. Savall, J. Wei, C. R. Butler, F. U. Axe, S. D. Bembenek, A. M. Fourie, P. J. Dunford, K. Lundeen, F. Coles, X. Xue, J. P. Riley, K. N. Williams, L. Karlsson and J. P. Edwards, *J. Med. Chem.*, 2008, **51**, 4150.

122. Y. Iizuka, T. Okuno, K. Saeki, H. Uozaki, S. Okada, T. Misaka, T. Sato, H. Toh, M. Fukayama, N. Takeda, Y. Kita, T. Shimizu, M. Nakamura and T. Yokomizo, *FASEB J.*, 2010, **24**, 4678.

123. T. Okuno, Y. Iizuka, H. Okazaki, T. Yokomizo, R. Taguchi and T. Shimizu, *J. Exp. Med.*, 2008, **205**, 759.

124. E. M. Davidson, S. A. Rae and M. J. Smith, *Ann. Rheum. Dis.*, 1983, **42**, 677.

125. J. Elmgreen, O. H. Nielsen and I. Ahnfelt-Ronne, *Ann. Rheum. Dis.*, 1987, **46**, 501.

126. A. Hashimoto, H. Endo, I. Hayashi, Y. Murakami, H. Kitasato, S. Kono, T. Matsui, S. Tanaka, A. Nishimura, K. Urabe, M. Itoman and H. Kondo, *J. Rheumatol.*, 2003, **30**, 1712.

127. S. Mathis, V. R. Jala and B. Haribabu, *Autoimmun. Rev.*, 2007, **7**, 12.

128. N. D. Kim, R. C. Chou, E. Seung, A. M. Tager and A. D. Luster, *J. Exp. Med.*, 2006, **203**, 829.

129. R. C. Chou, N. D. Kim, C. D. Sadik, E. Seung, Y. Lan, M. H. Byrne, B. Haribabu, Y. Iwakura and A. D. Luster, *Immunity*, 2010, **33**, 266.

130. J. Cowden, M. Sablad and Y. Zhang, unpublished results.

131. G. D. Anderson, K. L. Keys, P. A. De Ciechi and J. L. Masferrer, *Inflamm. Res.*, 2009, **58**, 109.

132. R. J. Aiello, P. A. Bourassa, S. Lindsey, W. Weng, A. Freeman and H. J. Showell, *Arterioscler. Thromb. Vasc. Biol.*, 2002, **22**, 443.

133. R. De Caterina, A. Mazzone, D. Giannessi, R. Sicari, W. Pelosi, G. Lazzerini, A. Azzara, R. Forder, F. Carey, D. Caruso, G. Galli and F. Mosca, *Biomed. Biochim. Acta*, 1988, **47**, S182–185.

134. R. Spanbroek, R. Grabner, K. Lotzer, M. Hildner, A. Urbach, K. Ruhling, M. P. Moos, B. Kaiser, T. U. Cohnert, T. Wahlers, A. Zieske, G. Plenz, H. Robenek, P. Salbach, H. Kuhn, O. Radmark, B. Samuelsson and A. J. Habenicht, *Proc. Natl Acad. Sci. USA*, 2003, **100**, 1238.

135. M. Back, A. Sultan, O. Ovchinnikova and G. K. Hansson, *Circ. Res.*, 2007, **100**, 946.

136. L. Huang, A. Zhao, F. Wong, J. M. Ayala, M. Struthers, F. Ujjainwalla, S. D. Wright, M. S. Springer, J. Evans and J. Cui, *Arterioscler. Thromb. Vasc. Biol.*, 2004, **24**, 1783.

137. J. H. Dwyer, H. Allayee, K. M. Dwyer, J. Fan, H. Wu, R. Mar, A. J. Lusis and M. Mehrabian, *N. Engl. J. Med.*, 2004, **350**, 29.

138. A. Helgadottir, S. Gretarsdottir, D. St. Clair, A. Manolescu, J. Cheung, G. Thorleifsson, A. Pasdar, S. F. Grant, L. J. Whalley, H. Hakonarson,

U. Thorsteinsdottir, A. Kong, J. Gulcher, K. Stefansson and M. J. MacLeod, *Am. J. Hum. Genet.*, 2005, **76**, 505.

139. S. H. Shah, E. R. Hauser, D. Crosslin, L. Wang, C. Haynes, J. Connelly, S. Nelson, J. Johnson, S. Gadson, C. L. Nelson, D. Seo, S. Gregory, W. E. Kraus, C. B. Granger, P. Goldschmidt-Clermont and L. K. Newby, *Atherosclerosis*, 2008, **201**, 148.

140. D. R. Crosslin, S. H. Shah, S. C. Nelson, C. S. Haynes, J. J. Connelly, S. Gadson, P. J. Goldschmidt-Clermont, J. M. Vance, J. Rose, C. B. Granger, D. Seo, S. G. Gregory, W. E. Kraus and E. R. Hauser, *Hum. Genet.*, 2009, **125**, 217.

141. J. Hartiala, D. Li, D. V. Conti, S. Vikman, Y. Patel, W. H. Tang, M. L. Brennan, J. W. Newman, C. B. Stephensen, P. Armstrong, S. L. Hazen and H. Allayee, *Hum. Genet.*, 2011, **129**, 617.

142. X. Z. Ding, M. S. Talamonti, R. H. Bell, Jr. and T. E. Adrian, *Anticancer Drugs*, 2005, **16**, 467.

143. D. Wang and R. N. Dubois, *Nat. Rev. Cancer*, 2010, **10**, 181.

144. D. Steinhilber, A. S. Fischer, J. Metzner, S. D. Steinbrink, J. Roos, M. Ruthardt and T. J. Maier, *Front. Pharmacol.*, 2010, **1**, 143.

145. E. Y. Kim, J. M. Seo, C. Kim, J. E. Lee, K. M. Lee and J. H. Kim, *Free Radic. Biol. Med.*, 2010, **49**, 1072.

146. J. M. Seo, K. J. Cho, E. Y. Kim, M. H. Choi, B. C. Chung and J. H. Kim, *Exp. Mol. Med.*, 2011, **43**, 129.

147. P. A. Janne, *J. Clin. Oncol.* (2006 ASCO Annual Meeting Proceedings Part I), 2006, A7024.

148. P. A. Janne, L. P.-A. Rodriguez, M. Gottfried, M. N. Reaume, T. Kaukel, Y. W. Oh, A. Sykes, N. Enas, L. H. Brail and J. V. Pawel, *J. Clin. Oncol.* (2006 ASCO Annual Meeting Proceedings Part I), 2006, 24, 7024.

149. M. W. Saif, H. Oettle, W. L. Vervenne, J. P. Thomas, G. Spitzer, C. Visseren-Grul, N. Enas and D. A. Richards, *Cancer J.*, 2009, **15**, 339.

150. L. Orning, G. Krivi and F. A. Fitzpatrick, *J. Biol. Chem.*, 1991, **266**, 1375.

151. T. D. Penning, L. J. Askonas, S. W. Djuric, R. A. Haack, S. S. Yu, M. L. Michener, G. G. Krivi and E. Y. Pyla, *Bioorg. Med. Chem. Lett.*, 1995, **5**, 2517.

152. F. Tsuji, Y. Miyake, H. Enomoto, M. Horiuchi and S. Mita, *Eur. J. Pharmcol.*, 1998, **346**, 81.

153. A. P. Ayscough and M. Whittaker, British Biotech Pharmaceuticals Limited, *PCT Pat. Appl.*, WO 1999/40910, 1999.

154. M. M. Thunnissen, P. Nordlund and J. Z. Haeggstrom, *Nat. Struct. Biol.*, 2001, **8**, 131.

155. M. M. Thunnissen, B. Andersson, B. Samuelsson, C. H. Wong and J. Z. Haeggstrom, *FASEB J.*, 2002, **16**, 1648.

156. A. D. Ferguson, B. M. McKeever, S. Xu, D. Wisniewski, D. K. Miller, T. T. Yamin, R. H. Spencer, L. Chu, F. Ujjainwalla, B. R. Cunningham, J. F. Evans and J. W. Becker, *Science*, 2007, **317**, 510.

157. H. Ago, Y. Kanaoka, D. Irikura, B. K. Lam, T. Shimamura, K. F. Austen and M. Miyano, *Nature*, 2007, **448**, 609.

158. D. Martinez Molina, A. Wetterholm, A. Kohl, A. A. McCarthy, D. Niegowski, E. Ohlson, T. Hammarberg, S. Eshaghi, J. Z. Haeggstrom and P. Nordlund, *Nature*, 2007, **448**, 613.

159. N. C. Gilbert, S. G. Bartlett, M. T. Waight, D. B. Neau, W. E. Boeglin, A. R. Brash and M. E. Newcomer, *Science*, 2011, **331**, 217.

160. T. D. Penning, S. S. Yu, J. Malecha, C. Liang and M. Russell, G. D. Searle & Co., *PCT Pat. Appl.*, WO 1998/40354, 1998.

161. S. A. Gregory, P. C. Isakson and G. Anderson, G. D. Searle & Co., *PCT Pat. Appl.*, WO 1997/29774, 1997.

162. B. B. Chen, H. Chen, M. Russell, J. M. Miyashiro, J. Malecha and T. D. Penning, G. D. Searle & Co., *PCT Pat. Appl.*, WO 1998/40364, 1998.

163. B. B. Chen, H. Chen and M. Russell, G. D. Searle & Co., WO 199840370, 1998.

164. N. S. Chandrakumar, B. B. Chen, M. Clare, B. N. Desai, S. W. Djuric, S. H. Docter, A. F. Gasiecki, R. A. Haack, C. Liang, J. M. Myashiro, T. D. Penning, M. A. Russell and S. Yu, G. D. Searle & Co., *PCT Pat. Appl.*, WO 1996/10999, 1996.

165. N. S. Chandrakumar, B. B. Chen, M. Clare, B. N. Desai, S. W. Djuric, S. H. Docter, A. F. Gasiecki, R. A. Haack, C. Liang, J. M. Myashiro, T. D. Penning, M. A. Russell and S. Yu, G. D. Searle & Co., *PCT Pat. Appl.*, WO 1996/11192, 1996.

166. T. D. Penning, N. S. Chandrakumar, B. B. Chen, H. Y. Chen, B. N. Desai, S. W. Djuric, S. H. Docter, A. F. Gasiecki, R. A. Haack, J. M. Miyashiro, M. A. Russell, S. S. Yu, D. G. Corley, R. C. Durley, B. F. Kilpatrick, B. L. Parnas, L. J. Askonas, J. K. Gierse, E. I. Harding, M. K. Highkin, J. F. Kachur, S. H. Kim, G. G. Krivi, D. Villani-Price, E. Y. Pyla and W. G. Smith, *J. Med. Chem.*, 2000, **43**, 721.

167. T. D. Penning, M. A. Russell, B. B. Chen, H. Y. Chen, C. D. Liang, M. W. Mahoney, J. W. Malecha, J. M. Miyashiro, S. S. Yu, L. J. Askonas, J. K. Gierse, E. I. Harding, M. K. Highkin, J. F. Kachur, S. H. Kim, D. Villani-Price, E. Y. Pyla, N. S. Ghoreishi-Haack and W. G. Smith, *J. Med. Chem.*, 2002, **45**, 3482.

168. T. D. Penning, N. S. Chandrakumar, B. N. Desai, S. W. Djuric, A. F. Gasiecki, C. D. Liang, J. M. Miyashiro, M. A. Russell, L. J. Askonas, J. K. Gierse, E. I. Harding, M. K. Highkin, J. F. Kachur, S. H. Kim, D. Villani-Price, E. Y. Pyla, N. S. Ghoreishi-Haack and W. G. Smith, *Bioorg. Med. Chem. Lett.*, 2002, **12**, 3383.

169. T. D. Penning, N. S. Chandrakumar, B. N. Desai, S. W. Djuric, A. F. Gasiecki, J. W. Malecha, J. M. Miyashiro, M. A. Russell, L. J. Askonas, J. K. Gierse, E. I. Harding, M. K. Highkin, J. F. Kachur, S. H. Kim, D. Villani-Price, E. Y. Pyla, N. S. Ghoreishi-Haack and W. G. Smith, *Bioorg. Med. Chem. Lett.*, 2003, **13**, 1137.

170. F. U. Axe, S. D. Bembenek, C. R. Butler, J. P. Edwards, A. M. Fourie, C. A. Grice, B. M. Savall, K. L. Tays and J. Wei, Janssen Pharmaceutica NV, *PCT Pat. Appl.*, WO 2005/012296, 2005.

171. F. U. Axe, S. D. Bembenek, C. R. Butler, J. P. Edwards, A. M. Fourie, C. A. Grice, B. M. Savall, K. L. Tays and J. Wei, Janssen Pharmaceutica NV, *PCT Pat. Appl.*, WO 2005/012297, 2005.

172. C. R. Butler, J. P. Edwards, A. M. Fourie, C. A. Grice, L. Karlsson, B. M. Savall, K. L. Tays and J. Wei, Janssen Pharmaceutica NV, *PCT Pat. Appl.*, WO 2006/105304, 2006.

173. G. M. Bacani, S. D. Bembenek, W. Eccles, J. P. Edwards, M. T. Epperson, L. Gomez, C. A. Grice, A. Kearney, A. M. Landry-Bayle, A. Lee-Dutra, K. J. McClure, T. Mirzadegan and A. Santillan, Janssen Pharmaceutica NV, US2009/0111794, 2009.

174. G. M. Bacani, D. Broggini, E. Y. Cheung, C. C. Chrovian, X. Deng, W. Eccles, A. M. Fourie, L. Gomez, C. A. Grice, A. Kearney, A. M. Landry-Bayle, A. Lee-Dutra, J. T. Liang, S. Lochner, N. S. Mani, A. Santillan, K. C. Sappey, K. Sepassi, V. M. Tanis, A. T. Wickboldt, J. J. M. Wiener and H. Zinser, Janssen Pharmaceutica NV, *PCT Pat. Appl.*, WO 2009/126806, 2009.

175. G. M. Bacani, C. C. Chrovian, W. Eccles, A. M. Fourie, L. Gomez, C. A. Grice, A. Kearney, A. M. Landry-Bayle, A. Lee-Dutra, A. Santillan, V. M. Tanis and J. J. M. Wiener, Janssen Pharmaceutica NV, *PCT Pat. Appl.*, WO 2010/132559, 2010.

176. V. Sandanayaka, J. Singh, L. Zhou and M. E. Gurney, deCODE Chemistry, Inc., *US Pat. Appl.*, US 2007/0149544, 2007.

177. V. Sandanayaka, J. Singh, L. Zhao and M. E. Gurney, deCODE genetics ehf, US 2008/0033013, 2008.

178. V. Sandanayaka, J. Singh, L. Zhao and M. E. Gurney, deCODE Chemistry, *US Pat. Appl.*, US 20070142434, 2007.

179. V. Sandanayaka, J. Singh, D. Sullins and M. E. Gurney, deCODE genetics ehf, *US Pat. Appl.*, US2008/0033024 2008.

180. V. Sandanayaka, J. Singh, M. Gurney, B. Mamat, P. Yu, L. Bedel and L. Zhao, deCODE Chemistry, *US Pat. Appl.*, US 2007/006820, 2007.

181. V. Sandanayaka, P. Chandrasekar and M. E. Gurney, deCODE genetics ehf, *PCT Pat. Appl.*, WO 2008/019284, 2008.

182. D. R. Davies, B. Mamat, O. T. Magnusson, J. Christensen, M. H. Haraldsson, R. Mishra, B. Pease, E. Hansen, J. Singh, D. Zembower, H. Kim, A. S. Kiselyov, A. B. Burgin, M. E. Gurney and L. J. Stewart, *J. Med. Chem.*, 2009, **52**, 4694.

183. V. Sandanayaka, B. Mamat, N. Bhagat, L. Bedell, G. Halldorsdottir, H. Sigthorsdottir, T. Andresson, A. Kiselyov, M. Gurney and J. Singh, *Bioorg. Med. Chem. Lett.*, 2010, **20**, 2851.

184. V. Sandanayaka, B. Mamat, R. K. Mishra, J. Winger, M. Krohn, L. M. Zhou, M. Keyvan, L. Enache, D. Sullins, E. Onua, J. Zhang, G. Halldorsdottir, H. Sigthorsdottir, A. Thorlaksdottir, G. Sigthorsson, M. Thorsteinnsdottir, D. R. Davies, L. J. Stewart, D. E. Zembower, T. Andresson, A. S. Kiselyov, J. Singh and M. E. Gurney, *J. Med. Chem.*, 2010, **53**, 573.

185. M. Congreve, R. Carr, C. Murray and H. Jhoti, *Drug Disc. Today*, 2003, **8**, 876.
186. D. R. Davies, B. Mamat, O. T. Magnusson, J. Christensen, M. H. Haraldsson, R. Mishra, B. Pease, E. Hansen, J. Singh, D. Zembower, H. Kim, A. S. Kiselyov, A. B. Burgin, M. E. Gurney and L. J. Stewart, *J. Med. Chem.*, 2010, **53**, 2330.
187. V. Sandanayaka, B. Mamat, P. Yu, L. Zhao, L. Bedell, N. Bhagat, J. Winger, M. Keyvan, B. Bock, M. Krohn, P. Chandrasekar, X. Mo, L.-M. Zhou, R. Mishra, E. Onua, J. Zhang, M. Porsteinsdottir, G. Halldorsdottir, H. Sigporsdottir, M. Friedman, D. Zembower, P. Andresson, J. Singh and M. Gurney, The 233rd ACS National Meeting, Chicago, IL, USA, 2007.
188. deCode Announces Positive Topline Results for Phase I Study of DG051 for the Prevention of Heart Attack. Available from: http://www.decode.com/news/news.php?story = 81
189. Positive Results from Phase IIa Study Pave Way for Phase IIb Trial of DG051 for the Prevention of Heart Attack. Available from: http://www.decode.com/news/news.php?story = 66
190. M. Chen, E. Claret, A. Cleve, D. Davey, W. Guilford, S.-K. Khim, T. Kirkland, M. J. Kochanny, A. Liang, D. Light, J. Parkinson, D. Vogel, G. P. Wei, B. Ye and H. Ye, Schering AG, *US Pat. Appl.*, US 2007/0155727, 2007.
191. D. Arnaiz, G. Brown, E. Claret, A. Cleve, D. Davey, W. Guilford, S.-K. Khim, T. Kirkland, M. J. Kochanny, A. Liang, D. Light, J. Parkinson, D. Vogel, G. P. Wei and B. Ye, Schering AG, *US Pat. Appl.*, US 2007/0155726, 2007.
192. T. A. Kirkland, M. Adler, J. G. Bauman, M. Chen, J. Z. Haeggstrom, B. King, M. J. Kochanny, A. M. Liang, L. Mendoza, G. B. Phillips, M. Thunnissen, L. Trinh, M. Whitlow, B. Ye, H. Ye, J. Parkinson and W. J. Guilford, *Bioorg. Med. Chem.*, 2008, **16**, 4963.
193. W. P. Jencks, *Adv. Enzymol. Relat. Areas Mol. Biol.*, 1975, **43**, 219.
194. B. Ye, J. Bauman, M. Chen, D. Davey, S. K. Khim, B. King, T. Kirkland, M. Kochanny, A. Liang, D. Lentz, K. May, L. Mendoza, G. Phillips, V. Selchau, S. Schlyer, J. L. Tseng, R. G. Wei, H. Ye, J. Parkinson and W. J. Guilford, *Bioorg. Med. Chem. Lett.*, 2008, **18**, 3891.
195. S. K. Khim, J. Bauman, J. Evans, B. Freeman, B. King, T. Kirkland, M. Kochanny, D. Lentz, A. Liang, L. Mendoza, G. Phillips, J. L. Tseng, R. G. Wei, H. Ye, L. Yu, J. Parkinson and W. J. Guilford, *Bioorg. Med. Chem. Lett.*, 2008, **18**, 3895.
196. H. Enomoto, Y. Morikawa, Y. Miyake, F. Tsuji, M. Mizuchi, H. Suhara, K. Fujimura, M. Horiuchi and M. Ban, *Bioorg. Med. Chem. Lett.*, 2008, **18**, 4529.
197. H. Enomoto, Y. Morikawa, Y. Miyake, F. Tsuji, M. Mizuchi, H. Suhara, K. Fujimura, M. Horiuchi and M. Ban, *Bioorg. Med. Chem. Lett.*, 2009, **19**, 442.

198. Positive Results from Phase IIa Study Pave Way for Phase IIb Trial of DG051 for the Prevention of Heart Attack. Available from: http://www.decode.com/news/news.php?story = 66

199. Prescribing information as posted on www.zyflocr.com

200. E. Israel, P. Rubin, J. P. Kemp, J. Grossman, W. Pierson, S. C. Siegel, D. Tinkelman, J. J. Murray, W. Busse, A. T. Segal, J. Fish, H. B. Kaiser, D. Ledford, S. Wenzel, R. Rosenthal, J. Cohn, C. Lanni, H. Pearlman, P. Karahalios and J. M. Drazen, *Ann. Intern. Med.*, 1993, **119**, 1059.

201. M. C. Liu, L. M. Dube and J. Lancaster, *J. Allergy Clin. Immunol.*, 1996, **98**, 859.

202. NCT00418613 from www.clinicaltrials.gov

203. P. G. Woodruff, R. K. Albert, W. C. Bailey, R. Casaburi, J. E. Connett, J. A. Cooper, Jr., G. J. Criner, J. L. Curtis, M. T. Dransfield, M. K. Han, S. M. Harnden, V. Kim, N. Marchetti, F. J. Martinez, C. E. McEvoy, D. E. Niewoehner, J. J. Reilly, K. Rice, P. D. Scanlon, S. M. Scharf, F. C. Sciurba, G. R. Washko and S. C. Lazarus, *COPD*, 2009, **8**, 21.

204. NCT00404313 from www.clinicaltrials.gov

205. NCT00421278 from www.clinicaltrials.gov

206. B. Lehnigk, K. F. Rabe, G. Dent, R. S. Herst, P. J. Carpentier and H. Magnussen, *Eur. Respir. J.*, 1998, **11**, 617.

207. J. Van Schoor, G. F. Joos, J. C. Kips, J. F. Drajesk, P. J. Carpentier and R. A. Pauwels, *Am. J. Respir. Crit. Care Med.*, 1997, **155**, 875.

208. S. M. Nasser, G. S. Bell, R. J. Hawksworth, K. E. Spruce, R. MacMillan, A. J. Williams, T. H. Lee and J. P. Arm, *Thorax*, 1994, **49**, 743.

209. E. M. de Jong, I. M. van Vlijmen, J. C. Scholte, A. Buntinx, B. Friedman, W. Tanaka and P. C. van de Kerkhof, *Skin Pharmacol.*, 1991, **4**, 278.

210. W. G. Roberts, T. J. Simon, R. G. Berlin, R. C. Haggitt, E. S. Snyder, W. F. Stenson, S. B. Hanauer, J. E. Reagan, A. Cagliola, W. K. Tanaka, S. Simon and M. L. Berger, *Gastroenterology*, 1997, **112**, 725.

211. H. Hakonarson, S. Thorvaldsson, A. Helgadottir, D. Gudbjartsson, F. Zink, M. Andresdottir, A. Manolescu, D. O. Arnar, K. Andersen, A. Sigurdsson, G. Thorgeirsson, A. Jonsson, U. Agnarsson, H. Bjornsdottir, G. Gottskalksson, A. Einarsson, H. Gudmundsdottir, A. E. Adalsteinsdottir, K. Gudmundsson, K. Kristjansson, T. Hardarson, A. Kristinsson, E. J. Topol, J. Gulcher, A. Kong, M. Gurney and K. Stefansson, *JAMA*, 2005, **293**, 2245.

212. GSK2190915 at www.clinicaltrials.gov

213. R. Alten, E. Gromnica-Ihle, C. Pohl, J. Emmerich, J. Steffgen, R. Roscher, R. Sigmund, B. Schmolke and G. Steinmann, *Ann. Rheum. Dis.*, 2004, **63**, 170.

214. F. Diaz-Gonzalez, R. H. E. Alten, W. G. Bensen, J. P. Brown, J. T. Sibley, M. Dougados, S. Bombardieri, P. Durez, P. Ortiz, G. de-Miquel, A. Staab, R. Sigmund, L. Salin, C. Leledy and S. H. Polmar, *Ann. Rheum. Dis.*, 2007, **66**, 628.

215. NCT00060801 at www.clinicaltrials.gov

216. D. J. Evans, P. J. Barnes, S. M. Spaethe, E. L. van Alstyne, M. I. Mitchell and B. J. O'Connor, *Thorax*, 1996, **51**, 1178.
217. J. M. Mommers, M. M. Van Rossum, M. E. Kooijmans-Otero, G. L. Parker and P. C. van de Kerkhof, *Br. J. Dermatol.*, 2000, **142**, 259.
218. From IDdb3.
219. NCT00424294 at www.clinicaltrials.gov.
220. L. Gronke, K. M. Beeh, R. Cameron, O. Kornmann, J. Beier, M. Shaw, O. Holz, R. Buhl, H. Magnussen and R. A. Jorres, *Pulm. Pharmacol. Ther.*, 2008, **21**, 409.
221. B. J. Gitlitz, P. Langecker, D. Blanchett, W. Lang, C. Baenziger, M. Moran, N. Moldawer and R. Figlin, *Proc. Am. Soc. Clin. Oncol.*, 2003, A1627.
222. Positive Results from Phase IIa Study Pave Way for Phase IIb Trial of DG051 for the Prevention of Heart Attack. Available from: http://www.decode.com/news/news.php?story = 66
223. C. A. Grice and L. Gomez, *Expert Opin. Ther. Pat.*, 2008, **18**, 1333.

CHAPTER 4
CRTH2 Antagonists

L. NATHAN TUMEY

Pfizer Global R&D, Worldwide Medicinal Chemistry, MS 8220-3563,
445 Eastern Point Rd, Groton, CT 06340, USA
Email: Nathan.Tumey@Pfizer.com

4.1 Role of PGD2 in Inflammation and Asthma

The modulation of prostanoid levels *via* blockage of the cyclooxygenase (COX) pathway is one of the oldest pharmacological tools known. The COX family of enzymes mediates the production of prostaglandin H2 (PGH2) from arachidonic acid. PGH2, in turn, is the key precursor to a wide variety of biologically important prostanoids such as prostaglandin D2 (PGD2, **1**), prostaglandin E2 (PGE2), prostaglandin F2a (PGF2a), prostacyclin (PGI2) and thromboxane A2 (TXA2). While these molecules are typically considered to be pro-inflammatory, they mediate a variety of sometimes opposing biological effects such as vasodilation, vasoconstriction, bronchodilation, bronchoconstriction, cell migration and platelet aggregation by binding to their cognate receptors. While the COX pathway has been a fruitful area of exploration, most recent efforts to modulate prostaglandin action have instead focused on the development of pharmacologic tools that specifically interact with the target receptors. It is hoped that specific targeting of a given prostanoid receptor will result in a more defined mode of action while limiting the possible side-effects. A variety of reviews have provided excellent overviews of prostaglandin receptors and modulators.[1,2]

One prostanoid in particular, PGD2, has been widely recognized for nearly a quarter of a century to be a key hallmark of allergic response (frequently known as atopy). Following an allergen challenge, patients with a variety of

RSC Drug Discovery Series No. 26
Anti-Inflammatory Drug Discovery
Edited by Jeremy I Levin and Stefan Laufer
© The Royal Society of Chemistry 2012
Published by the Royal Society of Chemistry, www.rsc.org

atopic disorders (asthma, rhinitis, dermatitis) have been observed to have increased levels of PGD2 in broncho-alveolar lavage (BAL) fluid,[3] nasal mucosa[4] and skin.[5] PGD2 is rapidly metabolized and its metabolites such as $9\alpha,11\beta$-PGF2 have been observed at increased levels in plasma and urine following allergen exposure.[6–8] Moreover, transgenic mice engineered to produce increased levels of PGD2 show increased leukocyte lung infiltration as compared to wild-type mice after allergen challenge.[9] Conversely, a selective inhibitor of PGD2 synthase (PGDS) has been reported to suppress inflammation in a murine allergen-induced asthma model.[10] Macrophages isolated from PGDS knockout mice and mice treated with a PGDS inhibitor have been shown to be resistant to lipopolysaccharide (LPS) induced cell migration, showing that the activity of this important immunostimulant is at least partially due to its effects on PGD2 production, rather than from LPS itself.[11]

While many cell types are known to produce PGD2, it is widely recognized that activated mast cells are responsible for the majority of its production during the early stages of an atopic response. Following the initial allergen exposure, the adaptive immune system elicits the production of a specific IgE that targets the relevant antigen. Upon subsequent exposure the allergen binds IgE and the complex becomes immobilized on the surface of mast cells located in various tissues including lung, skin, digestive tract, conjunctiva and mucosa. IgE is tightly bound to mast cells *via* a tight interaction between the Fc domain of the antibody and the cell-surface FcɛRI receptor.[12] The binding of antigen to IgE initiates an intra-cellular signalling cascade that results in mast cell degranulation thereby releasing large amounts of prostaglandins, along with histamine, cytokines and leukotrienes. The most abundant prostaglandin released from activated mast cells is PGD2.[13,14] As described above, this molecule proceeds to play a pivotal role in atopic pathology by eliciting many of the underlying physiological markings of allergic inflammation including vasodilation, bronchoconstriction, vascular permeability and lymphocyte recruitment.[15–17] While it is unclear if PGD2 plays an actual regulatory role in the allergic response, it is clear that PGD2 plays a major role in many of the clinical symptoms associated with allergic disease.

4.2 Receptors for PGD2: DP and CRTH2

As early as 1985 it was recognized that structural analogues of PGD2 could recapitulate some, but not all, of the functionality of PGD2 itself.[18] This suggested the possibility that multiple receptors may be involved in the transduction of PGD2 signalling. Today it is well understood that there are two high-affinity receptors of PGD2, the D prostanoid receptor (DP, also known as DP1) and the chemoattractant receptor homologous molecule expressed on T helper type 2 cells (CRTH2, also occasionally known as DP2). Additionally, there is some evidence that the TXA2 receptor (TP) and the PGE2 receptor (EP3) may have a weak affinity for PGD2. Nevertheless, it has become clear in recent years that the vast majority of the biological responses elicited by PGD2 can be attributed to the activation of either DP or CRTH2 (See Figure 4.1).

CRTH2 Activation	DP Activation
$G\alpha_i$ coupled	$G\alpha_s$ coupled
Decreased cAMP	Increased cAMP
Promotes chemotaxis in Th2 lymphocytes, eosinophils, and basophils	Inhibits chemotaxis
	Inhibits platelet aggregation
Promotes release of histamine, IL4, IL5, and IL13	Increases mucin secretion
	Promotes vasodilation and bronchodilation
Promotes eosinophilia/leukocytosis	**DP Expression**
CRTH2 Expression	Vasculature
Th2 lymphocytes	Smooth muscle
Eosinophils	Small intestine
Basophils	Nasal mucosa
CNS	Dendritic cells
	CNS

Figure 4.1 Comparing and contrasting CRTH2 and DP.

DP and CRTH2 are G-protein coupled receptors (GPCRs) that have different expression profiles and distinct, and sometimes opposing, biological roles. DP is a $G\alpha_s$ coupled receptor,[19] thereby increasing intra-cellular cAMP, while CRTH2 is a $G\alpha_i$ coupled receptor,[20] thereby decreasing intra-cellular cAMP. Activation of DP blocks chemotaxis of eosinophils and dendritic cells,[21] inhibits platelet aggregation,[22] promotes mucin secretion,[23] lowers intra-ocular pressure[24] and induces vasodilation.[22] It is expressed on smooth muscle, in the CNS,[25] retina,[19] small intestine,[19] vasculature,[26] osteoblasts,[27] platelets, dendritic cells and nasal mucosa.[6,23] In contrast, CRTH2 is expressed primarily on motile cells such as Th2 cells, eosinophils, monocytes and basophils where it plays a central role in PGD2-dependent chemotaxis.[28,29] CRTH2 has also been shown to be expressed in the stomach, heart, small intestine and thymus.[20] Consistent with its well-established role in chemotaxis, CRTH2 bears little sequence homology to DP and other prostanoid receptors but rather is highly homologous to various chemotactic receptors such as leukotriene B4 receptors (BLT1/BLT2) and the formyl peptide receptor.[30,31] In addition to its role in chemotaxis, CRTH2 activation leads to potentiation of histamine release from basophils,[32] production of inflammatory cytokines (IL-4, IL-5 and IL-13) by Th2 cells,[33] eosinophil degranulation[34] and prevention of Th2 cell apoptosis.[35] Collectively, these activities clearly play an important role in atopic pathology and provide a molecular basis for many of the physiological observations associated with PGD2 release.

In addition to its role in lymphocytes, CRTH2 may play a role in a variety of other PGD2-responsive cell types. Durand *et al.* have reported that CRTH2 is expressed on osteoblasts and osteoclasts where it appears to play an important role in PGD2-driven bone remodelling.[36] CRTH2 is also expressed on keratinocytes where its activation is reported to lead to up-regulation of the anti-microbial peptide hBD-3.[37]

The various cellular effects of CRTH2 activation appear to be divided into two mechanistic pathways, phosphatidylinositol 3-kinase (PI3K) activation of

AKT, leading to actin polymerization and cytokinesis, and calcium-mediated activation of calcineurin, leading to NFAT-promoted cytokine expression.[38] Altogether, the picture that emerges from the above studies is that an early stage atopic response leads to PGD2 release from mast cells, which, in turn, exerts its activity on leukocytes *via* CRTH2 to induce chemotaxis to the site of inflammation. (In fact, many of the *in-vivo* models used for CRTH2 antagonism involve blockage of eosinophilia and leukocytosis, *vide infra*.) Once at the site of inflammation, PGD2 further exerts its activity on leukocytes through activation of CRTH2 by promoting cytokine and histamine release. In summary, CRTH2 and DP are distinct and complementary receptors that mediate a great number of the observed phenotypes associated with PGD2. Modulation of one or both receptors may have a meaningful impact on the progression of an inflammatory episode.

4.3   Pharmacology of CRTH2

The elucidation of the role that CRTH2 plays in chemotaxis, cytokine release and other inflammatory processes has been greatly facilitated by the discovery of various molecular tool compounds that specifically agonize or antagonize the receptor. Numerous PGD2 metabolites such as 13-14 di-hydro,15-keto PGD2 (DK-PGD, **2**),[39] 9α,11β-PGF2,[40] PGF2α,[40] 15-deoxy-PGD2[41] and 15-deoxy-PGJ2[41] are known to be selective agonists of CRTH2, having little activity against DP. It has been extensively speculated that the activity of these metabolites specifically against CRTH2 combined with the short serum half-life of PGD2 (approximately 30 minutes) may imply that much of the observed PGD2-response *in vivo* may actually be driven by PGD2 metabolites such as those mentioned above.[42] CRTH2 is also known to be activated by several non-PGD2 derived ligands, such as PGH2[43] (**3**) and 11-dehydro-thromboxane B2[44] (**4**). Given that 11-dehydro-thromboxane B2 is the major metabolite of TXA2 (it is used as a biomarker for TXA2 in both plasma and urine), it has been suggested that this thromboxane-derived activity may be an important modulator of lymphocyte activation alongside PGD2-derived activity. In fact, it has been recently reported that CRTH2 also plays a critical role in cell migration in response to several non-PGD2 derived prostaglandins including PGH2 and 11-dehydro-thromboxane B2.[43,44] In addition to the natural ligands discussed above, numerous synthetic agonists of CRTH2 are also known, including the NSAID indomethacin[45] (**5**) and the selective agonist L888,607 (**6**),[46] both containing an indolic core linked to a carboxylic acid.

4.3.1   Role of CRTH2 in Animal Disease Models

CRTH2 selective agonists such as DK-PGD2 along with selective DP agonists (See Figure 4.2) such as BW245c[29,47] (**7**) have been indispensable tools in untangling the role that DP and CRTH2 play in allergic inflammation. For instance, Gervais showed in 2002 that CRTH2-selective agonist DK-PGD2

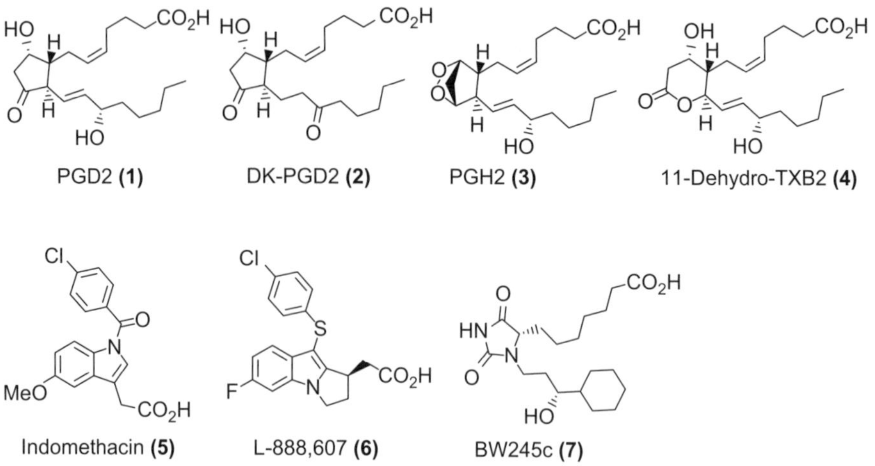

Figure 4.2 Agonists of CRTH2 and DP.

promotes eosinophil migration, degranulation and shape change while the DP-selective agonist BW245c does not.[34] Following on from these studies, Almishri showed in 2005 that these results translate to *in-vivo* models. Inhaled PGD2 and DK-PGD2 potently induced eosinophil infiltration into the lungs of brown Norway rats, while inhaled BW245c showed no effect.[48] Other studies have shown that DK-PGD2 (but not BW245c) exacerbates inflammation and lymphocyte recruitment in murine OVA-induced asthma and dermatitis models.[49] Consistent with these results, CRTH2-deficient mice have been shown to have reduced lymphocyte infiltration into their lungs following sensitization and chronic OVA exposure.[50] (Although, unexpectedly, cytokine production was actually *increased* as compared to wild-type mice.) Other studies have shown that CRTH2-deficient mice have decreased skin irritation, decreased dermal cytokine levels and decreased lymphocyte infiltration in an allergen-induced murine dermatitis model.[51]

Perhaps most importantly, numerous CRTH2 antagonists (*vide infra*) have been reported to ameliorate the symptoms associated with various allergen challenge animal models including murine antigen induced lung inflammation,[52–57] murine cigarette-smoke-induced lung inflammation,[58,59] murine allergic rhinitis,[53,57,58,60,61] guinea pig PGD2-induced airflow obstruction,[53] guinea pig airway hyper-responsiveness,[62] sheep airway hyper-responsiveness[63,64] and murine contact hypersensitivity.[65,66] The mechanism of action of these effects can be understood through studies showing that antagonists of CRTH2 block eosinophil release from bone marrow[67] and prevent eosiniphilia and leukocytosis.[59] These studies and the compounds associated with them will be discussed in more detail below. Collectively, the *in-vivo* results observed with CRTH2 agonists, CRTH2 antagonists and CRTH2-knockout animals clearly point to a significant role for modulation of this receptor in the treatment of allergic disease.

4.3.2   Role of CRTH2 in Human Disease

As most medical researchers are keenly aware, there is often a large disconnect between validation of a disease target in animal models and validation in human disease. One complicating factor that often arises is the differential expression of targets between various animal species and humans. One particular difference that may confound some of the animal model data described in Section 4.3.1 is that murine CRTH2 is expressed on both Th1 and Th2 cells, while in humans it is expressed solely on Th2 cells.[30,68] With this in mind, there is a variety of evidence that has accumulated over the past decade that strongly implicates CRTH2's role in human atopic disorders. Eosinophil shape change, a commonly used marker for CRTH2 activation, has been observed in human whole blood following PGD2 challenge. As would be expected, the effects of PGD2 are blocked by pre-treatment with a CRTH2 antagonist.[69] For example, a CRTH2 antagonist (ramatroban) blocked IgE/anti-IgE induced Th2 cell chemotaxis in human explants of nasal polyp tissue.[70] These studies both demonstrate that animal disease model data can be recapitulated in an *ex-vivo* human model. More importantly, several studies have shown that single nucleotide polymorphisms (SNPs) in the 3′-untranslated region (UTR) of the CRTH2 gene are associated with increased severity of asthma and, in some cases, food allergies in Chinese, German and African-American children,[71,72] although, interestingly, not Japanese children.[73] The SNPs are associated with a mutation that results in increased mRNA stabilization, presumably resulting in higher receptor density.

Perhaps the most compelling observation that suggests a therapeutic role for CRTH2 antagonists is that the TP-receptor antagonist ramatroban, marketed for allergic rhinitis in Japan by Bayer AG, is also a potent CRTH2 antagonist[74] and has shown clinical activity against bronchial asthma.[75] Interestingly, it was shown in the early 1990s that in addition to blocking TXA2-induced bronchoconstriction, ramatroban also blocked PGD2-induced bronchoconstriction.[76] While not recognized at the time, this activity is likely due to its activity against the CRTH2 receptor rather than its intended activity against the TP receptor. More recently, *in-vivo* administration of ramatroban has been shown to block eosinophil recruitment in blood[77] and in lungs[78] induced by the selective CRTH2 agonist DK-PGD2, whereas a TP-specific antagonist had no such effect. Additionally FITC-induced contact hypersensitivity has been shown to be ameliorated by pre-treatment with ramatroban, but not with a selective TP antagonsist.[66] While not conclusive, the weight of this evidence strongly suggests that the human efficacy of ramatroban is at least partially driven by its activity against the CRTH2 receptor.

4.4   CRTH2 Antagonists

The increasing interest in the biology of CRTH2 as a potential target for various inflammatory diseases has, of course, prompted significant efforts around the design and synthesis of small molecules to modulate this target.

A number of excellent reviews have been written in recent years that provide overviews of the various classes of CRTH2 antagonists that have been described in both peer-reviewed literature and patent literature.[79–85] While not trying to be comprehensive, the goal of this report is to give the reader a broad overview of the structural classes of molecules used to modulate CRTH2, provide a historical context for work going on across multiple organizations and give a snapshot of the current state of the art with regard to the therapeutic use of antagonists of this receptor. The discussion of these antagonists will largely follow a somewhat arbitrary deconstruction of the antagonists into three structural classes: tricyclic ramatroban analogues, indole acetic acids and phenyl/phenoxy acetic acids. The focus of this synopsis will be on structures and activity reported in peer-reviewed literature, although patent filings will be cited in so far as they support or supplement published literature. Those interested in a more thorough summary of CRTH2 patent literature and competitive intelligence are invited to refer to three excellent overviews of the area.[82,86,87]

4.4.1 Tricyclic Ramatroban Antagonists

The earliest reported CRTH2-selective antagonists were derived from the structure of the tricyclic dual TP/CRTH2 antagonist ramatroban (**8**) (See Figure 4.3). In early 2005 groups from Athersys and 7TM Pharma nearly simultaneously published SAR around this tricyclic scaffold showing that shortening the propionic acid side chain by one carbon increases the potency against CRTH2 by approximately 10-fold while decreasing potency against TP by 100-fold or more.[88,89] This SAR is observed in the case of both the parent tricyclic scaffold (**9**) and a variant of the parent scaffold in which the aromatic ring is transposed from one side of the scaffold to the other (**10**). Both compounds were shown to have transformed ramatroban, which is ~10× selective towards TP, to compounds that show >1000-fold selectivity for CRTH2. Moreover, compound **10** was shown to have little or no activity against the other PGD2 receptor, DP. This is perhaps not surprising given the lack of sequence homology between DP and CRTH2. The activity of compound **10**, which lacks a stereocentre, and the fact that compound **9** was tested as a racemic mixture strongly suggests that the stereochemistry of the sulfonamide is of little importance in the selectivity of these molecules. Additionally, these early SAR studies show that minor variations around the carboxylic acid portion of the molecule result in dramatic changes in activity, and that the sulfonamide region can be modified with retention of activity.

Subsequent to these early reports, scientists from Merck[63,90] and Amira[61] reported closely related variants of the above tricyclic scaffolds in which the indolic nitrogen is transposed resulting in the reverse-indole derivatives **11–13** and indolizine **14**, and aza-indole **15**, respectively. Highlights of the extensive SAR studies[90] around compound **11** (also known as MK-7246) have shown that affinity and selectivity of the (*R*) enantiomer are superior to the (*S*) enantiomer, the 4-fluorophenyl sulfonamide offers a 2×–10× boost in potency

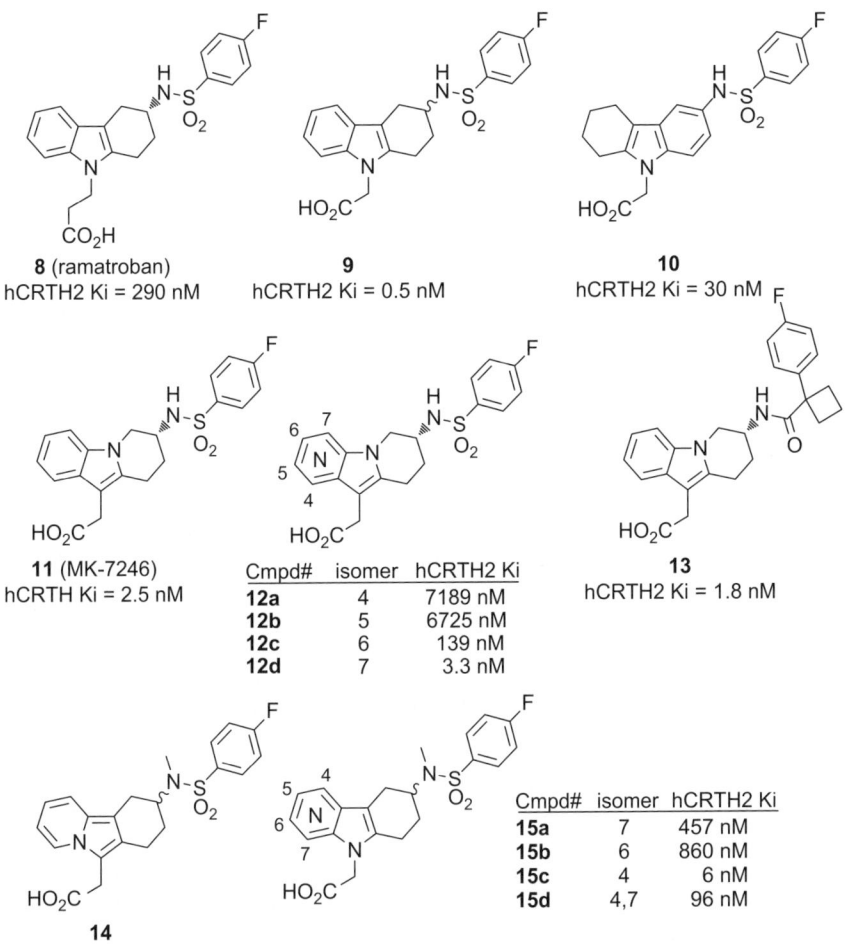

Figure 4.3 Tricyclic antagonists of CRTH2.

as compared to various other substituted phenyl sulfonamides, addition of a small alkyl group to the sulfonamide nitrogen offers a 2×–5× boost in potency and addition of an electron-withdrawing group (halogens in particular) around the 6-membered ring of the indole core offers a slight boost in potency. Compound **11** (MK-7246) was the lead compound identified from these studies and was advanced into PK studies where it exhibited excellent oral bioavailability (57–100%) in rat, dog and primate species. This compound is a modest inhibitor of CYP2C9 (9 μM) and 3A4 (34 μM) and is subject to a small amount of photodegradation. In spite of this, MK-7246 has shown promising activity in a sheep antigen-induced airway hyper-responsiveness model[63] and has been advanced into the clinic (see Section 4.7). The same team more recently described a back-up series of compounds that incorporates an aza-indole core (**12a–d**)[91] and replacements for the sulfonamide moiety (**13**).[92] Interestingly, the direct replacement of the sulfonyl group in **11** with a carbonyl resulted in a significant loss of potency. In order to more closely mimic the tetrahedral geometry of the sulfonamide, the team introduced a spacer between the carbonyl and 4-fluorobenzene ring, ultimately resulting in compound **13**.[92] This compound showed excellent pharmacokinetic properties along with similar *in-vitro* potency to the parent compound, **11**. Scientists from Amira reported on the SAR of a closely related compound, indolizine **14**, where they show that a variety of aza-indole cores (**15a–d**) are well tolerated as an indole replacement. Specifically, they report the identification of 4-aza-indole **15c** is marginally superior to various competitor compounds in terms of CYP inhibition, CYP-3A4 induction and DP selectivity. Interestingly, this compound showed single-digit nM activity in a whole-blood eosinophil shape change assay and was shown to be active in a murine OVA-induced lung inflammation model. SAR from both the Amira and Merck scaffolds reveal an important structural constraint around the carboxylic acid side. Both teams observed a dramatic reduction in activity when a nitrogen was placed in close proximity to the acetic acid side chain (compounds **12a,b** and **15a,b**). It is tempting to speculate that this loss in activity (for **12a** and **15a**) may be partially due to the formation of an intra-molecular H-bond that must be broken upon receptor binding.

4.4.2 Indole Acetic Acids and Related Antagonists

The indole acetic acid core (See Figure 4.4) was initially viewed as an agonist framework, due largely to early work by Hata and colleagues demonstrating that indomethacin (**5**) and closely related analogues such as sulindac sulfide (**16**) and *p*-bromobenzyl indomethacin (**17**) blocked intra-cellular cAMP production and induced chemotaxis in a HEK293 cell line transfected with mCRTH2.[93] However, an interesting report emerged in 2005 by Armer and colleagues curiously showing that the agonist Indomethacin (**5**) could be transformed to a potent antagonist by simply replacing the aryl ketone with an aryl sulfonamide (**18a–e**), particularly aryl sulfonamides containing electron withdrawing groups (**18b,c**).[94] The functional switch of agonist to antagonist was clearly demonstrated by showing that the compounds block PGD2 mediated Ca^{2+} flux in Chinese hamster ovary (CHO) cells and block PGD2-induced eosinophil shape change and Th2 cell chemotaxis. The IC_{50} for this functional activity was closely in line with the hCRTH2 binding activity. Additionally, the compounds were highly selective over DP and compound **18c** was shown to have acceptable pharmacokinetic properties after oral administration. A closely related series of compounds containing a 7-aza-indole core, as represented by compound **19**, was reported by Novartis in 2009.[95] This compound was broadly profiled against a panel of kinases, GPCRs and enzymes and found to have > 100-fold selectivity over every target evaluated. Especially pertinent to the receptor at hand, the compound was shown to have > 10 µM activity against a variety of prostanoid receptors including DP, EP2, EP3, EP4, FP and TP. Bonafoux and colleagues at Novartis also reported a related thienopyrrole series of

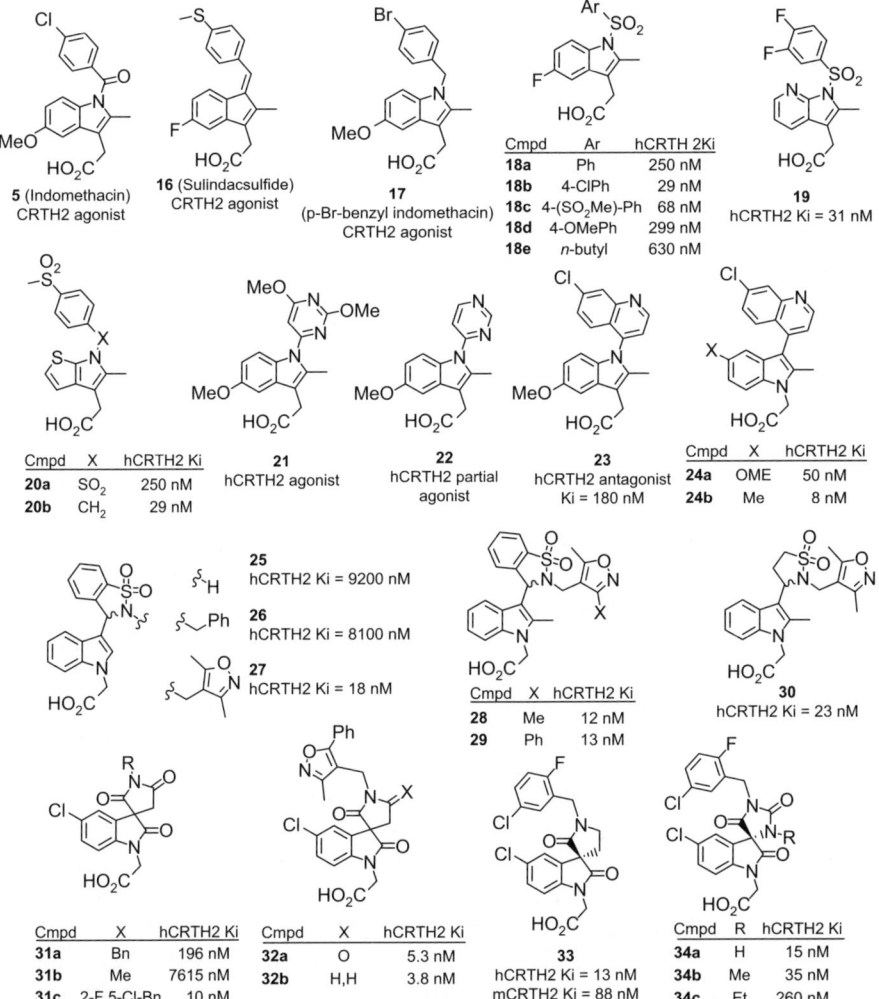

Figure 4.4 Indole acetic acid antagonists of CRTH2.

antagonists in which they demonstrated that sulfonyl moiety (**20a**) could be converted to a methylene (**20b**) with retention of activity.[96] Once again, these compounds were shown to be antagonists of the receptor by a Ca^{2+} flux assay. The antagonistic activity of these analogues, compared with the agonistic activity of indomethacin and sulindac sulfide (**16**) suggests that the precise placement of the pendant aromatic ring is important in determining the functional consequences of receptor binding. The antagonistic activity of *N*-benzyl thienopyrrole acetic acid **20b** is rather surprising in light of the previous report showing that *p*-bromobenzyl indomethacin (**17**) is a CRTH2 agonist.[93]

Further insight into the dissection of antagonism from agonism in the indole acetic acid core was provided by a thought-provoking publication by

Birkinshaw and colleagues from AstraZeneca[97] showing that *N*-aryl indole acetic acids (such as **21** and **22**) could be transformed from agonists into antagonists by changing the pendant aromatic group to a quinoline, as illustrated by compound **23**. Being structurally derived from indomethacin, it is not surprising that these compounds retained significant levels of COX-1 activity. Reversing the indole geometry (**24a,b**) significantly reduced the COX-1 activity with a concomitant improvement in potency against CRTH2. Interestingly, the team reports that the 2-methyl group on the indole ring offers approximately a 5-fold boost in potency as compared to the 2-H indole. This same SAR was also reported for the agonist indomethacin (**5**)[93] and has been subsequently reported for a variety of indole-like scaffolds (*vide infra*), perhaps not surprising given the similarity of this core to the ramatroban-like antagonists (**8–15**). Tumey and colleagues at Athersys have reported an interesting variant of the AstraZeneca motif in which the quinoline moiety is replaced by a reduced saccharin derivative (**25–29**).[98] Interestingly, a dramatic gain in potency is seen through the incorporation of a strategically placed isoxazole (**27**). This moiety seems to be the major contributor to the potency of these compounds, as changes on the indole core and saccharin "headpiece" have little effect on activity (**28–30**). As is typical for the indole acetic acid class of compounds, no significant activity against either DP or TP was observed.

Merck Serono has also reported an indole-inspired scaffold, represented by spirocyclic structures **31a–c**.[99] The series originated with HTS hit **31a**. Benzylic substituents on the amide/imide are the most potent, particularly those with halogens (**31c**). Interestingly, the most potent group on the amide/imide nitrogen proved to be an isoxazole, as illustrated in compounds **32a,b**. It is quite interesting to note the same substituent provided a dramatic boost in potency in the previously discussed indole series from Athersys (**28, 29**).[98] Systematic SAR around the remainder of the scaffold is not described; however, hints of the SAR emerge around various replacements for the succinimide ring such as **32b**. The authors present two interesting pharmacological observations that should be considered in any medicinal chemistry strategy pursuing this and related targets. First, they describe a GTPγS binding assay that allows for a functional evaluation of agonism/antagonism/inverse agonism across a broad range of compounds and they show that the lead compound **31a** is an inverse agonist of CRTH2. Unfortunately, no subsequent reference is made to this observation and its impact upon the further development of this series is unclear. Secondly, the authors show good correlation between murine-CRTH2 binding and human-CRTH2 binding for a subset of the described compounds. However, one compound (**33**) showed a significant difference between binding to the two isoforms. Most therapeutic programs rely upon animal disease models, particularly murine models, for compound advancement and therefore it is critical to understand the impact that differential binding between homologues may have upon the interpretation of biological data. In most cases, equivalent binding to human and animal homologues would be significantly advantageous in understanding efficacy and safety data that emerge from various *in-vivo* studies.

In a follow-up publication, a stability issue related to hydrolysis of the succinimide ring at pH > 6 is disclosed.[52] In order to address this liability, an additional nitrogen was incorporated into the scaffold, resulting in spirocyclic hydantoins such as **34a–c**. Compound **34a** showed improved aqueous stability as compared to **31c**; however, the compound still showed poor bioavailability. An elegant series of *in-vivo* experiments conclusively demonstrated that the problems associated with the compound were related to its poor permeability across intestinal membranes. Satisfyingly, this issue was predicted by poor Caco-2 permeability. The addition of a methyl group to the hydantoin nitrogen (**34b**) resulted in a modest improvement in Caco-2 permeability and a concomitant improvement in mouse oral bioavailability (37% for **34b** *vs.* 6% for **34a**). Compound **34b** was shown to block murine lung eosinophilia at 30 mpk (PO) following an aerosolized ovalbumin challenge.

4.4.3 Phenoxyacetic Acids, Phenyl Acetic Acids and Related Antagonists

The natural progression of SAR around the early lead ramatroban led from tricyclic analogues (Section 4.4.1) to bicyclic analogues (indolic derivatives, Section 4.4.2) and finally to monocyclic derivatives (See Figure 4.7). The earliest monocyclic CRTH2 antagonists were reported by Ulven (7TM Pharma) and were conceptually derived from a pharmacophore model based on indomethacin that included a negatively charged site attached to three hydrophobic regions.[100] This pharmacophore was used in an *in-silico* screen to identify approximately 600 compounds from a set of 1 million commercially available compounds. These 600 compounds were screened against CRTH2 and resulted in a ~10% hit rate for compounds with an $IC_{50} < 10 \mu M$. One such hit from this campaign is phenoxyacetic acid **35**. Replacement of the hydrazone with a more stable ketone linkage (**36**) ultimately resulted in compound **37** with excellent ADME properties that was advanced into mouse efficacy studies where it showed excellent activity in an OVA-induced lung inflammation model. At 5 mg/kg p.o. compound **37** blocked eosinophil recruitment and goblet cell proliferation after challenge with aerosolized OVA.

Sandham and colleagues (Novartis) describe the identification of a closely related scaffold derived from HTS hit **38a**.[101] An SAR progression is reported in which carboxylic acid replacements (**38a–f**) are explored followed by optimization of the aromatic substituent. Similar to SAR around related scaffolds, the acetic acid moiety is highly sensitive to replacement with isosteres (**38b,c**) and small electron withdrawing groups, halogens in particular, are favoured on the aromatic ring (**39c,d**). Yet another series of phenoxyacetic acids has been reported by AstraZeneca, with a focus on zwitterionic compounds such as **40a**.[102] Interestingly, a specific geometry of the amine-containing ring seems to be essential for activity. Replacement of the piperazine with a piperidine abolishes activity (**40a** *vs.* **40b**), while addition of a methyl group to the piperazine gives a significant boost in potency (**40a** *vs.* **41**).

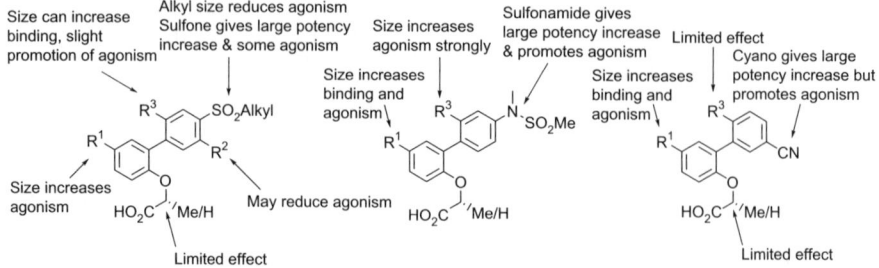

Figure 4.5   Summary of agonist/antagonist SAR for a series of phenoxyacetic acids. Reprinted from T. Luker, R. Bonnert, J. Schmidt, C. Sargent, S. W. Paine, S. Thom, G. Pairaudeau, A. Patel, R. Mohammed, E. Akam, I. Dougall, A. M. Davis, P. Abbott, S. Brough, I. Millichip and T. McInally, *Bioorg. Med. Chem. Lett.*, **21**, 3616–3621. © 2011, with permission from Elsevier.

Replacement of the phenyl sulfonamide with an amide (**42**) results in a nearly 100-fold loss in activity, which can be entirely regained by changing the benzamide to a phenyl acetamide (**43**). Several compounds from this series were shown to have acceptable PK in rat and good oral bioavailability.

Perhaps the most interesting observation from this series is that compound **44** was unexpectedly found to be a partial agonist of CRTH2. The team from AstraZeneca wisely screened all compounds of interest in an eosinophil shape change assay. As expected, the vast majority of compounds showed no activity in this assay. Compound **44**, however, showed ∼40% efficacy as compared to DK-PGD2. In a more recent publication the same group showed that minor changes in this scaffold can result in a functional switch from a very potent antagonist (**45**) to an exceptionally potent agonist (**46**).[103] A fascinating set of SAR work is presented, summarized in Figure 4.5.

The *in-silico* screen by 7TM Pharma described above also resulted in another series of compounds that has been recently disclosed.[104–106] The series originated with 2-aminothiazole acetic acid **47**, a structurally unique antagonist of CRTH2 with relatively modest potency. Extensive SAR studies explored substitution on both phenyl rings as well as various linkers between the thiazole and the 4-chlorophenyl moiety. A variety of potent compounds were identified, including 2-benzylthiazole **48**. Docking studies were performed on analogue **49**, giving a glimpse into a potential binding mode for CRTH2 antagonists. In agreement with earlier docking studies,[93] Lys210 forms the key interaction with the carboxylic acid. The thiazole orients the benzhydryl group into a large hydrophobic pocket lined by aromatic residues including Phe111, Phe87, Trp259 and Tyr262 (Figure 4.6). While Glu269 appears to make a key interaction with the pyridine nitrogen, the importance of this interaction is questionable given that replacement of the pyridine with groups such as phenyl and 4-Cl-phenyl result in only a slight loss in potency.

Amira Pharmaceuticals has reported extensive SAR and pharmacology of a phenylacetic acid class of inhibitors, represented by structures **50–54**.[57–60] These compounds, derived from an HTS effort, bear a resemblance to the

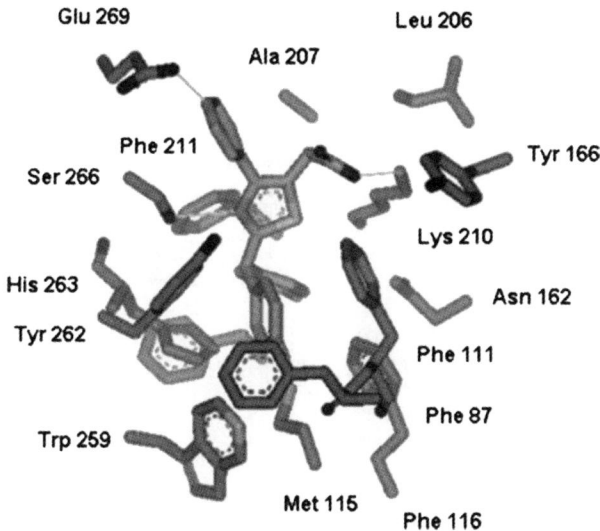

Figure 4.6 Docking of compound **49** to CRTH2. Reprinted from M. Grimstrup, Ø. Rist, J.-M. Receveur, T. M. Frimurer, T. Ulven, J. M. Mathiesen, E. Kostenis and T. Högberg, *Bioorg. Med. Chem. Lett.*, **20**, 1181–1185. © 2010, with permission from Elsevier.

AstraZeneca scaffold (**45–46**). Early structure optimization led to compound **50**, which proved to be very potent in both binding and functional assays. However, significant *in-vivo* metabolism of the carbamate was observed leading the team to more stable analogues such as urea **51** and cyclopropyl amides **52–53**. The detailed reports from Amira allow a unique opportunity to view the screening funnel for an active CRTH2 program, which has advanced into human clinical studies with AM211 (**51**). In addition to the standard CRTH2 binding analysis, the team employed a strategy of screening all leads of interest in a human whole-blood eosinophil shape change assay (hESC) and a serum-shift binding assay in human, mouse and guinea pig isoforms of CRTH2. Interestingly, compounds **50–54** have differential activity against the three homologues, and perhaps not surprisingly the compound with nearly equal activity against the various isoforms (**51**, AM211) was chosen for advanced animal studies and was advanced into clinical trials. Prior to undergoing testing in a disease model, compound **51** was evaluated for blockage of PGD2-induced leukocytosis at 2h post-dose and 18h post-dose in mice. This allowed a preliminary pharmacodynamic understanding of the antagonist's action without the various complexities associated with artificial disease models. Detailed pharmacology of several analogues has been also reported. Compound **53** (10 mpk/PO) was shown to reduce sneezing and nasal rubs following intranasal administration of OVA to sensitized mice, and was also shown to reduce cigarette smoke-induced influx of neutrophils, lymphocytes and macrophages into the lungs in a dose-dependent manner.[58]

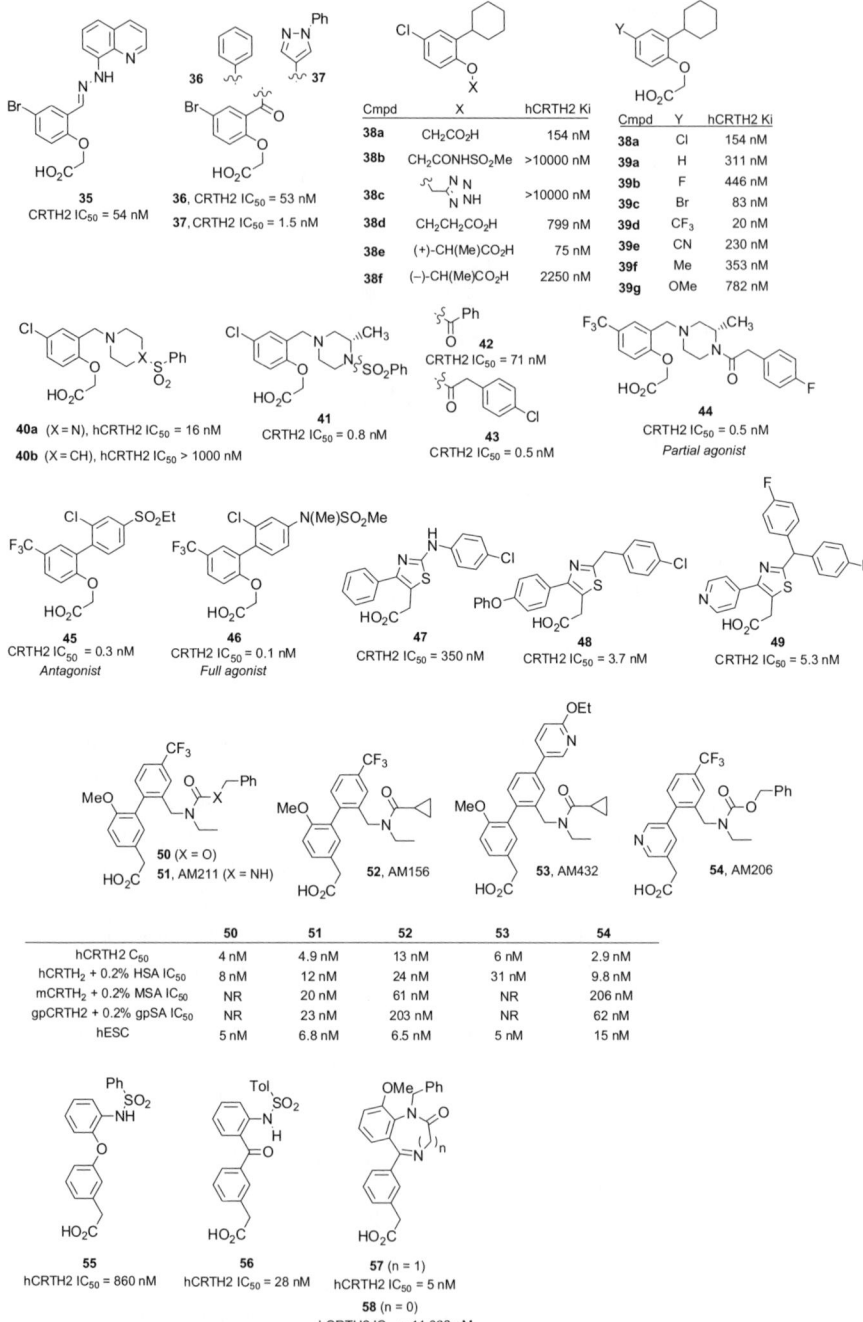

Figure 4.7 Phenoxy and phenyl acetic acid antagonists of CRTH2.

Finally, two additional series of phenylacetic acid CRTH2 antagonists have been described by teams from AstraZeneca,[107] both derived from an HTS hit closely related to compound **55**. Building on the observation that benzophenone **56** showed a significant improvement in potency compared to **55** it was found that benzodiazepinones such as **57** were exceptionally potent antagonists. The deletion of a single methylene group (**58**) results in a 1000-fold loss of activity, which is attributed to a lack of overlapping low-energy conformations between benzophenone **56** and quinazolinone **58**. Thus, compound **58** would have to adopt a higher energy conformation in order to bind in the same proposed binding mode as **56**. The same team reported that this series of phenylacetic acids could be moved from the *meta* position to the *para* position resulting in dual CRTH2/DP antagonists[53,108] (Figure 4.9 and Section 4.4.5). This will be discussed in more detail below.

4.4.4 Tetrahydroquinoline CRTH2 Antagonists

The most common structural element essential to all of the previously described scaffolds is a carboxylic acid or acid isostere. Presumably this moiety mimics the carboxylic acid of the natural ligand for the receptor, PGD2. One widely reported scaffold, however, defies this trend. In 2005 Mimura reported a small set of 4-amino tetrahydroquinolines to be antagonists of CRTH2[109] (See Figure 4.8). Scientists from Amgen followed up on this report and described the SAR of this scaffold in some detail.[110] They reported compound **59a** to be a 43 nM antagonist of CRTH2, and devoid of measurable activity against DP. Removal of the "northern" amide (**59b,c**) resulted in complete loss of activity, while a variety of larger amides were tolerated and even preferred (**59d,e,f**). A carboxylic acid side chain (**59f**) gives a meaningful boost in potency, possibly hinting at its binding mode in comparison to the previous antagonists. The addition of *ortho* substituents to the "southern" benzamide has a deleterious

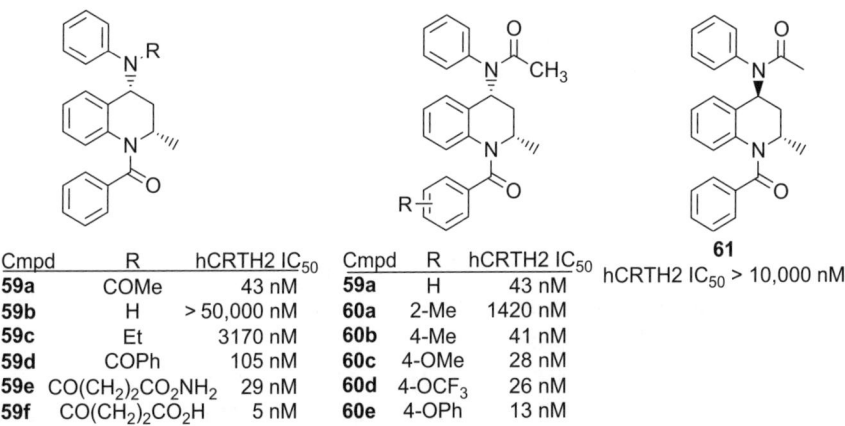

Cmpd	R	hCRTH2 IC_{50}
59a	COMe	43 nM
59b	H	> 50,000 nM
59c	Et	3170 nM
59d	COPh	105 nM
59e	$CO(CH_2)_2CO_2NH_2$	29 nM
59f	$CO(CH_2)_2CO_2H$	5 nM

Cmpd	R	hCRTH2 IC_{50}
59a	H	43 nM
60a	2-Me	1420 nM
60b	4-Me	41 nM
60c	4-OMe	28 nM
60d	$4\text{-}OCF_3$	26 nM
60e	4-OPh	13 nM

61
hCRTH2 IC_{50} > 10,000 nM

Figure 4.8 Tetrahydroquinoline CRTH2 antagonists.

effect on activity (**60a**) while particular *para* substituents provide a meaningful boost in potency (**60d,e**). Additional variations of this scaffold have been reported in the patent literature.[111–113]

4.4.5 Dual CRTH2/DP Antagonists

While there has been much speculation about the possibility of molecules that block the action of both CRTH2 and DP simultaneously, there are scant reports of such molecules. By far the most developed series of dual CRTH2/DP has been reported by Amgen[53,108] (See Figure 4.9). The series originated with the phenylacetic acids exemplified by **55–56**. In an intriguing series of SAR, it is reported that moving the acetic acid moiety from the *meta* position, as in **55**, to the *para* position, as in **62**, results in an increase in activity against DP. They further optimize this activity resulting in a clinical candidate, compound **63** (AMG009). However, relatively modest DP activity in plasma and perhaps other undisclosed issues led the team to abandon this compound in favour of the more potent analogue, **64** (AMG853). This compound has proceeded to Phase II clinical studies and will be briefly discussed in Section 4.7. It is interesting to note that a selective DP antagonist from Shionogi has been shown to be efficacious in a sheep model of asthma.[114]

As compelling as a dual DP/CRTH2 antagonist may seem, it is worth noting that many studies have shown that DP mediates some of the anti-inflammatory responses to PGD2 while CRTH2 mediates many of the pro-inflammatory responses. Indeed, selective *agonists* of DP have been shown to alleviate inflammatory symptoms in various animal models of atopic dermatitis.[115,116]

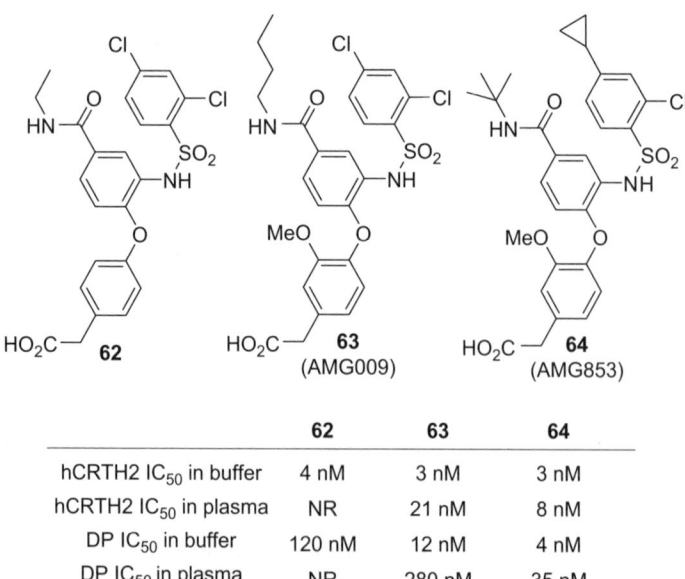

	62	63	64
hCRTH2 IC$_{50}$ in buffer	4 nM	3 nM	3 nM
hCRTH2 IC$_{50}$ in plasma	NR	21 nM	8 nM
DP IC$_{50}$ in buffer	120 nM	12 nM	4 nM
DP IC$_{50}$ in plasma	NR	280 nM	35 nM

Figure 4.9   Dual CRTH2/DP antagonists.

Spik *et al.* demonstrated in 2005 that the selective DP agonist BW245c modestly blocks OVA-induced eosinophil infiltration in murine asthma model.[49] In contrast, other reports have shown that selective DP antagonists have an anti-inflammatory effect in models of allergic rhinitis[117] and PGD2-induced vascular permeability.[118] Therefore, it could be argued that an ideal profile would be either a dual DP/CRTH2 antagonist or a DP-agonist/CRTH2-antagonist, the latter being a challenging goal in the near-term.

4.5 Activity of CRTH2 Antagonists in Animal Efficacy Studies

Animal efficacy studies form the very foundation of clinical pharmaceutical development. A sampling of animal efficacy data that have been generated for CRTH2 antagonists, in particular focusing on the variety of models in which CRTH2 antagonists have shown activity, is presented below.

During lead optimization, one of the most commonly used functional assays to assess CRTH2 antagonists is blockade of PGD2 (or DK-PGD2) induced migration of eosinophils or basophils. This *in-vitro* assay has been translated into an *in-vivo* model by measuring inhibition of DK-PGD2 induced leukocytosis in guinea pigs. Figure 4.10 shows an example of one compound (AM211, **51**) with good activity in this model.[60] While not an efficacy model

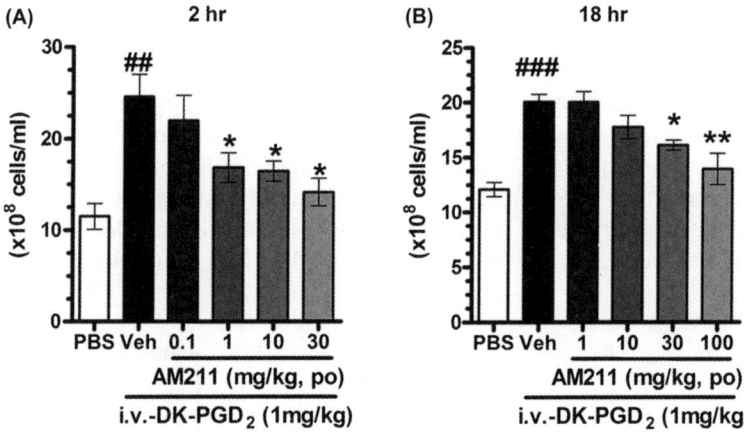

Figure 4.10 Effect of AM211 (**51**) on DK-PGD2-induced leukocytosis in guinea pigs. A and B, total cells/mL ($\times 10^6$) recovered in blood 30 min. after injection of DK-PGD2. Animals were orally dosed with vehicle (Veh) or 0.1, 1, 10 or 30 mg/kg AM211 (**51**) either 2 h (A) or 18 h (B) before the injection of DK-PGD2. PBS animals did not receive PD-PGD2. *P < 0.05; **P < 0.01 *versus* vehicle response. Reproduced from G. Bain, D. S. Lorrain, K. J. Stebbins, A. R. Broadhead, A. M. Santini, P. Prodanovich, J. Darlington, C. D. King, C. Lee, C. Baccei, B. Stearns, Y. Troung, J. H. Hutchinson, P. Prasit and J. F. Evans, *J. Pharmacol. Exp. Ther.*, 2011, **338**, 290–301. © 2011, the American Society for Pharmacology and Experimental Therapeutics. Used with permission.

per se, this system allows for an unambiguous assessment of functional blocking of CRTH2 in an *in-vivo* setting. Moreover, monitoring the activity of a drug in this model over time allows for a meaningful understanding of the pharmacodynamic properties of a compound.

Beyond a functional assessment, one of the first true efficacy models for evaluation is ovalbumen (OVA) induced lung inflammation. This model typically involves OVA sensitization (by IP injection) approximately 2 weeks prior to the study. On the day of the study, animals are typically dosed with the test compound 1 h prior to an aerosolized OVA or saline challenge. The aerosolized OVA induces a strong peribronchial inflammatory response manifested as an increase in lymphocyte cell count. Quantitation of lymphocyte recruitment is typically done by either histochemical visualization of lung tissue or by flow-cytometry of bronchiolar lavage fluid (BALF). An example of a compound that shows significant activity in a mouse model of OVA-induced lung inflammation is shown in Figure 4.11.[100] Thus, treatment with CRTH2 antagonist **37** results in a significant reduction of eosinophil recruitment and reduced proliferation of mucous-producing goblet cells. A variety of other compounds have shown activity in closely related model studies.[52–57] CRTH2 antagonists **51**, **52** and **54** have been reported to be active in a model system that induces airway inflammation with cigarette smoke rather than the classically used OVA.[58,59] Another model system uses DK-PGD2 induced reduction in enhanced pause (Penh) as a measure of airway resistance.[53] However, the interpretation of Penh measurements in animal models is controversial and data from these models should be used with caution.

A commonly used model of allergic rhinitis involves intra-nasal challenge with OVA on sensitized mice. Amelioration of allergic symptoms such as sneezing and nasal rubs is observed following a 10 mg/kg dose of **53** (AM432)[58] (Figure 4.12). A variety of other CRTH2 antagonists have also shown activity in this model.[53,57,60,61] Efficacy in lung and airway inflammation models has also been demonstrated by showing that CRTH2 antagonists block mucous cell hyperplasia and reduce levels of IL-13 and IL-17 in lung homogenate and BALF.[56,57]

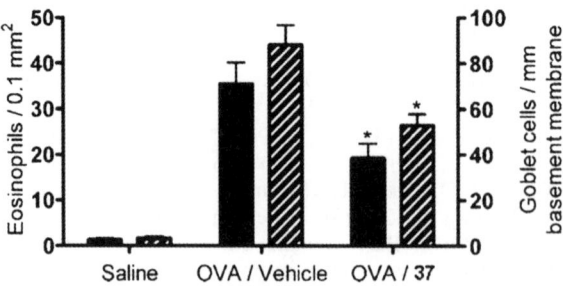

Figure 4.11   Inhibition of peribronchial eosinophilia (solid bars) and goblet cell hyperplasia (striped bars) in mice after oral administration of 5 mg/kg **37**. Reprinted with permission from *J. Med. Chem.*, 2006, **49**, 6638. © 2006, American Chemical Society.

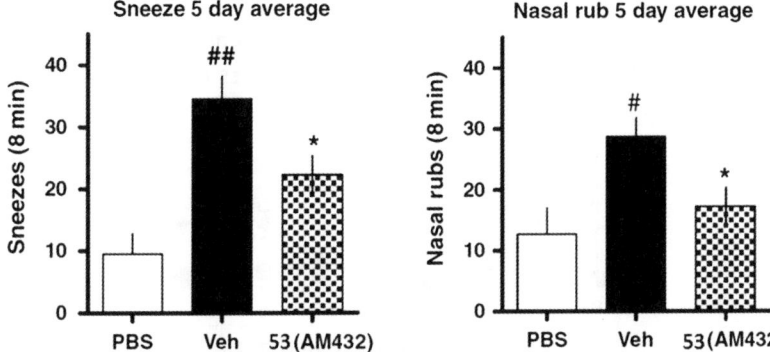

Figure 4.12   Mouse allergic rhinitis data for **53** (AM432) (10 mg/kg) *versus* PBS treated unsensitized mice and vehicle treated sensitized mice. Bars represent $+/-$ SEM of $n = 7$ mice per group. (Left) number of sneezes, 5-day average; (right) nasal rub counts, 5-day average. Reprinted from N. Stock, D. Volkots, K. Stebbins, A. Broadhead, B. Stearns, J. Roppe, T. Parr, C. Baccei, G. Bain, C. Chapman, L. Correa, J. Darlington, C. King, C. Lee, D. S. Lorrain, P. Prodanovich, A. Santini, J. F. Evans, J. H. Hutchinson and P. Prasit, *Bioorg. Med. Chem. Lett.*, 2011, **21**, 1036–1040. © 2011, with permission from Elsevier.

Probably the most widely recognized pre-clinical efficacy model for asthma is sheep model of experimental asthma (see Abraham[119] and references therein). This model is particularly relevant to clinical studies for three reasons. The model relies on natural sensitivity of sheep to *Ascaris sum* (a roundworm) and does not require any sort of artificial sensitization protocol, similar to humans, sheep exhibit both a quantifiable early-stage (1–3 h) and late-stage (3–8 h) bronchoconstriction response to antigens and sheep and humans have comparable lung size and physiology. Multiple CRTH2 antagonists, including MK-7246 (**11**) and AMG009 (**63**), have been shown to block the late-stage bronchoconstrictive response in the sheep model of experimental asthma.[63,64] The results of a 15 mg/kg dose of AMG009 are shown in Figure 4.13, clearly illustrating that this compound is effective primarily in the late-stage response (LAR).

The above models largely focus on lung and airway inflammation. However, it has become exceedingly clear that CRTH2 antagonism is also a viable therapeutic approach for atopic dermatitis (AD) and related skin disorders. As such it is worth a brief summary of one of the key AD models in use, fluorescein isothiocyanate (FITC) induced contact hypersensitivity. The murine version of this model begins with sensitization to FITC by multiple dermal exposures over an approximately 2-week period. Roughly 4 weeks after the initial exposure, the mice are challenged with FITC on one ear and vehicle on the other ear. After 24 h, ear thickness is measured by digital caliper. The difference in thickness between the two ears is the measure of FITC-induced inflammation. Frequently, a secondary measurement is also evaluated, such as serum IgE levels. Figure 4.14 shows an example of an undisclosed CRTH2 antagonist in

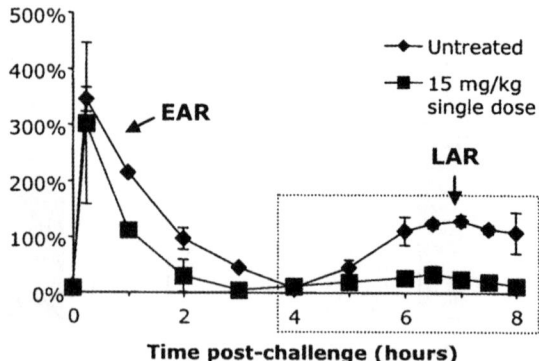

Figure 4.13   A 15 mg/kg dose of AMG-009 (**63**) blocks the late stage bronchoconstrictive response (LAR) to allergen in sheep but has little effect on the
early response (EAR). Reproduced from WO2009085177 (Amgen).

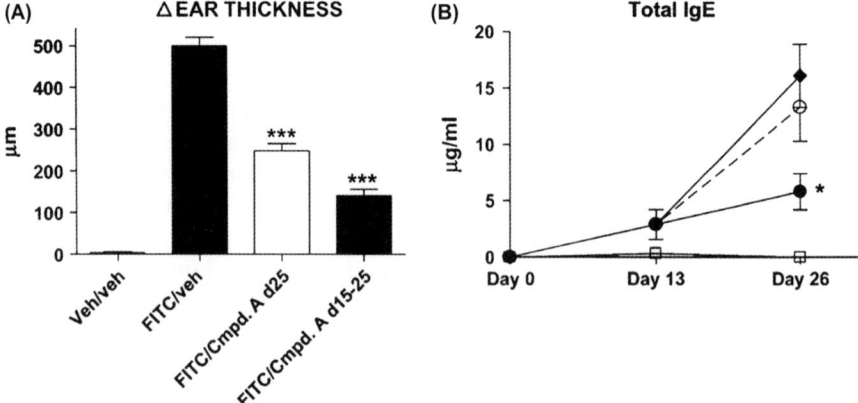

Figure 4.14   Ear swelling and serum IgE levels from BALB/c mice challenged with
FITC. Mice were sensitized to FITC on days 1, 2, 14 and 15 and challenged with FITC on both sides of their right ear on day 25. (A) 24 h
post-FITC challenge, ear thickness was determined by using digital
calipers and the change in ear thickness was determined by subtracting
the width of the right (FITC challenged) ear minus the width of the left
(vehicle challenged) ear. Each column shown is the mean 6 ± SEM. (B)
Serum was isolated from the mice on days 13 and 26, and the total IgE
serum levels were determined. Open squares – vehicle/vehicle-treated
cohort; closed diamonds – FITC/vehicle cohort; open circles – FITC/
Compound A day 25-treated cohort; and closed circles – FITC/Com-
pound A days 14–25-treated cohort. Each point shown is the average
6 ± SEM. ***$P < 0.001$; *$P < 0.05$. S. A. Boehme, E. P. Chen, K.
Franz-Bacon, R. Sásik, L. J. Sprague, T. W. Ly, G. Hardiman and
K. B. Bacon, *Int. Immunol.*, 2009, **21**(1), 1–17, by permission of Oxford
University Press.

this model.[120,121] It can be seen that drug treatment on the day of FITC challenge results in an approximately 50% decrease in ear swelling. Treatment over a more prolonged period prior to the challenge (days 15–25) results in an even more dramatic reduction in ear swelling (Figure 4.14, panel A). Importantly, this effect is even more pronounced when circulating IgE levels are compared between groups (Figure 4.14, panel B). CRTH2 antagonists have also shown promising activity in other related models of AD.[65,66]

4.6 CRTH2 Biomarkers for Clinical Development

The increasing number of clinical studies of CRTH2 antagonists has given rise to an interest in pharmacodynamic markers, which may assist in understanding the relationship between *in-vivo* drug exposure and physiological response (*i.e.* a PK/PD relationship). Several such markers have been identified, all relying on attenuation of signals of CRTH2-mediated cell activation. Historically, the most commonly used whole blood markers of CRTH2 activation have been eosinophil shape change and Th2 cell chemotaxis. Both of these phenotypes have been observed to be blocked by CRTH2 antagonists in whole blood.[94,101] More recently, researchers at Roche used gene profiling to identify mRNA that may be up-regulated in the presence of a CRTH2 agonist. They found that CRTH2 activation by DK-PGD2 in human whole blood from six donors resulted in increased expression of Charcot–Leyden crystal protein/galectin-10 (CLC/Gal-10) mRNA. As expected, this expression was largely attenuated by pre-treatment with a CRTH2 antagonist, CAY10471.[69] While the effect was relatively modest ($\sim$2–4-fold activation), the authors propose that this marker may prove more useful in a clinical setting than the more traditional markers of CRTH2 activation that rely upon expensive flow-cytometry equipment. Interestingly, CLC/Gal-10 has been known for many years to be expressed at elevated levels in patients with acute asthma, aspirin-induced asthma and allergic rhinitis.[122,123] It is tempting to speculate that the increased levels of this protein are due to over-activation of the CRTH2 receptor and that treatment with a CRTH2 antagonist may return levels of this protein to a more normal level.

4.7 CRTH2 Antagonists in Clinical Development

While numerous CRTH2 antagonists have entered clinical development, the results from only a handful of clinical trials have been reported. The status of various clinical CRTH2 programs as of 2010 has been reviewed.[87] Table 4.1 outlines the current status of various publicly known CRTH2 antagonists as of September 2011. By far the most detailed clinical report of CRTH2 antagonism comes from Oxygen's OC000459[124] (NCT01057927). In a study comparing OC000459 (200 mg, twice daily, N = 65) to placebo (N = 67), the authors report statistically significant improvements in pre-bronchodilator forced expiratory volume in 1 second (FEV$_1$) and in night-time symptom scores.

Adverse events were comparable to placebo, but respiratory infections were notably less common in the treatment group. The only additional report of human efficacy for a CRTH2 antagonist comes from Actelion's ACT-129968 (Setipiprant). This compound has been reported to meet the primary efficacy and safety end-points in studies of seasonal allergic rhinitis and mild asthma.[128]

Early safety and PK data have been reported for a variety of other clinical CRTH2 programs. Array BioPharma has reported that the Phase I antagonist ARRY502 was well tolerated in healthy subjects with a history of seasonal allergies. Moreover, the compound showed low inter-subject variability and produced a dose-dependent inhibition of PGDD2-stimulated *ex vivo* eosinophil shape change.[125] AstraZeneca has put multiple CRTH2 antagonists into the clinic. A Phase I trial of AZD5985 was discontinued due to tolerability issues (NCT00967356) and a Phase I trial of AZD8075 was discontinued after turbidity in urine was reported in 4 out of 8 healthy volunteers. However, AZD1981 has completed multiple Phase I and Phase II trials and is advancing as an experimental treatment for asthma and COPD. Amgen's dual CRTH2/ DP antagonist AMG853 (**64**) was well tolerated in a Phase I trial and dose-proportional PK was observed over a dose range of 5 mg to 400 mg. Moreover, this compound attenuated niacin-induced flushing and suppressed PGD2-induced responses *ex vivo*.[126] AMG853 advanced into a Phase II clinical trial which completed in early 2011 (NCT01018550). While no study results have been released for this trial, Amgen announced in April 2011 that the development of this compound would be discontinued.[127]

Other companies that have reported CRTH2 antagonists in their clinical pipeline include Merck, Novartis, Amira and Actimis (Table 4.1).

Table 4.1   Status of CRTH2 compounds in the clinic (as of September 2011).

Company	Compound ID	Status	Reference[a]
Actelion	ACT-129968 (Setipiprant)	Phase IIb	NCT01225315
Actimis	AP768	Phase I	Company website
Amira	AM211 (**51**)	Phase I	Company website
Amira	AM461	Phase I	Company website
Amgen	AMG853 (**64**)	Discontinued after Phase II (2011)	NCT01018550
Array BioPharma	ARRY-502	Phase I	NCT01349725
AstraZeneca	AZD1981	Phase II	NCT01197794
AstraZeneca	AZD8075	Discontinued after Phase I (2010)	NCT00787072
AstraZeneca	AZD5985	Discontinued after Phase I (2010)	NCT00967356
Merck	MK-7246 (**11**)	Phase I	Company website
Novartis	QAV680	Phase II completed	NCT00814216
Oxagen	OC000459	Phase IIb	NCT01057927
Oxagen	OC002417	Back-up candidate	Company website
Pulmagen	ADC3680B	Phase I	NCT01173770
Shionogi	S-555739	Phase IIb (Japan)	Company website

[a]NCT numbers can be searched at the website: www.clinicaltrials.gov

Unfortunately, few of these structures are known. However, a careful reading of the patent literature and some deductive reasoning has allowed for intelligent speculation about the structures of many of these candidates.[87]

4.8 Summary

In conclusion, the rapid proliferation of patent applications and early clinical candidates gives testament to the therapeutic promise of CRTH2 antagonism. The role that this target plays in atopic disorders has become increasingly clear, in particular over the past five years. Our understanding of CRTH2 biology has been principally enabled by the proliferation of tool compounds that selectively antagonize or agonize the two major PGD2 receptors, CRTH2 and DP. The primary function of CRTH2 appears to be induction of chemotaxis resulting in the recruitment of Th2 leukocytes, eosinophils and basophils to sites of activated mast cells. The blockade of this function by CRTH2 antagonists leads to the amelioration of a variety of inflammatory pathologies such as allergic rhinitis, asthma and atopic dermatitis. To date, over a dozen CRTH2 antagonists have been advanced into human clinical studies. While the promising activity reported for OC000459 bodes well for the field, it should also be pointed out that the attrition rate for Phase II clinical trials is notoriously high and no CRTH2 antagonist has yet advanced to Phase III trials. One of the most interesting pieces of clinical data that will emerge in the coming years is the efficacy of selective CRTH2 antagonists *versus* the efficacy of dual CRTH2/DP antagonists (principally from Amgen). While there is considerable evidence that selective CRTH2 antagonism is sufficient to elicit an anti-inflammatory response, it is also clear that DP also plays an important role in the PGD2 action. The untangling of these roles in a clinical setting will be a major advance for the field.

References

1. A. N. Hata and R. M. Breyer, *Pharmacol. Ther.*, 2004, **103**, 147.
2. P. Benoit, X. De Leval, B. Pirotte and J.-M. Dogne, *Expert Opin. Ther. Pat.*, 2002, **12**, 1225.
3. A. Miadonna, A. Tedeschi, C. Brasca, G. Folco, A. Sala and R. C. Murphy, *J. Allergy Clin. Immunol.*, 1990, **85**, 906.
4. R. M. Naclerio, H. L. Meier, A. Kagey-Sobotka, N. F. Adkinson, Jr., D. A. Meyers, P. S. Norman and L. M. Lichtenstein, *Am. Rev. Respir. Dis.*, 1983, **128**, 597.
5. R. M. Barr, O. Koro, D. M. Francis, A. K. Black, T. Numata and M. W. Greaves, *Br. J. Pharmacol.*, 1988, **94**, 773.
6. G. Bochenek, K. Nagraba, E. Nizankowska and A. Szczeklik, *J. Allergy Clin. Immunol.*, 2003, **111**, 743.
7. L. J. Roberts, 2nd, B. J. Sweetman, R. A. Lewis, K. F. Austen and J. A. Oates, *N. Engl. J. Med.*, 1980, **303**, 1400.

8. G. Bochenek, E. Nizankowska, A. Gielicz, M. Swierczynska and A. Szczeklik, *Thorax*, 2004, **59**, 459.

9. Y. Fujitani, Y. Kanaoka, K. Aritake, N. Uodome, K. Okazaki-Hatake and Y. Urade, *J. Immunol.*, 2002, **168**, 443.

10. K. Aritake, Y. Kado, T. Inoue, M. Miyano and Y. Urade, *J. Biol. Chem.*, 2006, **281**, 15277.

11. T. Tajima, T. Murata, K. Aritake, Y. Urade, H. Hirai, M. Nakamura, H. Ozaki and M. Hori, *Title J. Pharmacol. Exp. Ther.*, 2008, **326**, 493.

12. R. P. Siraganian, *Curr. Opin. Immunol.*, 2003, **15**, 639.

13. R. P. Schleimer, C. C. Fox, R. M. Naclerio, M. Plaut, P. S. Creticos, A. G. Togias, J. A. Warner, A. Kagey-Sobotka and L. M. Lichtenstein, *J. Allergy Clin. Immunol.*, 1985, **76**, 369.

14. R. A. Lewis, N. A. Soter, P. T. Diamond, K. F. Austen, J. A. Oates and L. J. Roberts, II, *J. Immunol.*, 1982, **129**, 1627.

15. R. J. Flower, E. A. Harvey and W. P. Kingston, *Br. J. Pharmacol.*, 1976, **56**, 229.

16. X. Norel, L. Walch, C. Labat, J.-P. Gascard, E. Dulmet and C. Brink, *Br. J. Pharmacol.*, 1999, **126**, 867.

17. L. Walch, C. Labat, J.-P. Gascard, V. De Montpreville, C. Brink and X. Norel, *Br. J. Pharmacol.*, 1999, **126**, 859.

18. S. Narumiya and N. Toda, *Br. J. Pharmacol.*, 1985, **85**, 367.

19. Y. Boie, N. Sawyer, D. M. Slipetz, K. M. Metters and M. Abramovitz, *J. Biol. Chem.*, 1995, **270**, 18910.

20. N. Sawyer, E. Cauchon, A. Chateauneuf, R. P. G. Cruz, D. W. Nicholson, K. M. Metters, G. P. O'Neill and F. G. Gervais, *Br. J. Pharmacol.*, 2002, **137**, 1163.

21. V. Angeli, C. Faveeuw, O. Roye, J. Fontaine, E. Teissier, A. Capron, I. Wolowczuk, M. Capron and F. Trottein, *J. Exp. Med.*, 2001, **193**, 1135.

22. B. J. Whittle, S. Moncada, K. Mullane and J. R. Vane, *Prostaglandins*, 1983, **25**, 205.

23. D. H. Wright, A. W. Ford-Hutchinson, K. Chadee and K. M. Metters, *Br. J. Pharmacol.*, 2000, **131**, 1537.

24. D. F. Woodward, C. S. Spada, S. B. Hawley, L. S. Williams, C. E. Protzman and A. L. Nieves, *Eur. J. Pharmacol.*, 1993, **230**, 327.

25. D. Gerashchenko, C. T. Beuckmann, Y. Kanaoka, N. Eguchi, W. C. Gordon, Y. Urade, N. G. Bazan and O. Hayaishi, *J. Neurochem.*, 1998, **71**, 937.

26. F. Nantel, C. Fong, S. Lamontagne, D. H. Wright, A. Giaid, M. Desrosiers, K. M. Metters, G. P. O'Neill and F. G. Gervais, *Prostaglandins Other Lipid Mediators*, 2004, **73**, 87.

27. M. A. Gallant, R. Samadfam, J. A. Hackett, J. Antoniou, J.-L. Parent and A. J. de Brum-Fernandes, *J. Bone Miner. Res.*, 2005, **20**, 672.

28. P. Gosset, F. Bureau, V. Angeli, M. Pichavant, C. Faveeuw, A.-B. Tonnel and F. Trottein, *J. Immunol.*, 2003, **170**, 4943.

29. H. Hirai, K. Tanaka, O. Yoshie, K. Ogawa, K. Kenmotsu, Y. Takamori, M. Ichimasa, K. Sugamura, M. Nakamura, S. Takano and K. Nagata, *J. Exp. Med.*, 2001, **193**, 255.
30. H. Abe, T. Takeshita, K. Nagata, T. Arita, Y. Endo, T. Fujita, H. Takayama, M. Kubo and K. Sugamura, *Gene*, 1999, **227**, 71.
31. R. Pettipher, *Br. J. Pharmacol.*, 2008, **153**, S191.
32. C. Yoshimura-Uchiyama, M. Iikura, M. Yamaguchi, H. Nagase, A. Ishii, K. Matsushima, K. Yamamoto, M. Shichijo, K. B. Bacon and K. Hirai, *Clin. Exp. Allergy*, 2004, **34**, 1283.
33. L. Xue, S. L. Gyles, F. R. Wettey, L. Gazi, E. Townsend, M. G. Hunter and R. Pettipher, *J. Immunol.*, 2005, **175**, 6531.
34. F. G. Gervais, R. P. G. Cruz, A. Chateauneuf, S. Gale, N. Sawyer, F. Nantel, K. M. Metters and G. P. O'Neill, *J. Allergy Clin. Immunol.*, 2001, **108**, 982.
35. L. Xue, A. Barrow and R. Pettipher, *J. Immunol.*, 2009, **182**, 7580.
36. M. Durand, M. A. Gallant and A. J. de Brum-Fernandes, *J. Bone Miner. Res.*, 2008, **23**, 1097.
37. N. Kanda, T. Ishikawa and S. Watanabe, *Biochem. Pharmacol.*, 2010, **79**, 982.
38. L. Xue, S. L. Gyles, A. Barrow and R. Pettipher, *Biochem. Pharmacol.*, 2007, **73**, 843.
39. C. J. Whelan, *Inflammation Res.*, 2009, **58**, 103.
40. H. Sandig, D. Andrew, A. A. Barnes, I. Sabroe and J. Pease, *FEBS Lett.*, 2006, **580**, 373.
41. G. Monneret, H. Li, J. Vasilescu, J. Rokach and W. S. Powell, *J. Immunol.*, 2002, **168**, 3563.
42. R. Schuligoi, R. Schmidt, G. Geisslinger, M. Kollroser, B. A. Peskar and A. Heinemann, *Biochem. Pharmacol.*, 2007, **74**, 107.
43. R. Schuligoi, M. Sedej, M. Waldhoer, A. Vukoja, E. M. Sturm, I. T. Lippe, B. A. Peskar and A. Heinemann, *J. Leukocyte Biol.*, 2009, **85**, 136.
44. E. Bohm, J. Sturm Gunter, I. Weiglhofer, H. Sandig, M. Shichijo, A. McNamee, E. Pease James, M. Kollroser, A. Peskar Bernhard and A. Heinemann, *J. Biol. Chem.*, 2004, **279**, 7663.
45. H. Hirai, K. Tanaka, S. Takano, M. Ichimasa, M. Nakamura and K. Nagata, *J. Immunol.*, 2002, **168**, 981.
46. F. G. Gervais, J.-P. Morello, C. Beaulieu, N. Sawyer, D. Denis, G. Greig, A. D. Malebranche and G. P. O'Neill, *Mol. Pharmacol.*, 2005, **67**, 1834.
47. N. A. Sharif, J. Y. Crider, S. X. Xu and G. W. Williams, *Title J. Pharmacol. Exp. Ther.*, 2000, **293**, 321.
48. W. Almishri, C. Cossette, J. Rokach, J. G. Martin, Q. Hamid and W. S. Powell, *Title J. Pharmacol. Exp. Ther.*, 2005, **313**, 64.
49. I. Spik, C. Brenuchon, V. Angeli, D. Staumont, S. Fleury, M. Capron, F. Trottein and D. Dombrowicz, *J. Immunol.*, 2005, **174**, 3703.

50. S. Kagawa, K. Fukunaga, T. Oguma, Y. Suzuki, T. Shiomi, K. Sayama, T. Kimura, H. Hirai, K. Nagata, M. Nakamura and K. Asano, *Int. Arch. Allergy Immunol.*, 2011, **155**, 6.

51. T. Satoh, R. Moroi, K. Aritake, Y. Urade, Y. Kanai, K. Sumi, H. Yokozeki, H. Hirai, K. Nagata, T. Hara, M. Utsuyama, K. Hirokawa, K. Sugamura, K. Nishioka and M. Nakamura, *J. Immunol.*, 2006, **177**, 2621.

52. S. Crosignani, C. Jorand-Lebrun, P. Page, G. Campbell, V. Colovray, M. Missotten, Y. Humbert, C. Cleva, J.-F. Arrighi, M. Gaudet, Z. Johnson, P. Ferro and A. Chollet, *ACS Med. Chem. Lett.*, 2011, **2**, 644.

53. J. Liu, Z. Fu, Y. Wang, M. Schmitt, A. Huang, D. Marshall, G. Tonn, L. Seitz, T. Sullivan, H. Lucy Tang, T. Collins and J. Medina, *Bioorg. Med. Chem. Lett.*, 2009, **19**, 6419.

54. Y. Shiraishi, K. Asano, K. Niimi, K. Fukunaga, M. Wakaki, J. Kagyo, T. Takihara, S. Ueda, T. Nakajima, T. Oguma, Y. Suzuki, T. Shiomi, K. Sayama, S. Kagawa, E. Ikeda, H. Hirai, K. Nagata, M. Nakamura, T. Miyasho and A. Ishizaka, *J. Immunol. (Baltimore, Md.: 1950)*, 2008, **180**, 541.

55. N. W. Lukacs, A. A. Berlin, K. Franz-Bacon, R. Sasik, L. J. Sprague, T. W. Ly, G. Hardiman, S. A. Boehme and K. B. Bacon, *Am. J. Physiol.*, 2008, **295**, L767.

56. L. Uller, M. Mathiesen Jesper, L. Alenmyr, M. Korsgren, T. Ulven, T. Hogberg, G. Andersson, G. A. Persson Carl and E. Kostenis, *Respir. Res.*, 2007, **8**, 16.

57. K. J. Stebbins, A. R. Broadhead, L. D. Correa, J. M. Scott, Y. P. Truong, B. A. Stearns, J. H. Hutchinson, P. Prasit, J. F. Evans and D. S. Lorrain, *Eur. J. Pharmacol.*, 2010, **638**, 142.

58. N. Stock, D. Volkots, K. Stebbins, A. Broadhead, B. Stearns, J. Roppe, T. Parr, C. Baccei, G. Bain, C. Chapman, L. Correa, J. Darlington, C. King, C. Lee, D. S. Lorrain, P. Prodanovich, A. Santini, J. F. Evans, J. H. Hutchinson and P. Prasit, *Bioorg. Med. Chem. Lett.*, 2011, **21**, 1036.

59. K. J. Stebbins, A. R. Broadhead, C. S. Baccei, J. M. Scott, Y. P. Truong, H. Coate, N. S. Stock, A. M. Santini, P. Fagan, P. Prodanovich, G. Bain, B. A. Stearns, C. D. King, J. H. Hutchinson, P. Prasit, J. F. Evans and D. S. Lorrain, *J. Pharmacol. Exp. Ther.*, 2010, **332**, 764.

60. G. Bain, D. S. Lorrain, K. J. Stebbins, A. R. Broadhead, A. M. Santini, P. Prodanovich, J. Darlington, C. D. King, C. Lee, C. Baccei, B. Stearns, Y. Troung, J. H. Hutchinson, P. Prasit and J. F. Evans, *J. Pharmacol. Exp. Ther.*, 2011, **338**, 290.

61. B. A. Stearns, C. Baccei, G. Bain, A. Broadhead, R. C. Clark, H. Coate, J. F. Evans, P. Fagan, J. H. Hutchinson, C. King, C. Lee, D. S. Lorrain, P. Prasit, P. Prodanovich, A. Santini, J. M. Scott, N. S. Stock and Y. P. Truong, *Bioorg. Med. Chem. Lett.*, 2009, **19**, 4647.

62. T. Terasaka, S. Ito, T. Zenkoh, H. Hayashida, H. Matsuda, J. Miyata, H. Nagata, Y. Takasuna, M. Kobayashi, M. Takeuchi and M. Ohta, Abstracts of Papers, 241st ACS National Meeting & Exposition, Anaheim, CA, United States, March 27–31, 2011, MEDI.

63. F. G. Gervais, N. Sawyer, R. Stocco, M. Hamel, C. Krawczyk, S. Sillaots, D. Denis, E. Wong, Z. Wang, M. Gallant, W. M. Abraham, D. Slipetz, M. A. Crackower and G. P. O'Neill, *Mol. Pharmacol.*, 2011, **79**, 69.
64. M. Grillo, A. Li, J. Liu, J. Medina, Y. Su, Y. Wang, J. Jona, A. Allgeier, J. Milne, J. Murry, J. Payacka and T. Storz, WO 2009085177, 2009.
65. M. Oiwa, T. Satoh, M. Watanabe, H. Niwa, H. Hirai, M. Nakamura and H. Yokozeki, *Clin. Exp. Allergy*, 2008, **38**, 1357.
66. K. Takeshita, T. Yamasaki, K. Nagao, H. Sugimoto, M. Shichijo, F. Gantner and K. B. Bacon, *Int. Immunol.*, 2004, **16**, 947.
67. J. F. Royer, P. Schratl, S. Lorenz, E. Kostenis, T. Ulven, R. Schuligoi, B. A. Peskar and A. Heinemann, *Allergy (Oxford, United Kingdom)*, 2007, **62**, 1401.
68. E. Chevalier, J. Stock, T. Fisher, M. Dupont, M. Fric, H. Fargeau, M. Leport, S. Soler, S. Fabien, M.-P. Pruniaux, M. Fink, C. P. Bertrand, J. McNeish and B. Li, *J. Immunol.*, 2005, **175**, 2056.
69. T.-A. Lin, G. Kourteva, H. Hilton, H. Li, N. S. Tare, V. Carvajal, J. S. Hang, X. Wei and L. M. Renzetti, *Biomarkers*, **15**, 646.
70. C. A. Perez-Novo, G. Holtappels, S. L. Vinall, L. Xue, N. Zhang, C. Bachert and R. Pettipher, *Allergy (Oxford, United Kingdom)*, **65**, 304.
71. J.-L. Huang, P.-S. Gao, R. A. Mathias, T.-C. Yao, L.-C. Chen, M.-L. Kuo, S.-C. Hsu, B. Plunkett, A. Togias, K. C. Barnes, C. Stellato, T. H. Beaty and S.-K. Huang, *Hum. Mol. Genet.*, 2004, **13**, 2691.
72. L. Cameron, M. Depner, M. Kormann, N. Klopp, T. Illig, E. von Mutius and M. Kabesch, *Allergy (Oxford, United Kingdom)*, 2009, **64**, 1478.
73. Y. Maeda, N. Hizawa, D. Takahashi, Y. Fukui, S. Konno and M. Nishimura, *Int. Arch. Allergy Immunol.*, 2007, **142**, 51.
74. H. Sugimoto, M. Shichijo, T. Iino, Y. Manabe, A. Watanabe, M. Shimazaki, F. Gantner and K. B. Bacon, *Title J. Pharmacol. Exp. Ther.*, 2003, **305**, 347.
75. H. Aizawa, M. Shigyo, H. Nogami, T. Hirose and N. Hara, *Chest*, 1996, **109**, 338.
76. H. P. Francis, S. J. Greenham, A. M. Thompson and P. J. Gardiner, *Br. J. Pharmacol.*, 1991, **104**, 596.
77. M. Shichijo, H. Sugimoto, K. Nagao, H. Inbe, J. A. Encinas, K. Takeshita, K. B. Bacon and F. Gantner, *J. Pharmacol. Exp. Ther.*, 2003, **307**, 518.
78. Y. Shiraishi, K. Asano, T. Nakajima, T. Oguma, Y. Suzuki, T. Shiomi, K. Sayama, K. Niimi, M. Wakaki, J. Kagyo, E. Ikeda, H. Hirai, K. Yamaguchi and A. Ishizaka, *J. Pharmacol. Exp. Ther.*, 2005, **312**, 954.
79. R. Schuligoi, E. Sturm, P. Luschnig, V. Konya, S. Philipose, M. Sedej, M. Waldhoer, B. A. Peskar and A. Heinemann, *Pharmacology*, 2010, **85**, 372.
80. R. Pettipher and T. T. Hansel, *Drug News Perspect.*, 2008, **21**, 317.
81. T. W. Ly and K. B. Bacon, *Expert Opin. Invest. Drugs*, 2005, **14**, 769.
82. T. Ulven and E. Kostenis, *Expert Opin. Ther. Pat.*, 2010, **20**, 1505.
83. R. Pettipher and T. T. Hansel, *Progress in Respir. Res.*, 2010, **39**, 193.
84. *Expert Opin. Ther. Pat.*, 2005, **15**, 115.

85. P. Norman, *Expert Opin. Ther. Pat.*, 2005, **15**, 1817.
86. J. J. Chen and A. L. Budelsky, *Prog. Med. Chem.*, 2011, **50**, 49.
87. P. Norman, *Expert Opin. Invest. Drugs*, 2010, **19**, 947.
88. M. J. Robarge, D. C. Bom, L. Nathan Tumey, N. Varga, E. Gleason, D. Silver, J. Song, S. M. Murphy, G. Ekema, C. Doucette, D. Hanniford, M. Palmer, G. Pawlowski, J. Danzig, M. Loftus, K. Hunady, B. A. Sherf, R. W. Mays, A. Stricker-Krongrad, K. R. Brunden, J. J. Harrington and Y. L. Bennani, *Bioorg. Med. Chem. Lett.*, 2005, **15**, 1749.
89. T. Ulven and E. Kostenis, *J. Med. Chem.*, 2005, **48**, 897.
90. M. Gallant, C. Beaulieu, C. Berthelette, J. Colucci, M. A. Crackower, C. Dalton, D. Denis, Y. Ducharme, R. W. Friesen, D. Guay, F. G. Gervais, M. Hamel, R. Houle, C. M. Krawczyk, B. Kosjek, S. Lau, Y. Leblanc, E. E. Lee, J.-F. Levesque, C. Mellon, C. Molinaro, W. Mullet, G. P. O'Neill, P. O'Shea, N. Sawyer, S. Sillaots, D. Simard, D. Slipetz, R. Stocco, D. Sorensen, V. L. Truong, E. Wong, J. Wu, H. Zaghdane and Z. Wang, *Bioorg. Med. Chem. Lett.*, 2011, **21**, 288.
91. D. Simard, Y. Leblanc, C. Berthelette, M. H. Zaghdane, C. Molinaro, Z. Wang, M. Gallant, S. Lau, T. Thao, M. Hamel, R. Stocco, N. Sawyer, S. Sillaots, F. Gervais, R. Houle and J.-F. Levesque, *Bioorg. Med. Chem. Lett.*, 2011, **21**, 841.
92. H. Zaghdane, M. Boyd, J. Colucci, D. Simard, C. Berthelette, Y. Leblanc, Z. Wang, R. Houle, J. F. Levesque, C. Molinaro, M. Hamel, R. Stocco, N. Sawyer, S. Sillaots, F. Gervais and M. Gallant, *Bioorg. Med. Chem. Lett.*, 2011, **21**, 3471.
93. A. N. Hata, T. P. Lybrand, L. J. Marnett and R. M. Breyer, *Mol. Pharmacol.*, 2005, **67**, 640.
94. R. E. Armer, M. R. Ashton, E. A. Boyd, C. J. Brennan, F. A. Brookfield, L. Gazi, S. L. Gyles, P. A. Hay, M. G. Hunter, D. Middlemiss, M. Whittaker, L. Xue and R. Pettipher, *J. Med. Chem.*, 2005, **48**, 6174.
95. D. A. Sandham, C. Adcock, K. Bala, L. Barker, Z. Brown, G. Dubois, D. Budd, B. Cox, R. A. Fairhurst, M. Furegati, C. Leblanc, J. Manini, R. Profit, J. Reilly, R. Stringer, A. Schmidt, K. L. Turner, S. J. Watson, J. Willis, G. Williams and C. Wilson, *Bioorg. Med. Chem. Lett.*, 2009, **19**, 4794.
96. D. Bonafoux, A. Abibi, B. Bettencourt, A. Burchat, A. Ericsson, C. M. Harris, T. Kebede, M. Morytko, M. McPherson, G. Wallace and X. Wu, *Bioorg. Med. Chem. Lett.*, 2011, **21**, 1861.
97. T. N. Birkinshaw, S. J. Teague, C. Beech, R. V. Bonnert, S. Hill, A. Patel, S. Reakes, H. Sanganee, I. G. Dougall, T. T. Phillips, S. Salter, J. Schmidt, E. C. Arrowsmith, J. J. Carrillo, F. M. Bell, S. W. Paine and R. Weaver, *Bioorg. Med. Chem. Lett.*, 2006, **16**, 4287.
98. L. N. Tumey, M. J. Robarge, E. Gleason, J. Song, S. M. Murphy, G. Ekema, C. Doucette, D. Hanniford, M. Palmer, G. Pawlowski, J. Danzig, M. Loftus, K. Hunady, B. Sherf, R. W. Mays, A. Stricker-Krongrad, K. R. Brunden, Y. L. Bennani and J. J. Harrington, *Bioorg. Med. Chem. Lett.*, 2010, **20**, 3287.

99. S. Crosignani, P. Page, M. Missotten, V. Colovray, C. Cleva, J.-F. Arrighi, J. Atherall, J. Macritchie, T. Martin, Y. Humbert, M. Gaudet, D. Pupowicz, M. Maio, P.-A. Pittet, L. Golzio, C. Giachetti, C. Rocha, G. Bernardinelli, Y. Filinchuk, A. Scheer, M. K. Schwarz and A. Chollet, *J. Med. Chem.*, 2008, **51**, 2227.

100. T. Hoegberg, T. Ulven, J.-M. Receveur, M. Grimstrup, O. Rist, T. M. Frimurer, L.-O. Gerlach, J. M. Mathiesen, E. Kostenis and L. Uller, *J. Med. Chem.*, 2006, **49**, 6638.

101. D. A. Sandham, C. Aldcroft, U. Baettig, L. Barker, D. Beer, G. Bhalay, Z. Brown, G. Dubois, D. Budd, L. Bidlake, E. Campbell, B. Cox, B. Everatt, D. Harrison, C. J. Leblanc, J. Manini, R. Profit, R. Stringer, K. S. Thompson, K. L. Turner, M. F. Tweed, C. Walker, S. J. Watson, S. Whitebread, J. Willis, G. Williams and C. Wilson, *Bioorg. Med. Chem. Lett.*, 2007, **17**, 4347.

102. T. Luker, R. Bonnert, S. W. Paine, J. Schmidt, C. Sargent, A. R. Cook, A. Cook, P. Gardiner, S. Hill, C. Weyman-Jones, A. Patel, S. Thom and P. Thorne, *J. Med. Chem.*, 2011, **54**, 1779.

103. T. Luker, R. Bonnert, J. Schmidt, C. Sargent, S. W. Paine, S. Thom, G. Pairaudeau, A. Patel, R. Mohammed, E. Akam, I. Dougall, A. M. Davis, P. Abbott, S. Brough, I. Millichip and T. McInally, *Bioorg. Med. Chem. Lett.*, 2011, **21**, 3616.

104. O. Rist, M. Grimstrup, J.-M. Receveur, T. M. Frimurer, T. Ulven, E. Kostenis and T. Hoegberg, *Bioorg. Med. Chem. Lett.*, 2010, **20**, 1177.

105. M. Grimstrup, O. Rist, J.-M. Receveur, T. M. Frimurer, T. Ulven, J. M. Mathiesen, E. Kostenis and T. Hoegberg, *Bioorg. Med. Chem. Lett.*, 2010, **20**, 1181.

106. M. Grimstrup, J.-M. Receveur, O. Rist, T. M. Frimurer, P. A. Nielsen, J. M. Mathiesen and T. Hoegberg, *Bioorg. Med. Chem. Lett.*, 2010, **20**, 1638.

107. J. Liu, A. C. Cheng, H. L. Tang and J. C. Medina, *ACS Med. Chem. Lett.*, 2011, **2**, 515.

108. J. Liu, A.-R. Li, Y. Wang, M. G. Johnson, Y. Su, W. Shen, X. Wang, S. Lively, M. Brown, S. J. Lai, F. G. Lopez De Turiso, Q. Xu, B. Van Lengerich, M. Schmitt, Z. Fu, Y. Sun, S. Lawlis, L. Seitz, J. Danao, J. Wait, Q. Ye, H. L. Tang, M. Grillo, T. L. Collins, T. J. Sullivan and J. C. Medina, *ACS Med. Chem. Lett.*, 2011, **2**, 326.

109. H. Mimura, T. Ikemura, O. Kotera, M. Sawada, S. Tashiro, E. Fuse, K. Ueno, H. Manabe, E. Ohshima, A. Karasawa and H. Miyaji, *Title J. Pharmacol. Exp. Ther.*, 2005, **314**, 244.

110. J. Liu, Y. Wang, Y. Sun, D. Marshall, S. Miao, G. Tonn, P. Anders, J. Tocker, H. L. Tang and J. Medina, *Bioorg. Med. Chem. Lett.*, 2009, **19**, 6840.

111. S. Ghosh, A. Elder, K. Carson, K. Sprott and S. Harrison, Millennium Pharmaceuticals, Inc., *PCT Pat. Appl.*, WO 2004/032848, 2004.

112. C. Kuhn, F. Feru, M. Bazin, M. Awad and S. Goldstein, Warner-Lambert, *EP Pat.*, 1413306, 2004.

113. W. Inman, J. Liu, J. Medina, S. Miao and H. Tang, Tularik Inc., *PCT Pat. Appl.*, WO 2005/007094, 2005.

114. M. Shichijo, A. Arimura, Y. Hirano, K. Yasui, N. Suzuki, M. Deguchi and W. M. Abraham, *Clin. Exp. Allergy*, 2009, **39**, 1404.

115. V. Angeli, D. Staumont, A.-S. Charbonnier, H. Hammad, P. Gosset, M. Pichavant, B. N. Lambrecht, M. Capron, D. Dombrowicz and F. Trottein, *J. Immunol.*, 2004, **172**, 3822.

116. I. Arai, N. Takano, Y. Hashimoto, N. Futaki, M. Sugimoto, N. Takahashi, T. Inoue and S. Nakaike, *Eur. J. Pharmacol.*, 2004, **505**, 229.

117. A. Arimura, K. Yasui, J. Kishino, F. Asanuma, H. Hasegawa, S. Kakudo, M. Ohtani and H. Arita, *J. Pharmacol. Exp. Ther.*, 2001, **298**, 411.

118. K. Torisu, K. Kobayashi, M. Iwahashi, Y. Nakai, T. Onoda, T. Nagase, I. Sugimoto, Y. Okada, R. Matsumoto, F. Nanbu, S. Ohuchida, H. Nakai and M. Toda, *Bioorg. Med. Chem. Lett.*, 2004, **14**, 4891.

119. W. M. Abraham, *Pulmonary Pharmacol. Ther.*, 2008, **21**, 743.

120. S. A. Boehme, E. P. Chen, K. Franz-Bacon, R. Sasik, L. J. Sprague, T. W. Ly, G. Hardiman and K. B. Bacon, *Int. Immunol.*, 2009, **21**, 1.

121. S. A. Boehme, K. Franz-Bacon, E. P. Chen, R. Sasik, L. J. Sprague, T. W. Ly, G. Hardiman and K. B. Bacon, *Int. Immunol.*, 2009, **21**, 81.

122. B. Ghafouri, K. Irander, J. Lindbom, C. Tagesson and M. Lindahl, *J. Proteome Res.*, 2006, **5**, 330.

123. M. Bryborn, C. Hallden, T. Saell and L. O. Cardell, *Allergy (Oxford, United Kingdom)*, **65**, 220.

124. N. Barnes, I. D. Pavord, A. Chuchalin, J. Bell, M. Hunter, T. Lewis, D. Parker, M. Payton, L. P. Collins, E. R. Pettipher, J. Steiner and C. M. Perkins, *Clin. Exp. Allergy*, 2011, in press.

125. L. Burgess, L. Anderson, C. Nugent, N. Klopfenstein, C. Eberhardt, L. Kass, C. Kass, S. Rojas-Caro, B. Bergstrom, S. Miller and S. Bell, in *Inflammation Research. Conference: 16th International Conference of the Inflammation Research Association*, Chantilly, VA United States, 2010.

126. C. Banfiel, J. Parnes, M. Emery, L. Ni, N. Zhang and P. Hodsman, in *2010 Annual Meeting of the American Academy of Allergy, Asthma and Immunology*, New Orleans, LA, USA, 2010.

127. http://investors.amgen.com/phoenix.zhtml?c = 61656&p = irol-presentations, "2011 Business Review Progress in R&D (PDF)", accessed December 19, 2011.

128. http://www1.actelion.com/en/scientists/development-pipeline/phase-2/setipiprant.page, accessed December 19, 2011.

Section 2
Kinases

CHAPTER 5

Dual Inhibition of Phosphodiesterase-4 and p38 MAP Kinase: A Strategy for Treatment of Chronic Inflammatory Diseases

WOLFGANG ALBRECHT[a] AND STEFAN LAUFER*[b]

[a] c-a-i-r biosciences GmbH, Paul-Ehrlich Str. 15, 72076 Tübingen, Germany, Email: w.albrecht@cair-biosciences.de; [b] Pharmazeutisches Institut, Eberhard-Karls-Universität Tübingen, Auf der Morgenstelle 8, 72076 Tübingen, Germany
*Email: stefan.laufer@uni-tuebingen.de

5.1 Introduction

The anti-TNFα therapies represent (one of) the most successful pharmacotherapeutic options for treatment of diseases associated with chronic inflammation. Currently, TNFα antagonists (infliximab, adalimumab, etanercept, certolizumab and golimumab) are approved and widely employed for the management of moderately to severely active rheumatoid arthritis (RA), ankylosing spondylitis (AS), Crohn's disease (CD), plaque psoriasis, psoriatic arthritis and juvenile idiopathic arthritis (JIA). More than two million patients worldwide have received treatment with either one of the anti-TNFα biologic agents. Without disregarding the significant improvement for treatment of these severe and

RSC Drug Discovery Series No. 26
Anti-Inflammatory Drug Discovery
Edited by Jeremy I Levin and Stefan Laufer
© The Royal Society of Chemistry 2012
Published by the Royal Society of Chemistry, www.rsc.org

debilitating diseases by biological TNFα-inhibitors, there remain limitations, such as (a) a still large percentage (approx. 30%) of patients who do not respond to the treatment with these drugs, (b) high costs and (c) the route of administration. Furthermore, for treatment of severe chronic inflammatory airway diseases (asthma and chronic obstructive pulmonary disease (COPD)), anti-cytokine therapy was not successful. Therefore, the development of orally available, small-molecule drugs that already interfere in the regulation of expression of inflammatory cytokines would represent a highly attractive alternative treatment option to the biological drugs.

The stimulation of cultivated monocytes with bacterial lipopolysaccharide (LPS) results in a fulminant production of the pro-inflammatory cytokines TNFα and IL-1β. Even before the first biological TNFα-inhibitor entered the market, researchers at SmithKline Beecham identified the mechanism of action of anti-inflammatory pyridinylimidazoles, which inhibited the cytokine release when cells were pre-incubated with these molecules prior to LPS-stimulation.[1] These compounds, initially denominated as cytokine-suppressive anti-inflammatory drugs (CSAIDs), were found to bind to a single 38 kDa protein kinase implicating that the expression of pro-inflammatory cytokines can be modulated by inhibition of this unique enzyme, the p38 MAP kinase (p38 MAPK). The concept of development of an orally available anti-cytokine small molecule that could be used for treatment of various inflammatory auto-immune diseases was adopted by essentially all large pharmaceutical companies and many biotech enterprises. In parallel to these development programs, results from basic research regarding the physiological role have become publicly available. The commercial access to the first prototype selective p38 MAPK inhibitor, SB203580, facilitated the experimental elucidation of various biological functions (using SB203580 as a search term, PubMed lists > 5000 publications).

Several reviews summarized the continuously increasing knowledge on the biology of p38 MAPK and its function in physiological and patho-physiological processes.[2–6] The purpose of the present article is not to add another review on the current status of p38 MAPK inhibitors as potential anti-inflammatory drugs but to suggest that p38 MAPK inhibition could represent a beneficial anti-inflammatory strategy if this mode of action is combined with the inhibition of Phosphodiesterase-4 (PDE4), a so far not considered dual mechanism.

5.2 p38 MAPK

Four isoforms of p38 MAPK are known, which are encoded by different genes. p38α MAPK is being ubiquitously expressed whereas the other isoforms, p38β, p38γ and p38δ MAPK are expressed in a tissue-specific manner. Several splice variants of p38α MAPK have been identified such as Mxi2, which exhibits a reduced binding to p38α MAPK substrates but binds to ERK-1 and -2 and regulates their nuclear import,[7] Exip, a member of the NF-κB pathway, and CSBP1, whose biological function has not been elucidated.

Most data are available for p38α and p38β MAPK, mainly because of the key function of the α-isoform in inflammation and other patho-physiological processes but also because SB203580 and most other inhibitors that are used as pharmacological tools inhibit both p38α and p38β with no or negligible affinity to p38γ/δ MAPK. Expression of p38γ, which has 63% amino acid identity with p38α, is largely restricted to skeletal muscle.[8] p38δ MAPK, with 61% amino acid identity with p38α, is mainly expressed in the testes, pancreas, small intestine and CD4$^+$ T-cells.[9] Lately, *in-vitro* studies have demonstrated that p38δ might be involved in keratinocyte differentiation and PKCδ-dependent keratinocyte apoptosis,[10] as well as the progression of neurodegenerative disorders referred to as tauopathies.[11] A very important patho-physiological function related to diabetes was identified using p38δ MAPK knockout mice.[12] These mice displayed improved glucose tolerance due to enhanced insulin exocytosis from pancreatic β cells. p38δ null mice are protected against high-fat-feeding-induced insulin resistance and oxidative stress-mediated β cell failure. Inhibition of protein kinase D (PKD) reverses enhanced insulin secretion from p38δ -deficient islets and glucose tolerance in p38δ null mice as well as their susceptibility to oxidative stress. In conclusion, the p38δ-PKD pathway integrates regulation of the insulin secretory capacity and survival of pancreatic β cells, pointing to a pivotal role for this pathway in the development of overt diabetes mellitus.[12] Unfortunately, a selective p38δ inhibitor is not available to further validate this potential target in an experimental disease model.

p38 MAPKs are activated by dual phosphorylation of the activation loop sequence Thr – Gly – Tyr. All p38 MAPK isoforms are activated by upstream MAPK kinases (canonical activation pathway, *cf.* Figure 5.1). The MAP2K kinases MKK3 and MKK6 specifically phosphorylate all four p38 MAPK isoforms whereas MKK4 activates p38α MAPK as well as the isoforms of c-Jun *N*-terminal kinase (JNK-1, -2 and -3). In addition, for p38α MAPK, three non-canonical mechanisms have been described, which finally induce an auto-phosphorylation. In the canonical pathway, several MAP3K are involved in the p38 MAPK activation (ASK1, DLK1, TAK1, TAO, TPL2, MLK3, MEKK3/4 and ZAK) and the known complexity of MAP3K regulation implies that p38 MAPKs may be activated as a consequence of many different types of cell stimulations.[2] In inflammation, the cytokine-receptor induced cell signalling, triggered by the TNF-receptor-associated family of E3 ubiquitous ligases (TRAF) and mediated through TAK1 or MEKK, may be of particular relevance.

As an example of the cross-talk between the p38 MAPK and the ERK-1/-2 pathway, the activation of protein phosphatase 2A is illustrated, which catalyzes the dephosphorylation of MKK-1/-2 and, as a consequence, the deactivation of ERK-1/2.

One auto-phosphorylation pathway, which might have some relevance in inflammation, has been identified in T-lymphocytes. Following activation of the T-cell receptor, p38α MAPK is being phosphorylated on Tyr323 by the tyrosine kinase ZAP70, which induces the auto-phosphorylation of the activity loop.[4] This mechanism appears to be critical for the normal function of Th1 T helper cells.

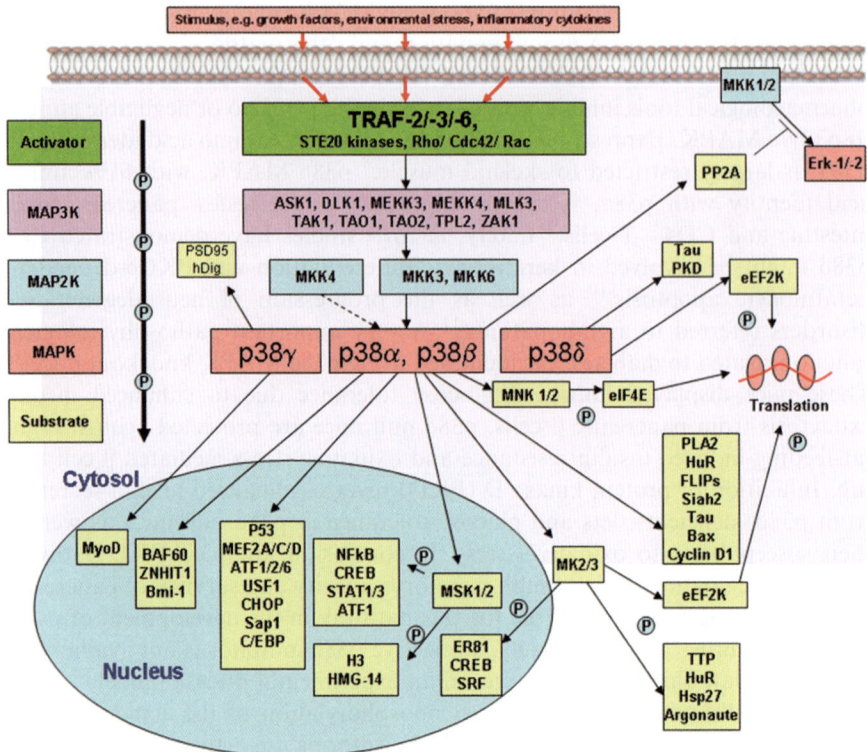

Figure 5.1 Illustration of the canonical activation pathway of p38 MAPK: environmental stimulus → activator → MAP3K → MAP2K → p38 MAPK. Downstream substrates of p38a MAPK include protein kinases such as MK2/MK3 or MSK 1/2, transcription factors like STAT 1/3 or cytosolic substrates like cytosolic phospholipase A_2 (cPLA$_2$).

As an example of the cross-talk between the p38 MAPK and the ERK-1/-2 pathway, the activation of protein phosphatase 2A is illustrated, which catalyzes the dephosphorylation of MKK-1/-2 and, as a consequence, the deactivation of ERK-1/2.

Two further auto-phosphorylation mechanisms have been described, which require the involvement of either TAB1 or Cdc7, but the exact physiological function is unclear and a relation to inflammation has not been mentioned.[2]

The identification of the various upstream MAP3Ks that mediate the activation of p38 MAPK implicate an important role of p38 MAPK in physiological and patho-physiological processes. Downstream, p38 MAPK isoforms phosphorylate a broad range of protein kinases, nuclear receptors and other cytoplasmatic proteins. MSK-1 and -2, two substrates of p38α MAPK, activate gene transcription by phosphorylation of several transcription factors (CREB,[13] ATF-1, the NF-kB subunit p65 (RelA) and STAT-1/-3). The p38α MAPK substrates MAPKAP-2 and -3 (MK2, MK3) phosphorylate the AU-rich element binding proteins tetratrisprolin and HuR, thereby controlling gene-expression at the post-transcriptional level.[14] The importance of MK-2 in inflammation was demonstrated with knockout mice that were insensitive to collagen-induced arthritis[15] and much less sensitive to oxazolone-induced skin inflammation.[16] Based on the hypothesis that inhibition of targets downstream of p38 MAPK would reduce the probability of clinical safety issues, a few companies initiated discovery programs but, based on publicly available data, no MK2-inhibitor is in an advanced stage of clinical development.

p38α MAPK phosphorylates the cytosolic phospholipase A$_2$ (cPLA$_2$), which results in an increased activity.[17] cPLA$_2$ liberates arachidonic acid from membrane phospholipids and triggers the formation of leukotrienes and prostaglandins, which in turn contribute to inflammatory pain and the maintenance of inflammation. Furthermore, the expression of cPLA$_2$ is also regulated *via* the p38 MAPK $\rightarrow$ MSK-1 $\rightarrow$ HuR signalling cascade.[18]

The cross-talk between p38 MAPK and the JNK or ERK-1/-2 pathway has been discussed as a potential reason or at least as a contributing factor to the failure of p38α MAPK inhibitors in clinical development. Activated p38α MAPK stimulates the interaction between protein phosphatase 2A (PP2A) and the [ERK-1/-2 – MKK-1/-2] complex leading to MKK-1/-2 dephosphorylation and, finally, to ERK-1/-2 deactivation.[19] This [p38 MAPK – ERK-1/-2] interaction plays an important role in the regulation of cell survival and apoptosis.

In IL-1β or sorbitol-stimulated epithelial cells and in LPS-stimulated macrophages, the inhibition of p38α MAPK by SB203580 resulted in a strong activation of JNK. Experimental data indicate that activated p38α MAPK phosphorylates TAB1, a subunit of the MAP3K kinase TAK1, which functions as a MAP3K upstream of p38 MAPKs and JNKs.[20]

The fact that so many companies underestimated the complexity of p38 MAPK biology may be surprising but it may be based on the hypothesis that inflammation, triggered by IL-1β and TNFα, is a result of a transient over-activity of p38α MAPK and an optimal inhibitor concentration, *i.e.* a clinically efficacious and safe dose, should modulate this patho-physiological mechanism without interfering with the homeostatic activity in non-inflamed tissue.

5.3   Binding Modes of p38 MAPK Inhibitors

SB203580 has been described as a p38α/β-selective MAPK inhibitor and, like most of the described inhibitors, it binds to the ATP pocket.[21] Analysis of pyridinylimidazole/p38α MAPK co-crystals revealed that the inhibitors exhibit a similar binding mode to that for ATP.[22] As the ATP binding site is highly conserved among the 518 human protein kinases, further structural features needs to be employed by ligands in order to achieve a reasonable kinase selectivity. Five distinct subareas within the ATP-binding site have been described, each differing in their chemical environments as well as in their local sequences (Figure 5.2).[23] The linker region between the N- and C-terminal domains ("hinge region"), surrounded by two lipophilic regions, serves as a highly significant anchor for the binding of ATP-site directed kinase inhibitors. The main sites of interaction for a number of investigated inhibitor molecules, which compete with the ATP co-substrate of p38, are a deep hydrophobic cavity and the amide proton of Met[109] (Figure 5.2). Therefore, a 4-fluorophenyl and a pyridin-4-yl group are considered as elemental for biological activity in this class of compounds. Moreover, the formation of a hydrogen bond with the side chain of Lys[53] and the absence of any negative interactions with this residue combined with a sterically and biologically favourable arrangement of its substituents in the 1-, 2-, 4- and 5-position have confirmed the efficacy of 'imidazole' for the pyridin-4-yl-/4-fluoro-phenyl pharmacophore. For SB203580, the inhibitor selectivity for p38 with respect to other kinases is provided by the 4-fluorophenyl substituent that fills the hydrophobic region I (also: hydrophobic back pocket) left unoccupied by ATP and, in the case of p38α and β, is more spacious than in the other p38 isoforms and related kinases. The selectivity of these compounds toward p38α has been attributed to this interaction, which is mediated by the presence of Thr[106] ("gatekeeper residue") in the ATP-binding site. About 20% of all human protein kinases share Thr as the gatekeeper residue.[24] In contrast, the second hydrophobic cavity below the linker region (hydrophobic region II, hydrophobic front region) is shallower and is more groove shaped, and is also left unoccupied by ATP and simple pyridin-4-yl imidazole inhibitors, thus making this hydrophobic groove a preferred objective for the development of novel potent and selective p38 inhibitors. In the p38 MAP kinase, the hydrophobic front region stretches above as well as beneath the plane in which the purine ring system of the bound ATP is situated. In these areas, the side chains of Val[30], Ala[40] and Leu[108] (above;

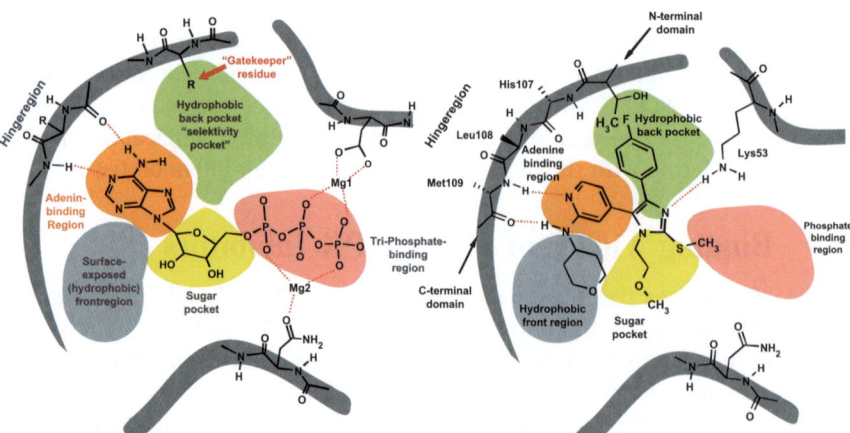

Figure 5.2   Binding mode of ATP (left drawing) and proposed binding mode of an exemplary inhibitor molecule to the ATP-binding site of p38α MAP kinase. The hinge part of the kinase polypeptide backbone is shown on the left. The hydrophobic back pocket and the surface-exposed front region are not utilized for interactions with the enzyme's native co-substrate ATP (reproduced from Andy Liedtke, PhD thesis, University of Tübingen, Germany).

numbering of p38α; more surface-exposed and thus accessible for different substituents) and of Ile[84], Met[109], Ala[157] and Leu[167] (beneath; deeper inside the ATP-cleft, thus not tolerating sterically demanding substituents) in each case constitute a hydrophobic surface. Possible substituents in the 2-position of the pyridine of SB-like inhibitors can principally orientate to the upper or the lower area depending on their conformation (a coplanar alignment of the 4-F-phenyl- and the pyridine ring is assumed). Basically such substituents can make interactions with both areas of the hydrophobic region II. The essential interactions of pyridinylimidazole inhibitors with the ATP-binding cleft are:

• hydrogen donor/acceptor functions within the hinge region (mainly gaining activity);
• space-filling lipophilic aryl residues,
• binding to the hydrophobic back pocket (mainly gaining selectivity);
• interactions with the hydrophobic front region (gaining both activity and selectivity).

Goldstein *et al.* described these motifs as "selectivity hotspots" and concluded that the clinical candidates Pamapimod, Talmapimod (= Scio 469) and VX-702, although belonging to different chemical classes, achieved their kinase selectivity by interactions with these regions.[24]

The clinical candidate of Boehringer Ingelheim, BIRB 0796 (Doramapimod) belongs to the class of diaryl urea p38 MAPK inhibitors, which bind spatially differently from the ATP pocket and with a different mode, when compared to ATP, the pyridinylimidazoles or the newer chemical classes of ATP-competitive inhibitors.[25] Kinases contain a highly conserved Asp-Phe-Gly (DFG-) motif in the active site with a conformation where the Phe-residue is buried in a hydrophobic pocket ("DFG-in conformation"). Binding of BIRB 0796 requires a conformational change of the Phe-residue to the "DFG-out conformation", which is incompatible with ATP-binding.[25] Kinetic binding and dissociation studies with BIRB 0796 and the pyridinylimidazole SK&F 86002 demonstrated that BIRB 0796 is a slow-binding compound that also dissociates very slowly. The association rate was some 500-fold slower and the dissociation rate some 1,000,000-fold slower when compared with those of SK&F 86002.[25]

As described above, activated p38α MAPK phosphorylates many different substrates and a common feature of all inhibitors is that the interaction with p38α MAPK results in an inhibition of all downstream mechanisms. CMPD1, a small molecule that is structurally unrelated to any of the inhibitors described above, has been described as a substrate-specific, ATP-non-competitive inhibitor of p38α MAPK.[26] The exact binding mode has not been elucidated but the CMPD1-p38α MAPK-interaction, which does not interfere with ATP-binding, inhibits the phosphorylation of MK2 but not of ATF-2. Except for its use as a pharmacological tool for mechanistic investigation, no data regarding the modulation of patho-physiological processes have been published.

5.4   Clinical Development of p38 MAPK Inhibitors

SB203580, although originally described as being selective for p38α/β MAPK, also interfered with other protein kinase at relevant, submicromolar concentrations.[27] Furthermore, the compound inhibited several cytochrome P450 (CYP) isoforms, which has been considered as being associated with the observed liver toxicity. Several clinical candidates, having the pyridinylimidazole scaffold with high potency, high selectivity and negligible CYP-interaction have been discovered and advanced to clinical trials. Published data are available for RWJ-67657, which was advanced to Phase I trials. In the human endotoxaemia model, the intravenous administration of LPS induces a rapid but only transient activation of the p38α MAPK and the ERK-1/-2 but not the JNK pathway.[28] In this model, single doses of 350 mg, 700 mg and 1400 mg RWJ-67657, administered 30 min. prior to LPS-infusion, resulted in a dose-dependent suppression of the pharmacodynamic markers TNFα, IL-6 and IL-8,[29] thus demonstrating an anti-inflammatory effect *in vivo*. Using a similar study design, the same effect was determined after single administration of 50 mg and 600 mg of BIRB-0796.[30] Clinical Phase II trials with BIRB-0796 were performed to investigate the efficacy in rheumatoid arthritis (doses between 5 and 30 mg b.i.d.), Crohn's disease (between 10 and 60 mg b.i.d.) and psoriasis (doses not disclosed). The results of the RA study have not been published but in the Crohn's disease[31] and the psoriasis study,[32] no treatment-related effects were observed.

As shown in Table 5.1, all p38 MAPK inhibitors, which were advanced to clinical Phase II trials, were tested for their efficacy in patients with RA and for most of the candidates the outcome of the respective clinical trial was the beginning of the termination of further development. During the 12-week treatment period, drug administration led to a rapid but only transient decrease of biomarkers of acute inflammation, such as C-reactive protein (CRP) or the soluble TNF-receptor. By week 12, these values returned to or near to baseline, which corresponded to the lack of clinical efficacy – irrespective of whether the drug was given with or without methotrexate (MTX) as concomitant basic therapy. Since the first comprehensive disclosure of clinical Phase II results obtained with Roche's p38 MAPK inhibitor pamapimod, similar results, observed with VX-702 and SCIO-469, have been published. The results of the clinical study with BIRB-796 in patients with rheumatoid arthritis have not been published, but in patients with Crohn's disease, the same transient decline of CRP was observed.[31] The molecular basis of this outcome is still unknown but several hypotheses have been discussed, including an adaptation of the signalling network to the specific blockade of the p38 MAPK pathway or the possibly underestimated relevance of anti-inflammatory features of activated p38 MAPK.[32,33] Irrespective of the real mechanisms that are triggered by p38α MAPK inhibition, the initial idea of a single drug/single target interaction as a straightforward anti-cytokine therapy represents – not for the first time – an over-simplification of apparently complex biological interactions. A recent perspective article summarized the outcomes of clinical Phase II studies performed with p38 MAPK inhibitors.[24] The authors concluded " … that the

Table 5.1 Chemical structures and development status of selected p38a MAPK inhibitors.

Substance	Chemical structure	Development status/ remark
RWJ-67657		Discontinued; no Phase II data reported; efficacy demonstrated in the human endotoxaemia pharmacodynamic model demonstrated[29]
AMG-548		Discontinued after Phase I; elevated liver enzymes reported
VX-745		Discontinued after a Phase II trial in RA; primary end-points were met but CNS adverse events expected based on non-clinical data
BIRB-0796 (Doramapimod)		Discontinued after Phase II; no efficacy in RA, CD and psoriasis; elevated liver enzymes after repeated administration
Pamapimod		Discontinued based on failure in Phase II trials in RA patients ($+/-$ co-medication with MTX)
SCIO-469 (Talmapimod)		Discontinued based on failure in Phase II trials in RA; outcome of studies in multiple myeloma unknown
VX-702		Discontinued based on failure in Phase II trials in RA patients ($+/-$ co-medication with MTX)

Table 5.1   (*Continued*)

Substance	Chemical structure	Development status/remark
PH-797804		Phase II; RA study completed but results not disclosed; clinical trials for treatment of COPD and neuropathic pain ongoing
BMS-582949		Phase II; clinical trials for treatment of RA, plaque psoriasis and atherosclerosis completed; results not reported
Losmapimod (GW-856553)		Phase II; clinical trials for treatment of RA, COPD and depression completed; results not reported
Dilmapimod (SB-681323)		Phase II; clinical trials for treatment of RA, COPD and neuropathic pain completed; results not reported

specific inhibition of p38a MAPK is unlikely to be a successful strategy toward treating chronic inflammatory disorders".[24] However, a conclusion, that p38 MAPK may not represent a druggable target for treatment of inflammatory diseases may also be a fallacy, because there is enough evidence that p38 MAPK plays an important role in RA but also in inflammatory bowel disease (IBD) or chronic obstructive pulmonary disease (COPD).

5.5   Evidences for p38 MAPK as Key Enzyme in Chronic Inflammation

Further to a general overview of the signalling pathways of all four p38 MAPK isoforms, Schett *et al.* discussed previous findings[34,35] regarding the relevance of these enzymes in rheumatoid arthritis.[36] Synovial tissue samples from patients with rheumatoid arthritis and osteoarthritis were analyzed by immuno- and histochemical methods. In synovial membranes from RA patients, not in those from OA patients, p38 MAPK was strongly activated. Active p38 MAPK was predominantly detected in the endothelium of synovial microvessels, in the cells of the lining layer as well as in macrophages and synovial fibroblasts.[36] Among the four different isoforms of p38 MAPK, the α- and γ-isoforms are the ones most abundantly expressed in the inflamed synovial tissue of patients with RA. Based on these findings, p38 MAPK is considered as a pivotal enzyme involved in the regulation of cytokine expression, which trigger and maintain synovial inflammation, cartilage damage and inflammatory bone loss.

Many p38 MAPK inhibitors demonstrated efficacy in the classical experimental models of adjuvant arthritis (AA) or collagen-induced arthritis (CIA). Furthermore, to investigate whether activation of p38 MAPK downstream of TNFα is a truly relevant factor for the induction of inflammatory – and especially destructive – joint disease *in vivo*, the efficacy of p38 MAPK inhibitors was tested in human TNF-transgenic mice.[37] This animal model of arthritis is based on stable transgenic over-expression of TNFα leading to inflammatory arthritis. Both tested specific inhibitors of p38 MAPK decreased the severity of synovial inflammation and cartilage damage and these effects were associated with reduced expression of pro-inflammatory cytokines, such as IL-1β, in the synovial membrane and a decrease of proteoglycan loss from articular cartilage, as well as an impaired differentiation of osteoclasts in inflamed synovial tissue.[37]

Both forms of IBD, ulcerative colitis (UC) and Crohn's disease (CD), are characterized by chronic relapsing inflammation of the gastro-intestinal tract with TNFα as a central mediator of initiation and amplification of mucosal inflammation.[38,39] A systematic analysis of tissue samples demonstrated a pronounced activity of p38α MAPK, JNK-1 and -2 as well as of ERK-1/-2.[40] Apparently, the activation of p38α MAPK was substantially higher when compared to that of the other MAP kinases. Furthermore, immunofluorescence analysis showed that p38α MAPK was strongly expressed by lamina propria macrophages and neutrophils.[40] Based on these findings, it has been concluded that p38α MAPK is of key importance for the maintenance of inflammation in inflamed tissue samples from both CD and UC patients.[40] Identical or essentially similar results confirmed these data[41–44] but the opposite, *i.e.* that the p38 MAPK pathway is not a major player in these diseases, has also been reported.[45] These controversial conclusions can be a result of the usually small populations that have been selected for these studies. In those studies, which demonstrated a pronounced activation of p38α MAPK, quantification of phosphorylated p38 MAPK demonstrated a substantial inter-individual variability within the analyzed population,[40,41] which underlines the complexity of this disease and reveals that p38 MAPK plays a key role in the regulation of inflammation in a significant subset but not in all patients.

Except for one study with BIRB-0796, no p38 MAPK selective inhibitor was tested in a clinical Phase II trial. In a small, open-labelled, Phase II trial, Semapimod (CNI-1493), earlier described as a dual p38 MAPK/JNK-inhibitor, showed good efficacy in Crohn's disease.[46] However, the beneficial effect of this compound, a guanyl hydrazone, which has to be administered intravenously, appears to be mainly associated with an inhibition of c-Raf and, downstream, of MEK.[47]

Over 50 cytokines have been identified in asthma and COPD, and Barnes, being aware that their role in the patho-physiology of these airway diseases has not been fully elucidated, depicted the complex networks and classified the cytokines into lymphokines (cytokines that are secreted by T-cells and regulate immune responses), pro-inflammatory cytokines (cytokines that amplify and perpetuate the inflammatory process), growth factors (cytokines that promote cell survival and result in structural changes in the airways), chemokines (cytokines that are chemotactic for inflammatory cells) and anti-inflammatory cytokines (cytokines that negatively modulate the inflammatory response).[48]

The pro-inflammatory cytokines TNFα, IL-1β and IL-6 are found in increased amounts in the sputum and BAL fluid in individuals with asthma and COPD and amplify inflammation, in part through the activation of NF-κB, which leads to the increased expression of multiple inflammatory genes. TNFα acts directly on human airway smooth muscle to increase the contractile response to spasmogens and may thus play a role in the AHR of asthma. TNFα is expressed in various cells in asthmatic airways. Treatment with biological TNFα improved lung function in a small study of patients with refractory asthma and reduced exacerbations in patients with moderate asthma. COPD patients, in contrast, did not benefit from treatment with infliximab when administrated at the doses that are effective in individuals with rheumatoid arthritis.[48] These findings suggest that in COPD, the role of TNFα is less important when compared to RA or CD.

Based on the failure of several specific anti-cytokine drugs, small molecules that interfere in cell signalling thereby modulating the release of several cytokines appear an attractive strategy in airway inflammation. Inhibitors of p38α MAPK showed a significant effect in experimental models airway inflammation[49–53] but *in-vitro* experiments, which addressed the direct effect of the p38 MAPK inhibitors SB-239063 and SD-282 on the release of different cytokines, were ambiguous. In LPS-stimulated macrophages from lung tissue and monocytes that were isolated from peripheral blood, p38 MAPK inhibitors significantly reduced the expression of TNFα. In contrast, the release of IL-8 was not modulated.[54] IL-8 appears to play a key role in COPD patho-physiology; the levels of this neutrophil chemoattractant are raised in the lungs of COPD patients[55] and are associated with the rate of disease progression.[56] The suppression of only a subset of inflammatory genes was observed when monocyte-derived macrophages and lung alveolar macrophages from COPD patients were stimulated with LPS in the presence of the p38 MAPK inhibitor SB-706504. In these experiments, the expression of several genes, including that of IL-8, remained resistant to p38 MAPK inhibition.[57]

Western blot analysis of both alveolar macrophages and the alveolar walls of smokers with COPD demonstrated an increased phospho-p38 expression in alveolar macrophages from COPD patients, in the absence of any change in total p38a levels.[58] Furthermore, phospho-p38 expression was related to the degree of lung function impairment and to the number of CD8 T-lymphocytes infiltrating the alveolar walls.[58] Thus, as described above for the analysis of biopsies from IBD patients, there was a significant inter-individual variability of the extent of p38α MAP activation.

5.6   New Indications for p38 MAPK Inhibitors

For the time being, no clinical results on the efficacy of p38 MAPK inhibitors in asthma or COPD have been reported but http://clinicaltrials.gov/, the official website of the FDA, where most sponsors announce planned and ongoing clinical studies, lists Phase II studies to investigate the efficacy of the selective p38 MAPK inhibitors Losmapimod (GW-856553) and Dilmapimod (SB-681323) in COPD patients. In a preliminary 5-way cross-over study, COPD patients received 7.5 mg and 25 mg SB-681323, 10 mg and 30 mg prednisolone or placebo. Serial blood samples were withdrawn and stimulated with either sorbitol ($\rightarrow$ hyperosmotic stress) or LPS.[59] After administration of either SB-681323 or prednisolone, TNFα-production in LPS-stimulated blood samples was significantly inhibited, with a stronger effect observed with prednisolone [59]. Furthermore, the phosphorylation of hsp27, a p38α MAPK specific substrate, was significantly inhibited after administration of SB-681323 while only a minor effect was observed with prednisolone.

In addition to these three major chronic inflammatory diseases, RA, IBD and asthma/COPD, p38α MAPK inhibitors may also be effective for treatment of neuropathic pain[60] or cardiovascular diseases.[61]

5.7   Therapeutic Potential of Drug Combinations with p38 MAPK Inhibitors

The failure of p38α MAPK inhibitors in clinical RA trials including the apparently mechanism-based transient reduction of biomarkers of acute inflammation raised the question whether p38α MAPK is a valid target for treatment of chronic inflammatory diseases. The central function of p38 MAPK with many different substrates and its influence on upstream MAP3Ks provide the basis for the hypothesis that inhibition of the p38 MAPK pathway results in an activation of alternative pathways.[32,33] To investigate the efficacy in RA, the p38 MAPK inhibitors were administered either as monotherapy or on top of a basic therapy with MTX, a standard therapy that is also used in combination with biological TNFα-inhibitors. Neither *in-vitro* studies nor results from experimental disease models suggest any beneficial effect of this combination.

p38 MAPK regulates TNFα-synthesis at the level of transcription (*via* MSK-1/-2[62]) and post-transcription (*via* MK2/MK3-mediated phosphorylation of tristertaprolin[63]). *In-vitro* experiments confirmed that inhibition of cytokine expression on the protein level is generally more pronounced than on mRNA expression. To efficiently block mRNA expression in LPS- or cytokine-stimulated cells, the combination of a p38α MAPK inhibitor with either an ERK-inhibitor[64] or a NF-κB-interacting substance[65] was proposed. Despite the fact that these experiments are far away from being in clinical practice, they underline the potential benefit of p38 MAPK inhibition as an add-on therapy.

As a more relevant application, the potentially synergistic effect of a combination of corticosteroids with p38 MAPK inhibitors was described.[57,66,67] Armstrong *et al.* used cultured alveolar macrophages that were isolated from bronchoalveolar lavage fluids of COPD patients.[65] The cells were cultured in the presence of different concentrations of BIRB-0796, dexamethasone and combinations thereof, stimulated with LPS, and cytokines were determined in the culture supernatants. At the lowest concentration (1 nM), BIRB-0796 mediated a synergistic cytokine suppressing potency. Corticoids and p38α MAPK inhibitors inhibit gene expression of pro-inflammatory cytokines, albeit by different mechanisms. Corticosteroids do not interfere with the p38α MAPK signalling cascade (monitored by quantification of phosphorylated hsp27), which explains the additional benefit of p38α MAPK inhibition.

An alternative approach to the development of drug combinations, which may require complex clinical dose-finding studies, is the development of drugs that target more than one enzyme or receptor, an approach that is denominated as polypharmacology. This strategy opposes the philosophy of rational drug design or, more specifically, the "one gene, one drug, one disease" paradigm, which has been challenged by several authors because it neglects the complex biology of patho-physiological processes.[68–70] This strategy, which has been considered as state-of-the-art in the pharmaceutical community, was also made responsible for the high rate of drugs failing in late-stage clinical development over the past decade.

5.8 Discovery and Development of PDE4 Inhibitors

A high attrition rate was also experienced with the class of PDE4 inhibitors. Like p38α MAPK inhibitors, the development of PDE4 inhibitors appeared to be straightforward as this target enzyme is primarily expressed in immune cells and, as an outcome of its inhibition, the intra-cellular concentration of the second messenger cyclic adenosine phosphate (cAMP) increased and induces anti-inflammatory signalling. Many PDE4 inhibitors were advanced into clinical development but, as of today, only one candidate, Roflumilast, has been approved. The drug is being marketed under the brand names Daxas® (EU) and Daliresp® (US) but may be prescribed only to a subpopulation of COPD patients, *i.e.* patients with severe COPD associated with chronic bronchitis and a history of exacerbations. The biology of PDE4 and the development of PDE4 inhibitors have been summarized in many review articles and only some characteristics that are considered relevant for the purpose of this article are briefly presented.

cAMP plays an important role in the immune system. It binds to the regulatory subunit of the cAMP-dependent protein kinase A (PKA), which modulates, by phosphorylation, the activity of target proteins and, as a consequence thereof, cellular function. The intra-cellular cAMP concentration varies between < 1 and 10 μM and is determined by the activities of adenylate cyclase (conversion of ATP into cAMP) and phosphodiesterase (conversion of cAMP into adenosine 5′-monophosphate). To modulate the cAMP concentration by pharmacologic intervention, PDE inhibition represents a straightforward strategy. There are 11 different families of PDE comprising 21 different gene products.[71] However, it has been estimated that, based on alternative transcriptional start sites and alternative splicing, the 21 genes might be translated into > 100 different mRNA products and proteins.[69] Any single cell type can express several different PDEs but each of these isoforms most likely plays a different and specific physiological role.[69]

In immune cells, PDE4 represents the most important cAMP-degrading enzyme and its involvement in pathological processes suggested a great potential for pharmacologic intervention in inflammatory diseases.[72] Since the discovery of rolipram as a selective PDE4 inhibitor, the complex biology of PDE4 enzymes and the therapeutic potential of its inhibitors has been extensively investigated. There are four PDE4 families (PDE4A, PDE4B, PDE4C and PDE4D) and the number of identified splice variants increased in the last years from 16[73] to over 20.[74] The expression of different isoforms in different cells, their protein structures and conformations, and the intra-cellular localization including association with binding proteins has been described in many original publications, review articles and book chapters.[75–77]

Rolipram was developed as an anti-depressant and, once in systemic circulation, crosses the blood-brain barrier and achieves high concentrations in the central nervous system. Rolipram causes nausea and emesis as side-effects and its clinical development was discontinued because of the absence of an acceptable therapeutic window. PDE4 inhibitors of the second generation were developed for treatment of inflammatory diseases and the primary objective was to increase the therapeutic window. Houslay *et al.*[70] and Spina[72] reviewed the status of clinical development of PDE4 inhibitors. At that time, Cilomilast and Roflumilast have been the only candidates that were advanced to clinical Phase III studies. As of today, the development of Cilomilast was discontinued while Roflumilast has been approved by the FDA and the European Medical Agency.

Celgene's orally administered PDE4 inhibitor Apremilast (lab code: CC-10004) is currently being tested in psoriatic arthritis (clinical Phase III) as well as in several other indications such as psoriasis, severe acne, atopic dermatitis, rheumatoid arthritis, osteoarthritis, ankylosing spondylitis, gout or uveitis.[78]

The efficacy and safety of another, highly potent, clinical candidate, GSK-256,066, is being tested in asthma and COPD (administration by inhalation) and in allergic rhinitis (intra-nasal administration).

Although results of non-clinical disease models suggest that PDE4 inhibitors may be effective in several indications, airway inflammation, in particular chronic obstructive airway disease (COPD), has been the most intensely investigated indication. This preference can be explained by the long experience with theophylline, an anti-asthmatic drug, and its, although very weak, PDEs inhibition potential as well as the findings that PDE4 is not only localized in inflammatory cell types that are involved in airway inflammation but is also responsible for cAMP-homeostasis and its effect on the contractibility of airway smooth muscles.[71,79]

5.9 Dual Inhibitors of p38 MAPK and PDE4

As an outcome of a search for new adenosine receptor antagonists, CGH2466 was discovered. This compound, however, is not only an antagonist of the adenosine receptor subtypes A_1, A_{2b} and A_3 but is also an inhibitor of p38α/β MAPK and of PDE4A, PDE4B and PDE4D (with highest potence against PDE4D). The compound inhibits p38α MAPK with an IC_{50} of 190 nM compared to 150 nM, which was determined for SB203580 in the same assay. In molecular modelling studies, CGH2466 was docked into p38 MAPK using the coordinates of ATP. According to this simulation, the overall orientation of CGH2466 was similar to that of the pyridinyl imidazoles. A hydrogen bond between the aminothiazole-NH_2 group and the carbonyl of Asp^{168} induced a twist of the molecule, which allowed for binding of the bulky dichlorophenyl group into the hydrophobic pocket. Similar docking experiments with CGH2466 into the active site of PDE4 resulted in a reasonable fit, which is based on the binding of the dichlorophenyl group into a hydrophobic region and a hydrogen bond between the pyridine-N and Thr^{345}/His^{387}, which is mediated *via* a water molecule.

In *in-vitro* experiments and experimental models of airway inflammation, the potency and efficacy of CHG2466 was superior when compared to SB203580, the PDE4 inhibitor Cilomilast or a selective adenosine receptor antagonist.

The pyridinylimidazole ML3403 has been described as a highly potent and selective inhibitor of p38α MAPK.[80,81] The 2-methylsulfanyl group of ML3403 and most of the structurally related derivatives are metabolically labile and extensively oxidized by several cytochrome P450 isoforms.[82] Further pharmacological characterization led to the discovery that sulfoxidation induced the affinity to PDE4-isoenzymes – without preference of any of the four isoforms. In U937 extracts, PDE4-activity was inhibited with an IC_{50} of $< 1 \mu M$. The configuration of the chiral sulfoxide group had no significant impact on p38 MAP kinase inhibition but influenced the PDE4 inhibition potency by a factor between 2 and 5.

These compounds showed no affinity to adenosine receptors and therefore can be denominated as dual inhibitors of p38α MAPK and PDE4. First results of experimental endotoxaemia (LPS-induced TNFα-release in the mouse and rat) and in experimental models of arthritis demonstrated a very high efficacy after oral administration. At identical doses, orally administered 30 min. prior to LPS-challenge, the inhibition of TNFα release was substantially more pronounced when compared with both tested reference drugs, Roflumilast and SB203580. This result cannot be explained by an additive effect of p38 MAPK and PDE4 inhibition because, on a molecular level, the potency of the lead dual inhibitor is too weak. Therefore, similar to what has been described for the combination of p38 MAPK inhibitors with corticosteroids, a synergistic effect is assumed.

5.10   Conclusion

Both inhibitors of p38 MAPK and PDE have originally been considered as promising drugs for treatment of inflammatory disorders. Based on the lack of efficacy in case of the p38 MAPK inhibitors, and an unfavourable ratio between clinical efficacy and side-effects in the case of the PDE4-inhibitors, both therapies did not meet the high expectations. Roflumilast has been approved for treatment of COPD but, due to its restricted use in a COPD sup-population, a high market penetration can no longer be expected.

The preliminary conclusion that inhibition of p38 MAPK is not a valid strategy in chronic inflammation is based on the results of clinical Phase II studies, in which the p38 MAPK inhibitor was administered either alone or in combination with MTX. Surprisingly, except for a very few *in-vitro* studies, the potential of p38 MAPK inhibition, when combined with a second mode of action, was never systematically investigated.

2-Methylsulfanylimidazoles have been well characterized as very potent p38α/β MAPK inhibitors with, based on the absence of interactions with cytochrome P450, low hepatotoxic potential. The 2-methylsulfanyl group of most of the prepared substances is metabolically unstable and the conversion into the corresponding 2-methylsulfinyl derivative resulted in metabolites that still inhibit p38 MAPK but also PDE4. First results regarding the anti-inflammatory effect of these compounds revealed an unexpectedly high potency *in vivo*. Further investigations are required to elucidate this potentially synergistic anti-inflammatory effect of dual p38α MAPK/PDE4-inhibition.

References

1. J. C. Lee, J. T. Laydon, P. C. McDonnell, T. F. Gallagher, S. Kumar, D. Green, D. McNulty, M. J. Blumenthal, J. R. Keys, S. W. Landvatter, J. E. Strickler, M. M. McLaughlin, I. R. Siemens, S. M. Fisher, G. P. Livi, J. R. White, J. L. Adams and P. R. Young, *Nature*, 1994, **372**, 739–746.
2. A. Cuadrado and A. R. Nebreda, *Biochem. J.*, 2010, **429**, 403–417.
3. A. Cuenda and S. Rousseau, *Biochim. Biophys. Acta*, 2007, **1773**, 1358–1375.
4. J. D. Ashwell, *Nat. Rev. Immunol.*, 2006, **6**, 532–540.
5. L. R. Coulthard, D. E. White, D. L. Jones, M. F. McDermott and S. A. Burchill, *Trends Mol. Med.*, 2009, **15**(8), 369–379.
6. J. Zhang, B. Shen and A. Lin, *Trends Pharmacol. Sci.*, 2007, **28**(6), 286–295.
7. B. Casar, V. Sanz-Moreno, M. N. Yazicioglu, J. Rodriguez, M. T. Berciano, M. Lafarga, M. H. Cobb and P. Crespo, *EMBO J.*, 2007, **26**, 635–646.

8. Z. Li, Y. Jiang, R. J. Ulevitch and J. Han, *Biochem. Biophys. Res. Commun.*, 1996, **228**(2), 334–340.

9. A. Risco and A. Cuenda, *J. Signal. Transduct.*, 2012, **2012**, 520289.

10. T. Efimova, A. M. Broome and R. L. Eckert, *Mol. Cell. Biol.*, 2004, **24**, 8167–8183.

11. C. Feijoo, D. G. Campbell, R. Jakes, M. Goedert and A. Cuenda, *J. Cell Sci.*, 2005, **118**, 397–408.

12. G. Sumara, I. Formentini, S. Collins, I. Sumara, R. Windak, B. Bodenmiller, R. Ramracheya, D. Caille, H. Jiang, KA. Platt, P. Meda, R. Aebersold, P. Rorsman and R. Ricci, *Cell*, 2009, **136**(2), 235–248.

13. J. A. Gustin, R. Pincheira, L. D. Mayo, O. N. Ozes, K. M. Kessler, M. R. Baerwald, C. K. Korgaonkar and D. B. Donner, *Am. J. Physiol. Cell Physiol.*, 2002, **286**(3), C547–C555.

14. N. Ronkina, M. B. Menon, J. Schwermann, C. Tiedje, E. Hitti, A. Kotlyarov and M. Gaestel, *Biochem. Pharmacol.*, 2010, **80**(12), 1915–1920.

15. M. Hegen, M. Gaestel, C. L. Nickerson-Nutter, L. L. Lin and J. B. Telliez, *J. Immunol.*, 2006, **177**(3), 1913–1917.

16. A. T. Funding, C. Johansen, M. Gaestel, B. M. Bibby, L. L. Lilleholt, K. Kragballe and L. Iversen, *J. Invest Dermatol.*, 2009, **129**(4), 891–898.

17. J. N. Walters, J. S. Bickford, D. E. Beachy, K. J. Newsom, J. D. Herlihy, M. V. Peck, X. Qiu and H. S. Nick, *Cell Signal.*, 2011, **23**(12), 1944–1951.

18. W. N. Lin, C. C. Lin, H. Y. Cheng and C. M. Yang, *Br. J. Pharmacol.*, 2011, **163**(8), 1691–1706.

19. M. R. Junttila, S. P. Li and J. Westermarck, *FASEB J.*, 2008, **22**(4), 954–965.

20. P. C. Cheung, D. G. Campbell, A. R. Nebreda and P. Cohen, *EMBO J.*, 2003, **22**(21), 5793–5805.

21. P. R. Young, M. M. McLaughlin, S. Kumar, S. Kassis, M. L. Doyle, D. McNulty, T. F. Gallagher, S. Fisher, P. C. McDonnell, S. A. Carr, M. J. Huddleston, G. Seibel, T. G. Porter, G. P. Livi, J. L. Adams and J. C. Lee, *J. Biol. Chem.*, 1997, **272**(18), 12116–12121.

22. L. Tong, S. Pav, D. M. White, S. Rogers, K. M. Crane, C. L. Cywin, M. L. Brown and C. A. Pargellis, *Nat Struct. Biol.*, 1997, **4**(4), 311–316.

23. S. A. Laufer, D. R. Hauser, D. M. Domeyer, K. Kinkel and A. J. Liedtke, *J. Med. Chem.*, 2008, **51**(14), 4122–4149.

24. D. M. Goldstein, A. Kuglstatter, Y. Lou and M. J. Soth, *J Med. Chem.*, 2010, **53**(6), 2345–2353.

25. C. Pargellis, L. Tong, L. Churchill, P. F. Cirillo, T. Gilmore, A. G. Graham, P. M. Grob, E. R. Hickey, N. Moss, S. Pav and J. Regan, *Nat. Struct. Biol.*, 2002, **9**(4), 268–272.

26. W. Davidson, L. Frego, G. W. Peet, R. R. Kroe, M. E. Labadia, S. M. Lukas, R. J. Snow, S. Jakes, C. A. Grygon, C. Pargellis and B. G. Werneburg, *Biochemistry*, 2004, **43**(37), 11658–11671.

27. M. W. Karaman, S. Herrgard, D. K. Treiber, P. Gallant, C. E. Atteridge, B. T. Campbell, K. W. Chan, P. Ciceri, M. I. Davis, P. T. Edeen, R. Faraoni, M. Floyd, J. P. Hunt, D. J. Lockhart, Z. V. Milanov, M. J.

Morrison, G. Pallares, H. K. Patel, S. Pritchard, L. M. Wodicka and P. P. Zarrinkar, *Nat. Biotechnol.*, 2008, **26**(1), 127–132.

28. B. van den Blink, J. Branger, S. Weijer, S. H. Deventer, T. van der Poll and M. P. Peppelenbosch, *Mol. Med.*, 2001, **7**(11), 755–760.

29. J. W. Fijen, J. G. Zijlstra, P. De Boer, R. Spanjersberg, J. W. Tervaert, T. S. Van Der Werf, J. J. Ligtenberg and J. E. Tulleken, *Clin. Exp. Immunol.*, 2001, **124**(1), 16–20.

30. J. Branger, B. van den Blink, S. Weijer, J. Madwed, C. L. Bos, A. Gupta, C. L. Yong, S. H. Polmar, D. P. Olszyna, C. E. Hack, S. J. van Deventer, M. P. Peppelenbosch and T. van der Poll, *J. Immunol.*, 2002, **168**(8), 4070–4077.

31. S. Schreiber, B. Feagan, G. D'Haens, J. F. Colombel, K. Geboes, M. Yurcov, V. Isakov, O. Golovenko, C. N. Bernstein, D. Ludwig, T. Winter, U. Meier, C. Yong and J. Steffgen, BIRB 796 Study Group, *Clin. Gastroenterol. Hepatol.*, 2006, **4**(3), 325–334.

32. M. C. Genovese, *Arthritis Rheum.*, 2009, **60**(2), 317–320.

33. S. E. Sweeney, *Nat. Rev. Rheumatol.*, 2009, **5**, 475–477.

34. G. Schett, M. Tohidast-Akrad, J. S. Smolen, B. Jahn-Schmid, C. W. Steiner, P. Bitzan, P. Zenz, K. Redlich, Q. Xu and G. Steiner, *Arthritis Rheum.*, 2000, **43**(11), 2501–2512.

35. A. Korb, M. Tohidast-Akrad, E. Cetin, R. Axmann, J. Smolen and G. Schett, *Arthritis Rheum.*, 2006, **54**(9), 2745–2756.

36. G. Schett, J. Zwerina and G. Firestein, *Ann. Rheum. Dis.*, 2008, **67**, 909–916.

37. J. Zwerina, S. Hayer, K. Redlich, K. Bobacz, G. Kollias, J. S. Smolen and G. Schett, *Arthritis Rheum.*, 2006, **54**(2), 463–472.

38. K. A. Papadakis and S. R. Targan, *Annu. Rev. Med.*, 2000, **51**, 289–298.

39. S. J. van Deventer, *Aliment. Pharmacol. Ther.*, 1999, **13**, 3–8.

40. G. H. Waetzig, D. Seegert, P. Rosenstiel, S. Nikolaus and S. Schreiber, *J. Immunol.*, 2001, **168**(10), 5342–5351.

41. G. Docena, L. Rovedatti, L. Kruidenier, A. Fanning, N. A. Leakey, C. H. Knowles, K. Lee, F. Shanahan, K. Nally, P. G. McLean, A. Di Sabatino and T. T. MacDonald, *Clin. Exp. Immunol.*, 2010, **162**(1), 108–115.

42. X. Zhao, B. Kang, C. Lu, S. Liu, H. Wang, X. Yang, Y. Chen, B. Jiang, J. Zhang, Y. Lu and F. Zhi, *J. Proteome Res.*, 2011, **10**(5), 2216–2225.

43. Y. J. Feng and Y. Y. Li, *Dig. Dis.*, 2011, **12**(5), 327–332.

44. F. Scaldaferri, M. Sans, S. Vetrano, C. Correale, V. Arena, N. Pagano, G. Rando, F. Romeo, A. E. Potenza, A. Repici, A. Malesci and S. Danese, *Eur. J. Immunol.*, 2009, **39**(1), 290–300.

45. G. Malamut, C. Cabane, L. Dubuquoy, M. Malapel, B. Derijard, J. Gay, C. Tamboli, J. F. Colombel and P. Desreumaux, *Dig. Dis. Sci.*, 2006, **51**, 1443–1453.

46. D. Hommes, B. van den Blink, T. Plasse, J. Bartelsman, C. Xu, B. Macpherson, G. Tytgat, M. Peppelenbosch and S. Van Deventer, *Gastroenterology*, 2002, **122**(1), 7–14.

47. M. Löwenberg, A. Verhaar, B. van den Blink, F. ten Kate, S. van Deventer, M. Peppelenbosch and D. Hommes, *J Immunol.*, 2005, **175**(4), 2293–2300.
48. P. J. Barnes, *J. Clin. Invest.*, 2008, **118**, 3546–3556.
49. E. B. Haddad, M. Birrell, K. McCluskie, A. Ling, S. E. Webber, M. L. Foster and M. G. Belvisi, *Br. J. Pharmacol.*, 2001, **132**(8), 1715–1724.
50. M. A. Birrell, S. Wong, K. McCluskie, M. C. Catley, E. L. Hardaker, S. Haj-Yahia and M. G. Belvisi, *J. Pharmacol. Exp. Ther.*, 2006, **316**(3), 1318–1327.
51. P. Nath, S. Y. Leung, A. Williams, A. Noble, S. D. Chakravarty, G. R. Luedtke, S. Medicherla, L. S. Higgins, A. Protter and K. F. Chung, *Eur. J. Pharmacol.*, 2006, **544**(1–3), 160–167.
52. J. Y. Ma, S. Medicherla, I. Kerr, R. Mangadu, A. A. Protter and L. S. Higgins, *J. Asthma Allergy.*, 2008, **1**, 31–44.
53. L. Munoz, E. E. Ramsay, M. Manetsch, Q. Ge, C. Peifer, S. Laufer and A. J. Ammit, *Eur. J. Pharmacol.*, 2010, **635**(1–3), 212–218.
54. S. J. Smith, P. S. Fenwick, A. G. Nicholson, F. Kirschenbaum, T. K. Finney-Hayward, L. S. Higgins, M. A. Giembycz, P. J. Barnes and L. E. Donnelly, *Br. J. Pharmacol.*, 2006, **149**, 393–404.
55. V. M. Keatings, P. D. Collins, D. M. Scott and P. J. Barnes, *Am. J. Respir. Crit. Care Med.*, 1996, **153**, 530–534.
56. T. M. Wilkinson, I. S. Patel, M. Wilks, G. C. Donaldson and J. A. Wedzicha, *Am. J. Respir. Crit. Care Med.*, 2003, **167**, 1090–1095.
57. L. M. Kent, L. J. Smyth, J. Plumb, C. L. Clayton, S. M. Fox, D. W. Ray, S. N. Farrow and D. Singh, *J. Pharmacol. Exp. Ther.*, 2009, **328**(2), 458–468.
58. T. Renda, S. Baraldo, G. Pelaia, E. Bazzan, G. Turato, A. Papi, P. Maestrelli, R. Maselli, A. Vatrella, L. M. Fabbri, R. Zuin, S. A. Marsico and M. Saetta, *Eur. Respir. J.*, 2008, **31**(1), 62–69.
59. P. Bhavsar, N. Khorasani, M. Hew, M. Johnson and K. F. Chung, *Eur. Respir. J.*, 2010, **35**(4), 750–756.
60. P. Anand, R. Shenoy, J. E. Palmer, A. J. Baines, R. Y. Lai, J. Robertson, N. Bird, T. Ostenfeld and B. A. Chizh, *Eur. J. Pain*, 2011, **15**(10), 1040–1048.
61. L. Sarov-Blat, J. M. Morgan, P. Fernandez, R. James, Z. Fang, M. R. Hurle, C. Baidoo, R. N. Willette, J. J. Lepore, S. E. Jensen and D. L. Sprecher, *Arterioscler. Thromb. Vasc. Biol.*, 2010, **30**(11), 2256–2263.
62. J. S. Arthur, *Front. Biosci.*, 2008, **13**, 5866–5879.
63. N. Ronkina, A. Kotlyarov, O. Dittrich-Breiholz, M. Kracht, E. Hitti, K. Milarski, R. Askew, S. Marusic, L. L. Lin, M. Gaestel and J. B. Telliez, *Mol. Cell Biol.*, 2007, **27**(1), 170–181.
64. K. Rutault, C. A. Hazzalin and L. C. Mahadevan, *J. Biol. Chem.*, 2001, **276**(9), 6666–6674.
65. J. M. Kuldo, J. Westra, S. A. Asgeirsdóttir, R. J. Kok, K. Oosterhuis, M. G. Rots, J. P. Schouten, P. C. Limburg and G. Molema, *Am. J. Physiol. Cell Physiol.*, 2005, **289**(5), C1229–1239.
66. P. Bhavsar, N. Khorasani, M. Hew, M. Johnson and K. F. Chung, *Eur. Respir. J.*, 2010, **35**(4), 750–756.

67. J. Armstrong, C. Harbron, S. Lea, G. Booth, P. Cadden, K. A. Wreggett and D. Singh, *J. Pharmacol. Exp. Ther.*, 2011, **338**(3), 732–740.
68. A. L. Hopkins, *Nat. Chem. Biol.*, 2008, **4**(11), 682–690.
69. Z. A. Knight, H. Lin and K. M. Shokat, *Nat. Rev. Cancer*, 2010, **10**(2), 130–137.
70. L. Xie, T. Evangelidis, L. Xie and P. E. Bourne, *PLoS Comput. Biol.*, 2011, **7**(4), e1002037.
71. A. T. Bender and J. A. Beavo, *Pharmacol. Rev.*, 2006, **58**, 488–520.
72. M. D. Houslay, P. Schafer and K. Y. J. Zhang, *Drug Dis. Today*, **10**, 1503–1519.
73. M. J. Sanz, J. Cortijo and E. J. Morcillo, *Pharmacol. Ther.*, 2005, **106**, 269–297.
74. D. Spina, *Br. J. Pharmacol.*, 2008, **155**, 308–315.
75. M. D. Houslay and D. R. Adams, *Biochem. J.*, 2003, **370**, 1–18.
76. K. Y. Zhang, P. N. Ibrahim, S. Gillette and G. Bollag, *Expert Opin. Ther. Targets*, 2005, **9**(6), 1283–1305.
77. A. T. Bender and J. A. Beavo, *Pharmacol. Rev.*, 2006, **58**(3), 488–520.
78. As of Feb 2012, http://clinicaltrials.gov listed 34 clinical studies with Apremilast (CC–10004).
79. C. Méhats, S.L. Jin, J. Wahlstrom, E. Law, D.T. Umetsu and M. Conti, *FASEB J.*, 2003, **200317**, 1831–1841.
80. S. A. Laufer, G. K. Wagner, D. A. Kotschenreuther and W. Albrecht, *J. Med. Chem.*, 2003, **46**(15), 3230–3244.
81. S. A. Laufer, D. R. Hauser, D. M. Domeyer, K. Kinkel and A. J. Liedtke, *J. Med. Chem.*, 2008, **51**(14), 4122–4149.
82. B. Kammerer, H. Scheible, W. Albrecht, C. H. Gleiter and S. Laufer, *Drug Metab. Dispos.*, 2007, **35**(6), 875–883.

CHAPTER 6

MAPKAP Kinase 2 (MK2) as a Target for Anti-inflammatory Drug Discovery

JEREMY J. EDMUNDS AND ROBERT V. TALANIAN*

Abbott Bioresearch Center, Worcester, MA 01605, USA
*Email: bob.talanian@abbott.com

6.1 Introduction

For many years the pharmacologic treatment of rheumatoid arthritis and related serious inflammatory diseases was limited to the use of prostaglandin synthesis inhibitors such as aspirin[1,2] and non-steroidal anti-inflammatory drugs (NSAIDS),[3] including the selective cyclooxygenase-2 inhibitors,[4,5] or compounds that had modest efficacy as a result of anti-proliferative effects, *i.e.* methotrexate[6] or leflunomide,[7] or by poorly understood mechanisms, *e.g.* gold salts.[8] Beginning in 1998, a dramatic breakthrough in the treatment of these diseases was realized with the approvals of the parenterally administered anti-cytokine biopharmaceutical drugs etanercept, infliximab and adalimumab.[9,10] These and more recent biologics bind and neutralize tumour necrosis factor-α (TNFα), a potent pro-inflammatory cytokine with a crucial signalling role in the cellular mechanisms of autoimmune disease propagation.[11,12] As such, they interfere with key molecular and cellular steps in the propagation of inflammatory disease, and clearly qualify as disease-modifying anti-rheumatic drugs (DMARDs). These compounds represent major improvements in efficacy and safety compared to previous generations of primarily palliative drugs.

RSC Drug Discovery Series No. 26
Anti-Inflammatory Drug Discovery
Edited by Jeremy I Levin and Stefan Laufer
© The Royal Society of Chemistry 2012
Published by the Royal Society of Chemistry, www.rsc.org

Current efforts to find improved follow-ons to the anti-TNFα biologics are focused primarily on achieving efficacy in the fraction of patients who do not respond adequately to them. For example, and typical of the anti-TNFα biologics, the ARMADA trial for Adalimumab showed that a 40 mg dose resulted in a 50% response according to the American College of Rheumatology clinical scoring (ACR50) in just over half of patients, while only a quarter achieved ACR70.[13] While such a degree of efficacy provides welcome benefit for many arthritis patients, there is clearly room for improvement. Also desirable for patient convenience and other benefits would be orally available drugs of at least similar clinical properties.[14,15] Although efforts to identify oral DMARDs have been pursued vigorously for decades, and at least one, the JAK kinase inhibitor tofacitinib has demonstrated efficacy in Phase III trials,[16] none have to date been approved. Multiple mechanisms continue to be pursued for the discovery and development of such compounds across the pharmaceutical industry. Selecting the mechanism(s) with the greatest promise for efficacy and therapeutic window is crucial for success. Herein, we will present support for Mitogen Activated Protein Kinase Activated Kinase 2 (MAPKAP-K2, or MK2) as a novel DMARD target, a summary of progress and difficulties in the discovery of MK2 inhibitors and our perspective on the promise of MK2 and related mechanisms for ongoing drug discovery work.

6.2   Oral Anti-cytokine Target Identification

6.2.1   p38 MAPK Signal Transduction Pathway

The p38 mitogen-activated protein kinase (MAPK) and its upstream and downstream signalling events arguably constitute the most explored set of mechanisms for inhibiting TNFα action with an orally administered drug.[14,17,18] There are four isoforms of p38; p38α, p38β, p38γ and p38δ, with p38α being most clearly associated with inflammation.[17] Cytokines and other extra-cellular stimuli such as growth factors, DNA damage and oxidative stress signal through multiple receptors and other mechanisms to activate a cascade of kinases starting with a MAP3K, for example MEKK3 or TAK1, then a MAP2K such as MKK3 or MKK6, and then a MAPK such as p38α (Figure 6.1).[19–21] By direct and indirect effects including the stabilization, translocation and translation of mRNAs, p38 plays a major role in the production of pro-inflammatory cytokines such as TNFα, IL-6 and IFNγ, as well as the induction of other pro-inflammatory mediators such as COX-2.[17] Enzymes at each level of the cascade have been explored for anti-cytokine drug discovery. It is difficult to generalize how upstream or downstream targets in such a pathway might vary in their potential for efficacy. For example, upstream targets might have multiple effects, enhancing efficacy, but might be bypassed by other signalling mechanisms, limiting the impact of inhibition. Undesirable side-effects are similarly difficult to predict. The specific properties of signalling mechanisms like that of the p38 pathway must be considered case by case to select the best targets, and empirical experience can be the best guide.

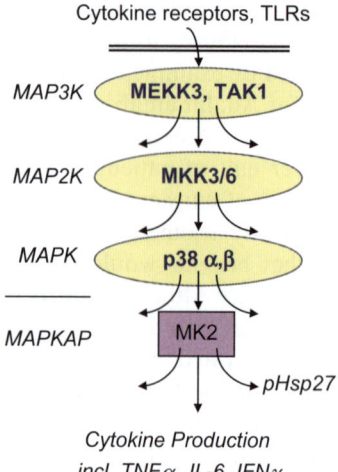

Figure 6.1 p38 – MK2 MAPK signalling cascade. Extra-cellular signals, for example from cytokine receptors or toll-like receptors (TLRs), activate a cascade of kinases resulting in p38 activation and further signalling. p38 has multiple substrates including MK2, which itself has multiple substrates including some that directly control cytokine production.

There have been many reports of p38 inhibitors with promising properties *in vitro* and in animal models of disease, but none have achieved clinical success.[12,22–24] Many targets beyond those related to cytokine production are regulated by p38, consistent with observed pleiotropic consequences of its inhibition and suggesting multiple mechanisms of toxicity[25] and even pro-inflammatory effects.[26] For example, in hepatocytes, p38 directly and indirectly down-regulates JNK, thereby modulating hepatocyte sensitivity to lipopolysaccharide (LPS) and TNFα-induced cell death; this may be an important mechanism of p38 inhibition-induced liver toxicity.[27,28] Activation of the kinases MSK1 and MSK2 by p38 may induce the anti-inflammatory cytokine IL-10,[29] and so inhibition of p38 might by this mechanism have a pro-inflammatory effect that contributes to the observed transient suppression of inflammatory markers by p38 inhibitors.[30,31] Thus, there are significant concerns that as an anti-inflammatory strategy p38 inhibition will not result in adequate efficacy or acceptable safety.

6.2.2 MK2 as an Oral Anti-inflammatory Drug Discovery Target

Many alternative mechanisms for the inhibition of cytokine production for the treatment of inflammatory disease are being pursued, several of which hold promise for avoiding the shortcomings of p38 inhibition.[32] One target of particular interest for anti-inflammatory drug discovery is the p38 substrate

MK2.[21,33] MK2 is a substrate of p38 whose activity may fully account for the anti-inflammatory effects of p38 inhibition in arthritis.[34] MK2 and p38α form a heterodimeric complex in cells,[35,36] which upon cellular stimulation trans-locates from the nucleus to the cytoplasm, where it is believed to have its primary pro-inflammatory effects. This occurs as a consequence of MK2 phosphory-lation by p38 on Thr334, exposing the otherwise buried MK2 nuclear export signal.[36–38] MK2 attracted wide attention as a potential drug discovery target when it was reported that MK2-deficient knockout mice are viable and fertile, and are defective in TNFα production.[39] Splenocytes derived from these animals are defective in the production of several pro-inflammatory cytokines, including TNFα, IL-6 and IFN-γ[39] and the animals themselves are resistant to collagen-induced arthritis, a mouse model of RA,[40] as well as in ovalbumin-induced airway inflammation, a mouse model of asthma.[41,42] Dosed orally, inhibitors of MK2 can block acute systemic induction of TNFα by LPS in rats,[43] and can reduce paw swelling in the rat SCW arthritis model.[44] This has led to the hypothesis that MK2 mediates most or all inflammatory signals of the p38 cascade while other p38 substrates regulate the pathways responsible for toxicity or attenuated efficacy. Thus, MK2 inhibition might deliver on the promise of p38 inhibition for anti-inflammatory efficacy while also giving a more favourable safety profile.

The mechanistically explained link between MK2 activity and TNFα pro-duction, coupled with the clinical validation of TNFα as a DMARD target, strongly support the use of MK2 inhibitors as oral DMARDs. MK2 inhibitors might find clinical application in other TNFα-driven diseases as well, such as psoriasis and Crohn's disease. For example, recent studies using the MK2 inhibitor lead "C23" (compound 7, Table 6.1) reduced puromycin aminonu-cleoside-induced podocyte injury *in vitro*, suggesting application in kidney disease.[43,45] Other potential applications have been suggested in breast cancer[46] and multiple myeloma.[47] Inhibitors of the p38 pathway have also been sug-gested for use in neurodegenerative disorders.[48,49]

6.2.3   Mechanism of Cytokine Regulation by MK2

Adenine- and uridine-rich element (ARE) binding proteins modulate mRNA stability and translation *via* their binding to the AREs present on the mRNA 3′ untranslated regions, thus regulating the production of pro-inflammatory cytokines.[50] A particularly well-studied ARE binding protein is the zinc finger protein tristetraprolin (TTP). In its unphosphorylated form, TTP binds to and promotes the degradation of TNFα mRNA, decreasing TNFα production.[51] TTP-deficient mice produce elevated TNFα and display a TNFα-driven disease phenotype that is fully blocked by anti-TNFα antibodies,[52] suggesting that TNFα in particular is the predominant pro-inflammatory consequence of TTP deficiency. Macrophages from mice deficient in both MK2 and TTP produce elevated amounts of TNFα similar to that of TTP knockouts, whereas in MK2 knockouts TNFα production was decreased.[53] MK2 phosphorylates and inac-tivates TTP,[54,55] consequently increasing TNFα production[53,56] by blocking

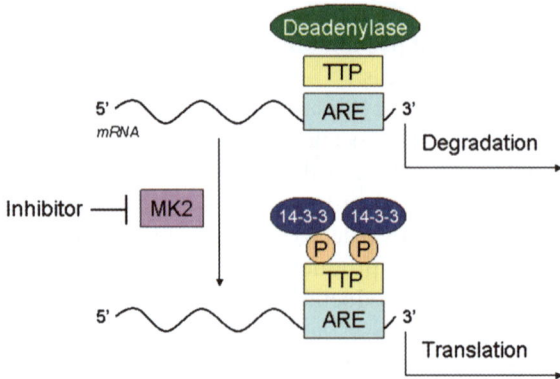

Figure 6.2   Model for the anti-TNFα consequences of MK2 inhibition. MK2 phosphorylation of TTP retards deadenylase recruitment, stabilizing cytokine
mRNA resulting in its increased translation. Inhibitors block MK2-
induced translation.

deadenylase recruitment.[57,58] According to the model that these studies suggest
(Figure 6.2), MK2 inhibition would restore down-regulation of TNFα production by TTP through its action on TNFα mRNA. In addition to TTP, MK2 and
p38 phosphorylate other mRNA-modulating proteins including poly(A)-binding
protein 1,[59] poly(A) ribonuclease,[60] hnRNP A0[61] and K-homology-type splicing
regulatory protein.[50] At least 100 mRNA transcripts are regulated by TTP,[62,63]
many of which contain ARE sequences and encode cytokines.

6.2.4   Functions of MK2 Beyond Cytokine Regulation

MK2 activity is implicated in many additional inflammatory mechanisms.
MK2 phosphorylates and activates 5-lipoxygenase,[64] and as a result may
promote leukotriene biosynthesis. MK2 also phosphorylates several transcription factors including CREB and ATF-1,[65] ER81[66] and SRF.[67] MK2 may
modulate sensitivity to TNFα-induced apoptosis at least in part by stabilizing
IL-1α mRNA.[68] MK2 phosphorylates heat shock protein 27 (Hsp27),[69] which
results in actin filament remodelling by releasing its inhibition of actin polymerization. This has an important role in multiple cellular effects including
migration, growth and differentiation.[70,71] Hsp27 phosphorylation may also
have a role in inflammatory signal transduction.[72] Incidentally, Hsp27 phosphorylation detected by phospho-specific antibodies is also a practical and
widely used marker of MK2 activity in cellular assays.

Recently emerging in the literature is an important role for p38 pathway
enzymes in cell cycle checkpoint control.[73,74] Thus, p38 is activated and
required for cell cycle arrest upon treatment of cells with topoisomerase inhibitors[75,76] and DNA damaging agents[77] or UV irradiation-induced DNA
damage.[78] In at least some settings, MK2 activity in particular may be
responsible for p38 pathway-induced cell cycle control. The mechanisms by

which MK2 regulates cell cycle checkpoint control include phosphorylation of isoforms of the cell cycle regulatory phosphatases Cdc25A, Cdc25B and Cdc25C[79,80] and repression of proto-oncogene c-Myc activity by activation of microRNA miR-34c.[81] While it has been suggested that MK2 inhibition might thus be useful in enhancing the anti-cancer activities of conventional chemotherapeutics,[80] it also implies that chronic use of MK2 inhibitors for inflammatory disease might elevate the risk of neoplastic transformations due to interference with cell cycle control. There is to our knowledge no direct evidence for such a risk, but it does suggest caution in advancing MK2 inhibitors for chronic, non-life-threatening indications and the advisability of monitoring of this specific risk in clinical and pre-clinical studies. Other biologic functions proposed for MK2 include the regulation of gut epithelial permeability through phosphorylation of cytoskeleton components[82] and regulation of skeletal muscle mass.[83] MK2 may also impact inflammatory and other mechanisms by raising cAMP levels as a consequence of attenuation of phosphodiesterase 4 (PDE4) isoform activation.[84,85] The long-term biologic consequences of modulation of multiple such systems by MK2 inhibitors are difficult to predict.

Related enzymes may have functions that overlap with those of MK2. It is known that p38 itself phosphorylates certain MK2 substrates, for example TTP.[26] MK3 is an enzyme highly homologous to MK2[86,87] with similar substrate specificity[88] and which is similarly transported to the cytoplasm as a result of cellular stress or mitogenic stimulation.[89] In macrophages from MK2-deficient mice, which are 90% defective in LPS-induced TNFα, the additional deletion of MK3 eliminates most of the residual TNFα production,[90] supporting at least partially redundant function between the two enzymes. Of course, high homology does not necessarily translate to similar potency of inhibitor leads, and so MK2 inhibitors may vary in their apparent biological properties due to independently varying MK3 potency. Recent evidence suggests distinct functions of MK2 and MK3, in particular, that MK2 blocks MK3 in its negative regulation of NF-κB and IRF3-dependent signalling.[91] A great deal of further work is clearly needed to fully understand this complex system and to predict with any confidence the consequences of pharmacologic inhibition of its key components.

6.3 Medicinal Chemistry

While it has been straightforward to design individual inhibitors that target MK2 specifically with at least modest selectivity with respect to other kinases, it has been unexpectedly difficult to create compounds with favourable solubility and permeability. As a result there are relatively few biochemically efficient MK2 inhibitors that have advanced to *in-vivo* pre-clinical studies and no compounds that have progressed through clinical studies. For an in depth review of a large variety of compounds with minimal to modest *in-vitro* activity against MK2 the reader is referred to an excellent recent review.[92] In this review the focus is on relatively recently disclosed compounds and those with reasonable cellular activity as depicted in Table 6.1.

The vast majority of disclosed MK2 inhibitors are classical type I inhibitors as revealed by crystallographic or biochemical studies. As such they bind to the ATP site of the kinase and thus compete with intra-cellular ATP (estimated concentration 1–5 mM) to inhibit phosphorylation and activation of the kinase.

Table 6.1   Representative MK2 inhibitors.

Compound #	Structure	MK2 IC_{50}, μM	Cellular EC_{50}, μM Inh. TNFα release
1		1.68	2.91 U937 cells
2		0.13	Not available
3		5.8	2.2 THP-1 cells
4		0.056	0.055 hPHM
5		0.29	80.8 U937 cells
6		10.1	0.83 U937 cells
7		0.126	4.8 U937 cells
8		0.002	0.300 THP-1 cells

Table 6.1   (*Continued*)

Compound #	Structure	MK2 IC$_{50}$, μM	Cellular EC$_{50}$, μM Inh. TNFα release
9		0.082	0.04 hPBMC
10		0.061	2.5 hPBMC
11		<0.003	0.1 hPBMC
12		0.05	0.3 hPBMC
13		0.007	0.45 hPBMC
14		0.020	0.75 THP-1 cells
15		0.005	0.160 hPBMC
16		0.11	4.4 THP-1 cells

Given the high homology of protein kinase active sites, selectivity profiles for therapeutic index discussions are generated in a variety of pharmaceutical company specific strategies that emphasize the importance of:

1) specific kinases that potentially contribute to TNFα production (*e.g.* JNK1, JNK2, p38, MK3, MK5, ERK2, IKK2);
2) those kinases with high sequence homology to MK2 (*e.g.* CDK2, MNK1, MSK1, MSK2);
3) kinases with demonstrated affinity for these types of compounds as revealed by broad kinome screening (*e.g.* Aurora A, GSK3β, JAK3, Kdr, c-Met, CLK1).

The earliest MK2 inhibitors disclosed in the literature included the amino-cyanopyridines, exemplified by compound **1**, that displayed relatively poor MK2 inhibitory activity and moderate cell potency. The most potent analogues displayed IC_{50}s in the low micromolar range, and inhibition of TNFα production from U937 cells upon LPS stimulation was rarely better than 1 μM. Encouragingly, however, compound **1** displayed *in-vivo* activity in a rat LPS-induced TNFα acute release assay. At a dose of 20 mg/kg p.o., this compound provided 60% inhibition of TNFα production at 1.5 h after LPS injection (a time point that corresponded to maximal TNFα production in control animals).[93] Compound exposure was not indicated.

Disclosures from Teijin[94,95] describe compounds such as TEI-I01800 (**2**), which is representative of a class of pyrazolo[1,5-*a*]pyrimidine MK2 inhibitors. Although these compounds have relatively modest *in-vitro* MK2 activity, their non-planar conformation distinguishes them from many other MK2 inhibitors. This series of compounds has subsequently been expanded by Bristol-Myers Squibb to include multiple 5,6-fused heterocycles such as imidazo-triazines and imidazo-pyrazines.[96,97]

At Wyeth a high-throughput screening campaign has identified a number of moderately active squarate MK2 inhibitors **3**.[98,99] These compounds benefit from a 4-aminopyridine interaction with the hinge Leu141, with the distal aryl ring interacting with the glycine rich loop. Analogues of **3** with improved MK2 potency did not exhibit a similar improvement in cellular potency, presumably due to a loss in cell permeability.

Several series of compounds were disclosed from Abbott, including indazole **4**,[100,101] as a result of MK2 crystal structure-based optimization of the high-throughput screening hit 4-(2-aminopyrimidin-4-ylamino) phenol. In this work it was discovered rather surprisingly that the diaminopyridine did not make the expected hinge interaction, but was in fact involved in a hydrogen bonding interaction with His106, Lys93, Asn191 and Glu190.

For Pfizer to achieve cellular activity within a series of zwitterionic carboline carboxylic acids exemplified by compound **5**, propyl ester **6** was prepared.[102] In a U937 cell assay an IC_{50} of 0.83 μM was observed suggesting a facile intra-cellular conversion to the parent acid, which inhibited MK2 with an IC_{50} of 0.29 μM. When the ester was evaluated in a rat LPS-induced TNFα model the prodrug, dosed at 40 mg/kg i.p. 1 hour prior to LPS stimulation, inhibited TNFα production by 84%.

A particularly fruitful area of research has been the pyrrolopyridones and related multi-cyclic lactam analogues. These series of compounds have been extensively investigated and optimized with the help of structure-based drug design. In fact there are over 20 crystal structures of MK2 inhibitors in the RCSB Protein Data Bank, over half of which belong to the pyrrolopyridone class. Many of these have been aligned and extracted in Table 6.2 and represented in Figure 6.3 for ease of viewing. The inhibitors are bound in the ATP pocket benefiting from a limited classical hinge interaction with Leu141 (N-H donor) or, with some inhibitors, an additional interaction with the backbone Glu139 (C=O acceptor) residue. In addition, Lys93 (catalytic lysine) and Asp207 (DFG) play predominant roles in dual interaction with the lactam moiety common to a number of inhibitors. As can be seen in the surface electrostatics plots, there is high complementarity between the inhibitors and the pockets created by the kinase residues. The glycine rich loop is typically observed with good electron density, suggesting close van der Waals interaction between the inhibitor and the protein residues (Leu193, Leu70, Val78, Met138 "gatekeeper" and Ala91). Compounds in this class often incorporate planar substituents that are accommodated by a large open hydrophobic flat surface of MK2 that is proximal to the hinge. Typically substituents that access this region of MK2 engender selectivity relative to CDK2 because of the smaller pocket of the latter.

An early example of a pyrrolo-pyridinone demonstrating *in-vivo* efficacy is compound **7** from Pfizer.[43] This fluorophenyl analogue inhibited TNFα production by greater than 80%, when dosed at 20 mg/kg orally 2 hours prior to LPS challenge. Typically TNFα production is maximal 90 minutes after the administration of LPS. Thus, the profound TNFα inhibition produced by compound **7** is reflective of its drug-like physical and pharmacokinetic properties. With a plasma concentration of 3.13 µg/mL 3.5 hours after dosing, the compound demonstrates excellent oral exposure and represents a suitable lead for further analoguing.

At Boehringer-Ingelheim a series of structurally related pyrazinoindolone inhibitors were modified to allow polar substituents on the lactam or polar thiazole carboxamide derivatives.[103–105] The piperidine derivative **8** displayed a good balance of physicochemical properties and cellular activity as revealed by an IC_{50} of 0.3 µM in THP-1 cells. This compound showed good PK properties in mice with high bioavailability and an AUC of 446 ng*h/mL when dosed orally at 1 mg/kg in PEG400. Oral activity was not reported for this analogue, although a closely related analogue did not inhibit TNFα when dosed orally at 60 mg/kg.

Pyrrolo-pyrimidones were modified with a styrene substituent proximal to the hinge binding pyridine functionality at Novartis.[106] This provided compounds such as **9** with modest kinase selectivity (IC_{50}: MK5 0.028 µM, Aurora A 0.28 µM, GSK3β 0.53 µM, JAK3 1 µM, JNK2 0.24 µM, KDR 0.19 µM and c-Met 0.14 µM) and remarkable cellular potency in inhibiting LPS-stimulated TFNα production with an IC_{50} of 0.04 µM in hPBMCs. In THP-1 cells, LPS

Table 6.2 Crystal structure of MK2 with inhibitors.

PDB	Name	ATP site and inhibitor
1NXK[123]	staurosporine	
3R30[110]	1-(2-aminoethyl)-3-[2-(quinolin-3-yl)pyridin-4-yl]-1H-pyrazole-5-carboxylic acid	
2P3G[43]	2-[2-(2-fluorophenyl) pyridin-4-yl]-1,5,6,7-tetrahydro-4H-pyrrolo[3,2-c]pyridin-4-one	
3R2Y[110]	2-(2-quinolin-3-ylpyridin-4-yl)-1,5,6,7-tetrahydro-4H-pyrrolo[3,2-c]pyridin-4-one	
3R2B[110]	2′-[2-(1,3-benzodioxol-5-yl)pyrimidin-4-yl]-5′,6′-dihydrospiro[piperidine-4,7′-pyrrolo[3,2-c] pyridin]-4′(1′H)-one	
3KGA[107]	6-{3-amino-1-[3-(1H-indol-6-yl) phenyl]-1H-pyrazol-4-yl}-3,4-dihydroisoquinolin-1(2H)-one	

Table 6.2   (*Continued*)

PDB	Name	ATP site and inhibitor
3M2W[109]	2′-(2-fluorophenyl)-1-methyl-6′,8′,9′,11′-tetrahydrospiro[azetidine-3,10′-pyrido[3′,4′:4,5]pyrrolo[2,3-f]isoquinolin]-7′(5′H)-one	
3FYK[115]	(3R)-3-(aminomethyl)-9-methoxy-1,2,3,4-tetrahydro-5H-[1]benzothieno[3,2-e][1,4]diazepin-5-one	
3FYJ[115]	(10R)-10-methyl-3-(6-methylpyridin-3-yl)-9,10,11,12-tetrahydro-8H-[1,4]diazepino[5′,6′:4,5]thieno[3,2-f]quinolin-8-one	
2PZY[103]	(4R)-N-[4-({[2-(dimethylamino)ethyl]amino}carbonyl)-1,3-thiazol-2-yl]-4-methyl-1-oxo-2,3,4,9-tetrahydro-1H-beta-carboline-6-carboxamide	
3KC3[100]	N4-[7-(1-benzofuran-2-yl)-1H-indazol-5-yl]pyrimidine-2,4-diamine	
3FPM[99]	3-{[(1R)-1-phenylethyl]amino}-4-(pyridin-4-ylamino)cyclobut-3-ene-1,2-dione	

Table 6.2   (*Continued*)

PDB	Name	ATP site and inhibitor
3A2C[94]	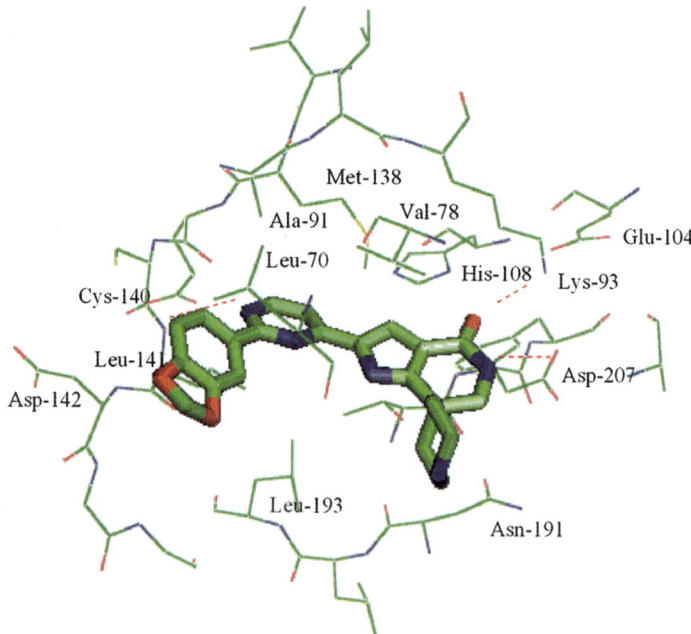N7-(4-ethoxyphenyl)-6-methyl-N5-[(3S)-piperidin-3-yl]pyrazolo[1,5-a]pyrimidine-5,7-diamine	

Figure 6.3   Cocrystal structure of pyrrolopyridinone derivative from PDB:3R2B bound to MK2 as an illustrative example of the pyridone class of MK2 inhibitors showing protein residues important for inhibitory activity.

stimulated phosphorylation of the MK2 substrate Hsp27 was dose-dependently inhibited from 0.5 μM to 5 μM, suggesting that cytokine release was mediated by MK2. A scaffold-hopping approach to modify the pyrrolo-pyrimidinone type MK2 inhibitors generated tetrahydroquinolones such as **10**.[107] The 3-aminopyrazole heterocycle was an effective moiety to replace the pyridine hinge interacting substituent of the classical pyrrolo-pyrimidinones. Furthermore, the indole of the preferred compound **10** introduced a hydrogen-bond interaction with the carbonyl of Phe90. This compound inhibited LPS-induced TNFα release in mice 68% at the high dose of 100 mg/kg p.o. with blood levels of 17.4 μM.

Compound **11** is from a series of spiroazetidines from Novartis with impressive MK2 kinase inhibitory activity that translates into good inhibition of TNFα in hPBMCs stimulated with LPS.[108,109] In particular, compound **11** benefits from a water-mediated interaction between the azetidine nitrogen and the carboxylate residue of Glu-145 of MK2 to afford an IC_{50} of $<0.003\,\mu M$, and $0.1\,\mu M$ in the hPBMC assay. However, the analogue **12** showed the most desirable composite properties. This compound exhibits an MK2 IC_{50} of $0.05\,\mu M$, and achieves 96% inhibition of TNFα, at $100\,mg/kg$ p.o. and achieves a plasma concentration of $30\,\mu M$ in LPS-treated mice. Furthermore, in rats a bioavailability of 24% and $t_{1/2}$ of 1.3 h is observed at $1\,mg/kg$ p.o. Thus, in a chronic collagen-induced arthritis model in DBA/1 mice compound **12** ($100\,mg/kg$, p.o., b.i.d.) reduced swelling and histological scores (joint damage and erosions, proteoglycan loss and inflammatory cell infiltrates), represented by a greater than 50% reduction in scoring of paws 40 days post immunization relative to control HPMC 603/Tween80. In a rat antigen induced arthritis model **12** ($100\,mg/kg$, p.o., b.i.d.) reduced knee swelling for 7 days of dosing, which can be represented by a reduced swelling ratio of 1.24× relative to a control vehicle value of 1.4×, or a value slightly greater than the positive control of dexamethasone $0.3\,mg/kg$ p.o with 1.14× at day 7. Due to the high plasma concentrations achieved in the *in-vivo* experiments the beneficial effects may also be attributed to other kinases, especially in the light of the fact that compound **12** is known to inhibit 14 kinases with at least 95% inhibition when tested at $1\,\mu M$ in a panel of 262 kinases.

Within a series of spiro-piperidinyl pyrrolo-pyridinones from Merck the initial absence of oral bioavailability was solved by moving the basic nitrogen of the spiro-4-piperidyl moiety towards the electron-deficient core. This created spiro-3-piperidyl analogues[110,111] with reduced pK_a and improved Caco-2 permeability. For example, compound **13**, as a racemate, exhibited a bioavailability of 48% in mice, at an undisclosed dose, and achieved an AUC of $3.5\,\mu M*h$. In hPBMCs the (*S*) isomer afforded an EC_{50} of $0.45\,\mu M$, reflective of its potent MK2 IC_{50} of $0.007\,\mu M$. When this analogue was profiled in an extensive 233 kinase panel at Millipore, both MK3 and MK5 were inhibited at least 95% at $1\,\mu M$. Other kinases that were inhibited at greater than 80% at $1\,\mu M$ include PIM1 and PIM3, DRAK1, ERK1 and 2 and CHK2.

Also of interest is a relatively unique series of thiourea MK2 inhibitors from Merck exemplified by compound **14**.[112] Through extensive analoguing, three compounds were selected for evaluation *in vivo*. Compound **14**, when dosed i.v. at $10\,mg/kg$ in an acute mouse model of inflammation, inhibited LPS-induced TNFα production by 75%. Compound exposures were not indicated.

Of particular interest is Pfizer's benzothiophene[113–117] MK2 inhibitor PF-3644022, **15**, with an MK2 IC_{50} of $0.005\,\mu M$. This compound is orally efficacious in the rat acute LPS-induced TNFα model and the chronic streptococcal cell wall-induced arthritis model with ED_{50} values of 6.9 and $20\,mg/kg$, respectively. It binds in the ATP binding pocket of both the phosphorylated and non-phosphorylated forms of MK2. When tested at the Km of ATP in a panel of 200 kinases compound **15** displays less than 20-fold selectivity to a number of kinases revealed by IC_{50}s of ASK1 0.06 μM, BrSK2 0.09 μM, CaMKII 0.07 μM, DRAK1 0.071 μM, Mer 0.076 μM, MK3 0.053 μM, MK5 0.005 μM and Pim1 0.088 μM. The selectivity of **15** is afforded principally by the methylpyridine substituent appended to the 2-position of the hinge binding pyridine. The lactam carbonyl is appropriately positioned to interact with Lys93 and Asp207 of the activation loop. While the position of the nitrogen in the lactam ring is important for potency it does not appear to make a direct hydrogen bond with the kinase. In a monocytic cell line U937, TNFα production was inhibited after LPS stimulation with an IC_{50} of 0.159 μM at the 4-hour time point. This concentration correlates well with the inhibition of phospho Ser82 of HSP27 (IC_{50} 0.2 μM), a substrate of activated MK2. More importantly, upon oral dosing in the Lewis rat model of LPS-stimulated TNFα production, **15** yielded an ED_{50} of 6.9 mg/kg, with an EC_{50} of 1.4 μM, or a total concentration similar to that required in the human whole blood LPS/TNFα assay (2 μM). Upon oral dosing b.i.d. for 12 days in the chronic rat streptococcal cell wall induced arthritis model compound **15** demonstrated an ED_{50} of 20 mg/kg and EC_{50} Cmin 0.91 μM, which likely reflects a need to suppress 50% of MK2 activity for the duration of dosing. The projected human dose of 300 mg twice a day coupled with the acute hepatotoxicity observed in dogs and monkeys failed to provide a sufficient therapeutic index to enable clinical development of PF-3644022.

In an effort to address the forecasted difficulty of identifying ATP-competitive compounds with sufficient biochemical efficiency, attention has also been directed toward uncompetitive[118] and non-ATP competitive compounds. At Merck[119] a high-throughput screen that relied upon affinity selection/mass spectrometry revealed a micromolar compound 5-(4-chlorophenyl)-*N*-(4-(piperazin-1-yl)phenyl)furan-2-carboxamide that displayed equivalent activity in an MK2 enzyme activity assay at either high (100 μM) or low (2 μM) ATP concentration. Through NMR analysis the lead was shown to localize with residues distinct from those observed upon binding of ATP, further reinforcing the unique characterization of the lead. The *N*-alkylation of the furan-2-carboxamide lead provided **16**, which displayed a MK2 IC_{50} of 0.11 μM, in the presence of 100 μM ATP and inhibited THP-1 LPS-stimulated TNFα production with an EC_{50} of 4.4 μM. This compound was further shown to inhibit MMP-13 production in chondrocytes and exhibited good selectivity against a panel of 150 kinases evaluated at 10 μM. This unique, non-ATP competitive binding mode and a favourable pharmacokinetic profile will likely provide an additional starting point for lead optimization efforts directed toward MK2 inhibitors.

6.4   Challenges in Translating MK2 Inhibition to Favourable *in-vivo* Effects

While MK2 drug discovery efforts have combined a simultaneous consideration of *in-vitro* potency, solubility, cell permeability and clearance to produce potentially low-dose compounds, it is clear that *in-vivo* activity is further hampered by limited inhibition of TNFα production in whole blood. This is presumably a result of the difficulty in achieving unbound plasma levels in excess of the cell-based assay EC_{50} values. These EC_{50}s from PBMCs or THP-1 cells tend to be relatively high, perhaps in part as a consequence of the strong affinity of ATP for the non-phosphorylated form of MK2 (MgATP $Km = 30 \mu M$,[100] compared, for example, with the value of $> 10 \text{mM}$ for non-phosphorylated p38α.[120]

In addition to the difficulties posed by the high ATP affinity of non-phosphorylated MK2, poor correlations have been observed between the inhibition of recombinant MK2[100] and cell assay potency within series of compounds suggesting further complexities.[43,101,121] For example, plotting the enzyme and cell assay potencies of a series of closely related 2-aryl pyridine MK2 inhibitors from Pfizer shows little correlation over a nearly 2 log range of potencies (Figure 6.4). Although this observation can be attributed to, for example, variations in analogue-specific properties that affect cell potency, such as membrane penetration, similarly poor correlations have been reported for other such series and seem to be typical for MK2 inhibitor series. A close correlation between inhibitor potency for recombinant MK2 and MK2 immunoprecipitated from human macrophages has been reported,[101] suggesting at least that recombinant enzyme is a good model for native enzyme. The poor enzyme-to-cell correlation for MK2 inhibitors is not well understood,

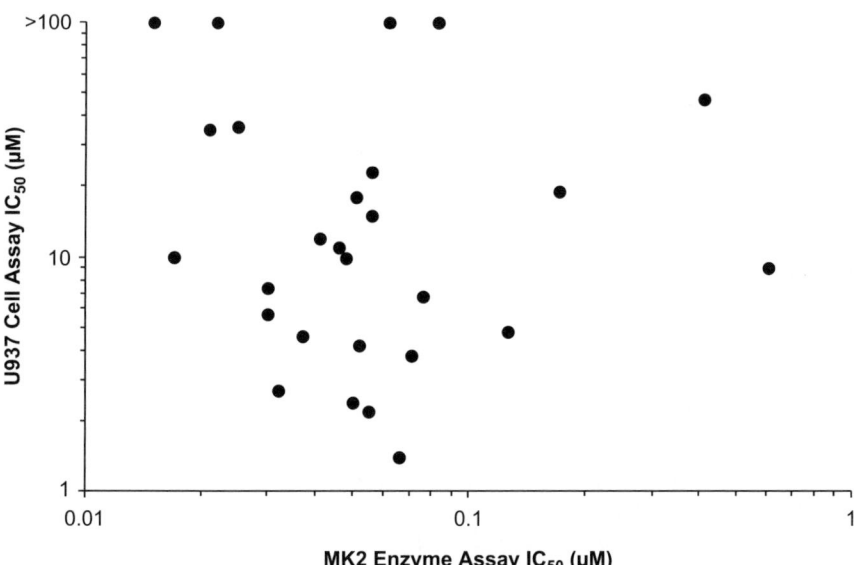

Figure 6.4 Replot of the data of Tables 1 and 2 from Anderson *et al.* comparing the potency of a series of 2-aryl pyridine analogues as inhibitors or recombinant MK2 enzyme and of LPS-induced TNFα in PMA-differentiated U937 cells.[43]

and may involve factors such as variable importance of other cellular targets that may influence the readout (for example, MK3). Given these uncertainties, a simple reductionist approach, *e.g.*, an MK2 enzyme to cell to *in-vivo* assay screening paradigm, may not be adequate to identify and advance inhibitor leads of the p38-MK2 pathway intended to block cytokine activity and have anti-inflammatory activity. More promising might be a phenotypic screening approach,[122] for example screening with cell assays for inhibitor leads that would yield leads of desirable properties first, and for which the specific molecular target(s) might be identified later.

6.5   Summary

The unmet medical need for a new generation of oral anti-inflammatory drugs has led to the thorough investigation of p38α and its downstream substrate MK2 as drug-discovery targets. However, MK2 inhibitor drug discovery has been hampered by poor *in-vivo* compound potency. Concerns about the promise of the efficacy and safety margin that might be expected from a nominally promising MK2 inhibitor lead stem from its role in important cellular processes such as cell cycle control and a poorly understood lack of correlation between enzyme and cell potency. At a minimum, thorough understanding of the pleiotropic biological roles of MK2 will be required to confidently predict its promise as a drug-discovery target. Unbiased phenotypic screening for inhibitors of cellular cytokine production might be a more productive avenue for anti-inflammatory drug discovery, at least until such understanding is achieved.

Acknowledgements

The authors gratefully acknowledge the expert assistance of Suzzanne Lunt in preparation of the manuscript.

References

1. J. R. Vane, *Nat. New Biol.*, 1971, **231**, 232.
2. J. B. Smith and A. L. Willis, *Nat. New Biol.*, 1971, **231**, 235.
3. B. N. Cronstein and G. Weissmann, *Annu. Rev. Pharmacol. Toxicol.*, 1995, **35**, 449.
4. C. J. Hawkey, *Lancet*, 1999, **353**, 307.
5. D. L. DeWitt, *Mol. Pharmacol.*, 1999, **55**, 625.
6. L. Genestier, R. Paillot, L. Quemeneur, K. Izeradjene and J. P. Revillard, *Immunopharmacology*, 2000, **47**, 247.
7. M. L. Herrmann, R. Schleyerbach and B. J. Kirschbaum, *Immunopharmacology*, 2000, **47**, 273.
8. R. L. Kaye, *J. Rheumatol., Suppl.*, 1982, **8**, 124.
9. C. Gabay, *Expert Opin. Biol. Ther.*, 2002, **2**, 135.
10. M. Feldmann and R. N. Maini, *Annu. Rev. Immunol.*, 2001, **19**, 163.

11. M. Feldmann, F. M. Brennan, B. M. Foxwell, P. C. Taylor, R. O. Williams and R. N. Maini, *J. Autoimmun.*, 2005, **25**(Suppl.), 26.
12. I. B. McInnes and G. Schett, *Nat. Rev. Immunol.*, 2007, **7**, 429.
13. M. E. Weinblatt, E. C. Keystone, D. E. Furst, L. W. Moreland, M. H. Weisman, C. A. Birbara, L. A. Teoh, S. A. Fischkoff and E. K. Chartash, *Arthritis Rheum.*, 2003, **48**, 35.
14. G. Wagner and S. Laufer, *Med. Res. Rev.*, 2006, **26**, 1.
15. J. S. Smolen and G. Steiner, *Nat. Rev. Drug Discovery*, 2003, **2**, 473.
16. K. Garber, *Nat. Biotechnol.*, 2011, **29**, 467.
17. G. L. Schieven, *Curr. Top. Med. Chem.*, 2005, **5**, 921.
18. G. Schett, J. Zwerina and G. Firestein, *Ann. Rheum. Dis.*, 2008, **67**, 909.
19. J. M. Kyriakis and J. Avruch, *Physiol. Rev.*, 2001, **81**, 807.
20. K. J. Cowan and K. B. Storey, *J. Exp. Biol.*, 2003, **206**, 1107.
21. M. Gaestel, A. Kotlyarov and M. Kracht, *Nat. Rev. Drug Discovery*, 2009, **8**, 480.
22. D. M. Goldstein and T. Gabriel, *Curr. Top. Med. Chem.*, 2005, **5**, 1017.
23. C. Dominguez, D. A. Powers and N. Tamayo, *Curr. Opin. Drug Discovery Dev.*, 2005, **8**, 421.
24. R. E. Alten, C. Zerbini, S. Jeka, F. Irazoque, F. Khatib, P. Emery, A. Bertasso, M. Rabbia and J. P. Caulfield, *Ann. Rheum. Dis.*, 2010, **69**, 364.
25. D. M. Dambach, *Curr. Top. Med. Chem.*, 2005, **5**, 929.
26. N. Ronkina, M. B. Menon, J. Schwermann, C. Tiedje, E. Hitti, A. Kotlyarov and M. Gaestel, *Biochem. Pharmacol.*, 2010, **80**, 1915.
27. L. Hui, L. Bakiri, A. Mairhorfer, N. Schweifer, C. Haslinger, L. Kenner, V. Komnenovic, H. Scheuch, H. Beug and E. F. Wagner, *Nat. Genet.*, 2007, **39**, 741.
28. J. Heinrichsdorff, T. Luedde, E. Perdiguero, A. R. Nebreda and M. Pasparakis, *EMBO Rep.*, 2008, **9**, 1048.
29. O. Ananieva, J. Darragh, C. Johansen, J. M. Carr, J. McIlrath, J. M. Park, A. Wingate, C. E. Monk, R. Toth, S. G. Santos, L. Iversen and J. S. Arthur, *Nat. Immunol.*, 2008, **9**, 1028.
30. S. E. Sweeney, *Nat. Rev. Rheumatol.*, 2009, **5**, 475.
31. D. Hammaker and G. S. Firestein, *Ann. Rheum. Dis.*, 2010, **69**, i77.
32. P. Cohen, *Curr. Opin. Cell Biol.*, 2009, **21**, 317.
33. S. Duraisamy, M. Bajpai, U. Bughani, S. G. Dastidar, A. Ray and P. Chopra, *Expert Opin. Ther. Targets*, 2008, **12**, 921.
34. G. Mbalaviele and J. B. Monahan, *Expert Opin. Drug Discovery*, 2008, **3**, 163.
35. A. White, C. A. Pargellis, J. M. Studts, B. G. Werneburg and B. T. Farmer, II, *Proc. Natl Acad. Sci. USA*, 2007, **104**, 6353.
36. E. ter Haar, P. Prabakhar, X. Liu and C. Lepre, *J. Biol. Chem.*, 2007, **282**, 9733.
37. R. Ben-Levy, S. Hooper, R. Wilson, H. F. Paterson and C. J. Marshall, *Curr. Biol.*, 1998, **8**, 1049.
38. K. Engel, A. Kotlyarov and M. Gaestel, *EMBO J.*, 1998, **17**, 3363.
39. A. Kotlyarov, A. Neininger, C. Schubert, R. Eckert, C. Birchmeier, H. D. Volk and M. Gaestel, *Nat. Cell Biol.*, 1999, **1**, 94.

40. M. Hegen, M. Gaestel, C. L. Nickerson-Nutter, L.-L. Lin and J.-B. Telliez, *J. Immunol.*, 2006, **177**, 1913.

41. M. M. Gorska, Q. Liang, S. J. Stafford, N. Goplen, N. Dharajiya, L. Guo, S. Sur, M. Gaestel and R. Alam, *J. Exp. Med.*, 2007, **204**, 1637.

42. M. M. Gorska, S. Stafford, N. Dharajiya, Q. Liang, S. Sur, M. Gaestel and R. Alam, *J. Allergy Clin. Immunol.*, 2004, **113**, S182.

43. D. R. Anderson, M. J. Meyers, W. F. Vernier, M. W. Mahoney, R. G. Kurumbail, N. Caspers, G. I. Poda, J. F. Schindler, D. B. Reitz and R. J. Mourey, *J. Med. Chem.*, 2007, **50**, 2647.

44. R. J. Mourey, G. Anderson, J. Hirsch, M. Meyers, S. Mnich, L. Stillwell, M. Thiede, W. Vernier, E. Webb, J. Zhang, M. Gaestel and J. Monahan, *14th International Conference of the Inflammation Research Association*, Cambridge, MD, 2006.

45. R. Pengal, A. J. Guess, S. Agrawal, J. Manley, R. F. Ransom, R. J. Mourey, R. Benndorf and W. E. Smoyer, *Am. J. Physiol. Renal Physiol.*, 2011, **301**, F509.

46. Q. Han, J. Leng, D. Bian, C. Mahanivong, K. A. Carpenter, Z. K. Pan, J. Han and S. Huang, *J. Biol. Chem.*, 2002, **277**, 48379.

47. T. Hideshima, K. Podar, D. Chauhan, K. Ishitsuka, C. Mitsiades, T. Tai Yu, M. Hamasaki, N. Raje, H. Hideshima, G. Schreiner, A. N. Nguyen, T. Navas, N. C. Munshi, P. G. Richardson, L. S. Higgins and K. C. Anderson, *Oncogene*, 2004, **23**, 8766.

48. G. D. Cuny, *Curr. Pharm. Des.*, 2009, **15**, 3919.

49. E. K. Kim and E.-J. Choi, *Biochim. Biophys. Acta*, 2010, **1802**, 396.

50. P. Anderson, *Nat. Immunol.*, 2008, **9**, 353.

51. H. Sandler and G. Stoecklin, *Biochem. Soc. Trans.*, 2008, **36**, 491.

52. G. A. Taylor, E. Carballo, D. M. Lee, W. S. Lai, M. J. Thompson, D. D. Patel, D. I. Schenkman, G. S. Gilkeson, H. E. Broxmeyer, B. F. Haynes and P. J. Blackshear, *Immunity*, 1996, **4**, 445.

53. E. Hitti, T. Iakovleva, M. Brook, S. Deppenmeier, A. D. Gruber, D. Radzioch, A. R. Clark, P. J. Blackshear, A. Kotlyarov and M. Gaestel, *Mol. Cell. Biol.*, 2006, **26**, 2399.

54. K. R. Mahtani, M. Brook, J. L. Dean, G. Sully, J. Saklatvala and A. R. Clark, *Mol. Cell. Biol.*, 2001, **21**, 6461.

55. C. A. Chrestensen, M. J. Schroeder, J. Shabanowitz, D. F. Hunt, J. W. Pelo, M. T. Worthington and T. W. Sturgill, *J. Biol. Chem.*, 2004, **279**, 10176.

56. G. Stoecklin, T. Stubbs, N. Kedersha, S. Wax, W. F. C. Rigby, T. K. Blackwell and P. Anderson, *EMBO J.*, 2004, **23**, 1313.

57. S. L. Clement, C. Scheckel, G. Stoecklin and J. Lykke-Andersen, *Mol. Cell. Biol.*, 2011, **31**, 256.

58. F. P. Marchese, A. Aubareda, C. Tudor, J. Saklatvala, A. R. Clark and J. L. E. Dean, *J. Biol. Chem.*, 2010, **285**, 27590.

59. F. Bollig, R. Winzen, M. Gaestel, S. Kostka, K. Resch and H. Holtmann, *Biochem. Biophys. Res. Commun.*, 2003, **301**, 665.

60. H. C. Reinhardt, P. Hasskamp, I. Schmedding, S. Morandell, V. M. A. T. M. van, X. Z. Wang, R. Linding, S.-E. Ong, D. Weaver, S. A. Carr and M. B. Yaffe, *Mol. Cell*, 2010, **40**, 34.

61. S. Rousseau, N. Morrice, M. Peggie, D. G. Campbell, M. Gaestel and P. Cohen, *EMBO J.*, 2002, **21**, 6505.

62. W. S. Lai, J. S. Parker, S. F. Grissom, D. J. Stumpo and P. J. Blackshear, *Mol. Cell. Biol.*, 2006, **26**, 9196.

63. G. Stoecklin, S. A. Tenenbaum, T. Mayo, S. V. Chittur, A. D. George, T. E. Baroni, P. J. Blackshear and P. Anderson, *J. Biol. Chem.*, 2008, **283**, 11689.

64. O. Werz, D. Szellas, D. Steinhilber and O. Radmark, *J. Biol. Chem.*, 2002, **277**, 14793.

65. Y. Tan, J. Rouse, A. Zhang, S. Cariati, P. Cohen and M. J. Comb, *EMBO J.*, 1996, **15**, 4629.

66. R. Janknecht, *J. Biol. Chem.*, 2001, **276**, 41856.

67. O. Heidenreich, A. Neininger, G. Schratt, R. Zinck, M. A. Cahill, K. Engel, A. Kotlyarov, R. Kraft, S. Kostka, M. Gaestel and A. Nordheim, *J. Biol. Chem.*, 1999, **274**, 14434.

68. K. A. Janes, H. C. Reinhardt and M. B. Yaffe, *Cell (Cambridge, MA, US)*, 2008, **135**, 343.

69. D. Stokoe, K. Engel, D. G. Campbell, P. Cohen and M. Gaestel, *FEBS Lett.*, 1992, **313**, 307.

70. S. Kostenko, M. Johannessen and U. Moens, *Cell. Signalling*, 2009, **21**, 712.

71. J. Landry and J. Huot, *Biochem. Soc. Symp.*, 1999, **64**, 79.

72. K. A. Alford, S. Glennie, B. R. Turrell, L. Rawlinson, J. Saklatvala and J. L. E. Dean, *J. Biol. Chem.*, 2007, **282**, 6232.

73. H. C. Reinhardt and M. B. Yaffe, *Curr. Opin. Cell Biol.*, 2009, **21**, 245.

74. T. M. Thornton and M. Rincon, *Int. J. Biol. Sci.*, 2009, **5**, 44.

75. T. Kurosu, Y. Takahashi, T. Fukuda, T. Koyama, T. Miki and O. Miura, *Apoptosis*, 2005, **10**, 1111.

76. A. Mikhailov, M. Shinohara and C. L. Rieder, *J. Cell Biol.*, 2004, **166**, 517.

77. Y. Hirose, M. Katayama, D. Stokoe, D. A. Haas-Kogan, M. S. Berger and R. O. Pieper, *Mol. Cell. Biol.*, 2003, **23**, 8306.

78. D. V. Bulavin, Y. Higashimoto, I. J. Popoff, W. A. Gaarde, V. Basrur, O. Potapova, E. Appella and A. J. Fornace, Jr., *Nature*, 2001, **411**, 102.

79. I. A. Manke, A. Nguyen, D. Lim, M. Q. Stewart, A. E. Elia and M. B. Yaffe, *Mol. Cell*, 2005, **17**, 37.

80. H. C. Reinhardt, A. S. Aslanian, J. A. Lees and M. B. Yaffe, *Cancer Cell*, 2007, **11**, 175.

81. I. G. Cannell, Y. W. Kong, S. J. Johnston, M. L. Chen, H. M. Collins, H. C. Dobbyn, A. Elia, T. R. Kress, M. Dickens, M. J. Clemens, D. M. Heery, M. Gaestel, M. Eilers, A. E. Willis and M. Bushell, *Proc. Natl Acad. Sci. USA*, 2010, **107**, 5375.

82. M. B. Menon, J. Schwermann, A. K. Singh, M. Franz-Wachtel, O. Pabst, U. Seidler, M. B. Omary, A. Kotlyarov and M. Gaestel, *J. Biol. Chem.*, 2010, **285**, 33242.

83. M. Norrby and S. Tagerud, *J. Cell. Physiol.*, 2010, **223**, 194.
84. K. F. MacKenzie, D. A. Wallace, E. V. Hill, D. F. Anthony, D. J. P. Henderson, D. M. Houslay, J. S. C. Arthur, G. S. Baillie and M. D. Houslay, *Biochem. J.*, 2011, **435**, 755.
85. J. E. Souness, D. Aldous and C. Sargent, *Immunopharmacology*, 2000, **47**, 127.
86. M. M. McLaughlin, S. Kumar, P. C. McDonnell, S. Van Horn, J. C. Lee, G. P. Livi and P. R. Young, *J. Biol. Chem.*, 1996, **271**, 8488.
87. G. Sithanandam, F. Latif, F. M. Duh, R. Bernal, U. Smola, H. Li, I. Kuzmin, V. Wixler, L. Geil and S. Shrestha, *Mol. Cell. Biol.*, 1996, **16**, 868.
88. A. D. Clifton, P. R. Young and P. Cohen, *FEBS Lett.*, 1996, **392**, 209.
89. V. Zakowski, G. Keramas, K. Kilian, U. R. Rapp and S. Ludwig, *Exp. Cell Res.*, 2004, **299**, 101.
90. N. Ronkina, A. Kotlyarov, O. Dittrich-Breiholz, M. Kracht, E. Hitti, K. Milarski, R. Askew, S. Marusic, L. L. Lin, M. Gaestel and J. B. Telliez, *Mol. Cell. Biol.*, 2007, **27**, 170.
91. C. Ehlting, N. Ronkina, O. Böhmer, U. Albrecht, K. A. Bode, K. S. Lang, A. Kotlyarov, D. Radzioch, M. Gaestel, D. Häussinger and J. G. Bode, *J. Biol. Chem.*, 2011, **286**, 24113.
92. A. Schlapbach and C. Huppertz, *Future Med. Chem.*, 2009, **1**, 1243.
93. D. R. Anderson, S. Hegde, E. Reinhard, L. Gomez, W. F. Vernier, L. Lee, S. Liu, A. Sambandam, P. A. Snider and L. Masih, *Bioorg. Med. Chem. Lett.*, 2005, **15**, 1587.
94. A. Fujino, K. Fukushima, N. Namiki, T. Kosugi and M. Takimoto-Kamimura, *Acta Crystallogr., Sect. D: Biol. Crystallogr.*, 2010, **D66**, 80.
95. G. Unoki, T. Kosugi, M. Takakuwa, H. Makino, K. Kataoka and Y. Yamakoshi, Teijin Pharma Limited, *US Pat. Appl.*, US 20070173519A1, 2007.
96. W. Vaccaro, Z. Chen, D. S. Dodd, T. N. Huynh, J. Lin, C. Liu, C. P. Mussari, J. S. Tokarski, D. R. Tortolani and S. T. Wrobleski, Bristol-Myers Squibb Company, USA, *PCT Int. Appl.*, WO 2007038314A2, 2007.
97. W. Vaccaro, Z. Chen, D. S. Dodd, T. N. Huynh, J. Lin, C. Liu, C. P. Mussari, J. S. Tokarski, D. R. Tortolani, S. T. Wrobleski and S. Lin, Bristol-Myers Squibb Company, USA, *US Pat. Appl.*, US 20080045536A1, 2008.
98. F. E. Lovering, S. J. Kirincich, W. Wang, J.-B. Telliez, L. Resnick, J. E. Sabalski, A. L. Banker, J. Butera and I. McFadyen, Wyeth, John, and Brother Ltd, USA, *PCT Int. Appl.*, WO 2009012375A2, 2009.
99. F. Lovering, S. Kirincich, W. Wang, K. Combs, L. Resnick, J. E. Sabalski, J. Butera, J. Liu, K. Parris and J. B. Telliez, *Bioorg. Med. Chem.*, 2009, **17**, 3342.
100. M. A. Argiriadi, A. M. Ericsson, C. M. Harris, D. L. Banach, D. W. Borhani, D. J. Calderwood, M. D. Demers, J. DiMauro, R. W. Dixon, J. Hardman, S. Kwak, B. Li, J. A. Mankovich, D. Marcotte, K. D. Mullen,

B. Ni, M. Pietras, R. Sadhukhan, S. Sousa, M. J. Tomlinson, L. Wang, T. Xiang and R. V. Talanian, *Bioorg. Med. Chem. Lett.*, 2010, **20**, 330.

101. C. M. Harris, A. M. Ericsson, M. A. Argiriadi, C. Barberis, D. W. Borhani, A. Burchat, D. J. Calderwood, G. A. Cunha, R. W. Dixon, K. E. Frank, E. F. Johnson, J. Kamens, S. Kwak, B. Li, K. D. Mullen, D. C. Perron, L. Wang, N. Wishart, X. Wu, X. Zhang, T. R. Zmetra and R. V. Talanian, *Bioorg. Med. Chem. Lett.*, 2010, **20**, 334.

102. J. I. Trujillo, M. J. Meyers, D. R. Anderson, S. Hegde, M. W. Mahoney, W. F. Vernier, I. P. Buchler, K. K. Wu, S. Yang, S. J. Hartmann and D. B. Reitz, *Bioorg. Med. Chem. Lett.*, 2007, **17**, 4657.

103. J.-P. Wu, J. Wang, A. Abeywardane, D. Andersen, M. Emmanuel, E. Gautschi, D. R. Goldberg, M. A. Kashem, S. Lukas, M. Wang, L. Martin, T. Morwick, N. Moss, C. Pargellis, U. R. Patel, L. Patnaude, G. W. Peet, D. Skow, R. J. Snow, Y. Ward, B. Werneburg and A. White, *Bioorg. Med. Chem. Lett.*, 2007, **17**, 4664.

104. Z. Xiong, D. A. Gao, D. A. Cogan, D. R. Goldberg, M.-H. Hao, N. Moss, E. Pack, C. Pargellis, D. Skow, T. Trieselmann, B. Werneburg and A. White, *Bioorg. Med. Chem. Lett.*, 2008, **18**, 1994.

105. D. R. Goldberg, Y. Choi, D. Cogan, M. Corson, R. DeLeon, A. Gao, L. Gruenbaum, M. H. Hao, D. Joseph, M. A. Kashem, C. Miller, N. Moss, M. R. Netherton, C. P. Pargellis, J. Pelletier, R. Sellati, D. Skow, C. Torcellini, Y. C. Tseng, J. Wang, R. Wasti, B. Werneburg, J. P. Wu and Z. Xiong, *Bioorg. Med. Chem. Lett.*, 2008, **18**, 938.

106. A. Schlapbach, R. Feifel, S. Hawtin, R. Heng, G. Koch, H. Moebitz, L. Revesz, C. Scheufler, J. Velcicky, R. Waelchli and C. Huppertz, *Bioorg. Med. Chem. Lett.*, 2008, **18**, 6142.

107. J. Velcicky, R. Feifel, S. Hawtin, R. Heng, C. Huppertz, G. Koch, M. Kroemer, H. Moebitz, L. Revesz, C. Scheufler and A. Schlapbach, *Bioorg. Med. Chem. Lett.*, 2010, **20**, 1293.

108. L. Revesz, A. Schlapbach, R. Aichholz, R. Feifel, S. Hawtin, R. Heng, P. Hiestand, W. Jahnke, G. Koch, M. Kroemer, H. Moebitz, C. Scheufler, J. Velcicky and C. Huppertz, *Bioorg. Med. Chem. Lett.*, 2010, **20**, 4715.

109. L. Revesz, A. Schlapbach, R. Aichholz, J. Dawson, R. Feifel, S. Hawtin, A. Littlewood-Evans, G. Koch, M. Kroemer, H. Moebitz, C. Scheufler, J. Velcicky and C. Huppertz, *Bioorg. Med. Chem. Lett.*, 2010, **20**, 4719.

110. T. Barf, A. Kaptein, S. de Wilde, R. van der Heijden, R. van Someren, D. Demont, C. Schultz-Fademrecht, J. Versteegh, M. van Zeeland, N. Seegers, B. Kazemier, B. van de Kar, M. van Hoek, J. de Roos, H. Klop, R. Smeets, C. Hofstra, J. Hornberg and A. Oubrie, *Bioorg. Med. Chem. Lett.*, 2011, **21**, 3818.

111. A. Kaptein, A. Oubrie, E. de Zwart, N. Hoogenboom, J. de Wit, B. van de Kar, M. van Hoek, G. Vogel, V. de Kimpe, C. Schultz-Fademrecht, J. Borsboom, M. van Zeeland, J. Versteegh, B. Kazemier, J. de Roos, F. Wijnands, J. Dulos, M. Jaeger, P. Leandro-Garcia and T. Barf, *Bioorg. Med. Chem. Lett.*, 2011, **21**, 3823.

112. S. Lin, M. Lombardo, S. Malkani, J. J. Hale, S. G. Mills, K. Chapman, J. E. Thompson, W. X. Zhang, R. Wang, R. M. Cubbon, E. A. O'Neill, S. Luell, E. Carballo-Jane and L. Yang, *Bioorg. Med. Chem. Lett.*, 2009, **19**, 3238.

113. R. J. Mourey, B. L. Burnette, S. J. Brustkern, J. S. Daniels, J. L. Hirsch, W. F. Hood, M. J. Meyers, S. J. Mnich, B. S. Pierce, M. J. Saabye, J. F. Schindler, S. A. South, E. G. Webb, J. Zhang and D. R. Anderson, *J. Pharmacol. Exp. Ther.*, 2010, **333**, 797.

114. D. R. Anderson, M. J. Meyers, R. G. Kurumbail, N. Caspers, G. I. Poda, S. A. Long, B. S. Pierce, M. W. Mahoney and R. J. Mourey, *Bioorg. Med. Chem. Lett.*, 2009, **19**, 4878.

115. D. R. Anderson, M. J. Meyers, R. G. Kurumbail, N. Caspers, G. I. Poda, S. A. Long, B. S. Pierce, M. W. Mahoney, R. J. Mourey and M. D. Parikh, *Bioorg. Med. Chem. Lett.*, 2009, **19**, 4882.

116. P.-C. Chiang, S. A. South, J. S. Daniels, D. R. Anderson, S. P. Wene, L. A. Albin, R. J. Mourey and J. G. Selbo, *J. Pharm. Sci.*, 2009, **98**, 248.

117. D. L. Morris, S. P. O'Neil, R. V. Devraj, J. P. Portanova, R. W. Gilles, C. J. Gross, S. W. Curtiss, W. J. Komocsar, D. S. Garner, F. A. Happa, L. J. Kraus, K. J. Nikula, J. B. Monahan, S. R. Selness, G. R. Galluppi, K. M. Shevlin, J. A. Kramer, J. K. Walker, D. M. Messing, D. R. Anderson, R. J. Mourey, L. O. Whiteley, J. S. Daniels, J. Z. Yang, P. C. Rowlands, C. L. Alden, J. W. Davis, II and J. E. Sagartz, *Toxicol. Pathol.*, 2010, **38**, 606.

118. H. Olsson, P. Sjoe, O. Ersoy, A. Kristoffersson, J. Larsson and B. Norden, *Bioorg. Med. Chem. Lett.*, 2010, **20**, 4738.

119. X. Huang, G. W. Shipps, C. C. Cheng, P. Spacciapoli, X. Zhang, M. A. McCoy, D. F. Wyss, X. Yang, A. Achab, K. Soucy, D. K. Montavon, D. M. Murphy and C. E. Whitehurst, *ACS Med. Chem. Lett.*, 2011, **2**, 632.

120. W. X. Zhang, R. Wang, D. Wisniewski, A. I. Marcy, P. LoGrasso, J.-M. Lisnock, R. T. Cummings and J. E. Thompson, *Anal. Biochem.*, 2005, **343**, 76.

121. J. P. Wu, J. Wang, A. Abeywardane, D. Andersen, M. Emmanuel, E. Gautschi, D. R. Goldberg, M. A. Kashem, S. Lukas, W. Mao, L. Martin, T. Morwick, N. Moss, C. Pargellis, U. R. Patel, L. Patnaude, G. W. Peet, D. Skow, R. J. Snow, Y. Ward, B. Werneburg and A. White, *Bioorg. Med. Chem. Lett.*, 2007, **17**, 4664.

122. D. C. Swinney and J. Anthony, *Nat. Rev. Drug Discovery*, 2011, **10**, 507.

123. K. W. Underwood, K. D. Parris, E. Federico, L. Mosyak, R. M. Czerwinski, T. Shane, M. Taylor, K. Svenson, Y. Liu, C.-L. Hsiao, S. Wolfrom, M. Maguire, K. Malakian, J.-B. Telliez, L.-L. Lin, R. W. Kriz, J. Seehra, W. S. Somers and M. L. Stahl, *Structure (Cambridge, MA, US)*, 2003, **11**, 627.

CHAPTER 7
Syk Kinase Inhibitors

NEELU KAILA,* MARK S. RYAN, ATLI THORARENSEN
AND EDDINE SAIAH

Inflammation Research Unit, Pfizer Research and Development, Cambridge,
MA 02140, USA
*Email: neelu.kaila@pfizer.com

7.1 Introduction

Rheumatoid arthritis (RA) is a chronic, autoimmune disease and is prevalent in approximately 1% of the worldwide population.[1] RA is a complex disease and involves multi-faceted pathways, cell types, inflammatory mediators and bone and tissue destruction. The pathogenesis of RA can be described as a cascade of pro-inflammatory cytokines and chemokines that help in the recruitment of inflammatory cells ultimately leading to bone and cartilage destruction. However, the whole inflammatory process in RA is not localized and includes multiple organs and tissues such as the bone marrow, spleen, lymph nodes, joints and cartilages. The process leading ultimately to RA can be categorized into three major steps: migration of inflammatory cells to the site of inflammation; inflammatory reactions with the release of various cytokines; and, finally, bone and cartilage degradation with the activation of osteoclasts and metalloproteases.[2] Understanding the signal transduction pathways that regulate gene expression in RA synovium has led to the exploration of novel therapeutic interventions. Among the various signalling molecules that regulate inflammatory pathways, protein kinases have an important role to play, as their abnormal or up-regulated expression is a common feature in the regulation of RA.[3] Due to their small and relatively conserved catalytic active site, kinases

RSC Drug Discovery Series No. 26
Anti-Inflammatory Drug Discovery
Edited by Jeremy I Levin and Stefan Laufer
© The Royal Society of Chemistry 2012
Published by the Royal Society of Chemistry, www.rsc.org

are considered "drugable" targets by small-molecule inhibitors. A number of small molecule kinase inhibitors have been launched primarily for the treatment of cancer and an increasing number of drug candidates are currently in the clinic for the treatment of chronic inflammatory indications.[4] Syk (Spleen tyrosine kinase) is a cytoplasmic tyrosine kinase involved in signalling in many of the cells that drive immune inflammation and plays a significant role in B-cell receptor signalling on auto-reactive B-cells and in FcγR signalling on mast cells, macrophages, osteoclasts and neutrophils. This supports the idea that Syk inhibition has the potential to interfere with key steps in the progression of RA and could be a broad acting anti-inflammatory agent.[5] As a result, significant drug-discovery efforts were initiated across the industry to discover and develop small-molecule Syk inhibitors. These molecules may change the way we treat disorders such as rheumatoid arthritis (RA), as well as a range of other inflammatory diseases. In the present chapter, we discuss the role of Syk and the effect of its inhibition on the outcome of inflammatory conditions. We also describe Syk inhibitors reported in literature, which consist mainly of traditional ATP competitive compounds. Many of these compounds are at the discovery stage with reported activity in cell-based assays and *in-vivo* models. We will also discuss the chemical equity in general or specific advanced compounds that progressed to the clinic.

7.2   The Role of Syk in Inflammation

Syk kinase is a non-receptor immune kinase expressed primarily by cells of the immune system. Syk has been characterized downstream of B-cell receptors (BCR) on B-cells, and Fc receptors that signal through the common Fc γ chain, such as FcεRI or FcγRI, FcγRIII and FcγRIV located on mast cells, basophils, monocytes, macrophages, NK cells, eosinophils and neutrophils.[6–11] An extracellular matrix receptor on platelets, GPVI, also signals through the Fc γ common chain and Syk.[12] In addition, recruitment of Syk through non-canonical hemi-ITAM (immunostimulatory tyrosine activation motif) has also been shown for integrins and c-type lectin signalling on dendritic cells and macrophages.[13]

The enzyme has been best characterized downstream of BCR signalling on B-cells (Figure 7.1) where, upon receptor activation by an antigen, src family kinases such as Lyn or Fyn phosphorylate the ITAM's of the intra-cellular tails of the receptor complex.[14] Syk recognizes tandem phosphorylated ITAMs and is recruited to the receptor to bind *via* its dual c-terminal SH_2 domains and becomes active, phosphorylating SLP-65, a scaffolding protein.[15] Similarly, Syk is recruited to an antibody-activated Fc receptor complex that has phosphorylated tandem ITAMs and becomes active, phosphorylating a related scaffold protein, SLP-76 and LAT.[16] These scaffold proteins nucleate a signalosome complex, recruiting other kinases, lipases and adapter molecules that link to discrete downstream signalling pathways. In B-cells, Btk (a kinase) and PLCγ2 (a lipase) mediate the activation of the Ca^{++} flux pathway that ultimately activates NFAT and NF-κB transcription factors, which transactivate

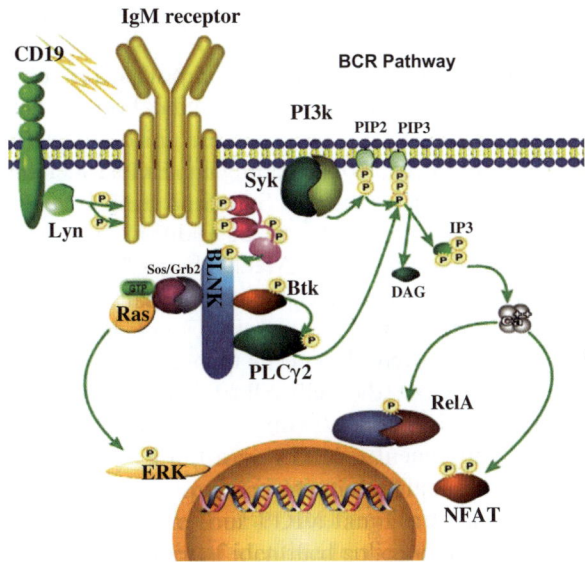

Figure 7.1 Crosslinking of the IgM receptor with antigen on B-cells results in recruitment of Src family kinases such as Lyn. These kinases phosphorylate ITAM motifs on co-receptor tails. This recruits Syk kinase, which phosphorylates a scaffold protein SLP-65 (BLNK). SLP-65 nucleates a complex of Btk and PLCγ2, which activates Ca^{++} flux, leading to NFAT and NF-κB activation. It also binds Grb2/Sos, which then activates the Ras/Mapk pathway leading to ERK activation.

the expression of antibody genes, among others.[17,18] Downstream of FcR, NFAT and NF-κB transactivate the expression of inflammatory genes, such as TNF and IL-8. In addition, in mast cells and basophils, the Ca^{++} flux mediates the immediate release of granules that store pro-inflammatory histamines from the cells. The signalosome complex also recruits other proteins such as guanine nucleotide exchange factors (GEFs) such as SOS and VAV. SOS activates the Ras/Raf/MEK/ERK pathway leading to AP-1 transcription factor activation, while VAV activates Rho family GTPases that mediate actin cytoskeletal changes such as receptor internalization.[19] In addition, in mast cells and basophils, Syk activates PLA_2 in the arachidonic acid pathway producing leukotrienes.

Syk plays a diverse pro-inflammatory role in each of the cell types where it is expressed. In B-cells it induces cellular activation and maturation, as well as antibody production. In mast cells, basophils, monocytes and neutrophils it promotes the release of inflammatory mediators such as histamine and leukotrienes and the secretion of downstream pro-inflammatory cytokines such as TNF and IL-8.[20] In platelets, GPVI signalling through Syk results in dense granule release and platelet activation. Syk is thought to regulate immunity to *Candida albicans* infection by signalling recognition of the yeast by Dectin-2 on dendritic cells and stimulating the development of IL-17 secreting T-cells to control the infection.[21] In each case, Syk is one of the initial events of a pro-inflammatory cascade.

Many cell types have been reported to be important for an inflammatory phenotype and, in order to determine the role of Syk, both total knockouts and cell type-specific knockouts have been generated. Two independent knockout strains of mice were produced, both yielding an almost identical phenotype, giving confidence in the conclusion that Syk expression is required for B-cell (and, hence, antibody) development.[22,23] In addition, Syk was required for FcγR mediated functions in these mice. Macrophages from one of these strains of knockout (KO) mice showed specific impairment of phagocytosis with opsonized antigens. Both the wild type ($syk^{+/+}$) and the KO ($syk^{-/-}$) could bind the Ig opsonized sheep red blood cells through their FcγRs, but only the wild type could phagocytose the cells, illustrating a requirement for Syk-mediated signal transduction downstream of FcγR leading to phagocytosis of antigens.[24] This was subsequently verified in a third genetic strain. Miller and colleagues generated an analogue sensitive kinase allele (ASKA) of Syk and produced transgenic macrophages from $syk^{-/-}$ animals. These macrophages not only had their competence for phagocytosis of opsonized antigens restored, but they could show a dose-dependent inhibition of phagocytosis and Ca^{++} flux with the inhibitor analogue designed for the allele.[25] Furthermore, the specific deletion of Syk in dendritic cells impaired disease generation in a murine OVA-induced diabetes model.[26] This specific knockout abolished downstream signalling of the Fc receptor upon antigen binding. Moreover, a B-cell specific knockout eliminated antigen presentation to T-cells, which also reduced disease in the same model.

Similarly, reconstitution of the immune compartment of a mouse following radio-ablation with $syk^{-/-}$ bone marrow completely blocked the developments of arthritis in a K/BxN serum transfer model, where the transferred auto-antibodies were unable to induce disease.[27] These data all suggest a strong role for Syk function in mediating inflammatory responses in multiple regulatory pathways.

Consistent with these results are those of small-molecule inhibitors of Syk. Fostamatinib or **R788** is the prodrug of **R406**, a small-molecule Syk inhibitor. Compound **R406** has been shown to reduce airway hyperresponsiveness (AHR) in an OVA induced allergen challenge model and decrease cytokine secretion (TNF, IL-13 and IL-6) *in vitro* in OVA-specific IgE challenged mast cells.[28] It has been shown to reduce auto-antibody formation in mice and attenuate collagen-induced arthritis (CIA) in rats.[29,30] Further, it reduced proteinuria and tissue injury through decreased macrophage cytokine secretion in a nephro-toxic rat model of lupus.[31]

Human studies have strongly correlated the expression of Syk enzyme (compared to FcεR or Lyn expression) in basophils to the magnitude of both the antigen induced Ca^{++} flux and also the subsequent histamine release.[32] Small-molecule inhibitors have shown efficacy in rheumatoid arthritis trials where a significant reduction in serum IL-6 was demonstrated, as well as an open label immune thrombocytopenic purpura (ITP) trial, where it was shown to increase the mean peak platelet count in more than half of the patients through the inhibition of anti-platelet antibodies.[33–35] In addition, in some B-cell lymphomas, inhibition of BCR signalling leads to apoptosis of the tumour cells.[36] A clinical trial of recurrent B-cell non-Hodgkins lymphoma patients showed a significant clinical benefit with an objective response rate of 55% for the small lymphocytic leukaemia/chronic lymphocytic leukaemia cohort.[37]

Small-molecule inhibitors of Syk inhibit signalling by competing with cellular ATP for binding to the kinase. The kinase normally cleaves the terminal phosphate group from ATP and attaches it to substrate proteins, which then become novel binding sites for downstream signalling events. By displacing ATP from Syk, the small molecule disrupts this signalling cascade. And while Syk kinase contributes to signalling in different immunoreceptor pathways, capable of binding to multiple ITAM bearing receptor tails as well as phosphorylating different target proteins in different signalling pathways, its biochemical mechanism of action is thought to be similar, suggesting that although an inhibitor may be selective it would still be capable of inhibiting multiple pathways. Studies to determine how Syk functions, biochemically and biophysically, to mediate membrane proximal signalling, may yield insights not only for better small-molecule design but also, as selective inhibitors develop, may decrease the potential field of targets in autoimmunity. Structurally, Syk is a 72 kDa protein that consists of 2 N-terminal SH2 domains separated by a linker denoted interdomain A, and connected to a C-terminal kinase domain by a second linker, interdomain B reviewed in reference 38. While there are no crystal structures of the full-length protein, crystal structures of a related protein, Zap-70, indicate that the two SH2 domains fold back on the protein, making contacts with tyrosine residues in the interdomain B linker.[39] This conformation of the protein is referred to as the resting or inactive state, where the level of enzymatic activity of the kinase domain is relatively low.[40] Much of the time the protein resides in the cytoplasm in this state.

Upon receptor activation, the intra-cellular tails of the receptor become phosphorylated, presenting a binding site for the tandem SH2 domains of Syk.[41] Binding of the SH2 domains to the receptor tails opens and extends the kinase, releasing its autoinhibition. This conformation is further stabilized by phosphorylation of the interdomain B tyrosines by src kinases such as Lyn or Fyn.[42] The fully active kinase bound to the receptor complex phosphorylates the N-terminal tyrosines of a scaffolding protein, such as SLP-65 in B-cells or SLP-76 in mast cells. These scaffolding proteins nucleate attachment and activation of other adaptor proteins, tyrosine kinases and lipases to transduce the signal from the membrane to a phosphorylation cascade through the cytoplasm that ends in transcription factor activation and gene expression program in the nucleus.

Thus far, Syk inhibition has been shown to be anti-inflammatory in multiple animal models, as well as in human studies. Further studies may elucidate which of the pathways described above are the most crucial for maximizing the anti-inflammatory benefit of Syk inhibition, while minimizing the risks of adverse effects for human therapies.

7.3 Structural Features of Syk

The SH2 domains of Syk selectively bind to ITAM upon tyrosine phosphorylation resulting in recruitment of Syk. This interaction has been evaluated by both NMR[42] and protein crystallography.[43] However, the design of small-molecule inhibitors of signalling originating through Syk has been exclusively focused on inhibition of the catalytic kinase domain. This effort has been aided by significant available structural information with the structures of six ligands co-crystallized with Syk available in the protein data bank (PDB) (Figure 7.2a).[44–46] The interactions of these small molecules within the ATP binding site of Syk are very well delineated by the broad structural diversity of these inhibitors. Furthermore, these compounds represent some of the key benchmark compounds that have been described in the literature as Syk inhibitors. The initial observation found by overlaying all of the structures in the same reference frame is that irrespective of whether the kinase structure is obtained with or without a small molecule inhibitor, very few conformational changes are observed among the many structures (Figure 7.2b). In general, these structures are composed largely of β-sheets in the N-terminal lobe, and largely α-helices in the C-terminal lobe. The structures are in the DFG-in conformation, representing the classical structural conformation found for most kinases. The only major movement that can be observed is in the P-loop where the loop movement is *ca.* 3.5 Å between Gly380 in 3FQQE *vs.* the location of the Gly380 in 3FQS keeping a fixed overlay of Met 448 and Ala 451 in the hinge region. This implies that the interaction of a small molecule does not have a significant effect on the overall tertiary structure of Syk. This is surprising considering that Gleevec binds to Abl kinase in a significantly different conformation compared to its binding mode in Syk. The ATP site of Syk can therefore be described as a fairly inflexible site requiring an inhibitor to fit into the pre-defined pocket.

Figure 7.3a illustrates that all the small-molecule Syk inhibitors adopt a shape that fits into the space that can be occupied by staurosporine. A common feature shared by these inhibitors is a hydrogen bond (HB) acceptor that interacts with the backbone NH of Ala451, anchoring it to the protein (Figure 7.3b). In compounds that contain the appropriate hydrogen bond donor as an NH-aryl moiety, an additional HB interaction is achieved with the Ala451 carbonyl. These interactions are nicely illustrated by **R406** binding to Syk. A rather unique but productive interaction is achieved by the basic amine found in **YM193306**, which makes multiple hydrogen bonds to Arg 498, ASN 499 and Asp 512 in the ribose pocket of Syk.

7.4 Known Chemotypes of Syk Inhibitors

Several pharmaceutical companies have been interested in the identification and development of Syk inhibitors for the treatment of RA and other indications. The work in this area has been summarized below; analogues have been classified based on chemotypes. In general Syk inhibitors reported in literature

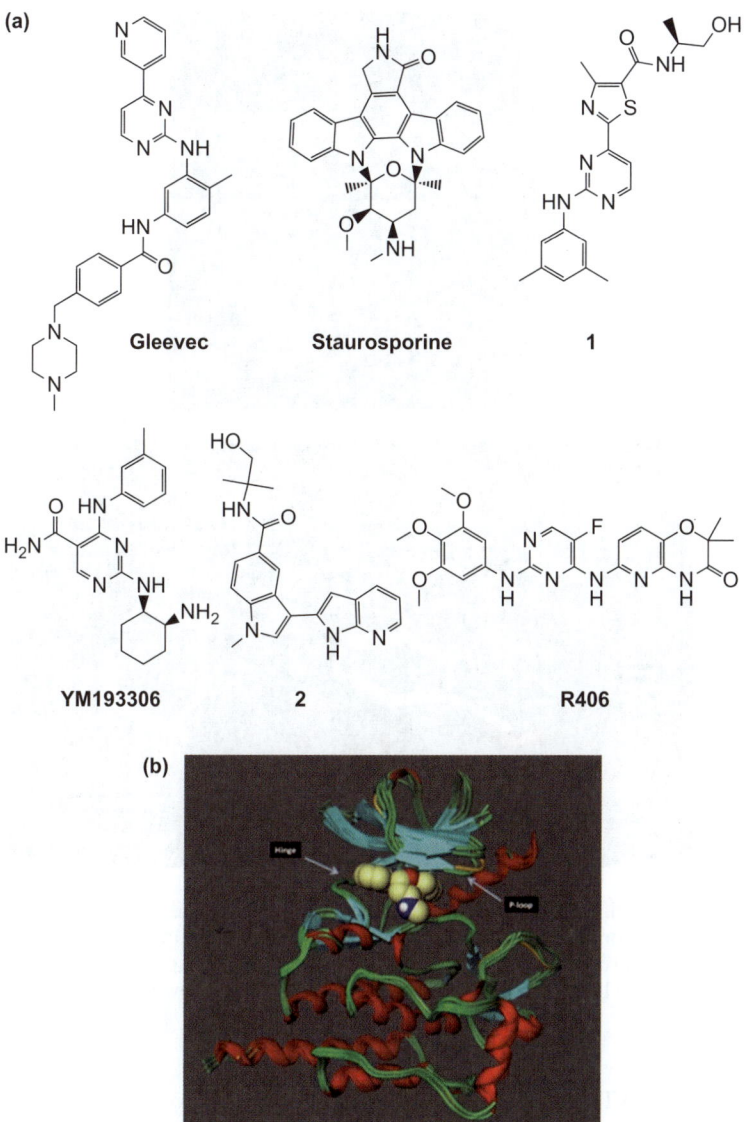

Figure 7.2 The small molecular inhibitors that have been co-crystallized with the SYK kinase domain in the ATP site. **a.** Six benchmark inhibitors co-crystallized with Syk available as PDB structures. **b.** The picture displays all PDB structures aligned in the same reference frame. The apo structure (1XBA) is highlighted in orange while structures (3FQE, 3FQH, 3FQS, 3EMG, 1XXBB and 1XBC) are in green. Staurosporin (1XBC) is highlighted in yellow (CPK model) to illustrate the interaction in the ATP site.

(a)

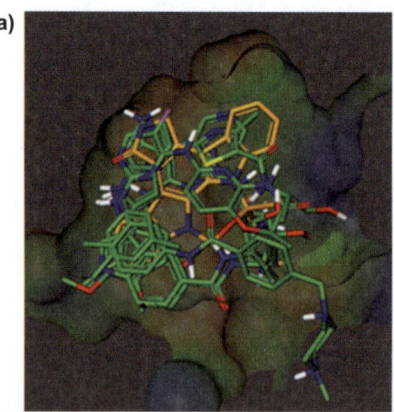

(b)

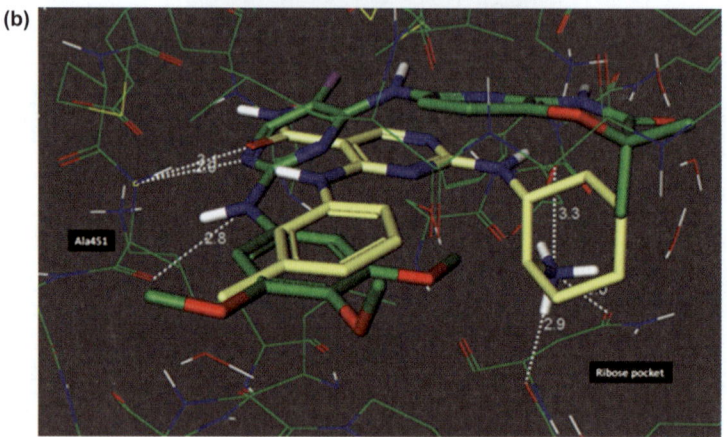

Figure 7.3   Overlays of reference compounds within the ATP-pocket of Syk.
a. Compounds from Figure 7.2 overlaid in the same reference frame with
Staurosporin (orange) and the surface of Staurosporin in the active site of
Syk (1XBC). **b.** R406 green and YM193306 yellow in the ATP site of Syk.

are traditional ATP competitive compounds, consisting of a hinge binder motif
substituted with an aryl group containing a solubilizing group. When comparing analogues from different organizations it is remarkable to see their
close structural similarity. In addition to the clinical compounds by Rigel and
Portola (discussed in the next section) several groups have identified compounds with good enzymatic potency derived from screening and structure
based drug design (SBDD) efforts. Some of these compounds are at the discovery stage with activity in cell based assays and *in-vivo* models. Some chemotypes have been reported as dual inhibitors of Syk with Zap-70, Btk, JAK-3,
Gsk-3, Flt3, Lyn, AurB and Ret, demonstrating that selectivity against these
kinases is an issue. A few advanced compounds have shown efficacy in
inflammatory disease models.

7.4.1 Aminopyrimidines

The presence of both hydrogen bond acceptor and donor capabilities in the diaminopyrimidine core has made this moiety a popular hinge binder in the kinase inhibitor field. Rigel was the first company to put a Syk inhibitor, **R112**, in the clinic (Figure 7.4). This symmetrical diaminopyrimidine, with modest potency in a Syk kinase enzyme assay (Syk $IC_{50} = 96$ nM), selectively inhibits the autophosphorylation of Syk and does not inhibit Lyn-dependent phosphorylation. In a Phase I intra-nasal dosing clinical study for the treatment of allergic rhinitis **R112** showed significant reduction (23% relative to placebo $p = 0.0005$) in global clinical symptoms. However, due to limited duration of action it could not be developed as a twice-a-day medication.[47] A second candidate put forward by Rigel is **R788** (fostamatinib), a prodrug of **R406**. It is the most advanced Syk inhibitor in the clinic and has demonstrated efficacy in Phase II trials for the treatment of RA. **R406** is potent in a Syk enzyme assay ($IC_{50} = 41$ nM) and specifically inhibits Syk-dependent FcR and BCR signalling in various cells with EC_{50}s ranging from 56 to 171 nM. It is efficacious in immune complex-mediated mouse inflammation models. **R406** showed reduction in clinical symptoms in CIA and K/BxN mouse models of RA. In the K/BxN serum transfer mouse model serum injected C57BL/6 mice with PO 10 mg/kg **R406** b.i.d. delayed the onset and reduced the severity of clinical arthritis. Paw thickening and clinical arthritis were reduced by approximately 50% in the reverse-passive Arthus mouse model on p.o. dosing.[29] Prophylactic treatment of mice with **R406** administered 1 h before immune complex challenge reduced the cutaneous reverse-passive Arthus reaction by approximately 72 and 86% at 1 and 5 mg/kg, respectively, compared with the vehicle control. A second compound from Rigel, **R343**, entered Phase I studies as an inhaled treatment of allergic asthma in December 2007.[48]

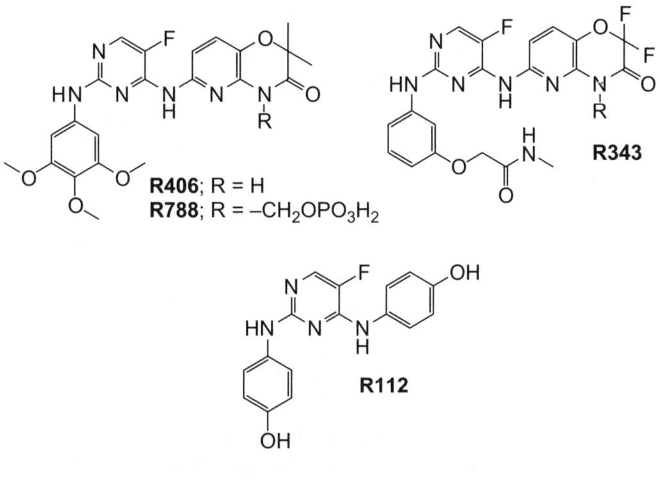

R406; R = H
R788; R = −CH₂OPO₃H₂

R343

R112

Figure 7.4 Compounds in clinical trials from Rigel.

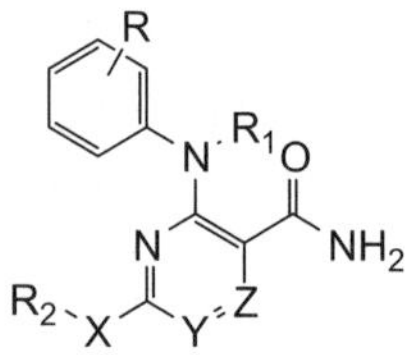

3a: R = 3-CF$_3$, Y = N, Z = CH, XR$_2$ = NH(CH$_2$)$_2$NH$_2$

Figure 7.5   Syk inhibitors disclosed by Yamanouchi/Astellas (generic structure).

Structures of type **3** were developed from an HTS hit and have been disclosed in a Yamanouchi/Astellas patent application (Figure 7.5).[49,50] A co-crystal of compound **YM193306** (K$_i$ = 10 nM, Figure 7.2) from this series bound to Syk shows the carboxamido group as the hinge binder with the diaminocyclohexyl group making polar interactions in the ribose pocket. The SAR observed shows that a carboxamido group substitution at the *meta*-position of the aniline and a diamine group at the 2-position of pyrimidine are required for optimal potency. For example, compound **3a** has an IC$_{50}$ of 41 nM in the Syk enzyme assay and inhibits serotonin (5-HT) release in rat basophilic leukemia (RBL) cells with an IC$_{50}$ of 460 nM. It showed good selectivity against other tyrosine kinases. This compound when administered subcutaneously also inhibited the anaphylaxis reaction in the PCA mouse model in a dose-dependent manner with an ID$_{50}$ of 13.2 mg/kg.[40,47,51]

Portola has filed a patent application on compounds of general structure **4**, structurally similar to the Yamanouchi aminopyrimidines.[52] Four different subclasses (**4a–d**) were disclosed. Based on the patent applications filed, some SAR can be deciphered. Compounds with structure **4a** (Syk IC$_{50}$ = 6 nM; pBlnk EC$_{50}$ = 500–700 nM) are the most potent, and the stereochemistry and ring size of the diamine appear crucial for potency (compare **5a**, **5b** and **6**). Compound **7** has been exemplified by Portola and has shown efficacy in the mouse CIA model when dosed p.o. at 30 mpk. GSK has also done some work on this class of compounds.[53] Their strategy focused on increasing HWB activity and selectivity over hERG. In a recent patent application[54] a very close analogue of **5** has been disclosed by the GSK group (**8**, **GSK143**; IC$_{50}$ = 31 nM; CD69 WB assay IC$_{50}$ = 251 nM). Compound **8** showed good efficacy in the rat Arthus model. It reduced the cutaneous reverse passive Arthus reaction in a dose-dependent manner by 50% and 70% when dosed p.o. at 10 and 30 mg/kg, respectively. The compound was terminated due to positive AMES test.

A second set of patent applications from Portola disclose compounds **9** wherein the 2-diamine substituent has been replaced with 2-amino-aryl group.[55] These compounds are similar to JAK inhibitors reported by Novartis.[56] Analogues **10** (Syk IC$_{50}$ = 4.5 nM; JAK-3 IC$_{50}$ = 2.7 nM) and **11** (Syk IC$_{50}$ = 9.4 nM; JAK-3 IC$_{50}$ = 4.1 nM) in this series are potent against JAK and Syk. In March 2010 Portola entered Phase I clinical trials with the Syk inhibitor **PRT062607** for chronic inflammatory diseases and certain cancers.[57]

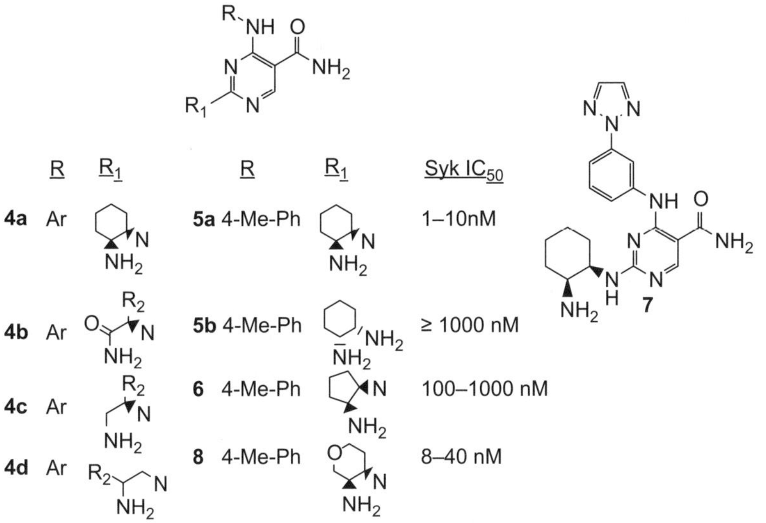

Figure 7.6 Diamino pyrimidine carboxamides as Syk inhibitors.

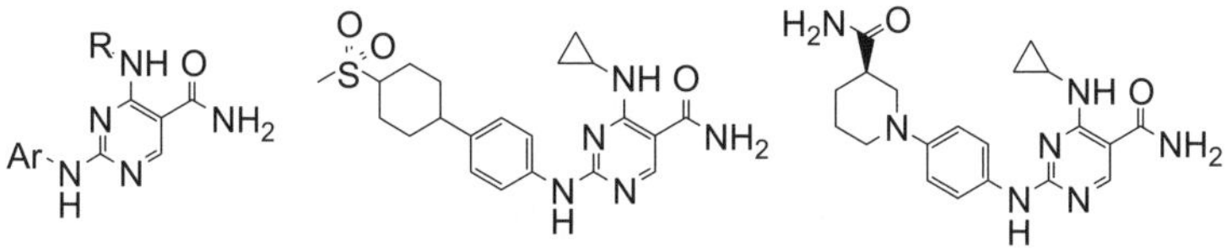

Figure 7.7 2-Aryl amino pyrimidine carboxamides as Syk inhibitors.

Novartis has also disclosed diaminopyrimidines (**12**) as Syk inhibitors.[58] These compounds have been evaluated for both anti-tumour and anti-inflammatory activity. The most recent Novartis Syk patent application includes specific reference to compounds **13–15**, though no biological data were given.[59]

GlaxoSmithKline (GSK) has also disclosed amino pyrimidine derivatives as Syk inhibitors. Patent applications have been filed on two pyrimidine cores. The initial application covers the indazolyl-diamino pyrimidines represented by general structure **16**.[60] Compound **17** is a specific analogue exemplified for this core. The second core claimed is shown in structure **18**, where the indazole group of **16** has been replaced with various solubilizing groups.[61]

Celltech has also done some work in the area of amino pyrimidine Syk inhibitors. In a 1998 patent application they disclosed compounds of structure **19** as inhibitors of Syk and Zap-70.[62] The majority of the examples are of general structure **20**, **21** with a 3,4,5-trimethoxy aniline substituent at the 2-position of the pyrimidine, similar to the Rigel chemotype. No biological data have been reported.

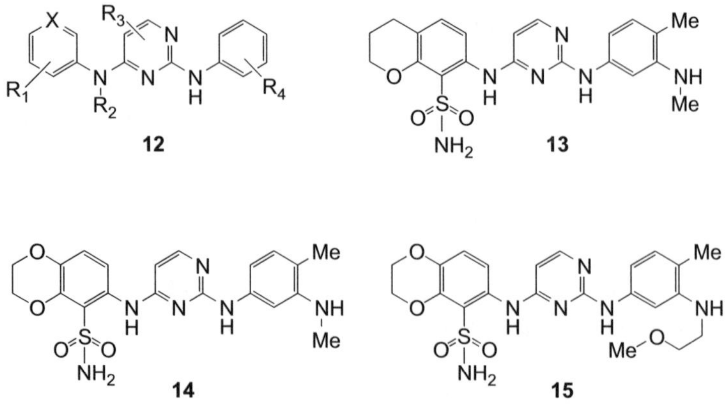

Figure 7.8 Diamino pyrimidine Syk inhibitors disclosed by Novartis.

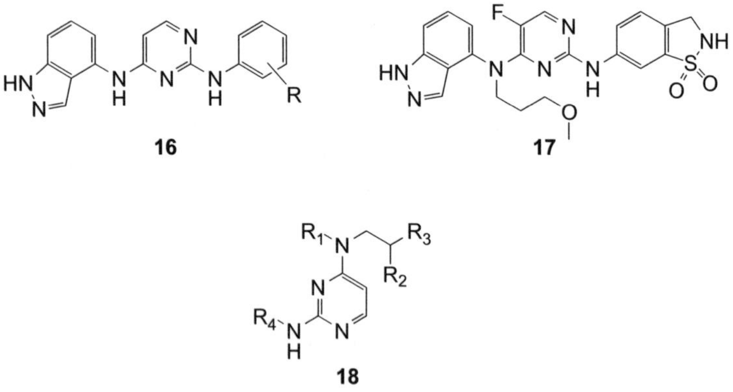

Figure 7.9 Diamino pyrimidine Syk inhibitors disclosed by GSK.

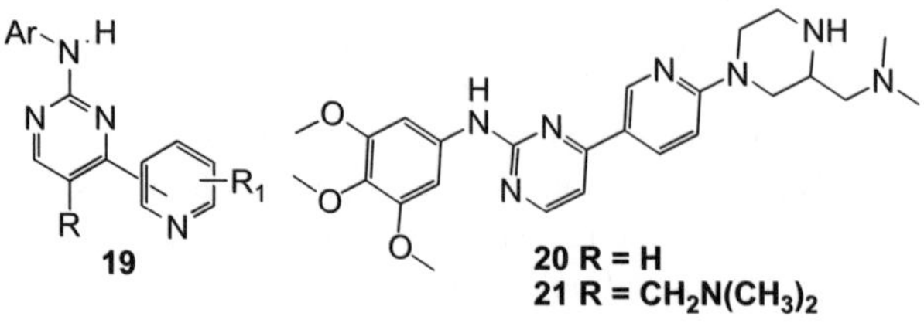

Figure 7.10 Celltech amino pyrimidine Syk inhibitors.

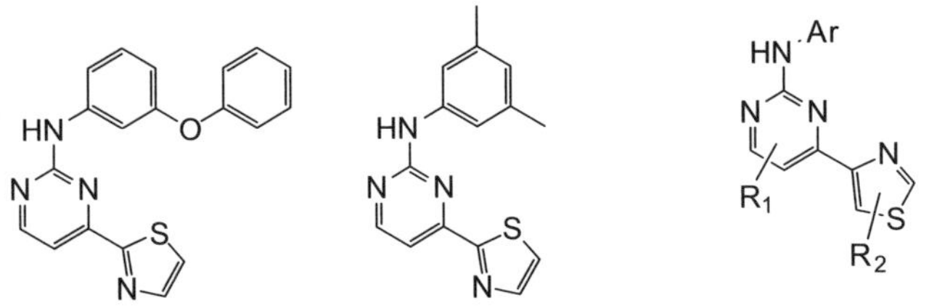

Figure 7.11 Amino pyrimidine Syk inhibitors disclosed by Vertex.

In an HTS screen, Vertex identified the 2-aminopyrimidine **22** as a lead.[45] Using an X-ray crystal structure with Syk as a guide, modifications were made to drive down potency to the nanomolar range. Compounds were tested for selectivity against a small panel of key kinases ROCK, SRC, ZAP70, CDK-2. In an effort to obtain selectivity, chiral groups were added distal to the hinge binding region. (*S*)-Alanol based compound **1** showed good inhibition of SYK with a K_i of 9 nM and an IC_{50} of 70 nM in the mast cell degranulation assay. Selectivity was good against ROCK, SRC, CDK-2 ($>250\times$) but poor against ZAP70 ($26\times$). The (*R*)-alanol analogue was 20-fold less active than enantiomer **1** for SYK inhibition. The co-crystal of compound **1** with the Syk kinase domain shows a hydrogen bond between the amide carbonyl of **1** and the side chain of Lys402. Efficacy data in the brown Norway rat asthma model for one of the analogues from this series, **VRT-124894**, have been presented but not published.[63,64] In addition, a patent application was filed disclosing the phenylamino pyrimidine-4-thiazoles of general structure **23** as Syk and Zap-70 inhibitors.[65,66] Selected analogues inhibited Syk activity in an enzymatic assay with IC_{50}s less than 0.5 μM.

7.4.2 Bicyclic 6/5 Cores

A number of companies have reported bicyclic hinge binding motifs. Portola[67,68] (**24**) and GSK[69,70] (**25**) filed patent applications disclosing pyrrolo-pyrimidine derivatives as Syk inhibitors. Some SAR can be deciphered from the Portola patents. Compounds where heterocycles or *para*-substituted anilines at the 2-position are combined with alkyl-amino substituents at the 4-position were found to be the most potent (Syk IC_{50}s = 1–10 nM) with modest selectivity against the JAK kinases (examples **26**, **27** and **28**). Substitution on the pyrrolo-pyrimidine ring is tolerated at the 5-position, as in compounds **27** and **30**, but the 6-methyl derivative **29** is inactive (compare **29** to **28**). GSK has filed two applications claiming 2-arylamino-4-alkylamino pyrrolo-pyrimidines. The most common substituents at the 4-position of the pyrrolo-pyrimidine are cyclobutyl or trifluoroethyl amine. No biological data have been provided.

The Bayer group has filed a patent application where the hinge binder is a 6/5 bicyclic scaffold, the imidazo[1,2-c]pyrimidine.[71] The most potent compounds have the general structure **31**, with R_1 being a methoxy or other solubilizing

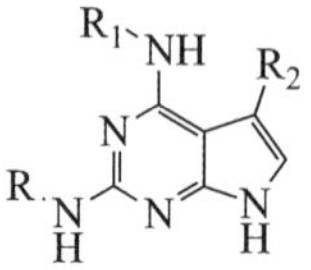

24

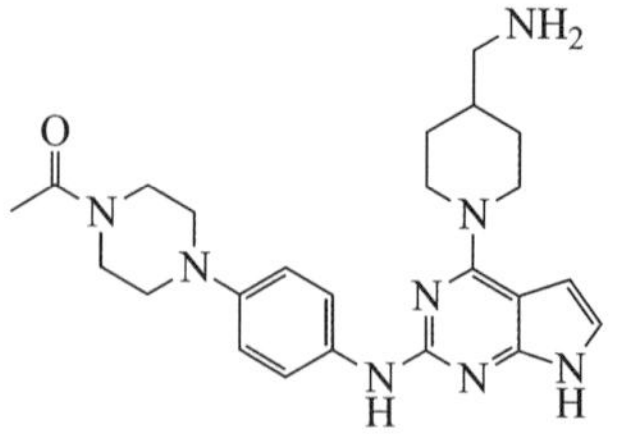

26
Syk IC50 = 1–10 nM
JAK 1,2,3 IC50 = 100–1000 nM

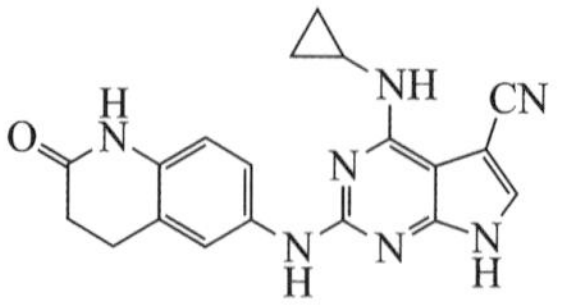

27
Syk IC50 = 1–10 nM
JAK 1,2,3 IC50 = 10–100 nM

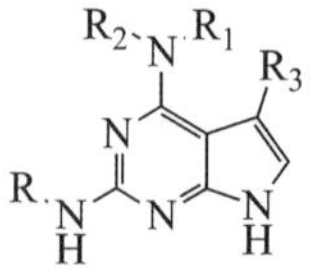

25

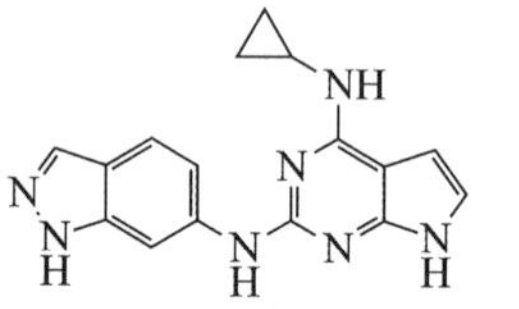

28
Syk IC50 = 43 nM
JAK 1,2,3 IC50 = 100–1000 nM

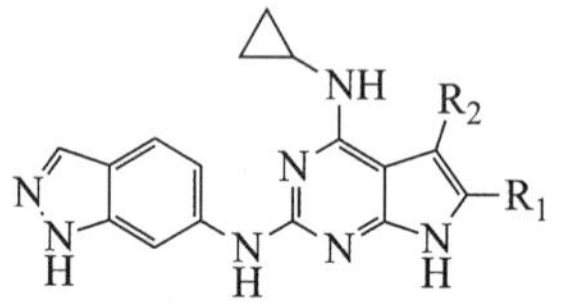

29; R_1 = Me; R_2 = H
Syk IC50 ≥ 1000 nM
30; R_1 = H; R_2 = Br
Syk IC50 = 1–10 nM

Figure 7.12 Pyrrolo pyrimidine containing Syk inhibitors.

group. The R group is most frequently thioethyl, aryl amino or nicotinamido. Two publications[72] have appeared showing data for their lead compound **32(BAY-61-3606)**, a potent (Syk K_i = 7.5 nM) inhibitor. It is selective against other tyrosine kinases, Lyn, Fyn, Src, Itk and Btk, which were not inhibited in concentrations up to 4.7 μM. This compound has been efficacious in suppression of allergic and asthmatic models *in vivo*. When dosed PO in rats **32** inhibited passive cutaneous anaphylactic reaction in an dose-dependent manner with an ED_{50} of 8 mg/kg. In an antigen-induced airway inflammation model in rats at 30 mg/kg p.o., **32** significantly inhibited eosinophil accumulation in the BAL fluid. In addition, when dendritic cells were pre-treated with **32**, the ability of dendritic cells to present immune complex antigens to Th2 cells was inhibited in a dose-dependent manner with an IC_{50} value of 1.08 μM paralleled by a reduced release of interleukin-4 production in Th2 cells, showing that Syk kinase inhibitors may be beneficial in selectively suppressing antibody-mediated antigen presentation in allergic diseases.[73]

The CGI group, now acquired by Gilead sciences, has filed two patent applications disclosing imidazo[1,2-a]pyrazines as Syk inhibitors.[74,75] One of these patent applications is broad and overlaps with the chemical matter disclosed in their Btk patent applications.[76] The general structure **33** summarizes the compounds disclosed. The second patent application is fairly narrow and includes general structure **34**. Some of the more refined structures are **35** and **36**. No biological data have been provided.

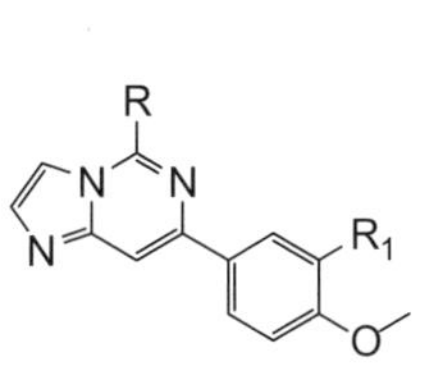

Figure 7.13   Imidazo pyrimidines from Bayer.

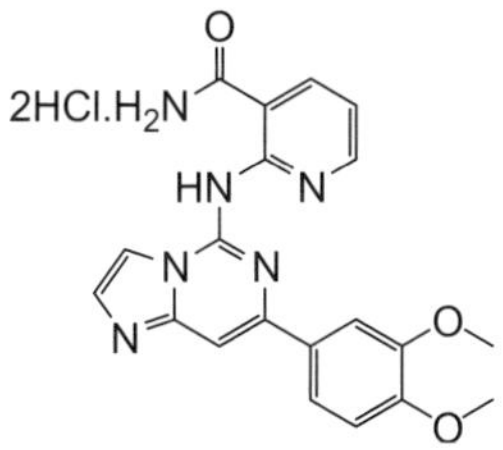

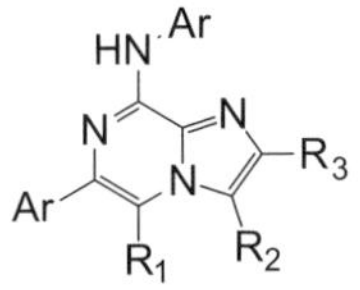

33

R_2, R_3 = small alkyl, halogen

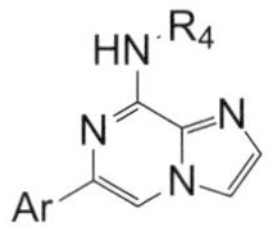

34

R_4 = pyridyl, pyridazinyl, pyrazole

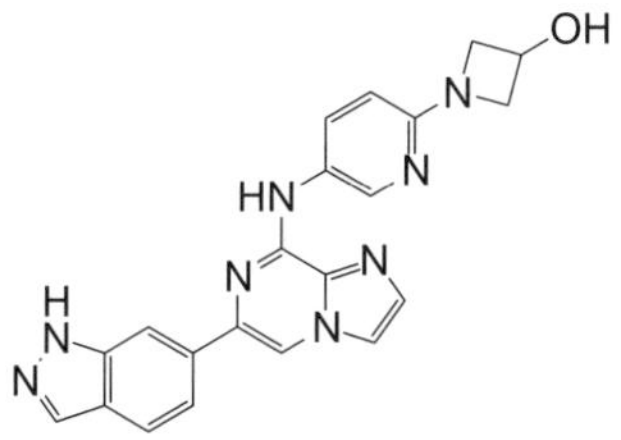

35

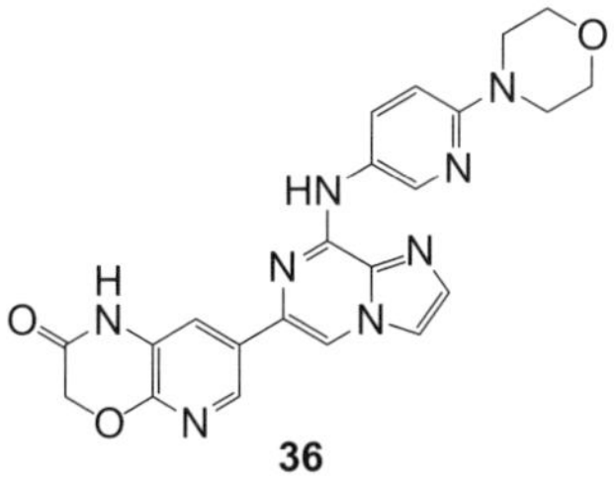

36

Figure 7.14   Imidazo pyrazines as Syk inhibitors.

A third 6/5 bicyclic scaffold that has been used as a hinge binder by workers at Celgene is the triazolopyridine exemplified in structure **37**. Celgene has filed a patent application[77] that includes Syk enzymatic data and describes some cell-based assays. Based on Syk enzymatic data the compounds have been divided into four subsets. The most potent compounds (IC_{50} = 0.005–250 nM) have a mono- or bicyclic arylamino substituent at the 2-position and an aryl or carbocyclic amine at the 5-position (examples **37a–d**).

Novartis has used the 6/5 bicyclic purines as hinge binders.[78] The claimed generic structure, **38**, defines X as N, O and S; however, in the exemplified compounds X is always NH. Enzymatic data are given for a few selected compounds. The *para* position of the aniline group at the 2-position is always substituted. For the majority of the compounds the R_1 group is cycloalkyl or alkyl. The most potent compound disclosed is **39** (Syk IC_{50} = 2 nM).

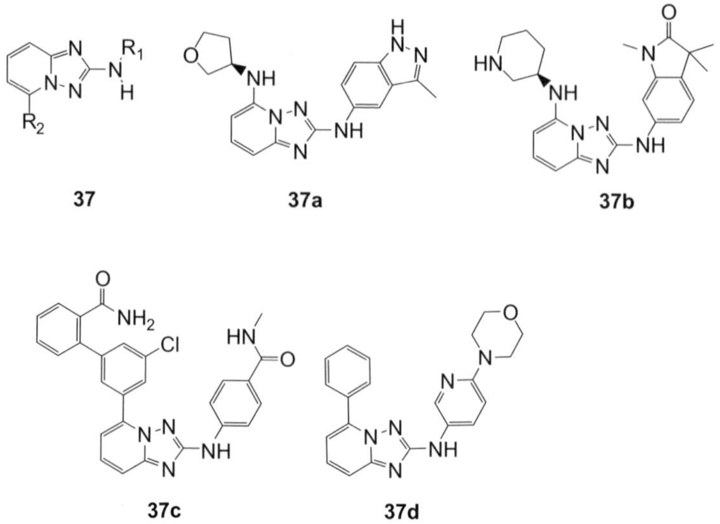

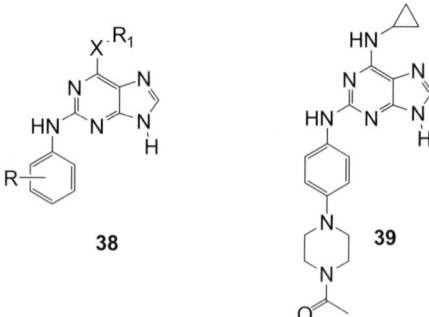

Figure 7.15   Triazolopyridines as Syk inhibitors.

Figure 7.16   Substituted purines as Syk inhibitors.

The majority of the work done by Kissei in the area of Syk inhibitors uses pyrimidine carboxamides as the hinge binder.[79–82] The carboxamide functionality can be placed on either the five- or the six-membered ring of the bicyclic core, as exemplified in compounds **40** and **41**, respectively. The aniline substituent adjacent to the carboxamido group, placed on either the pyrazole (**40**) or the pyrimidine (**41**) ring, occupies the same hydrophobic region making interactions with Leu377. In both cases there is a substituent (A) on the pyrimidine ring that is positioned to interact with the protein ribose pocket. SBDD work was based on literature reports and an approach utilizing structural insight from Zap-70 homology models constructed from a published Lck crystal structure.[83] Considerable work was done in order to identify the optimal

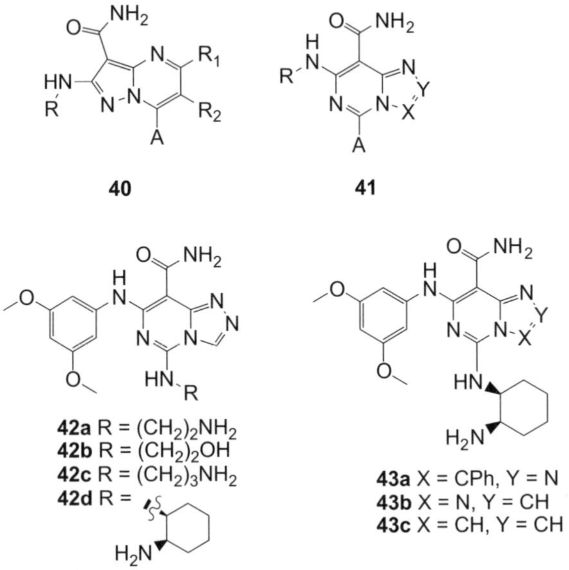

Figure 7.17 Bicyclic pyrimidine carboxamides from Kissei.

substitutuent for the ribose pocket. Thus, the ethylene diamino group of analogue **42a** provides good potency against Syk (IC$_{50}$ = 9 nM) and replacement with a hydroxyl group as in **42b**, or increasing the length of the tether to the primary amine (**42c**), results in a 700-fold loss of potency. The diaminocyclohexane analogue **42d** also shows good potency (IC$_{50}$ = 9 nM) and advanced to PBMC and human whole blood (HWB) assays where reduction of IL-2 formation was observed (PBMC IC$_{50}$ = 46 nM; WB IC$_{50}$ = 440 nM). Substitution in the triazole ring was not tolerated as compound **43a** was inactive (Syk IC$_{50}$ = 15 μM). A rearranged triazole group as shown in structure **43b** was found to be equipotent to 42d. No *in-vivo* data for these compounds have been reported due to poor bioavailability. Based on the hypothesis that lowering the PSA of the triazoles will improve Caco-2 permeability and bioavailability, the Kissei group designed a series of imidazo[1,2-c]pyrimidine derivatives.[84] Optimized compound **43c** showed strong inhibitory activities against Syk (IC$_{50}$ = 6 nM) and Zap-70 (IC$_{50}$ = 230 nM) kinases *in vitro*. When dosed p.o., **43c** showed *in-vivo* suppression of both the passive cutaneous anaphylaxis reaction at 10 mg/kg and concanavalin A-induced IL-2 production at 30 mg/kg in a mouse model.

Roche has filed several patent applications disclosing pyrrolo-pyrazines as dual inhibitors of JAK-3 and Syk.[85–88] Most of the compounds disclosed have a ketone at the 7-position of the core, as shown in structure **44** (R = alkyl) A range of substitutents have been exemplified at the 2-position; JAK-3 data are provided for a handful of compounds (JAK-3 IC$_{50}$ = 0.1–0.5 μM).

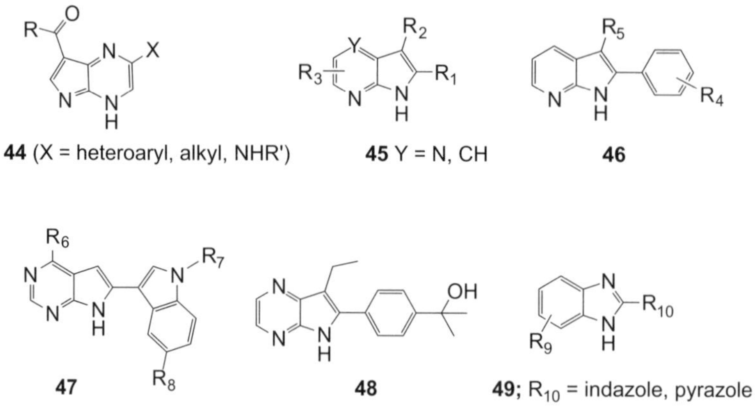

44 (X = heteroaryl, alkyl, NHR')   **45** Y = N, CH   **46**

47   **48**   **49**; R$_{10}$ = indazole, pyrazole

Figure 7.18   Bicyclic pyrroles and benzimidazoles as Syk inhibitors.

Pyrrolo-pyrazines (**45**), pyridines (**46**) and pyrimidines (**47**) have been disclosed as Syk inhibitors by researchers at Sanofi-Aventis.[89–91] More recent patent applications focus on compound **48** (Syk IC$_{50}$ = 1.7 nM) as a Syk inhibitor for the treatment of ocular disorders, macular degeneration and glomerulonephritis.[92–96] Compound **48** significantly delayed progression and severity of disease in the rat CIA when dosed p.o. b.i.d. at 10 mg/kg. However, there is no report of this compound reaching clinical trials. Sanofi-Aventis has filed another application that mentions inhibitors of Syk and other kinases such as EGFR and KDR.[97] This patent application discloses benzimidazoles with general structure **49** with SYK IC$_{50}$s ranging from 100 μM to 0.1 nM. No selectivity data are reported.

7.4.3   Bicyclic 6/6 Cores

Boehringer Ingelheim (BI) has filed patent applications disclosing 1,6-naphthyridines (**50**) as Syk inhibitors.[98–100] The substitution pattern on the naphthyridine core suggests that the N-1 nitrogen is probably the hinge binder, and the R$_2$ substituent occupies the ribose pocket. The most frequently exemplified R$_2$ substituent is a diaminopropane moiety. A variety of substituted aryl groups have been used at the 7-position (R$_3$). An additional patent application filed by BI restricts the R$_3$ substituent to the 3-methyl-4-dimethylamino phenyl group, but a variety of R$_2$ substituents where an alkyl or aryl group is connected to the naphthyridine through a linker have been disclosed (**51**).[101] In a follow-up application the R$_3$ substituent is defined as an aryl group bearing a cyclic amine, as shown in structure 52, and R$_2$ is defined similarly to earlier BI patent applications.[102] Based on the enzymatic data reported, the most potent compounds (>95% inhibition of Syk at 1 μM, **52a–c**) have a 4-morpholinophenyl group at the 7-position.

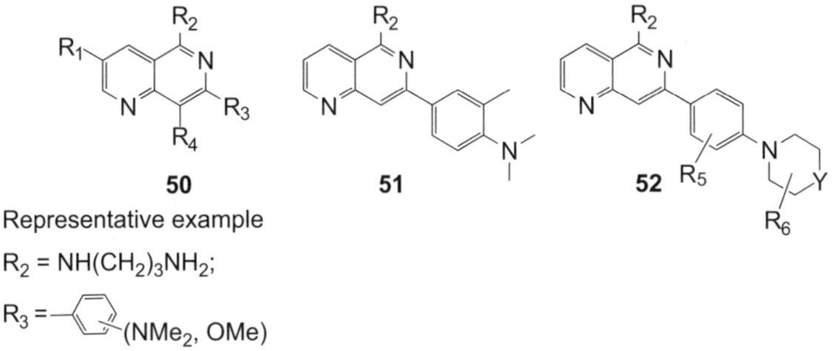

Figure 7.19 1,6- Naphthyridine as Syk inhibitors.

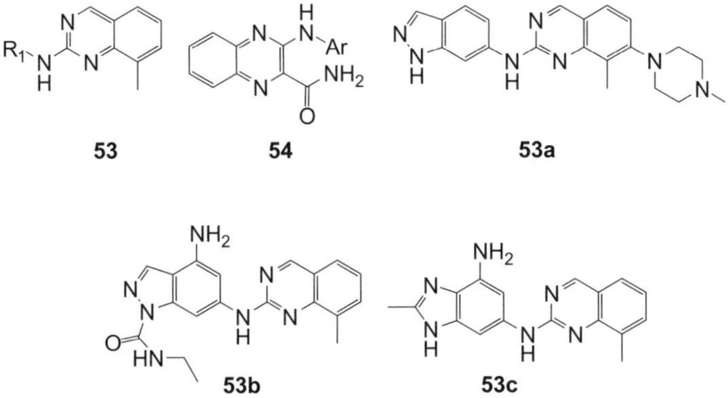

Figure 7.20 Quinazolines and quinoxalines as Syk inhibitors.

Carna Biosciences has filed a patent application disclosing 2-amino quinazolines (**53**) as general kinase inhibitors including Syk.[103] Enzymatic and cell-based data have been reported for Syk inhibition. The cell-based assay is an IgE-induced degranulation basophil assay. Structure **53a–c** describes compounds that were potent ($IC_{50} < 50$ nM) in the cell-based assay.

N-aryl-2-carboxamido quinoxalin-2-amine (**54**) compounds have been evaluated as JAK, Syk and GSK-3 inhibitors by Vertex, but no potency or selectivity data were provided.[104]

IRM LLC and Novartis, in collaboration, have disclosed 2,7-naphthyridinones, represented by generic structure **55a** and **55b**, as inhibitors of a range of kinase targets.[105] Syk is a target of importance in the application since specific enzymatic data for inhibition of this enzyme have been provided for 24 compounds with compound **55c** showing an IC_{50} of 3 nM in the Syk enzymatic assay and 120 nM in BAF3/Tel-Syk cellular assay. It is noteworthy that analogues where R_2 is a methyl group retain potency against Syk, signifying that this does not appear to be the hinge-binding motif.

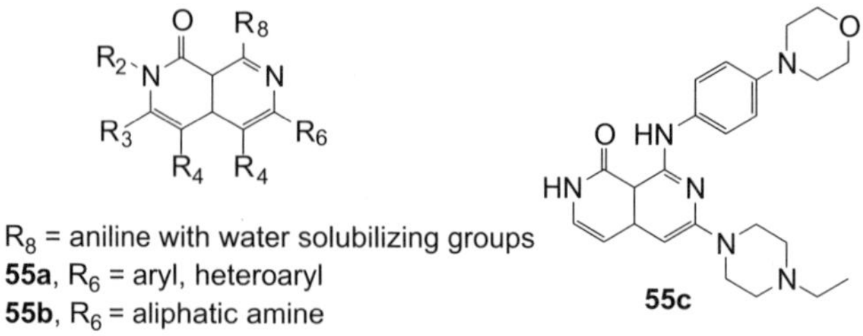

R8 = aniline with water solubilizing groups
55a, R6 = aryl, heteroaryl
55b, R6 = aliphatic amine

55c

Figure 7.21 Naphthyridinones as Syk inhibitors.

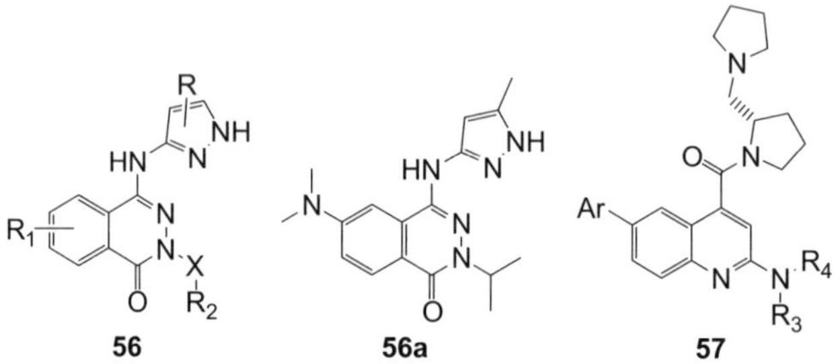

56   **56a**   **57**

Figure 7.22 Phthalazinones and quinolines as Syk inhibitors.

Aminopyrazole-phthalazinones (**56**) have been disclosed as dual Btk/Syk inhibitors in a patent application filed by researchers at Roche.[106] However, the data in the application show the compounds to be only weakly active. The most potent compound is **56a** with an IC_{50} of 300 nM against Syk.

Neogenesis has filed a patent application[107] centred on quinolines of general structure **57**. More than 400 analogues were prepared ranging in ZAP-70 potency. The primary assay used is ZAP-70 Delfia assay where the compounds were classified in three categories with IC_{50}s $< 1\,\mu M$, $1–10\,\mu M$ and $> 10\,\mu M$. No SYK potency has been reported.

7.4.4 Monocyclic Cores

Very few monocylic Syk inhibitors, excluding the amino-pyrimidines, have been reported.

Oxaprozin, its prodrugs and bioisosteres (**58**), have been disclosed for the topical treatment of eczema, dermatitis and pruritis by Astion.[108,109] The mode of action for these compounds is considered to be through the inhibition of Syk, Zap-70 and PDE-IV. It shows inhibition in ear swelling in the oxazalone induced mouse model when applied at doses of 250 μg/ear.

Scientists at the Centre National de la Recherche Scientifique have used tool compound **59** to develop an antibody displacement assay, which is used to identify Syk inhibitors. This assay uses two antibodies that bind to Syk at an

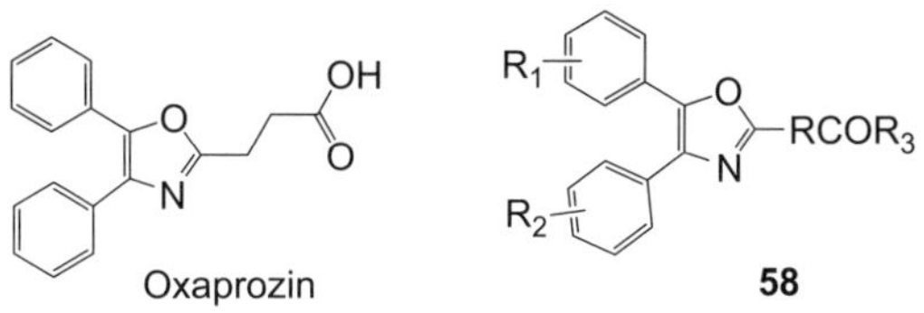

Figure 7.23 Oxaprozin and its analogues.

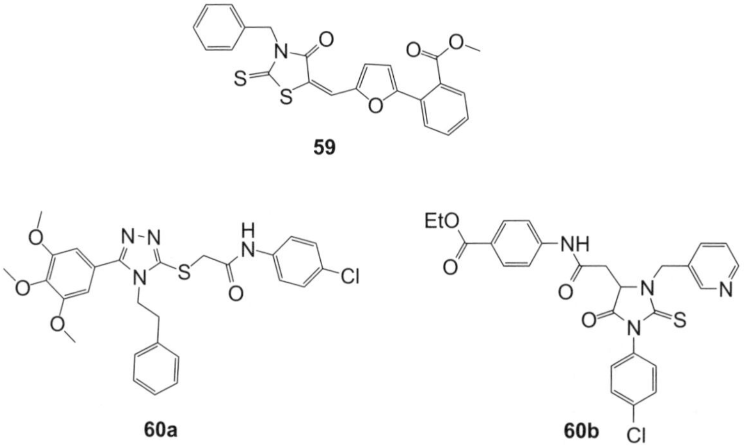

Figure 7.24 Syk inhibitors from Centre National de la Recherche Scientifique.

epitope that contains residues **65** to **74**. Using structural information derived from this assay a virtual screen was run on a library from Chembridge. One thousand compounds with good docking scores were selected and tested. Cell-based data have been included for selected compounds. The most potent compounds reported were **60a** and **60b** showing 50% inhibition of mast cell degrannulation at a concentration of 2 μg/mL.[110] Compound **59** was evaluated in an *in-vivo* passive cutaneous anaphylaxis mouse model where it inhibited increase of vascular permeability with an IC_{50} of 110 mg/kg.

Dipyridyl amines have been identified by the Japan Tobacco Company as Syk inhibitors.[111] Two subseries have been studied in detail, the thiazole (**61**) and thiophene (**62**) series. The most potent compounds reported are **61a–c** (only data reported for these compounds is $IC_{50} > 100$ nM in a Syk HTRF and human degrannulation assay).

7.5 Clinical Development of Syk Inhibitors

The level of interest in Syk has been amplified by the recent disclosure of **R788** and its efficacy in Phase II trials for RA, resulting in a significant commercial collaboration between Rigel and AstraZeneca.[112] Portola has initiated a Phase I

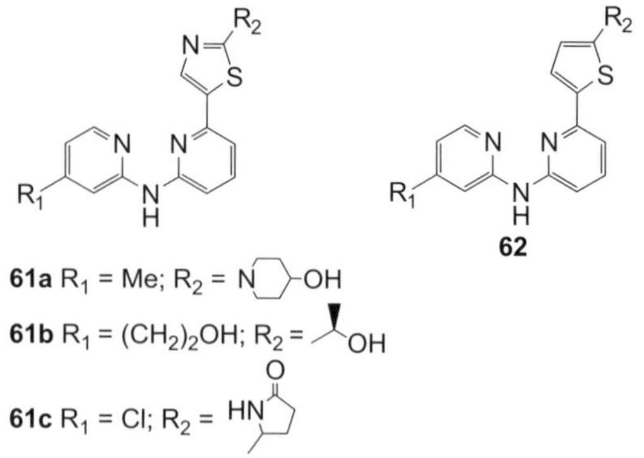

61a R_1 = Me; R_2 = N⟨ ⟩–OH

61b R_1 = $(CH_2)_2OH$; R_2 = ⌁$_{OH}$

61c R_1 = Cl; R_2 = HN⟨ ⟩=O

Figure 7.25 Dipyridyl amines as Syk.

clinical trial for its exclusive Syk inhibitor **PRT062607**.[57] Results to date suggest the compound is well tolerated and has a suitable profile for once-daily dosing. Phase IIA testing for rheumatoid arthritis is expected to start in the second half of 2012. In addition Biogen and Portola have entered into a licensing agreement to develop and commercialize **PRT062607** for treatment of RA and other inflammatory diseases.[113]

The initial disclosure of a Syk inhibitor in clinical development was compound **R112**. This compound is reported to have moderate affinity for Syk (K_i = 96 nM). **R112** was explored in a Phase II clinical trial for seasonal allergic rhinitis utilizing intra-nasal delivery with a metered spray pump.[114] The compound had efficacy in that trial, as measured by global symptom complex (GSC) with a statistical difference from placebo in the study, and was well tolerated with very few side-effects, mainly irritation around the site of delivery. Despite these results no further reports have appeared on the progression of **R112**. Recently, an inhaled formulation of **R343** was reported to have entered clinical trials for allergic asthma.[48] Very little information is available regarding the efficacy of this compound, but its development is a further illustration of the role Syk plays in allergic inflammatory events.

The most notable Rigel Syk inhibitor is **R406**, a compound that has been reported to be in various stages of clinical trials. **R406**, or its prodrug **R788**, is an orally bioavailable molecule with good affinity for the enzyme (K_i = 30 nM) and has demonstrated efficacy in pre-clinical rodent models of arthritis and inflammation.[30] The initial evaluation of **R406** in humans illustrated a dose-proportional increase in exposure of up to 400 mg, where the exposure plateaus. The effect of **R406** on basophil activation was evaluated in an *ex-vivo* assay by CD63 activation resulting in an EC_{50} of 1.06 µM.[29] In later clinical trials, **R406** was delivered as the methylenephosphate prodrug fostamatinib (**R788**), which demonstrated improved solubility and formulation properties over **R406**.[115] **R788** is reported to have good aqueous solubility (> 5 mg/mL), representing a 2000-fold improvement over that of **R406** (1–2 µg/mL). The improved aqueous solubility also translated into improved formulation solubility and facilitated oral administration and enhanced bioavailability. The metabolic fate of **R788** was recently evaluated in a human mass balance study complimented with *in-vitro* hepatic studies.[116] The conclusions from these studies suggest that the prodrug is converted into the parent (**R406**) in the intestine by alkaline phosphatases. The primary constituent identified in plasma was **R406**. Primary and secondary metabolism led to secretion into the bile. **R788** has been evaluated in a Phase II trial for immune thrombocytopenic purpura (ITP).[35] ITP is characterized as Fcγ receptor mediated destruction of platelets. The consequence of low platelet count is intra-cranial haemorrhage along with other bleeding complications. In this study a dose escalation from 75 mg to 175 mg twice daily (b.i.d.) was utilized with 16 enrolled subjects. Inhibition of Syk was efficacious in half of the patients at doses of 125 mg to 175 mg b.i.d. with an improvement in platelet count $> 100,000 /mm^3$. The major side-effects in this study were GI related such as diarrhoea. **R788** has been evaluated in a non-Hodgkin's lymphoma and chronic lymphocytic leukemia.[37] The initial dose evaluated in this trial was 200 mg or 250 mg b.i.d. for initial evaluation. Due to an increased frequency of dose-dependent toxicity at the 250 mg b.i.d. dose, such as hypertension, anaemia and nausea, the 200 mg b.i.d. dose was elected for further Phase II evaluation. **R788** provided a median progression free survival time of 4.2 months. There was no clear correlation between PK and clinical outcome or the number of $CD14^+$ cells in circulation. **R788** has also demonstrated clinical efficacy in Phase II trials of active RA in humans.[34,117] The initial 12-week trial of **R788** in RA utilized doses of 50 mg, 100 mg or 150 mg b.i.d. The 100 mg and 150 mg dose provided superior efficacy compared to placebo utilizing American College of Rheumatology criteria (ACR). The ACR50 responses for the 100 mg and 150 mg doses were 49% and 57%, respectively, *versus* 17% for placebo. The efficacy correlated with a decrease in IL-6 and MMP3 levels. The side-effect profile was similar to previous studies with predominant GI side-effects. In addition, hypertension occurred in *ca.* 5% of participants in the range of 4–7 mm Hg from baseline. **R788** was further evaluated at 100 mg b.i.d. or 150 once daily (q.d.) in a 6-month RA study with 457 enrolled patients. In this study the initial clinical improvement was observed after one week of dosing with ACR50 response rates of 43% and 32% *versus* 19% for placebo at the study end-point. The side-effects were similar to previous studies, but with a decreased frequency, which correlates with the reduced dose explored in this study compared with previous studies. Finally **R788** has been explored in RA patients that failed treatment with biologics.[33] In this trial **R788** was dosed at 100 mg b.i.d. in a 3-month study enrolling 229 patients and no significant differences with regard to ACR scores *versus* placebo were seen. The high placebo rate in this study along with the high medical need for RA patients that fail biologic treatment, along with questions raised by the study authors regarding the difference in patient composition in the placebo *vs.* the **R788** arm with regard to both disease severity and number of prior biological treatments surely will stimulate additional evaluation of this promising compound as a new oral RA medication.

In a recent *Nature Reviews* article Prof. Firestein points out that compounds from both Rigel and Portola are doing well in the clinical trials; however, **PRT062607** is claimed to be more selective than **R788** and therefore Portola data will provide direct support for Syk as a target, whereas **R788** is a promiscuous inhibitor and it is not certain whether efficacy and toxicity are coming primarily from Syk inhibition.[113]

7.6 Summary

In this review an overview of Syk biology and structural biology has been provided, along with a description of a wide variety of small-molecule inhibitors. A summary of several human clinical trials that have appeared in the literature with Syk inhibitors has also been presented. In general, the structural theme of most of the Syk inhibitors disclosed in the literature is that of a traditional ATP competitive inhibitor with different hinge-binding elements. In general the attachment to the hinge is an aryl amine that is decorated with a solubilizing group. The other substituents are broadly variable across the numerous scaffolds indicating that Syk has significant flexibility in its interactions in that portion of the kinase.

References

1. D. L. Scott, F. Wolfe and T. W. Huizing, *Lancet*, 2010, **376**, 1094.
2. M. Bajpai, P. Chopra, S. G. Dastidar and A. Ray, *Expert Opin. Investig. Drugs*, 2008, **17**, 641.
3. M. Gaestel, A. Mengel, U. Bothe and K. Asadullah, *Curr. Med. Chem.*, 2007, **14**, 2214.
4. S. Cohen and R. Fleischmann, *Curr. Opin. Rheumatol.*, 2010, **22**, 330.
5. M. Riccaboni, I. Bianchi and P. Petrillo, *Drug Discovery Today*, 2010, **15**, 517.
6. R. L. Geahlen, *Biochim. Biophys. Acta*, 2009, **1793**, 1115.
7. A. M. Gilfillan and J. Rivera, *Immunol. Rev.*, 2009, **228**, 149.
8. J. Abramson and I. Pecht, *Immunol. Rev.*, 2007, **217**, 231.
9. M. Simon, L. Vanes, R. L. Geahlen and V. L. Tybulewicz, *J. Biol. Chem.*, 2005, **280**, 4510.
10. S. Hida, S. Yamasaki, Y. Sakamoto, M. Takamoto, K. Obata, T. Takai, H. Karasuyama, K. Sugane, T. Saito and S. Taki, *Nat. Immunol.*, 2009, **10**, 214.
11. Z. Jakus, E. Simon, D. Frommhold, M. Sperandio and A. Mocsai, *J. Exp. Med.*, 2009, **206**, 577.
12. S. P. Watson, J. M. Herbert and A. Y. Pollitt, *J. Thromb. Haemost.*, 2010, **8**, 1456.
13. J. C. Spalton, J. Mori, A. Y. Pollitt, C. E. Hughes, J. A. Eble and S. P. Watson, *J. Thromb. Haemost.*, 2009, **7**, 1192.
14. M. Reth and J. Wienands, *Ann. Rev. of Immunol.*, 1997, **15**, 453.

15. J. G. Monroe, *Nat. Rev. Immunol.*, 2006, **6**, 283.
16. T. Kambayashi and G. A. Koretzky, *J. Allergy Clin. Immunol.*, 2007, **119**, 544.
17. J. B. Petro and W. N. Khan, *J. Biol. Chem.*, 2001, **276**, 1715.
18. S. Watanabe, M. Hashimoto, M. Ishiai, Y. Matsushita, T. Baba, T. Kishimoto, T. Kurosaki and S. Tsukada, *J. Biol. Chem.*, 2001, **276**, 38595.
19. G. A. Koretzky, F. Abtahian and M. A. Silverman, *Nat. Rev. Immunol.*, 2006, **6**, 67.
20. F. Nimmerjahn and J. V. Ravetch, *Immunol. Rev.*, 2010, **236**, 265.
21. A. Puel, S. Cypowyj, J. Bustamante, J. F. Wright, L. Liu, H. K. Lim, M. Migaud, L. Israel, M. Chrabieh, M. Audry, M. Gumbleton, A. Toulon, C. Bodemer, J. El-Baghdadi, M. Whitters, T. Paradis, J. Brooks, M. Collins, N. M. Wolfman, S. Al-Muhsen, M. Galicchio, L. Abel, C. Picard and J. L. Casanova, *Science*, 2011, **332**, 65.
22. A. M. Cheng, B. Rowley, W. Pao, A. Hayday, J. B. Bolen and T. Pawson, *Nature*, 1995, **378**, 303.
23. M. Turner, P. J. Mee, P. S. Costello, O. Williams, A. A. Price, L. P. Duddy, M. T. Furlong, R. L. Geahlen and V. L. Tybulewicz, *Nature*, 1995, **378**, 298.
24. M. T. Crowley, P. S. Costello, C. J. Fitzer-Attas, M. Turner, F. Meng, C. Lowell, V. L. Tybulewicz and A. L. DeFranco, *J. Exp. Med.*, 1997, **186**, 1027.
25. A. L. Miller, C. Zhang, K. M. Shokat and C. A. Lowell, *J. Immunol.*, 2009, **182**, 988.
26. L. Colonna, G. Catalano, C. Chew, V. D'Agati, J. W. Thomas, F. S. Wong, J. Schmitz, E. S. Masuda, B. Reizis, A. Tarakhovsky and R. Clynes, *J. Immunol.*, 2010, **185**, 1532.
27. Z. Jakus, E. Simon, B. Balazs and A. Mocsai, *Arthritis Rheum.*, 2010, **62**, 1899.
28. S. Matsubara, G. Li, K. Takeda, J. E. Loader, P. Pine, E. S. Masuda, N. Miyahara, S. Miyahara, J. J. Lucas, A. Dakhama and E. W. Gelfand, *Am. J. Respir. Crit. Care Med.*, 2006, **173**, 56.
29. S. Braselmann, V. Taylor, H. Zhao, S. Wang, C. Sylvain, M. Baluom, K. Qu, E. Herlaar, A. Lau, C. Young, B. R. Wong, S. Lovell, T. Sun, G. Park, A. Argade, S. Jurcevic, P. Pine, R. Singh, E. B. Grossbard, D. G. Payan and E. S. Masuda, *J. Pharmacol. Exp. Ther.*, 2006, **319**, 998.
30. P. R. Pine, B. Chang, N. Schoettler, M. L. Banquerigo, S. Wang, A. Lau, F. Zhao, E. B. Grossbard, D. G. Payan and E. Brahn, *Clin. Immunol.*, 2007, **124**, 244.
31. J. Smith, J. P. McDaid, G. Bhangal, R. Chawanasuntorapoj, E. S. Masuda, H. T. Cook, C. D. Pusey and F. W. Tam, *J. Am. Soc. Nephrol.*, 2010, **21**, 231.
32. D. W. MacGlashan Jr., *J. Allergy Clin. Immunol.*, 2007, **119**, 626.
33. M. C. Genovese, A. Kavanaugh, M. E. Weinblatt, C. Peterfy, J. DiCarlo, M. L. White, M. O'Brien, E. B. Grossbard and D. B. Magilavy, *Arthritis Rheum.*, 2011, **63**, 337.

34. M. E. Weinblatt, A. Kavanaugh, R. Burgos-Vargas, A. H. Dikranian, G. Medrano-Ramirez, J. L. Morales-Torres, F. T. Murphy, T. K. Musser, N. Straniero, A. V. Vicente-Gonzales and E. Grossbard, *Arthritis Rheum.*, 2008, **58**, 3309.
35. A. Podolanczuk, A. H. Lazarus, A. R. Crow, E. Grossbard and J. B. Bussel, *Blood*, 2009, **113**, 3154.
36. D. G. Efremov and L. Laurenti, *Expert Opin. Investig. Drugs*, 2011, **20**, 623.
37. J. W. Friedberg, J. Sharman, J. Sweetenham, P. B. Johnston, J. M. Vose, A. Lacasce, J. Schaefer-Cutillo, S. De Vos, R. Sinha, J. P. Leonard, L. D. Cripe, S. A. Gregory, M. P. Sterba, A. M. Lowe, R. Levy and M. A. Shipp, *Blood*, 2010, **115**, 2578.
38. J. M. Bradshaw, *Cell. Signal.*, 2010, **22**, 1175.
39. B. B. Au-Yeung, S. Deindl, L. Y. Hsu, E. H. Palacios, S. E. Levin, J. Kuriyan and A. Weiss, *Immunol. Rev.*, 2009, **228**, 41.
40. E. Papp, J. K. Tse, H. Ho, S. Wang, D. Shaw, S. Lee, J. Barnett, D. C. Swinney and J. M. Bradshaw, *Biochemistry*, 2007, **46**, 15103.
41. L. I. Pao, S. J. Famiglietti and J. C. Cambier, *J. Immunol.*, 1998, **160**, 3305.
42. T. D. Groesch, F. Zhou, S. Mattila, R. L. Geahlen and C. B. Post, *J. Mol. Biol.*, 2006, **356**, 1222.
43. K. Futterer, J. Wong, R. A. Grucza, A. C. Chan and G. Waksman, *J. Mol. Biol.*, 1998, **281**, 523.
44. S. Atwell, J. M. Adams, J. Badger, M. D. Buchanan, I. K. Feil, K. J. Froning, X. Gao, J. Hendle, K. Keegan, B. C. Leon, H. J. Muller-Dieckmann, V. L. Nienaber, B. W. Noland, K. Post, K. R. Rajashankar, A. Ramos, M. Russell, S. K. Burley and S. G. Buchanan, *J. Biol. Chem.*, 2004, **279**, 55827.
45. L. J. Farmer, G. Bemis, S. D. Britt, J. Cochran, M. Connors, E. M. Harrington, T. Hoock, W. Markland, S. Nanthakumar, P. Taslimi, E. Ter Haar, J. Wang, D. Zhaveri and F. G. Salituro, *Bioorg. Med. Chem. Lett.*, 2008, **18**, 6231.
46. A. G. Villasenor, R. Kondru, H. Ho, S. Wang, E. Papp, D. Shaw, J. W. Barnett, M. F. Browner and A. Kuglstatter, *Chem. Biol. Drug Des.*, 2009, **73**, 466.
47. E. S. Masuda and J. Schmitz, *Pulm. Pharmacol. Ther.*, 2008, **21**, 461.
48. P. Norman, *Expert Opin. Ther. Pat.*, 2009, **19**, 1469.
49. H. Hisamichi, R. Naito, S. Kawazoe, A. Toyoshima, K. Tanabe, E. Nakai, A. Ichikawa, A Orita and M. Takeuchi, Yamanouchi Pharmaceuticals, *PCT Int. Appl.*, WO 1999/31073, 1999.
50. H. Hisamichi, S. Kawazoe, K. Tanabe, A. Ichikawa, A. Orita, T. Suzuki, K. Onda and M. Takeuchi, Yamanouchi Pharmaceuticals, *PCT Int. Appl.*, WO 2000/075113, 2000.
51. H. Hisamichi, R. Naito, A. Toyoshima, N. Kawano, A. Ichikawa, A. Orita, M. Orita, N. Hamada, M. Takeuchi, M. Ohta and S. Tsukamoto, *Bioorg. Med. Chem.*, 2005, **13**, 4936.

52. Z. J. V. Jia Chandrasekar, W. Huang, M. Mehrotra, Y. Song, Q. Xu, S. M. Bauer and A. Pandey, Portola Pharmaceuticals, *PCT Int. Appl.*, WO 2009/136995, 2009.

53. J. Liddle, F. L. Atkinson, M. D. Barker, P. S. Carter, N. R. Curtis, R. P. Davis, C. Douault, M. C. Dickson, D. Elwes, N. S. Garton, M. Gray, T. G. Hayhow, C. I. Hobbs, E. Jones, S. Leach, K. Leavens, H. D. Lewis, S. McCleary, M. Neu, V. K. Patel, A. G. S. Preston, C. Ramirez-Molina, T. J. Shipley, P. A. Skone, N. Smithers, D. O. Somers, A. L. Walker, R. J. Watson and G. G. Weingarten, *Bioorg. Med. Chem. Lett.*, 2011, **21**, 6188.

54. F. L. Atkinson and V. K. Patel, Glaxo Group, *PCT Int. Appl.*, WO 2010/097248, 2010.

55. S. M. J. Bauer, J. Zhaozhong, Y. Song, Q. Xu, M. Mehrotra, J. W. Rose, W. Huang, C. Venkataramani and A. Pandey, Portola Pharmaceuticals, *PCT Int. Appl.*, WO 2009/145856, 2009.

56. M. R. G. Duthaler, P. Holzer, M. Streiff, G. Thoma, R. Waelchli and H. G. Zerwes, Novartis AG, *PCT Int. Appl.*, WO 2008/009458, 2008.

57. M. Dier and J. Bergan. Available from: http://www.portola.com/pdfs/Portola_syk_Ph_1_news_release_3.22.10.pdf (accessed July 8, 2010).

58. R. Baenteli, G. Zenke, N. G. Cooke, R. Duthaler, G. Thoma, A. Von Matt, T. Honda, N. Matsuura, K. Nonomura, O. Ohmori, I. Umemura, K. Hinterding and C. Papageoriou, Novartis AG, *PCT Int. Appl.*, WO 2003/078404, 2003.

59. R. Baenteli, M. C. Bernhard, P. Buehlmayer, N. G. Cooke, R. Duthaler, K. Hinterding, G. Thoma, M. Van Eis, A. Von Matt, L. Walliser and G. Zenke, Novartias AG, *US Pat. App.* US 2006/0247262, 2006.

60. F. L. Atkinson, M. D. Barker, S. A. Campos, L. A. Harrison, N. J. Parr and V. K. Patel, Glaxo group, *PCT Int. Appl.*, WO 2007/028445, 2007.

61. J. A. L. Linn, T. Stevens and K. Lawrence, SmithKline Beecham, *PCT Int. Appl.*, WO 2008/024634, 2008.

62. P. D. Davis, D. F. C. Moffat, M. J. Batchelor, M. C. Hutchings and D. M. Parry, Celltech Therapeutics, *PCT Int. Appl.*, WO 1998/18782, 1998.

63. L. J. Farmer, G. Bemis, S. D. Britt, J. Cochran, M. Connors, E. M. Harrington, T. Hoock, W. Markland, S. Nanthakumar, P. Taslimi, E. T. Haar, J. Wang, D. Zhaveri and F. G. Salituro, *American Chemical Society – 238th National Meeting*, 2009.

64. G. Brenchley, L. J. Farmer, E. M. Harrington, R. Knegtel, M. O'Donnell, F. G. Salituro, J. R. Studley and J. Wang, Vertex Pharmaceuticals, *PCT Int. Appl.*, WO 2004/087699, 2004.

65. L. J. Farmer, E. M. Harrington, F. G. Salituro and J. Wang, Vertex Pharmaceuticals, *PCT Int. Appl.*, WO 2004/087698, 2004.

66. E. Harrington, J. Wang, J. Cochran and S. Nanthakumar, Vertex Pharmaceuticals, *PCT Int. Appl.*, WO 2002/096905, 2002.

67. Y. Song, Q. Xu, S. M. Bauer, Z. J. Jia, M. Mehrotra and A. Pandey, Portola Pharmaceuticals, *PCT Int. Appl.*, WO 2009/131687, 2009.

68. Y. Song, Q. Xu and A. Pandey, Portola Pharmaceuticals, *PCT Int. Appl.*, WO 2009/026107, 2009.

69. R. A. Ancliff, F. L. Atkinson, M. D. Barker, P. C. Box, C. Daniel, P. M. Gore, S. B. Cuntrip, M. Hasegawa, G. G. A. Inglis, K. Kano, Y. Miyazaki, V. K. Patel, T. J. Ritchie, S. Swanson, A. L. Walker, C. R. Wellaway and M. Woodrow, Glaxo Group, *PCT Int. Appl.*, WO 2007/042299, 2007.

70. P. M. Gore, V. K. Patel, A. L. Walker and M. Woodrow, Glaxo Group, *PCT Int. Appl.*, WO 2007/042298, 2007.

71. T. Yura, A. B. Conception, K. H. Hahn, M. Hiraoka, H. Katsumada, N. Kawamura, T. Kokubo, H. Komura, Y. H. Lee, T. B. Lowinger, M. Motegi, T. Yamamoto and O. Yoshida, Bayer AG, *Jpn. Pat App.* JP 2001/302667, 2001.

72. N. Yamamoto, K. Takeshita, M. Shichijo, T. Kokubo, M. Sato, K. Nakashima, M. Ishimori, H. Nagai, Y. F. Li, T. Yura and K. B. Bacon, *J. Pharmacol. Exp. Ther.*, 2003, **306**, 11.

73. K. Nakashima, T. Kokubo, M. Shichijo, Y. F. Li, T. Yura and N. Yamamoto, *Eur. J. Pharmacol.*, 2004, **505**, 223.

74. S. A. Mitchell, K. S. Currie, P. A. Blomgren, J. E. Kropf, S. H. Lee, J. Xu, D. G. Stafford, J. P. Harding, A. J. Barbosa, Jr. and Z. Zhao, CGI Pharmaceuticals, *PCT Int. Appl.*, WO 2010/068257, 2010.

75. S. A. Mitchell, K. S. Currie, P. A. Blomgren, J. E. Kropf, S. H. Lee, J. Xu and D. G. Stafford, CGI Pharmaceuticals, *PCT Int. Appl.*, WO 2010/068258, 2010.

76. K. S. Currie, R. W. Desimone, S. A. Mitchell, D. A. I. Pippin, J. W. Darrow, X. Qian, M. Velleca and D. Qian, CGI Pharmaceuticals, *US Pat. Appl.*, US 2006/0183746, 2006.

77. S. Bahmanyar, R. J. Bates, K. Blease, A. A. Calabrese, T. O. Daniel, M. Delgado, J. Elsner, P. Erdman, B. Fahr, G. Ferguson, B. Lee, L. Nadolny, G. Packard, P. Papa, V. Plantevin-Krenitsky, J. Riggs, P. Rohane, S. Sankar, J. Sapienza, Y. Satoh, V. Sloan, R. Stevens, L. Tehrani, J. Tikhe, E. Torres, A. Wallace, B. W. Whitefield and J. Zhao, Signal Pharmaceuticals, *US Pat. App.* US 2010/0093698, 2010.

78. S. P. Collingwood, J. Hayler, D. M. Le Grand, H. Mattes, K. A. Menear, C. V. Walker and Z. Cockcroft, Novartis AG, *PCT Int. Appl.*, WO 2001/09134, 2001.

79. A. Hirabayashi, H. Mukoyama, H. Shiohara, H. Kobayashi, Y. Terao, K. Miyazawa, K. Misawa and H. Onoda, Showa Denko K. K., *Jpn. Pat. App.* JP2010/203748, 2010.

80. A. Hirabayashi, H. Shiohara, H. Kobayashi, Y. Terao, K. Miyazawa and K. Misawa, Kissei Pharmaceutical, *Jpn. Pat. App.* JP2004/238296, 2004.

81. H. Mukoyama, T. Nishimura, A. Nakayama, S. Kikuchi, Y. Komatsu and H. Onoda, Kissei Pharmaceutical, *Jpn. Pat. App.* JP2005/089352, 2005.

82. H. Mukoyama, H. Shiohara, T. Nishimura, A. Nakayama, S. Kikuchi, Y. Komatsu and H. Onoda, Kissei Pharmaceutical, *Jpn. Pat. App.* JP2005/008581, 2005.

83. A. Hirabayashi, H. Mukaiyama, H. Kobayashi, H. Shiohara, S. Nakayama, M. Ozawa, K. Miyazawa, K. Misawa, H. Ohnota and M. Isaji, *Bioorg. Med. Chem.*, 2008, **16**, 7347.

84. A. Hirabayashi, H. Mukaiyama, H. Kobayashi, H. Shiohara, S. Nakayama, M. Ozawa, E. Tsuji, K. Miyazawa, K. Misawa, H. Ohnota and M. Isaji, *Bioorg. Med. Chem.*, 2008, **16**, 9247.

85. D. J. Dubois, R. T. Hendricks, J. C. Hermann, R. K. Kondru, Y. Lou, T. D. Owens and C. W. Yee, Roche Palo Alto, *US Pat. App.* US 2009/0215724, 2009.

86. D. J. Dubois, T. R. Elworthy, R. T. Hendricks, J. C. Hermann, R. K. Kondru, Y. Lou, T. D. Owens and D. B. Smith, Roche Palo Alto, *US Pat. App.* US 2009/0215785, 2009.

87. T. R. Elworthy, R. T. Hendricks, R. K. Kondru, Y. Lou, T. D. Owens, M. Soth and H. Yang, Roche Palo Alto, *US Pat. App.* US 2009/0215788, 2009.

88. J. T. Bamberg, M. Bartlett, D. J. Dubois, T. R. Elworthy, R. T. Hendricks, J. C. Hermann, R. K. Kondru, R. Lemoine, Y. Lou, T. D. Owens, J. Park, D. B. Smith, M. Soth, H. Yang and C. W. Yee, Roche Palo Alto, *US Pat. App.* US 2009/0215750, 2009.

89. P. J. Cox, T. N. Majid, J. Y. Q. Lai, A. D. Morley, S. Amendola, S. Deprets and C. Edlin, Aventis Pharma, *PCT Int. Appl.*, WO 2001/047922, 2001.

90. P. J. Cox, T. N. Majid, S. Amendola, S. D. Deprets, C. Edlin, B. L. Pedgrift, F. Halley, M. Edwards, B. Baudoin, L. M. Mclay and D. J. Aldous, Aventis Pharma, *PCT Int. Appl.*, WO 2003/000695, 2003.

91. T. Oligino, C. Hahn, T. A. Gillespy, R. Dharanipragada and E. Matzkin, Sanofi-Aventis, *PCT Int. Appl.*, WO 2009/114373, 2009.

92. C. S. Hahn, Sanofi Aventis, *PCT Int. Appl.*, WO 2010/080563, 2010.

93. T. Oligino, T. A. Gillespy, R. Dharanipragada and E. M. Allen. *PCT Int. Appl.*, WO 2010/065571, 2010.

94. T. A. Gillespy, P. Eynott, E. M. Allen, K. T. Yu and A. Zilberstein, Sanofi Aventis, *PCT Int. Appl.*, WO 2008/033798, 2008.

95. G. E. Lee, F. L. Shrimp II and F. J. Weiberth, Sanofi Aventis, *PCT Int. Appl.*, WO 2010/107969, 2010.

96. K. J. Bordeau, B. C. Langevin, G. E. Lee, P. Logue, E. Secord and D. Sherer, Sanofi Aventis, *PCT Int. Appl.*, WO 2010/065690, 2010.

97. M. L. Edwards, P. J. Cox, S. Amendola, S. D. Deprets, T. A. Gillespy, C. D. Edlin, A. D. Morley, C. J. Gardner, B. Pedgrift, H. Bouchard, D. Babin, L. Gauzy, A. Le Brun, T. N. Majid, J. C. Reader, L. J. Payne, N. M. Khan and M. Cherry, Aventis Pharmaceuticals, *PCT Int. Appl.*, WO 2003/035065, 2003.

98. C. L. Cywin, S. E. Jakes, J. Heider, M. A. Bobko, R. L. Des Jarlais, M. Player, M. Winters and B.-P. Zhao, Boehringer Ingelheim, *US Pat. App.* US 2009/0171089, 2009.

99. C. L. Cywin, S. E. Jakes, J. Heider, M. A. Bobko, R. L. Des Jarlais, M. Player, M. Winters and B. Zhao, Boehringer Ingelheim, *US Pat. App.* US 2008/0114024, 2008.

100. C. L. Cywin, S. E. Jakes, J. Heider, M. A. Bobko, R. L. Des Jarlais, M. Player, J. Rinker, M. Winters and B. Zhao, Boehringer Ingelheim, *PCT Int. Appl.*, WO 2003/057695, 2003.

101. D. Fiegen, S. Handschuh, S. Hobbie, M. Hoffmann, T. Kono, Y. Sato, A. Schnapp and A. Schuler-Metz, Boehringer Ingelheim, *PCT Int. Appl.*, WO 2010/015518, 2010.

102. G. Dahmann, H. Engelhardt, D. Fiegen, S. Handschuh, S. Hobbie, M. Hoffmann, T. Kono, U. Reiser, Y. Sato and A. Schnapp, Boehringer Ingelheim, *PCT Int. Appl.*, WO 2010/015520, 2010.

103. M. Sawa, K. Yokota, H. Moriyama, M. Shin, S. Ro and J. M. Cho, Carna Biosciences and Crystal Genomics, *PCT Int. Appl.*, WO 2009/084695.

104. G. W. Bemis and J. P. Duffy, Vertex Pharmaceuticals, *PCT Int. Appl.*, WO 2005/056547, 2005.

105. B. Okram, T. Uno, Q. Ding, Y. Liu, Q. Jin, X. Wu, J. Che and F. S. Yan, IRM LLC, *PCT Int. Appl.*, WO 2009/097287, 2009.

106. D. M. Goldstein and M. Rueth, Roche Palo Alto, *US Pat. Appl.*, US 2007/0219195, 2007.

107. M. A. Siddiqui, D. Belanger, C. Dai and L. Zhao, Neogenesis Pharmaceuticals, *PCT Int. Appl.*, WO 2004/080463, 2004.

108. M. S. Weidner, Astion Development, P*CT Int. Appl.*, WO 2006/102898, 2006.

109. M. S. Weidner, Astion Development, *PCT Int. Appl.*, WO 2006/102899, 2006.

110. P. Dariavach, P. E. U. Martineau and B. Villoutreix, Centre National de la Recherche Scientifique, *PCT Int. Appl.*, WO 2009133294, 2009.

111. Y. Kodama, S. Noji, K. Imamura, R. Mizojiri, K. Aoki, H. Takagi, Y. Naka, G. Ito, K. Shinoda, A. Fujiwara, K. Kurihara and M. Tanaka, Japan Tobacco, *PCT Int. Appl.*, WO 2006093247, 2006.

112. M. Bajpai, *IDrugs*, 2009, **12**, 174.

113. M. H. Flight, *Nat. Rev. Drug Discovery*, 2012, **11**, nrd3631.

114. E. O. Meltzer, R. B. Berkowitz and E. B. Grossbard, *J. Allergy Clin. Immunol.*, 2005, **115**, 791.

115. R. Singh, S. Bhamidipati and E. Masuda, Rigel Pharmaceuticals, *PCT Int. Appl.*, WO 2006/078846, 2006.

116. D. J. Sweeny, W. Li, J. Clough, S. Bhamidipati, R. Singh, G. Park, M. Baluom, E. Grossbard and D. T. Lau, *Drug Metab. Dispos.*, 2010, **38**, 1166.

117. M. E. Weinblatt, A. Kavanaugh, M. C. Genovese, T. K. Musser, E. B. Grossbard and D. B. Magilavy, *N. Engl. J. Med.*, 2010, **363**, 1303.

CHAPTER 8

Janus Kinases – Just Another Kinase or a Paradigm Shift for the Treatment of Autoimmune Disease?

MICHAEL SKYNNER,* PHIL JEFFREY,
MICHAEL BINKS AND MICHAEL WOODROW

GlaxoSmithKline Research and Development Ltd, Epinova Discovery
Performance Unit, Medicines Research Centre, Gunnels Wood Road,
Stevenage, Hertfordshire, SG12NY, UK
Email: *michael.j.skynner@gsk.com; phillip.x.jeffrey@gsk.com;
michael.h.binks@gsk.com; michael.d.woodrow@gsk.com

8.1 Kinase Drug Discovery

Since the first kinase sequence (protein kinase A) was identified in 1981[1] multiple systematic approaches have uncovered a large protein family[2] comprising over 518 members. These kinases act as key signalling nodes within the cell, often in complex interrelated hierarchical networks, transducing extra-cellular signals to effect intra-cellular changes in gene transcription and protein activation.

Kinases have attracted considerable investment from drug-discovery companies over the last two decades due to their role as convergent points of signal transduction.[3–5] A quick snapshot of the level of investment is provided by the

RSC Drug Discovery Series No. 26
Anti-Inflammatory Drug Discovery
Edited by Jeremy I Levin and Stefan Laufer
© The Royal Society of Chemistry 2012
Published by the Royal Society of Chemistry, www.rsc.org

statistic that approximately 15% of all current drug-discovery activities are devoted to this target class (1,500 kinase projects in discovery/development out of a total of approximately 9,500 active programmes[6]) and kinases account for some 22% of available targets in the "druggable genome".[7]

Despite this intense activity only a handful of kinase inhibitors have successfully emerged from research and development pipelines and reached the market. Those that have are primarily for life-threatening indications in oncology and many of these medicines have adverse side-effects. This seems to represent a poor return on investment, but upon closer inspection there may be a number of confounding factors. Firstly, due to the large number of kinases and the conserved nature of their enzymatic site, *i.e.* the ATP binding pocket,[8] it has been difficult to identify molecules that bind the target with high affinity, yet with poor affinity to other "unwanted" kinases.[9–11] Secondly, kinase high-throughput screening campaigns have tended to identify similar chemotypes, primarily planar and lipophilic hinge binders with poor solubility, resulting in limited chemical diversity. The sheer size of the kinase family has also provided a challenge in assembling recombinant kinase panels and assays with sufficient family coverage to understand kinase selectivity[12–15] and even today this coverage is incomplete, particularly in the availability of cellular assays. These difficulties are compounded in lead optimization where the challenge is to maintain selectivity and cellular potency whilst also incorporating a range of additional drug-like properties (*e.g.* solubility, stability and safety). The net result of these collective efforts is a first generation of nine kinase drugs, many of which target multiple kinases often with fundamental cellular roles (see Table 8.1). Whilst this multi-kinase activity is potentially attractive for drugs to treat oncology, where tumours are prone to repeated mutational escape from selective target inhibition, such compounds have more limited application when the disease in question is both chronic and less life-threatening such as immune, metabolic or neurological diseases.

This first generation of kinase drugs is about to be supplemented by a second wave of emergent late-stage kinase inhibitors targeting a single or small cluster of kinase targets. These have been developed with the aid of recent advances in kinase drug discovery: rational drug design, structural biology and improved chemical library design. At the vanguard of these late stage drugs is a family of JAK inhibitors, the first of which is expected to reach the market in the next 2–3 years.

8.2   Janus Kinases (JAKs)

8.2.1   Identification and Structure of JAK Family Proteins

The Janus Kinase (JAK) proteins belong to a subfamily of tyrosine kinases first identified in 1989: JAK1 and 2,[16,17] TYK2[18] and JAK3.[19] They were named in reference to the Roman two-headed god, Janus, the mythical guardian of gates, doors, beginnings and ends. This name was adopted in reference to the paired

Table 8.1 Table of currently marketed kinase drugs (Pipeline Informa).

Inhibitor	Target	Company	Indication	Structure
Nilotinib (Tasigna AMN-107)	BCR-ABL, KIT, LCK, EPHA3, EPHA8, DDR1, DDR2, PDGFRB, MAPK11 & ZAK	Novartis	Drug-resistant chronic myelogenous leukaemia	
Dasatinib (Sprycel BMS-354825)	Multi- BCR/ABL and SRC family tyrosine kinases	BMS	Chronic myelogenous leukaemia, acute lymphocytic leukaemia	
Sunitinib (Sutent SU-11248)	PDGFRs, VEGFRs, multi-targeted receptor tyrosine kinase inhibitor	Pfizer	Gastrointestinal, stromal cancer, renal cancer	

Table 8.1 (*Continued*)

Inhibitor	Target	Company	Indication	Structure
Pazopanib (Votrient GW-2286)	VEGFR-1 VEGFR-2, VEGFR-3, PDGFR-α/β and C-KIT	GSK	Cancer, renal	
Lapatinib (Tykerb GW-2016)	EGFR, HER2	GSK	Cancer, breast	
Vandetanib (Zactima ZD6474)	VEGF, EGFR, RET	AstraZeneca	Cancer, thyroid	

Name	Structure	Targets	Company	Indication
Imatinib (STI571)		BCR-ABL, c-KIT, PDGFR	Novartis	Chronic myelogenous leukaemia, acute lymphocytic leukaemia, sarcoma, mastocytosis, hypereosinophilic syndrome
Sorafenib (Nexavar Bay 43-9006)		VEGFR, PDGFR, Raf	Bayer	Cancer, liver, renal
Erlotinib (Tarceva RG-1415)		EGFR	Astellas Roche	Cancer, non-small-cell lung and pancreatic

kinase JAK homology (JH-1), which contains the conserved tyrosine residues necessary for JAK activation and substrate binding and pseudo-kinase domains (JH-2) that characterizes the family. The JH-3 and JH-4 domain shares some homology with phosphate binding Src-homology-2 (SH2) domains and the final JH5 to JH7 domains contain a FERM domain (4.1 ezrin, radixin and moesin domain) involved in the association of JAKs with cytokine receptors and other kinases[20] (Figure 8.1).

When these proteins were first identified by PCR cloning it was in the heyday of kinase identification, when novel kinases were being described on an almost weekly basis. Therefore, it is no surprise that these were initially put to one side and referred to using a simple cloning acronym: "just another kinase".[21] This older nomenclature now seems highly incongruous given the status of this small kinase family as key targets for inflammatory and oncology drug research. As the JAK field encompasses a vast and rapidly expanding literature this chapter will not attempt to provide a comprehensive biological or chemical review but will instead identify key references for further reading. Here an overview of

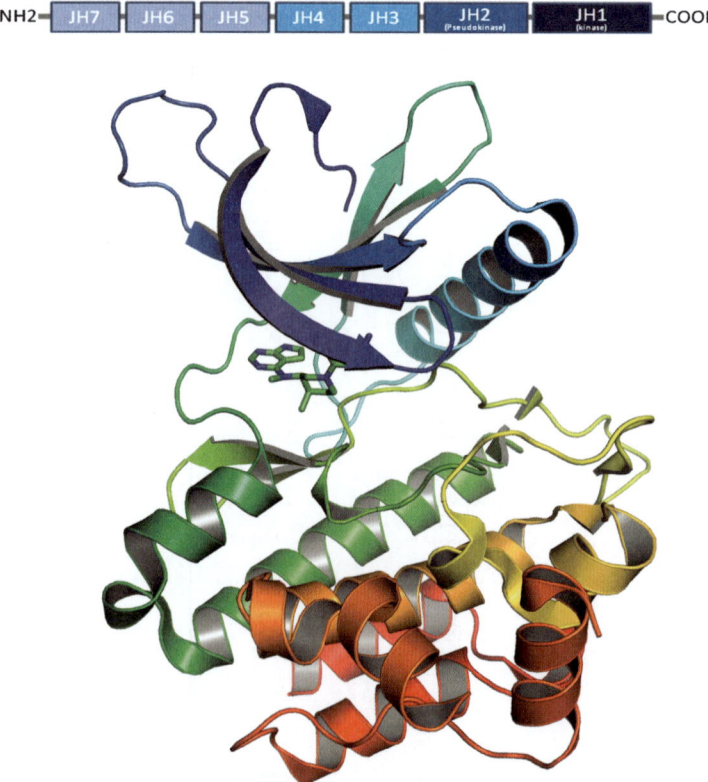

Figure 8.1   JAK3 protein domain structure and crystal structure of the kinase domain in complex with CP-690,550.

JAKs from a drug-discovery perspective, reviewing key headline aspects of their biology and target rationale, will be provided. Also reviewed will be data on late stage development compounds with an emphasis on their chemistry, drug-like properties and clinical progression and a commentary on future trends and perspectives.

8.2.2 Signalling Mechanisms of the JAKs

JAKs mediate signalling responses to over 38 different cytokines, which play a myriad of different roles in regulating processes such as growth, haematopoiesis and homeostasis. Most pertinently, the majority of these cytokines have distinct roles in the function and differentiation of both the innate and adaptive immune system and in the production of essential haematopoietic populations of such as leukocytes and reticulocytes.[22,23] JAK proteins are associated with intra-cellular cytoplasmic regions of cytokine receptors,[24] and auto-phosphorylate and self-activate when cytokines bind to their cognate receptors.[25] JAK activation initially leads to phosphorylation of tyrosine residues in the intra-cellular portion of cytokine receptors. These phosphorylation sites in turn serve as docking sites for the recruitment of STAT (signal transducer and activator of transcription) proteins, which associate with the receptor-JAK complex by virtue of paired phosphate binding SH2 domains. Once bound, the STATs in turn are phosphorylated by JAKs, dissociate from the receptor-JAK complex, form homo- or hetero-dimers and are translocated from the cytoplasm into the nucleus where they bind response elements within gene promoters activating as transcriptional activators to enact gene transcription[26,27] (Figure 8.2).

8.2.3 Feedback Inhibition of JAKs by SOCS Proteins

In parallel to activating gene transcription, STATs also activate a feedback inhibition loop through induction of a panel of otherwise poorly expressed genes, the Suppressor of Cytokine Signalling proteins (SOCS),[28] which are characterized by a common 40 amino acid module known as the SOCS box. This family of eight proteins, SOCS1–7 and the cytokine-induced SRC-homology 2 (SH2) protein CIS,[28–30] provides inhibition through complex multi-factorial interactions with activated JAK proteins, resulting in the suppression of JAK catalytic activity, inhibition of STAT recruitment and the induction of substrate degradation.[31,32] In contrast to the modest and largely viable phenotypes induced by deletion of the STAT proteins, deletion or over-expression of the negative regulatory SOCS proteins is largely deleterious.[31] Thus, SOCS1 null mice die neonatally due to fatty degeneration of the liver and haematopoietic organ infiltration[33,34] accompanied by lymphopoenia, lymphoid apoptosis and aberrant T-cell activation. SOCS2 deletion leads to gigantism and deregulated growth hormone signalling[35] and SOCS3 deletion results in mid-gestational embryonic lethality due to placental insufficiency and erythrocytosis.[36] Conditional knockout of SOCS1 or SOCS3 in immune cells

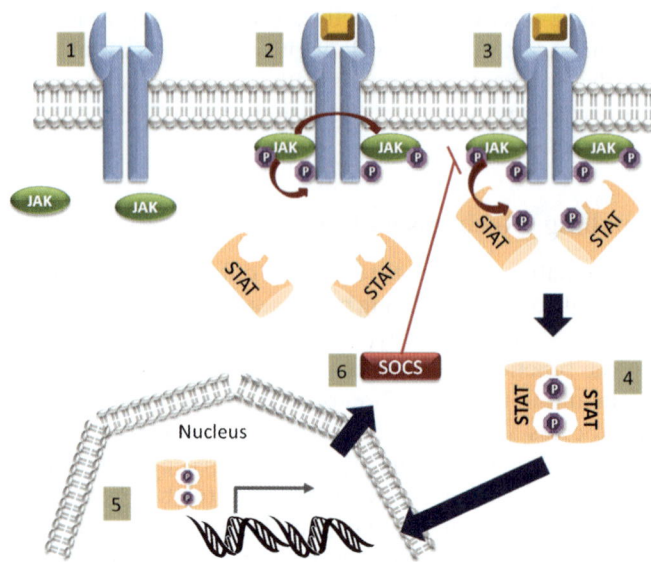

Figure 8.2   JAK STAT signalling. JAK proteins are initially disassociated from individual cytokine receptors (1), upon binding of a receptor ligand JAKs become associated with the individual subunits of the receptor and become activated by trans-phosphorylation and in turn phosphorylate amino acids in the intra-cellular portion of the receptor (2). These phosphorylation sites become recruitment sites for STAT proteins, which bind *via* SH2 domains and in turn become phosphorylated by the JAKs (3). Once phosphorylated STAT proteins dissociate from the receptor, dimerise (4) and are transported to the nucleus where they bind conserved response elements in the promoters of target genes to activate gene transcription (5). Negative feedback is provided *via* SOCS proteins, which are also under STAT mediated transcriptional control, which interact with JAK proteins and inhibit their activity (6).

leads to exacerbated responses in complex inflammatory disease models *e.g.* collagen induced arthritis (CIA),[37,38] and when over expressed SOCS3 represses CIA disease end-points[39] illustrating the key inhibitory role SOCS proteins play in inflammatory responses and cytokine mediated JAK signalling. There are limited data surrounding the function of the other SOCS proteins and more work is required to understand their function in systemic and immune function and their redundancy.

8.2.4   JAK Cytokine Signalling

Each JAK/STAT-dependent cytokine may engage and activate a different combination of both JAKs and STATs,[27] some promiscuously and some *via* discrete protein combinations (Table 8.2). So, for example, common gamma chain dependent cytokines such as IL2, IL4, IL7, IL9, IL15 and IL21 signal through JAK1 and JAK3 to activate STAT1, STAT3, STAT5 and STAT6,

Table 8.2 Cytokines bind and mediate their effects *via* different combinations of JAKs and STAT proteins. A qualitative indication of the degree of JAK usage by cytokines is shown using pluses and negatives.

Cytokine Receptor/Kinase	JAK-1	JAK-2	JAK-3	TYK-2	STAT
Type 1 Interferon signalling (IFNαβ and ILR-10, 20, 22)	+ + +	–	–	+	STAT1 STAT2 STAT3 STAT5
Type 2 Interferon signalling (IFNγ)	+ + +	+ + +	–	–	STAT1 STAT3 STAT5
Gp130 receptors (ILR-6, ILR-11, ILR-13, ILR-27 CNTFR, CTF1R, G-CSFR, LIFR, OSMR)	+	+ + +	–	+ +	STAT1 STAT3 STAT4 STAT5 STAT6
ILR-12, ILR-23	–	+ +	–	+ + +	STAT3 STAT4
Common β chain (ILR-3, 5, GM-CSFR)	–	+ + +	–	–	STAT5
Common γ chain(ILR-2, ILR-4, ILR-7, ILR-9, ILR-15, ILR-21)	+ + +	–	+ + +	–	STAT1 STAT3 STAT5 STAT6
Homodimeric Receptors (EPOR, TPOR, PRLR, GHR)	–	+ + +	–	–	STAT5

whilst cytokines dependent on homodimeric receptors activate a more discrete panel, signalling *via* JAK2 and STAT5 activation. The differential activation of JAK/STAT proteins provides an opportunity for therapeutic intervention, where individual cytokine signalling pathways can be inhibited through chemically targeting the individual JAK proteins. Inhibition of JAK3, for example, would eliminate IL2 mediated clonal expansion of T-cells, IL4 and IL21 mediated Th2 differentiation and B-cell function and the IL15 mediated development of NK cells. An additional level of selectivity may also be provided by the different cellular expression profiles of these JAKs, with JAK1 being widely expressed in immune and non-immune cell types whilst JAK3 is exclusively expressed in myeloid and activated lymphoid cells.[19,40] In contrast inhibition of JAK2, which is expressed primarily in bone marrow derived populations, would limit the numbers of erythrocytes, monocytes, neutrophils and platelets (Table 8.2).

8.2.5 Interrogating JAK Function through Loss of Function Studies

The pivotal role of JAKs can be highlighted through gene deletion studies in knockout mice, gain of function mutations and genome-wide gene-association studies (GWAS), with both JAK1 and JAK2 deletion proving to be embryonic lethal, whilst JAK3 and TYK2 knockouts are viable and fertile. JAK1 null mice die around birth from a profound deficiency in lymphoid development, especially early B-cell differentiation, and defects in cytokine signalling *via* the common gamma chain cytokines and type 1 interferons.[41] JAK2 null mice die *in utero* from severe defects in bone marrow and immune cell progenitor proliferation and a failure of erythrogenesis.[42,43] In contrast, JAK3 null mice are viable but exhibit a severe combined immune deficiency (SCID) phenotype affecting T-cells, B-cells and NK-cells.[44–46] This murine phenotype is mirrored in rare autosomally recessive human genetic mutations, which occur in approximately 10% of patients with B-cell positive, T/NK-cell negative, SCID.[47–49] Children with this condition present with life-threatening infections, which are often satisfactorily treated by bone marrow transplantations.[50] This same phenotype is replicated in individuals lacking a functional common gamma chain gene[51] and a related, but subtly different, phenotype in individuals carrying a mutation in IL7R. These patients show similar defects in lymphocyte populations but maintain functional NK-cells, illustrating the redundant function of IL7 in NK development.[52,53] The mildest phenotype is observed when TYK2 is ablated.[54] Mouse knockouts show no apparent phenotype, except for subtle immune effects such as a reduced propensity for $CD4^+$ T-cells to differentiate into IFNγ-producing Th1 cells in response to IL12 and a lowered NK-cell activity.[55–57] When challenged with a simple proinflammatory stimulus, such as LPS, these animals are protected, exhibit a restricted septic immune response compared to normal litter mates[58] and reduced disease severity in complex Th1/Th17 disease models including: experimental autoimmune encephalomyelitis (EAE), DSS colitis, delayed type hypersensitivity (DTH), TNBS colitis and imiquimod induced skin inflammation.[59,60] In humans an extremely rare genetic mutation has been described in which a premature stop codon is introduced in the TYK2 gene. This leads to Hyperimmunoglobulin E Syndrome, a primary immunodeficiency characterized by recurrent viral and staphylococcal skin abscesses and pneumonia and elevated serum immunoglobulin E probably caused by defective type 1 interferon, IL6 and IL23 signalling.[61,62]

Taken together these studies suggest that JAK3 and JAK1 are key signalling proteins in orchestrating the immune response and lymphocytic function, that JAK2 is a vital signalling intermediate in the production of key haematopoietic cells and TYK2 plays a more subtle role in fine tuning Th1/Th17 responses and immune cell activation.

8.2.6   JAK Proteins and STAT Responses

The complexity of the cytokine mediated JAK response is amplified downstream in the STAT response. JAKs can activate these six proteins in either a promiscuous or a selective manner, dependent on the cytokine/JAK pairing inducing the formation of STAT homo- or hetero-dimers. The role of these proteins has also been revealed by gene knockout studies.[63,64] STAT1 and STAT2 deficient mice show specific defects in type 1 and type 2 interferon responses and exhibit increased rates of microbial and viral infection.[65–67] Other STAT1 activating cytokines signal normally in these animals suggesting that other cytokines may have redundant signalling mechanisms.

In contrast, STAT3 ablation is early embryonic lethal illustrating a key non-redundant role in cytokine signalling and development.[68,69] To extend the viability of STAT3 null animals multiple groups have generated conditional and cell-specific knockouts that have shown multiple roles in processes such as growth, apoptosis and cell motility as well as discrete functions in various populations of immune cells.[70–72] STAT4 and STAT6 also play major roles in lymphocyte function, STAT4 nulls are defective in IL12 driven Th1 differentiation, whereas STAT6 nulls are defective in IL4 driven Th2 differentiation.[27,73–75]

STAT5 has a minor role outside the immune system. Thus, STAT5a-deficient mice are overtly normal but with defective mammary gland development,[76] whilst STAT5b-deficient mice show sexually dimorphic secretion of pituitary derived growth hormone[77] and females from dual STAT5a/b knockouts are infertile.[78] In the immune system STAT5a and STAT5b nulls are both defective in anti-CD3 induced T-cell proliferation, due to the failure of IL2 to induce IL2R expression on stimulation. However, whereas STAT5a knockouts can compensate through an exogenous supply of excess IL2, STAT5B nulls are unable to do so and exhibit additional effects on NK-cell proliferation due to defects in IL15 signalling.[79,80] As STATs provide an alternative druggable node, many companies have investigated these as therapeutic targets. These strategies have been largely unsuccessful to date since the STATs have proven difficult to target using small-molecule approaches due to their shallow SH2 binding sites and lack of enzyme function. However, some companies are attempting to target STATs using peptide and gene therapy approaches.[81–83]

8.3 Mutation of JAK Proteins – Gain of Function Studies

8.3.1 JAK2 Gain of Function

Gain of function mutations have now been described for three of the four JAKs, JAK1 and JAK3, and most notably dominant gain of function mutations in JAK2.[84–87] The most frequent JAK2 mutation is a guanine to thymidine mutation leading to a valine for phenylalanine substitution at amino acid 617 in the pseudokinase domain. This single amino acid change leads to constitutive phosphorylation and activation of JAK2 and results in a range of Philadelphia chromosome-negative myeloproliferative neoplasms (MPNs). These are diseases characterized by unchecked proliferation of terminally differentiated myeloid cells due to increased JAK2 drive initiated by haematopoietic cytokine signalling. The commonest of the MPNs are polycythaemia

vera (PV – 95% of patients carry the V617F mutation) and essential throm-
bocythaemia (ET – 60% of patients carry the V617F mutation).[88] Interestingly,
the same mutation gives rise to two clinically distinct diseases, probably
through gene dosage effects, biallelic expression of the mutated JAK gene in PV
versus mono-allelic expression in ET.[89,90] It is these MPN conditions that are
being targeted by the majority of JAK2 inhibitors in clinical development, the
most advanced of which is Incyte's dual JAK1/2 inhibitor INCB18424.

8.3.2   JAK1 and JAK3 Gain of Function

Rarer gain of function mutations have also been reported in JAK1 and JAK3
at autologous sites to the V617F site in JAK2. These have been associated with
leukaemias involving B-cells and T-cells. In the case of JAK1 up to 20% of
adult lymphoid leukaemias may be the result of increased JAK1 drive and the
presence of JAK1 mutations is correlated with an increased interferon gene
signature and poor survival.[91–94] Rare JAK3 mutations, such as the JAK3
alanine to valine mutation at amino acid 572, were first described in acute
megakaryoblastic leukaemia (AMKL)[95] and subsequently found to be present
in ~3% of all cases of cutaneous T-cell lymphoma (CTCL)[96] as well as a subset
of acute lymphocytic leukaemia (ALL) patients.[97,98] In addition to gain and
loss of function studies, genetic association with SNPs of unknown function
has implicated TYK2 in a range of inflammatory diseases such as lupus,
psoriasis, inflammatory bowel disease (IBD) and multiple sclerosis (MS).[99–101]

8.4   JAK Inhibitors for Immune Mediated Disease

8.4.1   Introduction

Mainstay treatments for severe chronic immune mediated diseases are steroids
and immune-suppressants such as methotrexate, azathioprine, mycophenylate
mofetil, cyclosporine and rapamycin. These may provide a level of disease
treatment for mild to moderate disease control, but are limited by adverse and
often serious side-effects, an increased cancer risk, dose-limiting toxicities and a
poor quality of life.[102] A treatment breakthrough was provided approximately
20 years ago with the approval of the first anti-TNFα therapies. However, as
the use of these and other biologic therapies has become more widespread it has
become apparent that these too have additional limitations including high
(20–30%) non-responder rates, infection rates, loss of efficacy over time, slow
reversibility in the event of infection and a requirement for intravenous or
subcutaneous administration with associated high treatment costs. The net
result is that there remains a significant unmet medical need for orally administered, safer, immune modifying drugs with more targeted mechanisms of
action and lower treatment costs. Kinase inhibitors are currently at the vanguard of these new emerging oral therapies with JAK, SYK and PI3Kinase
inhibitors at advanced stages of development. Below the most clinically

Table 8.3 Amino acid homology in the ATP binding pocket for the JAK kinases.

JAK1	JAK2	JAK3	TYK2	Consensus	Comment
HIS874	ASN859	ASN832	HIS907	POLAR	Turn in Gly-rich loop
VAL926	VAL911	VAL884	ILE960	ALIPHATIC	Back pocket
PHE946	TYR931	TYR904	TYR980	AROMATIC	Hinge
LEU947	LEU932	LEU905	VAL981	ALIPHATIC	Hinge donor and outer H-bond acceptor
SER949	TYR934	SER907	LEU983	–	Outer hinge
SER951	SER936	CYS909	SER985	SMALL	Solvent front & sugar pocket sidechain
LYS953	ARG938	ARG911	ARG987	POSITIVE	Surface helix
GLU954	ASP939	ASP912	ASP988	NEGATIVE	Surface helix
GLY1008	GLY993	ALA966	GLY1040	SMALL	Inner hydrophobe below purine

advanced of the JAK inhibitor molecules will be reviewed and their paths through initial discovery and development described.

8.4.2 JAK Drug Discovery

JAK inhibitors fall into two broad categories, for oncology targeting JAK2, and for inflammatory and autoimmune diseases targeting multiple members of the JAK family, either specifically or across the target class by exploiting differences in JAK homology at the ATP binding site. A particular challenge for the medicinal chemist has been the limited divergence between the family members, with only 9 amino acid differences across the 120 amino acid ATP binding kinase domain (Table 8.3), compared to 50% to 60% conservation between the JAKs outside this domain.[21] Despite this homology multiple JAK family selective series have been identified and patented with excellent selectivity against the general kinome. The first wave of these compounds is relatively unselective across the JAK family. However, many groups have since identified second-generation inhibitors, using information from crystal structures, with improved selectivity across the JAK family and to the wider kinome (Table 8.4).[103,104]

8.5 JAK Inhibitor Chemistry

8.5.1 Pan JAK Inhibitors

Many companies have identified pyrrolopyrimidines and similar structures as inhibitors of JAK kinases. Key exemplars of lead molecules are: Pfizer's tofacitinib (CP-690,550)[105] and Incyte's ruxolitinib (INCB18424).[106] Less advanced programs include: pyrrolopyrimidines from Targegen,[107] Novartis[108] and Portola,[109] pyrrolopyridines from Merck,[110] Astellas[111] and Vertex[112] and pyrrolopyrazines from Roche.[113] These vary in their substitution patterns,

Table 8.4 Summary of JAK inhibitor compounds.

Compound (Company)	Structure	Target activity (IC$_{50}$ unless indicated)	Status, Primary Indication	Reference
Tofacitinib CP-690,550 (Pfizer)		JAK1 3.2 nM, JAK2 4.1 nM, JAK3 1.6 nM, TYK2 34 nM	Phase III, RA	105
Ruxolitinib INCB-18424 (Incyte)		JAK1 3.3 nM, JAK2 2.8 nM, JAK3 428 nM, TYK2 19 nM	Pre-registration MFD	106
INCB-28050 LY3009104 (Incyte/Eli-Lilly)	Structure unknown	JAK1 5.9 nM, JAK2 5.7 nM, JAK3 >400 nM, TYK2 53 nM	Phase II, RA	165
VX-509 (Vertex)	Structure unknown	25–150-fold JAK3 selective in cell-based assays	Phase II, RA	171
R-348 (Rigel)	Structure unknown	JAK3 K$_i$ 16 nM	Phase I	115, 116

Compound	Structure	Selectivity	Phase	Ref.
Unnamed (Novartis)		JAK3 8 nM, JAK1 1017 nM, JAK2 2550 nM, TYK2 8055 nM	Discovery	117, 118
WYE-151650 (Pfizer)		JAK3 0.9 nM, 36-fold JAK1 selective, 14-fold JAK2 selective, 34-fold TYK2 selective	Discovery	119
PS-020613 (Pharmacopeia/Ligand)		JAK3 3 nM, JAK2 26 nM	Pre-clinical	120

Table 8.4 (Continued)

Compound (Company)	Structure	Target activity (IC$_{50}$ unless indicated)	Status, Primary Indication	Reference
AZD1480 (AstraZeneca)		JAK2 <3 nM, 16-fold JAK3 selective	Phase II, MFD	121, 122
CYT-387 (Cytopia)		JAK1 11 nM, JAK2 8 nM, JAK3 155 nM, TYK2 17 nM	Phase II, MFD	123, 124
SB-1518 Pacritinib (S*Bio)		JAK2 23 nM, FLT3 22 nM, JAK1 1280 nM, JAK3 520 nM	Phase II, MFD	126, 127
SB-1578 (S*Bio)	Structure unknown	JAK2 no published data	Phase I, RA/psoriasis	128

Name	Structure	Activity	Status	Ref.
NVP-BSK805 (Novartis)		JAK2 0.48 nM, JAK1 32 nM, JAK3 19 nM, TYK2 11 nM	Pre-clinical	129, 130
Unnamed (Merck)		JAK1 15 nM, JAK2 1 nM, JAK3 5 nM, TYK2 5 nM	Discovery	131
Unnamed (Vertex)		JAK2 K_i 1 nM, JAK3 K_i 6 nM	Tool molecule	132
AG490		JAK2 0.1 µM, JAK3 4.3 µM, plus multiple other kinases	Tool molecule	134
Lestaurtinib CEP-701 (Caephalon)		JAK2 0.9 nM, FLT3 2 nM, also TrkA, PDGFR	Phase III, AML	136

Table 8.4 (*Continued*)

Compound (Company)	Structure	Target activity (IC$_{50}$ unless indicated)	Status, Primary Indication	Reference
Pyridone-6 (Merck)		JAK1 15 nM, JAK2 1 nM, JAK3 5 nM, TYK2 nM	Tool molecule	137
Unnamed (Merck)		JAK2 0.8 nM, JAK3 300 nM	Discovery	138
SAR-302503 TG101348 (Sanofi-Aventis)		JAK2 3 nM, FLT3 15 nM, RET 48 nM	Phase II, MFD	169
GLPG0634 (Galapagos)	Structure not disclosed	JAK1 10 nM, JAK2 28 nM, JAK3 810 nM, TYK2 116 nM	Phase II, RA	173, 174
AC430 (Ambit)	Structure not disclosed	JAK2 63 nM	Phase I, RA	175

which may affect their binding modes, although all probably make a bi-dentate hinge-binding interaction through two nitrogen atoms in the bicyclic core.[114]

8.5.2 JAK3 Selective Molecules

Rigel's R-348 (prodrug of R-333), a dual JAK3 ($IC_{50} = 70–260$ nM in JAK3/1 cellular assays) and SYK kinase ($IC_{50} = 140$ nM in cellular assays) inhibitor,[115,116] is currently progressing as an oral therapy for rheumatoid arthritis (RA), psoriasis and transplantation and as a topical therapy for discoid lupus and Sjögren's syndrome. The structure has not been disclosed by Rigel but the scope of the patent covers the aminopyrimidine chemotype. Novartis have also reported a potent maleimide structure (JAK3 $IC_{50} = 8$ nM, JAK1 $IC_{50} = 1017$ nM, JAK2 $IC_{50} = 2550$ nM, TYK2 $IC_{50} = 8055$ nM) that shows exquisite (> 100-fold) JAK3 selectivity over other JAKs. Although the SAR is not described, the binding mode (from an X-ray structure co-crystallized with JAK3) shows the maleimide forming two hydrogen bonds to the hinge backbone and the CF_3 fitting into a small lipophilic pocket formed from protein side chains. Key to generating selectivity for JAK3 over JAK1 and JAK2 is a water-mediated hydrogen bond between the non-hinge binding maleimide carbonyl and the backbone amide of the catalytic asparagine 967 residue. This amino acid adopts a different conformation in JAK1 and JAK2 crystal structures. The maleimide shows only modest potency in cellular assays, possibly reflecting a more dominant role for JAK1 over JAK3 in common gamma chain cytokine signalling or a lesser role for JAK3 in initiating the trans-phosphorylation of JAK1.[117,118]

In contrast, WYE-151650, Wyeth's benzimidazoylpurinone JAK3 selective molecule (JAK3 $IC_{50} = 0.9$ nM),[119] which is potent in *ex vivo* JAK3 phospho-STAT assays, showed significant efficacy *in vivo* in DTH and CIA models, suggesting that JAK3 selective inhibition may deliver efficacy in autoimmune disease. Ligand has published the SAR of a structurally similar series of 2-benzimidazoylpurinones. This SAR illustrates a key role for electron-withdrawing substituents on the benzimidazole in defining JAK3 activity. Both benzyl and THP groups are tolerated at N-9 of the purinone core, with THP providing selectivity for JAK3 over JAK2. Combination of these in a chroman group led to the identification of PS-020613, which demonstrated moderate levels of JAK3 selectivity (JAK3 $IC_{50} = 3$ nM, JAK2 $IC_{50} = 26$ nM), good oral PK and *in-vivo* efficacy.[120]

8.5.3 JAK2 Selective Molecules

8.5.3.1 Aminopyrimidines

AstraZeneca has published the SAR and synthesis of AZD-1480, their lead JAK2 selective (JAK2 $IC_{50} = 3$ nM) aminopyrimidine molecule. The JAK2 selectivity achieved is presumed to be due to a steric clash between the fluoropyrimidine substituent and alanine 966 in the JAK3 ATP binding-site, which is better accommodated by a glycine substituent in the corresponding position in JAK2; JAK1 selectivity is not reported. Further, crystal structures show that the pyrazole rather than the pyrimidine is the hinge-binding functionality.[121,122] Cytopia are also progressing an aminopyrimidine JAK inhibitor, CYT-387 (JAK1 $IC_{50} = 11$ nM, JAK2 $IC_{50} = 18$ nM, JAK3 $IC_{50} = 155$ nM, TYK2 $IC_{50} = 17$ nM), for myeloproliferative disorders.[123,124] This compound was identified by the optimization of phenylaminopyrimidine hits from an HTS campaign and modelling suggests a conventional hinge-binding mode *via* the aminopyrimidine group.[125] Incorporation of a hydrogen bond donor such as OH, sulfonamide or carboxamide at the 4-position of the 4-aryl group increases JAK2 activity through the formation of a hydrogen bond to asparagine 994, whilst substitution of the carboxamide with a nitrile group further increases JAK2 potency. Both three and four substituted groups were tolerated at the 2-anilino position, although JAK2 activity is reduced by incorporation of such basic groups. CYT-387 derives JAK1 and JAK2 selectivity over JAK3 in a manner analogous to AZD-1480 through a steric clash between a morpholine and alanine 966 of JAK3.

8.5.3.2   Macrocycles

S*Bio have significantly improved the kinase selectivity and patentability of the ubiquitous 2-anilino-4-aryl pyrimidine kinase scaffold by constraining it into a olefin containing macrocyclic ring, with ether groups in the ring. These changes lead to reductions in logP and improvements in solubility. Addition of a pyrrolinine side chain, which formed a salt bridge with asparagine 939, also led to improvements in solubility, whilst substitution of the pyrimidine-ring 5-position improved JAK2 activity but reduced selectivity to JAK3. These changes led to the identification of SB-1518, a dual JAK2/FLT3 selective inhibitor (JAK2 $IC_{50} = 23$ nM, FLT3 $IC_{50} = 22$ nM, JAK1 $IC_{50} = 1280$ nM, JAK3 $IC_{50} = 520$ nM) that was active in cellular assays and demonstrated attractive oral pharmacokinetics.[126] S*Bio are developing SB-1518 (pacritinib) for MPN[127] and a second JAK2 selective molecule, SB-1578, for immune disease.[128]

8.5.3.3   Quinoxalines and Polycyclic Molecules

The published crystal structure of NVP-BSK805, Novartis' quinoxaline JAK2 inhibitor (JAK2 $IC_{50} = 0.48$ nM, JAK1 $IC_{50} = 32$ nM, JAK3 $IC_{50} = 19$ nM, TYK2 $IC_{50} = 11$ nM), shows the quinoxaline ring nitrogen as the hinge binder, the difluorobenzyl morpholine group extending into the hydrophobic pocket towards the glycine loop and the polar piperidinyl-pyrazole providing the solubilizing group.[129,130] Other companies pursuing similar templates include Merck with a series of tricyclic JAK2 inhibitors (JAK2 $IC_{50} = 2$ nM) discovered through optimization of a napthyridone hit from a virtual screen.[131]

Cyclization of a 2-amino-3-aryl-1,6-naphthyridin-5-one screening hit constrained the aryl rings as coplanar, allowing an improved fit into the narrow ATP binding pocket, which provided a thousand-fold increase in JAK2 potency. Although tertiary amines were tolerated at the 2-position and polar substituents were not, small lipophilic secondary amines were optimal.

Vertex demonstrated submicromolar JAK2 potency using both 3-phenyl substituted aza-indoles and 4-phenyl substituted deazapurines. JAK2 activity was further improved (JAK2 $IC_{50} = 1$ nM), with six-fold selectivity over JAK3, by cyclizing the three and four positions of an aza-indole into an 8-membered lactam ring. As expected the aza-indole forms two hydrogen bonds to the backbone NH and carbonyl of the hinge and the ring adopts a conformation allowing the lactam to form a hydrogen bond to the backbone NH of asparagines 994. Only the *S*-enantiomer is predicted to be active, as the phenol forms two additional hydrogen bonds, whilst the *R*-enantiomer cannot be docked into the active site.[132]

8.5.3.4 IP and Emerging Molecules

There are >130 JAK compound patents published. Approximately 35 of these specify JAK3 as the primary target, 25 specify JAK2 whilst the remainder are either pan-JAK or do not specify JAK selectivity. This illustrates the significant commercial interest in JAK inhibitors as molecules to treat both inflammation (JAK3 and pan JAK)[114] and oncology (JAK2).[133]

8.5.3.5 Non-specific Inhibitors and Tool Molecules

AG490, a screening hit originally identified as a JAK2 inhibitor in 1996,[134] is not specific for the JAK family but is frequently used as a commercially available tool inhibitor in mechanistic studies of the JAK/STAT pathway and multiple analogues have since been identified (LS-104).[135] Staurosporine analogues, for example Lestaurtinib (Caephalon), have also been progressed into advanced clinical trials in oncology, although it is unclear which of a myriad of kinases that are inhibited by these molecules is responsible for delivering clinical efficacy.[136] The first published JAK crystal structure was of JAK2[103] with Merck's "Pyridone-6",[137] a tool molecule with a pan-JAK profile (JAK1 $IC_{50} = 5$ nM, JAK2 $IC_{50} = 1$ nM, JAK3 $IC_{50} = 5$ nM, TYK2 $IC_{50} = 1$ nM). Subsequently, additional structures of JAK1, JAK3 and TYK2 have also been published, with "Pyridone-6" as well as CP-690,550 in the ATP binding site.[104]

Pyridone-6 is one of the earliest JAK inhibitors described. Whilst the SAR of this molecule has not been extensively described, JAK2 crystallography has shown the pyridine ring to orient toward the methionine 929 gatekeeper residue, the tert-butyl towards the glycine loop and the fluorophenyl to point outwards towards solvent. The carbonyl makes two hydrogen bonds to the hinge, whilst the flat, lipophilic tetracyclic ring system makes extensive Van der Waals interactions with the narrow pocket in which it resides.

Merck have also described a series of 1-amino-5H-pyrido[4,3-b]indol-4-car-boxamides that were obtained from work aimed at improving the physical properties of the tetracyclic "Pyridone-6" series through *de novo* design.[138] Small alkyl, and especially cycloalkyl, secondary amines at the 1-position gave improved JAK2 potency, whilst 6- and 9-position substituents were not toler-ated due to steric clash. Polar hetero-aryl groups were preferred side chains from the solvent exposed 7-position, which were used to modify the drug-like properties of the series. The net result was "Compound 65" (JAK2 $IC_{50} = 0.8$ nM, JAK3 $IC_{50} = 300$ nM).

8.6 Clinical Studies with JAK Inhibitors

8.6.1 Tofacitinib (CP-690,550)

8.6.1.1 Identification and Early Progression

Tofacitinib is the most advanced JAK inhibitor, currently in multiple Phase III trials for RA and psoriasis and in Phase II studies for inflammatory bowel disease and other conditions. An HTS hit (CP-352664 JAK3 $IC_{50} = 210$ nM) was identified using the recombinant catalytic domain of JAK3 and the project aspiration was to identify a potent JAK3 inhibitor with 100-fold selectivity over JAK2.[105] Replacement of the pyrrolopyrimidine 4-substituent with an *N*-methyl cycloalkyl group, as in CP-537,555, improved ligand efficiency and delivered improved JAK1 and cellular potency. Further optimization of the cycloalkyl group, using available natural products to easily access stereo-isomers, identified compounds, such as CP-634,558, with improved JAK3 potency. The cyclohexyl group was further optimized to an *N*-acyl piperidine to reduce lipophilicity, with the specific (3*R*,4*R*) stereochemistry important for JAK3 potency. This optimization resulted in CP-690,550 (Figure 8.3) ((3R,4R)-4-methyl-3-(methyl-1H-pyrrolo[2,3-d]pyrimidin-4-ylamino)-β-oxo-1-piperidinepropanenitrile).

Crystal structures of CP-690,550 bound in the kinase domains of JAK1, JAK2, JAK3 and TYK2 (Figure 8.4) show a binding mode with the pyrrolo-pyrimidine forming two hydrogen bonds to the hinge region of the ATP binding site, and the cyano group pointing towards the glycine loop, apparently

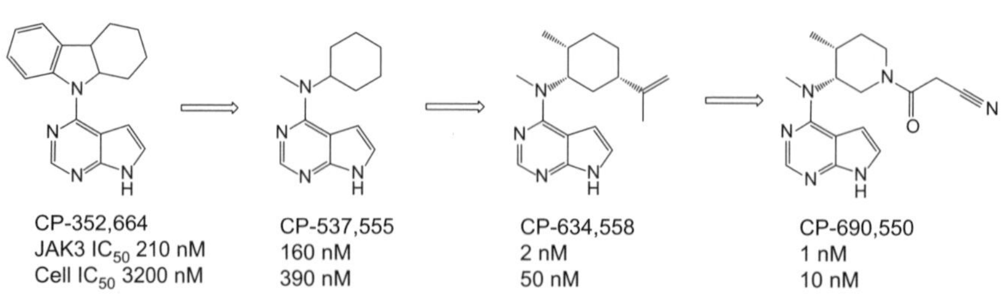

CP-352,664	CP-537,555	CP-634,558	CP-690,550
JAK3 IC_{50} 210 nM	160 nM	2 nM	1 nM
Cell IC_{50} 3200 nM	390 nM	50 nM	10 nM

Figure 8.3 Optimization scheme for CP-690,550.

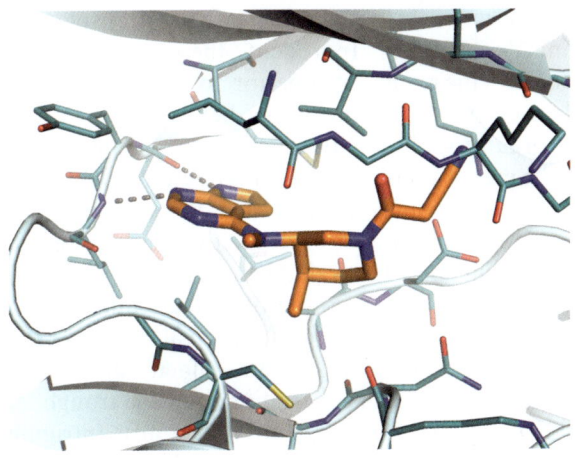

Figure 8.4 Crystal structure of CP-690,550 in the JAK3 active site showing hinge binding.

making several hydrogen bonds with residues in this loop.[104,139] The two stereo-centres and conformation of CP-690,550 have an important effect on the activity and kinase selectivity.[140] Preparation of the other three stereoisomers showed significantly reduced JAK3 activity (JAK3 Kd = 150 nM to 190 nM compared with $IC_{50} = 0.7$ nM for CP-690,550), similar JAK2 selectivity (JAK2 Kd = 270 nM to 600 nM compared with $IC_{50} = 2$ nM for CP-690,550) and modest changes in selectivity with activities at other kinases such as Map4K3, Map4K5 and Mst1 Mst2 that were not shown by CP-690,550. The reduced JAK activity of the stereoisomers is rationalized by molecular modelling, which shows that they need to adopt high energy conformations to dock into the JAK3 active site.

Tofacitinib was initially reported as a selective JAK3 inhibitor designed to selectively inhibit common gamma chain mediated cytokine signalling in lymphocytes *via* exploitation of a 20- and 100-fold selectivity against JAK2 and JAK1, respectively.[141] However, subsequent biochemical studies using recombinant isolated kinase domains have not upheld these early claims and now indicate that tofacitinib is "pan-JAK active" in biochemical assays with affinities across the JAK family differing by only 4–10-fold, although values vary dependent on assay conditions and operators.[9,139] In subsequent cellular assays, measuring cytokine induced STAT protein phosphorylation, tofacitinib has been reported to have similarly varying JAK selectivity with a cellular JAK2 selectivity window of 10–30-fold.

8.6.1.2 *Pre-clinical Development*

The early development path for tofacitinib reflected the initial hypothesis that this was a JAK3 selective molecule. Early pre-clinical data focused on animal models of organ rejection in both rodent cardiac transplant[142] and primate renal transplant[143,144] models. In rodent, these and other studies showed tofacitinib to be as efficacious or superior to current immune-suppressants (without co-therapies) (at concentrations between 44 ng/ml and 136 ng/ml) and to prevent organ rejection without significant depression of lymphocyte numbers, with the exception of NK- (93% reduction) and $CD3^+8^+$ cells (98% reduction). To achieve robust exposure in the rodent, Pfizer resorted to delivery *via* osmotic mini-pumps as tofacitinib has poor PK and a high clearance in the rodent (oral bioavailability *ca.* 30%, short terminal half-life 0.6 hours). In primate a similar profile of cellular inhibition and graft protection was observed with renal allograft survival of 53 to 83 days compared to 6 to 7 days in control, untreated animals (tofacitinib was given orally twice daily and dose adjusted through the study to achieve trough levels of drug in the ranges of 200 to 400 ng/ml or 50 to 100 ng/ml). Intriguingly, it wasn't for another two years that the first pre-clinical publications emerged[145,146] linking the potential utility of tofacitinib to rheumatoid arthritis where, as of writing, it is currently in multiple Phase III clinical trials. In earlier pre-clinical studies tofacitinib, again delivered by osmotic mini-pump for 28 days, dose-dependently reduced the clinical signs of RA (including inflammation, cartilage and bone damage) by up to 90% in the mouse CIA and rat adjuvant induced arthritis (AIA) with an ED_{50} of approximately 1.5 mg/kg/day based upon disease end-points in both models. This corresponds to serum levels of 5.8 ng/ml in mice (day 28) and 24 ng/ml in rats (day 24). Other supporting *in-vivo* studies have been performed, including DTH responses in SRBC (sheep red blood cells) sensitized mice, rodent allograft aorta transplantation vasculopathy and a murine model of allergic pulmonary inflammation where tofacitinib has shown efficacy.[147,148]

8.6.1.3 *Pharmacokinetics*

Clinical studies with tofacitinib administered orally to stable renal allograft recipients over the dose range of 5 mg, 15 mg and 30 mg (b.i.d., every 12 hours) for 29 days demonstrated that, at steady-state (day 29), tofacitinib is rapidly absorbed with maximum plasma concentrations (C_{max}) of 167 nM to 1125 nM and trough concentrations (12 h; C_{min}) of 25 nM to 82 nM.[149] Terminal half-life ($t_{1/2}$) ranged from 5.2 hours at 5 mg and 15 mg to 3.7 hours at the 30 mg dose. Inter-subject variability (percent coefficient of variation [CV%]) was low with corresponding values of 13% to 30% for C_{max}, 18% to 48% for the area under the plasma concentration-time curves (AUC_{0-12h}) and 39% to 55% for $t_{1/2}$.

Studies investigating the potential effects of tofacitinib (15 mg b.i.d.) on renal function in healthy volunteers[150] confirmed the earlier pharmacokinetic results including a median value for T_{max} (time to achieve C_{max}) of 1 hour (range 0.5 hour to 2 hours). There was no effect of tofacitinib on glomerulation filtration rate (GFR), renal plasma flow or creatinine clearance at this dose.

Co-administration of tofacitinib (30 mg b.i.d.) with methotrexate (15 mg to 25 mg/week) in RA patients was found to be safe and well tolerated with no clinically significant effect on the pharmacokinetic profile of either drug.[151] In healthy volunteers, co-administration of tofacitinib (30 mg q.d.) with fluconazole (200–400 mg), a mild inhibitor of CYP3A4 and CYP2C19, resulted in a reduction in tofacitinib $AUC_{0-\infty}$ and C_{max} of *ca.* 79% and *ca.* 27%, respectively, thus confirming the roles of CYPs 3A4 and 2C19 in the metabolic elimination of tofacitinib.[152]

8.6.1.4 Clinical Studies

Tofacitinib has completed multiple Phase I studies in both volunteer and patient settings at single doses spanning 0.1 mg to 100 mg and in repeat doses between 1 mg and 50 mg b.i.d. and 20 mg to 60 mg q.d. The therapeutic efficacy of tofacitinib has been evaluated in numerous Phase II studies in RA,[153–155] transplantation,[149,156] psoriasis,[157] ulcerative colitis (UC),[158] Crohn's disease,[159] dry eye and asthma. The compound has been trialled as a potential standalone therapy dosed once daily or twice daily, as a co-therapy in combination and referenced to "gold standard" comparator molecules. The detailed results of these trials have been extensively reviewed elsewhere and are beyond the scope of this chapter.[102,160] However, the outcomes from the most salient trials are summarized in Table 8.5. To date the most complete dataset is available in RA, where at the time of writing Pfizer are currently conducting multiple parallel Phase III studies.[155] These Phase III studies have been guided by the successful therapeutic outcomes from the Phase IIa/b studies where tofacitinib was shown to be highly efficacious in moderate to severe arthritis as determined by ACR scores (ACR20, 50 and 70) and comparable in efficacy to currently approved agents used as monotherapies.[153,154] However, what was also clear from early Phase II studies was that at higher doses (15 mg b.i.d.) there was an increased incidence of side-effects (anaemia, leukopoenia, thrombocytopenia, increased cholesterol LDL/HDL and increased creatinine), which were unlikely to be tolerated for long-term therapeutic dosing.[153,154] These early Phase II trials have narrowed the dosing regimen in the large Phase III trials to 5 mg and 10 mg twice daily. In IBD tofacitinib has shown impressive efficacy in UC, reaching the primary and secondary end-points on clinical response and remission rates,[158] but failed in a shorter duration study in Crohn's disease potentially because of the short four-week duration of the trial;[159] clinical progression is continuing in both areas. In longer term studies in transplantation, where tofacitinib is being trialled as an adjunct therapy in both stable and *de novo* renal transplant patients, again comparative efficacy has been observed to "gold standard" therapies, but as higher doses are being investigated (15 mg and 30 mg b.i.d.) a corresponding dose-dependent increase in side-effects has also been recorded.[161–163]

Clearly, the extensive clinical strategy being deployed around tofacitinib and the raft of encouraging clinical data generated to date is an indication that this

Table 8.5 Summary of JAK inhibitors in the clinic and headline clinical findings of on-going and completed clinical trials.

Compound and Indication	Trials	Summary of outcome and major findings
Tofacitinib (CP-690,550) Rheumatoid Arthritis	**Phase II** Multiple dose ranging Phase IIs in combinations with co-medications and in patients with inadequate treatment with anti-TNFs and MTX[153,154]	
	Phase IIa (Study 1019) 264 patients with moderate to severe RA; 5, 15 and 30 mg b.i.d., methotrexate or anti-TNFα failures 6-week dosing plus 6-week follow-up	Dose-dependent, clinically significant efficacy on disease activity (ACR20, 50 and 70) at 6 weeks, rapid onset of action within 2 weeks, 50% improvement in pain. Increase in mean LDL and HDL cholesterol and creatinine in all tofacitinib doses cohorts
	Phase IIb (Studies 1025 and 1035) Doses between 1 and 15 mg b.i.d. (and 20 mg q.d. 1025) as monotherapy or as dual therapy with anti-TNFα comparator, 24-week study	Significant effects ACR20, 50 and 70 at all doses above 1 mg. No dose response 3–15 mg. Efficacy in line with/superior to comparator anti-TNFα on ACR20–70 scores at 5 and 10 mg. Side-effects: decreased haemoglobin, anaemia, neutropoenia, increased LDL and HDL at 10 mg and above
	Phase III Multiple Phase III trials (400–900 patients) 5 or 10 mg doses as either monotherapy or in combination. Many with additional radiographic end-points and ACR scores[155]	
	Oral Solo Monotherapy in DMARD resistant patients – 5 and 10 mg b.i.d.	Study met 2/3 primary end-points. Positive ACR20 data in line with previous monotherapy and significant HAQ-DI benefits
	Oral Sync Dual therapy with background traditional DMARDS – 5 and 10 mg b.i.d.	All primary end-points met, at 6 months, limited dose discrimination 5 and 10 mg (ACR scores). Increase in LDL (5 mg = 15.7% and 10 mg = 18%) and modest increase in creatinine, opportunistic infections, neutropenia at 10 mg (2.8%) and increased anaemia

Table 8.5 (*Continued*)

Compound and Indication	Trials	Summary of outcome and major findings
	Oral Scan Dual therapy with background methotrexate – 5 and 10 mg b.i.d.	Significant improvement in structural damage inhibition at 10 mg but not 5 mg at 6 months. ACR20 end-points met for both 5 and 10 mg doses
	Oral Standard 5 and 10 mg b.i.d. dual therapy with background methotrexate. Humira comparator arm	ACR20 and DAS28 remission at 6 months
	Oral Step 5 and 10 mg b.i.d. dual therapy in anti-TNFα resistant patients with background methotrexate	Primary end-points met for 5 and 10 mg studies at 3 months (ACR20, DAS28)
Psoriasis	**Phase I** Small dose escalation study (n = 60) completed in psoriatic pastients with active lesions – 5, 10, 20, 30, 50 mg b.i.d. and 60 mg q.d.[157]	Significant improvement in physical global assessment at 30 and 50 mg b.i.d., dose-dependent effects on 14 day mPASI scores. Increase in cholesterol and LDL at 30 mg b.i.d.
	Phase II 12-week 1047 study in ~200 patients 2, 5 and 15 mg b.i.d.[155]	Statistically significant dose-dependent effects at 12 weeks (PASI 50/75/90)
Transplantation	**Phase I** 28 renal transplant patients >6 months post-transplantation 5, 15 and 30 mg b.i.d. with existing co-medications[149,156]	Significant reduction in NK cells in 15 and 30 mg b.i.d. with increased infections and decrease in haemoglobin and reticulocytes
	Phase IIa 61 new transplant recipients treated with 15 or 30 mg b.i.d. compared to tacrolimus over 6 months, with 12-month follow-up. MMF and prednisone as co-medications[162,163]	Reduction in NK cells and BPAR score superior to tacrolimus in 30 mg cohort. Total cholesterol increased 34 and 44% (15 and 30 mg b.i.d.). Decreased neutrophil counts and haemoglobin in both tofacitinib groups
Inflammatory Bowel Disease	**Phase II** Two Phase II studies in CD and UC, on-going Phase IIb studies in CD and Phase III in UC	
	Phase II (NCT00615199) - CD 140 patients with moderate/severe CD and co-medications. 1, 5 and 15 mg b.i.d. – 4-week dosing	Failed to achieve primary and secondary end-points at 4 weeks. Dose-dependent effects on CRP and FeC at 15 mg dose only

Table 8.5 (*Continued*)

Compound and Indication	Trials	Summary of outcome and major findings
	Phase II (NCT0078) – UC 194 patients with moderate/ severe UC and stable background medications. 0.5, 3, 10 and 15 mg b.i.d. 8-week dosing	Dose-dependent efficacy in clinical response and remission at 10 and 15 mg at 8 weeks
INCB28050 RA (NB INBC18424 progressing as a topical in Psoriasis)	**Phase II** One Phase II completed and one Phase II on-going **Phase IIa (Completed)** 125 patients with other stable oral therapies permitted. Three doses 4, 7 and 10 mg q.d., 6 months dosing	Clinically significant efficacy on ACR20, 50 and 70 scores at 12 and 24 weeks. No dose response, side-effects: dose-dependent decreases in haemoglobin and increases in LDL/HDL. 8 discontinuations due to AEs
	Phase IIb (Ongoing) 270 patients on stable co-medications. 2, 4 and 8 mg q.d. 1 mg q.d. with switch to 2 or 4 mg q.d. at 12 weeks, 6 months dosing	
VX-509RA	**Phase II** Single Phase II study in RA **Phase IIa** 200-patient study with stable co-medications. 25, 50, 100, 150 mg b.i.d. 12-week dosing	Interim report showed good signs and symptoms data for the highest two doses in methotrexate naive patients, meeting the primary end-points at 12 weeks for ACR20 and DAS28 improvement

compound is likely to have a significant role in the future treatment of auto-immune disease. A question that is as yet unresolved for the compound is how limiting the reported side-effects will be when large patient populations are exposed for long periods of treatments and what restriction, if any, the regulatory authorities may place on its use and dose selection. Clearly, the majority of the side-effects that have been recorded to date point to JAK2 inhibition being a major contributor to this profile (anaemia, neutropoenia, thrombocytopenia). However, a number of changes may not be directly attributable to this mechanism (cholesterol LDL/HDL and creatinine increases), which may implicate one or more of the other JAKs or some other compound-related effects. However, what is also unclear is what relative contribution

JAK2 inhibition delivers on clinical efficacy *versus* efficacy derived from inhibition of the other JAK family members.

8.6.2 INCB028050 and INCB018424 (ruxolitonib) for Immune Mediated Disease

8.6.2.1 Identification

Two compounds have been taken into the clinic by Incyte Corporation, INCB018424 (ruxolitonib) and INCB028050. Ruxolitonib is a dual JAK1 ($IC_{50} = 3.3$ nM) and JAK2 ($IC_{50} = 2.8$ nM) inhibitor, selective over JAK3 ($IC_{50} = 428$ nM) and TYK2 ($IC_{50} = 19$ nM)[106] in biochemical assays. Given that it shares structural features with tofacitinib, including the pyrrolopyrimidine core and pendant nitrile, it is likely to share the same binding mode although this has not been published. It is interesting that it shows a strikingly different selectivity profile in cellular assays (1000-fold selective for JAK1/2 over JAK3) from other structurally related pan-JAK inhibitors, illustrating that subtle changes in structure within the same chemotype can give rise to molecules that favour one or more of the JAKs. Incyte's second clinical molecule, INCB028050, has a similar biochemical profile to ruxolitonib being JAK1/2 selective: (JAK1 $IC_{50} = 5.9$ nM, JAK2 $IC_{50} = 5.7$ nM, JAK3 $IC_{50} = 400$ nM, TYK2 $IC_{50} = 53$ nM).[164] It is not known if this molecule is from the same structural class as ruxolitonib as Incyte is yet to release this structure.

8.6.2.2 Pre-clinical Development

Ruxolitonib was initially the Incyte lead molecule that was progressed into early Phase IIa trials in RA. Incyte shelved progression of ruxolitonib for RA in 2010, opting instead to switch to a combined in-house progression for topical use in psoriasis and a partnered approach with Novartis to prosecute higher doses in JAK2 gain of function myelofibrotic diseases (polythycemia vera, essential thrombocytopoenia). Ruxolitonib has no published clinical data in psoriasis; however, some recent pre-clinical data are available on the use of this compound as a topical agent.[165] Ruxolitonib was used successfully in a murine contact sensitivity model (twice daily topical application of INCB018424 (1.0% w/v)) where it was shown to inhibit tissue inflammation induced by IL23 or TSLP. In a parallel safety study the compound was well tolerated as a topical agent in mini-pigs where, significantly, there was no change in haematological parameters or blood counts after 28 days' administration. However, it is unclear from these studies whether significant levels of compound were available in the systemic circulation, despite high concentrations being reported locally in the skin. In parallel INCB028050, which has the advantage of once daily dosing over ruxolitonib's twice daily oral dosing regimen, is being progressed under a partnership with Eli Lilly for RA. INCB028050 has been shown to be efficacious on all measures of disease and disease score in pre-clinical rodent models of arthritis (AIA and CIA, at daily oral doses of 1 mg, 3 mg or 10 mg/kg). Furthermore in rat pharmacodynamic assays it has shown to inhibit IL6 and IL23 stimulated JAK1/2 mediated STAT3 phosphorylation with an IC_{50} of 128 nM and under serum free conditions in PBMCs with an IC_{50} of 48 nM, suggesting drug sequestering due to plasma protein binding. In contrast, in JAK3-mediated cellular assays INCB28050 was inactive at concentrations up to 10 µM. Moreover, Fridman *et al.* in an elegant study went on to correlate the dose used in these arthritis models with PD measures of target engagement. This suggested that constant target inhibition (whole blood IC_{50} determinations) was not required for efficacy, and indeed efficacy was achieved with as little as >55% pathway inhibition for 1 hour in a 24-hour period. This is in contrast to PK/PD predictions conducted with tofacitinib,[166] which suggested that for efficacy in RA this compound required constant target coverage above the JAK3 IC_{50}. This may be a reflection of the differential pharmacology of the two compounds or differences in assay format. However, it does raise the interesting question as to the optimal profile of target engagement to deliver efficacy and safety.

8.6.2.3 Pharmacokinetics of INCB018424

INCB018424 is a highly permeable molecule (P_{app} 215 nm/s determined in human Caco-2 monolayers) with high aqueous solubility (2.7 mg/L) and is categorized by the FDA under the BCS system as a Class 1 molecule.[167] Early clinical studies in healthy volunteers confirmed the promise of the pre-clinical studies where INCB018424 demonstrated rapid absorption and good oral bioavailability. In single oral ascending dosing studies in healthy volunteers using capsules (5 mg to 200 mg), INCB018424 showed high oral bioavailability (80%) and linear pharmacokinetics with no food effects. Oral dose clearance (CL/F) was low (*ca.* 20 L/h) with a corresponding moderate volume of distribution of the terminal phase ($V_z/F = 79$–87 L) resulting in a terminal half-life of 3 hours to 5 hours. Maximum plasma concentrations (C_{max}) ranged from 195 nM after a single 5 mg dose to 7010 nM after a single 200 mg dose occurring 1 hour to 2 hours post dose (T_{max}) for all doses with low inter-subject variability (percent coefficient of variation [CV%]) of 19% to 56% for C_{max}, and 9% to 34% for the area under the plasma concentration-time curves ($AUC_{0-\infty}$). Maximal inhibition of pSTAT3 corresponded with the C_{max} of INCB012484 and ranged from 40% at 5 mg to 90% at 200 mg. Based on population-based pharmacokinetic/pharmacodynamic (PK/PD) modelling, the IC_{50} was estimated at 254 nM with 95% confidence intervals (CI) ranges of 214 nM to 301 nM. After 10 days of dosing including both once (50 mg and 100 mg) and twice (15 mg, 25 mg and 50 mg) daily oral dose administrations, similar linear pharmacokinetics were observed on day 10 compared with day 1, with steady-state plasma levels achieved by day 2. Studies in man using [^{14}C]INCB018424 at a single oral solution dose of 25 mg[168] demonstrated that INCB018424 is cleared predominantly by metabolism followed by renal excretion with negligible renal clearance of the parent drug ($<1\%$ of dose).

8.6.2.4 INCB028050 Clinical Studies

Limited clinical information is available for INCB028050; however, a Phase II study has been completed with 125 patients in 6 months' dosing. Here, INCB028050 achieved the primary and ACR20 end-point (67% 4 mg q.d., 67% 7 mg q.d., 70% 10 mg q.d. at 24 weeks) as well as delivering efficacy at ACR50 and ACR70 scores and other secondary measures. The compound seemed to be broadly similar in terms of efficacy and side-effect profile (anaemia, increased LDL/HDL) to tofacitinib, however it has the potential to be given once daily compared to twice daily for tofacitinib. To date no published clinical data are available for review for ruxolitonib in psoriasis.

8.6.2.5 JAK2 Inhibitors for Myelofibrosis

Incyte is also prosecuting ruxolitonib as a treatment for myelofibrosis, where it recently achieved FDA approval and will be marketed under the trade name Jakafi. However, this is not the only JAK2 inhibitor that has been used in this patient population. A staurosporine analogue with JAK2 activity, CEP701, has been evaluated in myelofibrosis[136] where it met with limited success due to its poor tolerability and side-effects (GI toxicity, nausea, vomiting, diarrhoea). A second and more selective molecule, TG101348 from the aminopyrimidine structural class, has progressed to Phase II and shown efficacy on reducing peripheral blood counts (platelets and leukocytes) but with a similar toxicity profile to CEP-701.[169] In the initial Phase I study with ruxolitonib,[170] where doses up to 200 mg q.d. were explored, thrombocytopenia was observed in 8% of patients, which capped the maximum tolerated dose at 25 mg b.i.d. or 100 mg q.d. However, non-haematological toxicity was low. In this first study dose-dependent PD effects on JAK2 driven pSTAT3 were seen in both normal cells and cells carrying the JAK2 V617F mutation and significant clinical benefit was observed in reduction of splenomegaly and associated symptoms (weight loss, fatigue and pruritis). Since this first encouraging trial multiple Phase IIs and three Phase III trials have been initiated to test efficacy in myelofibrosis and in other haemoatopoietic leukaemias.

8.6.3 Vertex VX-509

Vertex are prosecuting VX-509, a JAK3 selective compound of unknown structure in RA. Vertex claim a high level of selectivity against the kinome and 25–100-fold selectivity for JAK3 over JAK2.[171] Vertex have recently completed a 14-day Phase I study and a 200-patient Phase II study in RA. Signs and symptoms data were good for the highest two doses (100 mg and 150 mg b.i.d.) in the RA study in methotrexate naive patients, meeting the primary end-points at 12 weeks for ACR20 and DAS28 improvement. For these cohorts ACR20/ 50/70 and DAS28 remission and improvement data are similar to those seen in tofacitinib monotherapy studies.[172]

8.6.4 Other JAK Inhibitors in Development

A number of other companies have JAK molecules at various stages of discovery and development. These include Galapagos with a JAK1 selective molecule (GLPG0634) for RA in Phase II.[173] This molecule recently completed a small (50-patient) four-week Phase II study where impressive efficacy was observed on ACR20/50/70 scores at 100 mg b.i.d. and 200 mg q.d. This efficacy was similar to that shown in trials with pan-JAK.[173,174]

Ambit are also developing a JAK2 selective molecule for RA in Phase II,[175] and Portola with a JAK3 molecule in pre-clinical development.[176] However, as limited published biological and clinical data are available on these molecules these will not be reviewed in any depth. These fall into a number of discrete classes based on their chemistry and selectivity profiles.

8.7   Conclusions and Future Perspectives

The JAK field has come a long way from the identification of "just another kinase" some 20 years ago. From these humble beginnings the small target class has emerged as one of the most promising therapies for the treatment of immune mediated diseases, acting as they do as nodal points to coordinate cytokine signalling. First generation drugs targeting the JAKs are delivering notable clinical efficacy across a broad spectrum of autoimmune diseases. However, these drugs are also demonstrating significant side-effects at clinically efficacious doses and these side-effects may ultimately restrict their clinical utility.

It is also clear that a second-generation approach is emerging, pioneered by JAK3 selective molecules, where the origin of these liabilities may be avoidable through the development of more selective inhibitors. Here, whilst there is much expectation that they may deliver an improved safety profile as compared to the "pan-JAKs" there also remains an unanswered question as to whether they will deliver improved or differentiated clinical benefit.

References

1. S. Shoji, D. C. Parmelee, R. D. Wade, S. Kumar, L. H. Ericsson, K. A. Walsh, H. Neurath, G. L. Long and J. G. Demaille, *Proc. Natl Acad. Sci. USA*, 1981, **78**, 848.
2. S. K. Hanks, A. M. Quinn and T. Hunter, *Science*, 1988, **241**, 42.
3. R. M. Eglen and T. Reisine, *Assay Drug Dev. Technol.*, 2009, **7**, 22.
4. J. Zhang, P. L. Yang and N. S. Gray, *Nat. Rev. Cancer*, 2009, **9**, 28.
5. L. L. Rokosz, J. R. Beasley, C. D. Carroll, T. Lin, J. Zhao, K. C. Appell and M. L. Webb, *Expert Opin. Ther. Targets*, 2008, **12**, 883.

6. Search carried out in *Pipeline Informa Healthcare*, 2011.
7. A. L. Hopkins and C. R. Groom, *Nat. Rev. Drug Discovery*, 2002, **1**, 727.
8. G. Manning, D. B. Whyte, R. Martinez, T. Hunter and S. Sudarsanam, *Science*, 2002, **298**, 1912.
9. M. W. Karaman, S. Herrgard, D. K. Treiber, P. Gallant, C. E. Atteridge, B. T. Campbell, K. W. Chan, P. Ciceri, M. I. Davis, P. T. Edeen, R. Faraoni, M. Floyd, J. P. Hunt, D. J. Lockhart, Z. V. Milanov, M. J. Morrison, G. Pallares, H. K. Patel, S. Pritchard, L. M. Wodicka and P. P. Zarrinkar, *Nat. Biotechnol.*, 2008, **26**, 127.
10. J. Baselga, *Science*, 2006, **312**, 1175.
11. J. S. Sebolt-Leopold and J. M. English, *Nature*, 2006, **441**, 457.
12. S. Bach, M. Knockaert, J. Reinhardt, O. Lozach, S. Schmitt, B. Baratte, M. Koken, S. P. Coburn, L. Tang, T. Jiang, D. C. Liang, H. Galons, J. F. Dierick, L. A. Pinna, F. Meggio, F. Totzke, C. Schaechtele, A. S. Lerman, A. Carnero, Y. Wan, N. Gray and L. Meijer, *J. Biol. Chem.*, 2005, **280**, 31208.
13. M. Bantscheff, D. Eberhard, Y. Abraham, S. Bastuck, M. Boesche, S. Hobson, T. Mathieson, J. Perrin, M. Raida, C. Rau, V. Reader, G. Sweetman, A. Bauer, T. Bouwmeester, C. Hopf, U. Kruse, G. Neubauer, N. Ramsden, J. Rick, B. Kuster and G. Drewes, *Nat. Biotechnol.*, 2007, **25**, 1035.
14. J. S. Melnick, J. Janes, S. Kim, J. Y. Chang, D. G. Sipes, D. Gunderson, L. Jarnes, J. T. Matzen, M. E. Garcia, T. L. Hood, R. Beigi, G. Xia, R. A. Harig, H. Asatryan, S. F. Yan, Y. Zhou, X. J. Gu, A. Saadat, V. Zhou, F. J. King, C. M. Shaw, A. I. Su, R. Downs, N. S. Gray, P. G. Schultz, M. Warmuth and J. S. Caldwell, *Proc. Natl Acad. Sci. USA*, 2006, **103**, 3153.
15. F. B. Sheinerman, E. Giraud and A. Laoui, *J. Mol. Biol.*, 2005, **352**, 1134.
16. A. F. Wilks, *Proc. Natl Acad. Sci. USA*, 1989, **86**, 1603.
17. A. F. Wilks, A. G. Harpur, R. R. Kurban, S. J. Ralph, G. Zuercher and A. Ziemiecki, *Mol. Cell. Biol.*, 1991, **11**, 2057.
18. I. Firmbach-Kraft, M. Byers, T. Shows, R. Dalla-Favera and J. J. Krolewski, *Oncogene*, 1990, **5**, 1329.
19. M. Kawamura, D. W. McVicar, J. A. Johnston, T. B. Blake, Y. Q. Chen, B. K. Lal, A. R. Lloyd, D. J. Kelvin and J. E. Staples, *Proc. Natl Acad. Sci. USA*, 1994, **91**, 6374.
20. N. L. Alicea-Velazquez and T. J. Boggon, *Curr. Drug Targets*, 2011, **12**, 546.
21. A. F. Wilks, *Semin. Cell Dev. Biol.*, 2008, **19**, 319.
22. J. L. Boulay, J. J. O'Shea and W. E. Paul, *Immunity*, 2003, **19**, 159.
23. P. J. Murray, *J. Immunol.*, 2007, **178**, 2623.
24. I. Behrmann, T. Smyczek, P. C. Heinrich, H. Schmitz-Van de Leur, W. Komyod, B. Giese, G. Mueller-Newen, S. Haan and C. Haan, *J. Biol. Chem.*, 2004, **279**, 35486.
25. I. Remy, I. A. Wilson and S. W. Michnick, *Science*, 1999, **283**, 990.
26. C. Schindler and I. Strehlow, *Adv. Pharmacol.*, 2000, **47**, 113.
27. J. J. O'Shea, M. Gadina and R. D. Schreiber, *Cell*, 2002, **109**, S121–S131.

28. T. Naka, M. Narazaki, M. Hirata, T. Matsumoto, S. Minamoto, A. Aono, N. Nishimoto, T. Kajita, T. Taga, K. Yoshizaki, S. Akira and T. Kishimoto, *Nature*, 1997, **387**, 924.

29. R. Starr, T. A. Willson, E. M. Viney, L. J. Murray, J. R. Rayner, B. J. Jenkins, T. J. Gonda, W. S. Alexander, D. Metcalf, N. A. Nicola and D. J. Hilton, *Nature*, 1997, **387**, 917.

30. T. A. Endo, M. Masuhara, M. Yokouchi, R. Suzuki, H. Sakamoto, K. Mitsui, A. Matsumoto, S. Tanimura, M. Ohtsubo, H. Misawa, T. Miyazaki, N. Leonor, T. Taniguchi, T. Fujita, Y. Kanakura, S. Komiya and A. Yoshimura, *Nature*, 1997, **387**, 921.

31. W. S. Alexander, *Nat. Rev. Immunol.*, 2002, **2**, 410.

32. M. Kubo, T. Hanada and A. Yoshimura, *Nat. Immunol.*, 2003, **4**, 1169.

33. J. C. Marine, D. J. Topham, C. McKay, D. Wang, E. Parganas, D. Stravopodis, A. Yoshimura and J. N. Ihle, *Cell*, 1999, **98**, 609.

34. R. Starr, D. Metcalf, A. G. Elefanty, M. Brysha, T. A. Willson, N. A. Nicola, D. J. Hilton and W. S. Alexander, *Proc. Natl Acad. Sci. USA*, 1998, **95**, 14395.

35. D. Metcalf, C. J. Greenhalgh, E. Viney, T. A. Willson, R. Starr, N. A. Nicola, D. J. Hilton and W. S. Alexander, *Nature*, 2000, **405**, 1069.

36. A. W. Roberts, L. Robb, S. Rakar, L. Hartley, L. Cluse, N. A. Nicola, D. Metcalf, D. J. Hilton and W. S. Alexander, *Proc. Natl Acad. Sci. USA*, 2001, **98**, 9324.

37. P. J. Egan, K. E. Lawlor, W. S. Alexander and I. P. Wicks, *J. Clin. Invest.*, 2003, **111**, 915.

38. P. K. K. Wong, P. J. Egan, B. A. Croker, K. O'Donnell, N. A. Sims, S. Drake, H. Kiu, E. J. McManus, W. S. Alexander, A. W. Roberts and I. P. Wicks, *J. Clin. Invest.*, 2006, **116**, 1571.

39. T. Shouda, T. Yoshida, T. Hanada, T. Wakioka, M. Oishi, K. Miyoshi, S. Komiya, K. I. Kosai, Y. Hanakawa, K. Hashimoto, K. Nagata and A. Yoshimura, *J. Clin. Invest.*, 2001, **108**, 1781.

40. D. C. Thomis and L. J. Berg, *Curr. Opin. Immunol.*, 1997, **9**, 541.

41. S. J. Rodig, M. A. Meraz, J. M. White, P. A. Lampe, J. K. Riley, C. D. Arthur, K. L. King, K. C. F. Sheehan, L. Yin, D. Pennica, E. M. Johnson, Jr. and R. D. Schreiber, *Cell*, 1998, **93**, 373.

42. H. Neubauer, A. Cumano, M. Muller, H. Wu, U. Huffstadt and K. Pfeffer, *Cell*, 1998, **93**, 397.

43. E. Parganas, D. Wang, D. Stravopodis, D. J. Topham, J. C. Marine, S. Teglund, E. F. Vanin, S. Bodner, O. R. Colamonici, J. M. Van Deursen, G. Grosveld and J. N. Ihle, *Cell*, 1998, **93**, 385.

44. T. Nosaka, J. M. A. van Deursen, R. A. Tripp, W. E. Thierfelder, B. A. Witthuhn, A. P. McMickle, P. C. Doherty, G. Grosveld and J. N. Ihle, *Science*, 1995, **270**, 800.

45. S. Y. Park, K. Saijo, T. Takahashi, M. Osawa, H. Arase, N. Hirayama, K. Miyake, H. Nakauchi, T. Shirasawa and T. Saito, *Immunity*, 1995, **3**, 771.

46. D. C. Thomis, C. B. Gurniak, E. Tivol, A. H. Sharpe and L. J. Berg, *Science*, 1995, **270**, 794.

47. P. Macchi, A. Villa, S. Giliani, M. G. Sacco, A. Frattini, F. Porta, A. G. Ugazio, J. A. Johnston and F. Candotti, *Nature*, 1995, **377**, 65.
48. S. M. Russell, N. Tayebi, H. Nakajima, M. C. Riedy, J. L. Roberts, M. J. Aman, T. S. Migone, M. Noguchi and M. L. Markert, *Science*, 1995, **270**, 797.
49. L. D. Notarangelo, P. Mella, A. Jones, G. de Saint Basile, G. Savoldi, T. Cranston, M. Vihinen and R. F. Schumacher, *Hum. Mutat.*, 2001, **18**, 255.
50. J. L. Roberts, A. Lengi, S. M. Brown, M. Chen, Y. J. Zhou, J. J. O'Shea and R. H. Buckley, *Blood*, 2004, **103**, 2009.
51. Y. Rochman, R. Spolski and W. J. Leonard, *Nat. Rev. Immunol.*, 2009, **9**, 480.
52. A. Puel, S. F. Ziegler, R. H. Buckley and W. J. Leonard, *Nat. Genet.*, 1998, **20**, 394.
53. C. M. Roifman, J. Zhang, D. Chitayat and N. Sharfe, *Blood*, 2000, **96**, 2803.
54. K. Shimoda, K. Kato, K. Aoki, T. Matsuda, A. Miyamoto, M. Shibamori, M. Yamashita, A. Numata, K. Takase, S. Kobayashi, S. Shibata, Y. Asano, H. Gondo, K. Sekiguchi, K. Nakayama, T. Nakayama, T. Okamura, S. Okamura, Y. Niho and K. I. Nakayama, *Immunity*, 2000, **13**, 561.
55. M. Karaghiosoff, H. Neubauer, C. Lassnig, P. Kovarik, H. Schindler, H. Pircher, B. McCoy, C. Bogdan, T. Decker, G. Brem, K. Pfeffer and M. Muller, *Immunity*, 2000, **13**, 549.
56. K. Shimoda, H. Tsutsui, K. Aoki, K. Kato, T. Matsuda, A. Numata, K. Takase, T. Yamamoto, H. Nukina, T. Hoshino, Y. Asano, H. Gondo, T. Okamura, S. Okamura, K. I. Nakayama, K. Nakanishi, Y. Niho and M. Harada, *Blood*, 2002, **99**, 2094.
57. N. Tokumasa, A. Suto, S. i. Kagami, S. Furuta, K. Hirose, N. Watanabe, Y. Saito, K. Shimoda, I. Iwamoto and H. Nakajima, *Blood*, 2007, **110**, 553.
58. M. Karaghiosoff, R. Steinborn, P. Kovarik, G. Kriegshaeuser, M. Baccarini, B. Donabauer, U. Reichart, T. Kolbe, C. Bogdan, T. Leanderson, D. Levy, T. Decker and M. Mueller, *Nat. Immunol.*, 2003, **4**, 471.
59. A. Oyamada, H. Ikebe, M. Itsumi, H. Saiwai, S. Okada, K. Shimoda, Y. Iwakura, K. I. Nakayama, Y. Iwamoto, Y. Yoshikai and H. Yamada, *J. Immunol.*, 2009, **183**, 7539.
60. M. Ishizaki, T. Akimoto, R. Muromoto, M. Yokoyama, Y. Ohshiro, Y. Sekine, H. Maeda, K. Shimoda, K. Oritani and T. Matsuda, *J. Immunol.*, 2011, **187**, 181.
61. Y. Minegishi, M. Saito, T. Morio, K. Watanabe, K. Agematsu, S. Tsuchiya, H. Takada, T. Hara, N. Kawamura, T. Ariga, H. Kaneko, N. Kondo, I. Tsuge, A. Yachie, Y. Sakiyama, T. Iwata, F. Bessho, T. Ohishi, K. Joh, K. Imai, K. Kogawa, M. Shinohara, M. Fujieda, H. Wakiguchi, S. Pasic, M. Abinun, H. D. Ochs, E. D. Renner, A. Jansson, B. H. Belohradsky, A. Metin, N. Shimizu, S. Mizutani, T. Miyawaki, S. Nonoyama and H. Karasuyama, *Immunity*, 2006, **25**, 745.

62. C. Woellner, A. A. Schaffer, J. M. Puck, E. D. Renner, C. Knebel, S. M. Holland, A. Plebani and B. Grimbacher, *Immunity*, 2007, **26**, 535.
63. S. Akira, *Stem Cells*, 1999, **17**, 138.
64. H. Yu, D. Pardoll and R. Jove, *Nat. Rev. Cancer*, 2009, **9**, 798.
65. M. A. Meraz, J. M. White, K. C. F. Sheehan, E. A. Bach, S. J. Rodig, A. S. Dighe, D. H. Kaplan, J. K. Riley and A. C. Greenlund, *Cell*, 1996, **84**, 431.
66. J. E. Durbin, R. Hackenmiller, M. C. Simon and D. E. Levy, *Cell*, 1996, **84**, 443.
67. C. Park, S. Li, E. Cha and C. Schindler, *Immunity*, 2000, **13**, 795.
68. K. Takeda, K. Noguchi, W. Shi, T. Tanaka, M. Matsumoto, N. Yoshida, T. Kishimoto and S. Akira, *Proc. Natl Acad. Sci. USA*, 1997, **94**, 3801.
69. S. A. Duncan, Z. Zhong, Z. Wen and J. E. Darnell, Jr., *Dev. Dyn.*, 1997, **208**, 190.
70. Z. Chen, A. Laurence, Y. Kanno, M. Pacher-Zavisin, B. M. Zhu, C. Tato, A. Yoshimura, L. Hennighausen and J. J. O'Shea, *Proc. Natl Acad. Sci. USA*, 2006, **103**, 8137.
71. I. Kinjyo, H. Inoue, S. Hamano, S. Fukuyama, T. Yoshimura, K. Koga, H. Takaki, K. Himeno, G. Takaesu, T. Kobayashi and A. Yoshimura, *J. Exp. Med.*, 2006, **203**, 1021.
72. N. Horiguchi, F. Lafdil, A. M. Miller, O. Park, H. Wang, M. Rejesh, P. Mukhopadhyay, X. Y. Fu, P. Pacher and B. Gao, *Hepatology*, 2010, **51**, 1724.
73. K. M. Murphy and S. L. Reiner, *Nat. Rev. Immunol.*, 2002, **2**, 933.
74. D. Agnello, C. S. R. Lankford, J. Bream, A. Morinobu, M. Gadina, J. J. O'Shea and D. M. Frucht, *J. Clin. Immunol.*, 2003, **23**, 147.
75. A. L. Wurster, T. Tanaka and M. J. Grusby, *Oncogene*, 2000, **19**, 2577.
76. X. Liu, G. W. Robinson, K. U. Wagner, L. Garrett, A. Wynshaw-Boris and L. Hennighausen, *Genes Dev.*, 1997, **11**, 179.
77. G. B. Udy, R. P. Towers, R. G. Snell, R. J. Wilkins, S. H. Park, P. A. Ram, D. J. Waxman and H. W. Davey, *Proc. Natl Acad. Sci. USA*, 1997, **94**, 7239.
78. S. Teglund, C. McKay, E. Schuetz, J. M. Van Deursen, D. Stravopodis, D. Wang, M. Brown, S. Bodner, G. Grosveld and J. N. Ihle, *Cell*, 1998, **93**, 841.
79. H. Nakajima, X. W. Liu, A. Wynshaw-Boris, L. A. Rosenthal, K. Imada, D. S. Finbloom, L. Hennighausen and W. J. Leonard, *Immunity*, 1997, **7**, 691.
80. K. Imada, E. T. Bloom, H. Nakajima, J. A. Horvath-Arcidiacono, G. B. Udy, H. W. Davey and W. J. Leonard, *J. Exp. Med.*, 1998, **188**, 2067.
81. J. Turkson, J. S. Kim, S. Zhang, J. Yuan, M. Huang, M. Glenn, E. Haura, S. Sebti, A. D. Hamilton and R. Jove, *Mol. Cancer Ther.*, 2004, **3**, 261.
82. J. Turkson, *Expert Opin. Ther. Targets*, 2004, **8**, 409.
83. S. Haftchenary, M. Avadisian and P. T. Gunning, *Anti-Cancer Drugs*, 2011, **22**, 115.
84. C. James, V. Ugo, J. P. Le Couedic, J. Staerk, F. Delhommeau, C. Lacout, L. Garcon, H. Raslova, R. Berger, A. Bennaceur-Griscelli, J. L. Villeval,

S. N. Constantinescu, N. Casadevall and W. Vainchenker, *Nature*, 2005, **434**, 1144.

85. E. J. Baxter, L. M. Scott, P. J. Campbell, C. East, N. Fourouclas, S. Swanton, G. S. Vassiliou, A. J. Bench, E. M. Boyd, N. Curtin, M. A. Scott, W. N. Erber and A. R. Green, *Lancet*, 2005, **365**, 1054.

86. R. L. Levine, M. Wadleigh, J. Cools, B. L. Ebert, G. Wernig, B. J. P. Huntly, T. J. Boggon, I. Wlodarska, J. J. Clark, S. Moore, J. Adelsperger, S. Koo, J. C. Lee, S. Gabriel, T. Mercher, A. D'Andrea, S. Froehling, K. Doehner, P. Marynen, P. Vandenberghe, R. A. Mesa, A. Tefferi, J. D. Griffin, M. J. Eck, W. R. Sellers, M. Meyerson, T. R. Golub, S. J. Lee and D. G. Gilliland, *Cancer Cell*, 2005, **7**, 387.

87. R. Kralovics, F. Passamonti, A. S. Buser, S. S. Teo, R. Tiedt, J. R. Passweg, A. Tichelli, M. Cazzola and R. C. Skoda, *N. Engl. J. Med.*, 2005, **352**, 1779.

88. P. J. Campbell and A. P. Green, *N. Engl. J. Med.*, 2006, **355**, 2452.

89. S. Dupont, A. Masse, C. James, I. Teyssandier, Y. Lecluse, F. Larbret, V. Ugo, P. Saulnier, S. Koscielny, J. P. Le Couedic, N. Casadevall, W. Vainchenker and F. Delhommeau, *Blood*, 2007, **110**, 1013.

90. L. M. Scott, M. A. Scott, P. J. Campbell and A. R. Green, *Blood*, 2006, **108**, 2435.

91. E. Flex, V. Petrangeli, L. Stella, S. Chiaretti, T. Hornakova, L. Knoops, C. Ariola, V. Fodale, E. Clappier, F. Paoloni, S. Martinelli, A. Fragale, M. Sanchez, S. Tavolaro, M. Messina, G. Cazzaniga, A. Camera, G. Pizzolo, A. Tornesello, M. Vignetti, A. Battistini, H. Cave, B. D. Gelb, J. C. Renauld, A. Biondi, S. N. Constantinescu, R. Foa and M. Tartaglia, *J. Exp. Med.*, 2008, **205**, 751.

92. E. G. Jeong, M. S. Kim, H. K. Nam, C. K. Min, S. Lee, Y. J. Chung, N. J. Yoo and S. H. Lee, *Clin. Cancer Res.*, 2008, **14**, 3716.

93. C. G. Mullighan, J. Zhang, R. C. Harvey, J. R. Collins-Underwood, B. A. Schulman, L. A. Phillips, S. K. Tasian, M. L. Loh, X. Su, W. Liu, M. Devidas, S. R. Atlas, I. M. Chen, R. J. Clifford, D. S. Gerhard, W. L. Carroll, G. H. Reaman, M. Smith, J. R. Downing, S. P. Hunger and C. L. Willman, *Proc. Natl Acad. Sci. USA*, 2009, **106**, 9414, S9414.

94. T. Hornakova, S. Chiaretti, M. M. Lemaire, R. Foa, R. Ben Abdelali, V. Asnafi, M. Tartaglia, J. C. Renauld and L. Knoops, *Blood*, 2010, **115**, 3287.

95. D. K. Walters, T. Mercher, T. L. Gu, T. O'Hare, J. W. Tyner, M. Loriaux, V. L. Goss, K. A. Lee, C. A. Eide, M. J. Wong, E. P. Stoffregen, L. McGreevey, J. Nardone, S. A. Moore, J. Crispino, T. J. Boggon, M. C. Heinrich, M. W. Deininger, R. D. Polakiewicz, D. G. Gilliland and B. J. Druker, *Cancer Cell*, 2006, **10**, 65.

96. M. G. Cornejo, M. G. Kharas, M. B. Werneck, S. Le Bras, S. A. Moore, B. Ball, M. Beylot-Barry, S. J. Rodig, J. C. Aster, B. H. Lee, H. Cantor, J. P. Merlio, D. G. Gilliland and T. Mercher, *Blood*, 2009, **113**, 2746.

97. R. Lai, G. Z. Rassidakis, Q. Lin, C. Atwell, L. J. Medeiros and H. M. Amin, *Hum. Pathol.*, 2005, **36**, 939.

98. M. G. Cornejo, T. J. Boggon and T. Mercher, *Int. J. Biochem. Cell Biol.*, 2009, **41**, 2376.

99. S. Sigurdsson, G. Nordmark, H. H. H. Goering, K. Lindroos, A. C. Wiman, G. Sturfelt, A. Joensen, S. Rantapaeae-Dahlqvist, B. Moeller, J. Kere, S. Koskenmies, E. Widen, M. L. Eloranta, H. Julkunen, H. Kristjansdottir, K. Steinsson, G. Alm, L. Roennblom and A. C. Syvaenen, *Am. J. Hum. Genet.*, 2005, **76**, 528.

100. P. R. Burton, D. G. Clayton, L. R. Cardon, N. Craddock, P. Deloukas, A. Duncanson, D. P. Kwiatkowski, M. I. McCarthy, W. H. Ouwehand, N. J. Samani, J. A. Todd, P. Donnelly, J. C. Barrett, D. Davison, D. Easton, D. M. Evans, H. T. Leung, J. L. Marchini, A. P. Morris, C. C. A. Spencer, M. D. Tobin, L. R. Cardon, A. P. Attwood, J. P. Boorman, B. Cant, U. Everson, J. M. Hussey, J. D. Jolley, A. S. Knight, K. Koch, E. Meech, S. Nutland, C. V. Prowse, H. E. Stevens, N. C. Taylor, G. R. Walters, N. M. Walker, N. A. Watkins, T. Winzer, R. W. Jones, W. L. McArdle, S. M. Ring, D. P. Strachan, M. Pembrey, G. Breen, D. Clair, S. Caesar, K. Gordon-Smith, L. Jones, C. Fraser, E. K. Green, D. Grozeva, M. L. Hamshere, P. A. Holmans, I. R. Jones, G. Kirov, V. Moskivina, I. Nikolov, M. C. O'Donovan, M. J. Owen, D. A. Collier, A. Elkin, A. Farmer, R. Williamson, P. McGuffin, A. H. Young, I. N. Ferrier, S. G. Ball, A. J. Balmforth, J. H. Barrett, T. D. Bishop, M. M. Iles, A. Maqbool, N. Yuldasheva, A. S. Hall, P. S. Braund, R. J. Dixon, M. Mangino, S. Stevens, J. R. Thompson, F. Bredin, M. Tremelling, M. Parkes, H. Drummond, C. W. Lees, E. R. Nimmo, J. Satsangi, S. A. Fisher, A. Forbes, C. M. Lewis, C. M. Onnie, N. J. Prescott, J. Sanderson, C. G. Matthew, J. Barbour, M. K. Mohiuddin, C. E. Todhunter, J. C. Mansfield, T. Ahmad, F. R. Cummings, D. P. Jewell, J. Webster, M. J. Brown, M. G. Lathrop, J. Connell, A. Dominiczak, C. A. Braga Marcano, B. Burke, R. Dobson, J. Gungadoo, K. L. Lee, P. B. Munroe, S. J. Newhouse, A. Onipinla, C. Wallace, M. Xue, M. Caulfield, M. Farrall, A. Barton, I. N. Bruce, H. Donovan, S. Eyre, P. D. Gilbert, S. L. Hilder, A. M. Hinks, S. L. John, C. Potter, A. J. Silman, D. P. M. Symmons, W. Thomson, J. Worthington, D. B. Dunger, H. E. Stevens, B. Widmer, T. M. Frayling, R. M. Freathy, H. Lango, J. R. B. Perry, B. M. Shields, M. N. Weedon, A. T. Hattersley, G. A. Hitman, M. Walker, K. S. Elliott, C. J. Groves, C. M. Lindgren, N. W. Rayner, N. J. Timpson, E. Zeggini, M. Newport, G. Sirugo, E. Lyons, F. Vannberg, A. V. S. Hill, L. A. Bradbury, C. Farrar, J. J. Pointon, P. Wordsworth, M. A. Brown, J. A. Franklyn, J. M. Heward, M. J. Simmonds, S. C. L. Gough, S. Seal, M. R. Stratton, N. Rahman, M. Ban, A. Goris, S. J. Sawcer, A. Compston, D. Conway, M. Jallow, M. Newport, G. Sirugo, K. A. Rockett, S. J. Bumpstead, A. Chaney, K. Downes, M. J. R. Ghori, R. Gwilliam, S. E. Hunt, M. Inouye, A. Keniry, E. King, R. McGinnis, S. Potter, R. Ravindrarajah, P. Whittaker, C. Widden, D. Withers, N. J. Cardin, T. Ferreira, J. Pereira-Gale, I. B. Hallgrimsdottir, B. N. Howie, Y. Y. Teo, D. Vukcevic, D. Bentley, M. A. Brown, M. Caulfield, A. Compston,

M. Farrall, A. S. Hall, A. T. Hattersley, A. V. S. Hill, M. Parkes, M. Pembrey, N. Rahman, M. R. Stratton, S. L. Mitchell, P. R. Newby, O. J. Brand, J. Carr-Smith, S. H. S. Pearce, R. McGinnis, A. Keniry, J. D. Reveille, X. Zhou, A. M. Sims, A. Dowling, J. Taylor, T. Doan, J. C. Davis, L. Savage, M. M. Ward, T. L. Learch, M. H. Weisman and P. Wordsworth, *Nat. Genet.*, 2007, **39**, 1329.

101. M. Ban, A. Goris, A. R. Lorentzen, A. Baker, T. Mihalova, G. Ingram, D. R. Booth, R. N. Heard, G. J. Stewart, E. Bogaert, B. Dubois, H. F. Harbo, E. G. Celius, A. Spurkland, R. Strange, C. Hawkins, N. P. Robertson, F. Dudbridge, J. Wason, P. L. De Jager, D. Hafler, J. D. Rioux, A. J. Ivinson, J. L. McCauley, M. Pericak-Vance, J. R. Oksenberg, S. L. Hauser, D. Sexton, J. Haines, S. Sawcer and A. Compston, *Eur. J. Hum. Genet.*, 2009, **17**, 1309.

102. L. Vijayakrishnan, R. Venkataramanan and P. Gulati, *Trends Pharmacol. Sci.*, 2011, **32**, 25.

103. I. S. Lucet, E. Fantino, M. Styles, R. Bamert, O. Patel, S. E. Broughton, M. Walter, C. J. Burns, H. Treutlein, A. F. Wilks and J. Rossjohn, *Blood*, 2006, **107**, 176.

104. J. E. Chrencik, A. Patny, I. K. Leung, B. Korniski, T. L. Emmons, T. Hall, R. A. Weinberg, J. A. Gormley, J. M. Williams, J. E. Day, J. L. Hirsch, J. R. Kiefer, J. W. Leone, H. D. Fischer, C. D. Sommers, H. C. Huang, E. J. Jacobsen, R. E. Tenbrink, A. G. Tomasselli and T. E. Benson, *J. Mol. Biol.*, 2010, **400**, 413.

105. M. E. Flanagan, T. A. Blumenkopf, W. H. Brissette, M. F. Brown, J. M. Casavant, S. P. Chang, J. L. Doty, E. A. Elliott, M. B. Fisher, M. Hines, C. Kent, E. M. Kudlacz, B. M. Lillie, K. S. Magnuson, S. P. McCurdy, M. J. Munchhof, B. D. Perry, P. S. Sawyer, T. J. Strelevitz, C. Subramanyam, J. Sun, D. A. Whipple and P. S. Changelian, *J. Med. Chem.*, 2010, **53**, 8468.

106. A. Quintas-Cardama, K. Vaddi, P. Liu, T. Manshouri, J. Li, P. A. Scherle, E. Caulder, X. Wen, Y. Li, P. Waeltz, M. Rupar, T. Burn, Y. Lo, J. Kelley, M. Covington, S. Shepard, J. D. Rodgers, P. Haley, H. Kantarjian, J. S. Fridman and S. Verstovsek, *Blood*, 2010, **115**, 3109.

107. G. Noronha, J. Cao, C. Chow, C. C. Mak, M. Palanki, E. Dneprovskaia, A. Mcpherson, V. P. Pathak, J. Renick and B. Zeng, Targegen Inc., USA, *PCT Int. Appl.*, WO, 2009/055674, 2009.

108. C. Gaul, M. Gerspacher, P. Holzer and C. Pissot Soldermann, Novartis AG, Switzerland, *PCT Int. Appl.*, WO 2009/098236, 2009.

109. Y. Song, Q. Xu and A. Pandey, Portola Pharmaceuticals, Inc., USA, *PCT Int. Appl.*, WO 2009/026107, 2009.

110. D. J. Guerin, J. Joon and E. Stanton, Merck & Co, Inc., USA, *PCT Int. Appl.*, WO 2009/054941, 2009.

111. T. Inoue, A. Tanaka, K. Nakai, H. Sasaki, F. Takahashi, S. Shirakami, K. Hatanaka, Y. Nakajima, K. Mukoyoshi, H. Hamaguchi, S. Kunikawa and Y. Higashi, Astellas Pharma Inc., Japan, *PCT Int. Appl.*, WO 2007/077949, 2007.

112. L. Farmer, G. Martinez-Botella, A. Pierce, F. Salituro, J. Wang, M. Wannamker and T. Wang, Vertex Pharmaceuticals Inc., USA, *PCT Int. Appl.*, WO 2007/084557, 2007.

113. J. T. Bamberg, M. Bartlett, D. J. Dubois, T. R. Elworthy, R. T. Hendricks, J. C. Hermann and R. Kondru, Roche Palo Alto LLC, USA, *US Pat. Appl.*, 20090215750, 2009.

114. L. J. Wilson, *Expert Opin. Ther. Pat.*, 2010, **20**, 609.

115. T. Deuse, J. B. Velotta, G. Hoyt, J. A. Govaert, V. Taylor, E. Masuda, E. Herlaar, G. Park, D. Carroll, M. P. Pelletier, R. C. Robbins and S. Schrepfer, *Transplantation*, 2008, **85**, 885.

116. B. Y. Chang, F. Zhao, X. He, H. Ren, S. Braselmann, V. Taylor, J. Wicks, D. G. Payan, E. B. Grossbard, P. R. Pine and D. C. Bullard, *J. Immunol.*, 2009, **183**, 2183.

117. G. Thoma, F. Nuninger, R. Falchetto, E. Hermes, G. A. Tavares, E. Vangrevelinghe and H. G. Zerwes, *J. Med. Chem.*, 2011, **54**, 284.

118. C. Haan, C. Rolvering, F. Raulf, M. Kapp, P. Drueckes, G. Thoma, I. Behrmann and H. G. Zerwes, *Chem. Biol.*, 2011, **18**, 314.

119. T. H. Lin, M. Hegen, E. Quadros, C. L. Nickerson-Nutter, K. C. Appell, A. G. Cole, Y. Shao, S. Tam, M. Ohlmeyer, B. Wang, D. G. Goodwin, E. F. Kimble, J. Quintero, M. Gao, P. Symanowicz, C. Wrocklage, J. Lussier, S. H. Schelling, A. G. Hewet, D. Xuan, R. Krykbaev, J. Togias, X. Xu, R. Harrison, T. Mansour, M. Collins, J. D. Clark, M. L. Webb and K. J. Seidi, *Arthritis Rheum.*, 2010, **62**, 2283.

120. A. G. Cole, A. C. Bohnstedt, V. Paradkar, C. Kingsbury, J. G. Quintero, H. Park, Y. Lu, M. You, I. Neagu, D. J. Diller, J. J. Letourneau, Y. Shao, R. A. James, C. M. Riviello, K. K. Ho, T. H. Lin, B. Wang, K. C. Appell, M. Sills, E. Quadros, E. F. Kimble, M. H. J. Ohlmeyer and M. L. Webb, *Bioorg. Med. Chem. Lett.*, 2009, **19**, 6788.

121. S. Ioannidis, M. L. Lamb, T. Wang, L. Almeida, M. H. Block, A. M. Davies, B. Peng, M. Su, H. J. Zhang, E. Hoffmann, C. Rivard, I. Green, T. Howard, H. Pollard, J. Read, M. Alimzhanov, G. Bebernitz, K. Bell, M. W. Ye, D. Huszar and M. Zinda, *J. Med. Chem.*, 2011, **54**, 262.

122. A. Scuto, P. Krejci, L. Popplewell, J. Wu, Y. Wang, M. Kujawski, C. Kowolik, H. Xin, L. Chen, Y. Wang, L. Kretzner, H. Yu, W. R. Wilcox, Y. Yen, S. Forman and R. Jove, *Leukemia*, 2011, **25**, 538.

123. A. Pardanani, T. Lasho, G. Smith, C. J. Burns, E. Fantino and A. Tefferi, *Leukemia*, 2009, **23**, 1441.

124. J. W. Tyner, T. G. Bumm, J. Deininger, L. Wood, K. J. Aichberger, M. M. Loriaux, B. J. Druker, C. J. Burns, E. Fantino and M. W. Deininger, *Blood*, 2010, **115**, 5232.

125. C. J. Burns, D. G. Bourke, L. Andrau, X. Bu, S. A. Charman, A. C. Donohue, E. Fantino, M. Farrugia, J. T. Feutrill, M. Joffe, M. R. Kling, M. Kurek, T. L. Nero, T. Nguyen, J. T. Palmer, I. Phillips, D. M. Shackleford, H. Sikanyika, M. Styles, S. Su, H. Treutlein, J. Zeng and A. F. Wilks, *Bioorg. Med. Chem. Lett.*, 2009, **19**, 5887.

126. A. D. William, A. C. H. Lee, S. Blanchard, A. Poulsen, E. L. Teo, H. Nagaraj, E. Tan, D. Chen, M. Williams, E. T. Sun, K. C. Goh, W. C. Ong, S. K. Goh, S. Hart, R. Jayaraman, M. K. Pasha, K. Ethirajulu, J. M. Wood and B. W. Dymock, *J. Med. Chem.*, 2011, **54**, 4638.

127. S. Hart, K. C. Goh, Y. C. Tan, C. Amalini and J. M. Wood, *Proceedings of the 101st Annual Meeting of the American Association for Cancer Research*, 2010, **70**(8), abstract number 2524.

128. Company website, http://www.sbio.com/main/pipeline.asp, 2011.

129. C. Pissot-Soldermann, M. Gerspacher, P. Furet, C. Gaul, P. Holzer, C. McCarthy, T. Radimerski, C. H. Regnier, F. Baffert, P. Drueckes, G. A. Tavares, E. Vangrevelinghe, F. Blasco, G. Ottaviani, F. Ossola, J. Scesa and J. Reetz, *Bioorg. Med. Chem. Lett.*, 2010, **20**, 2609.

130. F. Baffert, C. H. Regnier, A. De Pover, C. Pissot-Soldermann, G. A. Tavares, F. Blasco, J. Brueggen, P. Chene, P. Drueckes, D. Erdmann, P. Furet, M. Gerspacher, M. Lang, D. Ledieu, L. Nolan, S. Ruetz, J. Trappe, E. Vangrevelinghe, M. Wartmann, L. Wyder, F. Hofmann and T. Radimerski, *Mol. Cancer Ther.*, 2010, **9**, 1945.

131. T. Siu, E. S. Kozina, J. Jung, C. Rosenstein, A. Mathur, M. D. Altman, G. Chan, L. Xu, E. Bachman, J. R. Mo, M. Bouthillette, T. Rush, C. J. Dinsmore, C. G. Marshall and J. R. Young, *Bioorg. Med. Chem. Lett.*, 2010, **20**, 7421.

132. T. Wang, J. P. Duffy, J. Wang, S. Halas, F. G. Salituro, A. C. Pierce, H. J. Zuccola, J. R. Black, J. K. Hogan, S. Jepson, D. Shlyakter, S. Mahajan, Y. Gu, T. Hoock, M. Wood, B. F. Furey, J. D. Frantz, L. M. Dauffenbach, U. A. Germann, B. Fan, M. Namchuk, Y. L. Bennani and M. W. Ledeboer, *J. Med. Chem.*, 2009, **52**, 7938.

133. R. Kiss, P. P. Sayeski and G. M. Keseru, *Expert Opin. Ther. Pat.*, 2010, **20**, 471.

134. N. Meydan, T. Grunberger, H. Dadi, M. Shahar, E. Arpaia, Z. Lapidot, J. S. Leeder, M. Freedman, A. Cohen, A. Gazit, A. Levitzki and C. M. Roifman, *Nature*, 1996, **379**, 645.

135. T. Grunberger, P. Demin, O. Rounova, N. Sharfe, L. Cimpean, H. Dadi, A. Freywald, Z. Estrov and C. M. Roifman, *Blood*, 2003, **102**, 4153.

136. E. O. Hexner, C. Serdikoff, M. Jan, C. R. Swider, C. Robinson, S. Yang, T. Angeles, S. G. Emerson, M. Carroll, B. Ruggeri and P. Dobrzanski, *Blood*, 2008, **111**, 5663.

137. J. E. Thompson, R. M. Cubbon, R. T. Cummings, L. S. Wicker, R. Frankshun, B. R. Cunningham, P. M. Cameron, P. T. Meinke, N. Liverton, Y. Weng and J. A. DeMartino, *Bioorg. Med. Chem. Lett.*, 2002, **12**, 1219.

138. J. Lim, B. Taoka, R. D. Otte, K. Spencer, C. J. Dinsmore, M. D. Altman, G. Chan, C. Rosenstein, S. Sharma, H. P. Su, A. A. Szewczak, L. Xu, H. Yin, J. Zugay-Murphy, C. G. Marshall and J. R. Young, *J. Med. Chem.*, 2011, **54**, 7334.

139. N. K. Williams, R. S. Bamert, O. Patel, C. Wang, P. M. Walden, A. F. Wilks, E. Fantino, J. Rossjohn and I. S. Lucet, *J. Mol. Biol.*, 2009, **387**, 219.

140. J. K. Jiang, K. Ghoreschi, F. Deflorian, Z. Chen, M. Perreira, M. Pesu, J. Smith, D. T. Nguyen, E. H. Liu, W. Leister, S. Costanzi, J. J. O'Shea and C. J. Thomas, *J. Med. Chem.*, 2008, **51**, 8012.

141. P. S. Changelian, M. E. Flanagan, D. J. Ball, C. R. Kent, K. S. Magnuson, W. H. Martin, B. J. Rizzuti, P. S. Sawyer, B. D. Perry, W. H. Brissette, S. P. McCurdy, E. M. Kudlacz, M. J. Conklyn, E. A. Elliott, E. R. Koslov, M. B. Fisher, T. J. Strelevitz, K. Yoon, D. A. Whipple, J. Sun, M. J. Munchhof, J. L. Doty, J. M. Casavant, T. A. Blumenkopf, M. Hines, M. F. Brown, B. M. Lillie, C. Subramanyam, S. P. Chang, A. J. Milici, G. E. Beckius, J. D. Moyer, C. Su, T. G. Woodworth, A. S. Gaweco, C. R. Beals, B. H. Littman, D. A. Fisher, J. F. Smith, P. Zagouras, H. A. Magna, M. J. Saltarelli, K. S. Johnson, L. F. Nelms, S. G. Des Etages, L. S. Hayes, T. T. Kawabata, D. Finco-Kent, D. L. Baker, M. Larson, M. S. Si, R. Paniagua, J. Higgins, B. Holm, B. Reitz, Y. J. Zhou, R. E. Morris, J. J. O'Shea and D. C. Borie, *Science*, 2003, **302**, 875.

142. E. Kudlacz, B. Perry, P. Sawyer, M. Conklyn, S. McCurdy, W. Brissette, M. Flanagan and P. Changelian, *Am. J. Transplant.*, 2004, **4**, 51.

143. R. Paniagua, M. S. Si, M. G. Flores, G. Rousvoal, S. Zhang, O. Aalami, A. Campbell, P. S. Changelian, B. A. Reitz and D. C. Borie, *Transplantation*, 2005, **80**, 1283.

144. D. C. Borie, M. J. Larson, M. G. Flores, A. Campbell, G. Rousvoal, S. Zhang, J. P. Higgins, D. J. Ball, E. M. Kudlacz, W. H. Brissette, E. A. Elliott, B. A. Reitz and P. S. Changelian, *Transplantation*, 2005, **80**, 1756.

145. A. J. Milici, L. Audoly, G. E. Beckius, C. P. Gibbons, B. D. Perry, M. Fisher, K. Yoon, S. H. Zwillich and P. S. Changelian, Presentation 789, American College of Rheumatology Annual Scientific Meeting, 2006. Abstract: *Arthritis Rheum.*, 2006, **54**, S353.

146. A. J. Milici, E. M. Kudlacz, L. Audoly, S. Zwillich and P. Changelian, *Arthritis Res. Ther.*, 2008, **10**, R14.

147. G. Rousvoal, M. S. Si, M. Lau, S. Zhang, G. J. Berry, M. G. Flores, P. S. Changelian, B. A. Reitz and D. C. Borie, *Transplant Int.*, 2006, **19**, 1014.

148. E. Kudlacz, M. Conklyn, C. Andresen, C. Whitney-Pickett and P. Changelian, *Eur. J. Pharmacol.*, 2008, **582**, 154.

149. E. van Gurp, W. Weimar, R. Gaston, D. Brennan, R. Mendez, J. Pirsch, S. Swan, M. D. Pescovitz, G. Ni, C. Wang, S. Krishnaswami, V. Chow and G. Chan, *Am. J. Transplant.*, 2008, **8**, 1711.

150. N. Lawendy, S. Krishnaswami, R. Wang, D. Gruben, C. Cannon, S. Swan and G. Chan, *J. Clin. Pharmacol.*, 2009, **49**, 423.

151. S. Cohen, S. H. Zwillich, V. Chow, R. R. La Badie and B. Wilkinson, *Br. J. Clin. Pharmacol.*, 2010, **69**, 143.

152. V. Chow, G. Ni, R. LaBadie and G. Chan, *Clin. Pharmacol. Ther.*, 2008, **83**, PI-93.

153. J. M. Kremer, B. J. Bloom, F. C. Breedveld, J. H. Coombs, M. P. Fletcher, D. Gruben, S. Krishnaswami, R. Burgos-Vargas, B. Wilkinson, C. A. F. Zerbini and S. H. Zwillich, *Arthritis Rheum.*, 2009, 1895.

154. J. H. Coombs, B. J. Bloom, F. C. Breedveld, M. P. Fletcher, D. Gruben, J. M. Kremer, R. Burgos-Vargas, B. Wilkinson, C. A. F. Zerbini and S. H. Zwillich, *Ann. Rheum. Dis.*, 2010, **69**, 413.

155. Company website, www.pfizer.com, 2011.

156. E. A. F. J. van Gurp, W. Schoordijk-Verschoor, M. Klepper, S. S. Korevaar, G. Chan, W. Weimar and C. C. Baan, *Transplantation*, 2009, **87**, 79.

157. M. G. Boy, C. Wang, B. E. Wilkinson, V. F.-S. Chow, A. T. Clucas, J. G. Krueger, A. S. Gaweco, S. H. Zwillich, P. S. Changelian and G. Chan, *J. Invest. Dermatol.*, 2009, **129**, 2299.

158. W. J. Sandborn, S. Ghosh, J. Panes, I. Vranic, C. Su, J. Spanton and W. Niezychowski, *Digestive Disease Week*, 2011, Presentation number 594.

159. W. J. Sandborn, S. Ghosh, J. Panes, I. Vranic, J. Spanton and W. Niezychowski, *Digestive Disease Week*, 2011, Presentation number 745.

160. R. J. Riese, S. Krishnaswami and J. Kremer, *Best Pract. Res., Clin. Rheumatol.*, 2010, **24**, 513.

161. D. Wojciechowski and F. Vincenti, *Curr. Opin. Organ Transplant*, 2011, **16**, 614.

162. S. Busque, J. Leventhal, D. C. Brennan, S. Steinberg, G. Klintmalm, T. Shah, S. Mulgaonkar, J. S. Bromberg, F. Vincenti, S. Hariharan, D. Slakey, V. R. Peddi, R. A. Fisher, N. Lawendy, C. Wang and G. Chan, *Am. J. Transplant.*, 2009, **9**, 1936.

163. V. D. K. D. Sewgobind, M. E. Quaedackers, L. J. W. van der Laan, R. Kraaijeveld, S. S. Korevaar, G. Chan, W. Weimar and C. C. Baan, *Am. J. Transplant.*, 2010, **10**, 1785.

164. J. S. Fridman, P. A. Scherle, R. Collins, T. C. Burn, Y. Li, J. Li, M. B. Covington, B. Thomas, P. Collier, M. F. Favata, X. Wen, J. Shi, R. McGee, P. J. Haley, S. Shepard, J. D. Rodgers, S. Yeleswaram, G. Hollis, R. C. Newton, B. Metcalf, S. M. Friedman and K. Vaddi, *J. Immunol.*, 2010, **184**, 5298.

165. J. S. Fridman, P. A. Scherle, R. Collins, T. Burn, C. L. Neilan, D. Hertel, N. Contel, P. Haley, B. Thomas, J. Shi, P. Collier, J. D. Rodgers, S. Shepard, B. Metcalf, G. Hollis, R. C. Newton, S. Yeleswaram, S. M. Friedman and K. Vaddi, *J. Invest. Dermatol.*, 2011, **131**, 1838.

166. D. M. Meyer, M. I. Jesson, X. Li, M. M. Elrick, C. L. Funckes-Shippy, J. D. Warner, C. J. Gross, M. E. Dowty, S. K. Ramaiah, J. L. Hirsch, M. J. Saabye, J. L. Barks, N. Kishore and D. L. Morris, *J. Inflammation*, 2010, **7**, 41.

167. J. G. Shi, X. Chen, R. F. McGee, R. R. Landman, T. Emm, Y. Lo, P. A. Scherle, N. G. Punwani, W. Williams, V and S. Yeleswaram, *J. Clin. Pharmacol.*, 2011, **51**, 1644.

168. A. D. Shilling, F. M. Nedza, T. Emm, S. Diamond, E. McKeever, N. Punwani, W. Williams, A. Arvanitis, L. G. Galya, M. Li, S. Shepard, J. Rodgers, T. Y. Yue and S. Yeleswaram, *Drug Metab. Dispos.*, 2010, **38**, 2023.
169. A. Pardanani, J. R. Gotlib, C. Jamieson, J. E. Cortes, M. Talpaz, R. M. Stone, M. H. Silverman, D. G. Gilliland, J. Shorr and A. Tefferi, *J. Clin. Oncol.*, 2011, **29**, 789.
170. S. Verstovsek, H. Kantarjian, R. A. Mesa, A. D. Pardanani, J. Cortes-Franco, D. A. Thomas, Z. Estrov, J. S. Fridman, E. C. Bradley, S. Erickson-Viitanen, K. Vaddi, R. Levy and A. Tefferi, *N. Engl. J. Med.*, 2010, **363**, 1117.
171. Company website, http://investors.vrtx.com/releasedetail.cfm?ReleaseID= 603317, 2011.
172. T. Hoock, J. Hogan, S. Mahajan, D. Shlyakhter, L. Oh, L. Park, G. Ku, I. Catlett, M. Corbin, F. Salituro and M. Namchuk., Presentation 1136, American College of Rheumatology Annual Scientific Meeting, 2011. Abstract: *Arthritis Rheum.*, 2011, **63**(10), 1136.
173. Company website, http://www.glpg.com/pharmaceuticals/ra.htm, 2011.
174. Galapagos webcast, 22nd November 2011, available online at: http:// pulse.companywebcast.nl/playerv1_0/default.aspx?id=16993&bb=true& swf = true.
175. B. Belli, A. Dao, D. Brigham, R. Nepomuceno, E. Setti, E. Liu, G. Liu, M. J. Bhagwat, S. Hadd, W. Wierenga, M. Holladay and R. C. Armstrong, Poster, American College of Rheumatology General Meeting, 2010. Available at: http://www.ambitbio.com/posters_and_presentations.
176. Company website, http://www.portola.com/JAK3-Program, 2011.

CHAPTER 9

IKKβ as a Therapeutic Intervention Point for Diseases Related to Inflammation

ERICK R. R. YOUNG

Department of Medicinal Chemistry, Boehringer Ingelheim Pharmaceuticals, Inc., 900 Ridgebury, Rd/PO Box 368, Ridgefield, CT, USA
Email: erick.young@boehringer-ingelheim.com

9.1 Introduction

Inhibitor of I kappa B kinase beta (IKKβ also known as IKK2, is a ubiquitously expressed serine/threonine kinase which has received a great deal of attention in inflammatory, autoimmune, respiratory and oncology disease research.[1-3] This widespread interest springs from the tantalizing potential to leverage a small-molecule druggable target class to selectively disrupt activation of NF-κB-mediated transcription, a central mechanism of chronic inflammation.[4] The NF-κB transcription factor is a critical regulator of gene expression for multiple inflammatory pathways induced by cytokines, viral and bacterial infections, antigens, oxidative stress and DNA-damaging agents.[5] It has been hypothesized that this approach might provide the long sought after "safe steroid" profile through the discriminating modulation of a transcription factor pathway believed to be a primary source of steroidal anti-inflammatory properties.[5,6] IKKβ presents such an opportunity due to its critical importance for function of the IKK ternary complex. This complex is composed of IKKα (also known as IKK1), IKKβ and IKKγ (also known as NF-κB essential

RSC Drug Discovery Series No. 26
Anti-Inflammatory Drug Discovery
Edited by Jeremy I Levin and Stefan Laufer
© The Royal Society of Chemistry 2012
Published by the Royal Society of Chemistry, www.rsc.org

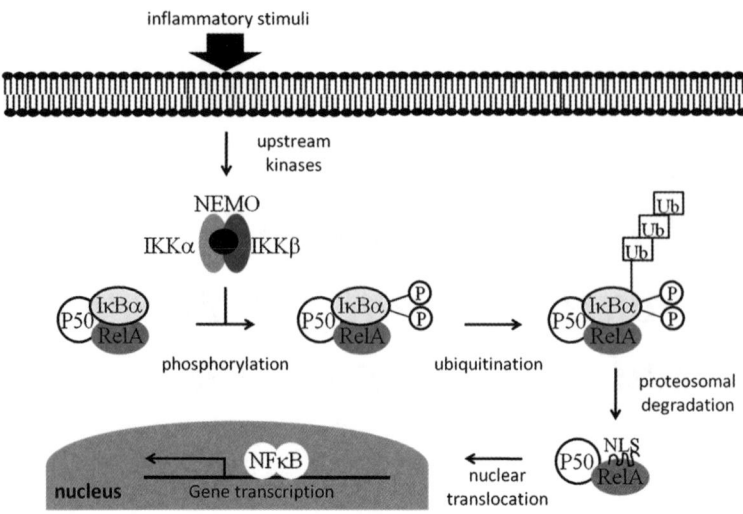

Figure 9.1   The role of IKKβ in the canonical NF-κB activation pathway.

modulator or NEMO). The IKKα and IKKβ catalytic subunits confer kinase activity while IKKγ is believed to serve as a non-enzymatic regulatory element important for scaffold-assisted activation.[7] The most well-characterized function of the IKK complex is its regulatory role in the canonical NF-κB pathway (see Figure 9.1). Once activated in response to inflammatory stimuli, the IKK complex carries out the site-specific phosphorylation of IκBα at the N-terminal Ser32 and Ser36 residues, which targets this protein for poly-ubiquitination and subsequent rapid degradation by the 26S proteasome.[8-10] Under non-inflammatory conditions, IκBα negatively regulates the transcriptional activity of the canonical NF-κB complex (RelA-p50) through cytoplasmic sequestration.[11,12] However, once this chaperone is cleared, nuclear localization sequences are exposed enabling translocation, DNA binding and transcription.[13] In some cases IκB proteins are gene products of the same NF-κB proteins they inhibit, creating a negative feedback loop, which contributes to the resolution of inflammation. In particular, IκBα competitively displaces Rel-A complexes from DNA and efficiently shuttles this NF-κB cargo to the cytoplasm through nuclear export reinitiating the cycle.[14,15] More recently it has been shown that the site of NF-κB sequestration for some IκB proteins may actually be nuclear rather than cytoplasmic.[11]

Typical high-level cartoon depictions, such as Figure 9.1, are helpful for conveying a foundational understanding of canonical NF-κB activation. However, these generalized representations of NF-κB, IκB and IKK also de-emphasize subtle complexities that could potentially confound the interpretation of functional data. "NF-κB" is a generic term referring to a family of homo- and heterodimeric proteins comprised of seven monomeric subunits (REL-A, REL-B, c-REL, p50, p52, p100 and p105). Each of these individual components influences unique gene transcription programs, some pro-inflammatory and others anti-inflammatory. Additionally, distinct NF-κB dimers can exhibit different binding specificities to the five IκB chaperones (IκBα, IκBβ, IκBγ, IκBε and BCL3).[1,16–18] Furthermore, "IKK" exists as an equilibrium distribution of ternary complexes, IKKα and IKKβ homo- and heterodimers, and individual monomers.[19–21] In some cases these various forms of IKK demonstrate differing IκB substrate specificity. Together, this complex network of external inflammatory stimuli, distribution and activities of IKK complexes, binding specificities of IκB proteins and NF-κB composition-dependent transcriptional programs underlies the highly specific regulation of inflammatory processes of the innate and adaptive immune responses.[9]

9.2   Target Validation

It is believed that the predominant NF-κB-dependent inflammatory pathway is the canonical mechanism described above. Evidence strongly supports the primary role played by IKKβ in this process, as it demonstrates 50-fold greater activity for IκBα phosphorylation than does IKKα due to a much lower Km for this substrate.[22] Additionally, TNFα and IL-1 induced NF-κB activation is greatly impaired in IKKβ$^{-/-}$ but not IKKα$^{-/-}$ mouse embryonic fibroblasts.[23] For these reasons IKKβ was historically pursued much more aggressively as a therapeutic target than other IKK isoforms. More recently, supporting evidence has emerged for involvement of IKKα activity in the canonical pathway.[24] Perhaps of greater significance, a critical inflammatory role of IKKα in the non-canonical, or alternative, activation of NF-κB has been elucidated, contributing to additional interest in this target.[9,25,26]

Aberrant NF-κB activity is implicated, by its over-expression, in multiple human disease pathologies. This over-expression has been observed in samples including, but not limited to, hematopoietic and solid tumours, inflamed synovium from rheumatoid arthritis (RA) patients, bronchial biopsies and airway epithelial cells from asthmatics, mucosal biopsy specimens from Crohn's disease and ulcerative colitis patients and in the microglia of active plaques from multiple sclerosis patients.[27–33] The therapeutic potential of IKKβ-dependent NF-κB pathway disruption in these diseases has been extensively studied by multiple biochemical approaches. Two of the more frequently employed technologies have been dominant negative IKKβ (mutant lacking a conserved lysine residue necessary for catalytic activity) and nondegradable IκBα dominant negative super repressor substrates (mutants lacking the serine phosphorylation site). These approaches allow for complementary blockade of the enzymatic reaction in question and, unlike genetic deletion, do not impact potential scaffolding roles of these two important proteins. Transfection of the IκBα dominant negative into synovial cells from RA patients substantially reduced the pro-inflammatory cytokines TNFα, IL-6, IL-8, IL-1β and MMP-1, -2, -3 and -13 expression, supporting the potential to reduce both disease severity and progression. In the same study, transfection of dominant negative IKKβ produced a very similar profile, while dominant negative IKKα had negligible effects.[34] It has also been shown that either direct intra-articular delivery of the dominant negative IKKβ transfectant or intact protein is efficacious in rat arthritis models as measured by swelling, histology and reduction of pro-inflammatory cytokines. Conversely, adenoviral transfection of wild type IKKβ exacerbates these disease states.[35,36] The transfection of IKKβ and IκBα dominant negatives also have profound anti-inflammatory effects on key pulmonary cell types, inhibiting the production of COX-2, IL-8, ICAM-1 and PGE-2 in A549 epithelial cells and IL-6, IL-8, GM-CSF, RANTES and MCP-1 in human airway smooth muscle cells in response to TNFα or IL-1β.[37,38] Once again, in both of these experiments IKKα dominant negative transfection had no effect. The dominant negative expression of IKKβ and IκBα have also been shown to sensitize a variety of tumour cells to apoptosis.[39–42] Selective expression of the IκBα dominant negative in renal fibroblasts inhibited fibrosis in a unilateral urethral obstruction model.[43]

9.3   IKKβ Therapeutic Target Considerations

While IKKβ inhibition holds great promise as a therapeutic strategy for a wide variety of diseases, its fundamental physiological role, coupled with the challenges of developing selective kinase inhibitors for chronic therapy and the interpretation and optimization of complex transcription factor-based functional effects, presents substantial hurdles for the successful prosecution of this target. The most formidable of these may well be safety.

9.3.1   Safety

NF-κB is known to be necessary for the production of a multitude of important gene products including cytokines, adhesion molecules, effector enzymes and anti-apoptotic factors central to inflammation, immunity, cell proliferation and survival.[44] Naturally, it has been a long-standing question whether the blockade of such a key pathway can be tolerated and these reservations were further heightened with observations of alarming mouse knockout phenotypes. For example, IKKβ$^{-/-}$ mice are embryonic lethal at 14 days of gestation due to extensive liver apoptosis.[23] This observation is mechanistically consistent with the known NF-κB-dependence of multiple anti-apoptotic survival genes and Rel-A or IKKγ deletion phenotypes.[45–47] In addition, IKKα$^{-/-}$ mice die shortly after birth, exhibiting severe skin and skeletal deformaties due to hyperproliferation and impaired differentiation of keratinocytes.[48] However, the demonstration that hepatocyte-specific ablation of IKKβ in conditional knockout mice caused no increased liver apoptosis in response to TNFα stimulation opened the possibility that these undesired effects might be restricted to developmental processes.[49] However, later reports raised additional, less obvious, toxicity risks due to the NF-κB dependent anti-inflammatory processes, which are also compromised upon IKKβ inhibition.[50] It was unexpectedly observed that mice with IKKβ deficient myeloid cells were actually more susceptible to endotoxin-induced shock despite reduced TNFα production. These animals showed notably increased mortality and exhibited higher levels of IL-1β due to enhanced processing of pro-IL-1β in the absence of NF-κB dependent protease inhibitory factors. These effects could be recapitulated in normal mice subjected to endotoxin challenge and prolonged administration of a small-molecule IKKβ inhibitor. Such "stress-induced" synergistic toxicities might escape detection in traditional single-agent pre-clinical and clinical safety assessments. It is also concerning that despite such aggressive pursuit of this target by the industry, no IKKβ inhibitor has successfully emerged from the clinic to date. Several investigators have specifically cited tolerability as the reason for termination of both pre-clinical and clinical programs as discussed later (see discussions in Sections 9.4.7 and 9.5).

9.3.2 NF-κB Pathway Complexities

Typically differences in functional effects of kinase inhibitors for a given target can be reasonably attributed to varying degrees of target coverage and/or off-target profiles. However, there is the potential that such interpretations may be less straightforward in the context of inhibiting NF-κB activation. As described earlier, each mechanistic step, from the initiating stimuli to transcription of differential gene products, introduces variables with respect to specificity and function. In theory this could contribute to scenarios in which IKK protein conformational differences, either as a function of complex composition (*i.e.* homo-/hetero- or ternary/dimer/monomer) or IκB substrate pairings, lead to a divergence of SAR and functional outcomes across inhibitor chemotypes. Full characterization of such relationships would be a substantial undertaking, if not time prohibitive, and has not been described as a component of program prosecution in the literature.

9.3.3 Isoform Selectivity

The IKK isoform dependence of the innate and adaptive immune responses provides a striking example of pharmacological outcomes being driven by the specificity of IKK/IκB chaperone/NF-κB networks.[8] The innate response is driven by canonical RelA-p50 NF-κB transcription, which is triggered by IKKβ-mediated phosphorylation and subsequent degradation of the chaperone IκBα. However, the adaptive immune response is driven by alternative RelB-p52 transcription, which is triggered by the IKKα-mediated phosphorylation of p100. Moreover, individual IKK isoforms play distinct additional roles in NF-κB activation such as phosphorylation-dependent nuclear import, transcriptional activation and HDAC activation.[51] Thus, the isoform selectivity profiles of small-molecule IKK inhibitors would be expected to impact pharmacological profiles in profound ways beyond simple differences in enzymatic potency against the IKK complex.

9.3.4 Broader Kinase Selectivity

Kinome selectivity is a standard consideration for dose-limiting toxicities of small molecule kinase inhibitors. Cross-reactivity with kinases in common, parallel or competing pathways of interest also presents an additional challenge for clear SAR interpretation. In the case of a program directed at the identification of selective inhibitors of IKKβ some obvious kinases to be monitored to avoid confounding functional data include: IKKα (influences several independent NF-κB activation processes as described above), MEKK3 (activates the ternary complex in response to LPS, IL-1 and TNF stimuli), NIK (activates IKKα in response to CD40L, BAFF, LTα1β2 stimuli), TAK-1 (activates both IKKα and IKKβ in response to IL-1), TBK1 and IKKε (both carry out inactivating phosphorylations of IKKα and IKKβ).[51–55] Lastly, the very low ATP Km of IKKβ may pose an intrinsic challenge for selective inhibition, with reported values of 100–600 nM, as compared to 10–500 μM values for the majority of the kinome.[22,56] This may also cause ATP competitive IKKβ inhibitors to suffer greater losses in potency when progressing from biochemical assays to physiologic conditions than is typically seen for other kinase targets.

9.4 Chemical Matter

Commensurate with the anticipated therapeutic and commercial potential of IKKβ inhibitors, a tremendous investment has been made in this target by both the pharmaceutical industry and academic research. At the time this manuscript was prepared 130 issued patents and published applications for small-molecule IKKβ inhibitors from 34 companies were identified. The associated chemical matter represents wide structural diversity, ranging from more common kinase inhibitor classes such as aminopyrimidines and β-carbolines to a number of novel chemotypes developed in the pursuit of this target.[1,57,58] It has been a common theme that most chemical series are capable of yielding selective IKKβ inhibitors and IKKα/β dual inhibitors, but potent IKKα isoform selective inhibitors have remained elusive. To date, opportunities for the rational design of potent and selective inhibitors of IKKβ have been limited to consensus pharmacophore and homology models. There are several examples in which these approaches have successfully guided scaffold hopping and engagement of specific molecular binding interactions with the target.[59–64] However, these tools have not been instructive in driving isoform selectivity, which appears to have been uniformly arrived at empirically. The first inhibitor X-ray co-crystal structures have been recently reported with Xenopus Laevis IKKβ, which has 74% identity to the human protein.[65] The potential impact of this disclosure on future inhibitor design remains to be determined. The IKKβ hinge region is comprised of the 3 amino acid sequence, Cys99-Tyr98-Glu97, adjacent to the Met96 gatekeeper residue. All of the binding models that have been proposed for IKKβ ATP competitive inhibitors to date share a common feature, a ligand H-bond acceptor engaging the backbone NH of the Cys99 hinge residue. One of the more interesting aspects of molecular design in this field is the context in which these H-bond acceptor atoms have been presented to the protein, with respect to both creative conformational restriction strategies and overall hinge region H-bond networks (*i.e.* mono-, di- or tridentate interactions with single or multiple amino acid residues). To assist the visualization of key molecular interactions described in the text, figures have been included throughout this chapter illustrating anticipated binding modes of representative IKKβ inhibitor chemotypes. These images were generated from an in-house homology model built from a 2.2-angstrom PHK/ATP crystal structure.[66] The development of potent inhibitors of IKKβ with suitable pharmacokinetic (PK) profiles, from multiple series, has enabled the timely pharmacological interrogation of the target. Demonstrations of efficacy in multiple disease relevant animal models with these compounds have further validated the broad therapeutic potential of this approach. Although the level of kinase selectivity achieved with early inhibitors, or the absence of such data, called into question conclusions regarding a discrete mechanism of action, a consistent picture has developed over time from the *in-vitro* and *in-vivo* effects of these initial tool compounds, biochemical pathway disruption techniques and now second-generation molecules with greatly improved selectivity profiles. While this review is not intended to capture the entire body of medicinal chemistry discoveries in this field, the major disclosed contributions will be presented, grouped by chemotype. The most recent advances will be highlighted, and insights will be offered into lessons learned across programs.

9.4.1 Aminopyrimidines

Aminopyrimidines are frequently identified in kinase inhibitor screens due to their ability to mimic the hydrogen bonding of adenine to the highly conserved hinge region of the protein. For this same reason they are among the most promiscuous classes of kinases inhibitors. Nonetheless, ease of synthesis and confidence in their binding mode has led to the ubiquitous use of aminopyrimidines for the development of kinase inhibitors. Not surprisingly, this has also been true for the case of IKKβ. Aminopyrimidines were among the earliest IKKβ inhibitors to be disclosed, with many companies contributing to a crowded intellectual property landscape for this class (Figure 9.2). However, relatively few scientific publications have emerged describing these programs given the volume of patent filings covering this chemotype.

Celltech has reported on the discovery of **1**, a dual inhibitor of IKKα ($IC_{50} = 154$ nM) and IKKβ ($IC_{50} = 146$ nM).[67] Homology models built from PKA, CAMK, ERK2, P38 and CDK2 predicted a "mode 2" binding interaction for this molecule, forming two hinge region hydrogen bonds from the aminopyrimidine to the backbone NH and carbonyl oxygen of Cys99. Exploration of SAR for the aniline ring resulted in compound **2**, with gains in IKKβ potency and isoform selectivity (IKKβ $IC_{50} = 11$ nM; 21-fold selective

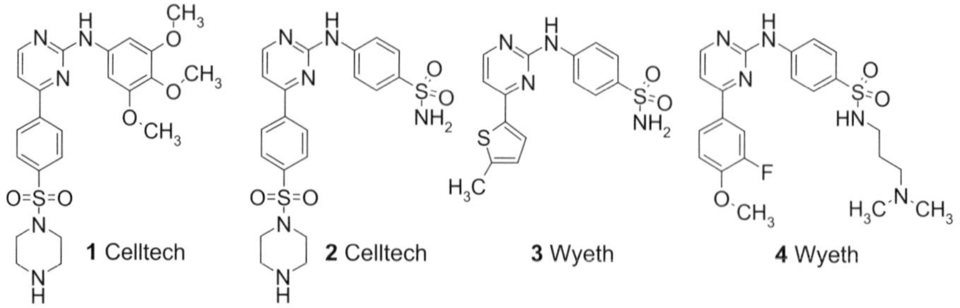

Figure 9.2 Aminopyrimidine derived IKKβ inhibitors.

over IKKα), although no broader kinome profiling was discussed. Researchers from Wyeth (now Pfizer) reported the discovery of the highly related lead structure **3**.[68] This compound shows comparable potency to **2** (IKKβ $IC_{50} = 19$ nM) and low micromolar potency in reporter gene and cell proliferation assays, but undesirable aqueous solubility, suggesting that the piperazinyl sulfonamide of **2** is beneficial for solubility but contributes minimally to potency in the Celltech series. The aminopyrimidine patent history from Wyeth culminated in a series of 2007 filings claiming specific salt forms of **4**, perhaps indicating it to be a more advanced compound from the series.[69] However, no further publications or announcements of clinical advancement have been reported.

Novartis has also published SAR around thiophene substituted aminopyrimidines (see Figure 9.3).[70] Initial optimization of weakly potent screening hits identified benzothiophene **5** as a reasonably potent IKKβ inhibitor ($IC_{50} = 600$ nM) with some selectivity over IKKα ($IC_{50} = 5$ μM). Curiously, these early analogues were significantly less potent against the IKK complex (compound **5** $IC_{50} = 18$ μM). Extending a strongly basic amine group from the benzoic acid handle provided analogues such as **6**, which were more potent and no longer suffered from an undesirable shift in the IKK complex assay (compound **6** IKKβ $IC_{50} = 25$ nM, IKKα $IC_{50} = 500$ nM, IKK complex $IC_{50} = 40$ nM). Compound **6** also showed mechanistically consistent inhibition of IκBα degradation and expression of NF-κB dependent adhesion molecules (E-selectin, ICAM, VCAM EC_{50} all ~ 200 nM). However, inhibition of non-NF-κB dependent pathways was also seen at just 10- to 20-fold higher concentrations. No kinase selectivity profiling data were disclosed for this compound. Also shown in Figure 9.3 are representative aminopyrimidines from Signal (**7**), Lilly (**8**) and Sanofi-Aventis (**9**). Compounds **7** and **8**, which bear strong structural similarities to the Novartis aminopyrimidines, are described in patent applications as having IKKβ IC_{50} values of less than 500 nM and 46 nM, respectively.[71–73] Compounds such as **9**, which incorporate the phenyl sulfonamide group common to the Wyeth and Celltech compounds (**2–4**) in the context of a diaminopyrimidine core, are included in a patent application from Sanofi-Aventis.[74]

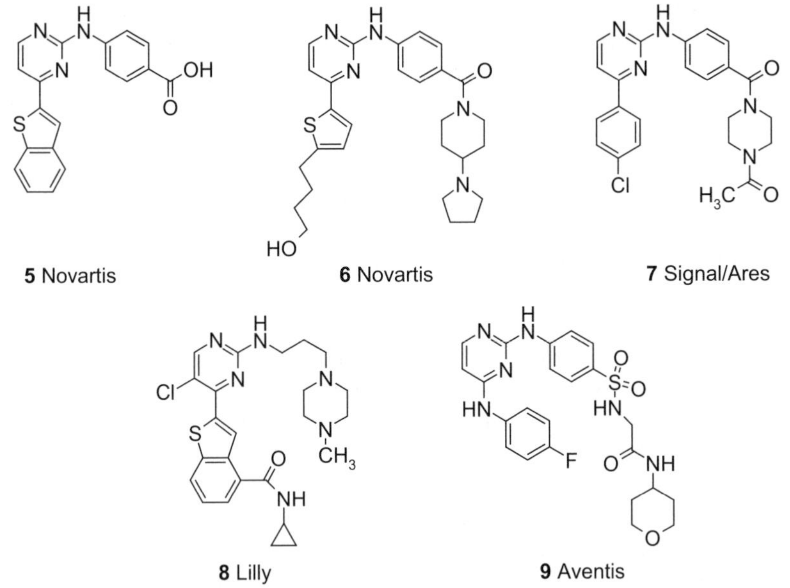

Figure 9.3   Additional aminopyrimidine based IKKβ inhibitors.

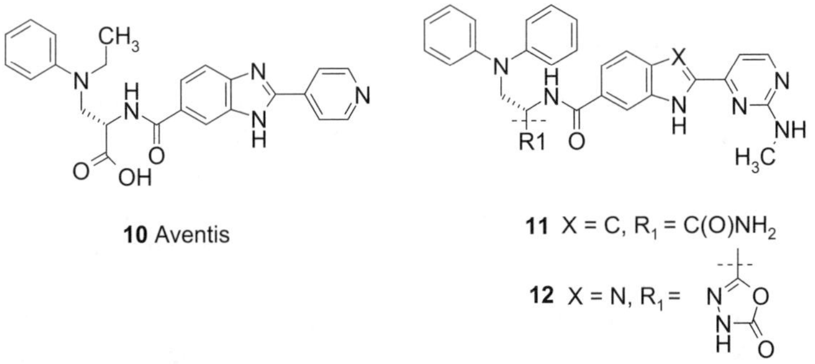

Figure 9.4   Benzimidazole and indole derived aminopyrimidine IKKβ inhibitors.

Sanofi-Aventis has also developed a structurally distinct series featuring an aminopyrimidine element which may have been the source of their clinical candidate (see Section 9.5). This chemical matter first appeared in a 2001 patent application describing IKKβ inhibitors featuring a benzimidazole core (Figure 9.4).[75]

All of the exemplified compounds were pyridyl-benzimidazoles, including a few examples of 2-amino substituted pyridines. Biological data were provided for 19 compounds, of which **10** was the most potent (IKKβ IC$_{50}$ = 70 nM) and showed selectivity over PKA, PKC, CKII (less than or equal to 50% inhibition at 100 µM). A closely related application published in the same year disclosed nine pyridyl-indole examples, but none of these indole analogues achieved the potency of **10**.[76] Two additional patent applications in 2004 replaced the pyridine head piece of **10** with aminopyrimidines. Optimization efforts appear to have focused on the left-hand side tertiary amine and heterocyclic variations thereof, as well as five-membered heterocycle replacements of the carboxylic acid and primary amide groups, resulting in a substantial improvement in potency (**11**, IKKβ IC$_{50}$ = 3 nM, **12**, IKKβ IC$_{50}$ = 12 nM).[77,78] Aminopyrimidine **12** was characterized in a number of *in-vivo* pain models in which it was found to have effects comparable to Celecoxib. Also in 2004, a compound of unknown structure from this series, referred to as I229, was reported to have a 1.9 nM IC$_{50}$ for the IKK complex and demonstrated less than 50% inhibition of a panel of 14 kinases at 10 μM concentration. In studies examining NF-κB-dependence of metabolic syndrome contributory mechanisms, I229 was shown to restore normal insulin signalling in adipocyte/muscle cell co-cultures and to inhibit CRP production in primary human hepatocytes in response IL-1 β, IL-6 or phorbol dibutyrate stimulation.[79,80] However, at the concentration tested, 10 μM, it is difficult to interpret these results in the absence of correlations to IκBα phosphorylation and broader specificity profiling data. Associated data from a 2006 patent on this same series also showed synergistic enhancement of apoptosis in both U937 and THP-1 cell lines when combined with TNF-related apoptosis inducing ligand (TRAIL), supporting their potential use as sensitizing agents for cytotoxic chemotherapies.[81]

9.4.2 Thiophenecarboxamides

Thiophenecarboxamides and related molecules inspired by them have been the most heavily explored class of chemical matter in the effort to develop IKKβ inhibitors (Figure 9.5). Multiple investigators have independently discovered these starting points *via* high-throughput screening and pursued them for lead identification due to their desirable ligand efficiency and straightforward synthesis amenable to SAR exploration. Perhaps more importantly, there was a high degree of confidence in the binding mode of these simple structures that could be used to guide molecular design in the absence of experimentally determined structural information. The first report of thiophenecarboxamides as IKKβ inhibitors appeared in a 2001 patent application from AstraZeneca.[82] This was an area of intense activity in subsequent years with numerous applications published by AstraZeneca, SmithKline Beecham (now Glaxo-SmithKline) and Pharmacia (now Pfizer) claiming very closely related chemical matter.[57,83–99] Collectively, these efforts blanketed the obvious conservative changes to the simple biaryl motif, including variations of the heterocycle, regiochemistry, substitution patterns and linkers.

AstraZeneca has reported on their hit to lead efforts, which initiated with the discovery that 5-aryl thiophene aminocarboxamides were low micromolar inhibitors of IKKβ (compound **13** IKKβ IC$_{50}$ = 3.2 μM).[100] Homologation of the primary aryl amine to urea **14** provided a remarkable 100-fold increase in IKKβ enzymatic potency (IKKβ IC$_{50}$ = 25 nM), which translated into

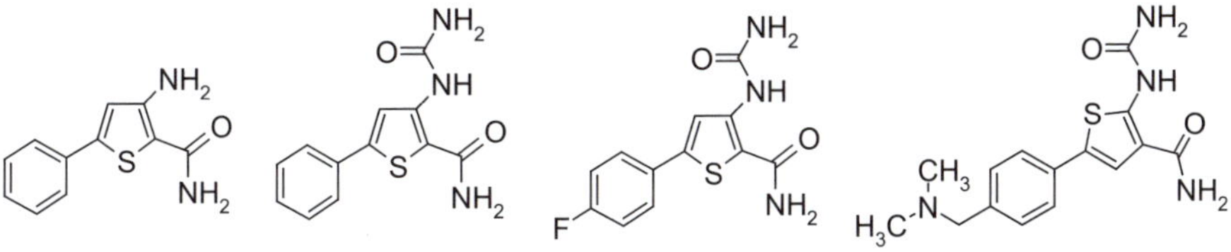

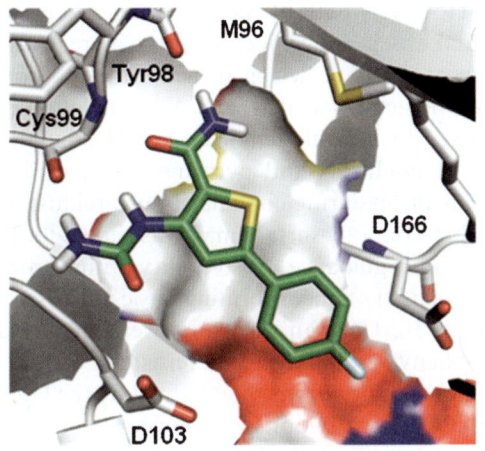

Figure 9.5 Evolution of AstraZeneca thiophene carboxamide series.

Figure 9.6 Proposed binding mode of compound **15**.

submicromolar cell-based potency as measured by the inhibition of TNFα production in LPS stimulated PBMCs ($IC_{50} = 250$ nM). Exploration of the SAR of the 5-aryl group identified 4-fluoro, as in compound **15**, as a preferred substituent which increased stability in rat hepatocytes and human microsomes and imparted desirable oral bioavailability in the rat (**14** F% = 27, **15** F% = 78) despite poor aqueous solubility (1.2 µg/mL). Representative molecules also demonstrated 25- to 100-fold selectivity over IKKα; no additional kinase selectivity profiling was described. A proposed binding mode for this class was developed using a human IKKβ homology model, which was further supported by an X-ray co-crystal structure of **15** bound to JNK-1, for which it has an IC_{50} of 1.6 µM.[83] In this model the vicinal urea and carboxamide groups are presented to the hinge region of the kinase as a coplanar array, which engages in a three-point "mode 1 + 2" hydrogen bonding network. Specifically, the carboxamide NH serves as an H-bond donor to the carbonyl of Glu97, the carboxamide oxygen serves as an H-bond acceptor for the Cys99 NH and the NH groups of the urea together form a bifurcated H-bond with the backbone carbonyl of Cys99. The PHK homology model shown in Figure 9.6 predicts the terminal amino group of the urea contributes more heavily to the third Cys99

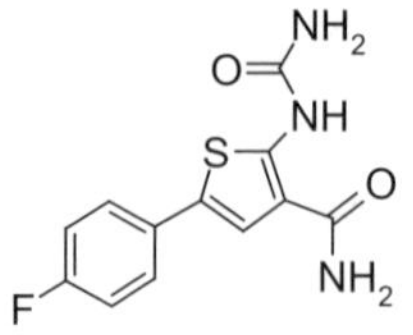

17 TPCA-1
GSK

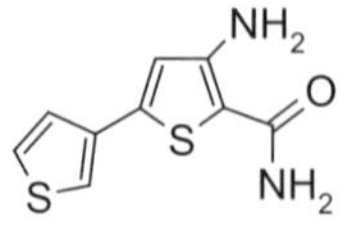

18 SC-514
Pharmacia

Figure 9.7 Thiophene carboxamide IKKβ inhibitors TPCA-1 and SC-514.

carbonyl H-bond. However, both of these binding models are consistent with multiple experimental observations, including the increase in potency observed upon incorporation of the urea, which provides an additional H-bond to the protein, the loss of potency associated with *N*-methylation of either the amide or urea functionality, which would eliminate key H-bond donors, and the loss of potency seen with imidazole and pyrazine cores, which would introduce repulsive interactions disfavouring coplanarity of ureido-carboxamide.

A solvent accessible Connolly surface map was shown for this binding model, which may have provided guidance for solubility optimization of **15**. Two later patent applications focused on incorporating basic amines into this chemotype, as shown for analogue **16**. These amine groups were linked *via* the *ortho* or *para* positions of the 5-phenyl group, vectors consistent with targeting polar and solvent exposed regions of the enzyme highlighted by the binding model.[85,86,101]

Compound **17** (TPCA-1), a thiophene regioisomer of **15** described by GSK, has been the most extensively studied thiophene carboxamide tool compound in the literature (Figure 9.7). It is reported to be an ATP competitive inhibitor with 18 nM and 400 nM IC_{50} values for IKKβ and IKKα, respectively, and demonstrates mechanistically consistent dose-responsive inhibition of NF-κB dependent pro-inflammatory mediators in multiple disease relevant cell types.[102] For example, **17** inhibits:

- TNFα, IL-6 and IL-8 production by LPS-stimulated human monocytes;[102]
- IL-6 and MCP1 production by LPS-stimulated human adipocytes;[103]
- B1 and B2 receptor expression by TNFα/IFNγ-stimulated human vascular endothelial cells and rat bronchia[104,105]
- GM-CSF, IL-8 and eotaxin release from human airway smooth muscle cells[106]
- TNFα and IL-6 production by choriodecidual cells.[107]

Consistent with these profiles, it has also proven efficacious in animal models of arthritis and asthma.[102,106] In a prophylactic mouse collagen induced arthritis (CIA) model TPCA-1 significantly reduced clinical score in a dose-responsive fashion, and at a 10 mg/kg i.p. b.i.d. dose showed efficacy comparable to the soluble TNF receptor fusion protein Enbrel®. It was also demonstrated in this experiment that TPCA-1 reduced levels of IL-1β, IL-6, TNFα and IFNγ in paw tissue. This compound was also efficacious in a therapeutic CIA protocol when dosed at 20 mg/kg i.p. b.i.d. In a rat allergic airway inflammation model, at a dose of 30 mg/kg, TPCA-1 inhibited lung tissue cytokine gene expression 6 h post antigen challenge (TNFα, IL-1β, eotaxin, IL-4, IL-5 and IL-13), decreased eosinophil and neutrophil influx into bronchoalveolar lavage fluid 24 h post antigen challenge and inhibited late asthmatic response. Compound **18** (SC-514) was also evaluated in a number of the experiments described above and showed similar profiles but of less magnitude, consistent with its weaker relative potency against the target (IKKβ $IC_{50} > 1 \mu M$).[108]

TPCA-1 has been frequently described as a selective inhibitor of IKKβ, with limited supporting data. However, a 2008 publication from GlaxoSmithKline requires the consideration of off-target activities for TPCA-1 and related analogues.[109] In this study 18 compounds, including TPCA-1, were tested against a 203-member kinase binding panel at 10 μM concentration. TPCA-1 was found to bind to 53 kinases at less than 10% of control, 12 of which were less than 1% of control. In particular, a K_i of 5 nM was determined for the binding of TPCA-1 to JAK2. Thiophene carboxamide urea IKKβ inhibitors such as **16** have now also been reported as potent (typical IC_{50} less than 100 nM) leads for Chk1.[110] It is not yet clear if a degree of kinome selectivity compatible with chronic therapy can be achieved with this class of molecules.

Boehringer Ingelheim has also reported on their IKKβ hit-to-lead program in which thiophenecarboxamides were identified *via* a high-throughput screen.[111] These, and other hit classes for which there were well-defined binding modes, were used to develop a pharmacophore model for scaffold hopping into new areas of opportunity. This approach led to the exploration of thienopyridine carboxamides, exemplified by generic structure **19** (Figure 9.8). Unlike the biaryl thiophenes, elaboration of this aryl amine core to the primary urea resulted in a loss of potency. This was attributed to a distortion of the urea from planarity by the newly introduced peri substituent ($\sim 30^\circ$ torsion angle predicted for $R_4 = H$), which would disrupt the mode of hinge region interaction proposed for **15**. However, this core did offer favourable microsomal stability and CYP inhibition profiles and two trajectories, from R_4 and R_6, to be explored to acquire additional interactions with the protein which were not present in the biaryl chemotype.

A follow-up publication described the SAR of the R_4 and R_6 positions of scaffold **19**, which led to substantial additive gains in potency.[112] Of particular interest was the empirical discovery that both IKKβ selective and dual IKKβ/IKKα inhibitors could be accessed through subtle modification of a left-hand side R_6 piperidine substituent. For example, the 4-aminopiperidine **20** displayed potent dual inhibition (IKKβ $IC_{50} = 41$ nM, IKKα $IC_{50} = 90$ nM) while the 4-hydroxypiperidine **21** was found to be 40-fold selective for IKKβ over IKKα (IKKβ $IC_{50} = 280$ nM, IKKα $IC_{50} = 11.2 \mu M$). The source of this differentiation is unfortunately difficult to understand in the absence of detailed experimental structural information. A homology model was shown illustrating

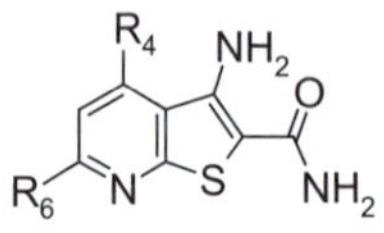

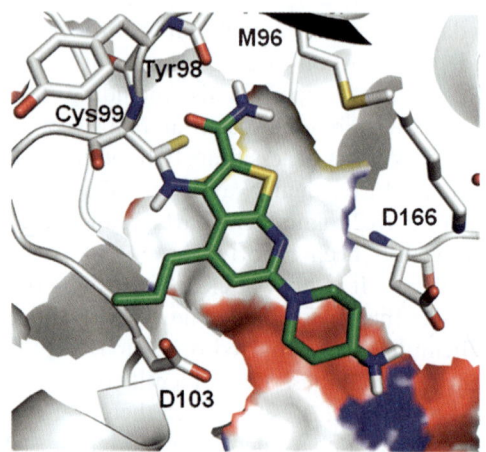

19 Boehringer Ingelheim 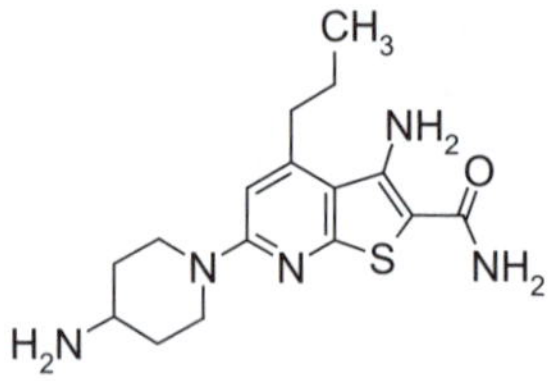   **20** Boehringer Ingelheim 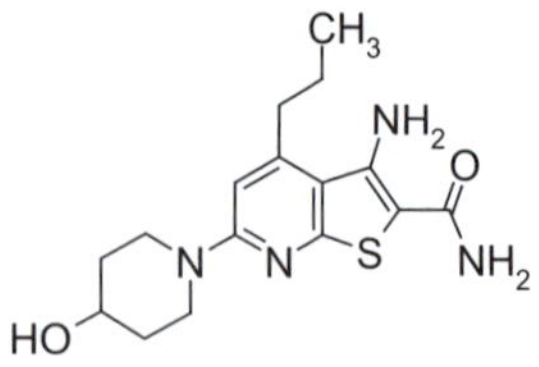   **21** Boehringer Ingelheim

Figure 9.8   Thienopyridine carboxamide IKKβ inhibitors.

Figure 9.9   Proposed binding mode of thienopyridine carboxamide **20**.

a classical "mode 1" hinge interaction in which the hydrogen bonding interactions of the carboxamide of **20** are the same as that described previously for **15**. In this case it is not clear that the lone amino group is capable of engaging the Cys99 carbonyl oxygen as predicted for the corresponding urea. However, the amino group is expected to serve a critical role in maintaining coplanar conformational restraint of the amide through intra-molecular hydrogen bonding. This model projects the R_4 and R_6 substituents into the ATP sugar binding region and more polar phosphate binding region, respectively (Figure 9.9). Early compounds from this series showed minimal inhibition across a panel of 43 additional kinases, with **20** showing a measurable IC_{50} for only HEK (7 μM).

Recently, researchers at the University of Dundee have reported the use of a selective inhibitor from this series in elucidating regulatory cross talk of the canonical IKKs and IKK-related kinases.[55] Compound **22** (Figure 9.10), at a concentration of 1 μM, was found to inhibit only one out of 124 additional kinases tested (IGF1, 20-fold selective). A US patent application describes the benefit of this compound in balancing both desirable potency (IKKβ $IC_{50} = 26$ nM; HeLa ICAM-1 expression $IC_{50} = 700$ nM) and sustained oral

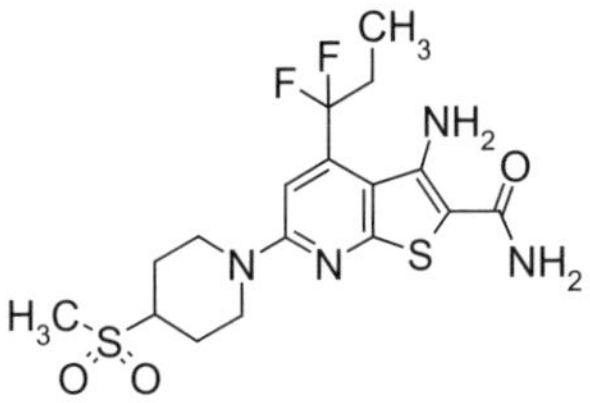

22 Boehringer Ingelheim 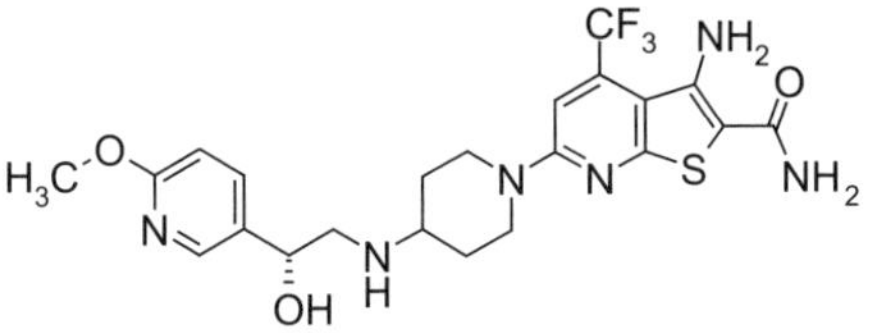   **23** Boehringer Ingelheim

Figure 9.10   Optimized thienopyridine analogues.

exposure in the rat (10 mg/kg p.o. plasma exposure 2 h = 1.59 μM, 6 h = 780 nM, 10 h = 650 nM), which translated into efficacy in a therapeutic rat CIA model (10 mg/kg b.i.d. p.o., 44% reduction in paw weight).[113]

Boehringer Ingelheim has filed additional thienopyridine carboxamide patent applications focused on molecules that extend further from the piperidine left-hand side.[114] One such compound is **23** (Figure 9.10), with increased molecular and cellular potency (IKKβ/α IC$_{50}$ = 9/595 nM; HeLa cell IκBα phosphorylation IC$_{50}$ = 290 nM; HeLa ICAM-1 expression IC$_{50}$ = 430 nM) while maintaining good kinase selectivity.[115] When tested against a panel of 60 kinases, compound **23** was greater than 100-fold selective against all but FLT3 (80-fold). This compound has been used to investigate the role of IKKβ in epithelial mesenchymal transition (EMT) mediated tumour growth and metastasis. Unlike other IKKβ inhibitors evaluated, **23** provided a large window between blockade of NF-κB activation (IC$_{50}$ = 1–2 μM) and general cytotoxicity (> 50 μM) in EpRas cells, possibly attributable to differences in kinase selectivity, enabling a clean interpretation of phenotypic cellular readouts.[115] In these studies IKKβ-dependent NF-κB activity was effectively blocked in EpRas cells, TGFβ-induced EMT prevented and EMT in mesenchymal CT26 cells was reversed. When compound **23** was dosed orally at 75 mg/kg b.i.d. in the mouse 4T1 xenograft model, only a 30% reduction in tumour volume was observed, but a pronounced effect on metastasis was achieved, with a 62% reduction in metastatic lung burden. Moreover the metastases that were present were found to be of smaller size and more highly differentiated compared to the vehicle-treated control group. Multiple patent applications have claimed variants of this thienopyridine class, including benzothiophenes, thienopyrimidines and 5-substituted thienopyridines as represented by compounds **24–26** (Figure 9.8).[116–118] However, there have not been public disclosures of this work translating into more advanced compounds (Figure 9.11).

9.4.3   2-Hydroxyphenylpyridines

An interesting class of IKKβ inhibitors appeared in patent applications from Bayer AG in 2002 featuring a novel electron-deficient 2-amino-6-hydroxy-phenylpyridine core (Figure 9.12).[119,120] Over the next two years three serial

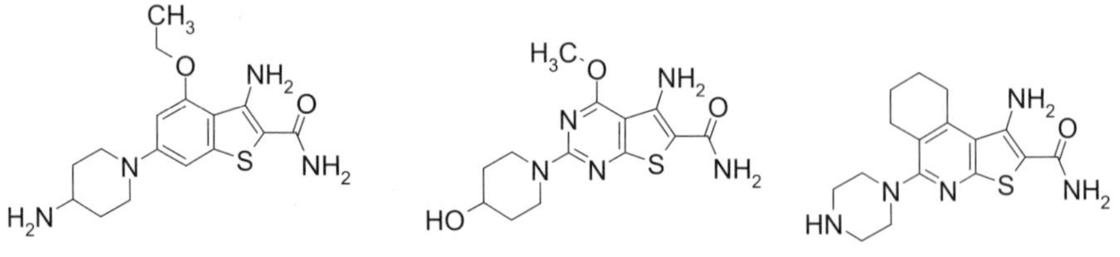

24 Boehringer Ingelheim[116] **25** Kyorin[117] **26** Astellas[118]

Figure 9.11 Representative thienopyridine follow-up patent applications.

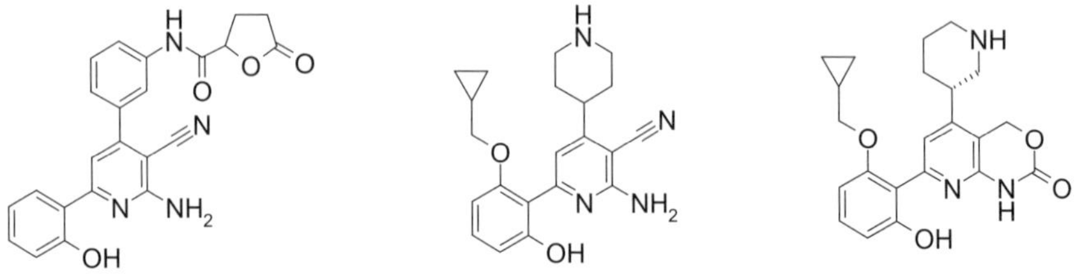

27 Bayer **28** ACHP Bayer **29** CHPD Bayer

Figure 9.12 2-Hydroxyphenylpyridine IKKβ inhibitors.

publications described enlightening SAR and promising cellular profiles, which led from screening hit **27** to the discovery and use of **28** (ACHP) and **29** (CHPD) as investigational tools.[121–123]

High-throughput screening identified **27** as a low micromolar potency hit with greater than 10-fold selectivity over IKKα, and measurable potency in cell-based assays (8–15 μM across a variety of cell types). A solid-phase, multi-component pyridine condensation methodology was developed and utilized to rapidly generate SAR around both phenyl rings.[124] The key discovery from this effort was the absolute requirement for the intra-molecular hydrogen bond between the phenolic hydroxyl group and the pyridine nitrogen atom. This was comprehensively demonstrated through loss of activity not only upon removal, migration or substitution of the phenol residue, but also upon introduction of an *ortho*-methyl group to either ring of the biaryl system, which precludes a coplanar geometry. The presence of this intra-molecular hydrogen bond in active compounds could be characterized by ^{1}H NMR spectroscopy, where a 3 ppm downfield shift of the acidic phenol proton was observed, relative to inactive compounds. An important role for the amino NH of **27** was also evident, as the corresponding 2-hydroxy and diethyl amino pyridines were inactive and *N*-acetylation resulted in weak activity. The subsequent publication introduced the piperidine substituents present in ACHP and CHPD, a modification that evolved out of the need to improve solubility and the realization that aromatic groups in this region were not necessary for potency.[122] Incorporation of these basic amine groups also provided a substantial potency increase. The nature of this interaction was highly specific as the 3-piperidinyl moiety was 10- to 20-fold more potent than the corresponding 2- and 4-piperidine isomers. In addition, later disclosures taught that the (3*S*)-3-piperidinyl fragment conferred 15-fold greater potency than the (3*R*)-3-piperidinyl enantiomer. The final paper in this series reports the incorporation of the pendant cyclopropylmethyl ether group of **28**, which is accompanied by further gains in potency.[123] Compound **28** is a very potent IKKβ inhibitor ($IC_{50} = 8.5$ nM) that is 30-fold selective over IKKα. Its desirable cell activity profile was shown through the inhibition of the production of a number of important pro-inflammatory mediators (TNFα, IL-6, IL-1β, RANTES, VCAM-1) across a panel of relevant cell types (Jurkat, huPBMC, mouse B-cell, HUVEC, A549) with IC_{50}s below 100 nM. This compound also showed modest to good oral bioavailability in mice (16%) and rats (60%) and at a 1 mg/kg p.o. dose demonstrated comparable efficacy to dexamethasone (0.3 mg/kg p.o.) in an arachadonic acid-induced mouse ear edema model. Very limited kinase selectivity data have been provided for **28**; however, the lack of activity in NFAT and AP-1 reporter gene assays supports these functional effects being mechanistically consistent with NF-κB signalling.

It appears that compound **29**, in which the amino and nitrile substituents of **28** are replaced by a ring-fused oxazinone, is a preferred compound from this inhibitor family based on a subsequent patent application claiming only this structure and salt forms thereof.[125] Presumably a key benefit of this modification is that it allows the incorporation of the more potent 3-piperidinyl fragment that had previously been shown to be incompatible with the nitrile group due to spontaneous intra-molecular cyclization. The pharmacological profile of **29** was disclosed in a later publication showing further improvement in potency and isoform selectivity (2 nM inhibitor of IKKβ, 135 nM inhibitor of IKKα). Compound **29** showed greater than 10 μM IC_{50}s against a panel of 10 kinases but more extensive kinome profiling was not described.[126] A battery of cell assays similar to those used to evaluate compound **28** were highlighted, including inhibition of IL-6 ($IC_{50} = 18$ nM) and IL-1β ($IC_{50} = 52$ nM) production by LPS-stimulated human PBMCs, and the inhibition of VCAM-1 production ($IC_{50} = 85$ nM) by TNFα-stimulated HUVECS. This compound has moderate clearance (2.31 l/h/kg) and high volume of distribution (6.91 l/kg) in the rat resulting in a 2.1-hour half-life and 69% oral bioavailability (2 mg/kg i.v., 10 mg/kg p.o. doses). Compound **28** has demonstrated efficacy in multiple rodent models of inflammation and asthma, including airway inflammation induced by ovalbumin or cockroach allergen (10 mg/kg racemate p.o. 24 h post challenge) and edema induced by arachadonic acid, phorbol ester or carrageenan challenges ($ED_{50} = 0.3$ mg/kg p.o.). Both compounds **28** and **29** have become tools used by independent investigators to examine IKKβ as a point of therapeutic intervention in a wide variety of NF-κB driven disease processes, including the suppression of TNFα-induced HIV replication,[127] neuroprotection through inhibition of the production of

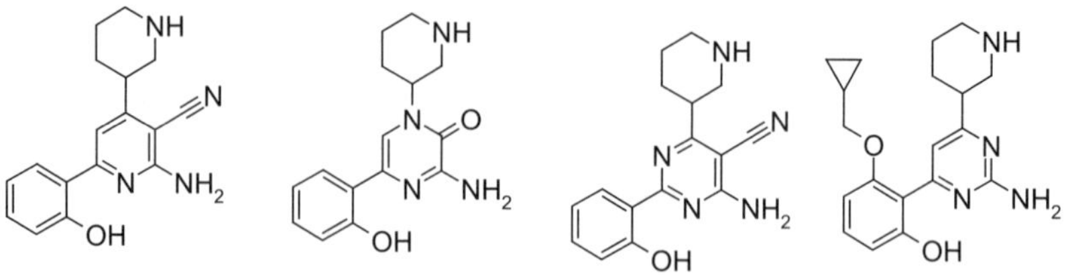

30 Bayer   **31** Pharmacia   **32** Pharmacia   **33** Astellas

Figure 9.13   2-Hydroxyphenylpyridine follow-up patent applications.

NF-κB-dependent pro-inflammatory mediators and NADPH-driven reactive oxygen species in microglia,[128] apoptotic sensitization in adult T-cell leukaemia and multiple myeloma cells,[129–130] protection against myocardial injury and dysfunction in an acute ischemia-reperfusion model[131] and the inhibition of pro-inflammatory cytokine production by TNFα-stimulated synovial fibroblasts from rheumatoid arthritis patients (compound **29** $EC_{50} = 510$ nM IL-6, 210 nM IL-8).[132]

Several other companies have also pursued opportunities within this IKKβ inhibitor class (Figure 9.13). For example, Pharmacia has filed patent applications on the amino-pyrazinone **31** ($IC_{50} = 1.08$ μM) and the aminopyrimidine nitrile derivative **32** ($IC_{50} = 633$ nM),[133,134] but both core modifications are significantly less potent than the analogous Bayer reference, **30** ($IC_{50} = 25$ nM). Perhaps a more attractive starting point is the regioisomeric pyrimidine **33** ($IC_{50} < 500$ nM) described by Astellas, which demonstrates efficacy in both the LPS mouse and rat CIA models at a 30 mg/kg p.o. q.d. dose.[135] Unlike **32**, incorporation of the second nitrogen into the heterocyclic core of **33** would not lead to non-productive competition for intra-molecular hydrogen bonding of the phenol, or electronic repulsion of the optimized 6-position ether substituent. Additionally, deletion of the reactive nitrile of **32** potentially provides greater compatibility with the preferred 3-piperidinyl group.

Although no binding model was proposed for the 2-hydroxyphenylpyridine inhibitor class in the initial Bayer publications, the disclosed SAR has been leveraged by others for the design of novel chemotypes. The anticipated binding mode is illustrated below in Figure 9.14.

9.4.4   Indole Carboxamides

Investigators from GSK recognized the potential of a shared hydrogen bonding pharmacophore between this rigidified biaryl motif and aryl carboxamides conformationally restricted by an adjacent hydrogen bond donor. Thus, a CDK2-based IKKβ homology model indeed predicted the phenolic oxygen of **34** and carboxamide oxygens of **35** and **36** (see Figure 9.15) serve as common hydrogen bond acceptor elements for Cys99.[136–138]

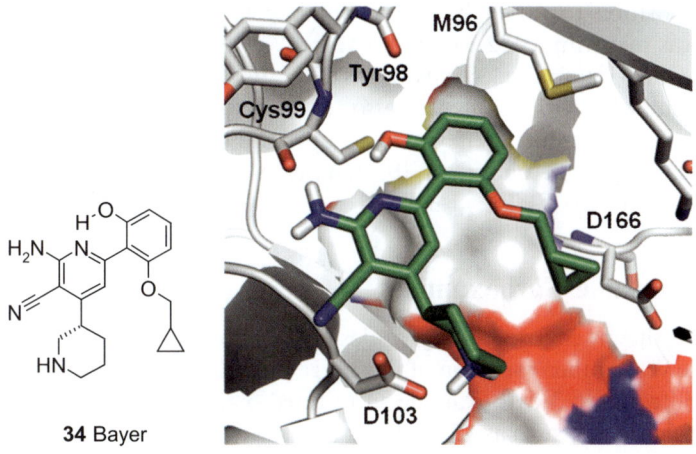

Figure 9.14   Proposed binding mode of 2-hydroxyphenylpyridine **34**.

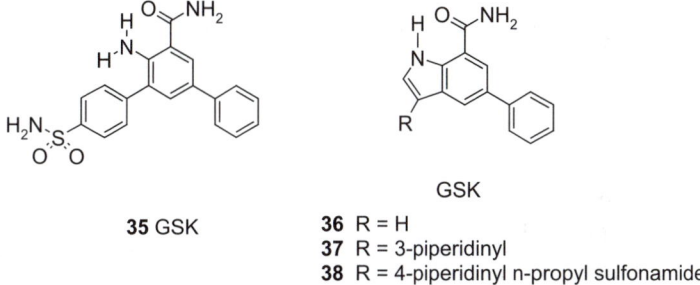

Figure 9.15   Optimization of indole carboxamide IKKβ inhibitors.

Given the demonstrated importance of the primary amino group of **34** for activity, it is likely that it also engages Cys99 as hydrogen bond donor for an overall "mode 2" binding interaction to the hinge. Another interesting feature of this binding model is the prediction that the potency increases associated with the 4-sulfonamide group of **35** (IKKβ IC$_{50}$ = 250 nM) and piperidinyl amine of **34** (IKKβ IC$_{50}$ = 3 nM, racemate) originate from interactions with a common residue, Asp103. This hypothesis served as the basis for targeting the 3-position of the weakly potent indole carboxamide **36** in an effort to optimize potency. Appending the 3-piperidinyl substituent to target electrostatic interactions with Asp103 resulted in a 10-fold improvement in both IKKβ potency (IC$_{50}$ = 158 nM) and IKKα selectivity (IC$_{50}$ = 10 μM) for **37**. Given the specific nature of this interaction in the Bayer series, the potential for further improvements through subtle structural modifications and chiral resolution may exist. Likewise, projecting the 4-piperidinyl *n*-propylsulfonamide from the

3-indole position to target hydrogen bonding interactions with Asp103 and Lys106 was met with a 122-fold improvement in IKKβ activity (IC$_{50}$ = 13 nM) and more than 150-fold selectivity over IKKα (IC$_{50}$ = 2.0 μM) for derivative **38**. An excellent selectivity profile was also reported for **38**, with greater than 100-fold selectivity observed in enzymatic assays against 45 additional kinases. In a binding assay format only 2 of 105 kinases gave Kd values less than 3 μM (CSNK1ε = 1.6 μM, FLT3 = 1.3 μM).

9.4.5   Pyrazole Carboxamides

Arguably, the fused tricyclic pyrazoles **39–42** from Pharmacia (now Pfizer) constitute yet another variation of this hinge binding motif (Figure 9.16). These molecules feature a 3-carboxamide substituted biaryl substructure common to the aryl thiophenes, such as **15**, and aryl indoles, exemplified by **38**. In this case, however, the carboxamide group is conformationally constrained *via* an intramolecular hydrogen bond between the carboxamide NH and the proximal pyrazole nitrogen rather than by engagement of the carboxamide oxygen with a neighbouring H-bond donor as for **15** and **38**.

An IKKβ homology model built from PKA and CHK-1 predicts the same anchoring hydrogen bond interaction between the Cys99 NH and the pyrazole carboxamide oxygen of the ligand, but can not confidently assign an overall binding mode since the predicted preference for interactions of the carboxamide NH with the backbone carbonyl of either Cys99 (mode 2) or Gly97 (mode 1) are ligand dependent in this series.[139] This class of molecules was first disclosed in a series of closely related 2003 patent applications[140–143] that included the characterization of tricycle **39** as a submicromolar inhibitor of IKKβ. This molecule appears to have been the starting point for optimization from the variety of tricyclic pyrazole cores described since the most heavily

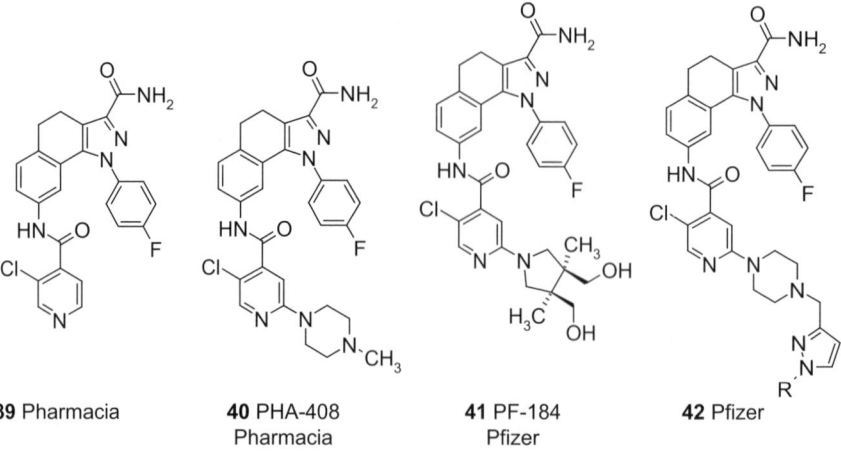

| **39** Pharmacia | **40** PHA-408 Pharmacia | **41** PF-184 Pfizer | **42** Pfizer |

Figure 9.16   Tricyclic pyrazole carboxamide IKKβ inhibitors.

characterized molecule emerging from this class is the closely related PHA-408 (**40**). PHA-408 is an ATP competitive, tight binding ($K_i = 6$ nM), potent IKKβ inhibitor (reported IC_{50}s from 1.6 to 40 nM) that is 350-fold selective over IKKα.[144] When profiled against a panel of 30 kinases, it was found to be 100-fold selective against all except PIM1, over which it is 15-fold selective. Broader kinome specificity has not been reported. PHA-408 demonstrates good oral rat PK properties (i.v. CL = 11 ml/min/kg, i.v. $t_{1/2} = 29$ h, F% = 60 @ 2 mg/kg i.v. and 5 mg/kg p.o. doses) and has been evaluated in a number of proof of principle and disease relevant models. For example, PHA-408 inhibits LPS-mediated TNFα production in the rat with an ED_{50} of 8 mg/kg p.o. t.i.d. In a streptococcal cell wall-induced chronic rat arthritis model (SCW), maximal efficacy for the inhibition of paw swelling and bone erosion was achieved at 10 mg/kg p.o. t.i.d.[145] Oral efficacy has also been demonstrated with this compound at 15 and 45 mg/kg p.o. doses in LPS and cigarette smoke induced rat airway inflammation models.[146] Perhaps the feature of PHA-408 that most distinguishes it from other IKKβ inhibitors is a very slow enzyme off rate ($K_{off} = 0.3$/h, $t_{1/2}$ of enzyme complex = 3h), which provides an extended duration of action and allows for co-immunoprecipitation with the IKK complex. This has enabled the development of detailed PK/PD relationships through the analysis of pharmacological responses as a function of target coverage, the latter measured by enzymatic inhibition of the pulled down IKK complex. Thus, PHA-408 has been used to establish *in-vitro* correlations between target coverage, IL-1β induced IKKβ activity, p65 phosphorylation, IκBα phosphorylation and degradation, and NF-κB DNA binding in synovial fibroblasts from rheumatoid arthritis patients. *In-vivo* correlations have also been examined in a number of animal models, including SCW induced chronic rat arthritis, in which relationships between PHA-408 plasma exposure, immunoprecipitate enzyme activity and paw volume were successfully established.[145] These PK/PD relationships were used in the design of rat safety studies to provide interpretable outcomes regarding target coverage.[147] In these experiments female Lewis rats were dosed at 3, 10, 15 and 20 mg/kg p.o. t.i.d. for 14 days. No adverse events were seen for the 10 mg/kg t.i.d. dose which was previously found to be fully efficacious in the SCW rat arthritis model. The two higher doses produced minor effects, specifically, mild liver inflammation and associated neutrophilia and lymphocyte depletion in the thymus, spleen and lymph nodes. The liver and neutrophilia effects were not seen with related compounds and so were not interpreted as being due to off-target activity. However, lymphoid depletion was believed to be IKKβ mechanism-dependent based upon similar results with additional structurally diverse IKKβ inhibitors. The most recent reports from Pfizer lay out strategies to expand the safety margin of this class of IKKβ inhibitor while leveraging the extended duration of action of PHA-408 in the development of inhaled therapeutics for inflammatory airway indications.[139] This optimization process was focused on maintaining desirable slow off rates and IKKβ potency profiles while reducing oral bioavailability through increased Phase I and II metabolism. It was hoped that the potential for any IKKβ mechanism-related toxicities, as well as potential off-target toxicities such as hERG inhibition, which was a general concern for the series, could be reduced through restricting systemic exposure. Through this effort PF-184 (**41**) was identified, demonstrating increased *in-vivo* clearance and decreased oral bioavailability in the rat (10% as compared to 60% for **40**). Intra-tracheal administration of PF-184 in an inhaled LPS rat model of airway inflammation showed comparable dose responsive inhibition of neutrophil influx and cytokine production relative to orally administered PHA-408.[148] Compound **41** also achieved efficacy comparable to intra-tracheally administered fluticasone, supporting the potential of IKKβ inhibition as a therapeutic strategy for pulmonary inflammation. Most recently, in an effort to incorporate alternative metabolic soft spots into the inhibitor to limit systemic exposure, substituted pyrazoles were appended to the methyl piperazine of PHA-408 to afford compounds of generic structure **42**.[139] A number of analogues were identified, which were rapidly metabolized and further increased potency and enzyme half-life; however, no additional pharmacology or safety data were reported for these compounds.

9.4.6 β-Carbolines

Prior to the disclosure of PHA-408, a patent application from Millennium Pharmaceuticals was published describing a quite different structural class of tricyclic IKKβ inhibitors, represented by PS-1145 (**44**, Figure 9.17).[149] Later reports disclosed that the natural product derivative 5-bromo-6-methoxy-β-carboline (**43**) was initially identified as an interesting, modestly potent ($IC_{50} = 1\,\mu M$), but poorly selective screening hit. Limited modification led to the discovery of **44**, a simply decorated compound with comparable potencies in both recombinant IKKβ ($IC_{50} = 100\,nM$) and endogenous IKK complex assays ($IC_{50} = 150\,nM$). PS-1145 was also shown to inhibit IκBα phosphorylation, NF-κB DNA binding and TNFα production in an LPS treated mouse model (60% inhibition, 50 mg/kg p.o.), strongly supporting the desired mechanism of action.[150] A publication from an independent lab characterized PS-1145 as having greater kinase selectivity than the IKKβ inhibitors BMS345541 (**46**) and SC514 (**18**), inhibiting just 2 out of 72 off-target kinases (PIM1 and PIM3) with similar potency.[151] A later report introduced an

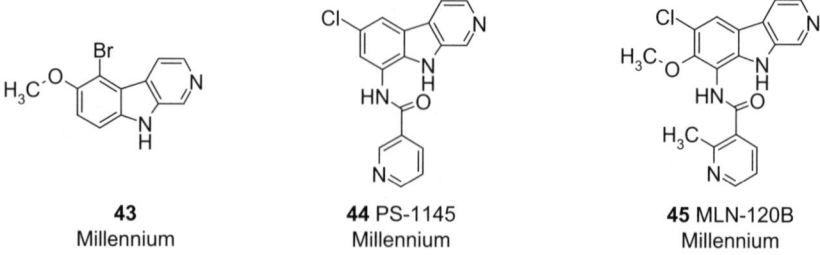

| **43** | **44** PS-1145 | **45** MLN-120B |
| Millennium | Millennium | Millennium |

Figure 9.17 β-Carboline IKKβ inhibitors.

improved analogue, MLN-120B (**45**), which incorporates 7-methoxy β-carbo-line and 2-methyl pyridyl benzamide substituents.[152] Like PS-1145, MLN-120B shows equipotent inhibition in IKKβ recombinant ($IC_{50} = 45$ nM) and IKK complex ($IC_{50} = 60$ nM) assay formats and mechanistically consistent pathway intervention. When tested against a panel of 30 kinases all IC_{50}s exceeded 50 μM, although the panel did not include PIM1 or PIM3. MLN-120B was also more selective than PS-1145 when profiled against a panel of transcription factor pathways.[152] PS-1145 showed significant inhibition of NFAT and AP-1 reporter gene assays, while MLN-120B did not. Compound **45** (MLN-120B) was also shown to inhibit the production of multiple NF-κB gene products involved in inflammation and joint destruction from disease relevant cell types, including IL-6, IL-8, MIP1 and RANTES in synoviocytes, MMP-1 and 13 in chondrocytes, and multiple cytokines in chord blood derived mast cells. Consistent with these properties, MLN-120B showed dose responsive inhibition of paw swelling, weight loss and bone loss in a rat adjuvant induced arthritis (AIA) model ($ED_{50} = 12$ mg/kg p.o. b.i.d.) with mechanistically consistent mRNA reduction of TNFα, IL-1β, iNOS and MMP13 at 30 mg/kg p.o. b.i.d.[153]

9.4.7 Imidazo Pyridines

Researchers from Bristol-Myers Squibb have developed an additional class of tricyclic IKKβ inhibitors for which a unique binding mode has been proposed.[154] The initial lead compound from this series, **46** (BMS-345541, Figure 9.18), is a modest inhibitor of IKKβ ($IC_{50} = 300$ nM), which is weakly active against IKKα ($IC_{50} = 4000$ nM) and inhibits LPS-stimulated production of NF-κB dependent cytokines in THP-1 cells, including TNFα, IL-1β, IL-8 and IL-6 at concentrations of 1000–5000 nM.

Detailed kinetic analysis of this compound revealed that it does not bind to the ATP site of either IKKα or IKKβ. Even more interestingly, it was found to compete with peptide substrate binding to IKKβ, but not IKKα. This led to a proposed model in which BMS-345541 binds to similar allosteric sites of both catalytically active subunits, but differentially perturbs their active sites. This compound possessed suitable PK properties in mouse (F% = 100, i.v. $t_{1/2} = 2.2$ h, Vss = 18.5 ml/kg, at doses of 2 mg/kg i.v. and 10 mg/kg p.o.) with respect to its potency profile to enable *in-vivo* studies and became the first

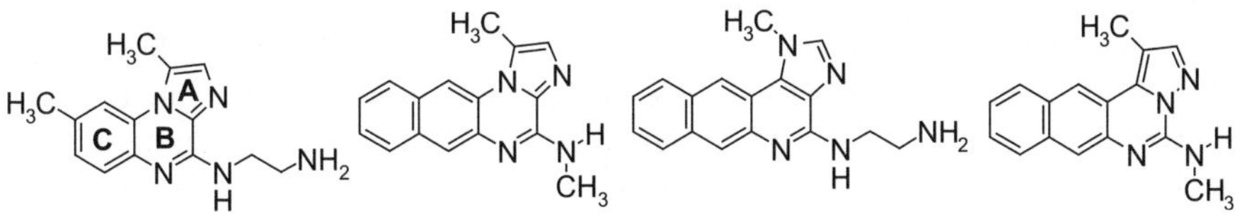

46 BMS345541   **47** BMS   **48** BMS   **49** BMS

Figure 9.18   BMS345541 and related analogues.

small-molecule IKKβ inhibitor reported to demonstrate pharmacological anti-inflammatory activity. In both prophylactic and therapeutic mouse collagen induced arthritis models, BMS-345541 showed dose responsive efficacy (10, 30 and 100 mg/kg p.o., q.d.) and provided essentially full remission of clinical score in the high dose group.[155] Profound effects on inflammation and histology were also seen in these studies. Efficacy was also demonstrated in the dextran sulfate sodium-induced colitis model with regard to weight retention, clinical score, inflammation and injury at doses of 30 and 100 mg/kg p.o., q.d.[156] These effects of BMS-345541 on intestinal inflammation were consistent with its *in-vitro* inhibition of adhesion molecule (ICAM-1 and VCAM-1) surface expression on TNFα-stimulated HUVECs. Compund **46** has also shown synergistic benefits in cardiac transplant models in combination with suboptimal doses of cyclosporin.[157] Additional *in-vitro* studies with BMS-345541 support potential applications to oncology, where it has been shown to cause sensitization of carboplatin resistant ovarian carcinomas and the arrest of multiple mitotic cell transitions, and metabolic disease, due to its ability to block insulin receptor substrate 1 and 2 degradation.[158–160]

The initial disclosure of BMS-345541 described minimal activity against a panel of 15 additional kinases and selective intra-cellular inhibition of IκBα phosphorylation over other pro-inflammatory signalling networks including c-Jun, STAT3 and MAPKAPK2. Independent investigators have shown this compound to have some measurable off-target activity in a broader panel of biochemical assays, significantly inhibiting 4 (ERK8, PKD1, CDK2, CK1) of 70 kinases approximately 2-fold less potently than IKKβ.[151] However, characterization in a battery of intra-cellular β-lactamase transcriptional reporter gene assays led others to conclude this compound was quite promiscuous, inhibiting multiple NF-κB independent transcriptional activities (AP1, ISRE, CRE, HRE, NFAT, SBE) with potency equal to on-target pathways.[161]

A series of subsequent publications described the systematic variation of the tricyclic core of BMS-345541 coupled with targeted opportunities amenable to rapid synthetic exploration. In the absence of an anticipated structural basis for allosteric binding, this empirical approach successfully led to substantial improvements in both potency and selectivity, culminating in the discovery of BMS-066 (**58**, *vide infra*). Initial improvements in potency were achieved through homologation of the tricyclic scaffold of **46** to a tetracyclic core, as shown for compounds **47–49** in Figure 9.18.[162] The tetracycle is absent from later analogues that alternatively access this space through a biaryl motif in which a fourth aromatic D-ring is appended to the tricyclic core as shown for analogues **50–54** in Figure 9.19.[163] Both of these modifications have been shown to be compatible with oral exposure and *in-vivo* activity in rodents. Pyrazolo and imidazo A-ring variants were equally tolerated and in later reports the imidazole regioisomer present in compound **48** is featured exclusively. The ethylenediamine side chain of **48** could be replaced with truncated alkyl amines, an important observation since it was discovered that general cytotoxicity was associated with some ethylenediamine containing analogues, attributed to their detergent-like properties. Later reports also taught that the

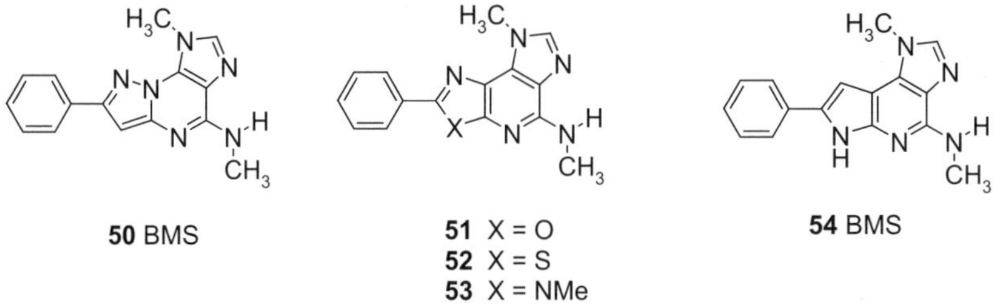

50 BMS **51** X = O
52 X = S
53 X = NMe **54** BMS

Figure 9.19 Imidazopyridine IKKβ inhibitors.

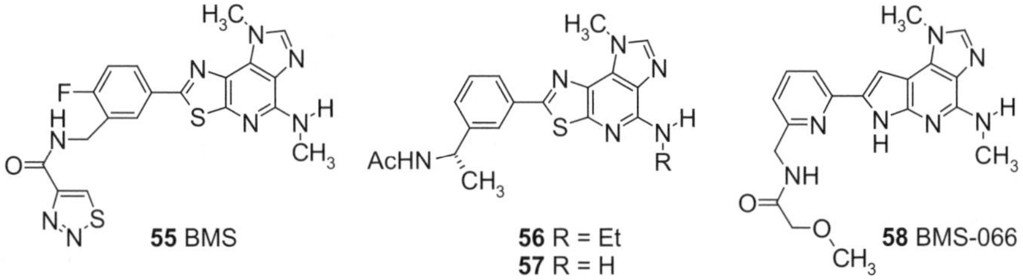

55 BMS **56** R = Et
57 R = H **58** BMS-066

Figure 9.20 Identification of BMS-066.

NH of the side chain is critical as the dimethylamine and methyl ether analogues were much less potent.[164] A variety of heterocycle replacements of the phenyl C-ring have also been examined. Initial pyrazole C-ring tricycles such as **50** were prone to oxidative metabolism of the remaining CH bond and therefore oxazole (**51**), thiazole (**52**), imidazole (**53**) and pyrrole C-ring analogues (**54**), all of which were accessible through a common synthetic intermediate, were targeted to remove this metabolic soft spot. Although this strategy was not successful in addressing clearance, it led to the serendipitous discovery of improved potency and isoform selectivity afforded by the thiazole core.

SAR of the left-hand side phenyl ring highlighted *meta*-substitution as the most profitable (Figure 9.20). The incorporation of polar groups at this position substantially reduced *in-vivo* clearance without negatively impacting potency, exemplified by **55** and **56**, the latter of which was also efficacious in a mouse CIA model at an oral dose of 60 mg/kg b.i.d.[165]

Unfortunately, it was later shown that **56** is converted *in vivo* to the Ames positive *N*-dealkylated metabolite **57** in appreciable concentration. However, the less potent pyrrole C-ring tricyclic core was found not to carry this intrinsic mutagenic liability. Further optimization of this chemotype through elaboration of the amide side chain and incorporation of a pyridine D-ring resulted in the discovery of BMS-066 (**58**).[166] Relative to the initial proof of concept compound, BMS-066 demonstrated greatly improved potency (IKKβ IC$_{50}$ = 9 nM), isoform selectivity (IKKα IC$_{50}$ = 4960 nM) and broader kinome selectivity, inhibiting just 4 kinases out of 150 with less than 100-fold selectivity (EphA1 = 76-fold, Ret = 68-fold, Flt4 = 40-fold, Brk = 34-fold). This compound was also found to have a desirable rat PK profile with very low clearance (6.4% Qh), good half-life (3.7 h) and high oral bioavailability (90%). It should be noted that the seemingly minor structural change to a 2-pyridyl D-ring in **58** exerts a profound effect on overall molecular rigidification. A recently described X-ray structure of this compound reveals a network of intra-molecular bifurcated hydrogen bonds relayed through the pyrrole NH, pyridyl nitrogen, benzyl amide NH and methyl ether oxygen atom, effectively freezing all rotatable bonds within the molecule. It is interesting to consider to what extent this reduced rotational freedom may contribute to the observed improvements in oral bioavailability and excellent off-target selectivity profile.

BMS-066 has been examined in mechanistic assays across multiple disease relevant cell types and shown to inhibit the immune function and proliferation of both B and T lymphocytes as well as RANKL-stimulated osteoclastogenesis. These effects were seen at concentrations consistent with cellular inhibition of IκBα phosphorylation. Extensive *in-vivo* characterization in rat adjuvant induced arthritis and mouse CIA pre-clinical models has been described.[167] In these experiments BMS-066 was found to be highly efficacious at lower oral doses than previous compounds from this series (5 and 10 mg/kg p.o. q.d. in rat; 20 mg/kg p.o. b.i.d. in mouse) and showed impressive effects with respect bone protection. This compound has also shown dose responsive efficacy in the TNBS-induced colitis model at oral doses of 3 and 10 mg/kg b.i.d., with effects comparable to dexamethasone on weight maintenance, colon length and histopathology scores.[166] Results from rodent pre-clinical safety studies with this chemotype were recently reported.[168] No adverse effects on liver or skin were observed, establishing clear differentiation between developmental knockout and pharmacological intervention phenotypes. However, a paradoxical finding of acute inflammation of the atrium with associated neutrophil influx was described. This cardiotoxicity was observed with both the initial lead, BMS-345541, and optimized candidate BMS-066.

9.4.8   Imidazo Pyridazines

Daichi has very recently disclosed an imidazo[1,2-b] pyridazine series of IKKβ inhibitors (Figure 9.21).[169,170] These compounds were developed from the screening hit **59**, which provided a desirable starting point for potency (IKKβ IC$_{50}$ = 1.1 µM) and good selectivity in kinase counter screens (GSK3β and CDK2 IC$_{50}$ > 100 µM). Optimization leading from **59** to **61** is detailed in two subsequent publications. Exploration of the SAR of the basic ethylenediamine group of **59** led to a 3.5-fold increase in molecular potency and measurable cell potency upon cyclization to a 2(S)-aminomethyl pyrrolidine side chain. Replacement of the dichlorobenzyl group of **59** with a cyclopropyl methyl, as in analogue **60**, provided a further additive improvement in enzymatic potency (IKKβ IC$_{50}$ = 55 nM). Key breakthroughs in cell potency were achieved

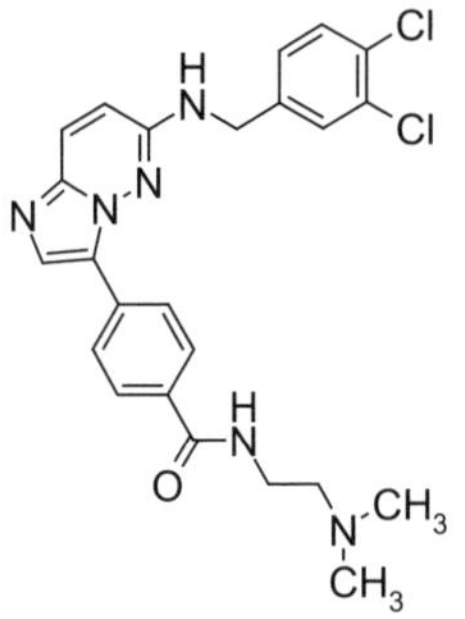

Figure 9.21 Imidazo pyridazine optimization.

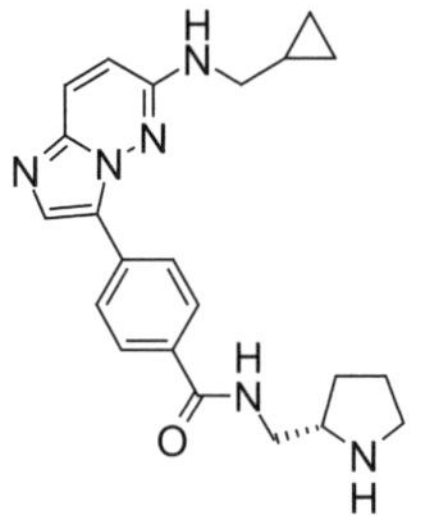

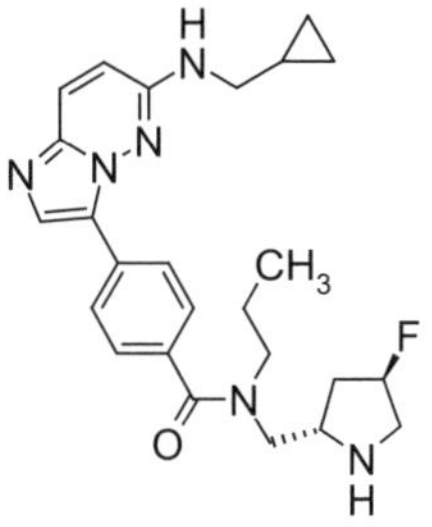

through improvements in permeability, as estimated by a PAMPA assay, accompanying the introduction of a fluorine substituent β to the pyrrolidine nitrogen atom and alkylation of the secondary amide. These effects on permeability were additive and correlated with reductions in potency shifts between enzymatic and cell-based assays. The resulting analogue **61** (IKKβ $IC_{50} = 18$ nM; TNFα inhibition in LPS-stimulated THP-1 cells $IC_{50} = 230$ nM) inhibited TNFα production by 52% in the LPS mouse model at a single oral dose of 30 mg/kg, which achieved plasma concentrations comparable to cell IC_{50} values 90 min. post administration. A homology model was generated from PKCα, DAPK1 and CDK2, which provided a rationalization for the observed SAR. It is predicted that the key hinge contact is made between the sp^2 imidazo nitrogen and the Cys99 backbone NH. The authors also proposed that these molecules must induce a significant disruption of the activation loop in order to occupy the ATP binding site, which results in favourable interactions of the basic amine with the perturbed Asp166 and Glu61 acidic residues. More extensive kinase profiling or disease model data have not yet been disclosed for this series.

9.5   Clinical Information

Clinical trial database queries for IKKβ inhibitors identify a number of compounds that likely function through other primary targets or multiple mechanisms of action. One must consider this information carefully when drawing conclusions regarding target-dependent safety and efficacy. To date, the clinical compounds for which there is the strongest confidence in IKKβ as the primary molecular target are MLN-0415 and SAR-113945. This opinion is based upon the reported characterization of the chemical series from which they are believed to have originated (see Figure 9.22). Other compounds advanced into clinical trials have either lacked accompanying data demonstrating direct inhibition of IKKβ at the molecular level (*i.e.* IKKβ enzymatic IC_{50}) or have had significant evidence indicating additional off-target activities, which confounds clear interpretations of IKKβ-dependent effects.

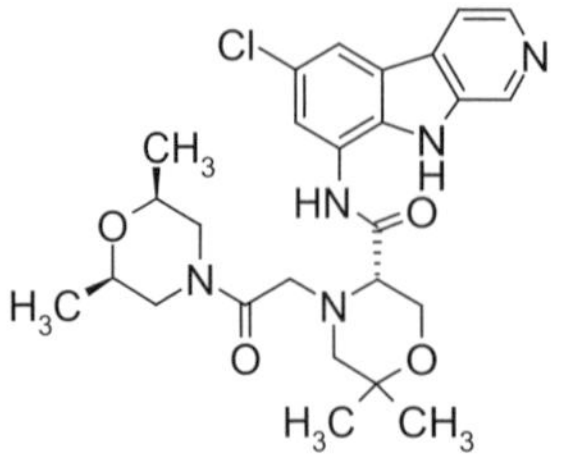

62 "MLN-0415"
Takeda

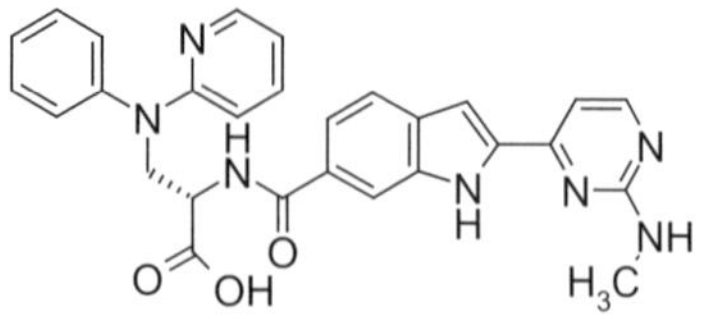

63 "SAR-113945"
Aventis

Figure 9.22 Potential structures of MLN-0415 and SAR-113945.

The first small-molecule IKKβ inhibitor to enter human trials was MLN-0415 in 2006.[171] Therapeutic indications under consideration were chronic obstructive pulmonary disorder, rheumatoid arthritis, inflammatory bowel disease and multiple sclerosis. The structure of this compound has not been disclosed but it is speculated to be the β-carboline **62** based upon a single compound patent filing of the corresponding mesylate salt by Millennium in 2009.[172] MLN-0415 successfully completed a single ascending dose trial but was later terminated due to a non-specified unfavourable safety profile in the multiple ascending dose study.[171]

Sanofi-Aventis advanced SAR-113945 to clinical trials in 2010 for the treatment of osteoarthritis and associated joint pain and a Phase I study has been completed. Forty patients were administered a single dose of SAR-113945 *via* intra-articular injection into the knee to assess safety, tolerability, pharmacokinetics and plasma concentrations of potential metabolites. Patient enrolment is currently underway for a two-part Phase I/II ascending single dose trial. The targeted patient enrolment is 24 with an estimated completion date of July 2012.[173] The structure of SAR-113945 has not been publicly disclosed. However, compound **63** may be a candidate based upon a single compound patent filing on salts and crystal forms.[174] In this patent application **63** is described as a subnanomolar IKKβ inhibitor with anticipated benefit for limited systemic exposure upon local administration, in line with the clinical route of administration of SAR-113945.

In 2007 SPC-839 (**64**), a compound originally developed by the Signal Research division of Celgene, was advanced into a Phase I trial for haematological malignancies by Merck Serono. However, by early 2008 this trial was terminated for portfolio repositioning considerations.[175] The structure of this compound has been disclosed (see Figure 9.23). It has been shown to be a potent ATP competitive reversible inhibitor of IKKβ ($IC_{50} = 62$ nM) with selectivity over IKKα ($IC_{50} = 13$ μM) and activity in cell-based assays (inhibition of ICAM-1 and VCAM-1 expression in HUVEC cells $IC_{50} = 1.0$ μM). Compound **64** has also shown efficacy in the rat AIA model, inhibiting paw swelling (65% at a 30 mg/kg p.o. q.d. dose) and providing near complete protection from bone destruction and ankle histopathology parameters

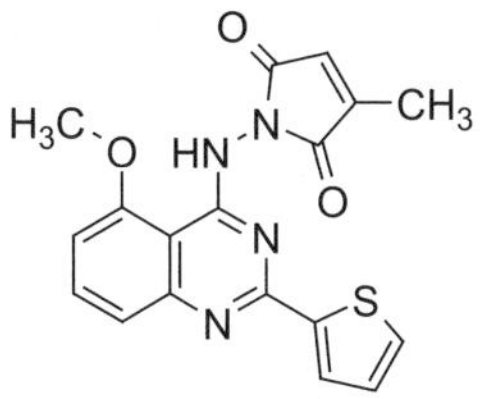

64 SPC-839
Signal/Merck Serono

Figure 9.23 Structure of SPC-839.

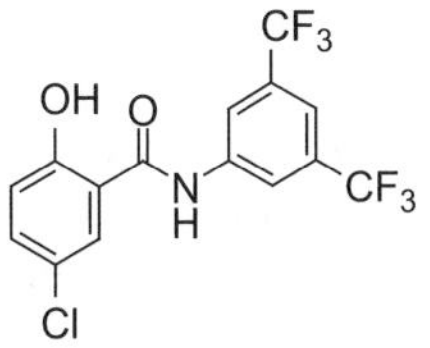

65 IMD-0354
Institute of Medical Molecular Design

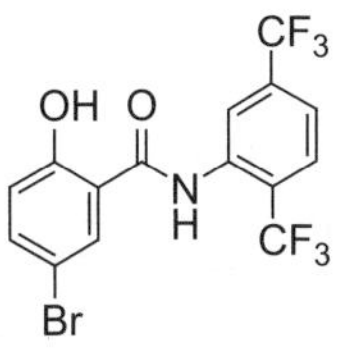

66 IMD-0560
Institute of Medical Molecular Design

Figure 9.24 Structures of IMD-0354 and IMD-0560.

(100 mg/kg p.o. q.d.).[176] However, there is evidence that SPC-839 also interacts with additional molecular targets, being originally developed as a potent dual inhibitor of both AP-1 and NF-κB transcription (IC_{50} in AP-1 Jurkat cell reporter gene assay = 8 nM).[177,178]

The salicylanilides IMD-0354, IMD-0560 (Figure 9.24) and their corresponding prodrugs IMD-1041 and IMD-2560 (structures unknown) developed by the Institute of Medicinal Molecular Design have been described in the literature as novel IKKβ inhibitors and included in discussions of the more advanced pharmacological agents for this target.[57,179–181] A Phase II study is currently being planned with IMD-0354 for atopic dermatitis and Phase I studies have been completed with IMD-1041 and IMD-2560, which are reportedly being positioned for COPD and RA indications, respectively.[182] However, the clinical progression of IMD-0354 and related molecules should be viewed with great caution regarding interpretation of direct IKKβ inhibition mechanism-related efficacy or tolerability. IMD-0354 demonstrates efficacy in multiple mechanistically relevant models and *in-vitro* data support that these effects are related to NF-κB pathway inhibition.[179,183–186] Nevertheless, its multi-target patenting history and the curious absence of data demonstrating direct inhibition of IKKβ in enzymatic assays strongly question whether IKKβ is the true molecular target. Prior to being described as the product of rational design based upon a specific binding mode of aspirin to IKKβ, IMD-0354 was patented not only as a general inhibitor of NF-κB activation, but also of the AP-1 and NFAT transcription factors.[179,187,188] Biological data supporting both IMD-0354 and IMD-0560 as inhibitors of all three of these pathways are provided in these applications. More recently IMD-0354 was claimed in a patent application for glutamate receptor modulation and specifically shown to have potent activity against mGluR1 ($EC_{50} = 111$ nM) and mGluR5 ($EC_{50} = 91$ nM).[189] Mechanistic studies question if disruption of NF-κB activation and inhibition of ensuing downstream functional effects by IMD-0354 can be fully accounted for by IKKβ inhibition. It has been shown that at the concentrations at which IMD-0354 achieves full blockade of LPS-stimulated TNFα production in cardiac myocytes (100 μM), IκBα phosphorylation is only partially inhibited (31%), suggesting an additional mechanism of action for IMD-0354.[190] In related experiments, at concentrations at which IMD-0354 strongly inhibited MCP1 and IL-1β production in this same cell type, IκBα phosphorylation was again only modestly effected, but inhibition of NF-kB translocation was nearly quantitative.[185] These observations have led some to suggest that perhaps the salicylanilides mechanistic point of NF-κB disruption may be at the level of NF-κB translocation or activation rather than kinase inhibition.[190]

The pyridyl-cyanoguanidine CHS-828, **67**, and more soluble prodrug EB-1627 (**68**, teglarinad chloride) from Leo Pharmaceuticals are another class of molecules historically appearing in IKKβ clinical update reviews, which now require reconsideration (Figure 9.25). The source of the antineoplastic activity of these molecules, which have progressed into Phase I/II clinical trials for oncology indications, was initially unknown.[191] With the discovery of potent inhibition of the pulled down IKK complex ($IC_{50} = 8$ nM), these effects were reasonably hypothesized to be IKKβ-dependent.[192] Interestingly, characterization of these molecules in typical, less complicated, IKKβ homodimer enzymatic assay was not described. It has now been shown by two groups *via* different approaches

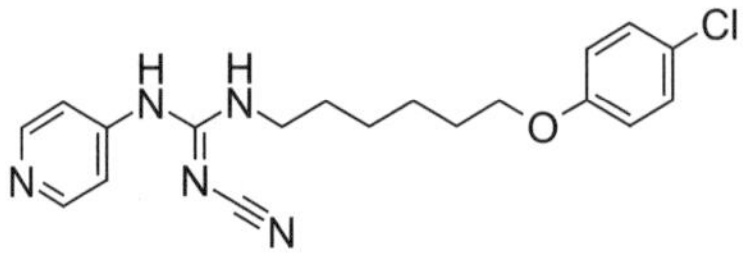

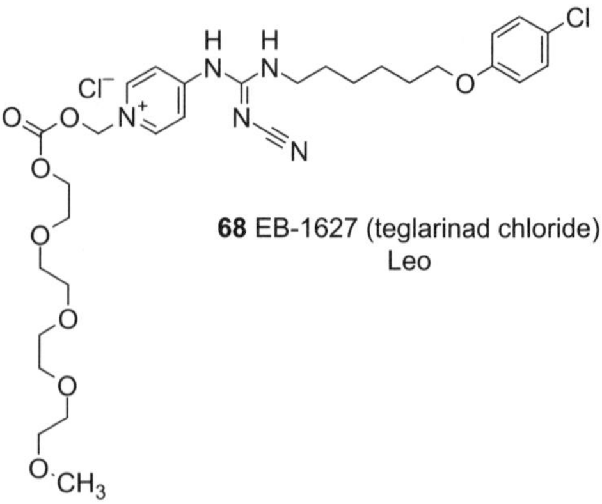

Figure 9.25 Structures of CHS-828 and teglarinad chloride.

that CHS-828 is a potent inhibitor of nicotinamide phosphoribosyltransferase ($K_i = 2.6$ nM) and that this is likely the primary cytotoxic mechanism, with NF-κB inhibition being a secondary consequence of NAD depletion.[193,194]

9.6 Future Outlook

The potential of IKKβ inhibition as a therapeutic strategy remains unrealized and it is not clear that chronic systemic inhibition of this target can be tolerated. Lingering safety concerns may limit the most suitable application of IKKβ inhibitors to subtherapeutic combination chemotherapeutic strategies for enhanced cytotoxicity and metastasis prevention. Previous optimism for this target was fueled by multiple compounds advancing into human trials in recent years. However, all but one of the former clinical candidates have now either been terminated or shown unlikely to operate solely by the intended mechanism. Additionally, development activities have not been reported for the second generation compounds with improved kinome selectivity, with some programs reporting termination due to pre-clinical toxicity observations. Patenting activity indicates a decline in investment from the pharmaceutical industry. From 2003 to 2009 an average of 15 IKKβ inhibitor applications were published per year. This dropped sharply to 6 applications appearing in 2010 and only a single application surfacing through September of 2011. Yet the pending SAR-113945 Phase II trial results have the potential to provide very interesting insight. If this compound is well tolerated in multi-dosing regimens it will be enlightening to determine if this could be related to differences in kinase selectivity (data not reported) or the limiting of systemic exposure through intra-articular administration. If the former scenario, clearer optimization paths may emerge. If the latter, innovative targeting strategies for selective tissue, disease site or cell type delivery may yet allow for exploitation of this target.

Acknowledgement

The author would like to thank Larry Anderson for his assistance with patent searches and analyses, Ming-Hong Hao for providing the homology model binding illustrations included in this text, Jun Li and John Ginn for manuscript reviews, and the Boehringer Ingelheim IKKβ project team for creating an enthusiastic environment in which to study this fascinating target, in particular co-project leader Jeff Madwed.

References

1. M. Karin, Y. Yamamoto and Q. M. Wang, *Nat. Rev. Drug Discovery*, 2004, **3**, 17.
2. J.-L. Luo, H. Kamata and M. Karin, *J. Clin. Invest.*, 2005, **115**, 2625.
3. T. D. Gilmore and M. Herscovitch, *Oncogene*, 2006, **25**, 6887.

4. P. J. Barnes and M. Karin, *New Engl. J. Med.*, 1997, **336**, 1066.
5. A. S. Baldwin Jr., *Annu. Rev. Immunol.*, 1996, **14**, 649.
6. P. J. Barnes, *Clin. Sci.*, 1998, **94**, 557–572.
7. L. A. Solt, L. A. Madge and M. J. May, *J. Biol. Chem.*, 2009, **284**, 27596.
8. M. S. Hayden and S. Ghosh, *Genes Dev.*, 2004, **18**, 2195.
9. G. Bonizzi and M. Karin, *Trends Immunol.*, 2004, **25**, 280.
10. A. Yaron, H. Gonen, I. Alkalay, A. Hatzubai, S. Jung, S. Beyth, F. Mercurio, A. M. Manning, A. Ciechanover and Y. Ben-Neriah, *EMBO J.*, 1997, **16**, 6486.
11. M. S. Hayden and S. Ghosh, *Cell*, 2008, **132**, 344.
12. T. T. Huang, N. Kudo, M. Yoshida and S. Miyamato, *Proc. Natl Acad. Sci. USA*, 2003, **100**, 1014.
13. C. F. Cervantes, S. Bergqvist, M. Kjaergaard1, G. Kroon, S.-C. Sue, H. J. Dyson and E. A. Komives, *J. Mol. Biol.*, 2011, **405**, 754.
14. P. Viatour, S. Legrand-Poels, C. van Lint, M. Warnier, M-P. Merville, J. Gielen, J. Piette, V. Bours and A. Chariot, *J. Biol. Chem.*, 2003, **278**, 46541.
15. T. T. Huang and S. Miyamoto, *Mol. Cell. Biol.*, 2001, **21**, 4737.
16. S. Ghosh, M. J. May and E. B. Kopp, *Annu. Rev. Immunol.*, 1998, **16**, 225.
17. Z. Li and G. J. Nabel, *Mol. Cell Biol.*, 1997, **17**, 6184.
18. M. Spieker, H. Darius and J. K. Liao, *J. Immunol.*, 2000, **164**, 3316.
19. M. Karin, *J. Biol. Chem.*, 1999, **274**, 27339.
20. E. Zandi, Y. Chen and M. Karin, *Science*, 1998, **281**, 1360.
21. D. M. Rothwarf, E. Zandi, G. Natoli and M. Karin, *Nature*, 1998, **395**, 297.
22. J. Li, G. W. Peet, S. S. Pullen, J. Schembri-King, T. C. Warren, K. B. Marcu, M. R. Kehry, R. Barton and S. Jakes, *J. Biol. Chem.*, 1998, **273**, 30736.
23. Z.-W. Li, W. Chu, Y. Hu, M. Delhase, T. Deerink, M. Ellisman, R. Johnson and M. Karin, *J. Exp. Med.*, 1999, **189**, 1839.
24. M. Adili, E. Merkhofer, P. Cogswell and A. S. Baldwin, *Plos ONE*, 2010, **5**, e9428.
25. S.-C. Sun, *Cell Res.*, 2011, **21**, 71.
26. E. Dejardin, *Biochem. Pharmacol.*, 2006, **72**, 1161.
27. B. Rayet and C. Gelinas, *Oncogene*, 1999, **18**, 6938.
28. M. L. Handel, L. B. McMorrow and E. M. Gravellesse, *Arthritis Rheum.*, 1995, **38**, 1762.
29. H. Asahara, M. Asanuma, N. Ogawa, S. Nishibayashi and H. Inoue, *Biochem. Mol. Biol. Int.*, 1995, **37**, 827.
30. L. A. Hart, V. L. Krishnan, I. M. Adcock, P. J. Barnes and K. F. Chung, *Am. J. Respir. Crit. Care Med.*, 1998, **138**, 1585.
31. L. Andersen, V. L. Jorgensen, A. Perner, A. Hansen, J. Eugen-Olsen and J. Rask-Madsen, *Gut*, 2005, **54**, 503.
32. S. Schreiber, S. Nikolaus and J. Hampe, *Gut*, 1998, **42**, 477.
33. D. Gveric, C. Kaltschmidt, L. Cuzner and J. Newcombe, *J. Neuropath. Exp. Neur.*, 1998, **57**, 168.

34. E. Andreakos, C. Smith, S. Kiriakidis, C. Monaco, R. de Martin, F. Brennan, E. Paleolog, M. Feldman and B. Foxwell, *Arthritis Rheum.*, 2003, **48**, 1901.

35. P. P. Tak, D. M. Gerlag, K. R. Aupperle, D. A. van de Geest, M. Overbeek, B. L. Bennett, D. L. Boyle, A. M. Manning and G. S. Fierstein, *Arthritis Rheum.*, 2001, **46**, 1897.

36. S. W. Tas, N. Hajji, D. J. Stenvers, G. S. Firestein, M. H. Vervoordeldonk and P. P. Tak, *Arthritis Rheum.*, 2006, **51**, 821.

37. M. C. Catley, M. B. Sukkar, K. F. Chung, B. Jaffee, S.-M. Liao, A. J. Coyle, E.-B. Haddad, P. J. Barnes and R. Newton, *Mol. Pharm.*, 2006, **69**, 697.

38. M. C. Catley, J. E. Chivers, N. S. Holden, P. J. Barnes and R. Newton, *Br. J. Pharmacol.*, 2005, **146**, 114.

39. S. Sanlioglu, G. Luleci and K. W. Thomas, *Cancer Gene Ther.*, 2001, **8**, 897.

40. H. Ni, M. Ergin, Q. Huang, J.-Z. Qin, H. H. Amin, R. L. Martinez, S. Saeed, K. Barton and S. Alkan, *Br. J. Haematol.*, 2001, **113**, 279.

41. C. Matteucci, A. Minutolo, E. Balestrieri, F. Marino-Merlo, P. Bramanti, E. Garaci, B. Macchi and A. Mastino, *Cell Death & Disease*, 2010, **1**, e81.

42. J. U. Ammann, C. Haag, H. Kasperczyk, K.-M. Debatin and S. Fulda, *Int. J. Cancer*, 2009, **125**, 1301.

43. T. Inoue, T. Takenaka, M. Hayashi, T. Monkawa, J. Yoshino, K. Shimoda, E. G. Neilson, H. Suzuki and H. Okada, *J. Am. Soc. Nephr.*, 2010, **21**, 2047.

44. P. A. Baeurle and V. R. Baichwal, *Adv. Immunol.*, 1997, **65**, 111.

45. A. A. Beg, W. C. Shah, R. C. Bronson, S. Ghosh and D. Baltimore, *Nature*, 1995, **6536**, 167.

46. D. Rudolph, W.-C. Yeh, A. Wakeham, B. Rudolph, D. Nallainathan, J. Potter, A. J. Elia and T. W. Mak, *Genes Dev.*, 2000, **14**, 854.

47. J. Dutta, Y. Fan, N. Gupta, G. Fan and C. Gelinas, *Oncogene*, 2006, **25**, 6800.

48. K. Takeda, O. Takeuchi, T. Tsujimura, S. Itami, O. Adachi, T. Kawai, H. Sanjo, K. Yoshikawa, N. Terada and S. Akira, *Science*, 1999, **284**, 316.

49. T. Luedde, U. Assmus, T. Wüstefeld, A. M. zu Vilsendorf, T. Roskams, M. Schmidt-Supprian, K. Rajewsky, D. A. Brenner, M. P. Manns, M. Pasparakis and C. J. Trautwein, *J. Clin. Invest.*, 2005, **115**, 849.

50. L. Eckmann, T. Nebelsiek, A. A. Figerle, S. M. Dann, J. Mages, R. Lang, S. Robine, M. F. Kagnoff, R. M. Schmidt, M. Karin, M. C. Arkan and F. R. Greten, *Proc. Natl Acad. Sci. USA*, 2008, **105**, 15058.

51. J. E. Hoberg, A. E. Popko, C. S. Ramsey and M. W. Mayo, *Mol. Cell. Biol.*, 2006, **26**, 457.

52. S. Wenjing, G. Ningling, Y. Yang, S. Burlingame, L. Xiaonan, M. Y. Zhang, F. Shenglong, F. Songbin and Y. Jianhua, *J. Biol. Chem.*, 2010, **285**, 7911.

53. L. Ling, C. Zhaodon and D. V. Goeddel, *Proc. Natl Acad. Sci. USA*, 1998, **95**, 3792.

54. S. Hiroaki, H. Miyoshi, W. Toriumi and S. Takahisa, *J. Biol. Chem.*, 1999, **274**, 10641.

55. K. Clark, M. Peggie, L. Plater, R. J. Sorcek, E. R. R. Young, J. B. Madwed, J. Hough, E. G. McIver and P. Cohen, *Biochem. J.*, 2011, **434**, 93.

56. K. Nandini, Q. K. Huynh, S. Mathialagan, T. Hall, S. Rouw, D. Creely, G. Lange, J. Caroll, B. Reitz, A. Donnelly, H. Boddupalli, R. G. Combs, K. Kretzmer and C. S. Trapp, *J. Biol. Chem.*, 2002, **277**, 13840.

57. P. D. G. Coish, P. L. Wickens and T. B. Lowinger, *Expert Opin. Ther. Pat.*, 2006, **16**, 1.

58. P. Bamborough, M. A. Morse and K. P. Ray, *Drug News and Perspect.*, 2010, **23**, 483.

59. S. M. Noha, A. G. Atanasov, D. Schuster, P. Markt, N. Fakhrudin, E. H. Heiss, O. Schrammel, J. M. Rollinger, H. Stuppner, V. M. Dirsch and G. Wolber, *Bioorg. Med. Chem. Lett.*, 2011, **21**, 577.

60. E. Sala, L. Guasch, J. Iwaszkiewicz, M. Mulero, M-J. Salvado, M. Pinent, V. Zoete, A. lien Grossdidier, S. Garcia-Vallve, O. Michielin and G. Pujadas, *PLOS One*, 2011, **6**, e16903.

61. S. Nagarajan, H. Choo, Y. S. Cho, K. J. Shin, K-S. Oh, B. H. Lee and A. N. Pae, *BMC Bioinf.*, 2010, **11**, S15.

62. S. Nagarajan, H. Choo, Y. S. Cho, K.-S. Oh, B. H. Lee, K. J. Shin and A. N. Pae, *Bioorg. Med. Chem.*, 2010, **18**, 3951.

63. S. Nagarajan, M. R. Doddareddy, H. Choo, Y. S. Cho, K.-S. Oh, B. H. Lee and A. N. Pae, *Bioorg. Med. Chem.*, 2009, **17**, 2759.

64. W. Long, P. Liu, Q. Li, Y. Xu and J. Gao, *QSAR Comb. Sci.*, 2008, **27**, 1113.

65. X. Guozhou, Q. Li, G. Napolitano, X. Wu, X. Jiang, M. Dreano and M. Karin, *Nature*, 2011, **472**, 325.

66. IKKβ-PHK homology model built using PRIME® software (Schrodinger, Inc.), refined using MAESTRO®(Schrodinger, Inc.) protein preparation protocol and docking carried out using GLIDE® software (Schrodinger, Inc.).

67. A. H. Bingham, R. J. Davenport, L. Gowers, R. L. Knight, C. Lowe, D. A. Owen, D. M. Parry and W. R. Pitt, *Bioorg. Med. Chem. Lett.*, 2004, **14**, 409.

68. A. L. Crombie, F.-W. Sum, D. W. Powell, D. W. Hopper, N. Torres, D. M. Berger, Y. Zhang, M. Gavriil, T. M. Sadler and K. Arndt, *Bioorg. Med. Chem. Lett.*, 2010, **20**, 2831.

69. M. Mirmehrabi, M. Sreenivasulu, T. Abdolsamad and D. Subodh, Wyeth, John, and Brother, Ltd., USA, *US Pat. Appl.*, US 2008/0275073, 2008.

70. R. Waelchli, B. Bollbuck, C. Bruns, T. Buhl, J. Eder, R. Feifel, R. Hersperger, P. Janser, L. Revesz, H.-G. Zerwes and A. Schlapbach, *Bioorg. Med. Chem. Lett.*, 2006, **16**, 108.

71. A. Kois, K. J. Macfarlane, Y. Satoh, S. S. Bhagwat, J. Parnes, M. S. Palanki and P. E. Erdman, Signal Pharmaceuticals, Inc., *PCT Int. Appl.*, WO 2002/046171, 2002.

72. S. Palmer and S. Nataraja, Applied Research Systems ARS Holding N.V., *PCT Int. Appl.*, WO 2007/030362, 2007.
73. K. R. Dahnke, H.-S. Lin, C. Shih, M. Q. Wang, B. Zhang and M. E. Richett, Eli Lilly and Company, *PCT Int. Appl.*, WO 07/092095, 2007.
74. J. Wagnon, J.-F. Nguefack, S. Jegham, M. Bosch, M. Bouaboula, P. Casellas, B. Tonnerre, J.-A. Olsen and S. Mignani, Sanofi-Aventis, *PCT Int. Appl.*, WO 2007/006926, 2007.
75. O. Ritzeler, H. U. Stilz, B. Neises, W. J. Bock, A. Walser and G. A. Flynn, Aventis Pharma Deutschland GMBH, *PCT Int. Appl.*, WO 2001/00610, 2001.
76. O. Ritzeler, H. U. Stilz, B. Neises, W. J. Bock, G. Jaehne and J. Habermann, Aventis Pharma Deutschland GMBH, *PCT Int. Appl.*, WO 2001/30774, 2001.
77. M. Michaelis, O. Ritzeler, G. Haehne, K. Rudolphi and H.-G. Schaible, Aventis Pharma Deutschland GMBH, *PCT Int. Appl.*, WO 2004/022057, 2004.
78. O. Ritzeler and G. Jaehne, Aventis Pharma Deutschland GMBH, *PCT Int. Appl.*, WO 2004/022553, 2004.
79. D. Dietzel, S. Ramrath, O. Ritzeler, N. Tennagels, H. Hauner and J. Ecke, *Int. J. Obes. Relat. Metab. Disord.*, 2004, **28**, 985.
80. Y. Ivashchenko, F. Kramer, S. Schafer, A. Bucher, K. Veit, V. Hombach, A. Busch, O. Ritzeler, J. Dedio and J. Torzewski, *Arterioscler. Thromb. Vasc. Biol.*, 2005, **25**, 186.
81. M. Lohn and Y. Ivashchenko, Sanofi-Aventis Deutschland GMBH, *DE Pat. Appl.*, DE 10 2004 034380.
82. A. Baxter, S. Brough, A. Faull, C. Johnstone and T. McInally, Astra Zeneca AB, *PCT Int. Appl.*, WO 2001/58890, 2001.
83. A. Faull, C. Johnstone, A. Morely and J. P. Poyser, Astra Zeneca AB, *PCT Int. Appl.*, WO 2003/010158, 2003.
84. D. Griffiths and C. Johnstone, Astra Zeneca AB, *PCT Int. Appl.*, WO 2003/010163, 2003.
85. A. D. Morley and J. P. Poyser, Astra Zeneca AB, *PCT Int. Appl.*, WO 2004/063185, 2004.
86. A. W. Faull, C. Johnstone, A. D. Morley and J. P. Poyser, Astra Zeneca AB, *PCT Int. Appl.*, WO 2004/063186, 2004.
87. S. Ashwell, T. Gero, S. Ioannidis, J. Janetka, P. Lyne, V. Oza, S. Springer, M. Su and D. Yu, Astra Zeneca AB, *PCT Int. Appl.*, WO 2005/016909, 2005.
88. J. F. Callahan and A. K. Roshak, SmithKline Beecham Corporation, *PCT Int. Appl.*, WO 2002/30353.
89. J. F. Callahan and A. K. Roshak, SmithKline Beecham Corporation, *PCT Int. Appl.*, WO 2002/30423, 2002.
90. C. A. Parrish, J. F. Callahan, Z. Wan, J. L. Burgess, R. A. Stavenger and D. A. Holt, SmithKline Beecham Corporation, *PCT Int. Appl.*, WO 2003/028731, 003.
91. C. A. Parrish, J. F. Callahan, Y. Li, R. A. Stavenger and D. A. Holt, SmithKline Beecham Corporation, *PCT Int. Appl.*, WO 2003/029241, 2003.

92. Z. Wan, J. L. Burgess and J. F. Callahan, SmithKline Beecham Corporation, *PCT Int. Appl.*, WO 2003/029242, 2003.

93. J. F. Callahan and Y. H. Li, SmithKline Beecham Corporation, *PCT Int. Appl.*, WO 2003/086309, 2003.

94. J. F. Callahan and Y. H. Li, Smithkline Beecham Corporation, *PCT Int. Appl.*, WO 2003/104218, 2003.

95. J. F. Callahan and Z. Wan, SmithKline Beecham Corporation, *PCT Int. Appl.*, WO 2003/104219, 2003.

96. J. F. Callahan and Y. H. Li, Smithkline Beecham Corporation, *PCT Int. Appl.*, WO 2004/053087, 2004.

97. M. Graneto, C. E. Hanau and T. D. Perry, Pharmacia Corporation, *PCT Int. Appl.*, WO 2003/037886, 2003.

98. T. J. Hagen, R. M. Weier, X. Xu, S. C. Houdek and M. Clare, Pharmacia Corporation, *PCT Int. Appl.*, WO 2004/009582, 2004.

99. M. Clare, T. R. Fletcher, B. C. Hamper, G. A. Hanson, R. F. Heier, H. Huang, P. J. Lennon, S. D. Oburn, M. T. Reding, S. A. Stealey, S. G. Wolfson and J. Xie, Pharmacia Corporation, *PCT Int. Appl.*, WO 2005/037797, 2005.

100. A. Baxter, S. Brough, A. Cooper, E. Floettmann, S. Foster, C. Harding, J. Kettle, T. McInally, C. Martin, M. Mobbs, M. Needham, P. Newham, S. Paine, S. Gallay, S. Salter, J. Unitt and Y. Xue, *Bioorg. Med. Chem. Lett.*, 2004, **14**, 2817.

101. *Expert Opin. Ther. Patents*, 2005, **15**, 343.

102. P. L. Podolin, J. F. Callahan, B. J. Bolognese, Y. H. Li, K. Carlson, T. G. Davis, G. W. Mellor, C. Evans and A. K. Roshak, *J. Pharm. Exp. Ther.*, 2005, **312**, 373.

103. J. J. O. Turner, K. M. Foxwell, R. Kanji, C. Brenner, S. Wood, B. M. J. Foxwell and M. Feldman, *Clin. Exp. Immunol.*, 2010, **162**, 487.

104. Y. Lei, Y. Zhang, Y. Cao, L. Edvinsson and C.-B. Xu, *Eur. J. Pharm.*, 2010, **634**, 149.

105. G. A. Koumbadinga, A. Désormeaux, A. Adam and F. Marceau, *Eur. J. Pharm.*, 2010, **647**, 117.

106. M. A. Birrell, E. Hardaker, S. Wong, K. McCluskie, M. Catley, J. De Alba, R. Newton, S. Haj-Yahia, K. T. Pun, C. J. Watts, R. J. Shaw, T. T. Savage and M. G. Belvisi, *Am. J. Crit. Care Med.*, 2005, **172**, 962.

107. D. De Silva, M. D. Mitchell and J. A. Keelan, *British J. Pharm.*, 2010, **160**, 1808.

108. N. Kishore, C. Sommers, S. Mathialagan, J. Guzova, M. Yao, S. Hauser, K. Huynh, S. Bonar, C. Mielke, L. Albee, R. Weier, M. Graneto, C. Hanau, T. Perry and C. S. Tripp, *J. Biol. Chem.*, 2003, **278**, 32861.

109. P. Bamborough, D. Drewry, G. Harper, G. K. Smith and K. Schneider, *J. Med. Chem.*, 2008, **51**, 7898.

110. J. Janetka, L. Almeida, S. Ashwell, P. J. Brassil, K. Daly, C. Deng, T. Gero, R. E. Glynn, C. L. Horn, S. Ioannidis, P. Lyne, N. J. Newcombe, V. B. Oza, M. Pass, S. K. Springer, M. Su, D. Toader, M. M. Vasbinger, D. Yu, Y. Yu and S. D. Zabludoff, *Bioorg. Med. Chem. Lett.*, 2008, **18**, 4242.

111. T. Morwick, A. Berry, J. Brickwood, M. Cardozo, K. Catron, M. DeTuri, J. Emeigh, C. Homon, M. Hrapchak, S. Jacober, S. Jakes, P. Kaplita, T. A. Kelly, J. Ksziazek, M. Liuzzi, R. Magolda, C. Mao, D. Marshall, D. McNeil, A. Prokopowicz III, C. Sarko, E. Scouten, C. Sledziona, S. Sun, J. Watrous, J.-P. Wu and C. Cywin, *J. Med. Chem.*, 2006, **49**, 2898.
112. J.-P. Wu, R. Fleck, J. Brickwood, A. Capolino, K. Catron, Z. Chen, C. Cywin, J. Emeigh, M. Foerst, J. Ginn, M. Hrapchak, E. Hickey, M.-H. Hao, M. Kashem, J. Li, W. Liu, T. Morwick, R. Nelson, D. Marshall, L. Martin, P. Nemoto, I. Potocki, M. Liuzzi, G. W. Peet, E. Scouten, D. Stefany, M. Turner, S. Weldon, C. Zimmitti, D. Spero and T. A. Kelly, *Bioorg. Med. Chem. Lett.*, 2009, **19**, 1547.
113. J. D. Ginn, R. J. Sorcek, M. R. Turner and E. R. R. Young, Boehringer Ingelheim Corporation, *US Pat. Appl.*, US 07/0293533, 2007.
114. Z. Chen, P. F. Cirillo, D. DiSalvo, W. Liu, D. R. Marshall, L. Wu and E. R. R. Young, Boehringer Ingelheim Pharmaceuticals, Inc., *PCT Int. Appl.*, WO 2005/056562, 2005.
115. M. A. Huber, H. J. Maier, A. Memetcan, E. Wiedemann, J. Braunger, G. Boehmelt, J. B. Madwed, E. R. R. Young, D. R. Marshall, H. Pehamberger, T. Wirth, N. Kraut and B. Hartmut, *Genes Cancer*, 2010, **1**, 101.
116. Z. Chen, J. D. Ginn, E. R. Hickey, W. Liu, C. Mao, T. M. Morwick, P. A. Nemoto, D. Spero and S. Sun, Boehringer Ingelheim Pharmaceuticals, Inc., *PCT Int. Appl.*, WO 2005/012283, 2005.
117. M. Wada, N. Sueda, T. Komine, K. Kastsuyama, H. Tsuchida, R. Kojima and O. Kawamura, Nisshin Kyorin Pharmaceutical Co., Ltd, *PCT Int. Appl.*, WO 2008/020622, 2008.
118. Y. Okamoto, K. Hattori, H. Kubota, I. Sato, T. Kanayama, K. Yokoyama, Y. Terai and M. Takeuchi, Astellas Pharma, Inc., *PCT Int. Appl.*, WO 2005/123745, 2005.
119. T. Murata, U. Masaomi, S. Sakakibara, T. Yoshino, H. Sato, T. Masuda, Y. Koriyama, M. Shimada, T. Shintani, H. Kadono, K. B. Ziegelbauer, K. Fuchikami and H. Komura, Bayer Aktiengesellschaft, *PCT Int. Appl.*, WO 2002/24679, 2002.
120. T. Murata, U. Masaomi, S. Sakakibara, T. Yoshino, Y. Ikegami, T. Masuda, M. Shimada, M. Shimazaki, T. B. Lowinger, K. B. Ziegelbauer, K. Fuchikami, H. Komura, M. Umeda and N. Yoshida, Bayer Aktiengesellschaft, *PCT Int. Appl.*, WO 2002/44153, 2002.
121. T. Murata, S. Mitsuyuki, S. Sakakibara, T. Yoshino, H. Kadono, T. Masuda, M. Shimazaki, T. Shintani, K. Fuchikami, K. Sakai, H. Inbe, K. Takeshita, T. Niki, M. Umeda, K. B. Bacon, K. B. Ziegelbauer and T. B. Lowinger, *Bioorg. Med. Chem. Lett.*, 2003, **13**, 913.
122. T. Murata, S. Mitsuyuki, S. Sakakibara, T. Yoshino, H. Kadono, T. Yoshino, T. Masuda, M. Shimazaki, T. Shintani, K. Fuchikami, K. B. Bacon, K. B. Ziegelbauer and T. B. Lowinger, *Bioorg. Med.Chem. Lett.*, 2004, **14**, 4013.
123. T. Murata, S. Mitsuyuki, S. Sakakibara, T. Yoshino, H. Kadono, T. Yoshino, T. Masuda, T. Shintani, H. Sato, Y. Koriyama, K. Fukushima,

N. Nunami, M. Yamauchi, K. Fuchikami, H. Komura, A. Watanabe, K. B. Bacon, K. B. Ziegelbauer and T. B. Lowinger, *Bioorg. Med. Chem. Lett.*, 2004, **14**, 4019.

124. T. Shintani, H. Kadono, T. Kikuchi, T. Schubert, Y. Shogase and M. Shimazaki, *Tetrahedron Lett.*, 2003, **44**, 5567.

125. T. Murata, S. Sasaki, T. Yoshino, H. Sato, Y. Koriyama, N. Nunami, M Yamamuchi, K. Fukushima, R. Grosser, K. Fuchikami, K. Bacon and T. Lowinger, Bayer Aktiengesellschaft, *PCT Int. Appl.*, WO 2003/076447, 2003.

126. K. Ziegelbauer, F. Gantner, N. W. Lucas, A. Berlin, K. Fuchikami, T. Niki, K. Sakai, H. Inbe, K. Takeshita, M. Ishimori, H. Komura, T. Murata, T. Lowinger and K. B. Bacon, *Br. J. Pharm.*, 2005, **145**, 178.

127. A. Florence, B. Victoriano, K. Asamitsu, Y. Hibi, K. Imai, N. G. Barzaga and T. Okamoto, *Antimicrob. Agents Chemother.*, 2006, **50**, 547.

128. F. Zhang, L. Qian, P. M. Flood, J.-S. Shi, J.-S. Hong and H.-M. Gao, *J. Pharm. Exp. Ther.*, 2010, **333**, 822.

129. T. Sanda, S. Iida, H. Ogura, K. Asamitsu, T. Murata, K. B. Bacon, R. Ueda and T. Okamoto, *Clin. Cancer Res.*, 2005, **11**, 1974.

130. T. Sanda, K. Asamitsu, H. Ogura, S. Iida, A. Utsunomiya, R. Ueda and T. Okamoto, *Leukemia*, 2006, **20**, 590.

131. N. C. Moss, W. E. Stansfield, M. S. Willis, R.-H. Tang and C. H. Selzman, *Am. J. Physiol. Heart Circ. Physiol.*, 2007, **293**, 2248.

132. A. Tsuchiya, K. Imai, K. Asamitsu, Y. Waguir-Nagaya, T. Otsuka and T. Okamoto, *J. Pharm. Exp. Ther.*, 2010, **333**, 236.

133. M. L. Boys, M. Clare and M. J. Mitton-Fry, Pharmacia Corporation, *PCT Int. Appl.*, WO 2005/035527, 2005.

134. M. Clare, T. J. Hagen, S. C. Houdek, P. J. Lennon, R. M. Weier and X. Xu, Pharmacia Corporation, *PCT Int. Appl.*, WO 2005/040133, 2005.

135. Y. Okamoto, H. Kubota, I. Sato, K. Hattori, T. Kanayama, K. Yokoyama, Y. Terai, M. Takeuchi, Astellas Pharma, Inc., *PCT Int. Appl.*, WO 2005/100341.

136. J. A. Christopher, B. G. Avitabile, P. Bamborough, A. C. Champigny, G. J. Cutler, S. L. Dyos, K. G. Grace, J. K. Kearns, J. D. Kitson, G. W. Mellor, J. V. Morey, M. A. Morse, C. F. O'Malley, C. B. Patel, N. Probst, W. Rumsey, C. A. Smith and M. J. Wilson, *Bioorg. Med. Chem. Lett.*, 2007, **17**, 3972.

137. J. A. Christopher, P. Bamborough, C. Alder, A. Campbell, G. J. Cutler, K. Down, A. M. Hamadi, A. M. Jolly, J. K. Kerns, F. S. Lucas, G. W. Mellor, D. D. Miller, M. A. Morse, K. D. Pancholi, W. Rumsey, Y. E. Solanke and R. Williamson, *J. Med. Chem.*, 2009, **52**, 3098.

138. D. D. Miller, P. Bamborough, J. A. Christopher, I. R. Baldwin, A. C. Champigny, G. J. Cutler, J. K. Kearns, T. Longstaff, G. W. Mellor, J. V. Morey, M. A. Morse, H. Nie, W. L. Rumsey and J. J. Taggart, *Bioorg. Med. Chem. Lett.*, 2011, **21**, 2255.

139. J. Xie, G. I. Poda, Y. Hu, N. X. Chen, R. F. Heier, S. G. Wolfson, M. T. Reding, P. J. Lennon, R. G. Kurumbail, S. R. Selness, X. Li,

N. N. Kishore, C. D. Sommers, L. Christine, S. L. Bonar, N. Venkatraman, S. Mathialagan, S. J. Brustkern and H.-C. Huang, *Bioorg. Med. Chem.*, 2011, **19**, 1242.

140. A. A. Bermanis, D. Bonafoux, M. Clare, J. Z. Crich, T. R. Fletcher, L. Geng, T. J. Hagen, B. C. Hamper, J. G. Hanson, S. C. Houdek, H. Huang, D. M. Iula, F. J. Koszyk, P. J. Lennon, S. Liao, S. Metz, M. Nguyen, D. S. Oburn, T. J. Owen, R. A. Partis, A. M. Scates, M. A. Stealey, M. B. Tollefson, M. L. Vazquez, R. M. Weier, S. Wolfson and X. Xu, Pharmacia Corporation, *PCT Int. Appl.*, WO 2003/024935, 2003.

141. S. Metz, M. Clare, J. Z. Crich, T. J. Hagen, G. J. Hanson, H. Huang, S. J. Houdek, M. A. Stealey, M. L. Vazquez, R. M. Weier and X. Xu, Pharmacia Corporation, *PCT Int. Appl.*, WO 2003/024936, 2003.

142. L. Geng, M. Clare, G. J. Hanson, H. Huang, D. Iula, S. Liao, M. A. Stealey, S. Metz, M. L. Vazquez and R. M. Weier, Pharmacia Corporation, *PCT Int. Appl.*, WO 2003/027075, 2003.

143. X. Xu, M. Clare, P. Lennon, S. Metz, M. Vazquez, R. M. Weier and S. G. Wolfson, Pharmacia Corporation, *PCT Int. Appl.*, WO 2003/070706, 2003.

144. D. F. Bonfoux, S. L. Bonar, M. Clare, A. M. Donnelly, J. L. Glaenzer, J. A. Guzova, H. Huang, N. N. Kishore, F. J. Koszyk, P. J. Lennon, A. Libby, S. Mathialagan, D. S. Oburn, S. A. Rouw, C. D. Sommers, C. S. Tripp, L. J. Vanella, R. Weier, S. G. Wolfson and H.-C. Huang, *Bioorg. Med. Chem.*, 2010, **18**, 403.

145. G. Mbalaviele, C. C. Sommers, S. L. Bonar, S. Mathialagan, J. F. Schindler, J. A. Guzova, A. F. Shaffer, M. A. Melton, L. J. Christine, C. S. Tripp, P.-C. Chiang, D. C. Thompson, Y. Hu and N. Kishore, *J. Pharm. Exp. Ther.*, 2009, **329**, 14.

146. S. Rajendrasozhan, J.-W. Hwang, H. Yao, N. Kishore and I. Rahman, *Pulm. Pharmacol. Ther.*, 2010, **23**, 172.

147. P.-C. Chiang, N. N. Kishore and D. C. Thompson, *J. Pharm. Sci.*, 2010, **3**, 1278.

148. C. D. Sommers, J. M. Thompson, J. A. Guzova, S. L. Bonar, R. K. Rader, S. Mathialagan, N. Venkatraman, V. W. Holway, L. E. Kahn, G. Hu, D. S. Garner, H.-C. Huang, P.-C. Chiang, J. F. Schneider, Y. Hu, D. M. Meyer and N. N. Kishore, *J. Pharm. Exp. Ther.*, 2009, **330**, 377.

149. J. Adams, Millennium Pharmaceuticals, Inc., *PCT Int. Appl.*, WO 2003/039545, 2003.

150. A. C. Castro, L. C. Dang, F. Soucy, L. Grenier, H. Mazdiyasni, M. Hottlelet, L. Parent, C. Pien, V. Palombella and J. Adams, *Bioorg. Med. Chem. Lett.*, 2003, **13**, 2419.

151. J. Bain, L. Plater, M. Elliott, N. Shpiro, C. J. Hastie, H. McLauchlan, I. Klevernic, S. C. Arthur, D. R. Alessi and P. Cohen, *Biochem. J.*, 2007, **408**, 297.

152. D. Wen, Y. Nong, J. G. Morgan, P. Gangurde, A. Bielecki, J. DaSilva, M. Keaveney, H. Cheng, C. Fraser, L. Schopf, M. Hepperle, G. Harriman, B. D. Jaffee, T. D. Ocain and Yajun Xu, *J. Pharm. Exp. Ther.*, 2006, **317**, 989.

153. L. Schopf, A. Savinainen, K. Anderson, J. Kujawa, M. DuPont, M. Silva, E. Siebert, S. Chandra, J. Morgan, P. Gangurde, D. Wen, J. Lane, Y. Xu, M. Hepperle, G. Harriman, T. Ocain and B. Jaffee, *Arthritis Rheum.*, 2006, **54**, 3163.

154. J. R. Burke, M. A. Pattoli, K. R. Gregor, P. J. Brassil, J. F. MacMaster, K. W. McIntyre, X. Yang, V. S. Iotzova, W. Clarke, J. Strnad, Y. Qiu and F. C. Zusi, *J. Biol. Chem.*, 2003, **278**, 1450.

155. K. W. McIntyre, D. J. Shuster, K. M. Gillooly, D. M. Dambach, M. A. Pattoli, P. Lu, X.-D. Zhou, Y. Qiu, F. C. Zusi and J. R. Burke, *Arthritis Rheum.*, 2003, **48**, 2652.

156. J. F. MacMaster, D. M. Dambach, D. B. Lee, K. K. Berry, Y. Qiu, F. C. Zusi and J. R. Burke, *Inflamm. Res.*, 2003, **52**, 508.

157. R. M. Townsend, J. Postelnek and V. Susulic, *Transplantation*, 2004, **77**, 1090.

158. N. Jinawath, C. Vasoontara, A. Jinawath, X. Fang, K. Zhao, K.-L. Yap, T. Guo, C. S. Lee, W. Wang, B. M. Balgley, B. Davidson, T.-L. Wang and I.-M. Shih, *PLOS One*, 2010, **5**, 11198.

159. H. Blazkova, C. von Schubert, K. Mikule, R. Schwab, N. Angliker, J. Schmuckli-Maurer, P. C. Fernandez, S. Doxsey and D. A. Dobbelaere, *Cell Cycle*, 2007, **6**, 2531.

160. T. Wada, S. Ohshima, E. Fujisawa, D. Koya, H. Tsuneki and T. Sasaoka, *Endocrinology*, 2009, **150**, 1662.

161. M. K. Hancock, C. S. Lebakken, J. Wang and Kun Bi, *Mol. BioSyst.*, 2010, **6**, 1834.

162. F. Beaulieu, C. Ouellet, E. H. Ruediger, M. Belema, Y. Qiu, X. Yang, J. Banville, J. R. Burke, K. R. Gregor, J. F. MacMaster, A. Martel, K. W. McIntyre, M. A. Pattoli, F. C. Zusib and D. Vyas, *Bioorg. Med. Chem. Lett.*, 2007, **17**, 1233.

163. J. Kempson, S. H. Spergel, J. Guo, C. Quesnelle, P. Gill, D. Belanger, A. J. Dyckman, T. Li, S. H. Watterson, C. M. Langevine, J. Das, R. V. Moquin, J. A. Furch, A. Marinier, M. Dodier, A. Martel, D. Nirschl, K. Van Kirk, J. R. Burke, M. A. Pattoli, K. Gillooly, K. W. McIntyre, L. Chen, Z. Yang, P. H. Marathe, D. Wang-Iverson, J. H. Dodd, M. McKinnon, J. C. Barrish and W. J. Pitts, *J. Med. Chem.*, 2009, **52**, 1994.

164. M. Belema, A. Bunker, V. N. Nguyen, F. Beaulieu, C. Ouellet, Y. Qiu, Y. Zhang, A. Martel, J. R. Burke, K. W. McIntyre, M. A. Pattoli, C. Daloisio, K. M. Gillooly, W. J. Clarke, P. J. Brassil, F. C. Zusia and D. M. Vyas, *Bioorg. Med. Chem. Lett.*, 2007, **17**, 4284.

165. A. J. Dyckman, C. M. Langevine, C. Quesnelle, J. Kempson, J. Guo, P. Gill, S. H. Spergel, S. H. Watterson, T. Li, D. S. Nirschl, K. M. Gillooly, M. A. Pattoli, K. W. McIntyre, L. Chen, M. McKinnon, J. H. Dodd, J. C. Barrish, J. R. Burke and W. J. Pitts, *Bioorg. Med. Chem. Lett.*, 2011, **21**, 383.

166. S. H. Watterson, C. M. Langevine, K. Van Kirk, J. Kempson, J. Guo, S. H. Spergel, J. Das, R. V. Moquin, A. J. Dyckman, D. Nirschl, K. Gregor, M. A. Pattoli, X. Yang, K. W. McIntyre, G. Yang, M. A. Galella,

H. Booth-Lute, L. Chen, Z. Yang, D. Wang-Iverson, M. McKinnon, J. H. Dodd, J. C. Barrish, J. R. Burke and W. J. Pitts, *Bioorg. Med. Chem. Lett.*, 2011, **21**, 7006.

167. K. M. Gillooly, M. A. Pattoli, T. L. Taylor, L. Chen, L. Cheng, K. R. Gregor, G. S. Whitney, V. Susulic, S. H. Watterson, J. Kempson, W. J. Pitts, H. Booth-Lute, G. Yang, P. Davies, D. W. Kukral, J. Strnad, K. W. McIntyre, C. J. Darienzo, L. Salter-Cid, Z. Yang, D. B. Wang-Iverson and James R. Burke, *J. Pharm. Exp. Ther.*, 2009, **331**, 349.

168. J. Kempson, "Optimization of a series of tricyclic inhibitors of IκB kinase, efficacy in chronic animal models of arthritis" oral presentation CHI Discovery on Target conference 11/03/2011 Boston Massachusetts.

169. H. Shimizu, S. Tanaka, T. Toki, I. Yasumatsu, T. Akimoto, K. Morishita, T. Yamasaki, T. Yasukochi and S. Iimura, *Bioorg. Med. Chem. Lett.*, 2010, **20**, 5113.

170. H. Shimizu, I. Yasumatsu, T. Hamada, Y. Yoneda, T. Yamasaki, S. Tanaka, T. Toki, M. Yokoyama, K. Morishita and S. Iimura, *Bioorg. Med. Chem. Lett.*, 2011, **21**, 904.

171. Information taken from Citeline TrialTrove commercial database (http://www.citeline.com) using either general mechanism of action search term "I kappa kinase inhibitor" or specific drugs tested search term "MLN-0415" [site accessed January 2012].

172. F. A. Hicks, I. M. Cooper, M. Langston and A. St. Clair Brown, Millennium Pharmaceuticals, Inc., *PCT Int. Appl.*, WO 2009/054970, 2009.

173. Information taken from Citeline TrialTrove commercial database (http://www.citeline.com) using either general mechanism of action search term "I kappa kinase inhibitor" or specific drugs tested search term "SAR-113945" [site accessed January 2012].

174. B. C. Langevin, D. Sherer, M. E. Buttrum and S. Rose, Aventis Pharmaceuticals, Inc., *PCT Int. Appl.*, WO 2006/076318, 2006.

175. Information taken from Citeline TrialTrove commercial database (http://www.citeline.com) using either general mechanism of action search term "I kappa kinase inhibitor" or specific drugs tested search term "SPC-839" [site accessed January 2012].

176. S. S. Bhagwat, B. L. Bennett, Y. Satoh, E. C. O'Leary, J. Leisten, G. S. Firestein, D. L. Boyle, M. Dreano, D. W. Anderson and C. E. Grimshaw, *Arthritis Rheum.*, 2001, **44**, S213.

177. M. Palanki, Signal Pharmaceuticals, Inc., *PCT Int. Appl.*, WO 1999/01441, 1999.

178. M. S. S. Palanki, P. E. Erdman, M. Ren, M. Suto, B. L. Bennett, A. Manning, L. Ranson, C. Spooner, S. Desai, A. Ow, R. Totsuka, P. Tsaob and W. Toriumib, *Bioorg. Med. Chem. Lett.*, 2003, **13**, 4077.

179. J.-I. Suzuki, M. Ogawa, S. Muto, A. Itai, M. Isobe, Y. Hirata and R. Nagai, *Expert Opin. Investig. Drugs*, 2011, **3**, 395.

180. J. Strnad and J. R. Burke, *Trends Pharmacol. Sci.*, 2007, **28**, 142.

181. M. R. Edwards, N. W. Bartlett, D. Clarke, M. Birrell, M. Belvisi and S. L. Johnston, *Pharmacol. Ther.*, 2009, **121**, 1.

182. Information taken from Citeline TrialTrove commercial database (http://www.citeline.com) using either general mechanism of action search term "I kappa kinase inhibitor" or specific drugs tested search terms "IMD-0354, IMD-0560, IMD-1041, and IMD-2560" [site accessed January 2012].

183. A. Sugita, H. Ogawa, M. Azuma, S. Muto, A. Honjo, H. Yanagawa, Y. Nishioka, K. Tani, A. Itai and S. Sone, *Int. Arch. Allergy Immunol.*, 2009, **148**, 186.

184. H. Ogawa, M. Azuma, S. Muto, Y. Nishioka, A. Honjo, T. Tezuka, H. Uehara, K. Izumi, A. Itai and S. Sone, *Clin. Exp. Allergy*, 2010, **41**, 104.

185. Y. Onaia, J.-I. Suzukia, T. Kakutaa, Y. Maejimaa, G. Haraguchia, H. Fukasawab, S. Mutob, A. Itaib and M. Isobea, *Cardiovascular Res.*, 2004, **63**, 51.

186. Y. Hayakawa, S. Maeda, H. Nakagawa, Y. Hikiba, W. Shibata, K. Sakamoto, A. Yanai, Y. Hirata, K. Ogura, S. Muto, A. Itai and M. Omata, *J. Gastroenterol.*, 2009, **44**, 935.

187. M. Susumu and I. Akiko, Institute of Medicinal Molecular Design, Inc., *PCT Int. Appl.*, WO 2003/103647, 2003.

188. M. Susumu and I. Akiko, Institute of Medicinal Molecular Design, Inc., *PCT Int. Appl.*, WO 2003/103654, 2003.

189. J. Welsh, N. Ai, S. Evans and R. D. Wood, *PCT Int. Appl.*, WO 2010/101648, 2010.

190. G. Hall, I. S. Singh, L. Hester, J. D. Hasday and Terry B. Rogers, *Am. J. Physiol. Heart Circ. Physiol.*, 2005, **289**, 2103.

191. Information taken from Citeline TrialTrove commercial database (http://www.citeline.com) using either general mechanism of action search term "I kappa kinase inhibitor" or specific drugs tested search terms "CHS-828 and EB-1627" [site accessed January 2012].

192. L. S. Olsen, P.-J. V. Hjarnaa, S. Latini, P. K. Holm, R. Larsson, E. Bramm, L. Binderup and M. W. Madsen, *Int. J. Cancer*, 2004, **111**, 198.

193. U. H. Olesen, M. K. Christensen, F. Bjorkling, M. Jaattela, P. B. Jensen, M. Sehested and S. J. Nielsen, *Biochem. Biophys. Res. Comm.*, 2008, **367**, 799.

194. H. Lövborg, R. Burman and J. Gullbo, *BMC Res. Notes*, 2009, **2**, 114.

CHAPTER 10

Bruton's Tyrosine Kinase (Btk)

MARK E. SCHNUTE,*[a] ADRIAN HUANG[b] AND
EDDINE SAIAH[a]

[a] Medicinal Chemistry, Pfizer Inc., 200 Cambridge Park Drive, Cambridge, MA 02140, USA; [b] Department of Chemistry, Wellesley College, 106 Central Street, Wellesley, MA 02481, USA
*Email: mark.e.schnute@pfizer.com

10.1 Introduction

Bruton's tyrosine kinase (Btk) is a non-receptor tyrosine kinase belonging to the Tec family of kinases, which constitutes the second largest family of cytoplasmic protein tyrosine kinases.[1] The Tec kinase family consists of five members: Btk, Interleukin-2 tyrosine kinase (Itk), Tec, Rlk and Bmx. These kinases (with the exception of Bmx) are highly expressed in haematopoietic cells. Each Tec family kinase shows a distinct pattern of cellular and tissue distribution. Btk is localized in bone marrow, spleen and lymph node tissue and is expressed primarily in B lymphocytes, myeloid and mast cells. Btk is specifically required for B-cell activation following engagement of the B-cell antigen receptor (BCR).[2] Mutations in the gene encoding Btk cause a primary immunodeficiency disease in humans known as X-linked agammaglobulinemia (XLA)[3,4] and X-linked immunodeficiency (*Xid*) in mice.

Btk is critical for B-cell development, differentiation and signalling and is activated by the upstream Src-family kinases. Btk has several known sites of phosphorylation including Tyr551 within the activation loop and Tyr223 within the SH3 domain. Once activated, Btk phosphorylates and activates phospholipase-Cγ (PLCγ), leading to calcium mobilization and activation of NF-κB and MAP kinase pathways.[5]

RSC Drug Discovery Series No. 26
Anti-Inflammatory Drug Discovery
Edited by Jeremy I Levin and Stefan Laufer
© The Royal Society of Chemistry 2012
Published by the Royal Society of Chemistry, www.rsc.org

Btk is present in specific cells of the myeloid lineage and contributes to the activation of the FcγR and FcεR signalling pathways in macrophages, neutrophils and mast cells. *Xid* mice have reduced FcεR dependent mast cell degranulation and impaired functioning of macrophages including TNFα production. The *xid* phenotype has also been shown to confer resistance to disease manifestations in pre-clinical arthritis models. Clinical studies using the anti-CD20 antibody rituximab to deplete mature B-cells support the role of B-cells in a number of inflammatory diseases such as rheumatoid arthritis,[6] systemic lupus erythematosus[7] and multiple sclerosis.[8]

Due to its key role as a mediator in multiple signal transduction pathways, Btk is considered a promising target for therapeutic intervention for several diseases including inflammatory diseases and cancer. Numerous research groups are actively investigating the role and function of Btk as well as identifying potent and selective Btk inhibitors as potential tool compounds and drug candidates. For the purpose of this chapter, we have divided the small molecule inhibitors section in two groups: non-covalent, reversible strategies and irreversible, covalent inhibitor strategies. Both approaches have benefited from the availability of structural biology information around this target in particular, and the Tec family in general. The first crystal structure of the Btk kinase domain was reported for the unphosphorylated form.[9] This structure captures Btk in an inactive conformational state where the C-helix is in an "out" conformation. Solution NMR and X-ray crystallographic structure information are also available for individual domains of the protein. Several inhibitor-bound X-ray co-crystal structures of Btk have been reported, including with dasatinib and the more selective Btk inhibitor PCI-29732. Interestingly, compound PCI-29732 binds to Btk in an inactive conformation in which the protein adopts a "C-helix out" conformation, which may offer opportunities for structure-based design of Btk selective inhibitors. Recent crystallography work indicates that Btk has the ability to adopt a variety of different inactive conformations upon binding to inhibitors,[10] including novel activation loop conformations, an atypical DFG conformation and a DFG-out conformation. Access to a DFG-out kinase inhibitor conformation for Btk may offer new opportunities to identify inhibitors with high kinome selectivity or compounds with slow off-rate kinetics.[11]

The first generation Btk inhibitors such as dasatinib[12] and LFM-A13[13] lacked the desired selectivity and had limited value as tool compounds to better understand and decipher the Btk pathway. Achieving selectivity for Btk *versus* the other 500 members of the kinome is challenging, especially for type I inhibitors. A shift towards inhibitors targeting inactive kinase conformations or covalent active site inhibitors is starting to address these challenges. Small-molecule Btk inhibitors with a new binding mode that stabilizes an inactive conformation have been reported and these compounds exhibit a remarkable profile in terms of potency and selectivity. The *in-vitro* and *in-vivo* profiles of these compounds will be discussed in the reversible inhibition strategy section.

Btk is well positioned for a covalent inhibition strategy. Btk contains a partially exposed Cys481, which is located in the active site near the ATP binding pocket. Only 11 kinases are known to have a cysteine at this position.

Thus, covalent targeting of Cys481 in Btk represents an attractive strategy to obtain general kinome selectivity. Several reports have disclosed the identification of potent and selective covalent inhibitors of Btk and these will be discussed in the covalent inhibitors section.

10.2 Btk Biochemistry

Btk is a multi-domain protein consisting of 659 amino acid residues (Figure 10.1). The protein is composed of an N-terminal pleckstrin homology (PH) domain, a Tec homology (TH) domain consisting of the Btk homology region (BH) and the polyproline (PPR) region, two src-homology domains (SH2 and SH3) and the C-terminal kinase domain (TK). The main function of the PH and TH domains is to recognize protein substrate and to provide a phosphoinositide binding site, which mediates the movement of Btk towards the cell membrane.[14,15] The SH3 region interacts with the PPR region in an intermolecular fashion to form a homodimer, which is believed to be the Btk resting state.[16] There are two tyrosine residues in Btk that undergo regulatory phosphorylation: Tyr551, which is located within the activation loop of the kinase domain, and Tyr223, which is located in the SH3 domain.[17] Phosphorylation of Btk at Tyr551 leads to disruption of the Tyr551/Arg544/Glu445 association cluster allowing Glu445 to form a salt bridge with Lys430 affording the catalytically active state.[18] Kinetic evaluation of Btk has indicated that Btk displays a ternary complex mechanism with either substrate or ATP being the first binding partner. However, there was a weak negative cooperativity between the ATP and substrate binding sites.[19] The ATP K_m value for Btk has been reported in the range of $84 \pm 20\,\mu M$ and $232 \pm 28\,\mu M$.[19,20] A very high degree of sequence homology exists for Btk between human, rat and mouse. Murine and human Btk share 99% sequence homology while only seven amino acid residues differ between murine and rat Btk, corresponding to 99% identity.[21,22]

10.3 Biological Function of Btk

10.3.1 Role in B-cell Development and Function

Btk is the only member of the Tec kinase family that is genetically associated with disease in humans. X-linked agammaglobulinemia (XLA) is a hereditary

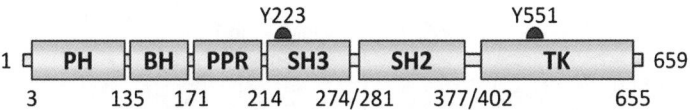

Figure 10.1 Representation of the Btk protein domains showing approximate sequence alignments and primary phosphorylation residues (Y223 and Y551). Domains shown: pleckstrin homology (PH), Btk homology region (BH), polyproline region (PPR), Src-homology 3 (SH3), Src-homology 2 (SH2) and tyrosine kinase (TK).

immunodeficiency disease that results from mutations in the gene encoding Btk. XLA patients have a profound decrease in the number of mature B-cells and very low immunoglobulin levels.[4,23] As a result, XLA patients are highly susceptible to bacterial infections and are usually not able to develop long-term anti-viral immunity. Immunoglobulin replacement therapy is required to prevent infections and allows patients to lead generally normal lives. The frequency of XLA has been estimated at 1 : 200,000 births and mutations are highly diverse within the protein, with more than 270 disease-causing missense and nonsense mutations having been documented.[24]

The normal development of B lymphocytes from stem cells in the bone marrow to circulating plasma cells or memory cells passes through several distinct phases (Figure 10.2).[25] In XLA patients, early B-cell development up to the pre-B-cell stage in the bone marrow is normal, at which point XLA patients generally have a ratio of pro-B to pre-B-cell numbers ten times higher than healthy individuals. In addition, XLA patients have almost no circulating plasma cells, representing the further developmental impairment or increased cell death at the transition from the pre-B to immature B-cell.

A similar but not identical phenotype to XLA has been observed in the CBA/ NM mouse strain, which carries a single point mutation in the Btk gene, changing residue 28 in the pleckstrin homology (PH) domain from an arginine to a cysteine (R28C). This mutation is referred to as the X-linked immunodeficiency (*xid*) mouse phenotype.[25] The *xid* phenotype in mouse is less severe than that observed for XLA in humans. The pre-B to immature B-cell transition is not completely blocked in the *xid* mouse leading to a reduction of peripheral B lymphocyte populations by only approximately 50%. The *xid* mouse does show greatly reduced IgG3 and IgM levels and is not able to produce antibodies to T-cell independent type 2 antigens, suggesting a defect in B-cell receptor (BCR) stimulation. The difference in phenotype between mouse *xid* and human XLA has been attributed to the ability of Tec kinase to compensate for Btk in the *xid* mouse during the B-cell developmental phase in the bone marrow. This proposal is supported by the observation that Btk$^{-/-}$Tec$^{-/-}$ mice show a severe defect in B-cell development similar to XLA in humans

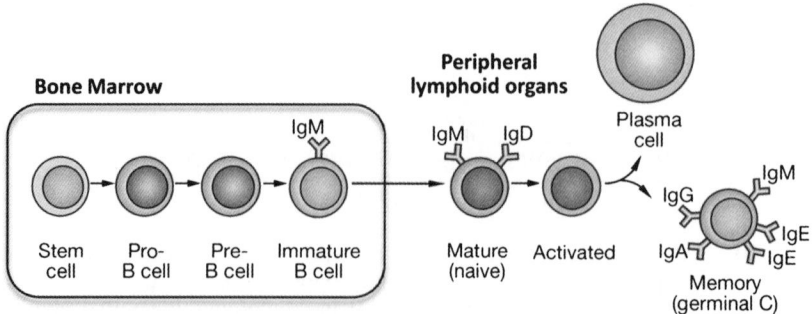

Figure 10.2   Stages of B lymphocyte development.

while Tec$^{-/-}$ mice alone exhibit phenotypically normal immune systems.[26] Based on observations from both the human XLA and mouse *xid* phenotypes, Btk plays an important role at several stages of B-cell development including (1) maturation and survival of immature B-cells in the bone marrow and (2) development of peripheral B-cell differentiation.

Signalling through the B-cell antigen receptor (BCR) is central to B-cell development, differentiation and response to antigen. Studies of the BCR signalling pathway have shown that Btk plays an important role, Figure 10.3.[27,28] BCR signalling is initiated upon binding of an antigen to the antigen-binding component membrane immunoglobulin (mIg), which leads to phosphorylation of tyrosine residues within the Ig-α and Ig-β by src family kinases such as Lyn, Blk and Fyn. These phosphorylation events create docking sites for two important kinases in the signalling process, phosphatidylinositol-3-kinase (PI3K) and spleen tyrosine kinase (Syk). Phosphatidylinositol-4,5-biphosphate is then phosphorylated by PI3K to form phosphatidylinositol-3,4,5-triphosphate (PIP$_3$). Production of PIP3 on the cytoplasmic plasma membrane is essential for recruitment of Btk through interactions in the pleckstrin homology (PH) domain. The resulting complex activates Btk towards phosphorylation at Tyr551. The process by which phosphorylation occurs remains unclear with evidence that PIP3 binding induces autophosphorylation[29] as well as evidence suggesting phosphorylation is mediated by Lyn and/or Syk kinases.[15,17] Once

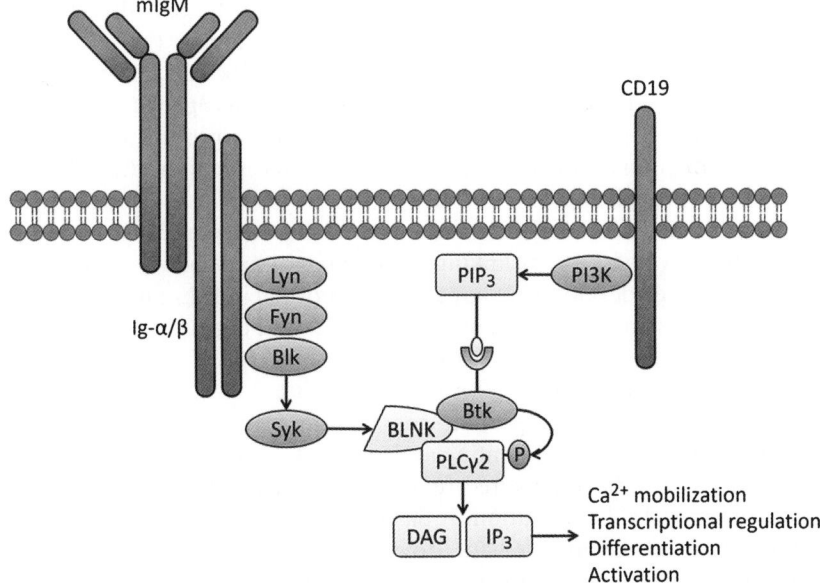

Figure 10.3   A simplified representation of the B-cell antigen receptor (BCR) signalling cascade. Abbreviations: PIP$_3$, phosphoinositol(3,4,5) triphosphate; BLNK, B-cell linker protein; PLCγ2, phospholipase Cγ2; DAG, diacylglycerol; IP$_3$, inositol triphosphate.

phosphorylated, Btk undergoes autophosphorylation at Tyr223 in the SH3 domain. Tyr223 phosphorylation may induce conformational changes that disfavour Btk homodimer formation allowing for interactions with other adapter molecules. BLNK, after phosphorylation by Syk, functions as a scaffold providing bindings sites for the Btk SH2 domain as well as the PLCγ2 SH2 domain bringing them into close proximity. Subsequent phosphorylation of PLCγ2 by Btk triggers a cascade of signalling events resulting in calcium mobilization and transcriptional activation.

10.3.2 Role in Myeloid and Other Cells

Btk is expressed in a range of haematopoietic cells in addition to B lymphocytes, including mast cells (basophils), macrophages, neutrophils and dendritic cells. All of these cell types play an important role in innate immunity and the response to inflammatory stimulus. As opposed to B-cells, the molecular role of Btk in these other cells is less well understood; however, evidence suggests Btk likewise functions in corresponding signalling cascades driving immune responses.[30] In general Btk appears to enable signalling primarily through two types of cell surface receptors including (1) the Toll-like receptors (TLR-3, -4, -6, -8 and -9) and (2) the high affinity Ig Fc receptors (FcεR and FcγR). The Toll-like receptors are transmembrane signal receptor proteins that play a role in innate immunity and especially recognition of bacterial pathogens. TLRs are highly expressed on mast cells, macrophages and dendritic cells. Recently, Liu has demonstrated that major histocompatibility complex (MHC) class II functions as an adapter protein for a complex with Btk and CD40 to enable TLR responses.[31] The Fc receptor is expressed in macrophages, neutrophils and mast cells. Fcγ receptors bind IgG antibodies triggering inflammatory cytokine production and phagocytic activity. Fcε receptors on the other hand are activated through IgE.

Mast cells are critical for the initiation of allergic responses. They are predominantly activated by crosslinking of the high-affinity IgE receptor FcεRI. Although mast cell development appears to be normal in Btk-null and *xid* mice, they demonstrate impaired function including delayed degranulation (release of histamine and leukotrienes) and release of cytokines (IL2, TNFα and IL6).[30,32] This suggests Btk is activated upon FcεRI ligation.

Macrophages play an important role in innate and adaptive immunity through phagocytosis and presenting antigens to T-cells. Macrophages isolated from XLA patients have impaired function and chemotaxis as well as TNFα production upon LPS stimulation. It has been suggested that Btk has an anti-apoptotic role in the macrophage downstream of inflammatory triggers.[33] Studies by Vuolteenaho have suggested that Btk is critical in TLR2 induced activation of signalling and inflammatory cytokine production in response to LTA stimulation.[34] Recently, Reif and Currie have reported that TNFα production in macrophages stimulated with immobilized immune complexes could be inhibited by the selective Btk inhibitor CGI1746 in a dose-dependent manner.[35] Subsequent studies suggested that Btk is activated downstream of FcγRIII and that Btk inhibition did not significantly reduce TLR-2 or TLR-4 mediated TNFα or IL-6 production. Btk may also regulate the survival of macrophages by enabling signalling through the M-CSF receptor.[36]

XLA patients are occasionally found to be neutropenic though this typically coincides with severe infection and is not characteristic of patients on adequate Ig replacement therapy. Based on the *xid* mouse, Btk-deficient neutrophils are impaired in maturation and function. Neutrophil migration into tissues, edema formation and haemorrhage are significantly reduced in *xid* mice after acute inflammatory response. Studies by Brunner looking at neutrophil development in *xid* mice have suggested that Btk is essential in GM-CSF and Toll-like receptor signalling.[37] In contrast, studies by Cunningham-Rurdles on XLA neutrophils have suggested that TLR signalling is Btk independent.[38] Dendritic cells from XLA patients have similar response to differentiation, maturation and antigen presentation compared to healthy counterparts. Sochorora has studied TLR mediated response of dendritic cells from XLA patients and found that TNFα and IL-6 responses to ssRNA that activate TLR-8 were markedly reduced while similar experiments with TLR2-5 were normal.[39]

Though the precise role Btk plays in haematopoietic cells other than B lymphocytes is still in the early stages of investigation, Btk appears to be involved in a wide range of anti-inflammatory processes and innate immunity driven events within these cell types.

10.4 Btk as an Anti-inflammatory Target

B-cells have a critical regulatory function of immune responses and are key drivers in the pathology of autoimmune and inflammatory diseases. Rheumatoid arthritis (RA) and systemic lupus erythematosus (SLE) are two representative autoimmune diseases where B-cell therapy has shown significant promise.[40–42] In RA, B-cells present antigens to T-cells resulting in their activation and subsequent production of inflammatory cytokines promoting disease. B-cells also produce autoreactive antibodies known as rheumatoid factors (RFs), which can initiate a cascade leading to local inflammation and tissue destruction. The benefit of B-cell targeted therapy in RA has been validated by clinical experience in B-cell depleting agents. Rituximab (RTX) is a chimeric monoclonal antibody that specifically depletes CD20⁺ B-cells resulting in a rapid and sustained peripheral depletion of B-cells.[40] RTX has been approved in the USA and Europe for the treatment of patients with RA who have had inadequate response to anti-TNFα therapy. A Btk inhibitor, however, would act through a different mechanism from RTX to influence B-cell population and function. Nonetheless, several small-molecule Btk inhibitors have demonstrated efficacy in pre-clinical animal models of RA (Section 10.6).[35,43–45] Since Syk is an upstream kinase to Btk in the BCR signalling pathway, recent promising Phase II clinical results for the Syk inhibitor R788 showing disease reduction in patients with RA is further suggestive of the importance of the pathway in disease.[46,47] The role of Syk in T-cell function in addition to that of B-cells, however, complicates the interpretation of these results in the context of a Btk selective inhibitor.

SLE is a complex autoimmune disease affecting multiple organs including the potential for life-threatening kidney and cardiovascular involvement. The off-label use of RTX for the treatment of SLE has also suggested potential benefit, especially in severe refractory cases; however, controlled clinical trials have been less conclusive.[40] The Phase II/III EXPLORE trial and the Phase III LUNAR trials both failed to meet their primary and secondary end-points for efficacy in SLE patients. More promising results have been seen with the B-cell therapy Belimumab, which was approved in 2011 by the FDA for the treatment of SLE.[48] Belimumab is a human monoclonal antibody specific for B lymphocyte stimulator (BLYS). It blocks the binding of BLYS to its receptor on B-cells resulting in compromised survival and differentiation. The Btk inhibitor PCI-32765 has shown the ability to inhibit antibody production and the development of kidney disease in the mouse MRL-Fas(lpr) lupus model.[44] Studies in mice that develop symptoms similar to SLE have suggested that basophils and mast cells in addition to B-cells may be involved in SLE pathology, suggesting that the broader cellular function Btk plays in myeloid cells could be advantageous towards developing SLE therapies.[49] A Btk inhibitor may also be beneficial in several other auto-antibody associated disorders including vasculitis, myasthenia gravis, systemic sclerosis, immune thrombocytopenia and Sjögren's syndrome.[40,41] There are also suggestions that dysregulation of B-cell trafficking across the blood-brain barrier is a contributing factor in multiple sclerosis.[50] *Xid* mice have shown reduced clinical disease in models of experimental autoimmune encephalomyelitis (EAE) compared to WT mice.[51] A controlled clinical trial in patients with relapsing remitting multiple sclerosis has shown that RTX therapy significantly reduced the number of new lesions and the number of patients with relapses.[52–54] B-cells are also believed to be contributors to islet inflammation in type 1 diabetes and may mediate autoimmune components of its pathology.[55] The development of B-cell directed therapeutics for the treatment of autoimmune and inflammatory disease remains in its infancy with only a handful of biological agents just recently reaching the market. The promising initial results, however, suggest that the unique mechanistic pathways impacted by Btk offer significant potential for therapeutic impact with an enzyme inhibitor.

10.5 Structural Biology

A high-resolution structure of the full-length Btk protein demonstrating the interconnectivity of the regulation domains is unfortunately not currently available. However, solution NMR and X-ray crystallographic structure information is available for individual domains of the protein (Table 10.1). The structure of the N-terminal pleckstrin homology (PH) and Btk homology (BH) domains has been determined by X-ray crystallography (Figure 10.4).[56–58]

Table 10.1 Representative solution and X-ray structural information for Btk.

Domain	Sequence	PDB	Description	Reference
PH, BH	3–170	1B55, 1BTK	X-ray	56, 57
SH3	212–275	1AWW, 1AWX	NMR	59
SH3	216–273	1QLY	NMR	60
SH2	270–387	2GE9	NMR	61
TK	397–659	1K2P	X-ray, mBtk	9
TK	382–659	3GEN	X-ray, PCI-29732	62
TK	382–659	3K54	X-ray, dasatinib	62
TK	393–659	3P08	X-ray, apo	35
TK	393–656	3OCS	X-ray, CGI1746	35
TK	387–659	3PJ3	X-ray, DFG out	10

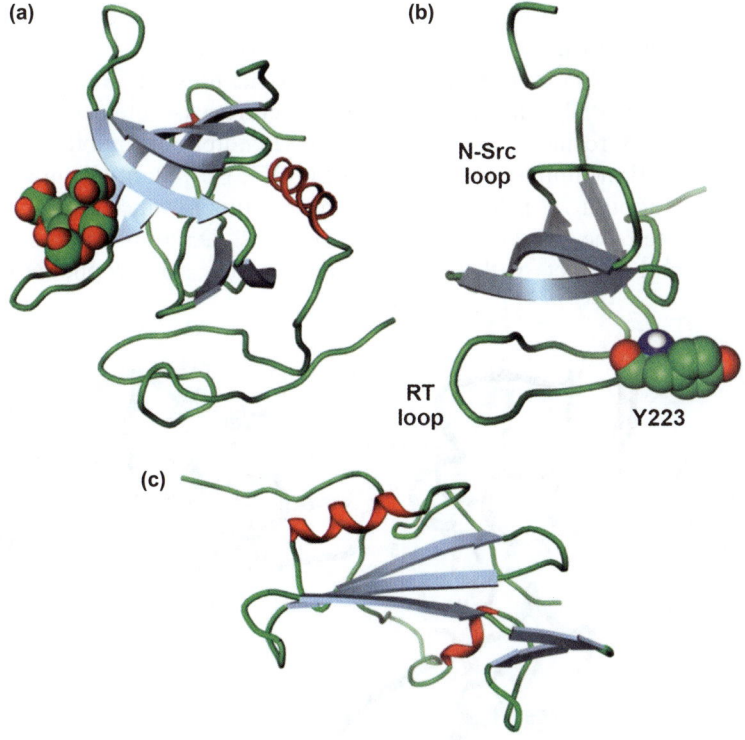

Figure 10.4 Structural representations of the Btk regulatory domains: (a) X-ray structure of PH and BH domains complexed with inositol-1,3,4,5-tetra-kisphosphate based on coordinates from 1B55 (Ref. 56), (b) NMR solution structure of SH3 domain based on coordinates from 1AWX (Ref. 59) and (c) NMR solution structure of SH2 domain based on coordinates from 2GE9 (Ref. 61).

The PH domain, involved in translocation of Btk to the plasma membrane through binding to PIP3, is occupied by inositol-1,3,4,5-tetrakisphosphate (IP4) in the reported structure, with the binding site formed by two β-loop regions. The Btk motif also forms a binding site for a zinc ion believed to be essential for structural stability. The SH3 and SH2 domains are small protein regions that mediate protein-protein interactions. In the case of Btk, BLNK is a likely partner for association with this region. The solution structures of both the Btk SH3 and SH2 domains have been solved by NMR spectroscopy (Figure 10.4).[59–61] The SH3 domain forms a well-defined structure of two short anti-parallel β sheets packed perpendicular to each other allowing for the side chain of Tyr223, which undergoes autophosphorylation during Btk activation, to be exposed.

The unique activation mechanism of Btk was first elucidated by Mao and co-workers with the aid of the X-ray crystal structure of unphosphorylated apo-murine Btk kinase domain (Figure 10.5).[9] Importantly, this structure captures Btk in an inactive conformational state where the C-helix is in an "out" conformation and the conserved salt bridge between Glu445 and Lys430 is absent. Instead, Glu445 forms a salt bridge with Arg544, which subsequently engages Tyr551 through a hydrogen bond with the hydroxyl group.

Several inhibitor-bound X-ray co-crystal structures of Btk have been reported (Figure 10.6). An early co-crystal structure was obtained with dasa-tinib, a Bcr/Abl tyrosine kinase inhibitor with serendipitous inhibitory activity against Btk, and the human Btk kinase domain Y551E mutant (Figure 10.5).[62] The protein adopts a "C-helix in" active conformation in which a salt bridge is formed between the Glu445 and Lys430 residues. Dasatinib makes several hydrogen bond contacts with the hinge and occupies a hydrophobic pocket

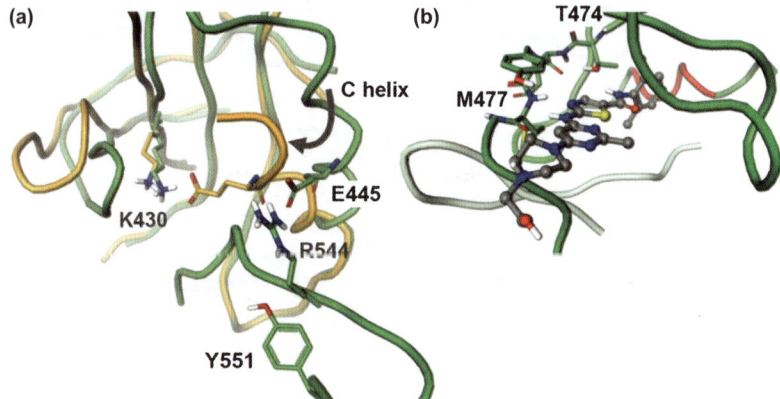

Figure 10.5   (a) Comparison of apo mBtk KD from coordinates 1K2P (Ref. 9) adopting inactive form (green) with Btk KD Y551E mutant from co-ordinates 3K54 demonstrating active conformation (orange). (b) X-ray structure of dasatinib co-crystallized with Btk KD Y551E mutant based on coordinates from 3K54 (Ref. 62).

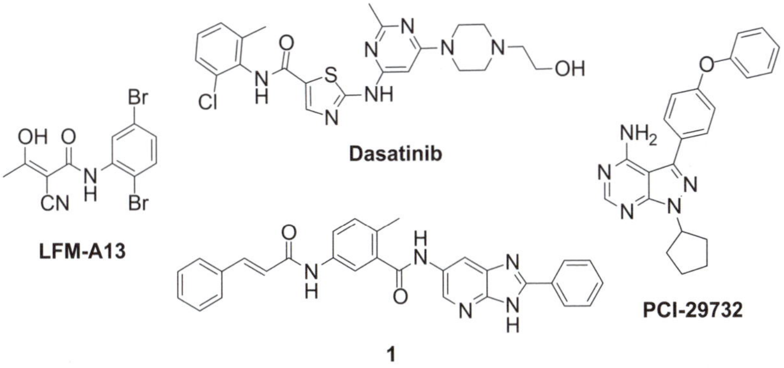

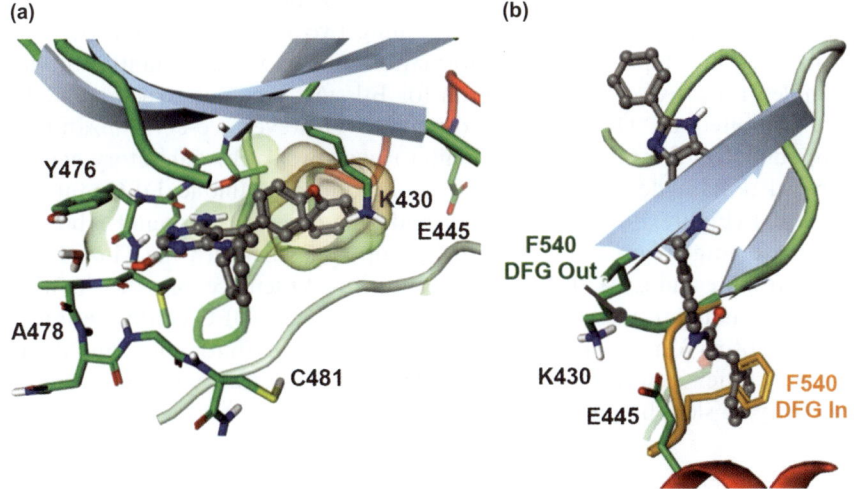

Figure 10.6 Btk inhibitors employed in co-crystallization studies and the leflonomide metabolite analogue LFM-A13.

Figure 10.7 (a) X-ray co-crystal structure of PCI-29732 with Btk KD based on coordinates from 3GEN (Ref. 62). (b) X-ray co-crystal structure of compound 1 with Btk KD based on coordinates from 3PJ3 (Ref. 10) demonstrating DFG out conformation (green). DFG in conformation described from coordinates 3K54 for Btk KD Y551E mutant (orange).

behind the Thr474 gatekeeper residue. Recently, structures of dasatinib co-crystallized with wild type Btk kinase domain have been reported (PDB ascension codes 3PIZ and 3PJ1).[35] These structures exhibit a very similar binding conformation to that described for the Y551E mutant.

Another X-ray co-crystal structure has been reported between Btk wild type kinase domain and the Btk inhibitor PCI-29732 (Figure 10.7).[60] The structure revealed that PCI-29732 binds to the ATP binding region with the

4-aminopyrazolopyrimidine moiety positioned in the site normally occupied by the adenine ring of ATP and consequently forming several interactions with the hinge region of Btk. The amino substituent of the ring forms a hydrogen bond contact with the Thr474 gatekeeper residue and the backbone carbonyl of Glu475. The pyrimidine N-3 atom accepts a hydrogen bond from the backbone amine of Met477 and the N-1 atom forms a water-mediated hydrogen bond network encompassing Tyr476 and Ala478. The cyclopentyl ring occupies the space where the ATP ribose would be expected. The phenoxyphenyl substituent enters a hydrophobic pocket behind the Thr474 gatekeeper residue. Compound PCI-29732 binds to Btk in an inactive conformation in which the protein adopts a "C-helix out" conformation and lacks the Glu445-Lys430 salt bridge. This conformation results in a large hydrophobic pocket behind the gatekeeper residue due to the shift in the C helix, which may offer opportunities for the structure-based design of Btk selective inhibitors.

   Work by Kuglstatter and co-workers has shown that Btk has the ability to adopt a variety of different inactive state conformations upon binding to inhibitors.[10] In addition to the previously known conformation, they identified four novel binding conformations, including two novel activation loop conformations, an atypical DFG conformation and, most importantly, the first example of a DFG-out conformation for Btk. All previous structures of Btk have exhibited a "DFG-in" conformation. However, co-crystallization of compound **1**, a Src family kinase inhibitor with weak Btk inhibitory activity, with Btk kinase domain was found to introduce a unique DFG out conformation (Figure 10.7). The azabenzimidazole core of compound **1** binds to the kinase hinge region while the tolyl group aligns between the Thr474 gatekeeper residue and Lys430. The styrene substituent then occupies the site where Phe540 normally resides, and consequently induces the DFG conformation. As seen with other kinases, access to a DFG out kinase inhibitor conformation for Btk may offer new opportunities to identify inhibitors with high kinome selectivity or slow binding kinetics.

10.6   Small-molecule Approaches to Btk Inhibition

Kinase-targeted drug discovery is a challenging field of research with two particularly confounding hurdles that must be confronted. The first of these is that a typical type I kinase inhibitor which binds at the kinase ATP pocket must compete with high intra-cellular ATP concentrations leading to overall poor biochemical efficiency[63] and high projected human dose. The second challenge is that the human kinome contains over 500 different kinases, many of which play critical roles in human physiology. As a result the necessary selectivity profile of any new kinase inhibitor is an important question. Both of these challenges are particularly relevant towards the development of kinase inhibitors for anti-inflammatory indications. In general, anti-inflammatory target indications require a chronic treatment regimen and, though debilitating, may not be immediately life threatening. As a result, the safety profile of any new agent is critical and therefore achieving a low dose, highly kinome-selective, agent is critical.

Early Btk inhibitor tool compounds lacked many of these attributes. The first reported inhibitor of Btk was LFM-A13, a leflonomide metabolite analogue (Figure 10.6).[13] LFM-A13 was a relatively weak inhibitor of Btk demonstrating an IC_{50} of 2.5 µM against recombinant Btk enzyme. Subsequently, it has been found also to inhibit JAK2 and Polo-like kinase (Plk).[64,65] Nonetheless, LFM-A13 has been extensively studied in oncology models and in inflammatory cell signalling.[66] More recently, by employing a chemical proteomics approach, Btk was found to be a major binding partner to the Bcr-Abl inhibitor dasatinib (Figure 10.6) currently approved for the treatment of chronic myelogenous leukaemia.[12] Dasatinib inhibited Btk with an IC_{50} value of 5 nM, demonstrating that highly potent Btk inhibitors were achievable. Unfortunately, dasatinib remains a fairly promiscuous kinase inhibitor showing significant inhibitory activity against over 60 other kinases. In order to address the above challenges of biochemical efficiency and kinome selectivity, drug discovery efforts to identify Btk inhibitors have moved in two directions: (1) ATP competitive inhibitors targeting inactive kinase conformations and (2) active site covalent inhibitors of the enzyme.

10.6.1 Non-covalent, Reversible Inhibitor Strategies

10.6.1.1 Discovery of CGI-1746 and GDC-0834

Workers at CGI Pharmaceuticals and Genentech have recently reported the discovery of CGI1746 (Figure 10.8), a small-molecule Btk inhibitor with a new

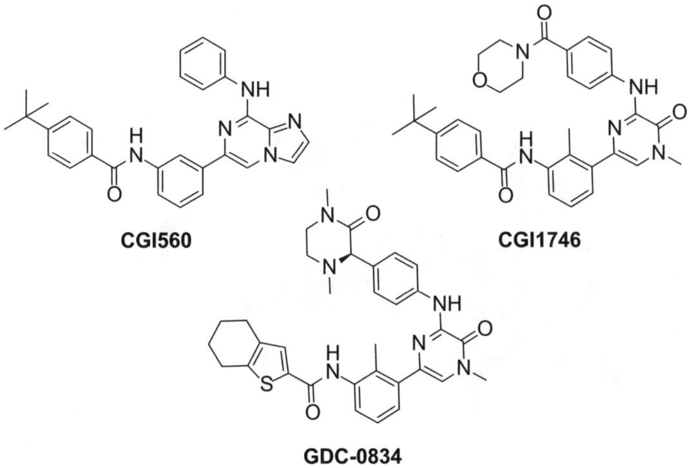

Figure 10.8   Representative Btk inhibitors described by CGI Pharmaceuticals and Genentech.

binding mode that stabilizes an inactive non-phosphorylated enzyme conformation.[35] CGI1746 was identified through optimization of the initial lead CGI560 (Btk $IC_{50} = 400$ nM). Initial SAR in this series suggested that the hydrogen bond donor/acceptor arrangement of the aminoimidazolopyrazine was an important structural element while the *para*-position of the aniline substituent was tolerant of substitution. Bioisosteric replacement of the imidazolopyrazine ring led to CGI1746 demonstrating an IC_{50} against Btk of 1.9 nM at the ATP K_m. The *tert*-butylphenylamide was found to be important for binding as the aniline alone was very weakly active ($IC_{50} = 5.4$ μM). Likewise, the acetate ($IC_{50} = 2.7$ μM) and benzylamine ($IC_{50} = 9.1$ μM) also showed poor activity suggesting the carbonyl and the substituted phenyl were necessary. Most importantly, CGI1746 showed exquisite kinome selectivity for Btk with 1000-fold selectivity over the closely related Tec and Src family kinases. Also, CGI1746 showed no significant binding to 385 non-mutant kinases when tested at 1 μM in an ATP free competition binding assay. A co-crystal structure of CGI1746 with human Btk kinase domain revealed that the inhibitor binds to an inactive conformation of Btk (Figure 10.9). In this enzyme conformation, Tyr551 is unphosphorylated as a result of shielding by CGI1746 and is instead engaged in a hydrogen bond with Asp521. CGI1746 interacts with the hinge region making two hydrogen bonds to Met477 as well as dipolar interactions with Glu475 and Thr474. A bifurcated hydrogen bond with Lys430 and a water molecule appears important to position the *tert*-butylphenyl group to occupy a large pocket that was designated the "H3" pocket formed from the Btk catalytic, phosphate binding and activation loops. Indeed CGI1746 was found to inhibit both the transphosphorylation step of Btk activation at Tyr551 as well as the autophosphorylation of Tyr223. Surface plasmon resonance demonstrated that CGI1746 binds to the inactive form of Btk with 32-fold greater affinity than the activated form ($K_d = 2.9$ nM and 94.1 nM, respectively). This difference was found to be driven by a difference in the off-rate

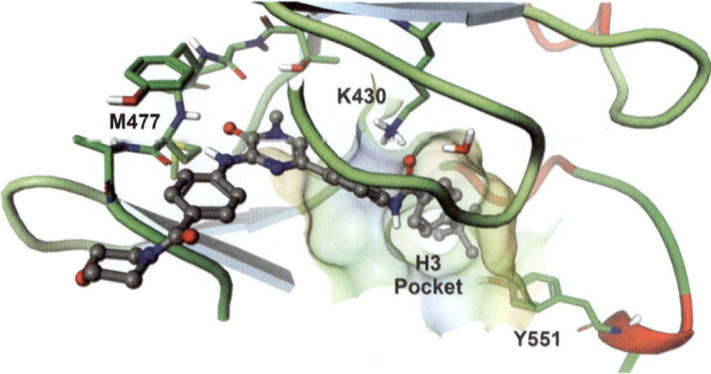

Figure 10.9 Structural representation for X-ray co-crystal of CGI1746 with Btk KD based on coordinates from 3OCS (Ref. 35).

kinetics for the compound between the active and inactive forms (dissociation $t_{1/2} = 19$ s and 888 s, respectively).

CGI1746 was found to be efficacious in the mouse collagen induced arthritis (mCIA) model in both a prophylactic and a therapeutic (treatment started three days after second collagen immunization) study design when dosed at 100 mg/kg (SC, b.i.d.) (Figure 10.10). In macrophages, Btk inhibition by CGI1746 was found to abolish FcγRIII induced TNFα, IL1β and IL-6 production. Similarly, CGI1746 was found to be efficacious when dosed at 100 mg/kg (SC, b.i.d.) in two lymphocyte independent models of arthritis including (1) the passive anti-collagen II antibody induced arthritis (CAIA) model, which requires FCγR signalling and (2) the antibody mediated K/BxN serum transfer model. These findings support the role of Btk in myeloid cell cytokine release and the potential of a Btk inhibitor to have a broader impact outside the B-cell compartment.

The same workers have also reported on a closely related analogue to CGI1746, GDC-0834 (Figure 10.8).[43] GDC-0834 was also a potent, selective Btk inhibitor with an enzyme IC_{50} of 5.9 nM and inhibited pBtk activity in rat spenocytes ($IC_{50} = 6.4$ nM). In mouse and rat, GDC-0834 inhibited pBtk-Tyr223 levels in the blood after correction for free plasma concentration with an IC_{50} of 7.7 nM and 5.6 nM, respectively. GDC-0834 also demonstrated a dose-dependent reduction of severity in arthritis markers when dosed orally in a prophylactic rat CIA model of arthritis. Utilizing disease progression from this model and pBtk-Tyr223 as a biomarker for target engagement, a pharmacokinetic/pharmacodynamic model was developed. This model suggested an overall 73% inhibition of pBtk was needed to decrease the rate of CIA disease progression by half.

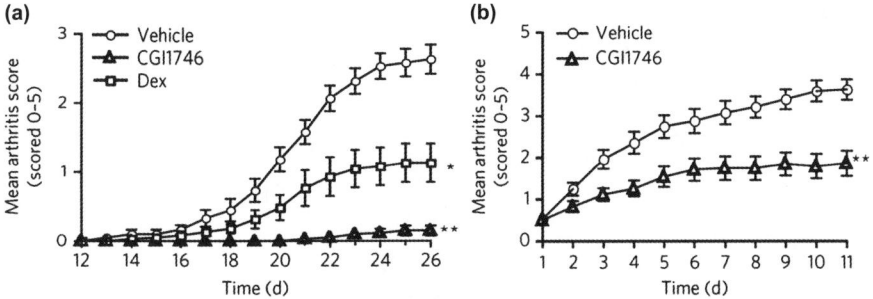

Figure 10.10 (a) Prophylactic treatment with CGI1746 protects from collagen-induced arthritis in B1ORIII mice. Mean clinical score (0–5 per paw, average per animal, n = 15 per group). Average daily clinical score in CGI1746 or dexathasone-treated mice compared with vehicle control: *P = 0.0002; **P < 0.0001. (b) Therapeutic treatment with CGI1746 protects from collagen-induced arthritis in B1ORIII mice (n = 15 per group. Terminal clinical score, **P = 0.0001. (Reprinted by permission from Macmillan Publishers Ltd: *Nature Chemical Biology*, Ref. 35, © 2010.)

Although the plasma clearance of GDC-0834 was low in the rat (4.7 mL/min/kg), clearance in the monkey was high (33 mL/min/kg) and *in-vitro* metabolite identification in human hepatocytes revealed the predominant formation of an inactive metabolite resulting from amide hydrolysis.[67] These findings prompted advancement of GDC-0834 into a single-dose human clinical trial to evaluate the pharmacokinetics. Unfortunately, in humans GDC-0834 plasma concentrations were below the limit of quantitation (< 1 ng/mL) in most samples from cohorts dosed at 35 and 105 mg while substantial concentrations of the amide hydrolysis metabolite were observed.

10.6.1.2 Other Reversible Inhibitors

Several companies are pursuing chemical series closely related to CGI1746 based on recent patent application filings (Figure 10.11). A general design strategy appears to be to mitigate the pharmacokinetic liabilities introduced by amide hydrolysis by (1) deactivation of the aniline nitrogen by substitution to the aryl ring or replacement of the aryl ring by piperidine or (2) replacement of the secondary amide by a cyclic tertiary amide or corresponding heterocycle. Representative examples of these strategies include compound **2** described by CGI pharmaceuticals (no biological data reported)[68] and compound **3** reported by Roche having an IC$_{50}$ of 1 nM in a human whole blood assay.[69] Recently, BMS has disclosed Btk inhibitors based on the latter strategy employing a carbazole amide hinge binding motif.[70] Compound **4** was reported to have Btk

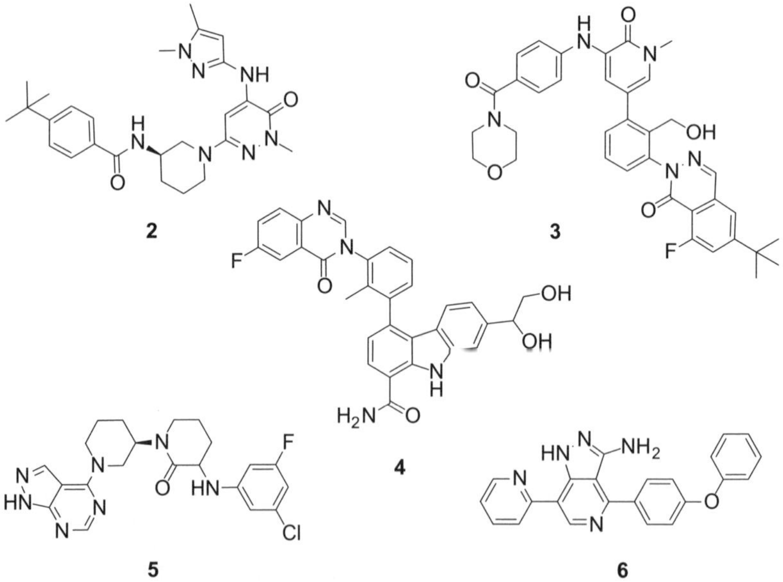

Figure 10.11 Representative non-covalent Btk inhibitors from the recent patent literature.

enzyme IC_{50} of 0.7 nM. Following the former strategy, Biogen Idec and Sunesis have disclosed inhibitors represented by compound **5** (Btk $IC_{50} < 100$ nM).[71,72] Pan and co-workers in 2007 described the Btk inhibitor PCI-29732 (Figure 10.6) as the starting point for their optimization efforts of a covalent inhibitor.[73] The inhibitor demonstrated good Btk enzyme inhibition ($IC_{50} = 8.2$ nM) but also inhibited numerous Tec and Src family kinases. Recently, BMS has disclosed structurally related aminopyrazolopyrimidines such as compound **6** (Btk $IC_{50} = 1$ nM).[74] Unfortunately, selectivity data for this series were not reported.

10.6.2   Covalent Inhibitor Strategies

10.6.2.1   Design of Covalent Inhibitors Targeting Btk

Drugs that employ a covalent inhibition mechanism form a covalent adduct with a targeted protein through bond formation with a specific amino acid residue.[75–77] This mechanism of drug action is common to many widely prescribed drugs including penicillin and the anti-platelet agent clopidogrel (Plavix®). As opposed to single-step enzyme kinetics observed with non-covalent inhibitors (eqn (10.1)), a covalent mechanism involves two distinct events (eqn (10.2)), namely (1) a reversible binding step where the inhibitor (I) and the target protein (E) form an initial EI complex and (2) a covalent bond formation step where the inhibitor modifies the protein forming the EI* complex. Depending on the nature of the bond forming partners, the EI* complex may form irreversibly such that the k_4 rate is negligible or there may be a slow reversible component.

$$E + I \underset{k_2}{\overset{k_1}{\rightleftharpoons}} EI \tag{10.1}$$

$$E + I \underset{k_2}{\overset{k_1}{\rightleftharpoons}} EI \underset{k_4}{\overset{k_3}{\rightleftharpoons}} EI^* \tag{10.2}$$

A covalent, irreversible inhibitor offers several drug design advantages over conventional equilibrium-based reversible inhibitors. Because the binding of the covalent inhibitor to the target protein is governed by both the non-covalent affinity and the subsequent inactivation step (k_3), very high potencies and ligand efficiencies can be achieved with this strategy. Another advantage is the potential for a prolonged pharmacodynamic effect of the drug beyond its pharmacokinetic residence time. Once the target protein reacts with the inhibitor to form the EI* complex, the enzyme remains inactive until dissociation, which would be negligible for an irreversible inhibitor. As a result, the intended pharmacological response of the drug may persist even though the drug itself has been cleared from the body. The dosing interval required for efficacy will therefore be dictated by the rate of new protein synthesis. In cases where biosynthesis of the target protein is slow, this strategy can allow for less frequent dosing and lower drug doses. The non-competitive nature of the enzyme kinetics can also allow for high biochemical efficiency for a covalent inhibitor in cases where there are high levels of endogenous enzyme substrate. This is particularly relevant for kinase inhibitors targeting the ATP binding pocket given the high physiological concentration of ATP. There are, however, potential risks associated with a covalent drug design strategy. Most important is the concern for potential idiosyncratic drug-related toxicities, which are rare adverse events generally caused by immunological response to a large molecule such as the EI* complex.[78]

Three design elements must be considered in the application of a covalent inhibition strategy: (1) the drug must have a reactive moiety either intrinsically or through mechanistic/metabolic activation, (2) the protein target must have a nucleophilic residue in or near the active site such as lysine, serine, or cysteine and (3) the enzyme resynthesis rate should be appreciably slow. Indeed Btk is well suited for this strategy. The enzyme turnover rate has been estimated on the order of 24 hours based on in-vivo inhibitor knockdown studies.[79] Also structural bioinformatics reveals that Btk has six cysteine residues in the kinase domain. Only two of these residues, Cys464 and Cys481, are partially exposed to solvent. A cysteine residue located near the solvent is believed to have a lower pK_a and therefore is more nucleophilic than a cysteine buried in a hydrophobic pocket. Of particular interest is Cys481, which is located in the active site near the ATP binding pocket (Figure 10.7). Only 11 kinases are known to have a cysteine at this position (Blk, Btk, Bmx, EGFR, ErbB2, ErbB4, Itk, JAK3, Tec, Txk and MAP2K7).[80] Consequently, covalent targeting of Cys481 in Btk could represent a strategy to gain general kinome selectivity. Within this set of kinases, the conceptual strategy of covalent inhibition has been clinically validated by EGFR inhibitors such as Neratinib.[81] As a result, several investigators have approached the design of Btk inhibitors with this strategy as well.

10.6.2.2 Discovery of PCI-32765

The identification of PCI-29732 (Figure 10.6) as a non-covalent Btk inhibitor by Pan and co-workers was the starting point for the discovery of the first reported covalent inhibitor of Btk.[44,71] PCI-29732 demonstrated an IC_{50} of 8.2 nM against Btk; however, it showed poor selectivity against many Tec and Src family kinases. By incorporating a covalent bond-forming group into this scaffold, it was hoped that selectivity against many of these kinases could be improved. Initial homology modelling based on Lck kinase, which was later corroborated by X-ray co-crystal structure analysis[60] (Figure 10.7), suggested that PCI-29732 binds in the ATP pocket with the aminopyrazine ring interacting with the hinge residues. This orientation places the cyclopentyl group in proximity to the covalent target residue, Cys481. Utilizing this structural information, a series of compounds was designed to position a Michael acceptor such as an acrylamide, vinyl sulfone or propionamide in proximity to the thiol for covalent bond formation. From this effort, the acrylamide PCI-32765 (Figure 10.12) was identified as a potent inhibitor of Btk with an IC_{50} of

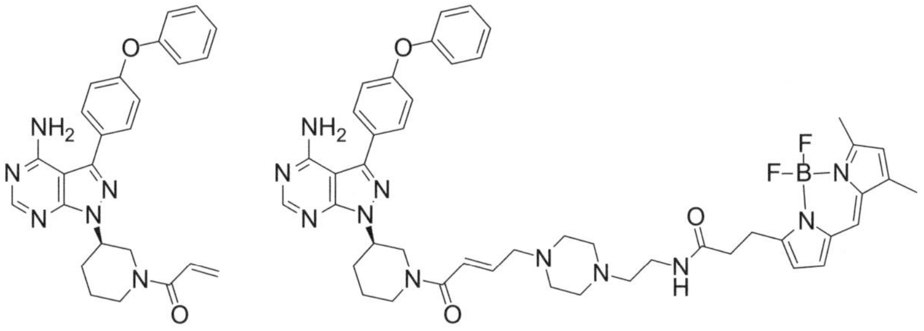

Figure 10.12 Covalent Btk inhibitor PCI-32765 and the fluorescent probe PCI-33380.

0.5 nM. In DOHH2 cellular assays, PCI-32765 inhibited autophosphorylation of Btk at Tyr223 ($IC_{50} = 11$ nM) and phosphorylation of the downstream substrate for Btk, PLCγ ($IC_{50} = 29$ nM). Irreversible binding of PCI-32765 to Btk was established by two approaches: (1) enzymatic activity of Btk pretreated with PCI-32765 was not recovered after multiple washings with inhibitor-free medium and (2) through mass spectroscopic characterization, the molecular weight corresponding to a 1:1 complex between PCI-32765 and Btk was observed.

The increased potency against Btk provided by the covalent mechanism of action provided improved selectivity against many Src family kinases not having a reactive cysteine, even though absolute IC_{50} magnitudes for them remained comparable (*e.g.* Lck, Src and Lyn; $IC_{50} = 2$ nM, 263 nM and 16.2 nM, respectively). The short apparent terminal plasma half-life of the compound in mouse (1.7–3.1 h) may provide a further pharmacodynamic selectivity enhancement for these kinases *versus* Btk. It should be noted, however, that PCI-32765 also potently inhibited several other kinases in enzyme assays that have a cysteine in the homologous location to Btk including Blk, Bmx, EGFR, Itk and JAK3 ($IC_{50} = 0.5$ nM, 0.8 nM, 5.6 nM, 10.7 nM and 16.1 nM, respectively). Nonetheless, PCI-32765 was found to be more than 1000-fold selective for inhibition of antigen receptor signalling in B-cells over T-cells, which do not express Btk, as determined by up-regulation of CD69.

In pre-clinical mouse efficacy models, PCI-32765 has shown promise as a therapeutic targeting rheumatoid arthritis and lupus.[44,45] In the murine collagen-induced arthritis model employing a therapeutic dosing regimen, orally administered PCI-32765 reduced levels of circulating antibodies and suppressed disease symptoms in a dose-dependent manner (Figure 10.13).[45] A dose of 12.5 mg/kg was found to completely reverse arthritic disease in this study. Target engagement *in vivo* was demonstrated employing a cell-permeable, fluorescently tagged derivative of PCI-32765, PCI-33380 (Figure 10.12). Utilizing the fluorescent probe, a dose-dependent level of Btk target occupancy was observed in the spleen in which the 12.5 mg/kg dose resulted in 85–90% occupancy when measured at 3 h after final dose in the mCIA study. The level

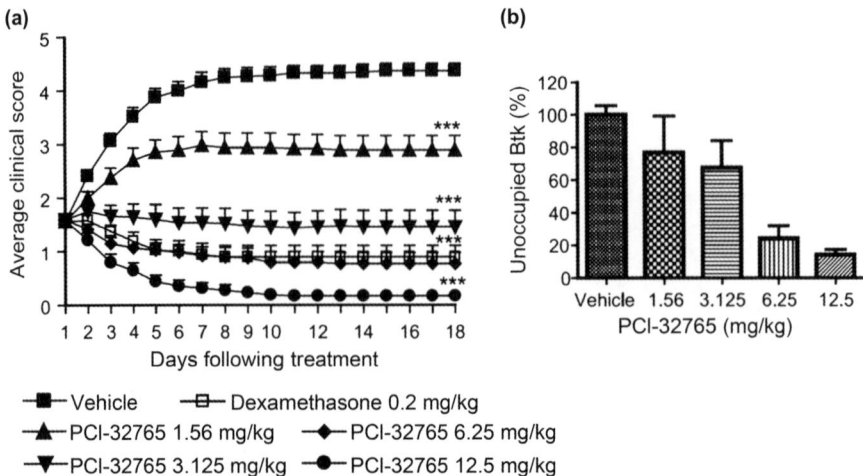

Figure 10.13   (a) PCI-32765 dose-dependently reduced clinical arthritis scores in mouse CIA model (n = 12 per dose group), ***p < 0.001 compared to vehicle (Mann–Witney U-test). (b) Using the fluorescent probe PCI-33380, the percent of Btk occupied by PCI-32765 in CIA mice was measured in splenocytes collected at 3 h (n = 6) following last dose (D18) of drug treatment. (Reproduced with permission from Biomed Central Ltd, © 2011, Chang *et al.*, Ref. 45.)

of occupancy subsequently diminished to baseline after 24 h. These results suggest that the level of Btk inhibition *in vivo* by PCI-32765 is correlated with the degree of *in-vivo* efficacy observed in the mCIA model. PCI-32765 also inhibited auto-antibody production and development of kidney disease in the MRL-Fas(lpr) lupus model (Figure 10.14).[44] Treatment with PCI-32765 reduced proteinuria, blood urea nitrogen (BUN) and serum anti-dsDNA levels in this model. Histological evaluation of the kidneys also showed a significant reduction in interstitial nephritis and perivascular inflammation as well as a trend towards reduced glomerulonephritis.

Compounds related to PCI-32765 with improved kinome selectivity, especially against EGFR, have been reported including 4-morpholinobut-2-enamide and but-2-ynamide derivatives.[82] Analogue PCI-45292 (structure not disclosed) was under investigation by Pharmacyclics as a Btk inhibitor for autoimmune indications based on its improved kinase selectivity and metabolic stability.[83] Unfortunately, development of PCI-45292 was suspended in 2011 due to preclinical toxicology findings.[84]

10.6.2.3   Imidazo[1,5-a]quinoxaline Inhibitors

Kim and co-workers have described a series of imidazo[1,5-a]quinoxalines which act as irreversible Btk inhibitors with good *in-vivo* efficacy in pre-clinical RA models.[85] The lead compound, **7** (Figure 10.15), was identified through

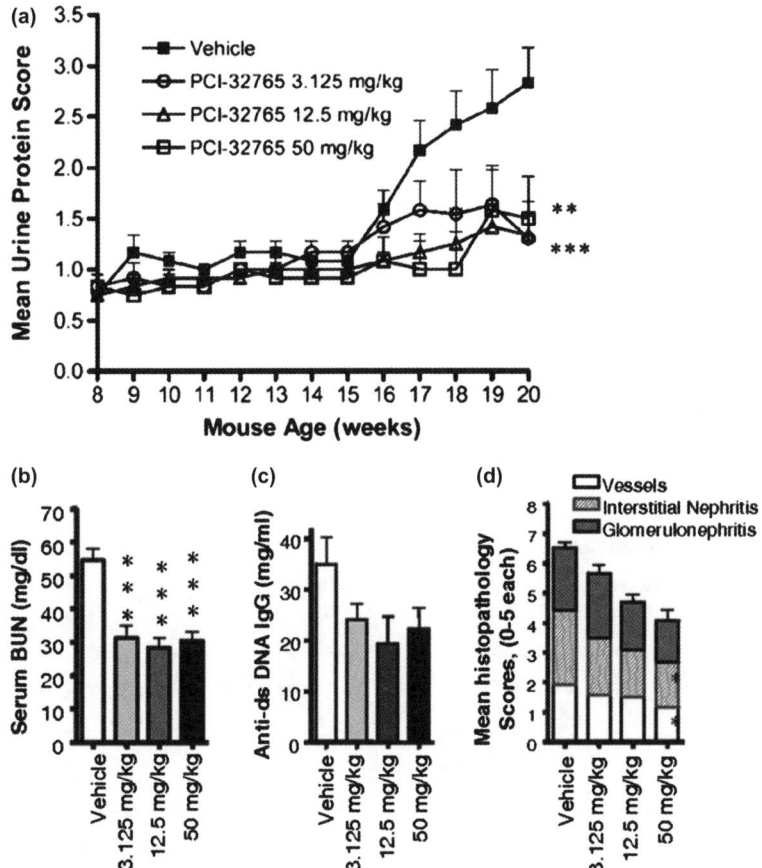

Figure 10.14 Inhibition of Btk reduces renal disease and autoantibody production in MRL-Fas(lpr) mice. Eight-week-old MRL-Fas(lpr) mice (n = 12) were randomized and treated orally with PCI-32765 or vehicle once daily for 12 weeks at different concentrations as indicated. (a) Range of urine protein concentration is calculated as a proteinurea score. (**P < 0.01 for 3.125 mg/kg and ***P < 0.001 for 12.5 and 50 mg/kg treatments; repeated measure ANOVA). (b) dsDNA-specific IgG in PCI-32765 treated and vehicle treated mice. (c) Serum BUN in PCI-32765 treated and vehicle treated mice (***P < 0.001). (d) Histopathology scores at week 20. Scores are mean ± SEM (n = 12; *P < 0.05). (Reproduced with permission, © 2010 National Academy of Sciences, USA, Ref. 44.)

modification of a known Lck inhibitor scaffold by incorporation of a Michael acceptor at the C8 position of the ring, thus enabling covalent modification of Cys481. Several electrophilic groups were examined, of which (*E*)-4-(dimethyl-amino)-*N*-methylbut-2-enamide showed the best combination of potency and physical properties. Compound **7** was a potent inhibitor of Btk in an enzyme assay ($IC_{50} = 1.9$ nM) and also showed good selectivity (> 100-fold) for

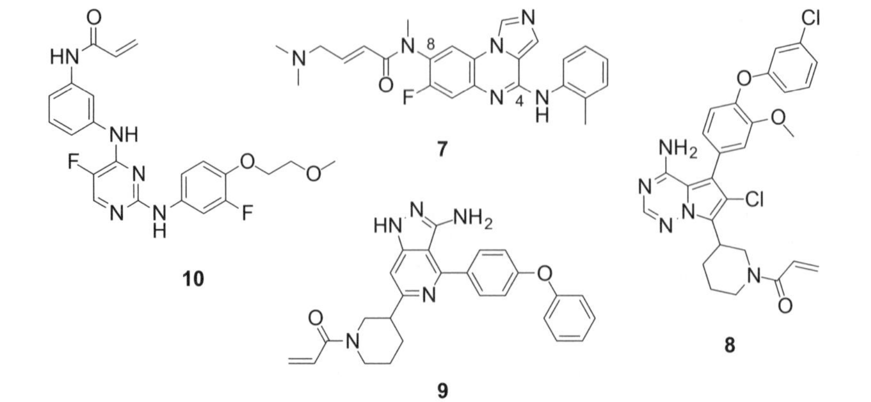

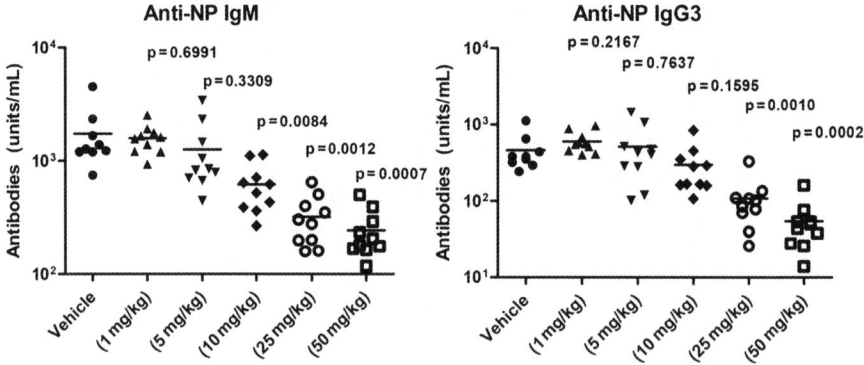

Figure 10.15   Representative covalent Btk inhibitors.

Figure 10.16   Inhibition of T-independent type 2 antibody response to NP-Ficoll by
compound **7** in female C57BL/6 mice. (Reprinted from Ref. 85, © 2011,
with permission from Elsevier.)

several Src family kinases (Lyn, Lck and Src). However, time-dependent kinetic
analysis suggested that compound **7** had comparable potency against EGFR
and Btk ($k_{inact}/K_I = 101,000 \, 1/M \cdot s$ and $53,200 \, 1/M \cdot s$, respectively). Early
compounds having a secondary amide substituent at C8 showed hyper-
proliferation of B-cells as opposed to inhibition even though they were potent
Btk enzyme inhibitors. Inhibition of B-cell proliferation was seen for *N*-methyl
tertiary amides such as compound **7** ($IC_{50} = 3.4 \, nM$). Substitution at the *ortho*-
position of the C4 aniline was found to be important for Btk inhibition with
methyl or chlorine being preferred. The pharmacokinetic profile of compound **7**
in the rat was promising with a short terminal plasma half-life (1.5 h) and
moderate oral bioavailability (34%). The *in-vivo* efficacy of compound **7** was
evaluated in a mouse NP-Ficoll model in which B-cell activation occurs through
a T-cell independent type 2 antigen response (Figure 10.16). Both anti-NP IgM

and anti-NP IgG3 antibody production were inhibited in a dose-responsive manner after oral dosing at 1–50 mg/kg daily. Similarly, compound **7** significantly inhibited progression of disease compared to vehicle control in a mouse collagen-induced arthritis model at oral doses of 3 and 10 mg/kg daily.

The covalent nature of Btk inhibition by compound **7** was supported by structural and analytical methods. Compound **7** was co-crystallized with a gatekeeper mutant (F435T) of Itk. The structure revealed that compound **7** binds in the ATP binding pocket engaging in hinge contacts to the Itk equivalents of the Thr474 gatekeeper and Met477 Btk residues. The 2-methyl-aniline substituent occupies a similar pocket to that observed for the 2-chloro-5-methyl substituent of dasatinib (Figure 10.5). A covalent bond was observed between the ATP binding pocket cysteine (Cys442, Itk) and the electrophilic moiety of compound **7**. Covalent adduct formation with Btk protein itself was confirmed through mass spectroscopic observation of a 1:1 stoichiometric complex between Btk and compound **7**.

10.6.2.4 Other Covalent Inhibitors

Two companies have reported covalent inhibitors structurally similar to the PCI-32765 chemotype (Figure 10.15). A pyrrolotriazine core has been reported by researchers at Locus pharmaceuticals.[86] The acrylamide **8** was described to be a potent inhibitor of Btk in an enzyme assay ($IC_{50} = 2$ nM) and also demonstrated appreciable selectivity against several Src family kinases including Lck, Lyn and Fyn (161, 1389 and 7222-fold, respectively). Compound **8** also inhibited CD69 up-regulation in B-cells ($IC_{50} = 130$ nM). Upon dosing in rats, favourable pharmacokinetics ($Cl = 0.74$ mL/hr/kg) and moderate oral bioavailability (24%) were seen. Researchers at Bristol-Myers Squibb have also combined electrophilic groups with an aminopyrazolopyridine scaffold.[72] The acrylamide **9** was reported to inhibit Btk with an IC_{50} of 4 nM.

Btk inhibitors based upon a diaminopyrimidine scaffold such as compound **10** (Figure 10.15) have been described by workers at Avila Therapeutics.[87] Compound **10** was a potent inhibitor of Btk in an enzyme assay ($IC_{50} < 10$ nM) and covalent modification of the protein was established by mass spectroscopic characterization of the adduct. It also inhibited calcium flux in Ramos cells ($IC_{50} < 10$ nM) and B-cell proliferation ($IC_{50} = 1$–10 nM). As was the case for PCI-32765, compound **10** was also a potent inhibitor of several other kinases with a cysteine residue at the same location as Btk including Itk, EGFR and JAK3 ($IC_{50} = 10$–100 nM, 10–100 nM and < 10 nM, respectively). Compound **10** was efficacious in the rat peptidoglycan-polysaccharide (PG-PS) arthritis model. Utilizing a biotinylated affinity probe, a 30 mg/kg dose of compound **10** resulted in 87% occupancy of Btk by drug at 2 hours after dosing, which persisted even to 24 hours after dosing (60%). From this work, the development candidate AVL-292 was identified; however, the structure has not been disclosed. AVL-292 is reported to be a potent inhibitor of Btk ($IC_{50} < 0.5$ nM) with good selectivity for Btk compared to several Src family kinases (Src, Fyn and Lck) in enzyme assays.[88] Inhibition of EGFR was poor in A431 cells

($IC_{50} > 1\,\mu M$) compared to inhibition of calcium flux mediated by Btk in Ramos cells ($IC_{50} = 1–10\,nM$) suggesting an improved level of selectivity within the family of covalently targeted kinases. AVL-292 was also efficacious in a semi-established mouse collagen induced arthritis model when dosed orally.

10.7 Clinical Experience with Btk Inhibitors

As of 2011, three Btk inhibitors had advanced to human clinical trials, namely PCI-32765, AVL-292 and GDC-0834. The last of these, GDC-0834, was discontinued due to poor pharmacokinetics in humans.[67] The most advanced inhibitor currently is PCI-32765. PCI-32765 is in Phase Ib/II studies targeting a number of B-cell driven non-Hodgkin's lymphomas including chronic lymphocytic leukaemia/small cell lymphocytic lymphoma (CLL/SLL), mantle cell lymphoma and diffuse large B-cell lymphoma.[89] Btk is also a promising oncology target in addition to its autoimmune therapeutic potential. Some B-cell lymphomas depend on chronic active BCR signalling and suppression of that signal can induce apoptosis and inhibit cell migration. Preliminary clinical results targeting these cancer cell types with PCI-32765 treatment has been very encouraging.[90,91] Unfortunately, human trials with PCI-32765 in patients with autoimmune disease have not been initiated. PCI-32765 has been generally well tolerated in clinical trials to date with some patients on continuous treatment for more than six months. Discontinuations of therapy due to serious adverse events have occurred in 3 of 83 patients, though this includes one patient that developed a drug hypersensitivity reaction.[85,92] Although the patient in question had a history of allergies to prescription medicines, drug hypersensitivity is a particular risk with covalent inhibitors because of the potential for hapten generation to the drug-protein conjugate. Otherwise, the most frequent side-effects were diarrhoea (a common side-effect of EGFR inhibition), nausea and dyspepsia. Severe neutropenia and thrombocytopenia were uncommon though more prevalent at higher doses. The application of the fluorescently tagged affinity probe PCI-33380 as an *ex-vivo* tool in the clinical setting has allowed for the measurement of PCI-32765 occupancy to Btk enzyme as a target biomarker. Indeed, full Btk occupancy by drug could be achieved by PCI-32765 at relatively low drug exposures even though the human pharmacokinetic half-life was only 6.3–10.8 h.[87] A good correlation was observed between Btk occupancy by PCI-32765 in PBMCs, drug plasma AUC and *ex-vivo* inhibition of basophil degranulation (a pharmacologic biomarker of Btk function).[93]

Less advanced is the covalent Btk inhibitor AVL-292.[84] A single ascending dose study in healthy volunteers has been completed evaluating doses in the range of 0.5–7.0 mg/kg. AVL-292 was reported to be generally safe and well tolerated in this study. Human pharmacokinetics was also favourable for a covalent inhibitor with the compound being rapidly absorbed and readily cleared ($t_{1/2} = 2–3\,h$). Maximum Btk target occupancy by drug was detected with doses as low as 2 mg/kg. A Phase Ib trial in patients with B-cell leukaemia began in June 2011.

At this time, the clinical experience with Btk inhibitors is limited in scope. Most obvious is the lack of clinical trials directed towards efficacy in auto-immune indications, which are essential to establish a linkage with the existing very promising pre-clinical data in animal efficacy models. The clinical safety profile of PCI-32765, though acceptable for continued studies in oncology, will need further scrutiny to move into less at-risk patient populations requiring chronic therapy. As demonstrated clinically with both compounds, the utilization of affinity probes to quantify drug occupancy to the target as a biomarker will greatly enable dose selection, especially as studies advance into the autoimmune population.

10.8 Conclusions

The critical role Btk plays as a mediator of signal transduction pathways in haematopoietic cells positions the target as an important linchpin in the regulation of autoimmune and inflammatory disease. Although Btk is most commonly associated with its role in B-cell development and signalling, the evolving understanding of its role in myeloid cells further expands the potential of Btk as a drug-discovery target. Common to many kinase targets, drug discovery of Btk inhibitors has faced many hurdles, not just potency and kinome selectivity. The structural biology understanding of the Tec kinase family and Btk especially has expanded significantly over the last few years and now offers valuable insights for the discovery of new inhibitors. The ability of Btk to adopt a variety of inactive kinase conformations including a DFG out conformation offers unique opportunities in structure-based drug design. Recent Btk inhibitors targeting an inactive conformation such as CGI1746 have demonstrated remarkable potency and kinome selectivity. Btk drug discovery has also been an excellent example of the potential for the strategy of covalent enzyme targeting. The ability of inhibitors such as PCI-32765 and compound **7** to inactivate the protein through adduct formation with the partially exposed Cys481 residue located near the ATP binding pocket demonstrates a valuable strategy to identify potent and selective inhibitors with potentially novel pharmacodynamic profiles. The availability of potent and orally bioavailable Btk inhibitors has allowed for the interrogation of the role Btk plays in disease. Indeed, Btk inhibitors have demonstrated efficacy in several pre-clinical models of rheumatoid arthritis including those driven not only by B-cells but also myeloid cell involvement, confirming the broad role of Btk in the immune system. The Btk inhibitor PCI-32765 has also been shown to inhibit auto-antibody production and development of kidney disease in a pre-clinical model of systemic lupus erythematosus. As of 2011, two Btk inhibitors have entered human clinical trials for oncology indications. Unfortunately, human trials in autoimmune disease have not been initiated. Nonetheless, the initial promising outcomes from the first Phase I/II trials is encouraging and opens the door for the ultimate advancement of Btk inhibitors to clinical trials for autoimmune and inflammatory indications.

Acknowledgement

The authors would like to thank Dr John Douhan for his critical review of this chapter.

References

1. A. J. Mohamed, L. Yu, C.-M. Bäckesjö, L. Vargas, R. Faryal, A. Aints, B. Christensson, A. Berglöf, M. Vihinen, B. F. Nore and C. I. Edvard Smith, *Immunol. Rev.*, 2009, **228**, 58.
2. W. N. Khan, *Immunol. Res.*, 2001, **23**, 147.
3. S. Tsukada, D. C. Saffran, D. J. Rawlings, O. Parolini, R. C. Allen, L. Klisak, R. S. Sparkes, H. Kubagawa, T. Mohandas, S. Duan, J. W. Belmont, M. D. Cooper, M. E. Conley and O. N. Witte, *Cell*, 1993, **72**, 279.
4. D. Vetrie, I. Vorechovskyt, P. Siderasj, J. Holland, A. Davies, F. Flinter, L. Hammarstrom, C. Kinnon, R. Levinsky, M. Bobrow, C. I. Edvard Smith and D. R. Bentley, *Nature*, 1993, **361**, 226.
5. L. A. Humphries, C. Dangelmaier, K. Sommer, K. Kipp, R. M. Kato, N. Griffith, I. Bakman, C. W. Turk, J. L. Daniel and D. J. Rawlings, *J. Biol. Chem.*, 2004, **279**, 37651.
6. J. C. Edwards, L. Szczepanski, J. Szechinski, A. Filipowicz-Sosnowska, P. Emery, D. R. Close, R. M. Stevens and T. Shaw, *N. Engl. J. Med.*, 2004, **350**, 2572.
7. C. Favas and D. A. Isenberg, *Nat. Rev. Rheumatol.*, 2009, **5**, 711.
8. S. L. Hauser, E. Waubant, D. L. Arnold, T. Vollmer, J. Antel, R. J. Fox, A. Bar-Or, M. Panzara, N. Sarkar, S. Agarwal, A. Langer-Gould, C. H. Smith and HERMES Trial Group, *N. Engl. J. Med.*, 2008, **358**, 676.
9. C. Mao, M. Zhou and F. M. Uckun, *J. Biol. Chem.*, 2001, **276**, 41435.
10. A. Kuglstatter, A. Wong, S. Tsing, S. W. Lee, Y. Lou, A. G. Villaseñor, J. M. Bradshaw, D. Shaw, J. W. Barnett and M. F. Browner, *Protein Sci.*, 2011, **20**, 428.
11. F. Zuccotto, E. Ardini, E. Casale and M. Angiolini, *J. Med. Chem.*, 2010, **53**, 2681.
12. O. Hantschel, U. Rix, U. Schmidt, T. Bürckstümmer, M. Kneidinger, G. Schütze, J. Colinge, K. L. Bennett, W. Ellmeier, P. Valent and G. Superti-Furga, *Proc. Natl Acad. Sci. USA*, 2007, **104**, 13283.
13. S. Mahajan, S. Ghosh, E. A. Sudbeck, Y. Zhengi, S. Downs, M. Hupke and F. M. Uckun, *J. Biol. Chem.*, 1999, **274**, 9587.
14. W. E. Lowry, J. Huang, M. Lei, D. Rawlings and X.-Y. Huang, *J. Biol. Chem.*, 2001, **276**, 45276.
15. K. Saito, A. M. Scharenberg and J.-P. Kinet, *J. Biol. Chem.*, 2001, **276**, 16201.
16. H. Hansson, M. P. Okoh, C. I. Edvard Smith, M. Vihinen and T. Härda, *FEBS Lett.*, 2001, **489**, 67.
17. T. Kurosaki and M. Kurosaki, *J. Biol. Chem.*, 1997, **272**, 15595.
18. R. E. Joseph, Q. Xie and A. H. Andreotti, *J. Mol. Biol.*, 2010, **403**, 231.

19. M. Dinh, D. Grunberger, H. Ho, S. Y. Tsing, D. Shaw, S. Lee, J. Barnett, R. J. Hill, D. C. Swinney and J. M. Bradshaw, *J. Biol. Chem.*, 2007, **282**, 8768.
20. L. Lin, R. Czerwinski, K. Kelleher, M. M. Siegel, P. Wu, R. Kriz, A. Aulabaugh and M. Stahl, *Biochemistry*, 2009, **48**, 2021.
21. J. M. Lindvall, K. Emelie, M. Blomberg, J. Väliaho, L. Vargas, J. E. Heinonen, A. Berglöf, A. J. Mohamed, B. F. Nore, M. Vihinen and C. I. Edvard Smith, *Immunol. Rev.*, 2005, **203**, 200.
22. S. Trukada, D. C. Saffran, D. J. Rawlings, O. Parolini, R. C. Allen, L. Klisak, R. S. Sparkes, H. Kubagawa, T. Mohandas, S. Duan, J. W. Belmont, M. D. Cooper, M. E. Conley and O. N. Witte, *Cell*, 1993, **72**, 279.
23. M. E. Conley, A. Broides, V. Hernandez-Trujillo, V. Howard, H. Kanegane, T. Miyawaki and S. A. Shurtleff, *Immunol. Rev.*, 2005, **203**, 216.
24. J. Väliaho, C. I. Edvard Smith and M. Vihinen, *Hum. Mutat.*, 2006, **27**, 1209.
25. A. Maas and R. W. Hendriks, *Dev. Immunol.*, 2001, **8**, 171.
26. W. Ellmeier, S. Jung, M. J. Sunshine, F. Hatam, Y. Xu, D. Baltimore, H. Mano and D. R. Littman, *J. Exp. Med.*, 2000, **192**, 1611.
27. T. Kurosaki, *Nat. Rev. Immunol.*, 2002, **2**, 354.
28. J. M. Dal Porto, S. B. Gauld, K. T. Merrell, D. Mills, A. E. Pugh-Bernard and J. Cambier, *Mol. Immunol.*, 2004, **41**, 599.
29. M. Dinh, D. Grunberger, H. Ho, S. Y. Tsing, D. Shaw, S. Lee, J. Barnett, R. J. Hill, D. C. Swinney and J. M. Bradshaw, *J. Biol. Chem.*, 2007, **282**, 8768.
30. U. Schmidt, N. Boucherona, B. Ungera and W. Ellmeiera, *Int. Arch. Allergy Immunol.*, 2004, **134**, 65.
31. X. Liu, Z. Zhan, D. Li, L. Xu, F. Ma, P. Zhang, H. Yao and X. Cao, *Nat. Immunol.*, 2011, **12**, 416.
32. M. Felices, M. Falk, Y. Kosaka and L. J. Berg, *Adv. Immunol.*, 2007, **93**, 145.
33. A. Khare, B. Viswanathan, R. Gund, N. Jain, B. Ravindran, A. George, S. Rath and V. Bal, *Apoptosis*, 2011, **16**, 334.
34. M. Liljeroos, R. Vuolteenaho, S. Morath, T. Hartung, M. Hallman and M. Ojaniemi, *Cell. Signalling*, 2007, **19**, 625.
35. J. A. Di Paolo, T. Huang, M. Balazs, J. Barbosa, K. H. Barck, B. J. Bravo, R. A. D. Carano, J. Darrow, D. R. Davies, L. E DeForge, L. Diehl, R. Ferrando, S. L Gallion, A. M. Giannetti, P. Gribling, V. Hurez, S. G. Hymowitz, R. Jones, J. E. Kropf, W. P. Lee, P. M. Maciejewski, S. A. Mitchell, H. Rong, B. L. Staker, J. A. Whitney, S. Yeh, W. B. Young, C. Yu, J. Zhang, K. Reif and K. S. Currie, *Nat. Chem. Biol.*, 2011, **7**, 41.
36. M. Melcher, B. Unger, U. Schmidt, I. A. Rajantie, K. Alitalo and W. Ellmeier, *J. Immunol.*, 2008, **180**, 8048.
37. K. Fiedler, A. Sindrilaru, G. Terszowski, E. Kokai, T. B. Feyerabend, L. Bullinger, H.-R. Rodewald and C. Brunner, *Blood*, 2011, **117**, 1329.
38. T. U. Marron, K. Rohr, M. Martinez-Gallo, J. Yu and C. Cunningham-Rundles, *Clin. Immunol.*, 2010, **137**, 74.

39. K. Sochorová, R. Horváth, D. Rožková, J. Litzman, J. Bartůňková, A. Šedivá and R. Špíšek, *Blood*, 2007, **109**, 2553.
40. P. Engel, J. A. Gómez-Puerta, M. Ramos-Casals, F. Lozano and X. Bosch, *Pharmacol. Rev.*, 2011, **63**, 127.
41. J. C. W. Edwards and G. Cambridge, *Nat. Rev. Immunol.*, 2006, **6**, 394.
42. I. Sanz and F. E.-H. Lee, *Nat. Rev. Rheumatol.*, 2010, **6**, 326.
43. L. Liu, J. D. Paolo, J. Barbosa, H. Rong, K. Reif and H. Wong, *J. Pharmacol. Exp. Ther.*, 2011, **338**, 154.
44. L. A. Honigberg, A. M. Smith, M. Sirisawad, E. Verner, D. Loury, B. Chang, S. Li, Z. Pan, D. H. Thamm, R. A. Miller and J. J. Buggy, *Proc. Natl Acad. Sci. USA*, 2010, **107**, 13075.
45. B. Y. Chang, M. M. Huang, M. Francesco, J. Chen, J. Sokolove, P. Magadala, W. H Robinson and J. J. Buggy, *Arthritis Res. Therapy*, 2011, **13**, R115.
46. M. Riccaboni, I. Bianchi and P. Petrillo, *Drug Discovery Today*, 2010, **15**, 517.
47. M. E. Weinblatt, A. Kavanaugh, M. C. Genovese, T. K. Musser, E. B. Grossbard and D. B. Magilavy, *N. Eng. J. Med.*, 2010, **363**, 1303.
48. I. Sanz, U. Yasothan and P. Kirkpatrick, *Nat. Rev. Drug Disc.*, 2011, **10**, 335.
49. N. Charles, D. Hardwick, E. Daugas, G. G. Illei and J. Rivera, *Nat. Med.*, 2010, **16**, 701.
50. O. Stüve, S. Cepok, B. Elias, A. Saleh, H.-P. Hartung, B. Hemmer and B. C. Kieseier, *Arch. Neurol.*, 2005, **62**, 1620.
51. A. Mangla, A. Khare, V. Vineeth, N. N. Panday, A. Mukhopadhyay, B. Ravindran, V. Bal, A. George and S. Rath, *Blood*, 2004, **104**, 1191.
52. S. L. Hauser, E. Waubant, D. L. Arnold, T. Vollmer, J. Antel, R. J. Fox, A. Bar-Or, M. Panzara, N. Sarkar, S. Agarwal, A. Langer-Gould and C. H. Smith, *N. Engl. J. Med.*, 2008, **358**, 676.
53. A. H. Cross and E. Waubant, *Biochim. Biophys. Acta*, 2011, **1812**, 231.
54. M. L. Kosmidis and M. C. Dalakas, *Ther. Adv. Neurol. Disord.*, 2010, **3**, 93.
55. J. L. Chamberlain, K. Attridge, C. J. Wang, G. A Ryan and L. S. K. Walker, *Expert Opin. Ther. Targets*, 2011, **15**, 703.
56. E. Baraldi, K. Djinovic Carugo, M. Hyvonen, P. L. Surdo, A. M. Riley, B. V. Potter, R. O'Brien, J. E. Ladbury and M. Saraste, *Structure*, 1999, **7**, 449.
57. M. Hyvonen and M. Saraste, *EMBO J.*, 1997, **16**, 3396.
58. K. Murayama, M. Kato-Murayama, C. Mishima, R. Akasaka, M. Shirouzu, Y. Fukui and S. Yokoyama, *Biochem. Biophys. Res. Commun.*, 2008, **377**, 23.
59. H. Hansson, P. T. Mattsson, P. Allard, P. Haapaniemi, M. Vihinen, C. I. Edvard Smith and T. Härd, *Biochemistry*, 1998, **37**, 2912.
60. S. R. Tzeng, Y. C. Lou, M. T. Pai, C. P. Chen, S. H. Chen and J. W. Cheng, *J. Biomol. NMR*, 2000, **16**, 303.
61. K.-C. Huang, H.-T. Cheng, M.-T. Pai, S.-R. Tzeng and J.-W. Cheng, *J. Biomol. NMR*, 2006, **36**, 73.
62. D. J. Marcotte, Y.-T. Liu, R. M. Arduini, C. A. Hession, K. Miatkowski, C. P. Wildes, P. F. Cullen, V. Hong, B. T. Hopkins, E. Mertsching,

T. J. Jenkins, M. J. Romanowski, D. P. Baker and L. F. Silvian, *Protein Sci.*, 2010, **19**, 429.

63. D. C. Swinney, *Nat. Rev. Drug Disc.*, 2004, **3**, 801.
64. E. van den Akker, T. B. van Dijk, U. Schmidt, L. Felida, H. Beug, B. Löwenberg and M. von Lindern, *Biol. Chem.*, 2004, **385**, 409.
65. F. M. Uckun, I. Dibirdik, S. Qazi, A. Vassilev, H. Ma, C. Mao, A. Benyumov and K. H. Emami, *Bioorg. Med. Chem.*, 2007, **15**, 800.
66. F. M. Uckun, H. E. Tibbles and A. O. Vassilev, *Anti-Cancer Agents Med. Chem.*, 2007, **7**, 624.
67. L. Liu, J. S. Halladay, Y. Shin, S. Wong, M. Coraggio, H. La, M. Baumgardner, H. Le, S. Gopaul, J. Boggs, P. Kuebler, J. C. Davis, Jr, X. C. Liao, J. W. Lubach, A. Deese, C. G. Sowell, K. S. Currie, W. B. Young, S. C. Khojasteh, Cornelis E. C. A. Hop and H. Wong, *Drug Metab. Dispos.*, 2011, **39**, 1840.
68. P. A. Blomgren, K. S. Currie, S. A. Mitchell and D. R. Brittelli, CGI Pharmaceuticals, USA, *PCT Int. Appl.*, WO 2010/068788, 2010.
69. S. Berthel, F. Firooznia, D. Fishlock, J.-B. Hong, Y. Lou, M. Lucas, T. D. Owens, K. Sarma, Z. K. Sweeney and J. P. G. Taygerly, Hoffmann-La Roche, USA, *US Pat. Appl.*, US 20100222325, 2010.
70. Q. Liu, D. G. Batt, G. V. Delucca, Q. Shi and A. J. Tebben, Bristol-Myers Squibb, USA, *US Pat. Appl.*, US 20100160303, 2010.
71. M. Bui, P. Conlon, D. A. Erlanson, J. Fan, B. Guan, B. T. Hopkins, A. Ishchenko, T. J. Jenkins, G. Kumaravel, D. Marcotte, N. Powell, D. Scott, A. Taveras, D. Wang and M. Zhong, Biogen Idec and Sunesis Pharmaceuticals, USA, *PCT Int. Appl.*, WO 2011/029043, 2010.
72. D. A. Erlanson, D. Marcotte, G. Kumaravel, J. Fan, D. Wang, J. H. Cuervo, L. Silvian, N. Powell, M. Bui, B. T. Hopkins, A. Taveras, B. Guan, P. Conlon, M. Zhong, T. J. Jenkins, D. Scott and A. A. Lugovskoy, Biogen Idec and Sunesis Pharmaceuticals, USA, *PCT Int. Appl.*, WO 2011/029046, 2011.
73. Z. Pan, H. Scheerens, S.-J. Li, B. E. Schultz, P. A. Sprengeler, L. C. Burrill, R. V. Mendonca, M. D. Sweeney, K. C. K. Scott, P. G. Grothaus, D. A. Jeffery, J. M. Spoerke, L. A. Honigberg, P. R. Young, S. A. Dalrymple and J. T. Palmer, *ChemMedChem*, 2007, **2**, 58.
74. J. Duan, B. Jiang and Z. Lu, Bristol-Myers Squibb, USA, *PCT Pat. Appl.*, WO 2011/019780, 2011.
75. J. Singh, R. C. Petter, T. A. Baillie and A. Whitty, *Nat. Rev. Drug Disc.*, 2011, **10**, 307.
76. M. H. Potashman and M. E. Duggan, *J. Med. Chem.*, 2009, **52**, 1231.
77. J. G. Robertson, *Biochemistry*, 2005, **44**, 5561.
78. J. Uetrecht, *Annu. Rev. Pharmacol. Toxicol.*, 2007, **47**, 513.
79. L. A. Honigberg, A. M. Smith, J. Chen, P. Thiemann and E. Verner, *Blood (ASH Annual Meeting Abstracts)*, 2007, **110**, 1592.
80. E. Leproult, S. Barluenga, D. Moras, J.-M. Wurtz and N. Winssinger, *J. Med. Chem.*, 2011, **54**, 1347.

81. H. J. Burstein, Y. Sun, L. Y. Dirix, Z. Jiang, R. Paridaens, A. R. Tan, A. Awada, A. Ranade, S. Jiao, G. Schwartz, R. Abbas, C. Powell, K. Turnbull, J. Vermette, C. Zacharchuk and R. Badwe, *J. Clin. Oncol.*, 2010, **28**, 1301.

82. L. Honigberg, E. J. Verner, J. J. Buggy, J. Joseph, D. J. Loury and W. Chen, Pharmacyclics, USA, *US Pat. Appl.*, US 20100254905, 2010.

83. B. Y. Chang, P. Thiemann, M. Francesco, M. M. Huang, M. Sirisawad, N. Purro, P. Jejurkar, C. Mani, D. Tonev, W. Chen, J. J. Buggy and D. J. Loury, *Arthritis Rheum.*, 2010, **62**(10), 286.

84. Pharmacyclics Press Release, March 2, 2011. http://ir.pharmacyclics.com/releasedetail.cfm?ReleaseID=553831

85. K.-H. Kim, A. Maderna, M. E. Schnute, M. Hegen, S. Mohan, J. Miyashiro, L. Lin, E. Li, S. Keegan, J. Lussier, C. Wrocklage, C. L. Nickerson-Nutter, A. J. Wittwer, H. Soutter, N. Caspers, S. Han, R. Kurumbail, K. Dunussi-Joannopoulos, J. Douhan III and A. Wissner, *Bioorg. Med. Chem. Lett.*, 2011, **21**, 6258.

86. K. J. Moriarty, Z. Konteatis, K. Moffett, Y. Lee and W. Chao, Locus Pharmaceuticals, USA, *PCT Int. Appl.*, WO 2010/126960, 2010.

87. J. Singh, R. Petter, R. W. Tester, A. F. Kluge, H. Mazdiyasni, W. F. Westlin III, D. Niu and L. Qiao, *PCT Int. Appl.*, WO, 2011/090760, 2011.

88. E. Evans, S. Aslanian, R. Karp, M. Sheets, P. Chaturvedi, H. Mazdiyasni, R. Tester, M. Nacht, R. Petter, J. Singh and W. Westlin, Keystone Symposium on Molecular and Cellular Biology, 2011. http://www.avilatx.com/news/2011_Keystone_Symposium_B_cell_meeting2.pdf

89. Pharmacyclics Press Release, June 6, 2011. http://ir.pharmacyclics.com/releasedetail.cfm?ReleaseID=583003

90. J. C. Byrd, K. A. Blum, J. A. Burger, S. E. Coutre, J. P. Sharman, R. R. Furman, I. W. Flinn, B. Grant, D. Richards, W. Zhao, N. Heerema, A. J. Johnson, R. Izumi, A. Hamdy and S. O'Brien, *J. Clin. Oncol. (ASCO Annual Meeting Proceedings)*, 2011, **29**(15), 6508.

91. R. Advani, J. P. Sharman, S. M. Smith, D. A. Pollyea, T. E. Boyd, B. W. Grant, K. S. Kolibaba, J. J. Buggy, A. Hamdy and N. H. Fowler, *J. Clin. Oncol. (ASCO Annual Meeting Proceedings)*, 2010, **28**(15), 8012.

92. Pharmacyclics Press Release, June 5, 2010. http://ir.pharmacyclics.com/releasedetail.cfm?ReleaseID=476851

93. D. A. Pollyea, S. Smith, N. Fowler, T. E. Boyd, A. M. Smith, M. Sirisawad, L. A. Honigberg, A. Hamdy and R. Advani, *Blood (ASH Annual Meeting Abstracts)*, 2010, **114**, 3713.

Section 3
GPCRs

CHAPTER 11
CCR1

J. ROBERT MERRITT*[a] AND ANNETTE GILCHRIST[b]

[a] Kean University, New Jersey Center for Science, Technology & Mathematics, 1000 Morris Ave., Union, NJ 07083, USA; [b] Midwestern University, Chicago College of Pharmacy, Department of Pharmaceutical Sciences, 551 31st St., Downers Grove, IL 60515, USA
*Email: jmerritt@kean.edu

11.1   Introduction

11.1.1   What is CCR1?

Chemokines are classified into four subfamilies based on the spacing between conserved N-terminal cysteines (CC, CXC, CX_3C and C). The CC chemokine receptor CCR1 was the first human CC chemokine receptor to be identified at the cDNA level.[1] Initially described as a receptor for chemokines CCL3 (macrophage inflammatory protein (MIP)-1α) and CCL5 (regulated upon activation, normal T-cell expressed and secreted; aka RANTES), the receptor has now been shown to be activated by a number of other chemokine ligands including CCL7 (monocyte chemotactic protein (MCP-3), CCL8 (MCP-2), CCL13 (MCP-4), CCL14 (hemofiltrate CC chemokine (HCC-1), CCL15 (HCC-2/MIP-5/leukotactin), CCL16 (liver specific CC chemokine (LCC)-1/HCC-4) and CCL23 (myeloid progenitor inhibitor factor (MPIF)-1/MIP-3).[2] Of importance is that several of the most potent ligands for CCR1 are generated from weak agonists by proteases that are up-regulated during inflammatory responses *in vivo*.[3]

Like other members of the G-protein coupled receptor (GPCR) superfamily, CCR1 mediates its biological effects through heterotrimeric G proteins acting

RSC Drug Discovery Series No. 26
Anti-Inflammatory Drug Discovery
Edited by Jeremy I Levin and Stefan Laufer
© The Royal Society of Chemistry 2012
Published by the Royal Society of Chemistry, www.rsc.org

as molecular switches to turn on/off downstream second messenger signalling pathways. Within the heterotrimer G-protein family there are 22 Gα subunits divided into four subfamilies ($G_{i/o}$, G_q, G_s and $G_{12/13}$), 5 Gβ subunits and 14 Gγ subunits.[4] While some of the subunits are expressed ubiquitously, others have temporal and/or differential expression.[5] This is important, as the pattern of G-proteins activated by a given ligand ultimately determines the cellular response. Early work suggested that signal transduction of CCR1 was mediated through pertussis toxin (PTx)-sensitive G-protein(s), suggesting a member of the $G_{i/o}$ family was involved,[6] but coupling to additional G-proteins ($Gα_{14}$, $Gα_{16}$) has been reported.[7,8] That $Gα_{16}$ may play a role in CCR1 signalling, as demonstrated by Tian *et al.*,[7] is interesting given its limited and overlapping distribution with CCR1. In addition, although $G_{12/13}$ signalling has not yet been noted following CCR1 activation, other chemokine receptors have been shown to utilize these G-proteins in their signalling profile.[9–12] Finally, it is important to point out that although numerous biochemical and physiological responses have been reported following CCR1 activation such as calcium flux, cell migration and CD11b up-regulation, less is known about which G-protein(s) is responsible for driving these events.

11.1.2   CCR1 Involvement in Cells

In humans, CCR1 is expressed on the cell surface of many cell types including monocytes, T-cells, basophils, eosinophils, neutrophils, platelets, mast cells, mesenchymal stem cells, dendritic cells, airway smooth muscle cells, neurons, astrocytes, osteoclasts and osteoblasts. A wealth of evidence indicates that CCR1 is intimately involved in monocyte and T-cell recruitment and activation. In addition, CCR1 may play a role in up-regulation of integrin CD11b (Mac-1), a protein involved in firm adherence of leukocytes to the endothelium, and protease secretion from monocytic cells.[13–16]

While the generation of CCR1 knockout mice[17] has allowed investigators to better define the role of CCR1 in a variety of diseases, it is important to point out that chemokine/receptor expression patterns differ between mouse and human. For example, in humans, CCR1 is highly expressed on circulating monocytes, and its ligands CCL3 and CCL5 have been shown to induce monocyte recruitment. However, in the mouse, CCR1 is expressed on neutrophils, and CCL3 appears to be the ligand principally responsible for recruiting these cells. Such diversity makes extrapolating data from murine models more challenging.

Chemokines aid in protecting the host from attack by dangerous pathogens and their actions are under tight control. The consequences of a breakdown in this stringent regulation of immune cell mobilization may result in autoimmune diseases like diabetes or systemic lupus erythematosus.[18] Given its importance in the trafficking of immune cells and Th1/Th2 cytokine balance, the dysregulation of CCR1 has been thought to contribute to the pathogenesis of several autoimmune diseases including rheumatoid arthritis (RA),[19] multiple sclerosis (MS)[20] and chronic obstructive pulmonary disease (COPD).[21] With this background, it is perhaps not surprising then that antagonists for CCR1 have undergone clinical testing for each of these diseases.[22] However, in addition to these indications, CCR1 antagonists have undergone *in-vivo* examination for transplant rejection[23] and colon cancer liver metastasis,[24] and have entered Phase II clinical trials for endometriosis associated pelvic pain,[25] and as a biomarker for Alzheimer's disease.[26] No doubt this wide range of indications reflects our recognition that CCR1 may function in a host of cellular responses.

11.1.3 CCR1 Involvement in Inflammatory Diseases

Together, chemokines and their receptors carefully orchestrate leukocyte movement, directing migration from blood to tissue. They often facilitate direct communication between the innate and adaptive immune responses and act as key mediators during the development of inflammatory responses. Because chemokines acting on CCR1 are essential mediators in the patho-physiology of a number of inflammatory diseases, the receptor has served as a popular candidate for therapeutic intervention. For example, a hallmark of inflammatory skin diseases such as irritant contact dermatitis (ICD) and contact hypersensitivity (CHS) is the over-expression of chemokines such as CCL5 with a subsequent resulting local accumulation of pro-inflammatory immune cells. As a result CCR1 antagonists have been evaluated following epicutaneous nickel challenge[27] and found to partially inhibit clinical manifestations of allergic contact dermatitis.

Clearly, the central hypothesis of a CCR1 antagonist as a therapeutic agent for autoimmune disease is that inhibition of monocyte and T-cell recruitment can ameliorate the underlying disease process. Pre-clinical studies with CCR1 deficient mice, anti-CCL3, anti-CCL5 or small-molecule inhibitors of CCR1 support the role of CCR1 in various immune related animal disease models, such as experimental autoimmune encephalomyelitis (EAE),[28] collagen induced arthritis (CIA),[29] adjuvant induced arthritis (AIA),[30] graft-*versus*-host disease (GVHD),[31] renal fibrosis,[32] radiation induced pulmonary fibrosis[33] and airway hyper-responsiveness and remodelling in response to infectious agents.[34]

A number of CCR1 antagonists have entered clinical trials. This includes AZD-4818 for chronic obstructive pulmonary disease,[21] BMS-817399 for rheumatoid arthritis,[35] BX-471 for multiple sclerosis[36] and endometriosis associated pelvic pain,[25] CCX354 for rheumatoid arthritis,[37] CP-481715 for allergic contact dermatitis[27] and rheumatoid arthritis[22,38] and MLN-3897 for rheumatoid arthritis.[39] The clinical failure of BX-471, CP-481,715, MLN3897 and AZD-4818 have tempered interest in CCR1 antagonists as immunomodulatory agents. However, it is conceivable that the clinical failure of these agents may be related to an inability to maintain high levels of receptor occupancy at all times, rather than poor target choice.[40]

11.1.4 CCR1 Involvement in Cancer

Tumour invasion and metastasis share many similarities with leukocyte trafficking, and CCR1 expression has been observed in carcinomas of the breast,[41] ovary,[42] prostate,[43] lung[44] and liver.[45] For ovarian and prostate cancer cells, the CCR1 ligand, CCL5, has been postulated to contribute to up-regulation of prostate cancer invasion.[46] Moreover, while CCR1 expression was observed in hepatoma cells and, to a lesser degree, endothelial cells in hepatoma tissues, it was not present in normal liver tissues.[45] Subsequently, the CCL3/CCR1 axis was shown to be involved in the progression of hepatocellular carcinoma.[46] Recently, Wang *et al.* observed that suppression of CCR1 expression (through use of miRNA) significantly attenuated the invasive ability of non-small-cell lung cancer cells.[44] In addition, CCR1 is also thought to be important in metastasis of cancer. Using a transplantable model of breast cancer (410.4 model), Robinson *et al.* showed that Met-CCL5 (an antagonist for both CCR1 and CCR5) slowed tumour growth and reduced macrophage infiltrate.[41] Likewise, using a mouse model for liver dissemination, Kitamura *et al.* demonstrated that mouse and human colon cancer cells secrete CCL9 and CCL15, and express CCR1.[24] Additionally, they found that lack of the *Ccr1* gene dramatically suppressed outgrowths of disseminated tumours in the liver. Finally, they demonstrated that the CCR1 antagonist BL5923 (Novartis) blocked immature myeloid cell accumulation and metastatic colonization, and significantly prolonged the survival of tumour-bearing mice. As a result, investigators have suggested that pharmacological intervention at CCR1 may serve as a novel therapeutic strategy to block tumour invasion at metastatic sites.[47]

In addition to solid neoplasms, CCR1 has also emerged as a potential target in multiple myeloma (MM), a clonal neoplasm of plasma cells.[48] Multiple myeloma is unique in its propensity to cause osteolysis, with 80% of patients suffering from progressive bone destruction. Myeloma cells require "homing" signals mediated by chemokines and chemokine receptors to enter the bone marrow (BM) where they interact with the stroma and receive proliferation/survival signals. Several investigators have shown that human MM cells express multiple chemokine receptors including CCR1.[49,50] Experiments by Lentzsch *et al.*[51] demonstrated the *in-vitro* chemotactic effect of CCL3 on myeloma cells. In addition, myeloma cells were shown to secrete high levels of CCL3 constitutively, and signalling through CCR1 enhanced adhesive interactions between myeloma and marrow stromal cells. The advent of DNA array studies showed CCL3 to be present at much higher levels in patients with MM than patients with other haematologic malignancies or normal control individuals.[52] Moreover, serum levels of CCL3 correlated with the extent of bone disease.[53] Important studies utilizing anti-sense and neutralizing antibodies to CCL3 and CCR1 showed that inhibition of either could alter disease progression in a mouse model of myeloma bone disease.[50,54] Finally, the use of CCR1 antagonists has been shown to inhibit bone lesions and/or tumour load *in vivo*.[54–56] Certainly, the data suggest that CCR1 plays a critical role in mediating the MM disease process, and hints that CCR1 antagonists may be an effective tool to block osteolytic bone disease in patients with MM. However, whether antagonists for CCR1 will provide a therapy for patients with MM or other forms of cancer remains to be seen given that the receptor is also an important driver of anti-tumour defence by attracting relevant cells to the tumour site[57] and its expression has been shown to correlate with patient survival.[58]

11.1.5 CCR1 Inhibitors

Since CCR1 appears to be involved in a variety of diseases, there has been considerable interest in developing small-molecule inhibitors. With the availability of large compound collections and high-throughput screening, a variety of novel chemotypes have been identified and optimized over the past dozen years. Many of these have demonstrated *in-vivo* animal efficacy, and at least six compounds have advanced as far as Phase II human clinical trials. These inhibitors are discussed in the subsequent sections.

11.2 CCR1 Inhibitors in Human Trials

11.2.1 Berlex

Researchers at Berlex were the first to publish examples of small-molecule CCR1 inhibitors.[59] These hydroxypiperidine antagonists were discovered by high-throughput screening of compound libraries (Figure 11.1). Compound **11.1**, an optimized compound from this series, is potent against recombinant human CCR1 ($K_i = 40$ nM, CCL3; $K_i = 60$ nM, CCL5) and selective *versus* other chemokine receptors, notably CCR5, which also binds to CCL3 and CCL5.

Further rounds of screening and optimization led to **11.2**. Over ten-fold more potent than **11.1**, **11.2** ($K_i = 1.0$ nM, CCL3; $K_i = 2.8$ nM, CCL5) was also a strong inhibitor of CCL3 and CCL5 induced calcium flux in CCR1 expressing HEK293 cells. Inhibition of CCL3 and CCL5 mediated leukocyte chemotaxis was achieved with 100 nM concentrations of **11.2**. In a human whole blood assay, pre-treatment with 1 μM **11.2** inhibited CCL3 induced CD11b up-regulation, an indicator of potential utility for multiple sclerosis. Compound **11.2** was more than 10,000-fold selective over other GPCRs tested, including CCR5.[20]

Although **11.2** was 100-fold less active against rat CCR1, with a K_i of 121 nM (CCL3) and less than 20% bioavailable in the rat, it was advanced into a rat disease model for multiple sclerosis. Dosed at 50 mg/kg (s.c., t.i.d.) in a rat EAE model, **11.2** achieved a 50% reduction in clinical score. Taking into account

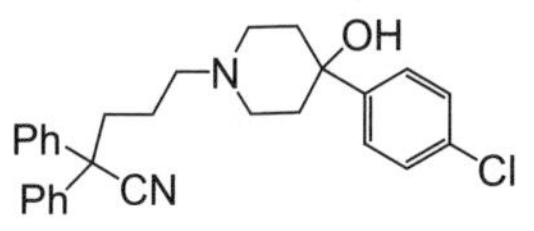

11.1, BX513 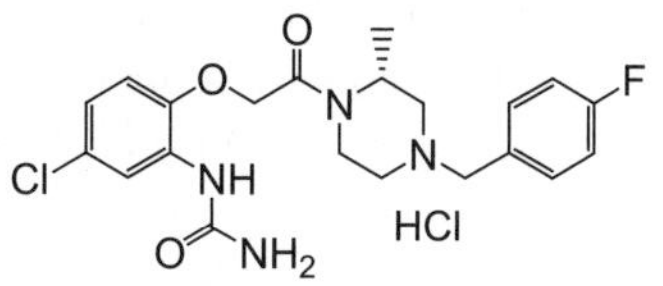   **11.2**, BX471

Figure 11.1   Berlex compounds.

greater inhibition for the human receptor, it was suspected that a dose of only 0.5 mg/kg might achieve the same effect in humans.[20] Furthermore, a p.o. dose of 4 mg/kg in dog afforded an improved pharmacokinetic profile ($t_{1/2} = 3$ h, $AUC_{0-6h} = 16$ µg × h/ml) offering hope for acceptable pharmacokinetics in human.

Compound **11.2** (BX-471) was the first CCR1 antagonist to advance into human clinical trials for multiple sclerosis. In a Phase II, randomized, double-blind, placebo-controlled trial with 105 relapsing/remitting multiple sclerosis patients, placebo or 600 mg of **11.2** was administered three times daily for 16 weeks. This dose achieved plasma concentrations greater than 100 ng/mL, more than 90% receptor coverage, and effectively reduced CD11b expression, as had been observed previously from human whole blood results. Despite this, MRI assessment of newly active brain lesions showed no difference between control and treated patients.[36] This trial provides strong evidence that CCR1 selective antagonism is not sufficient for remediation of multiple sclerosis. In 2008 Berlex was integrated into Bayer HealthCare, and there has been no further indication of clinical trial activity for BX-471.

Despite the lack of clinical trial efficacy, BX-471 continues to serve as a reference standard for the study of CCR1, and it has shown promise in a variety of *in-vitro* and *in-vivo* disease models. In a rat heart transplant rejection model, a combination of BX-471 and cyclosporine, at subtherapeutic doses, were more effective than either compound alone for the prolongation of cardiac graft survival.[23] In addition, BX-471 has demonstrated efficacy in a variety of rodent models for kidney diseases.[60–62] *In-vitro* multiple myeloma models showed that CCL5 stimulated osteoclast formation and adhesion of myeloma cells to stromal cells were significantly inhibited with BX-471.[50] In addition, osteolytic bone lesions were reduced by 40% in 5T2MM mice treated with BX-471.[54] Taken together, both these studies continue to support the therapeutic potential of small-molecule CCR1 antagonists such as BX-471.

11.2.2 Pfizer

Through collection screening, researchers at Pfizer identified quinoline carboxaldehyde **11.3** as a weak antagonist of CCR1 ($IC_{50} = 2.3$ µM, CCL3) (Figure 11.2). This compound also inhibited CCR1-mediated chemotaxis ($IC_{50} = 0.84$ µM, CCL3; $IC_{50} = 0.63$ µM, CCL5).[63] Further optimization afforded **11.4** ($IC_{50} = 28$ nM, CCL3). The quinoxaline replacement for quinoline provided a ten-fold improvement in potency. Likewise, the replacement of the cyclohexyl group with a phenyl ring afforded an additional ten-fold potency improvement. Compound **11.4** inhibited CCR1 mediated chemotaxis ($IC_{50} = 2$ nM, CCL3) and was 10–20 times more potent than **11.3** for inhibition of CCL3-induced up-regulation of CD11b in monocytes.[64]

Due to the poor microsomal stability of compounds like **11.4**, further modifications focused on the introduction of polar functional groups to improve bioavailability. Thus, introduction of a hydroxy group on the alkyl

PS031291 was roughly ten-fold more potent than BX-471 for inhibition of CCR1-mediated chemotaxis of THP-1 cells ($IC_{50} = 0.6$ nM). Furthermore, the compound was a much better inhibitor of CCL3-induced CCR1 internalization ($IC_{50} = 4$ nM) than BX-471 ($IC_{50} = 140$ nM) in a human whole blood assay. Like BX-471, PS031291 was moderately bioavailable in the rat ($F = 19\%$), and was significantly better in the dog ($F = 86\%$).[77]

In 2008, Pharmacopeia was acquired by Ligand Pharmaceuticals. There have been no further reported efforts for advancement of a pre-clinical candidate from this series.

11.3.2 Novartis

Researchers at Novartis discovered a novel series of cinnamides, which are structurally similar to BX-471. An early compound from this series, **11.13**, was potent against human CCR1 ($IC_{50} = 30$ nM, CCL3) (Figure 11.7). Furthermore, this compound was 95% bioavailable with a 16.5-hour half-life in the rat (3 mg/kg, p.o.). However, poor affinity for rodent CCR1 precluded its evaluation in disease models necessitating another round of optimization in search of a compound that would affect the rodent receptor.[29]

Installation of a bridged piperazine afforded compounds with activity for rat and mouse CCR1. An optimized compound from this series, **11.14**, was a potent inhibitor of CCR1 binding for human ($IC_{50} = 20$ nM, CCL3) and rat ($IC_{50} = 28$ nM, CCL3) receptors. This compound also inhibited calcium flux ($IC_{50} = 16$ nM, CCL3) and chemotaxis ($IC_{50} = 3$ nM, CCL3). Pharmacokinetic evaluation showed that the compound was 100% bioavailable with a 3.2-hour half-life in rats (3 mg/kg, p.o.).[78]

Since **11.14** was a potent inhibitor of rat CCR1 and possessed a good pharmacokinetic profile in rats, it was further evaluated in two rat acute experimental models. Compound **11.14**, when dosed at 30 mg/kg p.o., b.i.d., was slightly less effective than cyclosporine A in an experimental autoimmune encephalitis (EAE) model. The compound also inhibited knee-swelling when dosed at 5, 15 and 50 mg/kg (p.o., b.i.d.) in an antigen-induced arthritis (AIA) model but was less effective than dexamethasone.[78]

Recent publications indicate that compound **11.14** was further evaluated in a mouse model for colon cancer liver metastasis. Mice that were injected with colon cancer cells and then dosed at 50 mg/kg (p.o., b.i.d.) with CCR1 antagonist, BL5923 (structure not shown, but likely to be **11.14** based on the referenced

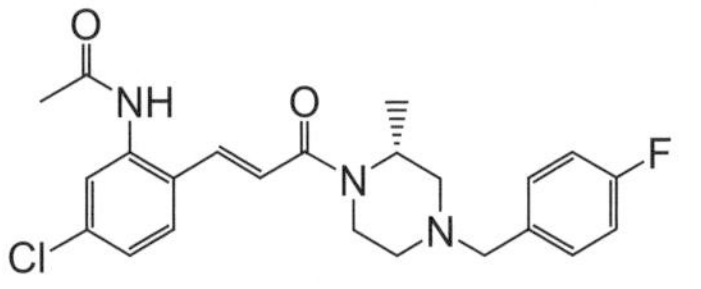

11.13 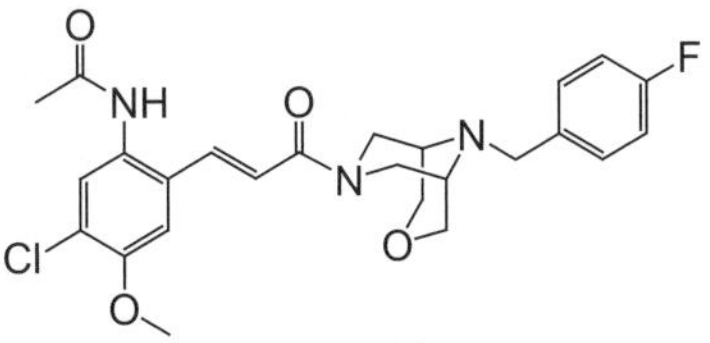   **11.14**

Figure 11.7 Novartis compounds.

manuscript),[78] showed reduced accumulation of immature myeloid cells at metastatic foci. Furthermore, mean survival time for these mice nearly doubled to 113 days. These results suggest that CCR1 antagonists could provide an adjuvant therapy for colorectal cancer patients after surgical resection of tumours.[24]

11.3.3 Banyu and Merck

Collaboration between researchers at Banyu Pharmaceuticals, Banyu Tsukuba Research Institute and Merck Research Laboratories resulted in a novel series of xanthenecarboxamide CCR1 antagonists. Collection screening afforded **11.15** ($IC_{50} = 510$ nM, CCL3) (Figure 11.8). Replacement of the *N*-hexyl group with cycloalkyl and arylmethyl groups improved potency.[79] However, lack of potency for rodent CCR1 precluded study of these compounds in rodent disease models. Fortuitously, transformation of the piperidine nitrogen into a quaternary ammonium ion afforded a compound, **11.16**, capable of modest inhibition of mouse CCR1 ($IC_{50} = 2100$ nM, CCL3) and improved inhibition of the human receptor ($IC_{50} = 14$ nM, CCL3). Further improvements were achieved through extension of the piperidine *N*-alkyl group and by substitution of the xanthene. The optimized compound, **11.17**, was isolated from a mixture of two isomers presumably resulting from the quaternary nitrogen stereocentre. The most potent isomer was an inhibitor of human ($IC_{50} = 0.9$ nM, CCL3) and rodent ($IC_{50} = 0.9$ nM, CCL3) CCR1. This potent **11.17** isomer also inhibited CCL3-induced calcium flux in human ($IC_{50} = 0.73$ nM) and mouse cells ($IC_{50} = 21$ nM).[80] Additionally, a compound from this series was found to have dual activity for the CCR3 receptor.[81] Compound **11.17**, owing to its poor oral bioavailability associated with the quaternary ammonium feature, was deemed

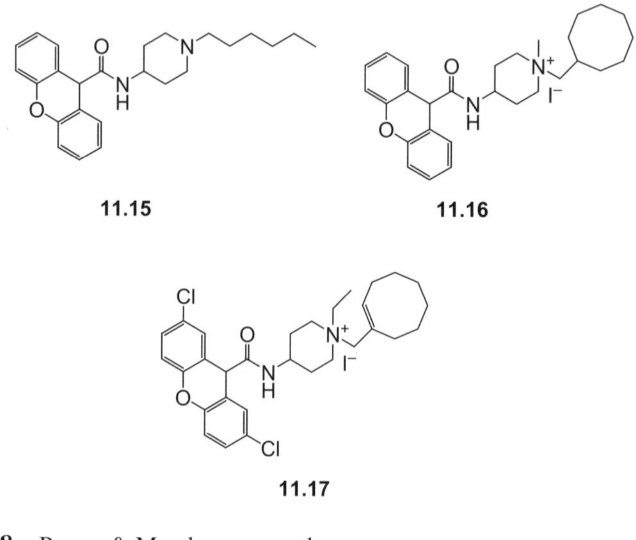

11.15   **11.16**

11.17

Figure 11.8 Banyu & Merck compounds.

unsuitable for further development. Optimization attempts that returned to the original piperidine tertiary amine motif afforded compounds with only moderate metabolic stability and affinity for only the human receptor.[82]

11.3.4 Bristol-Myers Squibb

Recent patent applications filed by Bristol-Myers Squibb (BMS) claim a series of hydroxy piperidine derivatives that are reminiscent of Millennium's CCR1 antagonists (Figure 11.9). Prodrugs of **11.18** are specifically claimed in one application.[83] A second application provides CCR1 binding data for specific isomer of **11.19** ($K_i = 2.4$ nM, CCL3).[84] BMS has recently reported advancement of BMS-817399, structure undisclosed, into Phase II trials with RA patients in combination with methotrexate.[85]

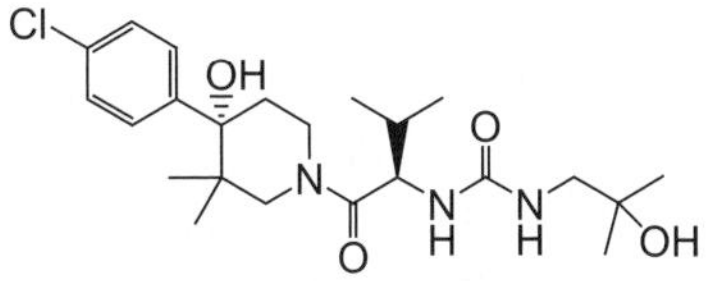

11.18 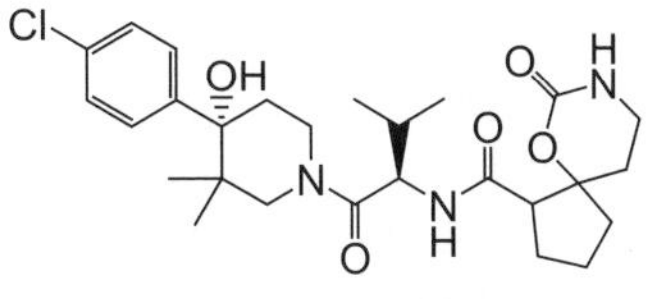   **11.19**

Figure 11.9   Bristol-Myers Squibb compounds.

11.3.5 Boehringer Ingelheim

Researchers at Boehringer Ingelheim recently filed patent applications for two related series of novel CCR1 antagonists (Figure 11.10). One application

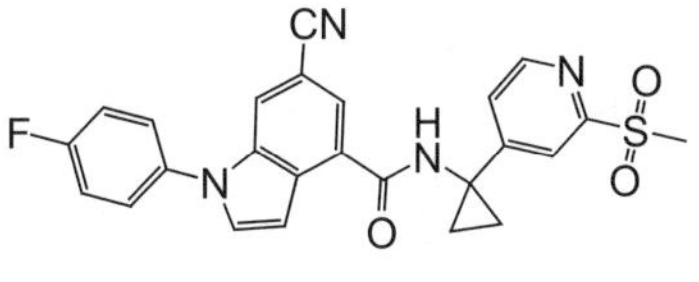

11.20 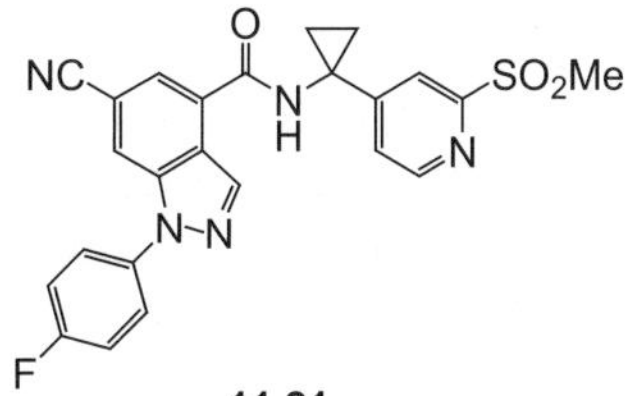   **11.21**

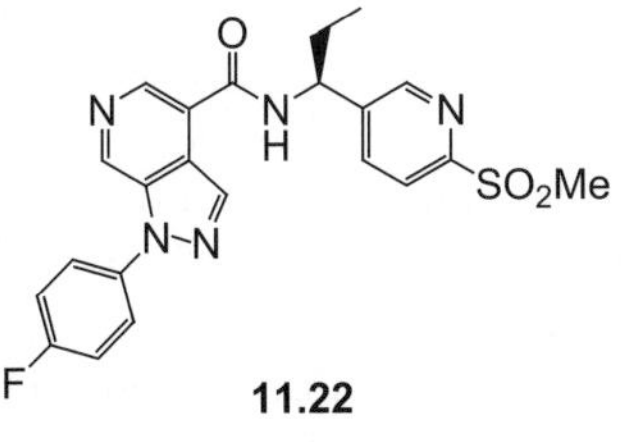

11.22

Figure 11.10   Boehringer-Ingelheim compounds.

identifies the indole, **11.20**, as a potent inhibitor of CCR1-mediated calcium flux ($IC_{50} = 0.8$ nM, CCL3).[86] A second application provides calcium flux data for indazole **11.21** ($IC_{50} = 0.1$ nM, CCL3).[87] More recent patents include multi-gram synthetic exemplifications for aza-indoles such as **11.22**.[88–90]

11.4 Strategies to Improve CCR1 Compounds

11.4.1 Increase Receptor Coverage

The *in-vivo* duration of drug action depends not only on pharmacokinetic properties like plasma half-life, it is also influenced by receptor coverage. Many investigators have speculated that, for effect, therapeutic use agents targeting CCR1 must provide high levels of target occupancy ($>95\%$). Long-lasting target-binding drugs are expected to be the most suitable agents for therapies that require continued, high levels of receptor occupancy. One method of increasing receptor coverage is the use of compounds with slow dissociation rates, of which there are a number of examples, including the CCR5 antagonists Maraviroc and Aplaviroc.[91] Yet, in most drug-screening programs, ligand-receptor interactions are traditionally quantified in terms of affinity and efficacy only. In the future, programs directed at the identification of potent CCR1 antagonists might benefit from the use of association/dissociation kinetic assays, in addition to the equilibrium binding assay, to focus on analogues with long residence times.[92] In addition to information gained on the offset rate, association/dissociation kinetic assays are more sensitive for detecting receptor conformational changes and can directly validate whether a test compound is acting allosterically.

11.4.2 Address Chemokine Promiscuity

The chemokine receptor CCR1 is activated by a plethora of ligands, yet the majority of CCR1 antagonists have been identified using CCL3, and little has been published on the effectiveness of CCR1 antagonists to block binding and/or receptor activation by alternative ligands. The exceptions are BX-471 with data published for CCL5, CCL7 and CCL15,[20,93] and CCX-354, which has published data for CCL15.[37] That many of the CCR1 antagonists are allosteric in nature,[94] and many allosteric inhibitors are "probe dependent",[95] highlights the importance of cooperativity between orthosteric and allosteric ligands. As investigators move from the screening process to structure optimization, it would be helpful to have an in-depth profile for the ability of CCR1 antagonists to inhibit binding or functional responses to alternative ligands.

Some allosteric modulators appear to work by altering the ability of the agonist to stabilize the receptor-G protein complex. For example, the small-molecule allosteric enhancers ATL525 and PD81,723 promote a slow ($t_{1/2} = 10$ min.) conversion of A_1AR from a mixture of low-/high-affinity state receptors to only high-affinity receptors (coupled to G-proteins), and reduce sensitivity of radioligand agonist binding to inhibition by GTPγS. Early binding work by other investigators has shown that CCL3 binding is sensitive to inhibition by GTP. Given the allosteric nature of several of the CCR1 compounds it would be interesting to know if GTP was able to have an effect on saturation binding studies. Unfortunately, the biological information available for most of the disclosed CCR1 antagonists is quite limited. For example, biological information on AZD-4818 and BMS-817399 is restricted to only what has been provided in the PCT patent applications.

11.4.3 Interrogate Signalling Using a Context-Dependent Approach

Antagonism of receptor activation can be the result of compounds that compete for the same binding site as the agonist, termed orthosteric antagonists, or they may be due to binding at a site distinct from the agonist binding site, termed allosteric antagonists. Allosteric antagonists may bind to a region with subsequent changes in receptor conformation that alter agonist affinity. While some CCR1 antagonists are orthosteric (CP481,715), others have been shown to be allosteric (BX-471) in nature. Evidence suggests that in addition to being probe-dependent (ligand bias), and pathway-dependent (functional selectivity), allosteric ligand behaviour may also be dependent on the cell type (context-dependent). Context-dependent signalling occurs when allosteric induced conformational changes in the receptor alter binding of interacting proteins that are present in different amounts in different cell lines.[96] Recent advancement in label-free technologies allows screening to be performed using cells with the most biologically relevant context, such as cell lines expressing endogenous receptors or primary cells.[97]

11.5 Conclusion

Pharmacodynamic parameters such as affinity, efficacy, orthosteric *vs.* allosteric binding and rate of dissociation from the biological target are ultimately reflected in the ability of a compound to alter biological activity. As a result, these parameters serve as the foundation on which chemical information on structure activity relationships (SARs) is established for target selectivity, therapeutic activity, pharmacokinetics and safety. In addition to quantifying effects in test systems, these parameters can also be used to predict subsequent activity in a therapeutic setting. However, to enhance the signal window, functional responses have frequently been generated in cell systems where CCR1 was heavily over-expressed, often alongside chimeric or promiscuous G-proteins. As a result, the increase in receptor density may ultimately influence receptor reserve, resulting in significant changes to efficacy and potency when a compound's effects on cellular responses are examined in native expression systems. Additionally, the milieu of downstream interacting proteins may differ from one cell line to another. Thus, recognition of phenomena such as biased agonism, probe-dependence and context-dependent signalling demands that SAR be performed, and published, using physiologically relevant systems with physiologically relevant end-points.

References

1. K. Neote, D. DiGregorio, J. Mak, R. Horuk and T. Schall, *Cell*, 1993, **72**, 415.
2. I. Charo and R. Ransohoff, *N. Engl. J. Med.*, 2006, **354**, 610.
3. R. Berahovich, Z. Miao, Y. Wang, B. Premack, M. Howard and T. Schall, *J. Immunol.*, 2005, **174**, 7341.
4. H. Hamm and A. Gilchrist, *Curr. Opin. Cell Biol.*, 1996, **8**, 189.
5. R. Eglen, A. Gilchrist and T. Reisine, *Comb. Chem. High Throughput Screen.*, 2008, **11**, 560.
6. B. Nardelli, H. Tiffany, G. Bong, P. Yourey, D. Morahan, Y. Li, P. Murphy and R. Alderson, *J. Immunol.*, 1999, **162**, 435.
7. Y. Tian, M. Lee, L. Yung, R. Allen, P. Slocombe, B. Twomey and Y. Wong, *Cell. Signal.*, 2008, **20**, 1179.
8. Y. Tian, D. New, L. Yung, R. Allen, P. Slocombe, B. Twomey, M. Lee and Y. Wong, *Eur. J. Immunol.*, 2004, **34**, 785.
9. R. Melnychuk, D. Streblow, P. Smith, A. Hirsch, D. Pancheva and J. Nelson, *J. Virol.*, 2004, **78**, 8382.
10. M. Rosenkilde, K. McLean, P. Holst and T. Schwartz, *J. Biol. Chem.*, 2004, **279**, 32524.
11. L. Shepard, M. Yang, P. Xie, D. Browning, T. Voyno-Yasenetskaya, T. Kozasa and R. Ye, *J. Biol. Chem.*, 2001, **276**, 45979.
12. W. Tan, D. Martin and J. Gutkind, *J. Biol. Chem.*, 2006, **281**, 39542.
13. C. Weber, K. Belge, P. von Hundelshausen, G. Draude, B. Steppich, M. Mack, M. Frankenberger, K. Weber and H. Ziegler-Heitbrock, *J. Leukoc. Biol.*, 2000, **67**, 699.
14. R. Gladue, L. Tylaska, W. Brissette, P. Lira, J. Kath, C. Poss, M. Brown, T. Paradis, M. Conklyn, K. Ogborne, M. McGlynn, B. Lillie, A. DiRico, E. Mairs, E. McElroy, W. Martin, I. Stock, R. Shepard, H. Showell and K. Neote, *J. Biol. Chem.*, 2003, **278**, 40473.
15. C. Weber, K. Weber, C. Klier, S. Gu, R. Wank, R. Horuk and P. Nelson, *Blood*, 2001, **97**, 1144.
16. C. Klier, E. Nelson, C. Cohen, R. Horuk, D. Schlöndorff and P. Nelson, *Biol. Chem.*, 2001, **382**, 1405.
17. J. L. Gao, T. A. Wynn, Y. Chang, E. J. Lee, H. E. Broxmeyer, S. Cooper, H. L. Tiffany, H. Westphal, J. Kwon-Chung and P. M. Murphy, *J. Exp. Med.*, 1997, **185**, 1959.
18. S. Ribeiro and R. Horuk, *Pharmacol. Ther.*, 2005, **107**, 44.
19. S. Hosaka, T. Akahoshi, C. Wada and H. Kondo, *Clin. Exp. Immunol.*, 1994, **97**, 451.
20. M. Liang, C. Mallari, M. Rosser, H. Ng, K. May, S. Monahan, J. Bauman, I. Islam, A. Ghannam, B. Buckman, K. Shaw, G. Wei, W. Xu, Z. Zhao, E.

Ho, J. Shen, H. Oanh, B. Subramanyam, R. Vergona, D. Taub, L. Dunning, S. Harvey, R. Snider, J. Hesselgesser, M. Morrissey and H. Perez, *J. Biol. Chem.*, 2000, **275**, 19000.

21. H. Kerstjens, L. Bjermer, L. Eriksson, K. Dahlström and J. Vestbo, *Respir. Med.*, 2010, **104**, 1297.

22. R. Gladue, M. Brown and S. Zwillich, *Curr. Top. Med. Chem.*, 2010, **10**, 1268.

23. R. Horuk, C. Clayberger, A. Krensky, Z. Wang, H. Grone, C. Weber, K. Weber, P. Nelson, K. May, M. Rosser, L. Dunning, M. Liang, B. Buckman, A. Ghannam, H. Ng, I. Islam, J. Bauman, G. Wei, S. Monahan, W. Xu, R. Snider, M. Morrissey, J. Hesselgesser and H. Perez, *J. Biol. Chem.*, 2001, **276**, 4199.

24. T. Kitamura, T. Fujishita, P. Loetscher, L. Revesz, H. Hashida, S. Kizaka-Kondoh, M. Aoki and M. Taketo, *Proc. Natl Acad. Sci. USA*, 2010, **107**, 13063.

25. NIH, *Study to Investigate the Efficacy of a Non-hormonal Drug Against Endometriosis Associated Pelvic Pain*, http://www.clinicaltrials.gov/ct2/show/NCT00185341, 2009, Accessed 01/15/2011.

26. M. Halks-Miller, M. Schroeder, V. Haroutunian, U. Moenning, M. Rossi, C. Achim, D. Purohit, M. Mahmoudi and R. Horuk, *Ann. Neurol.*, 2003, **54**, 638.

27. J. Borregaard, L. Skov, L. Wang, N. Ting, C. Wang, L. Beck, J. Sonne and A. Clucas, *Contact Dermatitis*, 2008, **59**, 211.

28. W. Karpus, N. Lukacs, B. McRae, R. Strieter, S. Kunkel and S. Miller, *J. Immunol.*, 1995, **155**, 5003.

29. L. Revesz, B. Bollbuck, T. Buhl, J. Eder, R. Esser, R. Feifel, R. Heng, P. Hiestand, B. Jachez-Demange, P. Loetscher, H. Sparrer, A. Schlapbach and R. Waelchli, *Bioorg. Med. Chem. Lett.*, 2005, **15**, 5160.

30. D. Barnes, J. Tse, M. Kaufhold, M. Owen, J. Hesselgesser, R. Strieter, R. Horuk and H. Perez, *J. Clin. Invest.*, 1998, **101**, 2910.

31. R. Horuk, S. Shurey, H. Ng, K. May, J. Bauman, I. Islam, A. Ghannam, B. Buckman, G. Wei, W. Xu, M. Liang, M. Rosser, L. Dunning, J. Hesselgesser, R. Snider, M. Morrissey, H. Perez and C. Green, *Immunol. Lett.*, 2001, **76**, 193.

32. H. Anders, V. Vielhauer, M. Frink, Y. Linde, C. Cohen, S. Blattner, M. Kretzler, F. Strutz, M. Mack, H. Gröne, J. Onuffer, R. Horuk, P. Nelson and D. Schlöndorff, *J. Clin. Invest.*, 2002, **109**, 251.

33. X. Yang, W. Walton, D. Cook, X. Hua, S. Tilley, C. Haskell, R. Horuk, A. Blackstock and S. Kirby, *Am. J. Respir. Cell. Mol. Biol.*, 2011, **45**, 127.

34. K. Blease, B. Mehrad, T. Standiford, N. Lukacs, S. Kunkel, S. Chensue, B. Lu, C. Gerard and C. Hogaboam, *J. Immunol.*, 2000, **165**, 1564.

35. NIH, *Proof-of-Concept Study With BMS-817399 to Treat Moderate to Severe Rheumatoid Arthritis (RA)*, http://clinicaltrials.gov/ct2/show/NCT01404585, 2011, Accessed 01/15/2011.

36. F. Zipp, H. Hartung, J. Hillert, S. Schimrigk, C. Trebst, M. Stangel, C. Infante-Duarte, P. Jakobs, C. Wolf, R. Sandbrink, C. Pohl and M. Filippi, *Neurology*, 2006, **67**, 1880.

37. D. Dairaghi, P. Zhang, Y. Wang, L. Seitz, D. Johnson, S. Miao, L. Ertl, Y. Zeng, J. Powers, A. Pennell, P. Bekker, T. Schall and J. Jaen, *Clin. Pharmacol. Ther.*, 2011, **89**, 726.
38. J. Haringman, M. Kraan, T. Smeets, K. Zwinderman and P. Tak, *Ann. Rheum. Dis.*, 2003, **62**, 715.
39. C. Vergunst, D. Gerlag, L. von Moltke, M. Karol, T. Wyant, X. Chi, E. Matzkin, T. Leach and P. Tak, *Arthritis Rheum.*, 2009, **60**, 3572.
40. M. Lebre, C. Vergunst, I. Choi, S. Aarrass, A. Oliveira, T. Wyant, R. Horuk, K. Reedquist and P. Tak, *PLoS One*, 2011, **6**, e21772.
41. S. Robinson, K. Scott, J. Wilson, R. Thompson, A. Proudfoot and F. Balkwill, *Cancer Res.*, 2003, **63**, 8360.
42. C. Scotton, D. Milliken, J. Wilson, S. Raju and F. Balkwill, *Br. J. Cancer.*, 2001, **85**, 891.
43. T. Akashi, K. Koizumi, O. Nagakawa, H. Fuse and I. Saiki, *Oncol. Rep.*, 2006, **16**, 831.
44. C. Wang, B. Sun, Y. Tang, H. Zhuang and W. Cao, *J. Cancer Res. Clin. Oncol.*, 2009, **135**, 695.
45. P. Lu, Y. Nakamoto, Y. Nemoto-Sasaki, C. Fujii, H. Wang, M. Hashii, Y. Ohmoto, S. Kaneko, K. Kobayashi and N. Mukaida, *Am. J. Pathol.*, 2003, **162**, 1249.
46. X. Yang, P. Lu, C. Fujii, Y. Nakamoto, J. Gao, S. Kaneko, P. Murphy and N. Mukaida, *Int. J. Cancer*, 2006, **118**, 1869.
47. T. Kitamura and M. Taketo, *Cancer Res.*, 2007, **67**, 10099.
48. A. Karash and A. Gilchrist, *Future Med. Chem.*, 2011, **3**, 1889.
49. C. Möller, T. Strömberg, M. Juremalm, K. Nilsson and G. Nilsson, *Leukemia*, 2003, **17**, 203.
50. Y. Oba, J. Lee, L. Ehrlich, H. Chung, D. Jelinek, N. Callander, R. Horuk, S. Choi and G. Roodman, *Exp. Hematol.*, 2005, **33**, 272.
51. S. Lentzsch, M. Gries, M. Janz, R. Bargou, B. Dörken and M. Mapara, *Blood*, 2003, **101**, 3568.
52. J. De Vos, G. Couderc, K. Tarte, M. Jourdan, G. Requirand, M. Delteil, J. Rossi, N. Mechti and B. Klein, *Blood*, 2001, **98**, 771.
53. E. Terpos, M. Politou, R. Szydlo, J. Goldman, J. Apperley and A. Rahemtulla, *Br. J. Haematol.*, 2003, **123**, 106.
54. E. Menu, E. De Leenheer, H. De Raeve, L. Coulton, T. Imanishi, K. Miyashita, E. Van Valckenborgh, I. Van Riet, B. Van Camp, R. Horuk, P. Croucher and K. Vanderkerken, *Clin. Exp. Metastasis*, 2006, **23**, 291.
55. B. Oyajobi, D. Dairaghi, A. Gupta, B. McCluskey, Y. Wang, L. Seitz, J. Powers, S. Miao, P. Zhang, T. Schall and J. Jaen, *Blood*, 2010, **116**, 3000.
56. S. Vallet and K. Anderson, *Expert Opin. Ther. Targets*, 2011, **15**, 1037.
57. N. Iida, Y. Nakamoto, T. Baba, K. Kakinoki, Y. Li, Y. Wu, K. Matsushima, S. Kaneko and N. Mukaida, *J. Leukoc. Biol.*, 2008, **84**, 1001.
58. C. Ohri, A. Shikotra, R. Green, D. Waller and P. Bradding, *BMC Cancer*, 2010, **10**, 172.

59. J. Hesselgesser, H. Ng, M. Liang, W. Zheng, K. May, J. Bauman, S. Monahan, I. Islam, G. Wei, A. Ghannam, D. Taub, M. Rosser, R. Snider, M. Morrissey, H. Perez and R. Horuk, *J. Biol. Chem.*, 1998, **273**, 15687.
60. V. Vielhauer, E. Berning, V. Eis, M. Kretzler, S. Segerer, F. Strutz, R. Horuk, H. Gröne, D. Schlöndorff and H. Anders, *Kidney Int.*, 2004, **66**, 2264.
61. H. Anders, V. Ninichuk and D. Schlöndorff, *Kidney Int.*, 2006, **69**, 29.
62. V. Ninichuk and H. Anders, *Am. J. Nephrol.*, 2005, **25**, 365.
63. J. Kath, A. DiRico, R. Gladue, W. Martin, E. McElroy, I. Stock, L. Tylaska and D. Zheng, *Bioorg. Med. Chem. Lett.*, 2004, **14**, 2163.
64. J. Kath, W. Brissette, M. Brown, M. Conklyn, A. DiRico, P. Dorff, R. Gladue, B. Lillie, P. Lira, E. Mairs, W. Martin, E. McElroy, M. McGlynn, T. Paradis, C. Poss, I. Stock, L. Tylaska and D. Zheng, *Bioorg. Med. Chem. Lett.*, 2004, **14**, 2169.
65. M. Brown, M. Avery, W. Brissette, J. Chang, K. Colizza, M. Conklyn, A. DiRico, R. Gladue, J. Kath, S. Krueger, P. Lira, B. Lillie, G. Lundquist, E. Mairs, E. McElroy, M. McGlynn, T. Paradis, C. Poss, M. Rossulek, R. Shepard, J. Sims, T. Strelevitz, S. Truesdell, L. Tylaska, K. Yoon and D. Zheng, *Bioorg. Med. Chem. Lett.*, 2004, **14**, 2175.
66. R. Gladue, S. Cole, M. Roach, L. Tylaska, R. Nelson, R. Shepard, J. McNeish, K. Ogborne and K. Neote, *J. Immunol.*, 2006, **176**, 3141.
67. A. Clucas, A. Shah, Y. Zhang, V. Chow and R. Gladue, *Clin. Pharmacokinet.*, 2007, **46**, 757.
68. R. Gladue, S. Zwillich, A. Clucas and M. Brown, *Curr. Opin. Investig. Drugs*, 2004, **5**, 499.
69. K. G. Carson and G. C. Harriman, Millennium Pharmaceuticals, Inc., *PCT Int. Appl.*, WO/2004/043965, 2004.
70. S. Vallet, N. Raje, K. Ishitsuka, T. Hideshima, K. Podar, S. Chhetri, S. Pozzi, I. Breitkreutz, T. Kiziltepe, H. Yasui, E. Ocio, N. Shiraishi, J. Jin, Y. Okawa, H. Ikeda, S. Mukherjee, N. Vaghela, D. Cirstea, M. Ladetto, M. Boccadoro and K. Anderson, *Blood*, 2007, **110**, 3744.
71. P. Norman, *Expert Opin. Ther. Pat.*, 2009, **19**, 1629.
72. ChemoCentryx, *Product Pipeline: CCR1 Program*, http://www.chemocentryx.com/product/CCR1.html, 2011, accessed 01/15/2011.
73. A. A. S. M. Presentation, *L11 - Safety and Efficacy of Oral Chemokine Receptor 1 Antagonist CCX354-C in a Phase 2 Rheumatoid Arthritis Study* http://acr.confex.com/acr/2011/webprogram/Paper24548.html#, 2011, accessed 01/15/2011.
74. P. Zhang, A. M. K. Pennell, J. K. Wright, J. W. Chen, M. R. Leleti, Y. Li, L. Li, Y. Xu, M. M. Gleason, Y. Zeng and K. L. Greenman, Chemocentryx, Inc., *PCT Int. Appl.*, WO/2008/147822, 2008.
75. J. Merritt, J. Liu, E. Quadros, M. Morris, R. Liu, R. Zhang, B. Jacob, J. Postelnek, C. Hicks, W. Chen, E. Kimble, W. Rogers, L. O'Brien, N. White, H. Desai, S. Bansal, G. King, M. Ohlmeyer, K. Appell and M. Webb, *J. Med. Chem.*, 2009, **52**, 1295.
76. J. Merritt, R. James, V. Paradkar, C. Zhang, R. Liu, J. Liu, B. Jacob, C. Chiriac, M. Ohlmeyer, E. Quadros, P. Wines, J. Postelnek, C. Hicks, W.

Chen, E. Kimble, L. O'Brien, N. White, H. Desai, K. Appell and M. Webb, *Bioorg. Med. Chem. Lett.*, 2010, **20**, 5477.

77. J. R. Merritt, R. Liu, M. Morris, M. Ohlmeyer, R. Zhang, K. Appell, S. Bansal, B. Jacob, G. King, J. Liu, M. Webb, W. Chen, H. Desai, C. Hicks, E. Kimble, J. Postelnek, E. Quadros, L. Rogers and P. Wines, 236th ACS National Meeting, Philadelphia, PA, United States, 2008.

78. L. Revesz, B. Bollbuck, T. Buhl, J. Dawson, R. Feifel, R. Heng, P. Hiestand, H. Sparrer, A. Schlapbach and R. Waelchli, *Lett. Drug Des. Discovery*, 2006, **3**, 689.

79. A. Naya, Y. Sagara, K. Ohwaki, T. Saeki, D. Ichikawa, Y. Iwasawa, K. Noguchi and N. Ohtake, *J. Med. Chem.*, 2001, **44**, 1429.

80. T. Saeki and A. Naya, *Curr. Pharm. Des.*, 2003, **9**, 1201.

81. I. Sabroe, M. Peck, B. Van Keulen, A. Jorritsma, G. Simmons, P. Clapham, T. Williams and J. Pease, *J. Biol. Chem.*, 2000, **275**, 25985.

82. A. Naya, M. Ishikawa, K. Matsuda, K. Ohwaki, T. Saeki, K. Noguchi and N. Ohtake, *Bioorg. Med. Chem. Lett.*, 2003, **11**, 875.

83. J. Hynes, P. H. Carter, L. A. M. Cornelius, T. G. M. Dhar, J. V. Duncia, J. B. Santella, H. Wu, S. Nair and J. S. Warrier, *Bristol-Myers Squibb, USA, WO*/, 2011/044309, 2011.

84. T. G. M. Dhar, J. V. Duncia, D. S. Gardner, W. Guo and J. Hynes, Bristol-Myers Squibb Company, *PCT Int. Appl.*, WO/2011/044197, 2011.

85. Bristol-Myers Squibb, *Study Connect*, http://www.bms.com/studyconnect/Pages/ProtocolPage.aspx?govid=IM126-004,NCT01404585, 2011, accessed 01/15/2011.

86. B. N. Cook and D. Kuzmich, Boehringer Ingelheim, *PCT Int. Appl.*, WO/2011/056440, 2011.

87. B. N. Cook, D. Kuzmich, C. Mao and H. Razavi, Boehringer Ingelheim, *PCT Int. Appl.*, WO/2011/049917, 2011.

88. B. N. Cook, D. DiSalvo, D. R. Fandrick, C. Harcken, D. Kuzmich, T. W.-H. Lee, P. Liu, J. Lord, C. Mao, J. Neu, B. C. Raudenbush, H. Razavi, J. T. Reeves, J. J. Song, A. D. Swinamer and Z. Tan, Boehringer Ingelheim, *US Pat. Appl.*, US 2011/7879873, 2011.

89. B. N. Cook, D. D. Salvo, C. Harcken, D. Kuzmich, T. W.-H. Lee, P. Liu, J. Lord, C. Mao, J. Neu, B. C. Raudenbush, H. Razavi and A. D. Swinamer, Boehringer Ingelheim, *US Pat. Appl.*, US 2011/8063065, 2011.

90. D. DiSalvo, D. Kuzmich, C. Mao, H. Razavi, C. R. Sarko, A. D. Swinamer, D. S. Thomson and Q. Zhang, Boehringer Ingelheim International GmbH, *US Pat.* US 8008327, 2011.

91. C. Watson, S. Jenkinson, W. Kazmierski and T. Kenakin, *Mol. Pharmacol.*, 2005, **67**, 1268.

92. G. Vauquelin and S. Charlton, *Br. J. Pharmacol.*, 2010, **161**, 488.

93. K. Furuichi, J. Gao, R. Horuk, T. Wada, S. Kaneko and P. Murphy, *J. Immunol.*, 2008, **181**, 8670.

94. N. Vaidehi, S. Schlyer, R. Trabanino, W. Floriano, R. Abrol, S. Sharma, M. Kochanny, S. Koovakat, L. Dunning, M. Liang, J. Fox, F. de

Mendonça, J. Pease, W. Goddard 3rd and R. Horuk, *J. Biol. Chem.*, 2006, **281**, 27613.

95. J. Jakubík, L. Bacáková, E. El-Fakahany and S. Tucek, *Mol. Pharmacol.*, 1997, **52**, 172.
96. C. Niswender, K. Johnson, N. Miller, J. Ayala, Q. Luo, R. Williams, S. Saleh, D. Orton, C. Weaver and P. Conn, *Mol. Pharmacol.*, 2010, **77**, 459.
97. C. Scott and M. Peters, *Drug Discov. Today*, 2010, **15**, 704.

CHAPTER 12

CCR2 Antagonists for the Treatment of Diseases Associated with Inflammation

CUIFEN HOU AND ZHIHUA SUI*

Johnson & Johnson Pharmaceutical Research and Development,
Welsh and McKean Roads, Spring House, PA 19477, USA
*Email: zsui@its.jnj.com

12.1 Introduction

Macrophages have been implicated in the pathogenesis of a variety of chronic inflammatory, pulmonary and autoimmune diseases. Therefore, a therapeutic strategy aimed at reducing the recruitment of monocytes/macrophages from the circulation to the site of inflammation would be a logical way to prevent the pathogenic effects of these cells. Since the cloning of the CCR2 receptor in 1994,[1] pre-clinical evidence has suggested that targeting the CCR2 pathway is a viable approach for drug discovery in inflammation-related diseases and conditions. This is based on the critical role this receptor plays in monocyte/marophage migration. In this chapter, we will review the role of CCR2 in disease and the medicinal chemistry efforts directed at the identification of antagonists, and briefly discuss pre-clinical and clinical challenges. Although there are still significant obstacles in the clinical development of CCR2 antagonists, from a medicinal chemistry perspective, CCR2 has become one of the most successful stories in small-molecule interventions of protein-protein interactions. The scientific activity around CCR2 has grown dramatically,

RSC Drug Discovery Series No. 26
Anti-Inflammatory Drug Discovery
Edited by Jeremy I Levin and Stefan Laufer
© The Royal Society of Chemistry 2012
Published by the Royal Society of Chemistry, www.rsc.org

especially in the past five years. A brief survey of the literature revealed that publications related to CCR2 alone (not including its ligands) have grown to more than 300 each year since 2006. There is a wide variety of chemical structures of CCR2 antagonists published and certainly many more to be disclosed or published. Several reviews on CCR2 antagonists have been published in the past few years.[2–11] While giving more attention to structures that were not covered in previous reviews, we will analyze the common structural features of the published CCR2 antagonists and summarize them based on the pharmacophores. There are three commonly used assays for evaluating CCR2 antagonists: 1. Binding assay that uses I^{125} labelled MCP-1 as the ligand; 2. Chemotaxis assay that measures the monocyte migration; 3. Calcium mobilization assay that measures the IP_3 signal transduction pathway (see Section 12.2.4). In most cases, only receptor binding potency is cited in this chapter, unless there is a major discrepancy between binding and functional potencies. Some drug discovery programs have used human recombinant CCR2 receptors in their assays while others used endogenous receptors, so potency comparisons between different labs are difficult. Even with the same protocol, such comparisons are inappropriate without detailed kinetic studies.

12.2 Biology of CCR2

12.2.1 Chemokines

Chemokines belong to the superfamily of cytokines, which induce chemotaxis (directed cell migration). These are small, water-soluble, heparin-binding proteins consisting of 68 to 120 amino acids residues with molecular weights of 8–12 kDa. More than 50 human chemokines and 20 chemokine receptors have been identified. Chemokines can be classified into four subfamilies (CXC, CC, XC and CX3C) based on the number and location of the conserved cysteine residues at the N-terminus of the molecule (Table 12.1). Among the four types, CXC and CC are the two major subfamilies while XC and CX3C are minor subfamilies.[12]

The CC family of chemokine contains two consecutive cysteines while in the CXC type the two cysteines are separated by another residue as illustrated in Figure 12.1, with IL-8 and MCP-1 (monocyte chemoattractant protein-1) as examples.[12,13]

Chemokines are also grouped into two main functional subfamilies: inflammatory and homeostatic chemokines. Inflammatory chemokines control the recruitment of leukocytes in inflammation and tissue injury. Homeostatic chemokines fulfil housekeeping functions such as navigating leukocytes to and

Table 12.1 Classification of chemokines.

Subfamily	CXC	CC	XC	CX3C
Examples	IL-8	MCP-1	lymphotactin	fractalkine
	GRO-α	RANTES		
	ENA-78	Eotaxin		

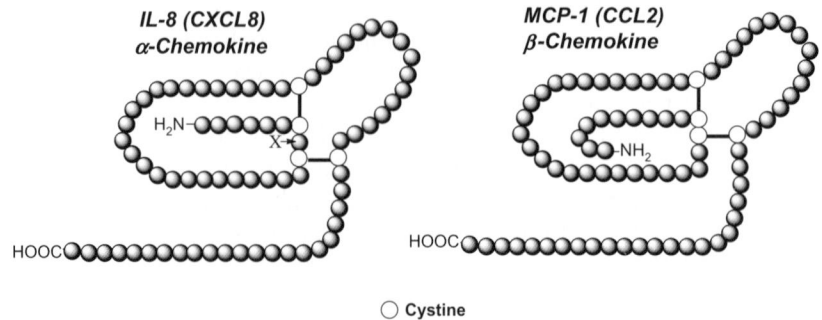

Figure 12.1

within secondary lymphoid organs as well as in bone marrow and thymus during haematopoiesis. Chemokines bind to specific cell surface transmembrane receptors, which are coupled to heterotrimeric G-proteins, whose activation leads to formation of intra-cellular signalling cascades that prompt migration toward the chemokine source. Directed migration of cells expressing the appropriate chemokine receptors occurs along a biochemical ligand gradient known as the chemokine gradient, allowing cells to move toward high local concentrations of chemokines at sites of inflammation. The regulation of leukocyte migration and activation and associated endothelial activation by chemokines are recognized as potentially important functions in the induction of acute and chronic inflammatory reactions.[14–17]

12.2.2   Monocyte Chemoattractant Protein-1 (MCP-1)

Monocyte chemoattractant protein-1 (MCP-1, CC chemokine ligand-2, CCL2) is a β or CC chemokine and a potent chemoattractant and activator for monocytes. MCP-1 attracts monocytes, macrophages, memory T-cells and mast cells to sites of chronic inflammation. MCP-1 plays a pivotal role in many inflammatory and autoimmune diseases, especially those characterized by a monocyte-rich infiltration.[18–21] MCP-1 has recently been implicated in various pathological conditions and diseases including atherosclerosis,[22–28] multiple sclerosis,[29–33] colitis,[34] inflammatory airway diseases,[35] cancer, uveitis,[36] obesity,[37 39] obesity-associated insulin resistance[40,41] and diabetic nephropathy.[42]

12.2.3   CC-Chemokine Receptor 2 (CCR2)

CCR2 is the only leukocyte MCP-1 receptor identified so far, and it mediates MCP-1 induced recruitment of circulating monocytes. CCR2 is a member of the CC chemokine receptors. It is predominantly expressed on monocytes, macrophages, basophils, immature dendritic cells, memory T-cells and mast cells. CCR2 is expressed in two slightly different forms (CCR2A and CCR2B) as a result of the alternative splicing of mRNA encoding the carboxy-terminal region that seems to determine their sensitivity to G-proteins. CCR2A consists of 360 amino acid (aa) residues. CCR2B (374 aa) represents alternative splicing at the carboxy terminal tail of the CCR2 protein.[47] The two have an identical sequence until residue 313, which is located in the C-terminal cytoplasmic region, and similar functional properties. Both RNAs are detectable in monocytes, blood derived dendritic cells (DC), natural killer (NK) cells and T lymphocytes but not in resting neutrophils or eosinophils. CCR2B appears to be the predominant form.[13] Several inflammatory/inducible chemokines are CCR2 agonists with similar affinity, including MCP-1/CCL2, MCP-3/CCL7, MCP-2/CCL8, MCP-4/CCL13 and mouse MCP-5. Only MCP-1 is exclusively selective for CCR2 *versus* other chemokine receptors. All of the other chemokines that interact with CCR2 also recognize other CC chemokine receptors.[13,43–46]

12.2.4 CCR2 Signal Transduction Pathways

MCP-1 exerts effects through binding to CCR2 on the surface of leukocytes targeted for activation and migration. As a GPCR, activated CCR2 catalyzes exchange of GTP for GDP on the $G\alpha$ subunit. As a result, conformational changes take place in CCR2, which leads to the dissociation of $G\beta\gamma$ dimer from $G\alpha$ and activation of multiple molecules of G-proteins ($G\alpha_i$, $G\alpha q$, $G\alpha_{12/13}$, $G\alpha_{14}$ and $G\alpha_{16}$). CCR2 can signal through different $G\alpha$ protein families, leading to distinct transduction pathways and biological effects. Activation of $G\alpha q$ causes the subsequent activation of phospholipase $C\beta2$ (PLC$\beta2$). PLC$\beta2$ cleaves phosphatidylinositol (4,5)-bisphosphate (PIP2) into two second messenger molecules known as inositol triphosphate (IP3) and diacylglycerol (DAG) that trigger intra-cellular signalling events. DAG activates protein kinase C (PKC) and IP3 triggers the release of calcium from intra-cellular stores. These events promote many signalling cascades such as the mitogen-activated protein kinase (MAPK) pathway that generate responses like chemotaxis, degranulation, release of superoxide anions and modification of integrin activity. The calcium mobilization assay is the most frequently used secondary assay for CCR antagonists. There is evidence that CCR2 can also activate several different intra-cellular effectors downstream of $G\alpha_i$ coupling, including: adenylyl cyclase, the low-molecular-weight proteins Ras, RhoA, Cdc42, phospholipase A2 (PLA2), phospholipase D, phosphatidylinositol-3-kinase (PI3K), tyrosine kinases Lyn, JAK2, cytoskeletal binding protein paxillin, downstream transcription factors STAT3/STAT5 and the MAP kinase pathways.[48–52] In addition, MCP-1 also induces CCR2B/$G\alpha_{12/13}$ interaction associated with increased RhoA GTPase activity and up-regulation of MCP-1. CCR2 induces $G\alpha_{12/13}$ activation resulting in actin-myosin contraction. $G\alpha_{12}$ and $G\alpha_{13}$ recruit the RhoA to the membrane to activate Rho-dependent kinase (ROCK) and myosin II, resulting in formation of contractile actomyosin complexes.[53] The $G\beta\gamma$ subunit was shown to be a positive regulator of a large number of effectors including adenylyl cyclase, PLC$\beta2$, PLA2, PI3Kγ and β-adrenergic receptor kinase. It is now clear that many effectors are regulated both by Gα and Gβγ subunits. The mitogen-activated protein kinases (MAP kinases) ERK1 and ERK2, Janus kinase JAK2, the stress-activated kinases JNK1 and p38 have all been implicated in MCP-1 signal transduction. In human monocytes, MCP-1 triggers tyrosine phosphorylation and activation of the JAK2/STAT3 pathway in a pertussis toxin (PTX)-independent manner.[18,54–58] MCP-1 has been shown to induce a rapid and transient activation of MAP-kinase in human monocytes, which was shown to be sensitive to H7.[54] This evidence indicates that serine/threonine kinases are required for MCP-1 stimulated chemotaxis.

CCR2 mediates MCP-1 induced recruitment of monocytes, macrophages and T lymphocytes through multiple signalling pathways. CCR2 is implicated in the trafficking of blood-borne monocytes to sites of inflammation and is involved in the pathogenesis of several chronic inflammatory, pulmonary and metabolic diseases.

One of the major challenges in the development of small-molecule CCR2 antagonists is the lack of pre-clinical cross-species activity. Sequence alignment indicates that human CCR2 shares 98%, 84% and 83% sequence homology with that of macaque, rat and mouse, respectively. MCP-1 shares 98%, 57% and 64% sequence homology with the corresponding ligand in macaque, rat and mouse, respectively.[59–62]

12.2.5 Roles of CCR2 and Its Ligands in Diseases

CCR2 and its ligands play key roles in the accumulation of inflammatory cells at the site of inflammation. They are involved in the patho-physiology of multiple disease states related to chronic inflammation.[63–67] An inhibitor of MCP-1 binding to CCR2 would be expected to block the migration of monocytes into the relevant tissues and potently suppress the inflammatory response. The literature linking MCP-1 and CCR2 to a variety of pathologies is reviewed below.

12.2.5.1 *Rheumatoid Arthritis*

Rheumatoid arthritis (RA) is a chronic disease that causes synovial inflammation and a progressive destruction of cartilage and bone in diarthrodial joints. One aspect of RA pathogenesis is the recruitment and activation of pro-inflammatory cells and secretion of cytokines and chemokines from macrophagic synoviocytes. Macrophages have been implicated in the disease progression of RA for many years. Macrophages are major producers of pro-inflammatory cytokines such as tumour necrosis factor α (TNFα) and interleukin-1β (IL-1β), and their excessive recruitment into the joint is believed to play a major role in the inflammatory process, ultimately leading to cartilage and bone destruction. MCP-1/CCR2 can attract monocytes, T-cells, NK cells and basophils. Therefore, using small molecules to target monocyte infiltration into joints is an interesting alternative to currently available therapies.

MCP-1 is highly expressed in synovial fluid in RA patients, and synovial tissue macrophages are the primary source of MCP-1 production.[68] The levels of MCP-1 in synovium correlate with swollen joint severity and CCR2 expressing cells are found in human RA joints.[69] Studies in different animal models of arthritis suggest that MCP-1 and CCR2 are involved in the pathogenesis of joint damage. An early report showed that injection of MCP-1 into rabbit joints resulted in macrophage infiltration of synovial tissues.[70] MCP-1 antibody significantly reduced the number of macrophages in the lesions and reduced ankle swelling in the rat collagen-induced arthritis (CIA) model.[71] Various laboratories have reported that small molecules as well as peptide CCR2 antagonists showed efficacy in this arthritis model. The blockade of CCR2 by a truncated form of MCP-1 also reduced the prevalence and severity of arthritis in an adjuvant-induced arthritis (AIA) model using MRL-lpr mice.[72,73] A well-known example of a small-molecule CCR2 antagonist is INCB3344 (Compound **5**).[74,75] Significant inhibition of severity in AIA in rats was achieved at 100 mg/kg, b.i.d., p.o., and this dosing schedule provided an IC_{50} at trough of $\sim 70\%$. Histologic analysis showed significant inhibition of joint damage and bone resorption in response to INCB3344 in the AIA model. In contrast to the studies described above, in which antagonists of MCP-1 were found to be effective treatments in RA disease models, one group has suggested that CCR2 KO mice are more susceptible to collagen-induced arthritis than are wild type (WT) mice.[76,77] Treatment of susceptible mice in the CIA model with a mAb against CCR2, early after immunization with collagen, inhibits the development of disease, but if treatment is delayed until day 21 the disease is more pronounced.[78] In spite of the conflicting reports, research and development activities focused on the use of CCR2 antagonists in RA continued. Many pharmaceutical companies have demonstrated the efficacy of small molecules in animal models of RA. However, enthusiasm for the development of CCR2 antagonists for RA has waned after recent reports of the lack of clinical efficacy in RA patients with the small-molecule CCR2 antagonist, MK-0812 (Compound **38**),[79] an MCP-1 antibody[80] and a CCR2 antibody.[81]

12.2.5.2 Multiple Sclerosis

Multiple sclerosis is an inflammatory demyelinating disease of the human central nervous system (CNS), which leads to axonal degeneration. The infiltration of leukocytes into the CNS is an essential step in the pathogenesis of MS. In humans, elevated expression of MCP-1 has been found in acute and chronic MS plaques. In MS, expression and cellular localization of MCP-1 and CCR2 have been described in the three compartments: brain, cerebrospinal fluid and blood.[82] Studies in a Th1-dependent animal model, experimental autoimmune encephalomyelitis (EAE), showed that MCP-1 was highly expressed in the CNS of affected rodents, and that MCP-1 antibodies could block disease relapse. Mice deficient for CCR2 did not develop EAE after active immunization, but did generate effector cells that could transfer the disease to naive wild-type recipients.[83] Deficiency in CCR2 confers resistance to

EAE induced with a peptide derived from myelin, oligodendrocyte glycoprotein peptide 35-55 (MOGp35-55). CCR2(−/−) mice immunized with MOGp35-55 failed to develop mononuclear cell inflammatory infiltrates in the CNS and did not show increased CNS levels of the chemokines RANTES, MCP-1, interferon (IFN)-inducible protein 10 (IP-10) or the chemokine receptors CCR1, CCR2 and CCR5. Additionally, T-cells from CCR2(−/−) immunized mice showed decreased antigen-induced proliferation and production of IFN-gamma compared with wild-type immunized controls, suggesting that CCR2 enhances the T helper cell type 1 immune response in EAE. These data indicate that CCR2 plays an important and potentially non-redundant role in the pathogenesis of EAE.[84] It was reported that CCR2(−/−) mice did not develop clinical EAE or characteristic CNS histopathology, and showed a significant reduction in T-cell- and CNS-infiltrating CD45(high)F4/80(+) monocyte sub-populations. Peripheral lymphocytes from CCR2(−/−) mice produced comparable levels of interferon-gamma (IFN-gamma) and interleukin-2 (IL-2) in response to antigen-specific restimulation when compared with control mice,[85] suggesting that the CCR2 pathway is involved in the pathogenesis of MS. Treatment of animals in the EAE model with small-molecule CCR2 antagonists has led to improvements in disease score and severity.[86,87]

12.2.5.3   Allergic Diseases and Asthma

Mounting evidence suggests that MCP-1 and its haematopoietic cell receptor CCR2 are involved in inflammatory disorders of the lung. In animal models of allergic asthma, idiopathic pulmonary fibrosis (IPF) and bronchiolitis obliterans syndrome (BOS), MCP-1 expression and protein production are increased and the disease process is attenuated by MCP-1 immunoneutralization. It is possible that MCP-1 acts *via* recruitment of regulatory and effector leukocytes, stimulation of histamine or leukotriene release from mast cells or basophils, induction of fibroblast production of transforming growth factor-beta (TGF-β) and procollagen, or enhancement of Th2 polarization. Recently, a polymorphism for MCP-1 has been described resulting in increased cytokine-induced release of MCP-1 by monocytes and increased risk of allergic asthma. These studies identify potentially important roles for MCP-1 in these lung inflammatory disorders. This presents both a challenge and an opportunity for targeting the MCP-1/CCL2/CCR2 axis in the disease.[88,89]

In asthma, MCP-1 is found in increased concentration in bronchoalveolar lavage (BAL) fluid from allergic asthmatic subjects and allergen challenge in stable asthmatics. MCP-1 has also been implicated in bronchial hyper-responsiveness. In an experimental model, instillation of MCP-1 into the lungs of normal mice has been shown to increase alveolar monocyte accumulation in the absence of lung inflammation. A combination of MCP-1 with LPS challenge provokes acute inflammation with early neutrophil migration and delayed monocyte influx into the lungs. In CCR2 knockout mice, or in normal mice treated with an anti-CCR2-blocking antibody, monocyte accumulation induced by MCP-1 or MCP-1 with LPS was inhibited by >90%, indicating the importance of the CCR2/CCL2 pathway in this model. That airway hyper-reactivity was mediated by MCP-1 through CCR2 was indicated by the finding that allergen-induced as well as direct MCP-1 instillation induced changes in airway hyper-reactivity were significantly attenuated in CCR2$^{-/-}$ mice. The neutralization of MCP-1 in allergic animals and instillation of MCP-1 in normal animals were correlated with changes in leukotriene C4 levels in the bronchoalveolar lavage. In addition, leukotriene C4 release was directly induced in pulmonary mast cells by exposure to MCP-1. These data identify MCP-1 and CCR2 as potentially important therapeutic targets for the treatment of hyper-reactive airway disease. Furthermore, neutralization of CCR2 by an anti-CCR2 monoclonal antibody reduces macrophage and eosinophil accumulation in the BAL of asthmatic monkeys. These results suggest that the selective blockade of a single chemokine receptor involved in the early stages of asthma can ameliorate later disease stages.[90–106] In light of the similarity between asthma and allergic rhinitis (AR), the use of CCR2 antagonists in the treatment of AR and other airway allergic diseases is potentially efficacious.

12.2.5.4 Obesity and Type 2 Diabetes

CCR2 and MCP-1 play a critical role in adipose macrophage accumulation. Chronic subacute inflammation is implicated in the pathogenesis of insulin resistance and type 2 diabetes (T2D). Components of the immune system are altered in obesity and T2D, with the most apparent changes occurring in adipose tissue, liver, pancreatic islets, the vasculature and circulating leukocytes. These immunological changes include altered levels of specific cytokines and chemokines, changes in the number and activation state of various leukocyte populations and increased apoptosis and tissue fibrosis. These changes suggest that inflammation participates in the pathogenesis of T2D. Preliminary results from clinical trials with salicylates and interleukin-1 antagonists support this notion and have opened the door for immunomodulatory strategies for the treatment of T2D that simultaneously lower blood glucose levels and potentially reduce the severity and prevalence of the associated complications of this disease.[107–109] Many reports have shown that obesity is associated with a low-grade, chronic inflammation, which may be a potential mechanism whereby obesity leads to insulin resistance.[110,111] The adipose tissue is an important endocrine organ that secretes many adipocytokines, such as leptin, adiponectin, tumour necrosis factor (TNFα) and MCP-1. The abnormal production of pro-inflammatory and anti-inflammatory adipokines in visceral fat is associated with metabolic syndrome. This suggests that inflammatory changes in the adipose tissue may contribute to the development of many aspects of the metabolic syndrome and result in T2D and atherosclerosis.[112,113] Adipose tissue inflammation induced by macrophage infiltration, through MCP-1/CCR2 pathways, has been identified as an important element in the development of visceral obesity and insulin resistance. CCR2 regulates monocyte and macrophage chemotaxis and local inflammatory responses. It has also been implicated in regulating glucose uptake in an insulin-responsive cell line. MCP-1 is highly expressed in the adipose tissue of obese subjects and rodents. CCR2 is expressed by adipocytes, and incubation of 3T3-L1 adipocytes with MCP-1 leads to a state of cellular insulin resistance, as evidenced by decreased insulin-stimulated glucose uptake. The addition of MCP-1 to differentiated adipocytes *in vitro* decreases expression of several adipogenic genes, suggesting a role for MCP-1 and CCR2 in adipogenesis.[114] It is hypothesized that MCP-1, acting through CCR2, might regulate obesity-induced inflammation in adipose tissue and systemic glucose homeostasis.[115–117] Mice fed a high-fat diet display CCR2 deficiency attenuated adipose tissue macrophage (ATM) accumulation, adipose tissue inflammation and systemic insulin resistance. In mice with pre-existing obesity, short-term pharmacologic antagonism of CCR2 reduced ATM content and improved insulin sensitivity. CCR2 antagonists also have direct inhibitory effects on the development of adiposity in animals fed with a high-fat diet.[114] These findings and significant internal research by pharmaceutical companies has stimulated interest in developing CCR2 antagonists for the treatment of diabetes and obesity. Several molecules are in clinical development for type 2 diabetes. The first report of clinical results of Phase II studies on CCX-140 (structure not disclosed), a CCR2 antagonist, has just been published. A statistically significant decrease in HbA1C was observed after only four weeks of treatment at 10 mg q.d., compared to placebo.[118]

12.2.5.5   Atherosclerosis

Atherosclerosis is recognized as an inflammatory disease. Macrophage recruitment and development play an important role in disease progression. It was reported as early as 1991 that expression of MCP-1 is up-regulated in human atherosclerotic plaques, indicating the role of MCP-1 in the pathogenesis of atherosclerosis.[65,119] It was shown in CCR2 KO mice crossed with apoE−/− mice that the selective absence of CCR2 markedly decreases lesion formation in apoE−/− mice. These data reveal an important role for MCP-1 in the development of early atherosclerotic lesions.[120] Correlations between polymorphisms in the MCP-1 locus that cause increased MCP-1 production and an increased risk for cardiovascular disease indicate the importance of MCP-1/CCR2.[121,122] The use of CCR2 antagonists and CCR2/MCP-1 antibodies for the treatment of atherosclerosis has attracted significant attention in the past ~15 years, although successful examples of small-molecule CCR2 antagonists in pre-clinical models of atherosclerosis have not been abundant. An anti-MCP-1 antibody, MLN1202, showed a potentially positive outcome in biomarkers related to atherosclerosis in recent clinical trials.[123]

12.2.5.6   Diabetic Nephropathy

Diabetic nephropathy is the leading cause of renal failure and accounts for significant morbidity and mortality in patients with diabetes. While hyperglycaemia and hypertension are well-established factors for the development of this complication, there is also evidence that low-grade inflammation plays an important role. MCP-1 is over-expressed in the kidneys of diabetic animals.[124] Within the glomeruli, MCP-1 is over-expressed in both crescent glomerulonephritis and other nephritic conditions. It has been suggested that macrophages may transiently infiltrate during the moderate stage of diffuse diabetic glomerulosclerosis, contributing to the irreversible structural damage of glomeruli.[125] MCP-1 can activate tubular epithelial cells *in vitro*, much like the action of MCP-1 on vascular epithelial cells and contribute to tubulointerstitial inflammation. Tubulointerstitial damage directly correlates with loss of renal function and the risk of progression to end-stage renal failure. MCP-1 can directly elicit an inflammatory response by inducing cytokine and adhesion molecule expression in the kidney.[126] CCR2 expression has been demonstrated in other cell types, both *in vitro* and *in vivo*, indicating that the MCP-1/CCR2 system has other effects beyond monocyte accrual.[18] Current therapies for diabetic nephropathy can only partially reduce the impact of MCP-1 on this disease, suggesting the need to directly block MCP-1 action *via* CCR2 antagonists or antibodies targeting CCR2 and/or MCP-1.[127] CCR2 antagonism with a CCR2 antagonist, RS504393 (Compound **12**), improves insulin resistance, lipid metabolism and diabetic nephropathy in type 2 diabetic mice.[128] In addition, treatment of type 2 diabetes in *db/db* mice with an orally active CCR2 antagonist, RO5234444 (Compound **24**), resulted in a higher total number of podocytes, less glomerulosclerosis, reduced albuminuria and a significantly improved glomerular filtration rate.[129] These results suggest that the inhibition of the MCP-1/CCR2 pathway may affect the development of diabetic nephropathy.

12.2.5.7 Cancer

A variety of recent studies have implicated MCP-1 and CCR2 in the formation and growth of tumours. It has been shown that MCP-1 is progressively over-expressed in tumour beds and may play a role in the clinical progression of solid tumours.[130] Cancer cells derived from several solid tumour types demonstrate functional receptors for MCP-1, suggesting that tumourigenicity may be achieved through direct effects of MCP-1 on malignant cells. However, a variety of normal host cells that co-exist with cancer cells in the tumour microenvironment also respond to MCP-1. These cells include macrophages, osteoclasts, endothelial cells, T-lymphocytes and myeloid-derived immune suppressor cells (MDSCs). MCP-1 mediated interactions between normal and malignant cells in the tumour microenvironment and plays a multi-faceted role in tumour progression.[130] MCP-1 is produced by cancer cells and multiple different host cells within the tumour microenvironment. MCP-1 mediates tumourigenesis in many different cancer types.[131] MCP-1 has been reported to promote prostate cancer cell proliferation, migration, invasion and survival, *via* binding to CCR2.[133–134] Furthermore, MCP-1 induces the recruitment of macrophages and promotes angiogenesis and matrix remodeling.[132] Targeting MCP-1 has been demonstrated as an effective therapeutic approach in pre-clinical prostate cancer models and, currently, a neutralizing monoclonal antibody against MCP-1 has entered into clinical trials for the treatment of prostate cancer.[131] In addition, MCP-1 recruits inflammatory monocytes to facilitate breast-tumour metastasis.[135–136] CCR2 antagonism may be of therapeutic benefit for the treatment of breast cancer.

12.2.5.8   Inflammatory Bowel Disease

The blockade of CCR2 also has the potential to provide a novel therapeutic alternative for the treatment of inflammatory bowel disease.[137–139] The expression of MCP-1 and MCP-3 has been found to be significantly increased in patients with ulcerative colitis.[137] Elevated MCP-1 expression was also detected in colonic mucosa from inflammatory bowel disease (IBD) patients and up-regulated chemokine expression correlated with increasing severity of the disease.[138] In addition, CCR2 expressing $CD4^+$ T lymphocytes are preferentially recruited to the ileum in Crohn's disease and CCR2 appears to be an important contributor to the accumulation of $CD4^+$ T lymphocytes in the ileum in small bowel Crohn's disease.[139]

12.2.5.9   Pulmonary Fibrotic Disease

There is evidence that pulmonary fibrosis, the end-point of a chronic inflammatory process characterized by some degree of lung inflammation and abnormal tissue repair and remodelling, may also be a therapeutic target for CCR2 antagonists. Thus, MCP-1 is induced in lung fibrosis that is accompanied by mononuclear cell recruitment and the activation of lung fibroblasts. Also, CCR2 knockout mice showed significantly reduced fibrosis as evidenced by decreased lung type I collagen gene expression, hydroxyproline content, lung TNFα and TGF-β1 expression relative to WT mice[140] and a deficiency of CCR2 improves bleomycin-induced pulmonary fibrosis by attenuation of both macrophage infiltration and production of macrophage-derived matrix metalloproteinases.[141] These findings suggest that the MCP-1/CCR2 functional pathway is involved in the pathogenesis of pulmonary fibrosis and that CCR2 antagonism may improve the outcome of this disease by regulating macrophage infiltration and macrophage-derived MMP-2 and MMP-9 production.[142]

12.2.5.10   Pain

Another role identified for the MCP-1/CCR2 system is as a regulator in the pathogenesis of neuropathic pain. Some of the studies indicative of this connection are summarized below. For example, mice lacking CCR2 have a marked attenuation of monocyte recruitment in response to various inflammatory stimuli and a reduction of inflammatory lesions in models of demyelinating disease.[143] In acute pain tests, responses were equivalent in CCR2 knockout and wild-type mice. In models of inflammatory pain, CCR2 knockout mice showed a 70% reduction in phase 2 of the intra-plantar formalin-evoked pain response but only a modest (20–30%) and non-significant reduction of mechanical allodynia after intra-plantar complete Freund's adjuvant (CFA). In a model of neuropathic pain, the development of mechanical allodynia was totally abrogated in CCR2 knockout mice. Administration of CFA induced marked up-regulation of CCR2 mRNA in the skin and a moderate increase in the sciatic nerve and dorsal root ganglia (DRG). In response to nerve ligation, persistent and marked up-regulation of CCR2 mRNA was evident in the nerve and DRG. To confirm that mRNA changes reported by PCR experiments were accompanied by changes in CCR2 protein as well as to determine which cell types express CCR2, immunohistochemistry was used to look at CCR2 distribution in various tissues in a neuropathic model. Consistent with the mRNA data, an increased number of CCR2-positive monocytes/macrophages was detected in neroma and perineurium as well as endoneurium.[143] Disruption of Schwann cells in response to nerve lesions resulted in the infiltration of CCR2-positive monocytes/macrophages not only to the neuroma but also to the DRG. Chronic pain also resulted in the appearance of activated CCR2-positive microglia in the spinal cord. Blocking spinal CCR2 with AZ889 reversed hyperalgesia in a model of neuropathic pain.[144] Furthermore, treatment with an MCP-1 neutralizing antibody prevented bone-marrow-derived microglia (BMDM) infiltration into the spinal cord after nerve injury,[145] confirming a key role of CCL2/CCR2 axis in pain pathways and suggesting that inhibition of this axis may result in novel pain therapies.

12.2.5.11   Stroke

Stroke is an example in which inflammation is now thought to exacerbate the tissue injury caused by the infarction. MCP-1 over-expression increases the infarct volume and the invasion of the ischemic area by monocytes and macrophages.[146] The inhibition of, or deficiency in, CCR2/MCP-1 has been associated with reduced injury.[147] Furthermore, MCP-1 deficient mice are resistant to permanent middle cerebral artery occlusion[148] and CCR2 KO mice are protected against ischemia-reperfusion injury.[149]

12.3   Medicinal Chemistry of CCR2 Antagonists

Reports of medicinal chemistry efforts on CCR2 began appearing in the literature a little more than a decade ago, with the first disclosures of small-molecule CCR2 antagonists published in 1998 and 1999.[150–153] At the time, there were not many successful examples of small-molecule inhibition of protein-protein interactions. Because protein-protein interactions often involve a large interaction surface with a large number of weak interactions, they are among the most challenging targets for inhibition by small molecules. Later studies revealed that small-molecule CCR2 antagonists may actually bind to similar regions as the biogenic amines instead of at the protein-protein interaction surfaces. As the result of intensive efforts of drug discovery scientists, chemokine receptors have now become one of the first few protein-protein interaction targets successfully prosecuted by medicinal chemists, and a summary of this work is provided herein. Throughout this section, we will highlight the efforts and successes in dialling out hERG liability as well as the opinions in developing selective CCR2 antagonists *versus* CCR2/CCR5 dual antagonists.

12.3.1 CCR2 Antagonist Pharmacophore

The known CCR2 antagonists are composed of diverse structural classes and pharmacophore analysis is challenging. In addition, X-ray crystal structures of CCR2 are unknown at this time, making the understanding of binding modes of small molecules difficult. The first CCR2 mutagenesis study and modelling analysis was done for a series of spiropiperidines and reported in 2000.[154] Many homology models[155,156] have since been developed and QSAR studies[157,158] as well as docking and scaffold hopping[159] have been performed to explain the SAR of CCR2 antagonists. It is now well accepted that one of the key interactions for the majority of small-molecule CCR2 antagonists is between the basic amine or quaternary ammonium ion of the ligand and Glu291 of the CCR2 receptor. Over the past decade, various structure types of CCR2 antagonists have been identified by research labs in industry and academia. Some of these arose through modifications of CCR5 antagonists and others were discovered *via* HTS screening. Yet the majority of the most potent CCR2 antagonists, with a few exceptions, contain a basic centre, which is often flanked by two aromatic rings or one aromatic and one aliphatic groups. More than 90% of CCR2 antagonists with low nanomolar potency in a receptor binding assay fit the cartoon model shown in Figure 12.2. The basic centre is important for potency, but not essential as long as a positive charge is present at physiological pH. In some cases, a basic nitrogen or a positive charge is absent at this position. This can be explained by the formation of strong hydrogen bonds (H-bonds) between the antagonists and Glu291 on the receptor either as a direct H-bond donor or *via* water bridge(s), or interactions with other nearby residues such as Thr292.[160] Such compounds are often less potent than their counterparts with a basic centre at this position. The aromatic ring on the right side is essential and substitution on the aromatic ring plays an

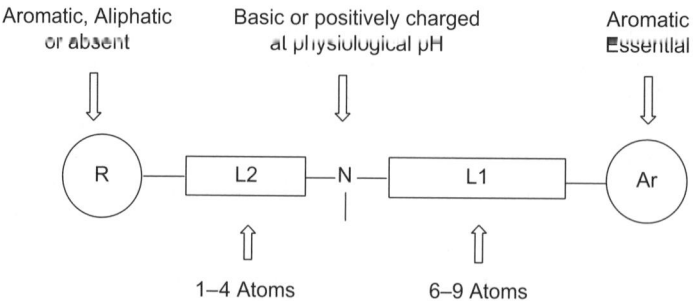

Figure 12.2 Pharmacophore model for the majority of CCR2 antagonists.

important role. This area is probably engaged in specific binding with the receptor. The linker L1 usually spans from 5 to 9 heavy atoms depending on the conformation. It can be lipophilic or polar, peptidic or as part of a saturated or partially unsaturated ring or even an aromatic ring. It appears that both the distance between AR_1 and N and the 3D relationship of the two moieties are important. As demonstrated in various reports, L1 is sensitive to modifications, indicating that it is engaged in specific binding as well as serving as a linker. In contrast, the left side of the antagonist tolerates more changes. Ring R can be aromatic or aliphatic. Linker L2 can be a short hinge or part of a ring or it can represent an aliphatic group in the absence of moiety R. It is worth mentioning that compounds where R is an aromatic ring in the models discussed above often suffer from a hERG liability. Such compounds frequently are linear in shape and share close similarity with the hERG pharmacophore.

It would not be surprising to find out in the future that CCR2 antagonists from different structural classes bind to different sites of the receptor. There is at least one report of potential allosteric binding of some CCR2 antagonists.[161] As with any small-molecule interactions with protein targets, CCR2 antagonists most certainly have different binding modes even for those that bind to the same site. This cartoon model certainly does not cover all CCR2 antagonists currently known. For example, some of the compact, non-basic structures might have a different binding site from the basic and linear compounds.

Throughout this chapter, the term "bisaryl" will be used to describe compounds where R is an aromatic group and "monoaryl" to describe compounds where R is an alkyl, saturated or a partially saturated ring. Current known CCR2 antagonists will be classified based on the putative central basic functional group and compounds that do not fit in the model will be discussed in the last two sections. Biological data will be cited directly from publications and potency comparisons between different laboratories will be avoided. The key SAR in the CCR2 literature will be covered, including all important lead compounds.

12.3.2   Pyrrolidines

Most pyrrolidine CCR2 antagonists reported so far belong to the bisaryl type (Figure 12.2, R = aryl). Among the first reported structures in this structural class are pyrrolidines represented by compound **1** (Figure 12.3). The first structures in this series were disclosed in 1999 in the patent literature[151,152] and published in 2004 and 2008.[162,163] It is interesting to note the sensitivity of the right-hand side of the molecule to modification. Thus, adding a trifluoromethyl group at the *meta*-position of the right-hand side aromatic ring increased potency by about 100-fold (**2** *vs.* **3**). Further modifications of these structures gave rise to compound (*R*)-**4** with low nanomolar potency in a receptor binding assay and antagonist activity in both chemotaxis assays (0.83 nM). It is worth noting that SAR studies revealed that the substituent effect on the right-side aromatic ring is much larger than that on the left-hand side aromatic ring.

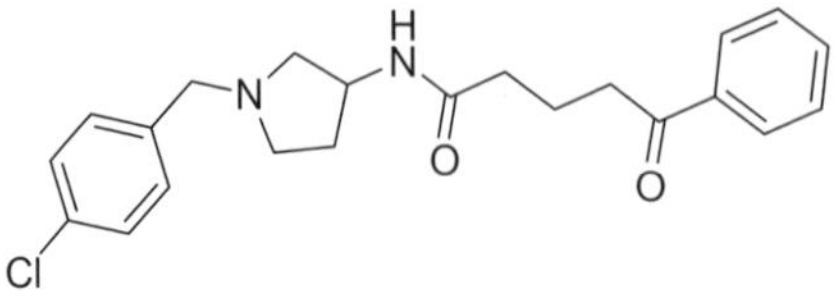

1

hCCR2 IC$_{50}$ (Binding) 11 μM

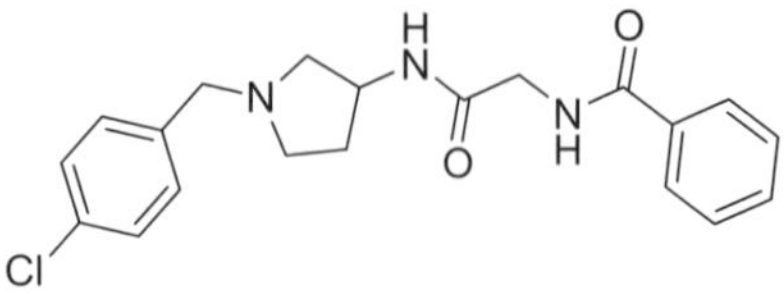

2

hCCR2 IC$_{50}$ (Binding) 42 μM

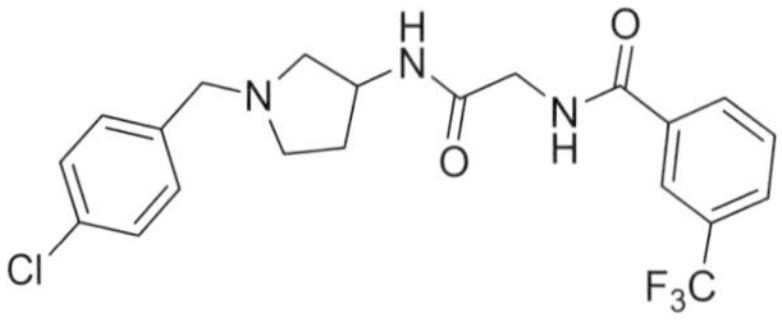

3

hCCR2 IC$_{50}$ (Binding) 0.37 μM

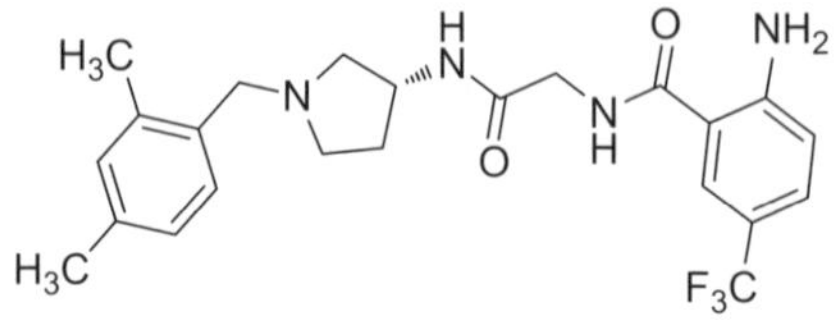

4

hCCR2 IC$_{50}$ (Binding) 3.2 nM

Figure 12.3

More elaborate pyrrolidine analogues such as INCB3344 (**5**) and INCB3284 (**6**) have also been reported (Figure 12.4).[164,165] The initial design idea for this series of compounds was to insert a carbocycle or a saturated heterocycle between the left-hand side aromatic ring and the tertiary amine in the above-mentioned pyrrolidine series in order to increase the basicity of the nitrogen. The authors noted that the hydroxyl group at the 4-position of the pyrrolidine of compound **7** was intended as a functional handle for further SAR exploration. Compound **7** has high affinity for hCCR2 while an analogue with the pyrrolidine and the phenyl at the 1- and 3-positions on the cyclohexane ring was inactive. Conversion of **7** into the ethoxy analogue **8** retains potency against hCCR2 while mCCR2 potency was improved. Unfortunately, compounds like **7** and **8** have a strong hERG binding liability. Further SAR studies on the left-hand side provided INCB3344. In pharmacokinetic studies in CD-1 mice (5 mg/kg i.v., 10 mg/kg p.o.), INCB3344 has a high clearance (CL) of 5.1 L/h/kg and a moderate volume of distribution (V$_{dss}$) of 2.8 L/kg. Consequently, the oral $t_{1/2}$ was short at 1.0 h. It has good oral exposure after a 10 mg/kg dose with an AUC of 2664 nM · h and a C$_{max}$ of 1886 nM. However, its hERG liability and CYP3A4 inhibition hindered its progression. Elimination of the ethoxy group on the pyrrolidine ring of **5** provided compounds **10** with similar potency against hCCR2 but lower potency against mCCR2 and no improvement in hERG liability. Twelve analogues were disclosed in this series with various pyridinyl, pyrimidinyl, pyradazinyl and pyrazinyl groups on the left-hand side. The potency in hCCR2 binding, chemotaxis assays and hERG binding are the same range while the hERG patch clamp showed differential potency. Balancing hERG liability of the series, as measured by patch clamp activity, with pharmacokinetic (PK) profiles led to the discovery of INCB3284. The clearance and volume of distribution for INCB3284 (**6**) are high in rats but low in the dog, monkey and chimpanzee. Oral exposure is higher in dogs and chimpanzees. The oral half-life is 2.9 h, 3.3 h, 5.1 h and 5.3 h in rats, dogs, monkeys and chimpanzees at 5 mg/kg, 5 mg/kg, 5 mg/kg and 1 mg/kg, respectively. Bioavailability is moderate at 20%, 31%, 13% and 23%, respectively. This compound progressed through pre-clinical safety and toxicity studies and entered clinical trials. Interestingly, in Phase I and Phase II clinical studies, INCB3284 had a $t_{1/2}$ of 15 hours.

A series of reverse pyrrolidines related to INCB3284, as exemplified by compound **11**, were disclosed in the patent literature. These compounds are reported to have binding IC_{50}s to CCR2 in the range of about 0.01 to about 500 nM.[166] It is likely that compound **11** is PF-4136309, currently in Phase II clinical trials for osteoarthritis pain and liver fibrosis (Figure 12.4).[220–223]

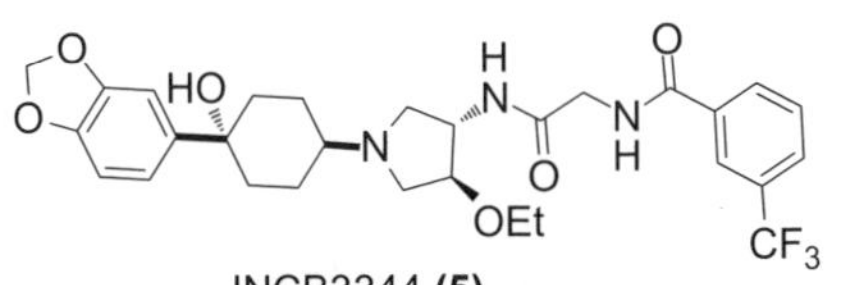

INCB3344 (**5**)
hCCR2 IC_{50} (Binding) 5.1 nM
mCCR2 IC_{50} (Binding) 9.5 nM

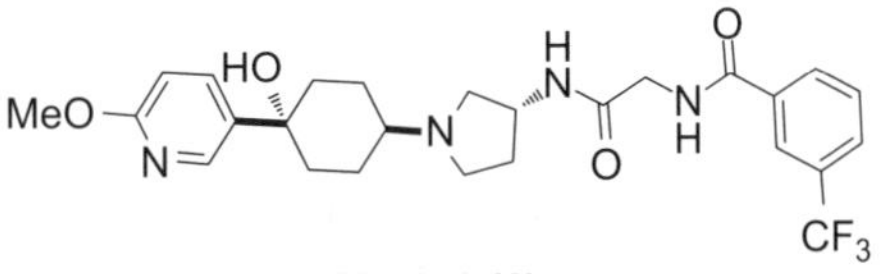

INCB3248 (**6**)
hCCR2 IC_{50} (Binding) 3.7 nM

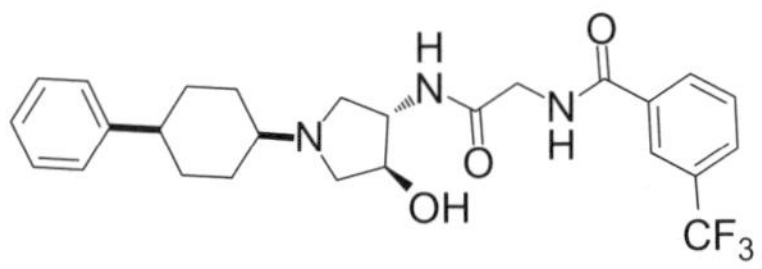

7
hCCR2 IC_{50} (Binding) 7.4 nM
mCCR2 IC_{50} (Binding) 186 nM
hERG IC_{50} (Binding) < 1 μM

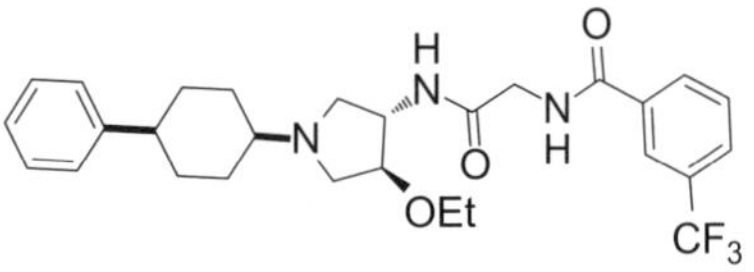

8
hCCR2 IC_{50} (Binding) 0.95 nM
mCCR2 IC_{50} (Binding) 1.9 nM
hERG IC_{50} (Binding) < 1 μM

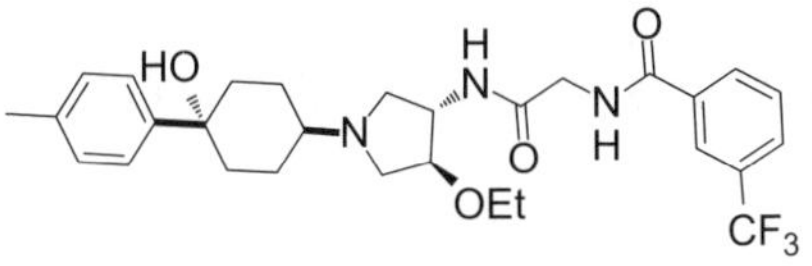

9
hCCR2 IC_{50} (Binding) 2.3 nM
mCCR2 IC_{50} (Binding) 2.1 nM
hERG IC_{50} (Binding) 11 μM

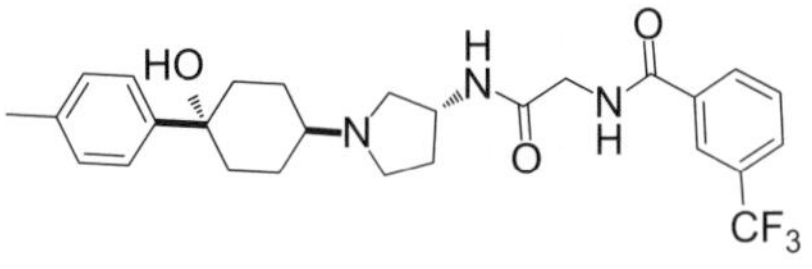

10
hCCR2 IC_{50} (Binding) 7.4 nM
mCCR2 IC_{50} (Binding) 186 nM
hERG IC_{50} (Binding) 13.2 μM

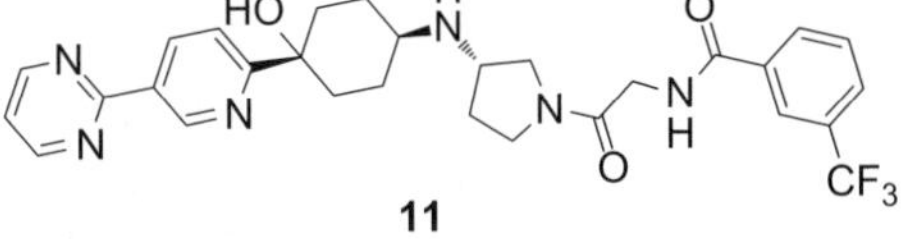

11

Figure 12.4

12.3.3   Piperidines

Both bisaryl and monoaryl piperidine CCR2 antagonists have been reported (Figure 12.5), with the majority of the advanced compounds of the bisaryl chemotype. One of the earliest CCR2 antagonists is spiropiperidine analogue (**12**).[154] It has moderate binding potency and weak agonist activity in chemotaxis assays. This compound was recently evaluated in a variety of animal models. In an LPS-induced acute lung injury model in mice, compound **12** at 5 mg/kg, i.p. provided significant protection against lung injury by attenuating the influx of leukocytes and protein into the bronchoalveolar space.[167] It was also reported that compound **12** improved insulin sensitivity in *db/db* mice and markedly decreased urinary albumin excretion and mesangial expansion, and suppressed pro-fibrotic and pro-inflammatory cytokine synthesis.[128] Urinary 8-isoprostane levels were also decreased. It is unclear why this compound was selected as a tool in these studies since compound **12** is not a potent functional antagonist and is not highly selective, though presumably its commercial availability was a factor. Compounds **13** and **14** are two other early piperidine analogues, derived from HTS screening. SAR studies were unable to increase its potency beyond mid double digit nanomolar.[168]

A series of dipiperidine analogues have been identified as functional CCR2 antagonists with high affinity for the receptor, as represented by compounds **15** and **16**.[169–172] Surprisingly, SAR studies led to the discovery that methyl analogue **17** is as potent as the corresponding acid (**15**) and alcohol (**16**). Presumably, substitution on the methylene between the two piperidine rings provides some advantageous conformational constraint. Replacing the methine that connects the two piperidine moieties in **17** with a nitrogen (hydrazine, **18**)[173] dramatically decreased CCR2 potency (hCCR2 binding: 59% at 25 µM) further indicating the importance of conformation and basicity of the piperidine. The *N*-oxide of the basic piperidine in **15** is fairly potent in the CCR2 binding assay (hCCR2 binding IC_{50}: 16 nM) and is a functional antagonist. This indicates that the proton on the basic amine at physiological pH does not play an essential role in the putative interaction with Glu291 of CCR2; the charge-charge interaction between the small molecule and Glu291 is more important.

A series of spiropiperidines with indolines (**19**), dihydrobenzofurans (**20**) and indanes (not shown) have been disclosed in the patent literature (there was no specific biological activity given, but it was noted that these compounds have $pK_i \geq 6$ in CCR2 binding assays).[174–177] Recently, a patent was published that claimed a process for the ketone reduction to form the secondary alcohol of **19** and **20**, with the left-hand side moiety of compound **19** exemplified, indicating that this analogue was interesting enough to be the focus of a chemical development effort.[178]

More recently, a series of reverse piperidines represented by analogues **21** and **22** (Figure 12.5) were disclosed in the patent literature.[179,180] Several compounds in this series have IC_{50}s in the subnanomolar range in receptor binding assays. Interestingly, a subset of the claimed compounds contained a carboxylic acid functional group on the phenyl ring, which is usually introduced into CCR2 antagonists in order to decrease hERG liability although there are no hERG data disclosed in the patents.

Two related series of ketopiperazine-containing spiropiperidines have been disclosed. Both series are potent antagonists as measured with calcium mobilization assays.[181,182] It is interesting to note that compound **24** (RO5234444)

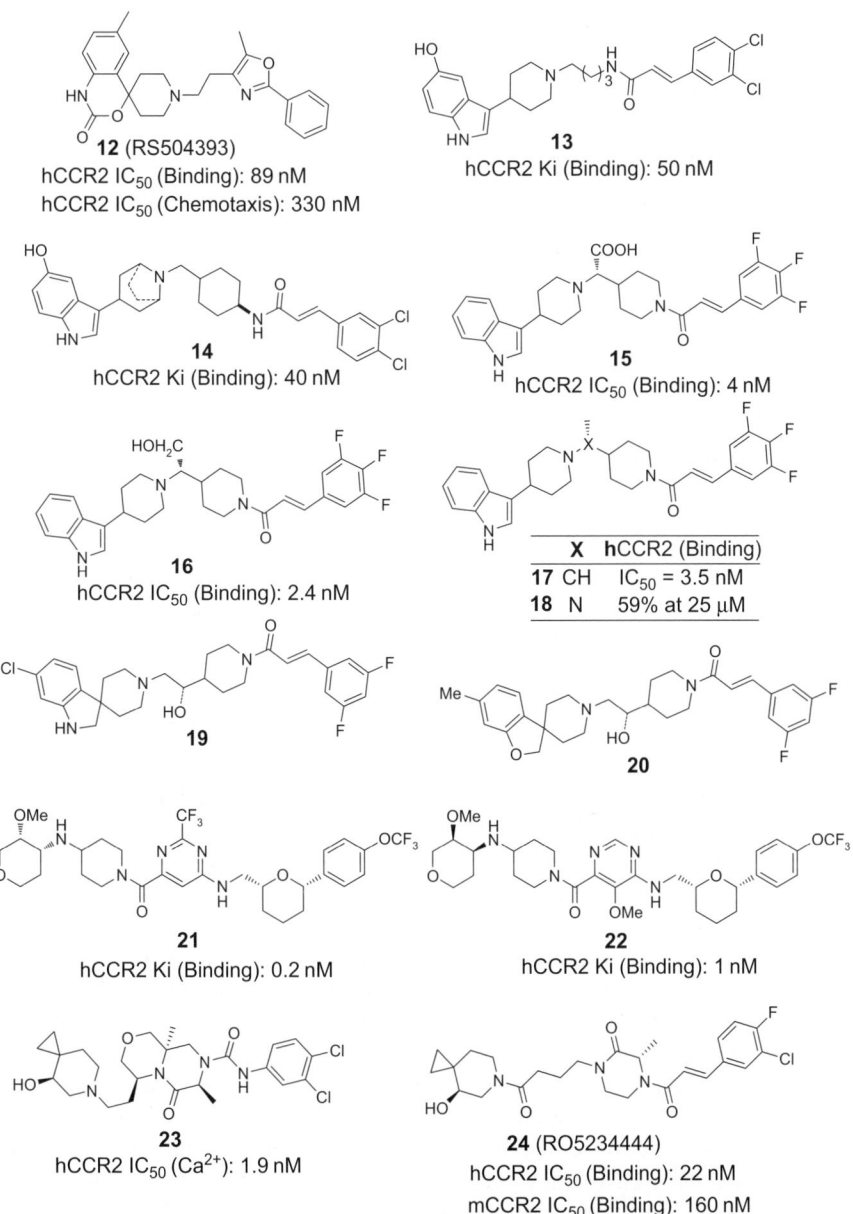

Figure 12.5

does not contain a basic nitrogen, yet it is a potent antagonist. This compound has an hCCR2 binding IC_{50} of 22 nM, mCCR2 binding of 160 nM and an IC_{50} of 50 nM in chemotaxis assays induced by MCP-1. The compound also blocked receptor internalization and has moderate activity in calcium mobilization assays (hCCR2 IC_{50}: 1.23 µM) based on the patent disclosure.[182] RO5234444 was evaluated in animal models for diabetic nephropathy in *db/db* mice *via* oral administration.[129] Male type 2 diabetic *db/db* mice were uninephrectomized to increase glomerular hyperfiltration to accelerate the development of glomerulosclerosis. Mice were chow-fed with or without admixed RO5234444 at 100 mg/kg to reach a trough plasma level of 1.23 µg/mL. The resulting CCR2 blockade reduced circulating monocyte levels, but did not affect total leukocyte or neutrophil numbers, and was associated with the reduction of macrophages and apoptotic podocytes in the glomerulus. The treatment produced a higher total number of podocytes, less glomerulosclerosis, reduced albuminuria and significantly improved glomerular filtration rate. There are no pharmacokinetic, safety or toxicology data reported for this compound at this time. Of note, the patent disclosed many analogues of **24**, with an aryl urea in place of the cinnamoyl moiety on the right side of the molecule and improved potency in calcium mobilization assays.[181,182]

12.3.4 Azetidines

Both bisaryl and monoaryl azetidine CCR2 antagonists have been identified. The azetidines are symmetrical and hence are achiral, which greatly simplifies their synthesis.

It is worth mentioning that in the pyrrolidine series the *R*-configuration at the 3-position of pyrrolidine is required for biological activity and the *S*-enantiomer is virtually inactive. In addition, replacing the pyrrolidine with a piperidine resulted in a loss of activity. Thus, it appeared that the pyrrolidine moiety was essential for activity until the azetidine analogues were discovered during SAR work on a series of piperidine-piperidine analogues (Figure 12.6).[183-188]

To eliminate the zwitterionic structures and improve the solubility of the earlier lead compound **15**, the piperidine moiety was replaced with various basic moieties. After screening of a large number of tertiary amines, azetidine emerged as a viable choice. To reduce the lipophilicity of the side chain, the cinnamoyl moiety was replaced by more polar fragments. Compound **25** was one of the first compounds to show promise. Further SAR studies improved potency to the low nanomolar range as shown with compound **26**. The effect of substitution on the cyclohexane ring on potency is shown for compounds **27a–27e**. Binding affinity decreases in the following order for the R group: $H > OH > F > NH_2 > COOH$. Further SAR studies identified analogues, such as dimethylamino analogue **28**, with a reduced hERG liability. The right-hand side phenyl ring is more sensitive to modifications. Among the many substituents scanned, an *o*-amino substituent showed acceptable potency. Since this amine could form an intra-molecular H-bond with the amide oxygen at the adjacent position on the aromatic ring, forming a pseudo 6-membered ring, various heterocyclic rings were explored, fused to the phenyl ring. Potent analogues like indazole **29** showed promise in balancing CCR2 potency with the hERG liability. As a comparison, its amide counterpart, compound **30**, has an IC_{50} of 37 nM. SAR studies have demonstrated that the left-hand side aromatic ring can be replaced with various functional groups including heteroaromatics such as pyridines, saturated heterocycles, carbocycles, alkyl groups and even simple esters (**31**). It is also apparent that small structural changes on the left side of the antagonist can significantly impact hERG binding, suggesting specific binding of this area to the hERG channel.

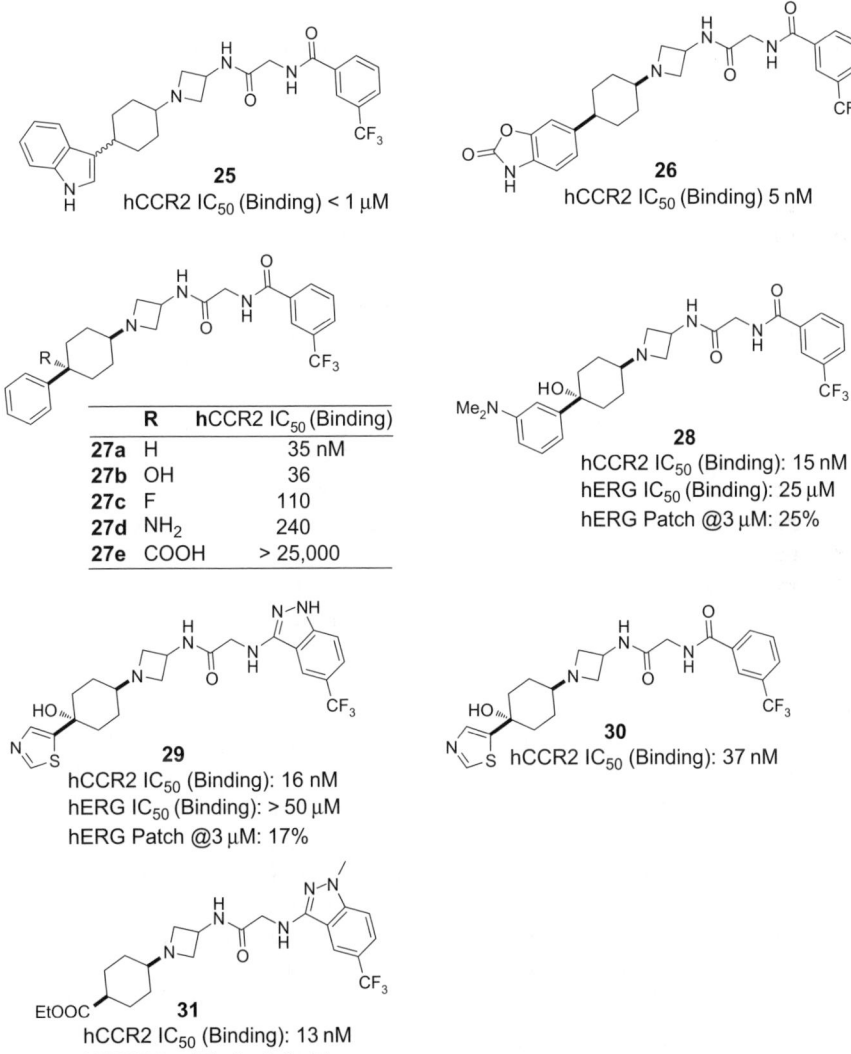

	R	hCCR2 IC_{50} (Binding)
27a	H	35 nM
27b	OH	36
27c	F	110
27d	NH₂	240
27e	COOH	> 25,000

25
hCCR2 IC_{50} (Binding) < 1 μM

26
hCCR2 IC_{50} (Binding) 5 nM

28
hCCR2 IC_{50} (Binding): 15 nM
hERG IC_{50} (Binding): 25 μM
hERG Patch @3 μM: 25%

29
hCCR2 IC_{50} (Binding): 16 nM
hERG IC_{50} (Binding): > 50 μM
hERG Patch @3 μM: 17%

30
hCCR2 IC_{50} (Binding): 37 nM

31
hCCR2 IC_{50} (Binding): 13 nM
hERG IC_{50} (Binding): 8 μM

Figure 12.6

Species-specific pharmacokinetic properties were also observed for the azetidines. The compounds generally have good oral exposure in higher species and lower exposure in rodents. For example, a derivative of compound **31** where 3-amino in the indazole moiety is substituted with methyl group (structure not shown) has an oral AUC of only 56 h*ng/mL in rats but 12,913 h*ng/mL in dogs at 10 mg/kg in both species.[188]

12.3.5   Cyclopentylamines

Both monoaryl and bisaryl types of cyclopentylamine CCR2 antagonists have been extensively reported in the literature and recently reviewed.[3] Cyclopentylamines such as **34**[189–192] arose from conformational constraint of the γ-amino-butyramides such as **33**, which in turn came from the modification of an early L-aryglycinamide screening hit (structure not shown).[193,194] Theoretically, this conformational constraint can increase potency by decreasing the entropy of binding and, in this case, there is a slight increase in potency in going from **33** to **34**. However, the cyclopentylamine scaffold offered more than just an increase in potency; they are potent CCR2 antagonists against both human and murine CCR2, which facilitated *in-vivo* evaluation. It is perhaps not surprising that a hERG liability was a major issue for the bisaryl chemotype in this series as well. For example, the left-hand side phenyl was substituted with various functional groups and replaced with a large number of heterocycles, but most of the compounds reported have an I_{Kr} binding IC_{50} in the nanomolar range and therefore have insufficient selectivity for CCR2. The only substitution that effectively decreases the hERG liability is the incorporation of a carboxylic acid (**35a** *vs.* **35b**). This finding is consistent with observations in several types of bisaryl CCR2 antagonists with diverse scaffolds. Because most CCR2 antagonists contain a basic amine, the addition of a carboxylic acid poses a challenge for achieving ideal oral pharmacokinetic properties. Therefore, detailed SAR of CCR2 binding, hERG binding and PK often are needed to identify a clinical candidate. This analysis was required for the identification of MK-0483 (**36**) as a clinical candidate.[194] Compound **36** is a potent functional antagonist of CCR2 with an IC_{50} of 0.3 nM in a chemotaxis assay.[195] Plasma protein binding was 98, 98, 97 and 97% in rat, dog, monkey and human, respectively. The hERG liability for this compound should be low, judging from the ratio of binding IC_{50} between I_{Kr} (33 µM) and CCR2 (4 nM). The data from hERG functional assay were not reported. MK-0483 is a CCR2/CCR5 dual antagonist with CCR5 potency at least 2.4-fold less than that of CCR2. It is interesting that this zwitterionic analogue has an acceptable PK profile in rat, dog and monkey. At 3 mg/kg, oral AUCs of 5.3, 33 and 4.7 µM · h are obtained with a bioavailability of 48, 63 and 66% in rat, dog and rhesus monkey, respectively. This advanced analogue, MK-0483, has been referred to as a backup clinical candidate, implying that it entered development after MK-0812 (**38**).[3] The current status and clinical indication(s) of this compound are unknown in the public domain.

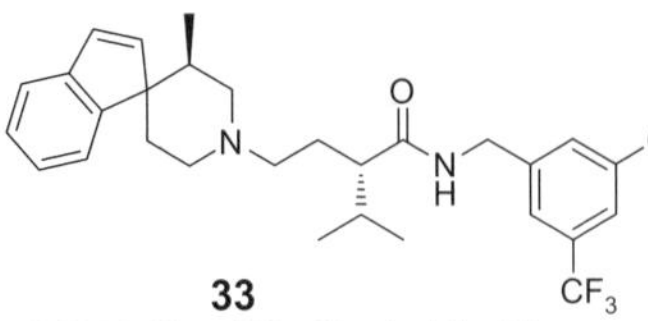

33
hCCR2 IC_{50} (Binding): 14 nM

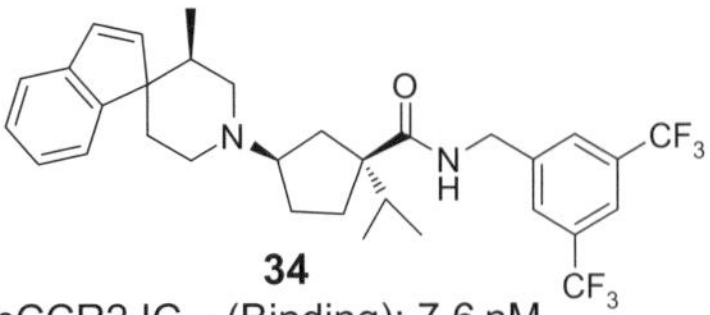

34
hCCR2 IC_{50} (Binding): 7.6 nM

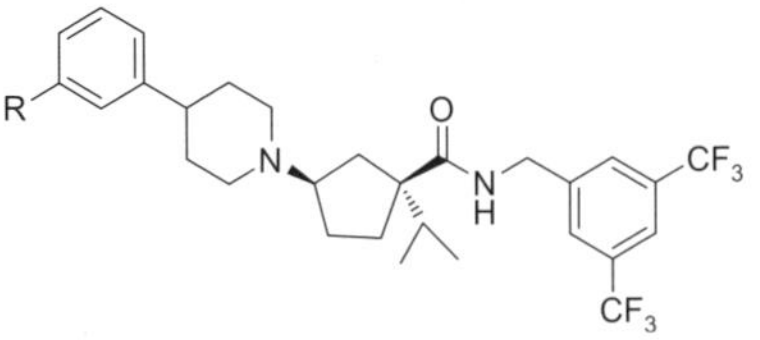

	R	hCCR2 IC_{50}	I_{Kr} IC_{50}
35a	F	4 nM	45 nM
35b	COOH	62 nM	2.8 μM

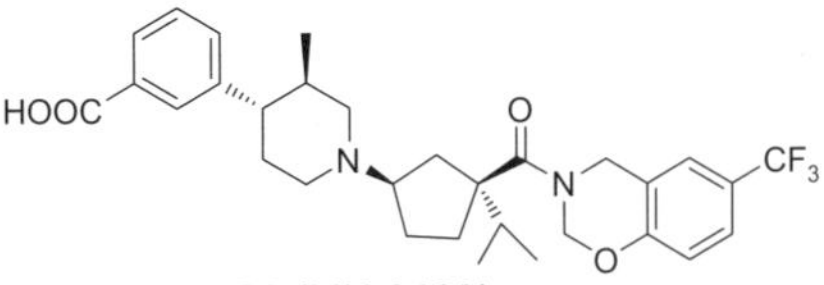

36 (MK-0483)
hCCR2 IC_{50} (Binding): 4 nM
I_{Kr} IC_{50} (Binding): 33 μM

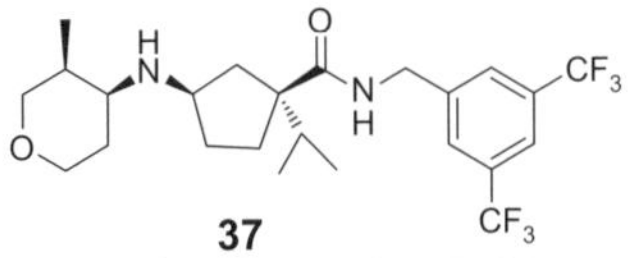

37
hCCR2 IC_{50} (Binding): 1.5 nM

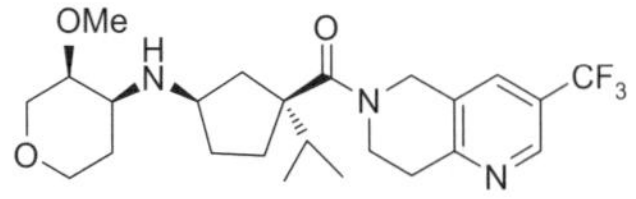

38 (MK-0812)
hCCR2 IC_{50} (Binding): 5 nM

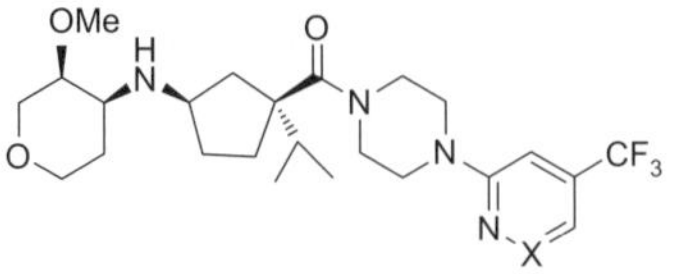

	X	hCCR2 IC_{50}	hCCR5 IC_{50}
39a	N	8.1 nM	--
39b	CH	3.0 nM	5.3 nM

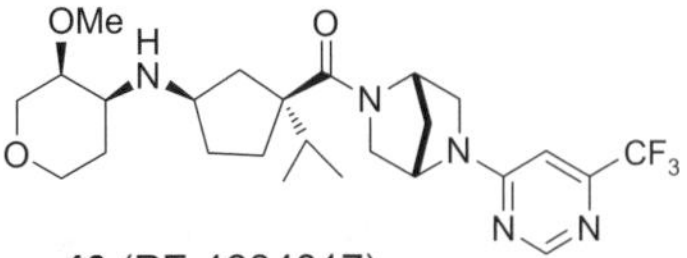

40 (PF-4634817)
hCCR2 IC_{50} (Binding): 1 nM
hERG IC_{50} (Binding): > 90 μM

Figure 12.7

The monoaryl class of cyclopentylamines is represented by MK-0812 (**38**), which entered Phase IIa clinical trials for the treatment of RA (Figure 12.7). This compound was the subject of much discussion prior to the disclosure of MK-0483. Actually, **37** is a hybrid of **34** with an early CCR2 analogue with a tetrahydropyran left-hand side.[192] It is worth noting that, unlike the bisaryl CCR2 antagonist SAR, a secondary amine is more active in the monoaryl series. Tertiary amines are generally less potent overall. The authors studied various carbocycles and saturated heterocycles on the secondary amine and noticed that 6-membered rings give the best potency and that the 4-tetra-hydropyran (THP) group provides greater potency than a cyclohexyl moiety. Inserting a methylene between the tetrahydropyran and amino group also resulted in a dramatic decrease in potency (not shown). Small substituents at the position adjacent to the amino group on the THP maintain binding potency but increase antagonist functional potency. The relative stereochemistry between this substituent and the amino group is important as *trans*-isomers are much less potent than *cis*-isomers.[192] In PK studies, **37** has moderate oral exposure in rat and dog but very low exposure in monkey. Further SAR studies of CCR2 binding and I_{Kr} binding, as well as efforts to improve the PK profile, led to the discovery of MK-0812.[3]

A series of piperazine analogues in the cyclopropylamine series, including two advanced compounds (**39a** and **39b**), have been reported recently (Figure 12.7).[196–198] The authors stated that **39b** (INCB10820/PF-4178903) was selected as a clinical candidate[197] and then was apparently replaced by **39a** (PF-4254196). Compound **39b** is a dual CCR2/CCR5 antagonist (CCR2 binding IC_{50}: 3.0 nM; CCR5 binding IC_{50}: 5.3 nM), and is a potent functional antagonist. It possesses good *in-vitro* metabolic stability in human liver microsomes as well as an acceptable *in-vivo* PK profile. However, as for many other CCR2 antagonists, it suffers from a low selectivity margin over hERG inhibition (binding IC_{50}: 1.7 µM). Exploration of SAR around the aromatic ring afforded compound **39a** with a reduced hERG liability (binding IC_{50}: 31 µM). It is a potent CCR2 antagonist with an IC_{50} of 7.7 nM in chemotaxis assays and does not inhibit CYPs (<3% at 3 µM against 3A4, 2C9, 2D6 and 1A2). In rat PK studies clearance is 2.8 mL/min/kg with 50% bioavailability. PF-4254196 was evaluated in selectivity, genetic toxicity and toxicology studies in rats and cynomolgus monkeys and was nominated for clinical development.[198] Another analogue in the series, **40** (PF-4634817), with a bridged piperazine on the right-hand side, was disclosed recently. This compound has entered Phase I clinical trials.[199]

12.3.6   Cyclohexylamines

Cyclohexylamine CCR2 antagonists are conceptually related to the early pyrrolidine antagonists, as shown in Figure 12.8. Both *cis*- and *trans*-cyclohexanediamine derivatives were evaluated. A series of non-basic cyclohexanediamines was also identified. Compound **42** was tested against mutant CCR2 receptors G291A and T292A and results showed that the key ligand-protein interaction was with Thr292, not G291 as is the case for most CCR2 antagonists.[200] Piperidine analogues such as **43** were studied later and it was determined with mutant receptor binding assays that these displayed significant reliance on an interaction with Glu291.[201] Later reports revealed that the nitrogen in the piperidine can be placed as an exocyclic amino group on the cyclohexane ring at the position corresponding to the 1- and 6-positions of the piperidine.[202] The benzamide bond of this chemotype can be replaced with a sulfonylmethyl moiety as exemplified by compound **44**.[203] Conformational constraint of the central amide into a γ-lactam afforded the pyrrolidone series. One of the first pyrrolidine compounds (**45**) showed excellent binding affinity and is a potent functional antagonist.[204] There are no reports of any of the pyrrolidones currently in clinical trials.

A series of indole-containing cyclohexylamines such as **46**, **47** and **48**, with good potency against CCR2 but undesirable selectivity over hERG activity, was discovered based on the piperidine leads. Studies on the SAR of this series

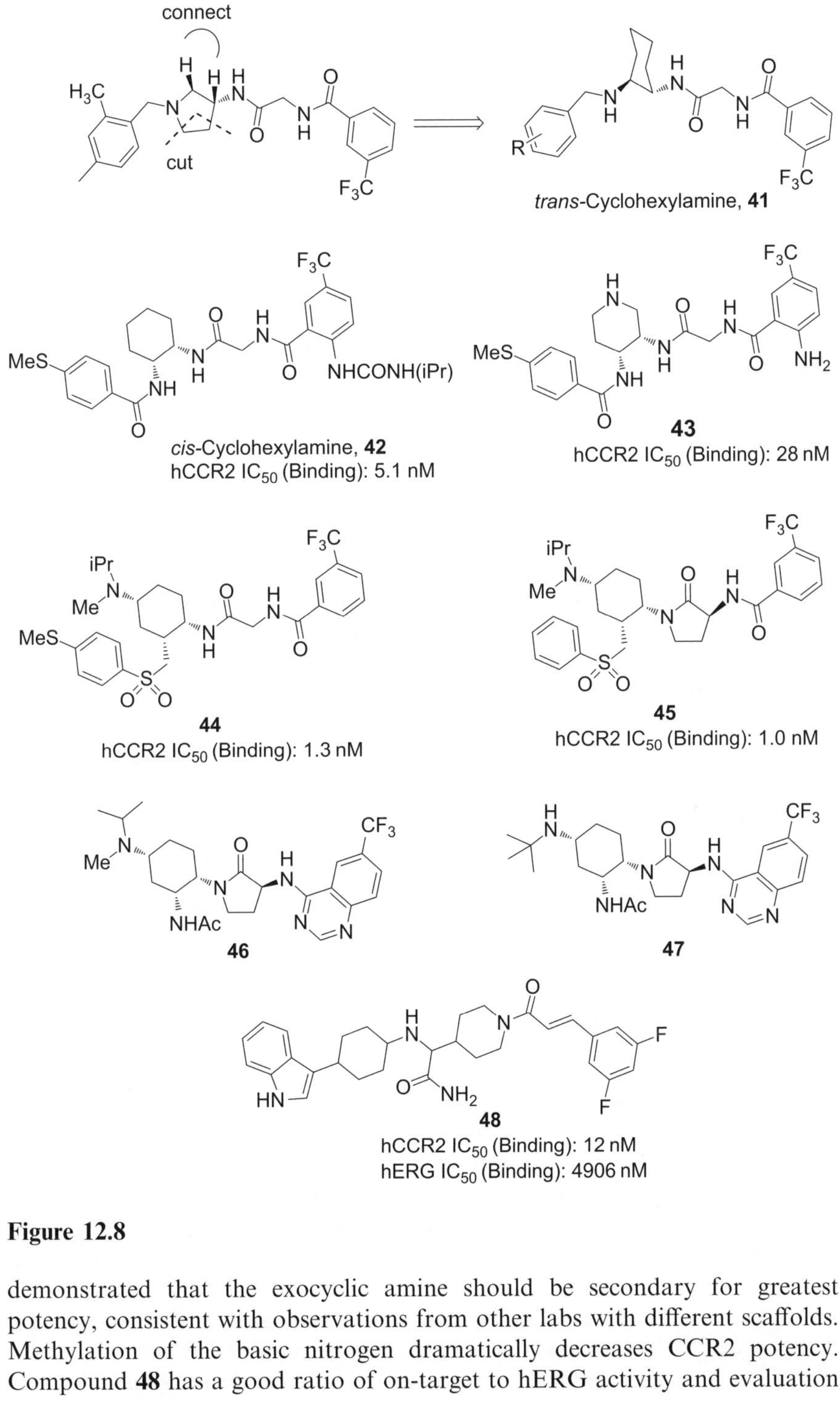

Figure 12.8

demonstrated that the exocyclic amine should be secondary for greatest potency, consistent with observations from other labs with different scaffolds. Methylation of the basic nitrogen dramatically decreases CCR2 potency. Compound **48** has a good ratio of on-target to hERG activity and evaluation

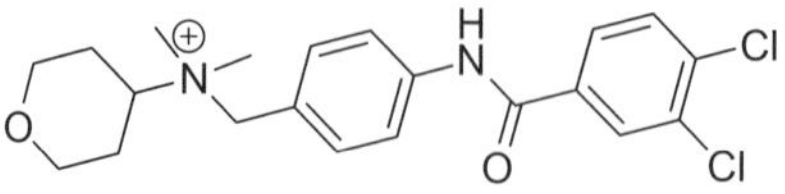

49

hCCR2 IC$_{50}$ (Binding): 20 nM
hCCR5 IC$_{50}$ (Binding): ~1 μM

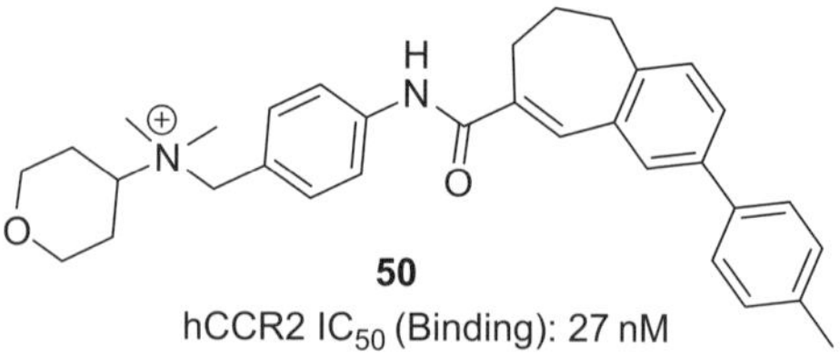

50

hCCR2 IC$_{50}$ (Binding): 27 nM
hCCR2 IC$_{50}$ (Binding): 1.5 nM

Figure 12.9

for cardiovascular safety in the guinea pig and did not show QT prolongation at doses up to 10 mg/kg, i.v.[205]

12.3.7 Quaternary Ammonium

Potent CCR2 antagonists containing a quaternary ammonium salt have also been reported (Figure 12.9).[206] Although there are no binding studies with G291A, T292A or any other mutants of CCR2, it is conceivable that the positively charged nitrogen may interact with Glu291 in the binding site of CCR2. These structures still fit in the pharmacophore model discussed earlier. SAR studies led to several compounds with human CCR2 binding IC$_{50}$s in the low double-digit nanomolar range as shown for compound **49**. This compound is selective over CCR5 and, as expected, has very good aqueous solubility. It is not orally bioavailable in pre-clinical models but has 74% and 100% bioavailability *via* intra-peritoneal and subcutaneous administration, respectively.[206] The compound was evaluated in ovalbumin-induced asthma model in mice *via* intra-peritoneal dosing and it showed good efficacy at 30 mg/kg. Other quaternary ammonium salts including **50** have been reported as CCR5/CCR2 dual antagonists.[206] With structural modifications on the right-hand side, these analogues can be converted from CCR5 selective to CCR2 selective compounds.[207]

12.3.8 Sulfonamides

It is worth noting that the structures of the sulfonamide-based CCR2 antagonists in this section and the next do not fit in the pharmacophore model discussed earlier (Figure 12.10). There are no reports of studies with mutant receptors, or any other studies, to elucidate the binding site of this class of compounds at this time. Sulfonamides have drawn a lot of recent attention due to the presentation of Phase IIa data of CCX-140 at the 71st Scientific Sessions of the American Diabetes Association Meeting in June of 2011.[118] Although the structure of CCX-140 has not been disclosed, many speculate that it might be a sulfonamide. There have been several patent filings on sulfonamide CCR2 antagonists.[208–216] The structures, in general, are quite similar and most of the modifications have been made on the right-hand side of the molecule. They can

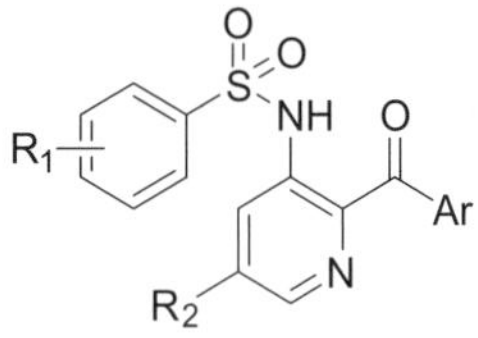

51

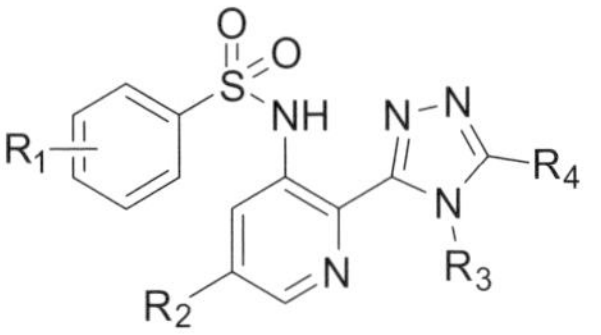

52

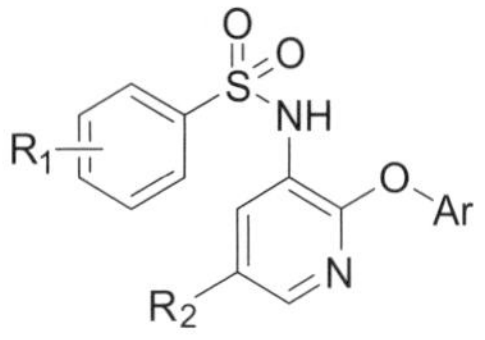

53

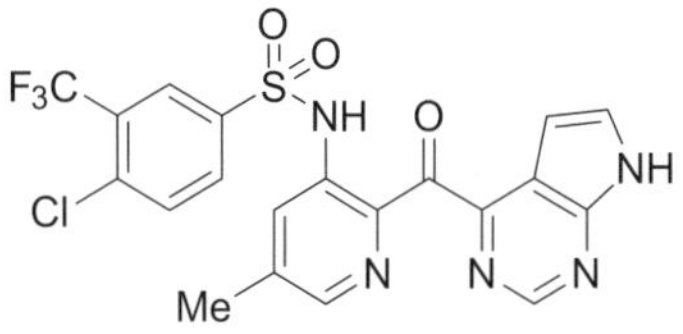

54

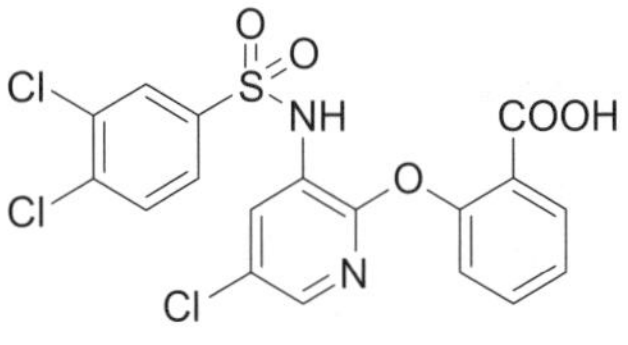

55
hCCR2 Binding, Ki = 1.26 nM (pKi = 8.9)

Figure 12.10

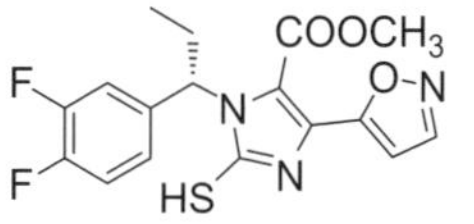

56
hCCR2 IC_{50} (Ca^{2+}): 15 nM

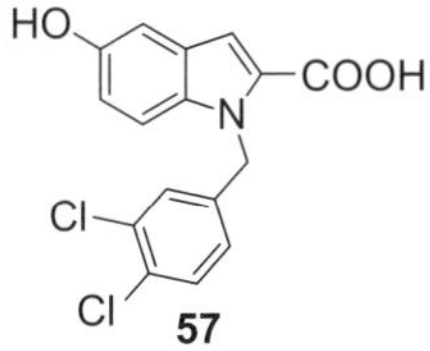

57
hCCR2 IC_{50} (Ca^{2+}): 15 nM

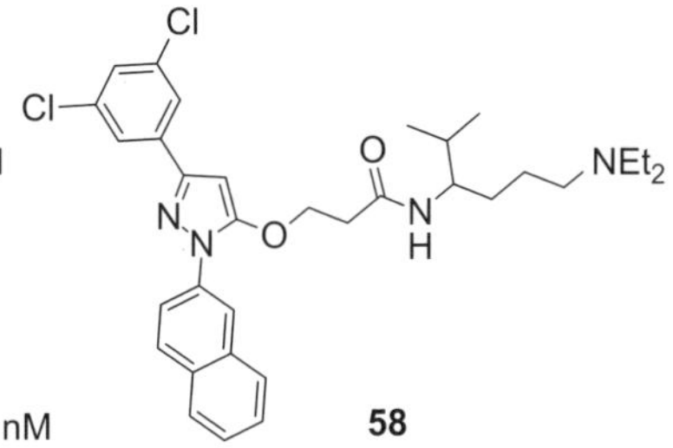

58
hCCR2 IC_{50} (binding): 6 nM

Figure 12.11

be loosely summarized into three general chemotypes: ketone linker (**51**), triazoles (**52**) and ether linker (**53**). The most frequently used R_1 groups are chloro and trifluoromethyl as well as small alkyl, and this phenyl ring is often disubstituted. The R_2 group is often a small substituent such as lower alkyl and chloro. The Ar group of **51** and **53** encompasses a large number of substituted phenyl and heterocycles including fused heterocycles. It seems that the right hand-side of the antagonist is more tolerant of modifications. Compound **54** has drawn particular attention because a specific crystalline form of its sodium salt was claimed, which suggests development activity for the compound. However, there is no process patent associated with this compound. Sulfonamides with oxygen linkers have also started appearing in the literature. Compound **55** is one of the most potent of these compounds reported in a recent publication.[217]

12.3.9 Other Structures

There have been a few reports of small heterocyclic CCR2 inhibitors (Figure 12.11) that do not fit in the known CCR2 pharmacophore model. Mercapto-imidazoles such as **56**[86,218] and indole carboxylic acids such as **57**[219] have been characterized extensively. There is also one report regarding a series of 1,3-bisaryl-pyrazoles such as **58**.[220]

12.4 Issues and Challenges

As with any drug discovery and development program, a CCR2 antagonist program has many challenges. The issues and challenges unique to CCR2 antagonists are discussed below.

12.4.1 Pre-clinical Challenges

The most obvious drug-development challenge facing CCR2 antagonists is the potential cardiovascular safety issue. As discussed throughout this chapter, many CCR2 antagonists share hERG pharmacophore characteristics and often have some hERG binding activity, which frequently translates into functional hERG blockade. Thus, subtle SAR studies on both CCR2 and hERG binding are required to separate the targets, especially for the bisaryl class of structures. However, after extensive medicinal chemistry efforts, most compounds have a low hERG liability in pre-clinical cardiovascular safety evaluations. Nevertheless, except for a few indications, cardiovascular safety margins, especially against QT prolongation, remain a concern for clinical development and careful pre-clinical cardiovascular evaluations are clearly warranted.

The second challenge is differences between species. As discussed in Section 12.2, the sequence homology of hCCR2 and rodent CCR2 is ~80% and the homology between MCP-1 and JE (mouse CCR2 ligand) is only 64%. This not only makes pharmacological evaluation of discovery compounds challenging, but also significantly impacts clinical translation. To overcome this difficulty, animal models of mCCR2 KO/hCCR2 KI (humanized) mice have been developed, greatly facilitating the pre-clinical evaluation of compounds. Still, challenges exist for certain indications. For example, there are no leptin-deficient mouse models with humanized CCR2 to use for the evaluation of compounds to treat diabetes. Non-human primate (NHP) models are available for some indications but do not work well for others. In addition, the high cost of NHP models makes them inappropriate for early discovery efficacy assessment.

The third challenge relates to achieving selectivity among chemokine receptors. The most common off-target activity occurs on CCR5, followed by CCR1, because of their close sequence homology with CCR2. Many CCR2 antagonists show cross-reactivity with CCR5, including some of the development candidates from Merck, BMS and Incyte/Pfizer. Initially, such cross-reactivity was viewed as a potential liability and efforts have been expended to eliminate this off-target activity. With the failure of selective CCR2 antagonists in recent clinical trials, discussion has shifted to consider whether targeting more than one chemokine receptor is necessary to achieve clinical efficacy. While the discussion is still ongoing and more clinical data are needed to draw conclusions, it is likely that in some indications dual antagonists may be more efficacious, and in other indications, selective CCR2 antagonists may work better clinically.

One advantage of known CCR2 antagonists is that they are relatively polar compounds so their pharmaceutical properties are in general quite good and it is possible to identify BCS Class I (high solubility, high permeability) development candidates.

12.4.2 Clinical Challenges

Table 12.2 summarizes current clinical compounds, both antibodies and small molecules, based on information in the public domain. This is probably not a complete list of all clinical compounds in the CCR2 field since many clinical trials are registered without information on the mechanism of action. CCR2 antagonists have been evaluated for a wide range of clinical indications including RA, MS, atherosclerosis, AR, pain, T2D and cancer. It is fair to say that, based on the published data and clinical outcomes with both small-molecule CCR2 antagonists and CCR2/MCP-1 antibodies, this therapeutic approach has not translated well with regard to pre-clinical predictions of efficacy. Even with the failed trials in RA and MS, there are signs of clinical proof of concept emerging from biomarkers in atherosclerosis studies to indications of efficacy in T2D. The question is whether the desired clinical efficacy end-points can be reached to differentiate CCR2 antagonists from the current standard of care for the particular indication. On a positive note, many of the clinical CCR2 antagonists do not seem to have target-related safety issues in Phase I. This, combined with the fact that the pharmaceutical properties of many of these compounds are excellent, is encouraging for clinical designs of indications requiring high drug exposure.

One of the first indications for CCR2 antagonists and CCR2/MCP-1 antibodies was RA. Three molecules were taken to Phase II for RA, including ABN912 (CCR2 antibody), MLN1202 (MCP-1 antibody) and MK0812 (**38**).[79–81] As all three molecules failed in RA trials, one of the most important questions in the CCR2 antagonist field and in chemokine receptors in general becomes whether blocking a single chemokine receptor can translate into sufficient clinical efficacy. There are at least two schools of thought. Some experts in the field argue that the lack of clinical efficacy is because of the lack of predictability of animal models and the redundancy of chemokines.[225] Lack of predictability of animal models is not unique to chemokine receptors as drug targets. Many biological targets face the same challenge because of the differences between human physiology and animal physiology, especially in lower species such as rodents. The redundancy of chemokines and their receptors is

Table 12.2 CCR2 antagonists in clinical development.

Drug	Company	Indication	Ph	Status	Source	Ref.
ABN912	Novartis	RA	II	Failed	MAb	81
MLN1202	Millennium	RA	II	Failed	MAb	80
		MS	II	Completed		
		Atherosclerosis	II	Completed		
		Cancer	I	Active		
CNTO888	J&J	Cancer	II	Ongoing	MAb	221
		Pulmonary fibrosis	II	Ongoing		
MK-0812	Merck	RA	II	Failed	SM	3, 79
		MS	II	Completed		
JNJ-17166864	J&J	AR	II	Completed	SM	221
Unknown	BMS	Cardiovascular	Pre-II	Unclear	SM	222
CCX-140	ChemoCentryx	Diabetes	II	Completed	SM	118
		Nephropathy	I-II	Active		223
INCB-3284	Incyte	Diabetes	I	Unclear	SM	
PF-04136309	Pfizer	OA pain	II	Completed	SM	221
		Liver fibrosis	II	Active		224

somewhat unique. Since there are 40 chemokines and 19 receptors, and many play similar roles in the immune system in the same tissues, it is surprising from an evolutionary perspective, for a complex biological system like the human body, to rely on one single pathway even for one aspect of the immune systems. Some scientists believe that targeting more than one chemokine receptor is the answer to the clinical efficacy challenge. We believe that dual chemokine receptor antagonists are a potential solution as long as no unmanageable additional side-effects arise from the second chemokine receptor for the target indications. The major challenges with such approaches are the ideal ratio of the activity on the two targets from a therapeutic perspective, and the medicinal chemistry efforts to achieve such a profile. Such medicinal chemistry efforts are much more difficult than optimizing a single target if a pre-determined specific ratio of the two targets is required for the molecule. Many of the current dual chemokine receptor antagonists were discovered by serendipity, not by design. Perhaps aggressive pharmacological screening in reliable animal models combined with medicinal chemistry efforts to design dual antagonists is a potential answer.

The second point of view is that the failure of clinical trials in the chemokine field, including CCR2 and CCR1 for RA, is because of target selection for the indication and clinical trial design.[226] We believe that targeting a single chemokine receptor can achieve the desired clinical outcome in certain diseases. For some diseases the complete blockade of multiple chemokine pathways might not be necessary. Perhaps indications that do not involve complicated cross-talk between chemokines are the best choices for CCR2 antagonists.

12.5 Conclusion

Significant progress has been made in understanding the complex biology of MCP-1 and CCR2. Their roles in inflammatory diseases have been intensively studied and to some extent elucidated. Tremendous efforts from the pharmaceutical industry have been spent on CCR2 antagonists with great success from a medicinal chemistry point of view. As a result, a large, diverse set of CCR2 antagonists with excellent potency and suitable pharmaceutical properties have been identified and a number of clinical compounds have been advanced. This at least confirms that small-molecule intervention of protein-protein interaction targets can be successful. Unfortunately, clinical development of CCR2 antagonists has not been as successful as the pre-clinical disease models would predict. Recent reports of lack of clinical efficacy in RA and MS cast some doubt on CCR2 antagonists as monotherapy for autoimmune diseases and inflammation. Whether this is due to the redundancy of chemokines and their receptors or due to a mismatch of indications with the target is still under active debate. Recent reports of clinical efficacy of a CCR2 antagonist in T2D and positive biomarker changes in the atherosclerosis trial of an MCP-1 antibody provided clinical evidence that therapeutic intervention of MCP-1/CCR2 can produce an efficacious response in human disease. The question remains as to whether CCR2 antagonists can reach the desired clinical end-points as single-agent therapy in some indications, or must be considered as either combination therapy or part of dual chemokine receptor antagonists. The ongoing clinical trials in additional indications with diverse structural classes of CCR2 antagonists may provide answers to the above questions.

Acknowledgement

The authors are grateful to Dr David Uhlinger for proofreading of the manuscript and helpful suggestions.

References

1. I. L. Charo, S. J. Myers, A. Herman, C. Franci, A. J. Connolly and S. R. Coughlin, *Proc. Natl Acad. Sci. USA*, 1994, **91**, 2752.
2. R. Horuk and W. Guilford, *Expert Opin. Ther. Pat.*, 2011, **21**, 1275.
3. M. Struthers and A. Pasternak, *Curr. Top. Med. Chem.*, 2010, **10**, 1278.
4. P. H. Carter, *Expert Opin. Ther. Pat.*, 2010, **20**, 283.
5. M. E. Sobhia, R. Singh, P. Kare and S. Chavan, *Expert Opin. Drug Discov.*, 2010, **5**, 543.
6. M. Xia and Z. Sui, *Expert Opin. Ther. Pat.*, 2010, **19**, 295.
7. J. E. Pease and R. Horuk, *Expert Opin. Ther. Pat.*, 2009, **19**, 39.
8. J. E. Pease and R. Horuk, *Expert Opin. Ther. Pat.*, 2009, **19**, 199.
9. P. H. Carter, R. J. Cherney and I. K. Mangion, *Annu. Rep. Med. Chem.*, 2007, **42**, 211.
10. P. J. Higgins, C. E. Schwartz and J.-M. Nicolas, in *Chemokine Biology – Basic Research and Clinical Application*, ed. K. Neote, G.L. Letts and B. Moser, Birkhäuser Verlag, Basel, 2007, p. 115.
11. J. J. Onuffer and R. Horuk, *Trends Pharmacol. Sci.*, 2002, **23**, 459.

12. B. J. Rollins, *Blood*, 1997, **90**, 909.
13. P. M. Murphy, M. Baggiolini, I. F. Charo, C. A. Hebert, R. Horuk, K. Matsushima, L. H. Miller, J. J. Oppenheim and C. A. Power, *Pharmacol. Rev.*, 2000, **52**, 145.
14. C. Gerard and B. J. Rollins, *Nat. Immunol.*, 2001, **2**, 108.
15. A. D. Luster, *N. Engl. J. Med.*, 1998, **338**, 436.
16. S. L. Deshmane, S. Kremlev, S. Amini and B. E. Sawaya, *J. Interferon Cytokine Res.*, 2009, **29**, 313.
17. I. F. Charo and R. M. Ransohoff, *N. Engl. J. Med.*, 2006, **354**, 610.
18. A. Yadav, V. Saini and S. Arora, *Clin. Chim. Acta.*, 2010, **411**, 1579.
19. I. F. Charo and R. M. Ransohoff, *N. Engl. J. Med.*, 2006, **354**, 610.
20. A. Viola and A. Luster, *Annu. Rev. Pharmacol. Toxicol.*, 2008, **48**, 171.
21. E. Melgarejo, M. Á. Medina, F. Sánchez-Jiménez and J. L. Urdiales, *Int. J. Biochem. Cell Biol.*, 2009, **41**, 998.
22. B. Coll, C. Alonso-Villaverde and J. Joven, *Clin. Chim. Acta*, 2007, **383**, 21.
23. I. Martinovic, N. Abegunewardene, M. Seul, M. Vosseler, G. Horstick, M. Buerke, H. Darius and S. Lindemann, *Circ. J.*, 2005, **69**, 1484.
24. J. Petrkova, J. Szotkowska, Z. Hermanova, J. Lukl and M. Petrek, *Mediators Inflammation*, 2004, **13**, 39.
25. O. Satiroglu, H. A. Uydu, A. Demir, M. Bostan, M. Atak and E. Bozkurt, *J. Exp. Med.*, 2011, **224**, 301.
26. I. F. Charo and M. B. Taubman, *Circ. Res.*, 2004, **95**, 858.
27. C. Gonzalez-Quesada and N. G. Frangogiannis, *Curr. Atheroscler. Rep.*, 2009, **11**, 131.
28. U. Ikeda, K. Matsui, Y. Murakami and K. Shimada, *Clin. Cardiol.*, 2002, **25**, 143.
29. D. J. Mahad, S. J. L. Howell and M. N. Woodroofe, *J. Neurol. Neurosurg. Psychiatry*, 2002, **72**, 498.
30. D. Mahad, M. K. Callahan, K. A. Williams, E. E. Ubogu, P. Kivisäkk, B. Tucky, G. Kidd, G. A. Kingsbury, A. Chang, R. J. Fox, M. Mack, M. B. Sniderman, R. Ravid, S. M. Staugaitis, M. F. Stins and R. M. Ransohoff, *Brain*, 2006, **129**, 212.
31. L. Izikson, R. S. Klein, A. D. Luster and H. L. Weiner, *Clin. Immunol.*, 2002, **103**, 125.
32. G. Conductier, N. Blondeau, A. Guyon, J. L. Nahon and C. Rovére, *J. Neuroimmunol.*, 2010, **224**, 93.
33. D. W. Holman, R. S. Klein and R. M. Ransohoff, *Biochim. Biophys. Acta*, 2011, **1812**, 220.
34. W. I. Khan, Y. Motomura, H. Wang, R. T. EI-Sharkawy, E. F. Verdu, M. Verma-Gandhu, B. J. Rollins and S. M. Collins, *Am. J. Physiol. Gastrointest. Liver Physiol.*, 2006, **291**, G803.
35. C. K. Chen, M. L. Kuo, K. W. Yeh, L. S. Ou, L. C. Chen, T. C. Yao and J. L. Huang, *J. Asthma*, 2009, **46**, 225.
36. M. A. Ahad, T. Missotten, A. Abdallah, P. A. Lympany and S. Lightman, *Mol. Vision*, 2007, **13**, 388.

37. A. Chen, S. Mumick, C. Zhang, J. Lamb, H. Dai, D. Weingarth, J. Mudgett, H. Chen, D. J. MacNeil, M. L. Reitman and S. Qian, *Obes. Res.*, 2005, **13**, 1311.
38. H. Kanda, S. Tateya, Y. Tamori, K. Hiasa, R. Kitazawa, S. Kitazawa, H. Miyachi, S. Maeda, K. Egashira and M. Kasuga, *J. Clin. Invest.*, 2006, **116**, 1494.
39. J. Huber, F. W. Kiefer, M. Zeyda, B. Ludvik, G. R. Silberhumer, G. Prager, G. J. Zlabinger and T. M. Stulnig, *J. Clin. Endocrine Metab.*, 2008, **93**, 3215.
40. H. Sell and J. Eckel, *Curr. Opin. Lipidol.*, 2007, **18**, 258.
41. N. Kamei, K. Tobe, R. Suzuki, M. Ohsugi, T. Watanabe, N. Kubota, N. Ohtsuka-Kowatari, K. Kumagai, K. Sakamoto, M. Kobayashi, T. Yamauchi, K. Ueki, Y. Oishi, S. Nishimura, I. Manabe, H. Hashimoto, Y. Ohnishi, H. Ogata, K. Tokuyama, M. Tsunoda, T. Ide, K. Murakami, R. Nagai and T. Kadowaki, *J. Biol. Chem.*, 2006, **281**, 26602.
42. C. Ruster and G. Wolf, *Front. Biosci.*, 2008, **13**, 944.
43. J. David and F. Mortari, *Clin. Appl. Immunol. Rev.*, 2000, **1**, 105.
44. G. Lazennec and A. Richmond, *Trends Mol. Med.*, 2010, **16**, 133.
45. D. D'Ambrosio, P. Panina-Bordignon and F. Sinigaglia, *J. Immunol. Methods*, 2003, **273**, 3.
46. S. J. Ono, T. Nakamura, D. Miyazaki, M. Ohbayashi, M. Dawson and M. Toda, *Mol. Allergy Clin. Immunol.*, 2003, **11**, 1185.
47. I. F. Charo, S. J. Myers, A. Herman, C. Franci, A. J. Connolly and S. R. Coughlin, *Proc. Natl Acad. Sci.*, 1994, **91**, 2752.
48. S. Sozzani, M. Locati, D. Zhou, M. Rieppi, W. Luini, G. Lamorte, G. Bianchi, N. Polentarutti, P. Allavena and A. Mantovani, *J. Leukocyte Biol.*, 1995, **57**, 788.
49. S. K. Biswas and A. Sodhi, *Int. Immumopharmacol.*, 2002, **2**, 1095.
50. Y. A. Berchiche, S. Gravel, M.-E. Pelletier, G. St-Onge and N. Heveker, *Mol. Pharmacol.*, 2011, **79**, 488.
51. J. Dawson, W. Miltz, A. K. Mir and C. Wiessner, *Expert Opin. Ther. Targets*, 2003, **7**, 35.
52. M. Sharma, *Crit. Rev. Biotechnol.*, 2010, **30**, 1.
53. M. E. Monzon, R. M. Forteza and S. M. Casalino-Matsuda, *Am. J. Physiol. Lung Cell Mol. Physiol.*, 2011, **300**, L204.
54. H. Yen, Y. Zhang, S. Penfold and B. J. Rollins, *J. Leukocyte Biol.*, 1997, **61**, 529.
55. M. Mellado and J. M. Rodriguez-Frade, *J. Immunol.*, 1998, **161**, 805.
56. S. J. Turner, J. Domin, M. D. Waterfield, S. G. Ward and J. Westwick, *J. Biol. Chem.*, 1998, **273**, 25987.
57. B. Cambien, M. Pomeranz and M. A. Millet, *Blood*, 2001, **97**, 359.
58. J. H. Wain, J. A. Kirby and S. Ali, *Clin. Exp. Immunol.*, 2002, **127**, 436.
59. S. Yamagami, Y. Tokuda, K. Ishii, H. Tanaka and N. Endo, *Biochem. Biophys. Res. Commun.*, 1994, **202**, 1156.
60. T. Kurihara and R. Bravo, *J. Biol. Chem.*, 1996, **271**, 11603.

61. D. Lu, X. J. Yuan, R. J. Evans Jr, A. T. Pappas, H. Wang, E. W. Su, C. Hamsouchi and C. Venkataraman, *BMC Immunol.*, 2005, **6**, 15.
62. B. J. Margulie, D. A. Hauer and J. E. Clements, *AIDS Res. Hum. Retroviruses*, 2001, **17**, 981.
63. T. Ellingsen, A. Buus and K. Stengaard-Pedersen, *J. Rheumatol.*, 2001, **28**, 41.
64. M. C. Grimm, S. K. O. Elsbury, P. Pavli and W. F. Doe, *J. Leukocyte Biol.*, 1996, **59**, 804.
65. N. A. Nelken, S. R. Coughlin, D. Gordon and J. N. Wilcox, *J. Clin. Invest.*, 1991, **88**, 1121.
66. A. R. Sousa, S. J. Lane, J. A. Nakhosteen, T. Yoshimura, T. H. Lee and R. N. Poston, *Am. J. Respir. Cell Mol. Biol.*, 1994, **10**, 142.
67. M. J. Verma, A. Lloyd, H. Rager, R. Strieter, S. Kunkel, D. Taub and D. Wakefield, *Curr. Eye Res.*, 1997, **16**, 1202.
68. A. E. Koch, S. L. Kunkel, L. A. Harlow, B. Johnson, H. L. Evanoff, G. K. Haines, M. D. Burdick, R. M. Pope and R. M. Strieter, *J. Clin. Invest.*, 1992, **90**, 772.
69. M. Harigai, M. Hara, T. Yoshimura, E. J. Leonard, K. Inoue and S. Kashiwazaki, *Clin. Immunol. Immunolpathol.*, 1993, **69**, 83.
70. T. Akahoshi, C. Wada, H. Endo, K. Hirota, S. Hosaka, K. Takagishi, H. Kondo, S. Kashiwazaki and K. Matsushima, *Arthritis Rheum.*, 1993, **36**, 762.
71. H. Ogata, M. Takeya, T. Yoshimura, T. Takagi and K. Takahashi, *J. Pathol.*, 1997, **182**, 106.
72. J. H. Gong and I. Clark-Lewis, *J. Exp. Med.*, 1995, **181**, 631.
73. J. H. Gong, L. G. Ratkay, J. D. Waterfield and I. Clark-Lewis, *J. Exp. Med.*, 1997, **186**, 131.
74. C. M. Brodmerkel, R. Huber, M. Covington, S. Diamond, L. Hall, R. Collins, L. Leffet, K. Gallagher, P. Feldman, P. Collier, M. Stow, X. M. Gu, F. Baribaud, N. Shin, B. Thomas, T. Burn, G. Hollis, S. Yeleswaram, K. Solomon, S. Friedman, A. L. Wang, C. B. Xue, R. C. Newton, P. Scherle and K. Vaddi, *J. Immunol.*, 2005, **175**, 5370.
75. N. Shin, F. Baribaud, K. Wang, G. Yang, R. Wynn, M. B. Covington, P. Feldman, K. B. Gallagher, L. M. Leffet, Y. Y. Lo, A. Wang, C. B. Xue, R. C. Newton and P. A. Scherle, *Biochem. Biophys. Res. Comm.*, 2009, **387**, 251.
76. M. P. Quinones, S. K. Ahuja, F. Jimenez, J. Schaefer, E. Garavito, A. Rao, G. Chenaux, R. L. Reddick, W. A. Kuziel and S. S. Ahuja, *J. Clin. Invest.*, 2004, **113**, 856.
77. M. P. Quinones, C. A. Estrada, Y. Kalkonde, S. K. Ahuja, W. A. Kuziel, M. Mack and S. S. Ahuja, *J. Mol. Med.*, 2005, **83**, 672.
78. H. Bruehl, J. Cihak, M. A. Schneider, J. Plachy, T. Rupp, I. Wenzel, M. Shakarami, S. Milz, J. W. Ellwart, M. Stangassinger, D. Schloendorff and M. Mack, *J. Immunol.*, 2004, **172**, 890.
79. A. Beaulieu, F. Hasler, E. M. Mola, K. Pavelka, J. DeMartino, M. Struthers, P. Chen, J. Koen and A. Melian, *Ann. Rheum. Dis.*, 2006, **65**, 175.

80. J. J. Haringman, D. M. Gerlag, T. J. M. Smeets, D. Baeten, F. Van den Bosch, B. Bresnihan, F. C. Breedveld, H. J. Dinant, F. Legay, H. Gram, P. Loetscher, R. Schmouder, T. Woodworth and P. P. Tak, *Arthritis Rheum.*, 2006, **54**, 2387.

81. C. E. Vergunst, D. M. Gerlag, L. Lopatinskaya, L. Klareskog, M. D. Smith, F. van den Bosch, H. J. Dinant, Y. Lee, T. Wyant, E. W. Jacobson, D. Baeten and P. P. Tak, *Arthritis Rheum.*, 2008, **58**, 1931.

82. D. J. Mahad and R. M. Ransohoff, *Semin. Immunol.*, 2003, **15**, 23.

83. D. R. Huang, J. Wang, P. Kivisakk, B. J. Rollins and R. M. Ransohoff, *J. Exp. Med.*, 2001, **193**, 713.

84. L. Izikson, R. S. Klein, I. F. Charo, H. L. Weiner and A. D. Luster, *J. Exp. Med.*, 2000, **192**, 1075.

85. B. T. Fife, G. B. Huffnagle, W. A. Kuziel and W. J. Karpus, *J. Exp. Med.*, 2000, **192**, 899.

86. M. Buntinx, B. Hermans, J. Goossens, D. Moechars, R. A. H. J. Gilissen, J. Doyon, S. Boeckx, E. Coesemans, G. Van Lommen and J. P. Van Wauwe, *J. Pharm. Exp. Therap.*, 2008, **327**, 1.

87. M. Braddock, *Expert Opin. Invest. Drugs*, 2007, **16**, 909.

88. C. E. Rose Jr., S. S. Sung and S. M. Fu, *Microcirculation*, 2003, **10**, 273.

89. C. Daly and B. J. Rollins, *Microcirculation*, 2003, **10**, 247.

90. C. Palmqvist, A. J. Wardlaw and P. Bradding, *Br. J. Pharmacol.*, 2007, **151**, 725.

91. S. T. Holgate, *Clin. Exp. Allergy*, 2008, **38**, 872.

92. T. R. Bai and D. A. Knight, *Clin. Sci.*, 2005, **108**, 463.

93. D. Vercelli, *Nat. Rev. Immunol.*, 2008, **8**, 169.

94. P. J. Barnes, *Nat. Rev. Immunol.*, 2008, **8**, 183.

95. S. T. Holgate and R. Polosa, *Nat. Rev. Immunol.*, 2008, **8**, 218.

96. S. J. Galli, M. Grimbaldeston and M. Tsai, *Nat. Rev. Immunol.*, 2008, **8**, 478.

97. E. M. Campbell, I. F. Charo, S. L. Kunkel, R. M. Strieter, L. Boring, J. Gosling and W. Lukacs, *J. Immunol.*, 1999, **163**, 2160.

98. M. Mallado, A. M. DeAna, L. Gomez, C. Martinez-A and M. Rodriguez-Frade, *J. Pharmcol. Exp. Ther.*, 2008, **324**, 769.

99. T. Okuma, Y. Terasaki, K. Kaikita, H. Kobayashi, W. A. Kuziel, M. Kawasuji and M. Takeya, *J. Pathol.*, 2004, **20**, 594.

100. D. Hartl, M. Griese, T. Nicolai, G. Zissel, C. Prell, D. Reinhardt, D. J. Schendel and S. Krauss-Etschmann, *Respir. Res.*, 2005, **6**, 93.

101. M. Sugar, H. Iyonaga, N. Saita, H. Yamasaki and M. Ando, *Eur. Respir. J.*, 1999, **1**, 376.

102. Y. K. Kim, H. B. OH, E. Y. Lee, J. E. Lee and Y. Y. Kim, *Allergy*, 2007, **62**, 207.

103. Z. Navratilova, *Biomed. Pap. Med. Fac. Univ. Palacky Olomouc Czech Repub.*, 2006, **150**, 191.

104. M. Toda, T. Nakamura, M. Ohbayashi, Y. Ikeda, M. Dawson, C. C. Aye, D. Miyazaki and S. J. Ono, *Expert Rev. Clin. Immunol.*, 2007, **3**, 351.

105. M. L. Castellani, K. Bhattacharya, M. Tagen, D. Kempuraj, A. Perrella, M. Delutis, W. Boucher and P. Conti, *Int. J. Immunopathol. Pharmacol.*, 2007, **20**, 447.

106. R. X. Zhang, S. Q. Yu, J. Z. Jiang and G. J. Liu, *J. Investig. Allergol. Clin. Immunol.*, 2007, **17**, 329.

107. M. Y. Donath and S. E. Shoelson, *Nat. Rev. Immunol.*, 2011, **11**, 98.

108. A. B. Goldfine, R. Silver, W. Aldhahi, D. Cai, E. Tatro, J. Lee and S. E. Shoelson, *Clin. Transl. Sci.*, 2008, **1**, 36.

109. M. M. Rumore and K. S. Kim, *Ann. Pharmacother.*, 2010, **44**, 1207.

110. A. W. Ferrante, *J. Intern. Med.*, 2007, **262**, 408.

111. J. G. Neels and J. M. Olefsky, *J. Clin. Investig.*, 2006, **116**, 33.

112. C. N. Lumeng, S. M. DeYoung, J. L. Bodzin and A. R. Saltiel, *Diabetes*, 2007, **56**, 16.

113. H. Kanda, S. Tateya, Y. Tamori, K. Kotani, K. I. Hiasa, R. Kitazawa, S. Kitazawa, H. Miyachi, S. Maeda, K. Egashira and M. Kasuga, *J. Clin. Investig.*, 2006, **116**, 1494.

114. S. P. Weisberg, D. Hunter, R. Huber, J. Lemieux, S. Slaymaker, K. Vaddi, I. Charo, R. Leibel and A. W. Ferrante, *J. Clin. Investig.*, 2006, **116**, 115.

115. A. Bouloumie, C. A. Curat, C. Sengenes, K. Lolmede, A. Miranville and R. Busse, *Curr. Opin. Clin. Nutr. Metab. Care*, 2005, **8**, 347.

116. J. M. Bruun, A. S. Lihn, S. B. Pederson and B. Richelsen, *J. Clin. Endocrinol. Metab.*, 2005, **90**, 2282.

117. T. Christiansen, B. Richelsen and J. M. Bruun, *Int. J. Obesity*, 2005, **29**, 146.

118. M. Hanefeld, E. Schell, I. Gouni-Berthold, M. Melichar, I. Vesela, T. Sullivan, S. Miao, D. Johnson, J. Jaen, P. Bekker and T. J. Schall, *Abstract Book*, *71st Scientific Sessions*, *American Diabetes Association*, 310 OR, San Diego, 2011 (*Diabetes*, 2011, **60**(1), A85).

119. S. Yiä-Herttuala, B. A. Lipton, M. E. Rosenfeld, T. Sarkioja, T. Yoshimura, E. J. Leonard, J. L. Witztum and D. Steinberg, *Proc. Natl Acad. Sci. USA*, 1991, **88**, 5252.

120. L. Boring, J. Gosling, M. Cleary and I. F. Charo, *Nature*, 1998, **394**, 894.

121. C. Szalai, G. T. Kozma, A. Nagy, A. Bojszko, D. Krikovszky, T. Szabo and A. Falus, *J. Allergy Clin. Immunol.*, 2001, **108**, 375.

122. C. Szalai, J. Duba, Z. Prohaszka, A. Kalina, T. Szabo, B. Nagy, L. Horvath and A. Csaszar, *Atherosclerosis*, 2001, **158**, 233.

123. M. Davidson, J. Lekstrom-Himes, J. Gilbert, D. Donaldson, Y. Lee, M. Hu and J. Xu, *Circulation*, 2007, **116**, II-172.

124. S. Giunti, F. Barutta, P. C. Perin and G. Gruden, *Curr. Vascul. Pharmacol.*, 2010, **8**, 849.

125. T. Furuta, T. Saito, T. Ootaka, J. Soma, K. Obara, K. Abe and K. Yoshinaga, *Am. J. Kidney Dis.*, 1993, **21**, 480.

126. C. Viedt and S. R. Orth, *Nephrol. Dial. Transplant*, 2002, **17**, 2043.

127. G. H. Tesch, *Am. J. Physiol. Renal Physiol.*, 2008, **294**, F697.

128. Y. S. Kang, M. H. Lee, H. K. Song, G. J. Ko, O. S. Kwon, T. K. Lim, S. H. Kim, S. Y. Han, K. H. Han, J. E. Lee, J. Y. Han, H. K. Kim and D. R. Cha, *Kidney Int.*, 2010, **78**, 883.

129. S. G. Sayyed, M. Ryu, O. P. Kulkarni, H. Schmid, J. Lichtnekert, S. Grüner, L. Green, P. Mattei, G. Hartmann and H. J. Anders, *Kidney Int.*, 2011, **80**, 68.

130. J. Zhang, L. Patel and K. J. Pienta, *Cytokine Growth Factor Rev.*, 2010, **21**, 41.

131. J. Zhang, L. Patel and K. J. Pienta, *Prog. Mol. Biol. Transl. Sci.*, 2010, **95**, 31.

132. K. Mizutani, S. Sud, N. A. McGregor, G. Martinovski, B. T. Rice, M. J. Craig, Z. S. Varsos, H. Roca and K. J. Pienta, *Neoplasia*, 2009, **11**, 1235.

133. R. D. Loberg, C. Ying, M. Craig, L. Yan, L. A. Snyder and K. J. Pienta, *Neoplasia*, 2007, **9**, 556.

134. J. Zhang, Y. Lu and K. J. Pienta, *J. Natl Cancer Inst.*, 2010, **102**, 522.

135. X. Lu and Y. Kang, *J. Biol. Chem.*, 2009, **284**, 29087.

136. B. Z. Qian, J. Li, H. Zhang, T. Kitamura, J. Zhang, L. R. Campion, E. A. Kaiser, L. A. Snyder and J. W. Pollard, *Nature*, 2011, **475**, 222.

137. M. Uguccioni, P. Gionchetti, D. F. Robbiani, F. Rizzello, S. Peruzzo, M. Campieri and M. Baggiolini, *Am. J. Pathol.*, 1999, **155**, 331.

138. C. Banks, A. Bateman, R. Payne, P. Johnson and N. Sheron, *J. Pathol.*, 2003, **199**, 28.

139. S. J. Connor, N. Paraskevopoulos, R. Newman, N. Cuan, T. Hampartzoumian, A. R. Lloyd and M. C. Grimm, *Gut*, 2004, **53**, 1287.

140. M. Gharaee-Kermani, R. E. McCullumsmith, I. F. Charo, S. L. Kunkel and S. H. Phan, *Cytokine*, 2003, **24**, 266.

141. T. Okuma, Y. Terasaki, K. Kaikita, H. Kobayashi, W. A. Kuziel, M. Kawasuji and M. Takeya, *J. Pathol.*, 2004, **204**, 594.

142. B. B. Moore, R. Paine, 3rd, P. J. Christensen, T. A. Moore, S. Sitterding, R. Ngan, C. A. Wilke, W. A. Kuziel and G. B. Toews, *J. Immunol.*, 2001, **167**, 4368.

143. C. Abbadie, J. A. Lindia, A. M. Cumiskey, L. B. Peterson, J. S. Mudgett, E. K. Bayne, J. A. DeMartino, D. E. MacIntyre and M. J. Forrest, *Proc. Natl Acad. Sci. USA*, 2003, **100**, 7947.

144. A. Serrano, M. Paré, F. McIntosh, S. J. Elmes, G. Martino, C. Jomphe, E. Lessard, P. M. Lembo, F. Vaillancourt, M. N. Perkins and C. Q. Cao, *Mol. Pain*, 2010, **6**, 90.

145. J. Zhang, X. Q. Shi, S. Echeverry, J. S. Mogil, Y. De Koninck and S. Rivest, *J. Neurosci.*, 2007, **27**, 12396.

146. D. Chen, Y. Ding, B. Schröppel, N. Zhang, S. Fu, D. Chen, H. Zhang and J. S. Bromberg, *Am. J. Transplant.*, 2003, **3**, 1216.

147. P. M. Hughes, P. R. Allegrini, M. Rudin, V. H. Perry, A. K. Mir, C. Wiessner and J. Cereb, *Blood Flow Metab.*, 2002, **22**, 308.

148. O. B. Dimitrijevic, S. M. Stamatovic, R. F. Keep and A. V. Andjelkovic, *Stroke*, 2007, **38**, 1345.

149. G. Conductier, N. Blondeau, A. Guyon, J. L. Nahon and C. Rovère, *J. Neuroimmunol.*, 2010, **224**, 93.
150. A. J. Barker, J. G. Kettle and A. W. Faull, Zeneca Limited, UK, *PCT Int. Appl.*, WO 1999/40913, 1999.
151. T. Shiota, F. Miyagi, T. Kamimura, T. Ohta, Y. Takano and H. Horiuchi, Teijin Limited, Japan, *PCT Int. Appl.*, WO 2000/69432, 2000.
152. T. Shiota, K. Kataoka, M. Imai, T. Tsutsumi, M. Sudoh, R. Sogawa, T. Morita, T. Hada, Y. Muroga, O. Takenouchi, M. Furuya, N. Endo, C. M. Tarby, W. A. Moree and S. L. Teig, Teijin Ltd, Japan; Combichem, Inc., *PCT Int. Appl.*, WO 1999/25686, 1999.
153. O. M. Z. Howard and T. Yoshimura, *Expert Opin. Ther. Pat.*, 2001, **11**, 1147.
154. T. Mirzadegan, F. Diehl, B. Ebi, S. Bhakta, I. Polsky, D. McCarley, M. Mulkins, G. S. Weatherhead, J.-M. Lapierre, J. Dankwardt, D. Morgans, Jr., R. Wilhelm and K. Jarnagin, *J. Biol. Chem.*, 2000, **275**, 25562.
155. X. F. Shi, S. Liu, J. Xiangyu, Y. Zhang, J. Huang, S. Liu and C.-Q. Liu, *J. Mol. Model.*, 2002, **8**, 217.
156. P. H. Carter and A. J. Tebben, *Methods Enzymol.*, 2009, **461**, 249.
157. K. Srikanth, P. C. Nair and M. E. Sobhia, *Bioorg. Med. Chem. Lett.*, 2008, **18**, 1450.
158. P. C. Nair, K. Srikanth and M. E. Sobhia, *Bioorg. Med. Chem. Lett.*, 2008, **18**, 1323.
159. P. C. Nair and M. E. Sobhia, *J. Chem. Inf. Model.*, 2008, **48**, 1891.
160. T. A. Berkhout, F. E. Blaney, A. M. Bridges, D. G. Cooper, I. T. Forbes, A. D. Gribble, P. H. E. Groot, A. Hardy, R. J. Ife, R. Kaur, K. E. Moores, H. Shillito, J. Willetts and J. Witherington, *J. Med. Chem.*, 2003, **46**, 4070.
161. P. H. Carter, *Curr. Opin. Chem. Biol.*, 2002, **6**, 510.
162. W. J. Moree, K. Kataoka, M. M. Ramirez-Weinhouse, T. Shiota, M. Imai, M. Sudo, T. Tsutsumi, N. Endo, Y. Muroga, T. Hada, H. Tanaka, T. Morita, J. Greene, D. Barnum, J. Saunders, Y. Kato, P. I. Myers and C. M. Tarby, *Bioorg. Med. Chem. Lett.*, 2004, **14**, 5413.
163. W. J. Moree, K. Kataoka, M. M. Ramirez-Weinhouse, T. Shiota, M. Imai, T. Tsutsumi, M. Sudo, N. Endo, Y. Muroga, T. Hada, D. Fanning, J. Saunders, Y. Kato, P. L. Myers, C. M. Tarby and M. Christine, *Bioorg. Med. Chem. Lett.*, 2008, **18**, 5413.
164. C.-B. Xue, A. Wang, D. Meloni, K. Zhang, L. Kong, H. Feng, J. Glenn, T. Huang, Y. Zhang, G. Cao, R. Anand, C. Zheng, M. Xia, Q. Han, D. J. Robinson, L. Storace, L. Shao, M. Li, C. M. Brodmerkel, M. Covington, P. Scherle, S. Diamond, S. Yeleswaram, K. Vaddi, R. Newton, G. Hollis, S. Friedman and B. Metcalf, *Bioorg. Med. Chem. Lett.*, 2010, **20**, 7473.
165. C.-B. Xue, H. Feng, G. Cao, T. Huang, J. Glenn, R. Anand, D. Meloni, K. Zhang, L. Kong, A. Wang, Y. Zhang, C. Zheng, M. Xia, L. Chen, H. Tanaka, Q. Han, D.J. Robinson, D. Modi, L. Storace, L. Shao, V. Sharief, M. Li, L.G. Galya, M. Covington, P. Scherle, S. Diamond, T. Emm, S. Yeleswaram, N. Contel, K. Vaddi, R. Newton, G. Hollis, S. Friedman and B. Metcalf, *ACS Med. Chem. Lett.*, 2011, **2**, 450.

166. C.-B. Xue, B. Metcalf, A. Q. Han, D. J. Robinson, C. Zheng, A. Wang and Y. Zhang, Incyte Corporation, *PCT Int. Appl. Pub.*, WO 2005/060665, 2005.

167. D. Yang, L. Tong, D. Wang, Y. Wang, X. Wang and X. Bai, *Respir. Physiol. Neurobiol.*, 2010, **170**, 253.

168. J. Witherington, V. Bordas, D. G. Cooper, I. T. Forbes, A. D. Gribble, R. J. Ife, T. Berkhout, J. Gohil and P. H. E. Groot, *Bioorg. Med. Chem. Lett.*, 2001, **11**, 2177.

169. M. Xia, C. Hou, D. DeMong, S. Pollack, M. Pan, M. Singer, M. Matheis, W. Murray, D. Cavender and M. Wachter, *Bioorg. Med. Chem. Lett.*, 2008, **18**, 6468.

170. M. Xia, C. Hou, D. DeMong, S. Pollack, M. Pan, J. Brackley, M. Singer, M. Matheis, D. Cavender and M. Wachter, *Bioorg. Med. Chem. Lett.*, 2008, **18**, 3562.

171. M. Xia, C. Hou, D. E. DeMong, S. R. Pollack, M. Pan, J. A. Brackley, N. Jain, C. Gerchak, M. Singer, R. Malaviya, M. Matheis, G. Olini, C. Cavender and M. Wachter, *J. Med. Chem.*, 2007, **50**, 5561.

172. M. Xia, C. Hou, S. Pollack, J. Brackley, D. DeMong, M. Pan, M. Singer, M. Matheis, G. Olini, D. Cavender and M. Wachter, *Bioorg. Med. Chem. Lett.*, 2007, **17**, 5964.

173. C. Cai and Z. Sui, Janssen Pharmaceutica NV, *US Pat. Appl.*, US 20090318498, 2009.

174. B. W. Budzik, H. S. Eidam, R. M. Fox, K. B. Goodman, D. B. Gotchev, P. A. Haile, T. V. Hughes, R. Liu, N. A. Miller, T. A. Miskowski, C. A. Sehon, A. Q. Viet, G. Z. Wang and J. Zhang, Glaxo Group Limited, *PCT Int. Appl.*, WO 2008/157741, 2008.

175. B. W. Budzik, M. J. Bury, M. Gu, R. Liu, F. Ren, C. A. Sehon, G. Z. Wang, J. Zhang, Glaxo Group Limited, *PCT Int. Appl.*, WO 2009/023754, 2009.

176. B. W. Budzik, P. A. Haile, T. V. Hughes, C. A. Sehon and G. Z. Wang, Glaxo Group Limited, *PCT Int. Appl.*, WO 2009/061881, 2009.

177. P. Eidam, P. A. Haile, T. V. Hughes, T. A. Miskowski and C. A. Sehon, Glaxo Group Limited, *PCT Int. Appl.*, WO 2010/093578, 2010.

178. P. A. Haile, C. A. Sehon and H. Wang, Glaxo Group Limited, *PCT Int. Appl.*, WO 2010/099098, 2010.

179. H. Ebel, S. Frattini, K. Gerlach, R. Giovannini, C. Hoenke, R. Mazzaferro, M. Santagostino, S. Scheuerer, C. Tautermann and T. Trieselmann, Boehringer Ingelheim GmbH, *PCT Int. Appl.*, WO 2011/073154, 2011.

180. H. Ebel, S. Frattini, K. Gerlach, R. Giovannini, C. Hoenke, M. Santagostino, S. Scheuerer and T. Trieselmann, Boehringer Ingelheim GmbH, *PCT Int. Appl.*, WO 2011/073155, 2011.

181. J. Aebi, A. Binggeli, L. Green, G. Hartmann, H. P. Maerki and P. Mattei, Hoffmann-La Roche, Inc., *US Pat. Appl.*, US 20110082294, 2011.

182. J. Aebi, A. Binggeli, L. Green, G. Hartmann, H. P. Maerki and P. Mattei, Hoffmann-La Roche, Inc., *US Pat. Appl.*, US 20110092698, 2011.

183. X. Zhang, H. R. Hufnagel, C. Hou, D. L. Johnson, Z. Sui, B. Fegely and D. Breslin, Janssen Pharmaceutica NV, *US Pat. Appl.*, US 20100144695, 2010.
184. X. Zhang, H. R. Hufnagel and Z. Sui, Janssen Pharmaceutica NV, *PCT Int. Appl.*, WO 2010121036, 2010.
185. X. Zhang, H. R. Hufnagel, C. Cai, J. Lanter, T. Markotan and Z. Sui, Janssen Pharmaceutica NV, *PCT Int. Appl.*, WO 2010/121011, 2010.
186. X. Zhang, H. R. Hufnagel, C. Hou, D. L. Johnson and Z. Sui, Janssen Pharmaceutica NV, *PCT Int. Appl.*, WO 2010/121046, 2010.
187. X. Zhang, H. Hufnagel, T. Markotan, J. Lanter, C. Cai, C. Hou, M. Singer, E. Opas, S. McKenney, C. Crysler, D. Johnson and Z. Sui, *Bioorg. Med. Chem. Lett.*, 2011, **21**, 5577.
188. X. Zhang, H. Hufnagel, C. Hou, E. Opas, S. McKenney, C. Crysler, J. O'Neill, D. Johnson and Z. Sui, *Bioorg. Med. Chem. Lett.*, 2011, **21**, 6042.
189. L. Yang, G. Butora, R. X. Jiao, A. Pasternak, C. Zhou, W. H. Parsons, S. G. Mills, P. P. Vicario, J. M. Ayala, M. A. Cascieri and M. MacCoss, *J. Med. Chem.*, 2007, **50**, 2609.
190. G. Butora, R. Jiao, W. H. Parsons, P. P. Vicario, H. Jin, J. M. Ayala, M. A. Cascieri and L. Yang, *Bioorg. Med. Chem. Lett.*, 2007, **17**, 3636.
191. A. Pasternak, S. D. Goble, P. P. Vicario, J. Di Salvo, J. M. Ayala, M. Struthers, J. A. DeMartino, S. G. Mills and L. Yang, *Bioorg. Med. Chem. Lett.*, 2008, **18**, 994.
192. S. Kothandaraman, K. L. Donnely, G. Butora, R. Jiao, A. Pasternak, G. J. Morriello, S. D. Goble, C. Zhou, S. G. Mills, M. MacCoss, P. P. Vicario, J. M. Ayala, J. A. DeMartino, M. Struthers, M. A. Cascieri and L. Yang, *Bioorg. Med. Chem. Lett.*, 2009, **19**, 1830.
193. L. Yang, C. Zhou, L. Guo, G. Morriello, G. Butora, A. Pasternak, W. H. Parsons, S. G. Mills, M. MacCoss, P. P. Vicario, H. Zweerink, J. M. Ayala, S. Goyal, W. A. Hanlon, M. A. Cascieri and M. S. Springer, *Bioorg. Med. Chem. Lett.*, 2006, **16**, 3735.
194. A. Pasternak, D. Marino, P. P. Vicario, J. M. Ayala, M. A. Cascierri, W. Parsons, S. G. Mills, M. MacCoss and L. Yang, *J. Med. Chem.*, 2006, **49**, 4801.
195. A. Pasternak, S. D. Goble, M. Struthers, P. P. Vicario, J. M. Ayala, J. Di Salvo, R. Kilburn, T. Wisniewski, J. A. DeMartino, S. G. Mills and L. Yang, *ACS Med. Chem. Lett.*, 2010, **1**, 14.
196. J. I. Trujillo, W. Huang, R. O. Hughes, D. J. Rogier, S. R. Turner, R. Devraj, P. A. Morton, C.-B. Xue, G. Chao, M. B. Covington, R. C. Newton and B. Metcalf, *Bioorg. Med. Chem. Lett.*, 2011, **21**, 1827.
197. C. Zheng, G. Cao, M. Xia, H. Feng, J. Glenn, R. Anand, K. Zhang, T. Huang, A. Wang, L. Kong, M. Li, L. Galya, R. O. Hughes, R. Devraj, P. A. Morton, D. J. Rogier, M. Covington, F. Baribaud, N. Shin, P. Scherle, S. Diamond, S. Yeleswaram, K. Vaddi, R. Newton, G. Hollis, S. Friedman, B. Metcalf and C.-B. Xue, *Bioorg. Med. Chem. Lett.*, 2011, **21**, 1442.
198. R. O. Hughes, D. J. Rogier, R. Devraj, C. Zheng, G. Cao, H. Feng, M. Xia, R. Anand, L. Xing, J. Glenn, K. Zhang, M. Covington,

P. A. Morton, J. M. Hutzler, J. W. Davis, II, P. Scherle, F. Baribaud, A. Bahinski, Z.-L. Mo, R. Newton, B. Metcalf and C.-B. Xue, *Bioorg. Med. Chem. Lett.*, 2011, **21**, 2626.

199. R. O. Hughes, D. J. Rogier, R. Devraj, C.-B. Xue, G. Cao, S. R. Turner, P. A. Morton, K. Keys, M. Covington, B. R. Bond, Y. Yu, H. Meade, W. F. Hood, S. Roeberds, R. Newton, B. Metcalf. *Abstracts of Papers, 242nd ACS National Meeting, American Chemical Association*, MEDI-26, Denver, 2011.

200. R. J. Cherney, R. Mo, D. T. Meyer, D. J. Nelson, Y. C. Lo, G. Yang, P. A. Scherle, S. Mandlekar, Z. R. Wasserman, H. Jezak, K. A. Solomon, A. J. Tebben, P. H. Carter and C. P. Decicco, *J. Med. Chem.*, 2008, **51**, 721.

201. R. J. Cherney, D. J. Nelson, Y. C. Lo, G. Yang, P. A. Scherle, H. Jezak, K. A. Solomon, P. H. Carter and C. P. Decicco, *Bioorg. Med. Chem. Lett.*, 2008, **18**, 5063.

202. R. J. Cherney, J. B. Brogan, R. Mo, Y. C. Lo, G. Yang, P. B. Miller, P. A. Scherle, B. F. Molino, P. H. Carter and C. P. Decicco, *Bioorg. Med. Chem. Lett.*, 2009, **19**, 597.

203. R. J. Cherney, R. Mo, D. T. Meyer, M. E. Voss, Y. C. Lo, G. Yang, P. B. Miller, P. A. Scherle, A. J. Tebben, P. H. Carter and C. P. Decicco, *Bioorg. Med. Chem. Lett.*, 2009, **19**, 3418.

204. R. J. Cherney, R. Mo, D. T. Meyer, M. E. Voss, M. G. Yang, J. B. Santella, III, J. V. Duncia, Y. C. Lo, G. Yang, P. B. Miller, P. A. Scherle, Q. Zhao, S. Mandlekar, M. E. Cvijic, J. C. Barrish, C. P. Decicco and P. H. Carter, *Bioorg. Med. Chem. Lett.*, 2010, **20**, 2425.

205. J. C. Lanter, T. P. Markotan, X. Zhang, C. Hou, M. Singer, E. Opas, S. McKenney, C. Crysler, D. Johnson, C. J. Molloy and Z. Sui, *Bioorg. Med. Chem. Lett.*, 2011, accepted.

206. B. Lagu, C. Gerchak, M. Pan, C. Hou, M. Singer, R. Malaviya, M. Matheis, G. Olini, D. Cavender and M. Wachter, *Bioorg. Med. Chem. Lett.*, 2007, **17**, 4382.

207. M. Shiraishi, Y. Aramaki, M. Seto, H. Imoto, Y. Nishikawa, N. Kanzaki, M. Okamoto, H. Sawada, O. Nishimura, M. Baba and M. Fujino, *J. Med. Chem.*, 2000, **43**, 2049.

208. S. Ungashe, ChemoCentryx, Inc., *PCT Int. Appl.*, WO 2006076644, 2006.

209. S. Ungashe, J. J. Wright, A. M. K. Pennell, Z. Wei and A. Melikian, ChemoCentryx, Inc., *US Pat. Appl. Publ.*, US 20070021466, 2007.

210. C. Brooks, P. A. Cleary, K. B. Goodman, S. Peace, J. Philp, C. A. Sehon, C. Smethurst and S. P. Watson, Glaxo Group Limted, *PCT Int. Appl.*, WO 2007/014054, 2007.

211. C. Brooks, S. Peace, C. Smethurst and S. P. Watson, Glaxo Group Limited, *PCT Int. Appl.*, WO 2007/014008, 2007.

212. S. Ungashe, Z. Wei, A. Basak, T. T. Charvat, W. Chen, J. Jin, J. Moore, Y. Zeng, S. Punna, D. Dairaghi, D. Hansen, A. M. K. Pennell and J. J. Wright, ChemoCentryx, Inc., *US Pat. Appl.*, US 20070037794, 2007.

213. K. B. Goodman, C. A. Sehon, P. A. Cleary, J. Philip and S. Peace, Glaxo Group Limted, *PCT Int. Appl.*, WO 2007/067875, 2007.

214. T. T. Charvat, C. Hu, J. Jin, Y. Li, A. Melikian, A. M. K. Pennell, S. Punna, S. Ungashe and Y. Zeng, ChemoCentryx, Inc., *PCT Int. Appl.*, WO 2008/008375, 2008.

215. A. Krasinski, S. Punna, S. Ungashe, Q. Wang and Y. Zeng, Chemocentryx, Inc., *PCT Int. Appl.*, WO 2009/009740, 2009.

216. S. Ungashe, Z. Wei, A. Basak, T. T. Charvat, W. Chen, J. Jin, J. Moore, Y. Zeng, S. Punna, D. Dairaghi, D. Hansen, A. M. K. Pennell, J. J. Wright, A. Krasinski and Q. Wang, *US Pat. Appl. Publ.*, US 20110118248, 2011.

217. S. Peace, J. Philp, C. Brooks, V. Piercy, K. Moores, C. Smethurst, S. Watson, S. Gaines, M. Zippoli, C. Mookherjee and R. Ife, *Bioorg. Med. Chem. Lett.*, 2010, **20**, 3961.

218. J. Doyon, E. Coesemans, S. Boeckx, M. Buntinx, B. Hermans, J. P. Van Wauwe, R. A. H. J. Gilissen, A. H. J. De Groot, D. Corens and G. Van Lommen, *ChemMedChem*, 2008, **3**, 660.

219. J. G. Kettle, A. W. Faull, A. J. Barker, D. H. Davies and M. A. Stone, *Bioorg. Med. Chem. Lett.*, 2004, **14**, 405.

220. A. B. Pinkerton, D. Huang, R. V. Cube, J. H. Hutchinson, M. Struthers, J. M. Ayala, P. P. Vicario, S. R. Patel, T. Wisniewski, J. A. DeMartino and J.-M. Vernier, *Bioorg. Med. Chem. Lett.*, 2007, **17**, 807.

221. http://www.clinicaltrials.gov/ (Aug. 2011).

222. http://www.bms.com/research/pipeline/Pages/default.aspx (Aug. 2011).

223. http://www.chemocentryx.com/product/CCR2.html (Aug. 2011).

224. http://www.pfizer.com/files/research/pipeline/2011_0811/pipeline_2011_0811.pdf (Aug. 2011).

225. R. Horuk, *Nature Rev. Drug. Dis.*, 2009, **8**, 23.

226. T. J. Schall and A. E. I. Proudfoot, *Nature Rev. Immunol.*, 2011, **11**, 355.

CHAPTER 13

Recent Advances in Selective CB₂ Agonists for the Treatment of Pain

E. J. GILBERT*[a] AND C. A. LUNN[b]

[a] Department of Medicinal Chemistry, Merck Research Laboratories, 2015 Galloping Hill Road, Kenilworth, NJ, 07033, USA; [b] Department of *In Vitro* Pharmacology, Merck Research Laboratories, 2015 Galloping Hill Road, Kenilworth, NJ, 07033, USA
*Email: eric.gilbert@merck.com

13.1 Introduction

The cannabinoid 2 (CB₂) receptor was first cloned from HL60 cells in 1993.[1] *In situ* hybridization experiments showed that CB₂ receptor mRNA was expressed in high levels in macrophage present in the marginal zone of the spleen, but was undetectable in any brain tissue. The association of CB₂ receptor expression with macrophage supported the idea that the CB₂ receptor may have a regulatory role in immune modulation. This result provided a convenient explanation for the numerous immune activities associated with cannabinoids,[2] and suggested a distinct target for the development of a new class of immune modulators free of potential CNS deficits.

Since then, many immunomodulatory activities have been ascribed to the cannabinoid CB₂ receptor, and many attempts have been reported seeking small-molecule effectors of these activities. However, the pharmacology

RSC Drug Discovery Series No. 26
Anti-Inflammatory Drug Discovery
Edited by Jeremy I Levin and Stefan Laufer
© The Royal Society of Chemistry 2012
Published by the Royal Society of Chemistry, www.rsc.org

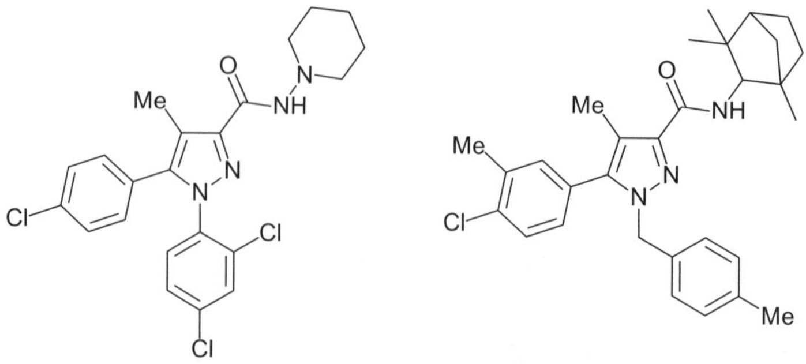

SR-141716   SR-144528

Figure 13.1 CB$_2$ inverse agonists.

required for this action has proved to be complex, with agonists and antagonists showing similar therapeutic benefit in experimental systems modelling immune diseases.[3,4] The discovery that the endogenous cannabinoid anandamide could modulate pain, while both receptor inverse agonists SR141716A (rimonabant) and SR144528 (Figure 13.1) reversed this response suggesting that CB$_2$ plays some role in the regulation of pain.[5] The consistent ability of cannabinoid agonists to modulate pain may be an exception to the pharmacological complexity within CB$_2$ biology, and has led to an extensive literature on the role of CB$_2$ agonists in pain.[6]

In this chapter, an update of the state of the art of cannabinoid CB$_2$ receptor-specific agonists, their chemistry and their potential therapeutic importance will be presented. Current thoughts on the biology of the CB$_2$ receptor, with particular emphasis on pain, the target area showing greatest promise for agonist intervention, will be summarized. The newest chemical advances in cannabinoid CB$_2$ receptor-specific agonists will be reviewed, focusing on those chemistries with a potential to lead to an orally active therapeutic. Finally, ongoing clinical trials featuring cannabinoid CB$_2$ receptor-specific agonists will be reviewed.

13.2   Finding Cannabinoid CB$_2$ Receptors in the CNS

The initial excitement surrounding the discovery of the cannabinoid CB$_2$ receptor was driven by the hope that this receptor represented a therapeutic target not present within the central nervous system. This offered the possibility that ligands specific to CB$_2$ would eliminate the psychoactivity that limited the clinical application of cannabinoids.[1] Subsequent studies have conclusively demonstrated the existence of functional cannabinoid CB$_2$ receptors within the CNS. Functional cannabinoid CB$_2$ receptors have been found in microglia cells,[7] which is not surprising as these cells can perform immunological functions within the brain.[8] Microglia are responsible for the ability of cannabinoid CB$_2$ receptor agonists to serve as neuroprotective agents in neurodegenerative disorders. Examples of this function are shown in the ability of HU-308 (Figure 13.2) administered intraperitoneally to protect striatal projection neurons from malonate-induced death, a model for Huntington's disease. This effect was diminished with co-administration of the cannabinoid CB$_2$ receptor antagonist SR144528. Sagredo and colleagues speculate that this effect is linked to a reduction in the generation of pro-inflammatory molecules like TNF-α.[9] Similar conclusions were reported by Palazuelos, who showed that cannabinoid CB$_2$ knockout mice showed exacerbated motor symptoms in the R6/2 transgenic mouse model for Huntington's disease. The investigators also showed that the cannabinoid CB$_2$ receptor-specific agonist HU-308 decreased striatal neurodegeneration after excitotoxicity caused by intra-striatal administration of quinolinic acid. The effect appears mediated by microglial cells and not astroglial cells.[10] Similarly, Ternianov and colleagues show that overexpression of CB$_2$ cannabinoid receptors reduced the vulnerability of nigrostriatal dopaminergic neurons to the neurotoxicity induced by intra-caudate administration of 6-hydroxydopamine.[11] This result, believed to be mediated by diminished recruitment of astrocytes and microglia to the insult site, may have implications for the treatment of Parkinson's disease.

Cannabinoid CB$_2$ receptor protein has also been found in the cerebellum, hippocampus, olfactory tubercule, islands of Calleja, cerebral cortex, striatum, thalamic nuclei, amygdala and some areas of the brainstem. Recent immunocytochemical studies have also found cannabinoid CB$_2$ receptors in adult rat retina, associated with retinal pigmentary epithelium, inner photoreceptor segments, horizontal and amacrine cells, cells localized in ganglion cell layer and in fibres of inner plexiform layer, although no function has yet to be ascribed to these receptors.[12] The existence of the receptor in these tissues is consistent with non-immune activities associated with cannabinoid CB$_2$ receptors. For example, Morgan and colleagues have presented pharmacological evidence of functional cannabinoid CB$_2$ receptors in synapses capable of suppressing GABAergic spontaneous inhibitory postsynaptic current amplitude in medial entorhinal cortex, as measured in tissue slices.[13] These studies showed that the current amplitude was suppressed using the CB$_2$-selective agonist JWH-133 (Figure 13.2), and the effect was reversed by the CB$_2$ receptor-specific inverse agonist AM-630 (Figure 13.2). Pre-addition of AM-630 blocked the effects of the CB$_2$ receptor agonist JWH-133. JTE-907 (Figure 13.2), a CB$_2$ receptor inverse agonist structurally unrelated to AM-630, also elicited increased GABAergic neurotransmission in this experimental system. The discovery of functional cannabinoid CB$_2$ receptors within neuronal synaptic tissue is an important foundation supporting a primary therapeutic target for cannabinoid CB$_2$ receptor-specific agonists – the treatment of neuropathic pain.

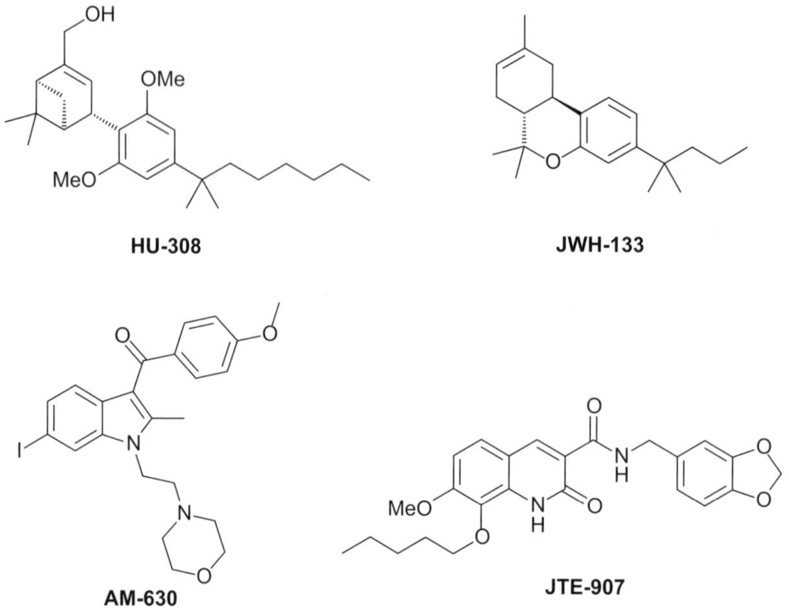

Figure 13.2   CB$_2$ agonists (HU-308 and JWH-133) and CB$_2$ inverse agonists (AM-630 and JTE-907).

13.3   Target Biology for Cannabinoid CB$_2$ Receptor Agonists

13.3.1   Cannabinoid CB$_2$ Receptor Involvement in Pain

The effective treatment of neuropathic pain continues to be a significant unmet medical need. Of particular interest are therapeutics that lack the often debilitating depressant side-effects associated with opioid-like compounds that target centres in the brain. A compound with the ability to generate an analgesic effect at a peripheral neuron *via* a peripheral receptor would therefore be highly desirable. Much of the work on the cannabinoid CB$_2$ receptor seeks to bolster this target as a peripheral modulator of the pain response. Sagar and colleagues describe endocannabinoid regulation of spinal nociceptive processing in a model of neuropathic pain.[14] In this study, a rat model of neuropathic pain employing spinal nerve ligation was interrogated for paw withdrawal following mechanical punctate stimulation using both innocuous (10 g) and noxious (15–60 g) Von Frey monofilaments. The investigators show that an antinociceptive effect of the fatty acid amide hydrolase inhibitor UCM707 (Figure 13.3) is blocked by the cannabinoid CB$_2$ antagonist SR144528 administered directly to the spinal cord, a result of modulating endocannabinoid interaction with CB$_2$ *in vivo*. Curto-Reyes and colleagues describe the spinal and peripheral analgesic effects of the CB$_2$ cannabinoid

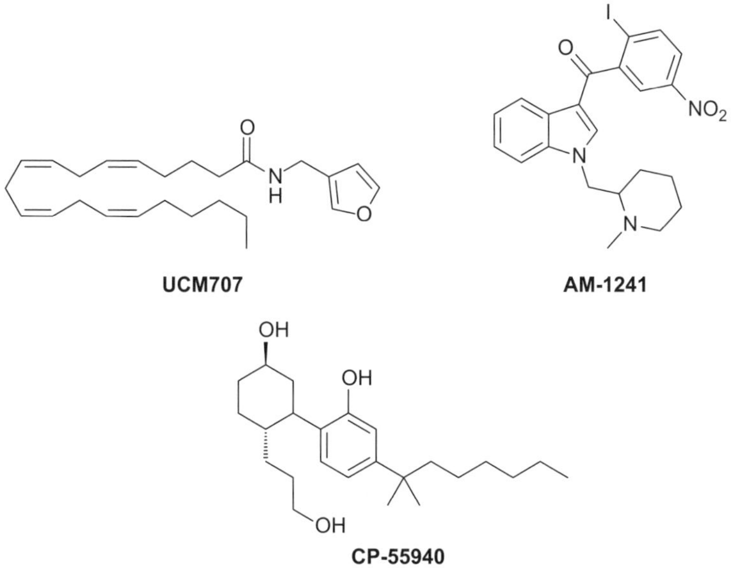

Figure 13.3   Fatty acid amide hydrolase inhibitor UCM707, selective CB$_2$ agonist
AM-1241 and non-selective CB$_2$ agonist CP-55,940.

receptor agonist AM1241 (Figure 13.3) in a bone cancer-induced pain model.[15]
To induce bone pain, the investigators inject osteosarcoma or melanoma cells
directly into the medullar cavity of the tibia. Thermal hyperalgesia and
mechanical allodynia were then determined following drug treatment in the
presence or absence of the cannabinoid CB$_2$ antagonist SR144528. The investigators conclude that spinal CB$_2$ receptors are involved in the anti-allodynic effect
induced by the CB$_2$ agonist AM1241, but that both peripheral and spinal
receptors participate in the anti-hyperalgesic effects of the compound.

Xu and colleagues describe the pharmacological characterization of a novel
N-alkyl isatin acylhydrazone, **13.21** (Figure 13.11).[16] This compound, MDA19,
showed 70-fold selectivity for the rat CB$_2$ receptor over the rat CB$_1$ receptor in
competitive binding experiments *versus* [^{3}H]-CP-55,940 (Figure 13.3). However,
the pharmacology of this compound is complex, exhibiting different pharmacologies dependent on the assay system used for the evaluation. Other
examples of cannabinoids behaving as "protean agonists" include AM1241[17],
L768242[18] and AM630.[19] MDA19 behaves as an inverse agonist, based on its
ability to modulate [^{35}S]-GTPγS binding to rat CB$_2$ receptors in recombinant
CHO membrane fractions. Using these same membrane fractions, the compound behaves as a neutral antagonist in a cAMP activation assay. Finally,
MDA19 induced phosphorylation of Erk1/2 in rat CHO cells stably expressing
rat CB$_2$ receptors, thus behaving as an agonist. Using both a spinal nerve
ligation model and a paclitaxel-induced neuropathy model for neuropathic
pain, MDA19 was shown to be effective in treating allodynia in rat and mice
models of neuropathic pain by activating CB$_2$ receptors, without affecting the

locomotor behaviour of the animals at the active dose. This result suggests that, though the pharmacology may be complex, cannabinoid CB_2 receptor-specific ligands can have significant beneficial therapeutic results.

Finally, investigations in a model of osteoarthritis found paradoxical nociceptive activities of the cannabinoid CB_2 receptor agonist GW405833 (**13.18**, Figure 13.10). Using intra-articular injection of sodium monoiodoacetate to induce osteoarthritis, Schuelert and colleagues found that the local application of GW405833 reduced joint afferent firing rate by up to 31% in control knees, and that the CB_2 receptor antagonist AM630 attenuated the sensitizing effect.[20] This antinociceptive effect was not seen in the arthritic knee, where the receptor agonist increased firings, consistent with an increased pain response. Behavioural studies confirmed the sensitizing effect of GW405833, as mice shifted more weight off of an osteoarthritic limb following drug administration.

The complex pharmacology associated with cannabinoid CB_2 receptor-specific compounds complicates the development of receptor-specific therapeutics. However, complex pharmacology is not unique to the cannabinoid receptor system. Protean agonists, ligands that exhibit different pharmacologies in different assay systems, have been found for the histamine H3 receptor,[21,22] the secretin receptors[23] and the α2A adrenoceptors.[24,25] Ligands have also been described that specifically target one of several possible signal transduction pathways mediated through one seven-transmembrane receptor. These compounds, called biased agonists, are predicted to have therapeutic importance in several medically important indications, including an ability to increase bone mass *via* the parathyroid hormone receptor,[26] an ability to modulate cardiac function *via* the beta2-adrenergic receptor[27] and an ability to maximize anti-psychotic efficacy *via* the dopamine D2 receptor.[28] The possible therapeutic value of biased agonists for the cannabinoid receptor system has also been discussed.[29] Interestingly, we have obtained preliminary data suggesting that the PPARα receptor agonist fenofibrate behaves as a biased agonist for the cannabinoid CB_2 receptor in several cell-based assay systems.[30] Experimental evidence has linked cannabinoids and PPAR receptors,[31] and one clinical study suggested fenofibrate reduces systemic inflammation independent of its effects on lipid and glucose metabolism.[32] Whether the anti-inflammatory activity of fenofibrate is linked to cannabinoid biology (or *via* another off-target activity) must await further experimentation.

13.4 CB_2 Agonist Chemotypes

The realization of the therapeutic potential of cannabinoid agonists for the modulation of pain requires a small molecule that targets this receptor with potency and high specificity.[33,34] Medicinal chemistry efforts have focused on selectivity over the centrally located CB_1 receptor in order to avoid psychotropic side effects. The main strategy the field has used to avoid CB_1-mediated side effects has been to build in CB_2/CB_1 at the molecular level. One could also imagine pursuing a non-brain penetrant CB_2 agonist, thus avoiding the central CB_1 receptor; however, the lipophilic nature of CB_2 agonists would make this a tremendous challenge. Furthermore, while there is no direct evidence of CB$_2$ involvement in central nociception pathways, empirical evidence suggests that CNS penetration of a CB$_2$ agonist may be important. Numerous scaffolds have been interrogated in the search for an efficacious CB$_2$ agonist that modulates the pain response. For instance, groups have pursued acyclic and mono-, bi- and tricyclic cores in addition to bi- and triaryl cores. The lipophilic nature of CB$_2$ agonists consistently leads to medicinal chemistry efforts focused on improving solubility, lowering ClogP and improving metabolic stability. Binding potency, CB$_1$ selectivity and functional activity of a substrate can vary across species, which has the effect of reducing the confidence in the ability of pre-clinical pain models to predict clinical outcomes. While several CB$_2$ agonists have reached the clinic, a proof of concept for the treatment of pain in humans has yet to be realized.

13.4.1 Acyclic Cores

13.4.1.1 α-Amido Sulfones/Sulfonamide

The structurally similar α-amidosulfonamide **13.1** and α-amidosulfone **13.2** (Figure 13.4) have been reported separately by Boehringer Ingelheim and Amgen, respectively. Compound **13.1** was reported to be a CB$_2$ agonist with an EC$_{50}$ of 1.3 nM (cAMP), although no further data were provided.[35] The sulfone **13.2** was reported to be a potent and selective CB$_2$ agonist (hCB$_2$ EC$_{50}$ = 29 nM in a GTP-Europium binding assay; CB$_1$/CB$_2$ = 86).[36] The 4-chlorophenyl substituent on the sulfone engendered the best combination of potency and selectivity whereas a smaller fluorine substituent led to an erosion of CB$_2$ selectivity. Sulfone **13.2** was reported to have low intrinsic clearance in human liver microsomes (HLM Cl$_{int}$ = 14 μl/min/mg), modest protein binding across species (human PPB = 94%) and good exposure (AUC$_{0-inf}$ = 3570 ng·h/mL, 10 mg/kg) and bioavailability in rat (F = 43%).

13.4.2 Monocyclic Cores

13.4.2.1 5-Membered Heterocycles

Optimization of a high-throughput screening hit by the Taisho Pharmaceutical Company resulted in the identification of the sulfonamide **13.3** (Figure 13.5) as a potent and selective CB$_2$ agonist (hCB$_2$ IC$_{50}$ = 16 nM, CB$_1$/CB$_2$ = 106 in the

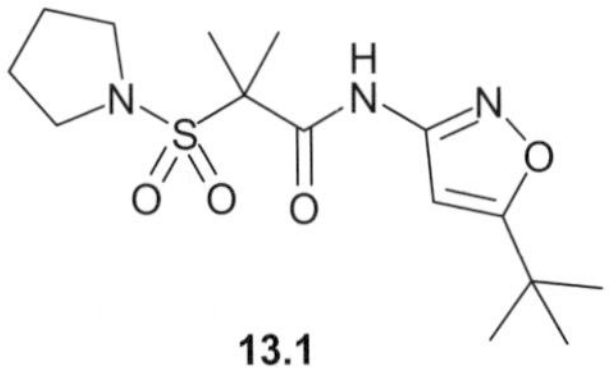

13.1 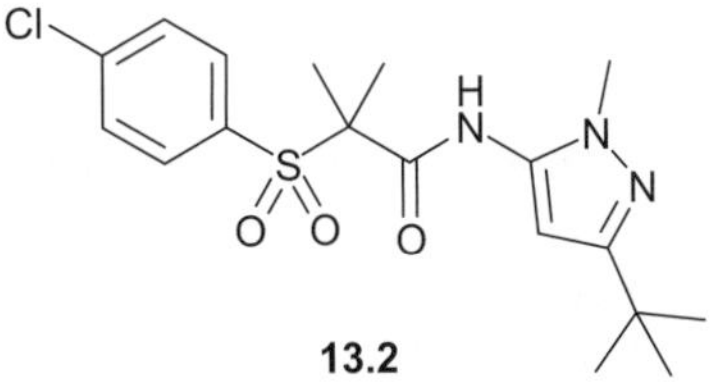   **13.2**

Figure 13.4   CB$_2$ agonists with an acyclic core.

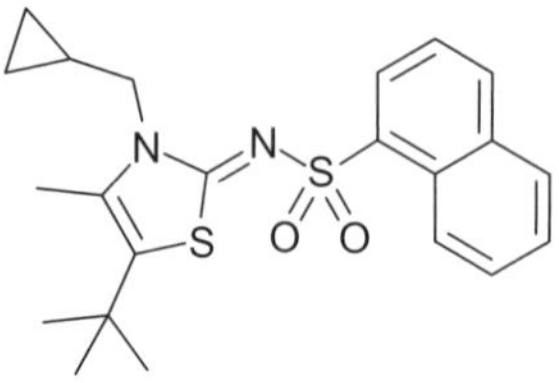

Figure 13.5   CB$_2$ agonists with a 5-membered heterocyclic core.

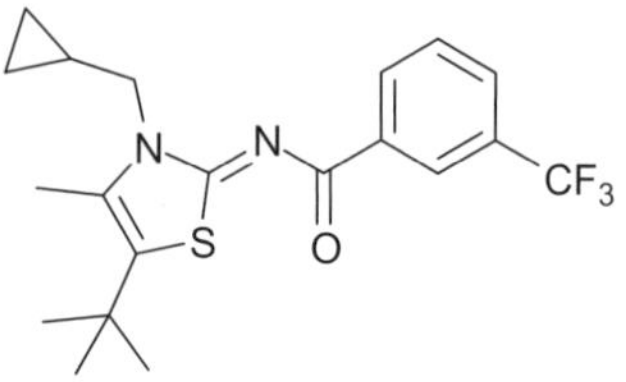

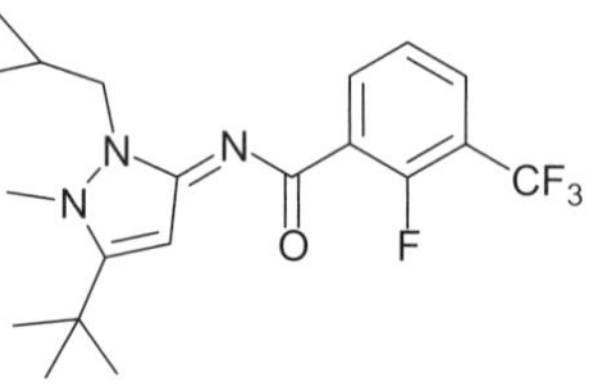

[^{3}H]-CP-55,940 binding assay).[37] In the [^{35}S]GTPγS functional assay, **13.3** showed an E_{max} of 100% relative to the full CB$_2$ agonist CP-55,940 with a CB$_2$ EC$_{50}$ of 7.2 nM. In an effort to improve the metabolic stability of the sulfonamide series, the analogous amide series was explored, leading to **13.4** (hCB$_2$ IC$_{50}$ = 13 nM, CB$_1$/CB$_2$ = 269 in the [^{3}H]-CP-55,940 binding assay).[38] It was found that the amide series did have improved metabolic stability relative to the sulfonamide series with **13.4** showing 93% remaining in human liver microsomes. Additional exploration of the SAR suggested that the *N*-cyclopropylmethyl group and the bulky lipophilic 5-*t*-butyl moiety were important for obtaining good CB$_2$ binding potency, selectivity over the CB$_1$ receptor and good agonist activity. Further optimization led to the pyrazole derivative **13.5** with improved potency and selectivity (hCB$_2$ IC$_{50}$ = 3 nM, CB$_1$/CB$_2$ = 1380 in the [^{3}H]-CP-55,940 binding assay).[39] Oral administration of **13.5** in rats resulted in good plasma exposure (10 mg/kg *p.o.*, AUC$_{0-8h}$ = 2160 ng · h/mL, C_{max} = 545 ng/mL, $t_{1/2}$ = 1.7 h). Compound **13.5** reversed mechanical hyperalgesia in a dose-dependent manner in the Randall–Sellitto model of inflammatory pain in rats. Notably, the pyrazolylidene **13.5** demonstrated much improved solubility in water as compared to **13.4** (5.9 mg/mL *vs.* <0.01 mg/mL at 25 °C).

13.4.2.2   6-Membered Heterocycles

13.4.2.2.1   Aminopyridine/pyrimidine.   Using a pharmacophore model as a guide, select compounds from the Glaxo SmithKline (GSK) corporate compound collection were tested for binding affinity at the human CB$_1$ and CB$_2$ receptor. This screen resulted in the identification of pyrimidinyl ester **13.6** (Figure 13.6) as a ligand for the CB$_2$ receptor (CB$_2$ EC$_{50}$ = 630 nM in the cAMP assay) with no activity at the CB1 receptor at concentrations up to 30 μM.[40] Optimization studies led to the identification of amide **13.7** (GW842166X),

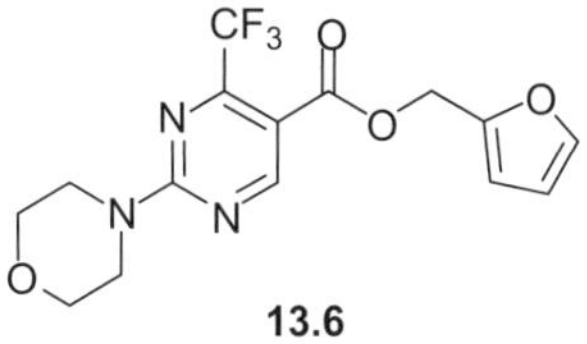

13.6 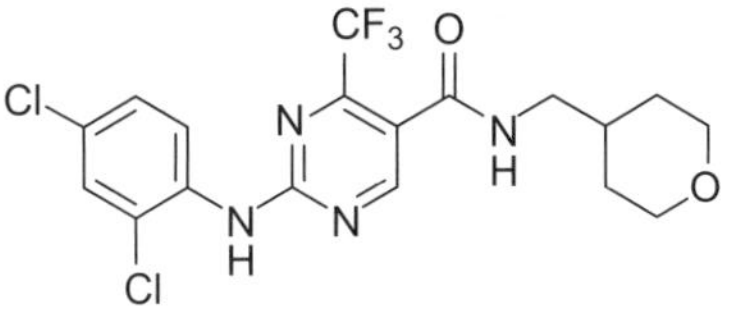   **13.7**

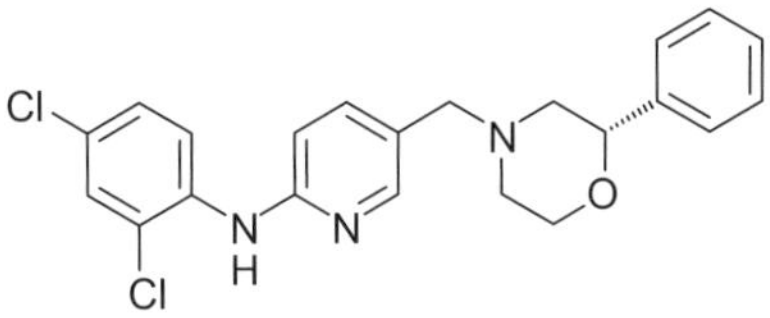

13.8

Figure 13.6   CB$_2$ agonists with a 6-membered heterocyclic core.

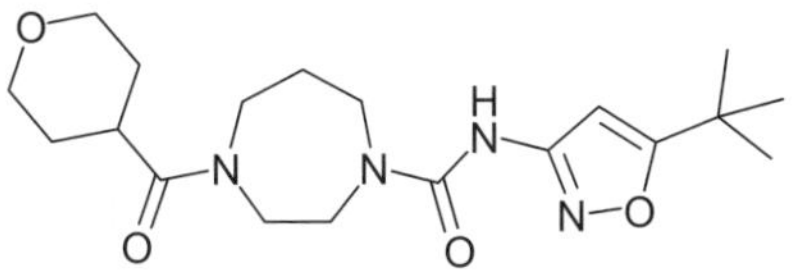

13.9

Figure 13.7   CB$_2$ agonist with a 7-membered heterocyclic core.

which was progressed to the clinic for the treatment of pain associated with osteoarthritis and rheumatoid arthritis. Compound **13.7** possesses good human CB$_2$ potency (EC$_{50}$ = 63 nM) and also shows no CB$_1$ agonist activity at concentrations up to 30 μM in both human and rat CB$_1$ recombinant assays. When dosed orally in rats, **13.7** had a half-life of 3 hours and a bioavailability of 58%. In addition, **13.7** had a clean CYP450 profile; however, it suffers from poor solubility (2 μg/mL at pH 7.4). It was efficacious in the CFA model of inflammatory pain with an oral ED$_{50}$ of 0.1 mg/kg. This effect could be reversed by the administration of a CB$_2$ antagonist, suggesting that the anti-hyperalgesia effects of **13.7** are mediated by the CB$_2$ receptor.

Compound **13.8**, which possesses an aminopyridine central ring and lacks the amide functionality of **13.7**, was shown to be a potent CB$_2$ agonist in a cellular assay (CB$_2$ cAMP EC$_{50}$ = 10 nM, 100% efficacy), with good selectivity over CB$_1$ (CB$_1$/CB$_2$ = >2000).[41] Examination of **13.8** in a zymosan-induced paw inflammation model showed an 85% decrease in paw inflammation 3 hours post dose (100 mpk, *p.o.*).

13.4.2.3   7-Membered Heterocyclic Core

A 7-membered agonist with a diazapane core ring system has been reported by Evotec and Boehringer Ingelheim, as a high-throughput screening hit.[42] Efforts

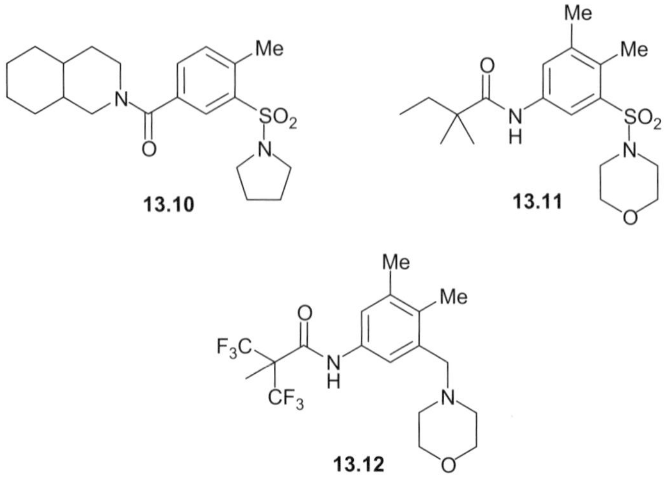

Figure 13.8 CB$_2$ agonists with a 6-membered non-heterocyclic aromatic core.

to improve the microsomal stability of the initial hit while maintaining CB$_2$ potency and selectivity over CB$_1$ resulted in the identification of **13.9** (Figure 13.7) (CB$_2$ EC$_{50}$ = 67 nM, cAMP (104% efficacy), CB$_1$ EC$_{50}$ = > 20,000; human liver microsome $t_{1/2}$ = > 120 minutes; Rat AUC (10 µmol/kg, *p.o.*) = 1380 nmol·h/L, F% = 29). However, compound **13.9** was found to have 3% bioavailability in the cynomologous monkey, which was attributed to intestinal efflux based on an efflux ratio of 6 in caco-2 cells.[43]

13.4.2.4 6-Membered Aromatic Core

The aryl sulfonamide **13.10** (Figure 13.8) was the product of the optimization of a non-selective high-throughput screening hit by Boehringer Ingelheim and Evotec AG. In a cellular assay, **13.10** was found to be a potent and selective CB$_2$ agonist (CB$_2$ cAMP EC$_{50}$ = 4 nm, CB$_1$/CB$_2$ = 560). Methyl substitution on the phenyl ring was found to be critical for binding potency as the des-methyl analogue had a CB$_2$ EC$_{50}$ > 20 µM.[44]

A series of reverse amide analogues of **13.10** were prepared at the Aldolor Corporation culminating in the identification of **13.11** (hCB$_2$ K$_i$ = 17 nM, CB$_1$/CB$_2$ = 150 in the [^{3}H]-CP-55,940 binding assay). As with **13.10**, methyl substitution alpha to the sulfonamide was critical for CB$_2$ binding potency. Compound **13.11** showed a robust anti-allodynic effect in a post-surgical pain *in vivo* model.[45] Deletion of the sulfonyl moiety of the sulfonamide led to a series exemplified by **13.12** (hCB$_2$ K$_i$ = 2.7 nM, CB$_1$/CB$_2$ = 190 in the [^{3}H]-CP-55,940 binding assay).[46] In a hind paw incision model of postoperative pain, **13.12** (10 mg/kg IP) showed an anti-allodynic effect comparable to morphine (3 mg/kg IP); however, no effect was observed when **13.12** was dosed orally.

13.4.3 Bicyclic Cores

13.4.3.1 Benzimidazole Core

The benzimidazole sulfone **13.13** (Figure 13.9) was the product of the optimization of a screening hit by researchers at Johnson and Johnson. A bulky hydrophobic group in the 2-position of the benzimidazole was found to improve binding potency at the human receptor while maintaining full agonism.[47] Substitution on the pyridyl sulfone was used to improve selectivity over CB$_1$ and to optimize metabolic stability. A 4-pyridyl sulfone suffered from poor metabolic stability in human microsomes; however, the 3-ethoxypyridyl sulfone **13.13** had an improved metabolic profile and excellent binding potency and selectivity (hCB$_2$ cAMP EC$_{50}$ = 0.3 nM; CB$_1$/CB$_2$ = 4266). Compound **13.13** showed limited brain penetration when dosed in male Sprague-Dawley rats (AUC$_{0-inf}$ = 2501 ng · h/mL, C_{max} = 731 ng/mL, $t_{1/2}$ = 3.4 h, %F = 43%, b/p = 0.1 at 10 mg/kg, *p.o.*), which was thought to mitigate the psychotropic side-effects observed with similar compounds when dosed in the rat. However, low solubility and inhibition of Cytochrome P450 2C9 and 2C19 remained liabilities for this compound.

Benzimidazole **13.14** was prepared by Pfizer researchers in the pursuit of a selective and brain penetrant CB$_2$ agonist.[48] Compound **13.14** was potent and selective (hCB$_2$ cAMP EC$_{50}$ = 2.7 nM, hCB$_1$ cAMP EC$_{50}$ = > 10 μM) and had excellent bioavailability (100%) upon oral dosing (2 mg/kg, *p.o.*) with a brain plasma ratio of 0.3. However, **13.14** also showed affinity for the hERG channel (hERG patchclamp IC$_{50}$ = 2.6 μM). Replacing the cyclopropyl group of **13.14** with a less lipophilic 2-tetrahydrofuranyl group led to **13.15**, which had an improved hERG profile (hERG patchclamp IC$_{50}$ = 63 μM). Compound **13.15**

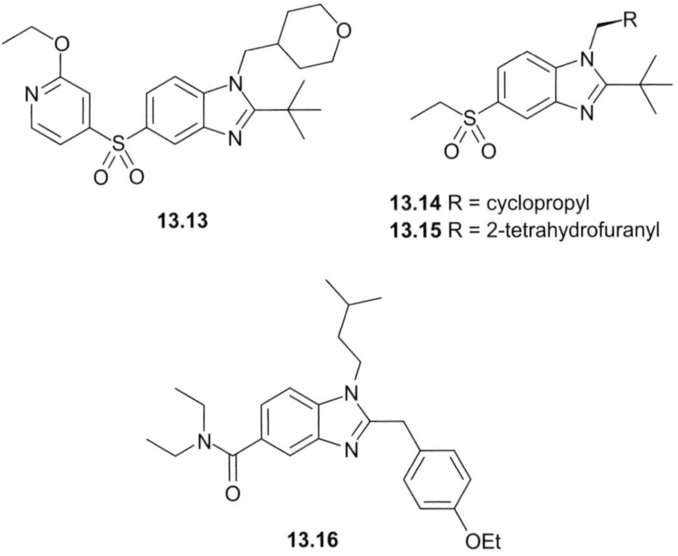

13.13

13.14 R = cyclopropyl
13.15 R = 2-tetrahydrofuranyl

13.16

OEt

Figure 13.9 CB$_2$ agonists with benzimidazole core.

also maintained a good *in vitro* profile (hCB$_2$ cAMP EC$_{50}$ = 11 nM, hCB$_1$ cAMP EC$_{50}$ = 15 μM) and oral bioavailability (100%) with improved brain penetration relative to **13.14** (b/p = 1). Studies to asses the *in vivo* pharmacology of **13.15** are reported to be underway.

Researchers at AstraZeneca also pursued CB$_2$ agonists possessing a benzimidazole core.[49] Optimization of a screening hit and extensive SAR studies led to a series exemplified by **13.16** (hCB$_2$ K$_i$ = 4.5 nM, CB$_1$/CB$_2$ = >1000 in the [^{3}H]-CP-55,940 binding assay; hCB1 GTPγ[^{35}S] EC$_{50}$ = 2.9 nM, E_{max} = 63% relative to WIN-55,212-2). Investigation of the benzimidazole *N*-substituent suggested that alkyl groups were preferred in terms of CB$_2$ binding potency; however, CB$_1$ selectivity could be improved with more hydrophilic groups. While the tertiary amide was preferred, a secondary amide was tolerated; however, a primary amide led to the loss of CB$_2$ affinity. Finally, the optimal benzyl linker was a methylene while a ketone or a secondary alcohol led to ~10X loss in binding potency.

13.4.3.2   Indole-based Cores

13.4.3.2.1   Indole Core.   The indole core can be found in the early days of cannabinoid receptor research as exemplified by WIN-55,212-2 (**13.17**, Figure 13.10) which was reported as a CB$_2$ agonist (K$_i$ = 3.3 nM in the [^{3}H]-CP-55,940 competitive binding assay, CB$_1$/CB$_2$ = 19) in 1992 by researchers at Sterling-Winthrop.[50,51] The acyl indole GW-405833 (**13.18**) was reported to be a selective and potent CB$_2$ modulator (hCB$_2$ K$_i$ = 3.9 nM; CB$_1$/CB$_2$ = 4772 in the [^{3}H]-CP-55,940 binding assay).[52] In a forskolin-mediated cAMP assay, **13.18** was found to be a partial agonist with a maximum inhibition of 45%. The intra-peritoneal administration of **13.18** in rats was reported to reduce mechanical hyperalgesia in a dose-dependent manner in neuropathic-, incision- and inflammation-induced pain models (ED$_{50}$ at 1 hour time point = 0.08 mg/kg, 2.6 mg/kg and 0.17 mg/kg, respectively).

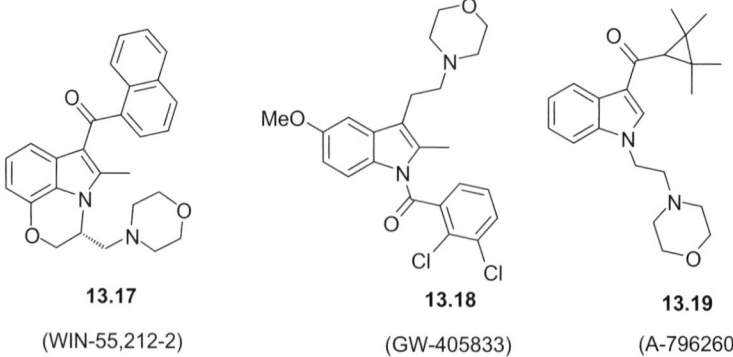

13.17	**13.18**	**13.19**
(WIN-55,212-2)	(GW-405833)	(A-796260)

Figure 13.10   CB$_2$ agonists with an indole core.

Abbott has reported the *N*-alkyl indole A-796260 (**13.19**, Figure 13.10).[17,53] In the [^{3}H]-CP-55,940 competitive binding assay **13.19** demonstrated good binding potency at the human (K$_i$ = 4.4 nM) and rat receptor (K$_i$ = 13 nM). The tetramethylcyclopropyl group led to superior human CB$_2$ binding potency and selectivity over the CB$_1$ receptor relative to other acyl carbocycles investigated. Selectivity *versus* the CB$_1$ receptor was found to be 193-fold for human and 30-fold for rat. Of note is the much diminished selectivity *versus* CB$_1$ in the rat relative to the human receptor. In cases such as this, care must be taken when interpreting *in vivo* efficacy as it may be difficult to decouple the analgesic effects resulting from binding to the CB$_1$ receptor. Functional activity was measured in a cyclase assay, which showed that **13.19** behaved as an agonist (hCB$_2$ EC$_{50}$ = 0.71 nM, E_{max} of 78%). In the complete Freund's adjuvant (CFA)-induced chronic inflammatory thermal hyperalgesia model, *i.p.* administration of **13.19** resulted in a dose-dependent attenuation of thermal hyperalgesia with an ED$_{50}$ of 2.8 mg/kg. The analgesic effects of **13.19** could be blocked by the pre-administration of either of the CB$_2$ antagonists SR144528 or AM630. Compound **13.19** also demonstrated analgesic effects in post-operative (ED$_{50}$ = 18 mg/kg, *i.p.*) and neuropathic pain (ED$_{50}$ = 15 mg/kg, *i.p.*) models.

13.4.3.2.2 Aza-indole Core. The aza-indole **13.20** (Figure 13.11) was the culmination of efforts to improve on the binding potency and aqueous solubility of the pyrimidine **13.7**.[40] Compound **13.20** showed good potency at the human CB$_2$ receptor and excellent selectivity over the CB$_1$ receptor (hCB$_2$ cAMP EC$_{50}$ = 5 nM, CB$_1$/CB$_2$ = > 1200). It had a clean CYP450 profile and low intrinsic clearance in human (< 0.6 (mL/min)/g) and rat (< 0.8 (mL/min)/g) microsomes. Solubility in the simulated intestinal fluid assay was ∼ 60-fold that of **13.7**. In the acute CFA model of inflammatory pain, **13.20** had an impressive ED$_{50}$ of 0.02 mg/kg (*p.o.*) with full reversal of hyperalgesia observed at 0.1 mg/kg (97 nM plasma concentration). This effect could be reversed by the co-administration of the CB$_2$ antagonist AM630. The compound was also efficacious in a chronic joint pain model (10 mg/kg, *p.o.*, *b.i.d.*, 5 days) with efficacy comparable to an oral dose of the COX-2 inhibitor rofecoxib. The efficacy of **13.20** in these pain models was attributed in part to its good CNS

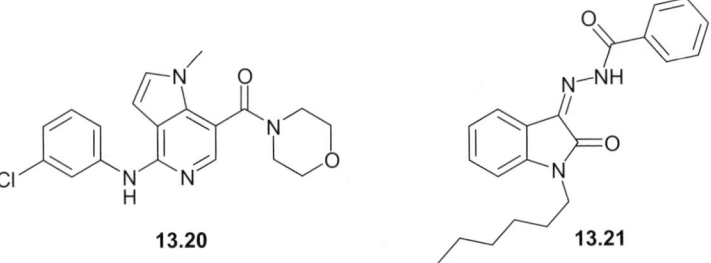

13.20   **13.21**

Figure 13.11 CB$_2$ agonists with an aza-indole and oxindole core.

penetration (130 mg/kg, *p.o.*, b/p = 1.1 : 1 at 1, 3 and 5 h) suggesting that brain penetration may be important for an efficacious CB_2 agonist.

13.4.3.2.3 Oxindole Core. Compound **13.21** (MDA19) (Figure 13.11) from the University of Texas M. D. Anderson Cancer Center was reported to be a protean agonist since it exhibits various functional activities depending on the assay.[16] In addition, it also exhibits different functional responses at the human and rat CB_2 receptors. It had modest affinity and relatively poor selectivity for the human CB_2 receptor ($K_i = 43$ nM, $CB_1/CB_2 = 4$ in the [^{3}H]-CP-55,940 competitive binding assay). It was, however, more potent and selective at the rat CB_2 receptor ($K_i = 16$ nM, $CB_1/CB_2 = 70$). When tested in a paclitaxel-induced neuropathy mouse model, **13.21** attenuated tactile allodynia at 10 mg/kg in $CB_2^{+/+}$ mice. No effect was observed in $CB_2^{-/-}$ mice, suggesting that the observed efficacy was not due to CB_1 off-target effects.

13.4.3.3 Quinoline-3-carboxamides

Compound **13.22** (Figure 13.12) exemplifies a series of quinoline-3-carboxamides bearing both a lipophilic adamantyl amide and *N*-alkyl substitution.[54] The *N*-pentyl moiety on the quinolone was superior to other *N*-alkyl chain lengths and *N*-benzyl, regardless of the amide substituent. Relative to aryl amides, the adamantyl amide provided better binding potencies ($> 20\times$) and improved selectivity over CB1. In the [^{3}H]-CP-55,940 competitive binding assay, **13.22** showed good affinity for the human CB_2 receptor ($K_i = 6.3$ nM) with 194-fold selectivity over the human CB_1 receptor. Compound **13.22** was deemed to be a CB_2 agonist based on its ability to demonstrate analgesic activity in the formalin test of acute peripheral and inflammatory pain in mice (3 mg/kg, *i.p.*).

13.4.3.4 Imidazopyridines

The discovery of the imidazopyridine **13.23** (Figure 13.12) stems from optimization efforts around core modifications of GW405833 (**13.18**).[55] Compound **13.23** possessed an hCB_2 cAMP IC_{50} of 5 nM with modest selectivity ($CB_1/CB_2 \sim 500$). The morpholino group was found to impart good CB_2 agonist activity along with high plasma free fraction ($f_u = 24\%$). Compound **13.23** was evaluated in a rat CFA hyperalgesia model along with an imidazopyridine analogue (**13.24**) possessing no detectable CB_1 agonist activity. Compound **13.23** exhibited dose-dependent reversals in paw withdrawal approaching naproxen-like effects at 100 mpk (*p.o.*). In contrast, the analogue devoid of CB_1 agonism showed no change in the paw withdrawal threshold at 100 mpk. Exposures for both compounds in the plasma, brain and CSF were well above their *in vitro* CB_2 IC_{50} values. In order to explain these results, it was suggested that the modest CB_1 agonism of **13.23** plays an important role in its *in vivo* efficacy. Therefore, it was suggested that CB_2 agonism alone may not be sufficient for analgesic activity.

13.4.3.5   Decahydroquinoline

The concept that some degree of CB$_1$ agonist activity needs to be present in order for a CB$_2$ agonist to demonstrate analgesia is bolstered by work from Merck on an entirely different chemotype. The decahydroquinoline **13.25** (Figure 13.12) (hCB$_2$ cAMP IC$_{50}$ = 7 nM, CB$_1$/CB$_2$ = 688) arose from efforts to optimize a high-throughput screening hit.[56] It was quickly determined that a *cis* ring fusion of the decahydroquinoline system was crucial for binding potency. Although the tertiary alcohol was not important for binding potency, retaining it simplified the synthesis and improved the physical properties of final targets. The presence of the fused cyclohexyl ring was found to be crucial for potency and, in the end, the unsubstituted phenyl substituent was found to be optimal. The amide moiety was optimized using an iterative library approach resulting in the identification of **13.25**. In the rat CFA hyperalgesia model, **13.25** demonstrated a dose-dependent reversal of paw withdrawal threshold (ED$_{50}$ = 30 mpk, subcutaneous). In contrast, the analogue **13.26**, which was devoid of CB$_1$ affinity (CB$_1$ cAMP IC$_{50}$ = >17,000 nM), was not efficacious in the rat CFA hyperalgesia model despite exposure in the plasma, brain and cerebral spinal fluid well above its CB$_2$ IC$_{50}$. This result was taken as further evidence that some degree of CB$_1$ agonist activity is needed for anti-nociception in animal models.

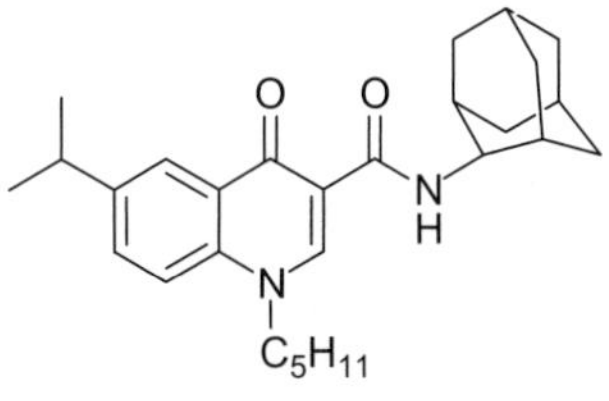

13.22 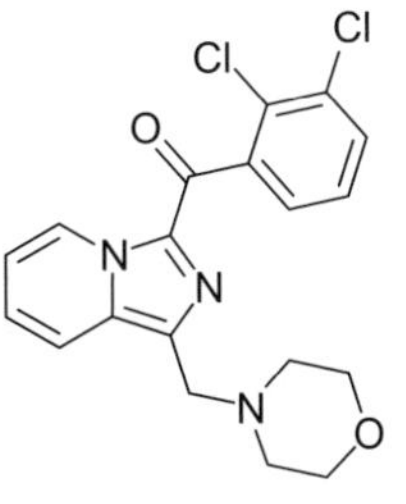   **13.23** 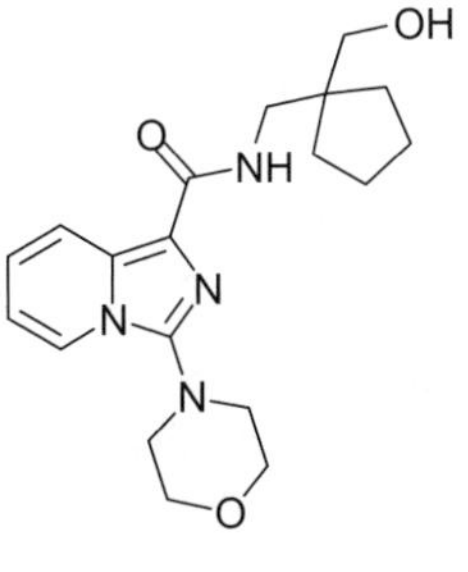   **13.24**

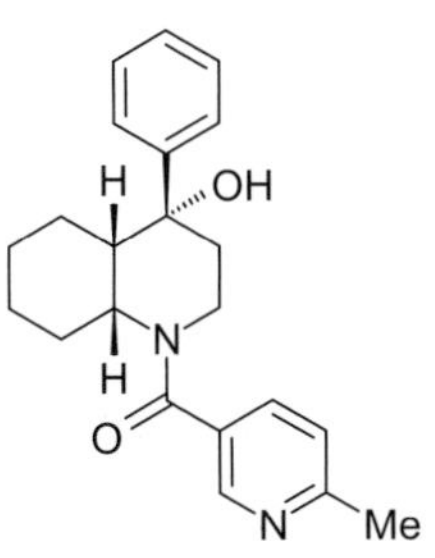

13.25 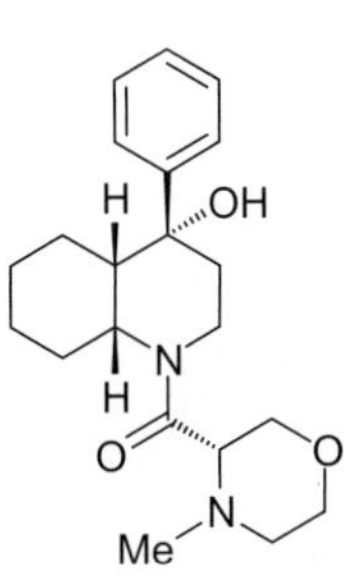   **13.26**

Figure 13.12   CB$_2$ agonists with a bicyclic core.

13.4.4 Tricyclic Cores

13.4.4.1 *Tetrahydropyrrole[3,4-b]indole Core*

The tetrahydropyrrole[3,4-b]indole core of **13.27** (Figure 13.13) was present in an internal screening hit at AstraZeneca.[57] SAR studies around the sulfonamide determined that this functionality was superior to both amides and *N*-alkyl substitution, with the ethyl sulfonamide superior to other alkyl and aryl sulfonamides. Deletion of the methyl group on the piperidine ring of the amide substituent resulted in >10-fold loss of hCB$_2$ affinity. Polar substituents such as a morpholine amide were not tolerated. Finally, the tetrahydropyran group on the tetrahydropyrrole afforded good potency along with selectivity over CB$_1$ (**13.27**: hCB$_2$ K$_i$ = 18 nM, CB$_1$/CB$_2$ = 351 in the [³H]-CP-55,940 competitive binding assay).

13.4.4.2 *Benzoquinolizinone Core*

Another CB$_2$ agonist with a tricyclic core is Sch-35966 (**13.28**). This di-pivaloyl ester was reported to be a CB$_2$ agonist with good binding affinities across species and good selectivity over the CB$_1$ receptor (hCB$_2$ K$_i$ = 6.8 nM CB$_1$/CB$_2$ = 387; monkey CB$_2$ K$_i$ = 5.4 nM, CB$_1$/CB$_2$ = 944; rat CB$_2$ K$_i$ = 2.4 nM, CB$_1$/CB$_2$ = 1250).[58]

13.4.5 Biaryl Cores

13.4.5.1 *Biaryl Oxadiazoles*

The biaryl oxadiazole system was discovered by researchers at Amgen following an HTS screening campaign that was biased by a pharmacophore model and *in-silico* physicochemical filters.[59] SAR studies around the oxadiazole isomer as exemplified in **13.29** (Figure 13.14) showed that deletion of either of the ring nitrogens resulted in a loss of potency. However, the tetrazole and the isomeric oxadiazole analogues showed comparable potency to the oxadiazole. The aryl ring attached to the oxadiazole ring was sensitive to substitution, with

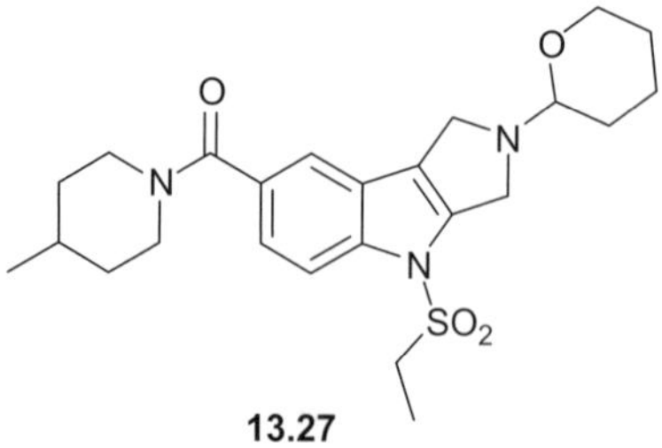

13.27 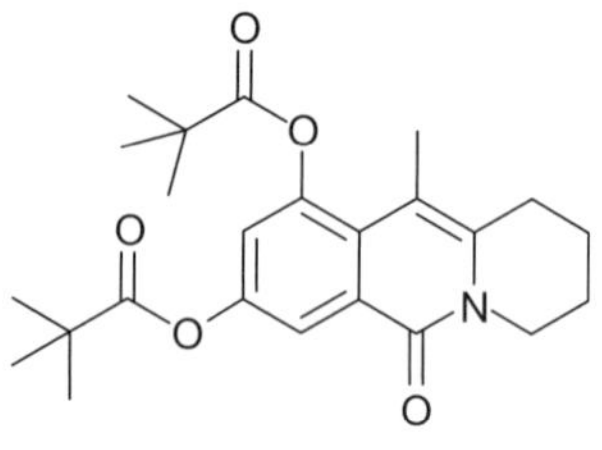   **13.28**

Figure 13.13   CB$_2$ agonists with a tricyclic core.

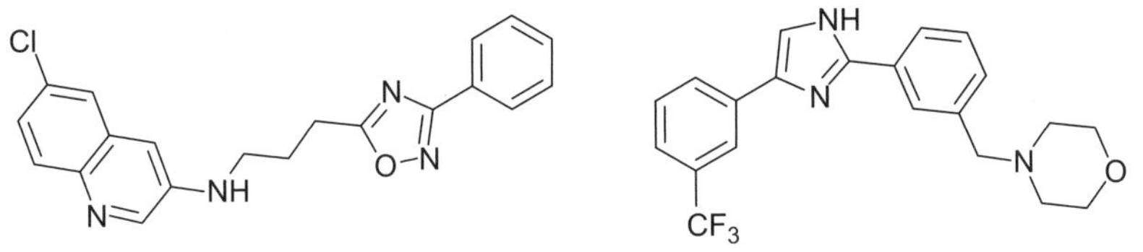

Figure 13.14 CB$_2$ agonists with a biaryl and a triaryl core.

2,4-disubstitution providing the best potency. Pharmacokinetic properties, water solubility and metabolic stability could be tuned with various *N*-linked aryl and heteroaryl groups, leading to **13.29** with >100-fold selectivity and excellent potency and oral exposure (hCB$_2$ cAMP EC$_{50}$ = 2.2 nM, CB$_1$/CB$_2$ = 255; AUC (10 mpk/*p.o.*) = 43 µg · h/mL, $t_{1/2}$ = 5.1 h, F = 100%).

13.4.6 Triaryl Cores

The triaryl core class of CB$_2$ agonist is exemplified by the initial lead imidazole **13.30** (hCB$_2$ cAMP EC$_{50}$ = 9 nM, E_{max} 80%; hCB$_1$ EC$_{50}$ > 10 µM; rat AUC (10 mpk, *p.o.*) = 866 nM · h).[60] SAR studies around this core found that *N*-1 substitution of the imidazole was not desirable, the morpholine ring could be replaced with a piperidine ring and substitution on the 4-aryl ring was tolerated although selectivity over CB$_1$ could be adversely affected. It was reported that several analogues from this 2,4-diphenyl-1H-imidazole series were active in rodent neuropathic pain models with the results to be disclosed in a future publication.

The modulation of the pain response by CB$_2$ agonists has been well validated in pre-clinical animal models with a diverse set of chemotypes. While selectivity over the CB$_1$ receptor is considered important to avoid unwanted psychotropic side-effects, Merck researchers have suggested that modest CB$_1$ agonism may be necessary in CB$_2$-mediated analgesia as exemplified in **13.23** and **13.25**. Implicit in this theory is that brain penetration would also be a necessity. The potential importance of brain penetration was put forth by Pfizer and GSK researchers in work leading to the brain penetrant benzimidiazole **13.15** and aza-indole **13.20**, respectively. Further work is needed to determine the importance of brain penetration and modest CB$_1$ activity in a CB$_2$ agonist for the treatment of pain.

13.5 Clinical Status

Five companies have recently been in the clinic or are entering the clinic with selective CB$_2$ agonists for the treatment of pain. Evotec, in collaboration with Boehringer Ingelheim, reportedly advanced a compound into Phase I development in 2010 for the treatment of neuropathic pain.[61] More recently it was

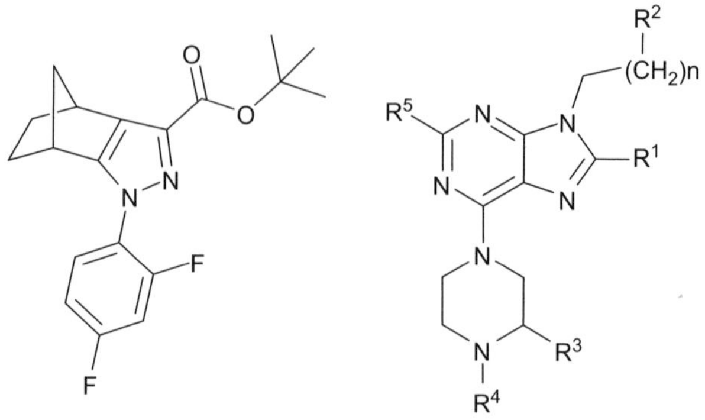

13.31   **13.32**

Figure 13.15   Tedalinab and Eli Lilly's piperazinyl purine core.

reported that a back-up compound had also entered into Phase I clinical trials for the treatment of neuropathic pain. Though not explicitly stated, it is assumed that these clinical compounds are CB_2 agonists based on their joint publication record in the CB_2 agonist field.[41–44] The structures of these clinical compounds have not been reported.

Glenmark's CB_2 agonist, GRC-10693 or tedalinab (**13.31**, Figure 13.15), completed a Phase I clinical trial in August of 2009 in Europe for neuropathic pain and osteoarthritis.[62] Improvements in various pain scores were observed at both 2 and 13 days (300 mg *q.d.*, 14 days), suggesting a proof of mechanism for the treatment of pain with a CB_2 agonist. However, two years after the completion of the Phase I study, the compound has not yet progressed to Phase II.

Eli Lilly has progressed their CB_2 agonist LY2828360 to Phase II clinical trials for the treatment of osteoarthritic knee pain.[63] Although the structure of LY2828360 has not been disclosed, recent patent applications suggest it possesses a piperazinyl purine core (**13.32**).[64–66]

GlaxoSmithKline completed a Phase IIa trial for subjects with osteoarthritis in October of 2007 with their CB_2 agonist **13.7** (GW842166X).[67] Two additional Phase II studies were conducted looking at pain following third molar tooth extraction and osteoarthritis of the knee.[68,69] Both trials were completed in October of 2008, and the results of the trial involving third molar tooth extraction have now been published.[70] The study was carried out with pre-operative doses of 100 mg and 800 mg of GW842166, 800 mg ibuprofen (plus 400 mg post-operative) and placebo. The investigators found that, in comparison to ibuprofen, GW842166 offered no clinically meaningful analgesia. They speculate that the high plasma protein binding of the drug (>99%) resulted in insufficient free drug to have a therapeutic effect, though the need for co-activation of the CB_1 receptor cannot be formally excluded. The compound is no longer mentioned in the company's pipeline and therefore development is presumed to be discontinued.

Finally, Abbott recently reported in a presentation to investors that they are in Phase 1 clinical trials with the CB_2 agonist ABT-521 for the treatment of pain. Although the structure of ABT-521 has not been reported, it is believed to be from the indole series that includes A-796260 (**13.19**).[71]

In this review, we have examined the role of the cannabinoid CB_2 receptor in the pain response, and discussed the diverse set of reported chemotypes that may lead to a clinical candidate capable of modulating this response. Though other activities have been implicated in the biology of the CB_2 receptor, pain modulation continues to be the best validated and most clinically significant use for cannabinoid CB_2 receptor-specific agonists. Future work will be required to realize this potential.

References

1. S. Munro, K. L. Thomas and M. Abu-Shaar, *Nature*, 1993, **365**, 61.
2. R. Tanasescu and C. S. Constantinescu, *Immunobiology*, 2010, **215**, 588.
3. B. K. Atwood and K. Mackie, *Br. J. Pharmacol.*, 2010, **160**, 467.
4. C. A. Lunn, *Current Topics in Medicinal Chemistry*, 2010, **10**, 768.
5. A. Calignano, G. La Rana, A. Giuffrida and D. Piomelli, *Nature*, 1998, **394**, 277.
6. D. Turcotte, J.-A. Le Dorze, F. Esfahani, E. Frost, A. Gomori and M. Namaka, *Expert Opin. Pharmacother.*, 2010, **11**, 17.
7. G. A. Cabral and F. Marciano-Cabral, *J. Leukocyte Biol.*, 2005, **78**, 1192.
8. M. B. Graeber, *Science*, 2010, **330**, 783.
9. O. Sagredo, S. Gonzalez, I. Aroyo, R. Pazos Maria, C. Benito, I. Lastres-Becker, P. Romero Juan, M. Tolon Rosa, R. Mechoulam, E. Brouillet, J. Romero and J. Fernandez-Ruiz, *Glia*, 2009, **57**, 1154.
10. J. Palazuelos, T. Aguado, M. R. Pazos, B. Julien, C. Carrasco, E. Resel, O. Sagredo, C. Benito, J. Romero, I. Azcoitia, J. Fernandez-Ruiz, M. Guzman and I. Galve-Roperh, *Brain*, 2009, **132**, 3152.
11. A. Ternianov, J. M. Perez-Ortiz, M. E. Solesio, M. S. Garcia-Gutierrez, A. Ortega-Alvaro, F. Navarrete, C. Levia, M. F. Galindo and J. Manzanares, *Neurobiol Aging*, 2012, **33**, 421.
12. E. M. Lopez, P. Tagliaferro, E. S. Onaivi and J. J. Lopez-Costa, *Synapse*, 2011, **65**, 388.
13. N. H. Morgan, I. M. Stanford and G. L. Woodhall, *Neuropharmacology*, 2009, **57**, 356.
14. R. Sagar Devi, D. Jhaveri Maulik, D. Richardson, A. Gray Roy, E. de Lago, J. Fernandez-Ruiz, A. Barrett David, A. Kendall David and V. Chapman, *Eur. J. Neurosci.*, 2010, **31**, 1413.
15. V. Curto-Reyes, S. Llames, A. Hidalgo, L. Menendez and A. Baamonde, *Br. J. Pharmacol.*, 2010, **160**, 561.
16. J. Xu Jijun, P. Diaz, F. Astruc-Diaz, S. Craig, E. Munoz and M. Naguib, *Anesth. Analg.*, 2010, **111**, 99.
17. B. B. Yao, G. C. Hsieh, J. M. Frost, Y. Fan, T. R. Garrison, A. V. Daza, G. K. Grayson, C. Z. Zhu, M. Pai, P. Chandran, A. K. Salyers, E. J.

Wensink, P. Honore, J. P. Sullivan, M. J. Dart and M. D. Meyer, *Br. J. Pharmacol.*, 2008, **153**, 390.

18. I. Mancini, R. Brusa, G. Quadrato, C. Foglia, P. Scandroglio, L. S. Silverman, D. Tulshian, A. Reggiani and M. Beltramo, *Br. J. Pharmacol.*, 2009, **158**, 382.
19. D. Bolognini, M. G. Cascio, D. Parolaro and R. G. Pertwee, *Br. J. Pharmacol.*, 2011, May 26. doi: 10.1111/j.1476-5381.2011.01503.x [Epub ahead of print].
20. N. Schuelert, C. Zhang, A. J. Mogg, L. M. Broad, D. L. Hepburn, E. S. Nisenbaum, M. P. Johnson and J. J. McDougall, *Osteoarthritis Cartilage*, 2010, **18**, 1536.
21. F. Gbahou, A. Rouleau, S. Morisset, R. Parmentier, S. Crochet, J.-S. Lin, X. Ligneau, J. Tardivel-Lacombe, H. Stark, W. Schunack, C. R. Ganellin, J.-C. Schwartz and J.-M. Arrang, *Proc. Natl Acad. Sci. USA*, 2003, **100**, 11086.
22. M. B. Henry, S. Zheng, C. Duan, B. Patel, G. Vassileva, C. Sondey, J. Lachowicz and J. J. Hwa, *Endocrinology*, 2011, **152**, 828.
23. S. C. Ganguli, C.-G. Park, M. H. Holtmann, E. M. Hadac, T. P. Kenakin and L. J. Miller, *J. Pharmacol. Exp. Ther.*, 1998, **286**, 593.
24. C. C. Jansson, J. P. Kukkonen, J. Nasman, G. Huifang, S. Wurster, R. Virtanen, J.-M. Savola, V. Cockcroft and E. O. Akerman, *Mol. Pharmacol.*, 1998, **53**, 963.
25. P. J. Pauwels, I. Rauly and T. Wurch, *J. Pharmacol. Exp. Ther.*, 2003, **305**, 1015.
26. D. Gesty-Palmer and L. M. Luttrell, *Br. J. Pharmacol.*, 2011, **164**, 59.
27. K. Chakir, C. Depry, V. L. Dimaano, W.-Z. Zhu, M. Vanderheyden, J. Bartunek, T. P. Abraham, G. F. Tomaselli, S.-B. Liu, Y. K. Xiang, M. Zhang, E. Takimoto, N. Dulin, R. P. Xiao, J. Zhang and D. A. Kass, *Sci. Transl. Med.*, 2011, **3**, 100ra88.
28. J. A. Allen, J. M. Yost, V. Setola, X. Chen, M. F. Sassano, M. Chen, S. Peterson, P. N. Yadav, X. P. Huang, B. Feng, N. H. Jensen, X. Che, X. Bai, S. V. Frye, W. C. Wetsel, M. G. Caron, J. A. Javitch, B. L. Roth and J. Jin, *Proc. Natl Acad. Sci. USA*, 2011, **108**, 18488.
29. B. Bosier, G. G. Muccioli, E. Hermans and D. M. Lambert, *Biochem. Pharmacol.*, 2010, **80**, 1.
30. D. McGuinness, A. Malikzay, R. Visconti, K. Lin, M. Bayne, F. Monsma and C. A. Lunn, *J. Biomol. Screening*, 2009, **14**, 49.
31. Y. Sun, S. P. H. Alexander, M. J. Garle, C. L. Gibson, K. Hewitt, S. P. Murphy, D. A. Kendall and A. J. Bennett, *Br. J. Pharmacol.*, 2007, **152**, 734.
32. R. Belfort, R. Berria, J. Cornell and K. Cusi, *J. Clin. Endocrinol. Metab.*, 2010, **95**, 829.
33. G. A. Thakur, R. Tichkule, S. Bajaj and A. Makriyannis, *Expert Opin. Ther. Pat.*, 2009, **19**, 1647.
34. S. Han, J. Thatte and R. M. Jones, *Annu. Rep. Med. Chem.*, 2009, **44**, 227.

35. A. Berry, R. Betageri, D. Riether, E. R. Hickey, L. Wu, R. M. Zindell and S. Khor, Boehringer Ingelheim International GMBH, *PCT Int. Appl.,* WO 2010/077836, 2010.
36. I. E. Marx, E. F. DiMauro, A. Cheng, R. Emkey, S. A. Hitchcock, L. Huang, M. Y. Huang, J. Human, J. H. Lee, X. Li, M. W. Martin, R. D. White, R. T. Fremeau and V. F. Patel, *Bioorg. Med. Chem. Lett.,* 2009, **19**, 31.
37. H. Ohta, T. Ishizaka, M. Yoshinaga, A. Morita, Y. Tomishima, Y. Toda and S. Saito, *Bioorg. Med. Chem. Lett.,* 2007, **17**, 5133.
38. H. Ohta, T. Ishizaka, M. Tatsuzuki, M. Yoshinaga, I. Iida, Y. Tomishima, Y. Toda and S. Saito, *Bioorg. Med. Chem. Lett.,* 2007, **17**, 6299.
39. H. Ohta, T. Ishizaka, M. Tatsuzuki, M. Yoshinaga, I. Iida, T. Yamaguchi, Y. Tomishima, N. Futaki, Y. Toda and S. Saito, *Bioorg. Med. Chem.,* 2008, **16**, 1111.
40. G. M. P. Giblin, A. Billinton, M. Briggs, A. J. Brown, I. P. Chessell, N. M. Clayton, A. J. Eatherton, P. Goldsmith, C. Haslam, M. R. Johnson, W. L. Mitchell, A. Naylor, A. Perboni, B. P. Slingsby and A. W. Wilson, *J. Med. Chem.,* 2009, **52**, 5785.
41. R. Zindell, D. Riether, T. Bosanac, A. Berry, M. J. Gemkow, A. Ebneth, S. Loebbe, E. L. Raymond, D. Thome, D.-T. Shih and D. Thomson, *Bioorg. Med. Chem. Lett.,* 2009, **19**, 1604.
42. D. Riether, L. Wu, P. F. Cirillo, A. Berry, E. R. Walker, M. Ermann, B. Noya-Marino, J. E. Jenkins, D. Albaugh, C. Albrecht, M. Fisher, M. J. Gemkow, H. Grbic, S. Loebbe, C. Moeller, K. O'Shea, A. Sauer, D.-T. Shih and D. S. Thomson, *Bioorg. Med. Chem. Lett.,* 2011, **21**, 2011.
43. R. Zindell, E. R. Walker, J. Scott, P. Amouzegh, L. Wu, M. Ermann, D. Thomson, M. B. Fisher, C. L. Fullenwider, H. Grbic, P. Kaplita, B. Linehan, M. Patel, M. Patel, S. Loebbe, S. Block, C. Albrecht, M. J. Gemkow, D.-T. Shih and D. Riether, *Bioorg. Med. Chem. Lett.,* 2011, **21**, 4276.
44. M. Ermann, D. Riether, E. R. Walker, I. F. Mushi, J. E. Jenkins, B. Noya-Marino, M. L. Brewer, M. G. Taylor, P. Amouzegh, S. P. East, B. W. Dymock, M. J. Gemkow, A. F. Kahrs, A. Ebneth, S. Loebbe, K. O'Shea, D.-T. Shih and D. Thomson, *Bioorg. Med. Chem. Lett.,* 2008, **18**, 1725.
45. I. Sellitto, B. Le Bourdonnec, K. Worm, A. Goodman, M. A. Savolainen, G.-H. Chu, C. W. Ajello, C. T. Saeui, L. K. Leister, J. A. Cassel, R. N. DeHaven, C. J. LaBuda, M. Koblish, P. J. Little, B. L. Brogdon, S. A. Smith and R. E. Dolle, *Bioorg. Med. Chem. Lett.,* 2010, **20**, 387.
46. K. Worm, D. G. Weaver, R. C. Green, C. T. Saeui, D.-M. S. Dulay, W. M. Barker, J. A. Cassel, G. J. Stabley, R. N. De Haven, C. J. La Buda, M. Koblish, B. L. Brogdon, S. A. Smith and R. E. Dolle, *Bioorg. Med. Chem. Lett.,* 2009, **19**, 5004.
47. B. M. P. Verbist, M. A. J. De Cleyn, M. Surkyn, E. Fraiponts, J. Aerssens, M. J. M. A. Nijsen and H. J. M. Gijsen, *Bioorg. Med. Chem. Lett.,* 2008, **18**, 2574.

48. C. Watson, D. R. Owen, D. Harding, K. Kon-I, M. L. Lewis, H. J. Mason, M. Matsumizu, T. Mukaiyama, M. Rodriguez-Lens, A. Shima, M. Takeuchi, I. Tran and T. Young, *Bioorg. Med. Chem. Lett.*, 2011, **21**, 4284.

49. D. Page, E. Balaux, L. Boisvert, Z. Liu, C. Milburn, M. Tremblay, Z. Wei, S. Woo, X. Luo, Y.-X. Cheng, H. Yang, S. Srivastava, F. Zhou, W. Brown, M. Tomaszewski, C. Walpole, L. Hodzic, S. St-Onge, C. Godbout, D. Salois and K. Payza, *Bioorg. Med. Chem. Lett.*, 2008, **18**, 3695.

50. T. E. D'Ambra, K. G. Estep, M. A. Bell, M. A. Eissenstat, K. A. Josef, S. J. Ward, D. A. Haycock, E. R. Baizman, F. M. Casiano, N. C. Beglin, S. M. Chippari, J. D. Grego, R. K. Kullnig and G. T. Daley, *J. Med. Chem.*, 1992, **35**, 124.

51. C. C. Felder, K. E. Joyce, E. M. Briley, J. Mansouri, K. Mackie, O. Blond, Y. Lai, A. L. Ma and R. L. Mitchell, *Mol. Pharmacol.*, 1995, **48**, 443.

52. K. J. Valenzano, L. Tafesse, G. Lee, J. E. Harrison, J. M. Boulet, S. L. Gottshall, L. Mark, M. S. Pearson, W. Miller, S. Shan, L. Rabadi, Y. Rotshteyn, S. M. Chaffer, P. I. Turchin, D. A. Elsemore, M. Toth, L. Koetzner and G. T. Whiteside, *Neuropharmacology*, 2005, **48**, 658.

53. J. M. Frost, M. J. Dart, K. R. Tietje, T. R. Garrison, G. K. Grayson, A. V. Daza, O. F. El-Kouhen, B. B. Yao, G. C. Hsieh, M. Pai, C. Z. Zhu, P. Chandran and M. D. Meyer, *J. Med. Chem.*, 2010, **53**, 295.

54. S. Pasquini, L. Botta, T. Semeraro, C. Mugnaini, A. Ligresti, E. Palazzo, S. Maione, V. Di Marzo and F. Corelli, *J. Med. Chem.*, 2008, **51**, 5075.

55. B. W. Trotter, K. K. Nanda, C. S. Burgey, C. M. Potteiger, J. Z. Deng, A. I. Green, J. C. Hartnett, N. R. Kett, Z. Wu, D. A. Henze, K. Della Penna, R. Desai, M. D. Leitl, W. LeMaire, R. B. White, S. Yeh, M. O. Urban, S. A. Kane, G. D. Hartman and M. T. Bilodeau, *Bioorg. Med. Chem. Lett.*, 2011, **21**, 2354.

56. P. J. Manley, A. Zartman, D. V. Paone, C. S. Burgey, D. A. Henze, K. Della Penna, R. Desai, M. D. Leitl, W. LeMaire, R. B. White, S. Yeh, M. O. Urban, S. A. Kane, G. D. Hartman, M. T. Bilodeau and B. W. Trotter, *Bioorg. Med. Chem. Lett.*, 2011, **21**, 2359.

57. D. Page, H. Yang, W. Brown, C. Walpole, M. Fleurent, M. Fyfe, F. Gaudreault and S. St-Onge, *Bioorg. Med. Chem. Lett.*, 2007, **17**, 6183.

58. W. Gonsiorek, C. A. Lunn, X. Fan, G. Deno, J. Kozlowski and R. W. Hipkin, *Br. J. Pharmacol.*, 2007, **151**, 1262.

59. Y. Cheng, B. K. Albrecht, J. Brown, J. L. Buchanan, W. H. Buckner, E. F. DiMauro, R. Emkey, R. T. Fremeau, J.-C. Harmange, B. J. Hoffman, L. Huang, M. Huang, J. H. Lee, F.-F. Lin, M. W. Martin, H. Q. Nguyen, V. F. Patel, S. A. Tomlinson, R. D. White, X. Xia and S. A. Hitchcock, *J. Med. Chem.*, 2008, **51**, 5019.

60. S.-W. Yang, J. Smotryski, J. Matasi, G. Ho, D. Tulshian, W. J. Greenlee, R. Brusa, M. Beltramo and K. Cox, *Bioorg. Med. Chem. Lett.*, 2010, **21**, 182.

61. http://www.evotec.com/archive/en/Press-releases/2011/Milestone-received-as-Boehringer-Ingelheim-starts-Phase-I-clinical-trial-in-Pain/2167/1

62. http://www.glenmarkpharma.com/GLN_NWS/pdf/Glenmarks_molecule_ neuropathicpain_osteoarthritis_GRC10693_succes.pdf
63. http://clinicaltrials.gov/ct2/show/NCT01319929?term=LY2828360&rank=1
64. P. C. Astles, R. Guidetti, M. W. Tidwell and S. P. Hollinshead, Eli Lilly and Company, *US Pat. Appl.*, US 2010/0160288, 2010.
65. P. C. Astles, R. Guidetti, A. J. Sanderson and S. P. Hollinshead, Eli Lilly and Company, *PCT Int. Appl.*, WO 2011/123372, 2011.
66. S. P. Hollinshead, Eli Lilly and Company, *PCT Int. Appl.*, WO 2011/ 123482, 2011.
67. ClinicalTrials.gov Identifier NCT00479427.
68. ClinicalTrials.gov Identifier NCT00444769.
69. ClinicalTrials.gov Identifier NCT00447486.
70. T. Ostenfeld, J. Price, M. Albanese, J. Bullman, F. Guillard, I. Meyer, R. Leeson, C. Costantin, L. Ziviani, P. F. Nocini and S. Milleri, *Clin. J. Pain*, 2011, **27**, 668.
71. http://media.corporate-ir.net/media_files/irol/94/94004/presentations/ Abbott_Investor_Day_2011.pdf

Section 4
Sphingolipids

CHAPTER 14

S1P Receptor Agonists

CRAIG A. MILLER

Boehringer Ingelheim RCV GmbH & Co KG, Dr. Boehringer-Gasse 5-11,
Vienna, Austria, 1121
Email: craig.miller@boehringer-ingelheim.com

14.1 S1P Biology and Relevance to the Pathological Features of Multiple Sclerosis

14.1.1 S1P Biosynthesis, Receptor Expression and Function

Sphingosine-1-phosphate (S1P) is a secondary signalling lysophospholipid produced *via* the biosynthesis of ceramide, a key component in the sphingo-myelin pathway. Ceramidase converts ceramide to sphingosine, which exists in dynamic equilibrium with the phosphorylated form, S1P.[1] The equilibrium is established between competing phosphorylation by ubiquitously expressed sphingosine kinase type 1 (SphK1) and type 2 (SphK2) and the reverse process of dephosphorylation by phospholipid phosphatases.[2] S1P can also undergo terminal degradation by sphingosine-1-phosphate *lyase* to 2-hexadecanal and phosphoethanol (Figure 14.1).

All cells can produce S1P, which is present in submicromolar levels in various biological fluids and tissues. S1P exerts diverse biological functions under both physiological and pathological conditions, including regulation of cell pro-liferation, differentiation, survival, adhesion, migration, morphogenesis and cytoskeletal rearrangements.[3]

S1P can act in both an autocrine and a paracrine fashion by high-affinity engagement with five known cell surface G-protein-coupled receptors (GPCRs)

RSC Drug Discovery Series No. 26
Anti-Inflammatory Drug Discovery
Edited by Jeremy I Levin and Stefan Laufer
© The Royal Society of Chemistry 2012
Published by the Royal Society of Chemistry, www.rsc.org

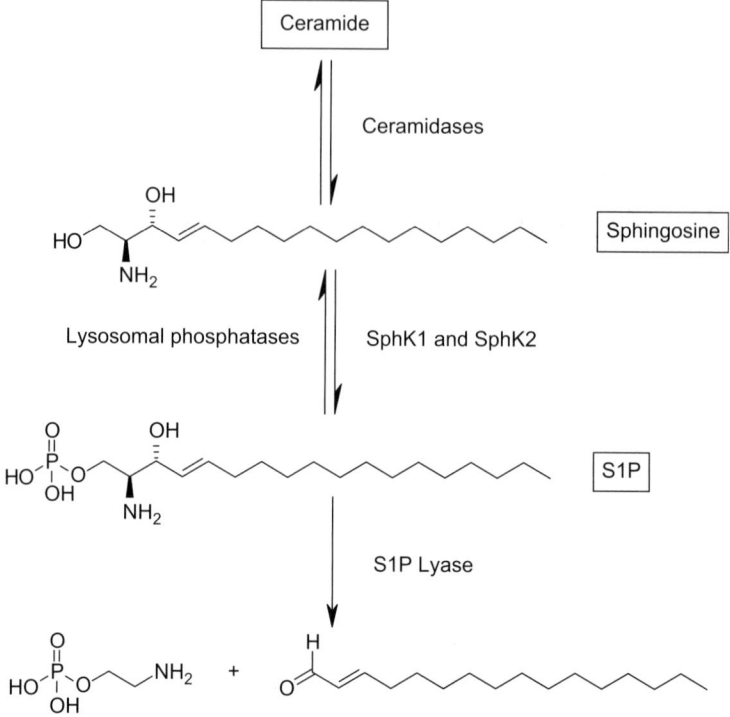

Figure 14.1   S1P biosynthetic pathway.

S1P$_{1,2,3,4,5}$, which were originally discovered as endothelium differentiation gene (Edg) receptors.[4,5] The various receptors are coupled with a diverse array of G-protein subunits resulting in complex and differentiated functions from each receptor.

The S1P$_1$ receptor is widely expressed by many cell types including immune (T and B lymphocytes), neural, smooth muscle and endothelial cells.[6] Within the context of the adaptive immune system, the S1P$_1$ receptor acts as a key recirculation regulator of T-cells.[7] Lymphocytes and thymocytes have been shown to undergo chemotactic egress from lymph nodes and the thymus along an S1P concentration gradient which is higher in the blood than in tissue.[8] Naive T-cells transiently down-regulate the S1P$_1$ receptor in lymph nodes once a suitable antigen is presented, allowing the T-cell to remain in contact with antigen-presenting cells and become activated. After clonal expansion of the newly activated T-cell, cell surface expression of the S1P$_1$ receptor is up-regulated allowing cells to respond to the S1P gradient and migrate back into circulation.[9]

In the cardiovascular system, atrial cardiomyocytes and cardiac endothelial cells express high levels of the S1P$_1$ receptor, which may play a role in heart rate regulation through activation of the IK$_{Ach}$ G-protein-gated channel in

Table 14.1 Distribution and key functions of the S1P receptor subtypes.

Receptor	Distribution (messenger RNA)	Key functions
$S1P_1$	Ubiquitously expressed with high expression on lymphocytes, neural cells and vasculature	– Lymphocyte egress from secondary lymphoid organs – Neural cell migration/function – Embryonic development of cardiovascular and nervous systems – blood vessel formation – endothelial barrier function
$S1P_2$	Ubiquitous	– Vascular tone – Endothelial barrier function – Inner ear maintenance affecting hearing and balance
$S1P_3$	Ubiquitous including the CNS and the endothelium	– Endothelial barrier function – Neural cell migration/function
$S1P_4$	Lymphocytes (low levels)	– Unknown
$S1P_5$	Brain/white matter, oligodendrocytes	– Oligodendrocyte function – Natural killer cell migration

humans.[10] This expression and functional pattern is believed to be reversed in the rodent where $S1P_3$ may play a more dominant role. Vascular smooth muscles express the $S1P_{1-3}$ receptors and are involved in the regulation of vascular tone and blood pressure.

In the central nervous system (CNS), virtually all cell lineages express S1P receptors,[11] including astrocytes ($S1P_3 > S1P_1 > S1P_2 > S1P_5$), oligodendrocytes ($S1P_5 > S1P_1 = S1P_2 > S1P_3$), neurons ($S1P_1 = S1P_3 > S1P_2 = S1P_5$) and microglia ($S1P_1 > S1P_2 > S1P_3 = S1P_5$). The $S1P_1$ receptor on neural stem cells facilitates migration along an S1P gradient to sites of injury in the CNS. Both $S1P_5$ and $S1P_1$ contribute to the control of oligodendrocyte survival and function, as well as modulation of myelination following injury. The various roles of the S1P receptors is beyond the scope of this work and has been the subject of a comprehensive review.[7] A summary of the key physiological functions of the receptors can be found in Table 14.1.[12]

14.1.2 Pathological Features of Multiple Sclerosis

Multiple sclerosis (MS) is a chronic autoimmune disease of the CNS[13] and is the most common cause of neurological impairment of young adults in the western world, with a median age of onset of 28 years. Eighty-five percent of patients initially present with the relapsing-remitting (RRMS) form of the disease which is characterized by an acute relapse phase of neurological impairment followed by a remittent period.[14] Disability in patients with MS is caused by incomplete recovery following the relapse phase resulting in progressive, irreversible neurodegeneration. The pathological features of the disease are inflammation of the CNS resulting in demyelination, astrogliosis and loss of oligodendrocytes and neurons, culminating in neurological deficiencies and impairment. While the pathogenesis of the disease is poorly understood, a key aspect is activation of lymphocytes in the periphery, which subsequently migrate to the CNS where they elicit their auto-aggressive effect.

Until recently, first-line treatment for patients suffering from RRMS was interferon-β (IFNβ) or glatiramer acetate (GA), both of which are believed to function exclusively on the immune system. Unfortunately, these disease-modifying maintenance therapies produce only a minor effect on relapse rate (29%–34% *versus* placebo) and have a modest impact on disease progression after two years of treatment. Second-line therapy includes the more effective VLA-4 monoclonal antibody natalizumab (Tysabri®).[15] However, treatment with this agent is associated with the potential for patients to develop progressive multi-focal leukoencephalopathy (PML), a rare though usually fatal infectious demyelinating disease of the brain.

In 2010 the United States Food and Drug Administration (FDA) approved fingolimod (FTY720/Gilenya®; Novartis), a prodrug S1P$_{1,3,4,5}$ receptor agonist, as the first orally bioavailable, first-line treatment option for patients with RRMS. Fingolimod demonstrates improved clinical efficacy relative to established therapies and is believed to derive its function by operating simultaneously in both the periphery and CNS. The approval of fingolimod marks a critical milestone for those suffering from RRMS.

14.2   Discovery and Development of Fingolimod (FTY720/Gilenya®; Novartis)

14.2.1   Discovery of Fingolimod and S1P Receptor Binding Mode

Fingolimod[16,17] (FTY720/Gilenya®; Novartis) was first identified during a chemical derivatization program of the fungal metabolite myriocin in the mid 1990s. Though the mode of action of fingolimod was not initially elucidated, it was noted that the molecule induced lymphopenia in both animals and in humans. Lymphopenia is characterized by reduction in circulating lymphocytes, predominantly affecting CD4$^+$ T-cells, CD8$^+$ T-cells and B-cells with fingolimod. In animal models of bone marrow transplantation, fingolimod showed efficacy in preventing graft-*versus*-host disease further implying an immunological function for the compound.[18]

Fingolimod was later discovered to undergo bioconversion *in vivo* to the phosphate metabolite fingolimod-P.[19] Phosphorylation occurs *via* SphK2 and produces exclusively (S)-fingolimod-P (Figure 14.2). Unlike fingolimod, this metabolite is a potent agonist of four of the five known S1P receptors in *in-vitro* biochemical assays (S1P$_{1,4,5}$ EC$_{50}$ = 0.3–0.6 nM, S1P$_3$ EC$_{50}$ = 3 nM).

The S1P receptors share high sequence homology to each other and belong to the rhodopsin-like subfamily of GPCRs. A variety of high-resolution crystal structures are available for these class A GPCRs. The crystal structure of

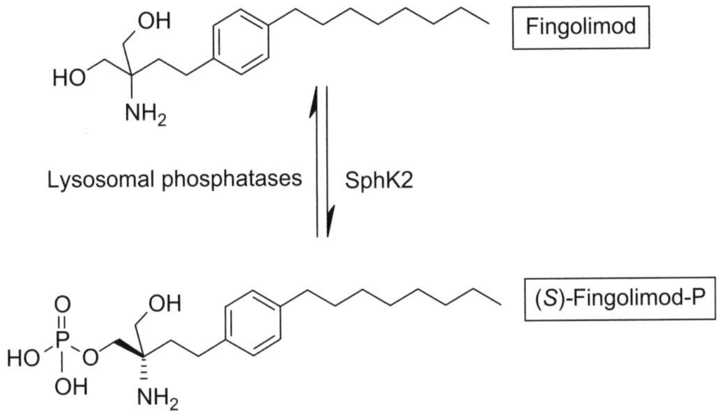

Figure 14.2 Bioconversion of fingolimod to the active S1P receptor agonist (*S*)-fingolimod-P.

rhodopsin has allowed the generation of homology models for both the $S1P_1$ and $S1P_3$ receptors. These models have been used to explain the binding potencies of fingolimod and second-generation agonists to these two receptors.[20] In all receptors, fingolimod is predicted to bind in a narrow cavity located in the upper half of the seven-transmembrane domain (7TM) helical bundle in a linear extended mode interacting with TM 3, 5 and 6. The anionic phosphate group is expected to interact with Arg102 and the amine with Glu164, which are conserved in all S1P receptors. The second group of interactions consists of nine aromatic or hydrophobic amino acids, which define the so-called hydrophopic pocket. Of these nine amino acids only one is divergent in comparing $S1P_1$ to $S1P_3$ (Leu276 in $S1P_1$ and Phe263 in $S1P_3$). This structural difference has often been exploited in second-generation agonists to make selective $S1P_1$ agonists (*vide infra*) and may contribute to the ten-fold $S1P_1$/$S1P_3$ selectivity observed with fingolimod. Thus, the small Leu276 forms productive van der Waals interactions with the phenyl ring of fingolimod while the bulkier Phe263 residue in $S1P_3$ reduces the size of the binding pocket, resulting in a steric clash with the ligand and a loss of binding affinity (Figure 14.3).

14.2.2 *In-vitro* and *in-vivo* **Pharmacology of Fingolimod**

Agonism of the S1P receptors in transfected cell lines with fingolimod-P induces receptor internalization and degradation, allowing the compound to act as a "functional antagonist"[21,22] and to elicit the immunomodulatory activity observed *in vivo*. Treatment of mice with fingolimod causes internalization of the $S1P_1$ receptor in lymph node T-cells, which occurs with a concomitant decrease in circulating lymphocytes and an increase in lymphostatic T-cells in the lymph nodes. Conditional $S1P_1^{-/-}$ knockout mice show a phenotype similar to fingolimod treatment[23] indicating that the $S1P_1$ receptor is required for egress of certain lymphocytes from the secondary lymphoid organs.

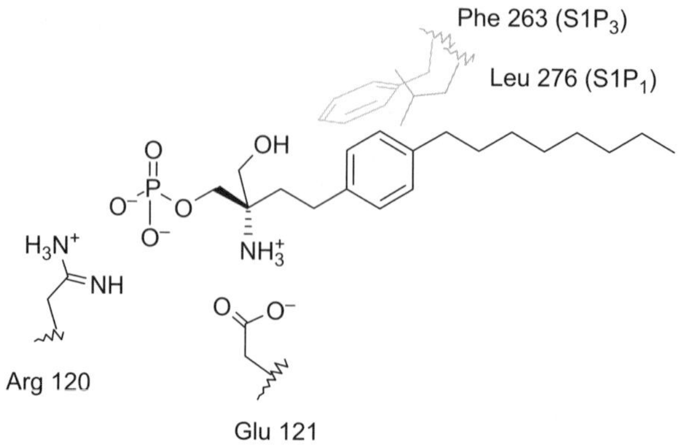

Figure 14.3 Homology model of fingolimod with polar interactions and divergent hydrophobic amino acids in S1P$_1$ and S1P$_3$.

Both clinically[24] and pre-clinically, fingolimod has been shown to selectively reduce CCR7$^+$ naive and central memory T-cells (T$_N$ and T$_{CM}$) in the blood while not effecting CCR7$^-$ effector memory T-cells (T$_{EM}$). The T$_{CM}$ lineage includes the pro-inflammatory helper T 17 cells (T$_h$17).[25] Within the context of MS pathology, more than 90% of accumulated CNS T-cells are of the T$_{CM}$ phenotype, while being void of T$_N$ cells and involving relatively few T$_{EM}$ cells.[26]

Although the CNS effects of fingolimod are poorly understood, significant evidence exists for a CNS component to efficacy.[27] In support of that assertion, it has been shown that rodents that receive a therapeutic dose of fingolimod (0.3 mg/kg p.o.) display picomolar concentration levels of fingolimod-P in cerebrospinal fluid (CSF) indicating that fingolimod can access the CNS compartment. This dose in rodents results in fingolimod concentrations of 5.4 ng/mL in plasma and 0.07 ng/mL in the CSF, while fingolimod-P concentration levels reach 7.4 ng/mL and 0.23 ng/mL in the plasma and CSF, respectively. Also, fingolimod and S1P have both been shown to promote the survival and extension of oligodendrocytes[28] (myelin forming cells) *via* S1P$_5$-mediated activation of the Akt and/or extra-cellular signal-regulated kinase (ERK) signalling pathways. Fingolimod may also help prevent the inflammatory cytokine-mediated death of oligodendrocyte precursor cells (OPCs) in an S1P$_1$-dependent fashion.[14] In addition, enhanced migration, survival and proliferation of astrocytes[29] has been documented with *in-vitro* treatment of fingolimod and in knockout mice in an S1P$_1$- or S1P$_3$-dependent manner.[30] Thus, pre-treatment of cerebellar slices with fingolimod prior to lysolecithin-induced demyelination results in enhanced remyelination, astrogliosis and microgliosis.[31] Furthermore, treatment with fingolimod in the murine demyelinating cuprizone model has been shown to lead to a reduction in toxic demyelination and an increase in the number of oligodendrocytes in the corpus callosum.[32] Therefore, the preponderance of both *in-vitro* and *in-vivo* evidence, taken together, supports a direct role of fingolimod in enhanced remyelination or glioprotective effects.

The pre-clinical validation supporting fingolimod as a potential treatment for MS includes its activity in the induced experimental autoimmune encephalitis (EAE) animal model of MS where a 0.3 mg/kg p.o. dose of fingolimod is sufficient to prevent the onset of neurodegeneration and paralysis, and this occurs in conjunction with a reduction in the number of T-cells and macrophages in the spinal cord. In addition, treatment with fingolimod is more efficacious than IFN-β, the established standard of care for MS, in this model. Initiation of fingolimod treatment four weeks after the onset of disease in the dark agouti rat model of EAE reversed paralysis and normalized electrophysiological function.[33] This correlated with a decrease in demyelination in the brain and spinal cord and a reversal of breakdown in blood-brain barrier integrity as measured by immunoglobulin precipitation.[34]

14.2.3 Clinical Efficacy of Fingolimod

The compelling pre-clinical pharmacology of fingolimod has been recapitulated in a placebo-controlled six-month Phase II clinical trial.[35] This study included 287 patients from Canada and Europe with RRMS and fingolimod was administered once daily at a dose of 1.25 mg or 5.0 mg. Both doses significantly reduced not only the cumulative number of gadolinium enhanced lesions by up to 80%, but also reduced annualized relapse rate by more then 50% compared to placebo (relapse rate: 1.25 mg = 0.35, 5.0 mg = 0.36, placebo = 0.77). An open-label extension of this study demonstrated that continuous treatment with fingolimod for up to 48 months maintains suppression of clinical disease activity with 63% to 70% of patients being relapse-free.[36,37]

A pivotal two-year double-blind placebo-controlled Phase III study of fingolimod at a dose of 0.5 or 1.25 mg daily (FREEDOMS study)[38] evaluated safety and efficacy in 1,272 patients with RRMS. The annual relapse rate was reduced by more than half relative to placebo (0.5 mg = 0.18, 1.25 mg = 0.16, placebo = 0.40) and the probability of disability after three months was significantly reduced (0.5 mg = 17.7%, 1.25 mg = 16.6%, placebo = 24.1%). An additional one-year double-blind, double-dummy Phase III trial of fingolimod at a daily dose of 0.5 mg or 1.25 mg *versus* established weekly IFN-β1a treatment (TRANSFORMS study)[39] demonstrated superiority to the current standard of care. The annualized relapse rate was significantly lower in the fingolimod treatment groups (0.5 mg = 0.16, 1.25 mg = 0.20, IFN-β1a = 0.33) although disability progression was the same in all groups, potentially as a result of the short duration of the study. The median change in brain volume and the median number of new or enlarged T2 lesions over 24 months were the same with each dose of fingolimod and improved as compared to IFN-β1a.

Adverse events from the TRANSFORM study were higher in the fingolimod 1.25 mg dose group (10.7%) *versus* the 0.5 mg dose (7%) or IFN-β1a (5.8%). Two deaths resulting from herpes infection were reported in the high-dose group, although the extent to which fingolimod may have contributed to these deaths is unclear as both cases involved confounding factors. Other adverse events included mild dose-dependent decreases in pulmonary function, macular edema (<1% of patients), slight increase in blood pressure (1–2 mm Hg average increase) and elevated liver enzyme counts. The most common adverse event was transient acute bradycardia often observed following the first dose treatment and not with subsequent daily doses. It is believed that the observed bradycardia is mechanistically linked to $S1P_1$ agonism and activation of the IK_{Ach} G-protein-gated channel in atrial cardiomyocytes.[10] Subsequent receptor internalization/desensitization may abrogate this response with chronic dosing.

The improved safety profile of the 0.5 mg dose of fingolimod *versus* the 1.25 mg daily dose, and the compelling improvement in efficacy over the current standard of care, led the FDA to approve the 0.5 mg dose of fingolimod as a first-in-line treatment option for patients with RRMS in 2010.

14.3 Medicinal Chemistry Approaches to Second-generation S1P Receptor Agonists

14.3.1 Overview of Second-generation Agonists

The clinical success of fingolimod has generated intense interest within the pharmaceutical industry to develop second-generation S1P receptor agonists with improved safety, efficacy or convenience, relative to fingolimod. This effort has been challenged by the nature of this lipophilic phospholipid receptor, which is difficult to target with compounds possessing drug-like properties. Also impacting S1P receptor agonist drug-discovery programs has been the rapidly evolving understanding of the various mechanisms responsible for the efficacy and side-effects associated with fingolimod, which has shaped the selectivity profile of virtually all second-generation agonists. For example, based on pre-clinical observations in the rodent, it was initially believed that the major side-effect associated with fingolimod treatment, transient asymptomatic bradycardia, was mediated by agonism of the $S1P_3$ receptor. This belief has guided the optimization strategy of numerous pharmaceutical companies to generate $S1P_1$ selective agonists. Also, as the CNS effects of fingolimod have become more apparent, BBB permeable derivatives with different profiles of receptor agonism ($S1P_1$ selective, dual $S1P_{1/5}$ or $S1P_{1/3/5}$) have been targeted.

Medicinal chemistry approaches to second-generation analogues can broadly be divided into two efforts: 1) direct agonists of the target and 2) prodrugs that require bioactivation to a phosphate metabolite in analogy to fingolimod. An overview of the compounds and compound classes discussed in this chapter, which have been selected to demonstrate the challenges and strategies used to develop small-molecule S1P receptor agonists, is given in Table 14.2.

14.3.2 Direct Agonists – Triaryl Scaffold

The majority of published S1P receptor agonists share a common structural motif derived from the triaryl core scaffold initially disclosed by Merck in 2004.[40] The commercially available compound **SEW2871** (Table 14.3) was

Table 14.2 Overview of selected second-generation S1P receptor agonists.

Compound/ parent company	Structure	Agonist type	S1P receptor profile	Optimization parameters (comment)
SEW2871/ Merck		Direct	$S1P_1$	– Potency, selectivity, PK – (first reported direct agonist of $S1P_1$)
19/GSK		Direct	$S1P_{1/5}$	– Potency, selectivity, BBB permeability, off-target toxicity
ACT-128800/ Actellion		Direct	$S1P_1$	– Potency, selectivity, PK (novel chemotype, long $t_{1/2} \sim 30\,h$ in humans)
AMG 369/ Amgen		Direct	$S1P_1$	– Selectivity, PK, off-target toxicity (rationally designed scaffold)
53/Amgen		Direct	$S1P_1$	– Potency, selectivity, solubility, permeability (lacks polar head group)
CS-0777/ Daiichi Sankyo		Prodrug	$S1P_{1/5}$	– PK ($S1P_3$ selective, human PK of phosphate metabolite)

identified in a high-throughput screening (HTS) campaign as a potent agonist of the $S1P_1$ receptor with exquisite selectivity over the other S1P receptors. It also had a potential advantage over many other known agonists in that it does not function as a prodrug and require bioconversion for activity, eliminating an additional optimization parameter and potentially allowing faster development times. Hybridization of this compound with previously identified amino acid containing derivatives[41] resulted in analogue **1** (Table 14.3), which maintains a favourable selectivity profile and potency. Selectivity for the $S1P_3$ receptor could be further improved through substitution on the central phenyl residue *ortho* to the ether linkage as shown for compounds **2** and **3**.

The *in-vivo* activity of compounds **1**, **2** and **3** was assessed in a model of peripheral blood lymphocyte (PBL) lowering. Maximal PBL lowering at 24 hours was achieved in both the beagle dog and rhesus monkey following a single 1 mg/kg p.o. dose for both compound **2** and **3**. In the rodent, a 3 mg/kg p.o. dose of **1** resulted in maximal PBL lowering, with the lowest counts achieved 3 to 4 hours post dose. Rebound of lymphocyte levels starts to occur after 24–32 hours and returns to pre-dose levels in 72 hours. A sequential 3 mg/kg challenge with oxadiazole **1** recapitulates the initial PBL response.

Table 14.3 S1P receptor EC_{50} (nM)[a] for SEW2871 analogues.

Compound	Structure	$S1P_1$	$S1P_3$	$S1P_4$	$S1P_5$
SEW2871		37	> 10,000	> 10,000	4600
1		1.2	530	1600	23
2		3.2	2000	1300	18
3		1.8	7500	1500	13

[a]EC_{50} values were determined by displacement of [^{33}P]-labelled S1P by test compounds from human S1P receptors expressed on CHO cell membranes.

While **1** was a potent and selective $S1P_1$ receptor agonist, its selectivity *versus* $S1P_3$ was moderate at 500-fold and an additional increase in receptor selectivity was desired, driven by the hypothesis that the transient bradycardia observed with fingolimod was due in part to $S1P_3$ agonism. Screening of additional commercially available compounds related to **SEW2871** and initial SAR produced **4** (Table 14.4), which demonstrates a high level of $S1P_3$ selectivity.[42] Modification of the central oxadiazole ring arrangement results in compounds **5** and **6**, with improved $S1P_1$ agonist activity. Further modification to the substituent on the terminal phenyl ring afforded isobutyl analogue **7**, which achieved improved $S1P_3$ selectivity by increasing $S1P_1$ activity.

While compound **7** has 20,000-fold selectivity for $S1P_1$ over the $S1P_3$ receptor it only has moderate selectivity over $S1P_4$ and is not selective for $S1P_5$. To further refine this selectivity profile and generate exclusive $S1P_1$ agonists the amino acid polar head group was replaced with a straight chain carboxylic acid and a variety of substituents were explored on the phenyl fragments, as exemplified by compounds **8** and **9** (Table 14.5).[43] While carboxylate **8** is potent and reasonably selective for $S1P_1$, the absolute potency could be increased with the addition of an isopropyl group to yield **9**, which maintains high levels of selectivity, ~ 1200-fold over $S1P_3$ and 440-fold over $S1P_5$. Substitution of the trifluoromethyl group of **9** for a cyano moiety provides analogue **10** and improves selectivity over $S1P_3$ by 10-fold, although selectivity over $S1P_5$ erodes.

Table 14.4 S1P receptor EC_{50} $(nM)^a$ for oxadiazole derivatives with improved S1P$_3$ selectivity.

Compound	Structure	SIP$_1$	SIP$_3$	SIP$_4$	SIP$_5$
4		100	>10,000	4300	1000
5		8.2	>10,000	640	20
6		3.8	12,000	370	5.2
7		0.6	12,000	70	1.0

^{a}EC$_{50}$ values were determined by displacement of [^{33}P]-labelled S1P by test compounds from human S1P receptors expressed on CHO cell membranes.

On testing *in vivo*, compound **9** demonstrated maximal PBL lowering in rats following a 0.3 mg/kg p.o. dose, a 10-fold lower dose than that required for **7**. The rat pharmacokinetic parameters for this series of agonists with improved selectivity have low clearance, low volume of distribution at steady state, but also very short half-lives (Table 14.5).

Substitution on the ethyl linker to the carboxylic acid, as shown in analogues **11**, **13** and **14**, or nitrogen incorporation into the pendant phenyl ring, as exemplified by pyridyl analogue **12**, significantly lengthen half-life for each of these examples without erosion of the S1P receptor selectivity profile (Table 14.5).

Numerous examples appear in the literature that builds from the general triaryl template with various S1P receptor selectivities. Utilizing this template GSK has generated S1P$_1$/S1P$_5$ dual agonists, which penetrate the CNS compartment and therefore offer the potential to capture the putative CNS effects of fingolimod.

The initial compounds from GSK featured analogues of **7** in which the methyl-azetidine head group was constrained in a tetrahydroisoquinoline (Figure 14.4).[44] Compound **15** is a potent agonist of the S1P$_1$ receptor with greater than 1000-fold selectivity over S1P$_3$. It induces full lymphopenia at a 0.1 mg/kg p.o. dose in rats and has excellent cross-species PK with, for example, low clearance (5 mL/min/kg) and a $t_{1/2}$ of 3 hours after a 1 mg/kg i.v. dose in rats. However, permeability in an MDCKII-MDR1 assay was increased for **15** in the presence of a Pgp inhibitor (efflux ratio for **15** = 6.0; efflux ratio for

Table 14.5   S1P receptor EC_{50} $(nM)^a$ for oxadiazole derivatives with improved $S1P_{3/5}$ selectivity and rat pharmacokinetic parameters.

Compound	Structure	$S1P_1$	$S1P_3$	$S1P_5$	Rat $Cl_p{}^b$	Rat $t_{1/2}{}^b$
8		3.2	>10000	2600	–	–
9		0.08	110	40	8.7	1.1
10		0.08	1100	6.5	2.3	0.7
11[c]		0.12	184	3.9	2.9	3.1
12		0.44	2760	28.6	1.2	2.4
13[c]		0.45	>10000	13.6	1.0	5.9
14[c]		0.8	>1000	37.7	1.2	7.4

$^a EC_{50}$ values were determined by displacement of $[^{33}P]$-labelled S1P by test compounds from human S1P receptors expressed on CHO cell membranes.
$^b Cl_p$ was measured following 2.0 mpk i.v. dose of the respective compounds to male Sprague-Dawley rats and reported in units of mL/min/kg. $t_{1/2}$ is reported in units of hours.
^{c}Data reported for the racemic mixture.

15 + Pgp inhibitor $= 0.5$), indicating that it is a substrate for active transport *in vitro* and unlikely to cross the blood-brain barrier (BBB). It is likely that the *in-vitro* active transport is in part a result of the zwitterionic nature of **15**.

In order to design CNS penetrable analogues, the carboxylic acid of the head group was abandoned and replaced with neutral derivatives, as shown for compounds **16–19** (Table 14.6). These derivatives were selected to decrease the

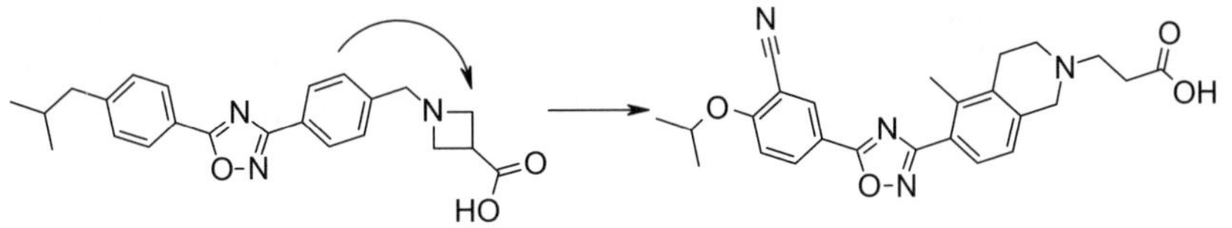

Figure 14.4 Modification of **7** to **15**.

Table 14.6 *In-vitro* and *in-vivo* profile of oxadiazole agonists.

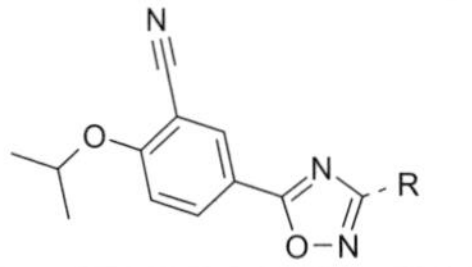

Compound	16	17	18	19
R	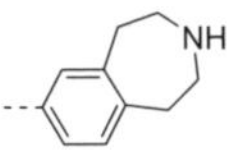	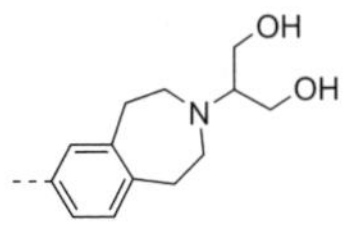	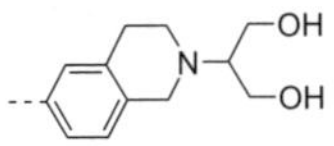	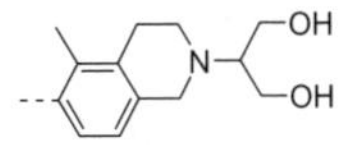
S1P$_1$ pEC$_{50}$[a]	7.7	7.7	8.1	8.5
S1P$_3$ pEC$_{50}$[b]	<4.5	4.8	5.2	<4.5
Measured pK$_a$	9.7	7.8	7.3	7.3
hERG pIC$_{50}$[c]	5.5	4.7	4.7	4.9
Rat V$_{ss}$ (L/kg)[d]	9.7	7.6	3.7	4.1
Rat $t_{1/2}$ (h)[d]	6.3	5.3	0.6	1.0
MW, cLogP, PSA	375, 3.6, 97	449, 3.3, 116	434, 2.5, 116	449, 3.0, 116

[a]β-Arrestin functional assay of human S1P$_1$ receptor expressed on CHO cell membranes.
[b]EC$_{50}$ values were determined by compound induced binding of [^{35}S]-GTPγS to human S1P receptors expressed on CHO cell membranes.
[c]max inhibition % @ 30μM, PatcheXpress.
[d]1 mg/kg i.v. DMSO/10% kleptose HPB 0.9% saline.

pK$_a$ of the tetrahydroisoquinoline amine in order to minimize off-target risks including hERG inhibition, phospholipidosis and high V$_{ss}$, the latter potentially resulting in drug accumulation following chronic dosing and prolonged pharmacodynamic half-lives.[45] It was noted during this optimization effort that rat V$_{ss}$ correlated strongly with measured pK$_a$, with V$_{ss}$ less than 5 L/kg observed when the amine pK$_a$ was less than 7.5, although little correlation was seen with hERG Patchexpress IC$_{50}$s.

Diol derivative **17** retains S1P$_1$ potency relative to the secondary amine **16** but also high V$_{ss}$ and $t_{1/2}$ in rodents. Ring contraction to the tetrahydroisoquinoline, which displays the amine in the benzylic position, further reduces the amine pK$_a$ resulting in lower V$_{ss}$ and $t_{1/2}$ for compounds **18** and **19** (Table 14.6).

When analogue **19** was tested in MDCKII-MDR1 cell lines for permeability, the basolateral-apical (B-A) to A-B ratio was unaffected by the presence or absence of a Pgp inhibitor indicating that the compound is not a substrate for active transport by Pgp. A CNS penetration study in rats demonstrates that **19** can enter the CNS compartment with a brain to plasma ratio of 1.85 : 1. The ability of **19** to enter the CNS is surprising given the compound properties (MW = 449, PSA = 116 Å^2), which are a significant deviation from the accepted CNS drug-like space (MW < 305, PSA < 70 Å^2).[46] This may be due to the presence of an internal hydrogen bond between the pendant diols, which can improve permeability.

Tetrahydroisoquinoline **19** is a partial agonist of the S1P$_5$ receptor (S1P$_5$ pEC$_{50}$ = 7.7, efficacy = 65%) with excellent solubility in simulated intestinal fluid (FeSSIF = > 1000 ng/mL). The compound induces full lymphopenia in rats at a 0.3 mg/kg p.o. dose with lymphocyte levels returning to baseline within 24 hours.

14.3.3   Direct Agonists – Additional Chemotypes

The use of the triaryl template has resulted in a crowded intellectual property landscape for this agonist series. The identification of chemotypes with improved freedom to operate is a natural advantage in any drug-discovery program. Acetlion identified iminothiazolidinone **20** (Figure 14.5) as a potent agonist of the S1P$_1$ and S1P$_3$ receptors that does not require bioconversion to a phosphate metabolite for activity.

A systematic SAR effort to identify S1P$_1$ selective compounds based on this core, with appropriate efficacy and PK properties suitable for q.d. dosing centred on modifications at four regions of the scaffold (Table 14.7).

Conversion of the hydrazinyl group of **20** to *n*-propyl or *iso*-propyl (**22** and **23**) can retain potency and slightly improve S1P$_3$ selectivity, while the methyl analogue **21** is less active. Modifications at the R$_1$ position are limited to methyl incorporation with larger groups losing potency (**24** *vs.* **25**). At R$_3$ a chloro substituent provides optimal potency and selectivity as compared to hydrogen, methyl or methoxy (**23** *vs.* **26–28**). A series of analogues, **29–32**, was also prepared replacing the phenol of compounds **20–28** with a linker to an alcohol moiety. Introduction of a chiral diol head group further improves potency and selectivity ultimately producing **ACT-128800**. This compound has moderate S1P$_1$ selectivity over the other S1P receptors as measured in a GTPγS

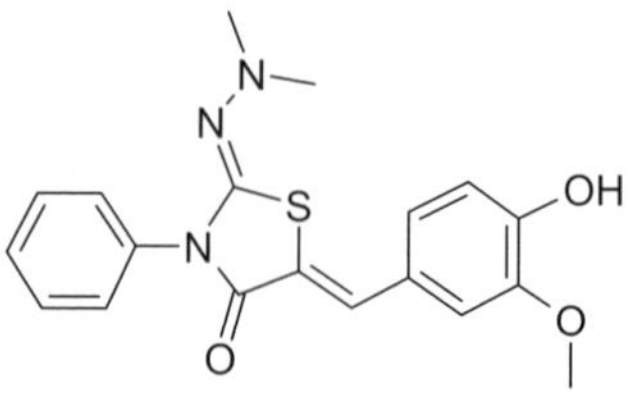

20

Figure 14.5   Iminothiazolidinone S1P$_1$ agonist **20**.

Table 14.7 Receptor EC_{50} (nM) for thiazolidinone derivatives.

Compound	R_1	R_2	R_3	R_4	$S1P_1{}^a$	$S1P_3{}^a$
20	H	dimethylamino	Cl	OH	62	140
21	H	methyl	Cl	OH	990	8810
22	H	n-propyl	Cl	OH	67	189
23	H	isopropyl	Cl	OH	47	120
24	Me	isopropyl	Cl	OH	34	139
25	OMe	isopropyl	Cl	OH	106	428
26	H	isopropyl	Me	OH	37	50
27	H	isopropyl	OMe	OH	200	335
28	H	isopropyl	H	OH	122	95
29	H	isopropyl	H	˙˙O⌒⌒OH	58	68
30	H	isopropyl	H	˙˙O⌒⌒OH (OH)	143	240
31	Me	n-propyl	Cl	˙˙O⌒⌒OH	11	124
32	Me	n-propyl	Cl	˙˙O⌒⌒OH (OH)	9.7	109
ACT-128800	Me	n-propyl	Cl	˙˙O⌒⌒OH (OH)	9.1	123

^{a}EC_{50} values were determined by compound induced binding of [^{35}S]-GTPγS to human S1P receptors expressed on CHO cell membranes.

Table 14.8 Pharmacokinetic parameters of **ACT-128800** in rats and dogs.^{a}

Species	% F^a	C_{max} (ng/mL)	$t_{1/2}$ (h)	Cl (mL/min/kg)	V_{ss} (L/kg)
Wistar rat	35	409	1.3	25	3.3
Beagle dog	69	1360	10	1.3	0.9

^{a}Oral administration of 3 mg/kg or i.v. administration of 0.3 mg/kg.

functional assay ($S1P_2/S1P_1 = >10,000$, $S1P_3/S1P_1 = 14$, $S1P_4/S1P_1 = 122$, $S_1P_5/S1P_1 = 6.5$). Dose response experiments in male Wistar rats demonstrated maximal efficacy and lymphocyte reduction at a 3 mg/kg p.o. dose (65 to 75% reduction in blood lymphocytes at 3 hours post-dose). The duration of action was extended at higher doses, with sustained reduction of blood lymphocytes at 24 hours achieved with a 100 mg/kg p.o. dose.

The bioavailability, C_{max} and $t_{1/2}$ for **ACT-128800** are higher in the beagle dog than the rodent, and predicted to be suitable for a once-daily dosing regimen in humans (Table 14.8). In a single-ascending dose study in humans the compound showed good exposure but a long half-life of 22–33 h.

The extensive SAR data published in the peer-reviewed literature have fuelled rational drug design approaches to generate novel chemical series

Table 14.9 Receptor EC_{50} (μM) for benzofuran derivatives.

Compound	R	Aryl	$GTP\gamma S^a$ $S1P_1$	$S1P_3{}^b$ Ca^{2+}	$S1P_1$ RI^c
33	*n*-Bu		0.35	>25	2.48
34	O-*n*-Bu		0.52	>7.2	NA
35	CH_2Ph		0.44	>4.3	2.07
36	O-*n*-Bu		3.2	>25	>1.0
37	CH_2Ph		0.62	3.43	0.253
38	CH_2Ph		0.16	27	0.057

[a]EC_{50} values were determined by compound induced binding of [^{35}S]-GTPγS to human S1P receptors expressed on CHO cell membranes.
[b]Ca^{2+} flux measured in human $S1P_3$ receptor in $G_{q/i5}$ transfected CHO-K1 cells.
[c]Receptor internalization assay of human $S1P_1$ receptor-GFP fusion protein in U2O cells.

outside of the traditional triaryl motif. Amgen Inc. utilized *in-silico* models of the $S1P_1$ receptor and PREDICT methodology[47] to identify novel scaffolds. These scaffolds were designed to contain the Merck amino acid polar head group (*supra infra*), a moderately polar core that lacks an H-bond donor, and a hydrophobic tail, along with standard drug-like property considerations of logP and molecular weight (Table 14.9).[48] Benzofuran derivative **33** was subsequently identified by this methodology as a moderately potent but selective, agonist of $S1P_1$. Modification of the hydrophobic tail portion of this molecule, including the butyl ether derivative **34** and benzyl substituted analogue **35**, maintain selectivity but do not improve potency. Replacement of the central phenyl linker with a thiazole, exemplified by **36**, erodes potency while subtle modifications including the introduction of a fluorine dramatically improve $S1P_1$ potency (**37** and **38**).

The benzyl-substituted benzofuran **38** produces lymphopenia activity comparable to fingolimod in Lewis rats at a 3 mg/kg p.o. dose, with sustained reduction in lymphocyte levels at 24 hours post-dose. A 10 mg/kg p.o. dose of **38** completely reverses the course of disease in the EAE mouse model of human multiple sclerosis. This compound has low clearance (4–15% of liver blood flow), a moderate volume of distribution at steady state (1.0–2.2 L/kg) and a long half-life (5–21 h) in the rat, non-human primate and canine. Its oral bioavailability is consistently good across species (51–88%). In a four-day repeat dose toxicity study in rats the exposure was proportional at doses up to 60 mg/kg with a NOAEL of 20 mg/kg. To evaluate $S1P_3$-related cardiovascular toxicity, a bolus dose of **38** was delivered to telemeterized female SD rats. No effect was observed on heart rate at any dose, although moderately elevated mean arterial pressure (MAP) was observed at doses of 20 and 40 mg/kg from 4–6 h post dosing, with peak effects corresponding to C_{max}. MAP increases have been reported in clinical trials of fingolimod and may be linked mechanistically to $S1P_3$ agonism in humans.[14]

Toxicology assessment of **38** at a 40 mg/kg oral dose produced pro-convulsive activity in the rat, which had not previously been reported for an S1P agonist, indicating that it is most likely an off-target toxicity. Subsequent SAR modifications aimed at eliminating this issue focused on identifying close structural analogues of **38** that lacked the pro-convulsive effect.

In this effort, the screening of heterocycles bioisosteric with the benzofuran of **38** identified benzothiazole derivative **39** as having equivalent $S1P_1$ potency and $S1P_3$ selectivity to **38**.[49] The incorporation of fluorine at the 2-position of the benzyl moiety to provide compound **40** further improves selectivity and did not show pro-convulsive effects in SD rats at a daily p.o. dose of 100 mg/kg for four days. In addition, benzothiazole derivative **40** reduced lymphocyte levels to near maximum levels after a 3 mg/kg p.o. dose. Further modifications to **40** focused on reducing the logP (clogP = 3.9) and increasing PSA (57 Å²)[50] to a range predicted to reduce the probability of off-target toxicity (clogP < 3, PSA > 75 Å²). Thus, incorporation of a single nitrogen, resulting in a thiazolopyridine core, decreases clogP by 1.1 units and increases PSA by 12 Å² and analogues of **40** and the benzylic regioisomer **42** bearing this ring system were therefore screened. Potency was eroded for each of the aza analogues of **40**, such as **41**, while thiazolopyridine **43** retained $S1P_1$ potency relative to the analogous benzothiazole **42**. Efforts to "shield" this polar atom with either benzylic methyl or *geminal*-dimethyl groups, as in compounds **44–46**, improve potency and selectivity, while incorporation of a cyclopropyl moiety was superior in both aspects, resulting in **AMG 369** (Table 14.10).

This improvement in potency has been rationalized by presuming favourable entropic pre-organization of the ligand, as the cyclopropyl group favours a 180° dihedral angle about the benzylic bond while the *gem*-dimethyl substitution favours a 120° and 240° angle.

The high level of *in-vitro* $S1P_1$ potency of **AMG 369** translates into increased *in-vivo* activity with maximal lymphocyte lowering in rats observed following a 0.1 mg/kg p.o. dose. In a rat EAE model **AMG 369** was fully efficacious at a 0.1 mg/kg q.d. p.o. dose. **AMG 369** has similar pharmacokinetic properties to **38** with a long half-life observed across species ($t_{1/2}$ = 5.9 h rat, 38 h dog, 24 h non-human primate), a common profile for S1P receptor agonists. Cardiovascular assessment of **AMG 369** in telemeterized rats established a no-effect level for heart rate and MAP changes at 10 mg/kg p.o., indicating a wide safety margin for $S1P_3$-associated cardiovascular toxicity.

Table 14.10   Receptor EC_{50} (µM) for benzothiazole and thiazolopyridine derivatives.

Compound	R_1	R_2	Y	$S1P_1$ RI^a	$S1P_3^b$ Ca^{2+}
39	(benzyl)	H	CH	0.042	1.21
40	(fluorobenzyl)	H	CH	0.042	3.47
41	(benzyl)	H	N	0.138	2.89
42	H	(phenyl)	CH	0.221	3.47
43	H	(phenyl)	N	0.199	0.478
44	H	(1-phenylethyl)	N	0.035	0.877
45	H	(phenethyl)	N	0.038	0.996
46	H	(2-phenylpropyl)	N	0.033	2.51
AMG 369	H	(phenylcyclopropyl)	N	0.002	0.888

^{a}Receptor internalization assay of human $S1P_1$ receptor-GFP fusion protein in U2O cells.
^{b}Ca^{2+} flux measured in human $S1P_3$ receptor in $G_{q/i5}$ transfected CHO-K1 cells.

Contrary to the majority of S1P receptor agonists described in the literature, potent and selective agonists of the $S1P_1$ receptor can be generated without the presence of a polar head group, as demonstrated by **SEW2871**. The polar group is often required to enhance the solubility of the triaryl scaffold, as well as other chemotypes, due to the limited tolerance for polarity in the hydrophobic pocket of the S1P receptors. However, a polar head group often introduces additional liabilities such as isoform selectivity, oxidative metabolism and Phase II conjugation, among others, and therefore a core scaffold with greater intrinsic solubility would be advantageous. One such core was discovered by Amgen when an HTS campaign identified carbamothioylbenzamide **47**, lacking a polar head group, as a potent and selective agonist, albeit with poor solubility (Table 14.11).[51] The potential toxicity associated with the carbamothioylbenzamide moiety prompted

Table 14.11 Receptor EC$_{50}$ (μM)^a, solubility and permeability of carbamoylbenzamide derivatives.

Compound	X	Y	R$_1$	R$_2$	S1P$_1$	Sol^b	P$_{app}$^{c}
47	S	CH	Cl	⟨N⟩--	0.43	<1.0	1.8
48	O	CH	Cl	⟨N⟩--	0.23	<1.0	<1.0
49	O	CH	CF$_3$	⟨N⟩--	0.013	<1.0	<1.0
50	O	CH	CF$_3$	⟨⟩--	0.068	<1.0	<1.0
51	O	CH	CF$_3$	⟨=N⟩--	0.013	19	7.4
52	O	N	CF$_3$	⟨⟩--	0.035	6.6	2.3

^{a}Receptor internalization assay of human S1P$_1$ receptor-GFP fusion protein in U2O cells.
^{b}Solubility (μg/mL) measured in phosphate-buffered saline (PBS).
^{c}Apparent permeability ($\times 10^{-6}$ cm/s) through porcine proximal tubule cells (LLC-PK1 cell line).

conversion of the sulfur atom to oxygen to give **48**. The potency of **48** could be further improved by the replacement of the R$_1$ chlorine atom with a CF$_3$ group, as in **49**, or by exchange of the piperidinyl R$_2$ group with a phenyl, exemplified by **50**, although solubility remains low for both analogues. Nitrogen incorporation in either the left- or right-side phenyl rings improves solubility and permeability for analogues **51** and **52** without eroding potency or selectivity.

Compound **52** is more than 1000-fold selective over the S1P$_5$ receptor and greater than 10,000-fold selective over the other S1P receptor family members. The pharmacokinetic profile of **52** in rats includes moderate clearance (0.25 L/kg/h), high V$_{ss}$ (6.3 L/kg) and long $t_{1/2}$ (19 h). Circulating lymphocyte levels could be maximally reduced over 24 hours in rats following a single 1 mg/kg p.o. dose.

14.3.4 Prodrug Agonists

Despite the success of fingolimod, relatively few examples appear in the literature of prodrug derived compounds. This may be due to the issues associated with the discovery and development of compounds that require bioconversion *in vivo* to form an active agonist. Not only are the challenges with single agent, direct agonists present, such as isoform selectivity, but also the added challenge of insertion into the S1P biosynthetic pathway and phosphorylation by SphK1 or SphK2. Despite these challenges second-generation prodrug agonists of the S1P receptors have been generated and experienced clinical success. Three compounds that are prodrugs for the corresponding phosphate ester, **KRP-203**, **CS-0777** and **53**, are shown in Figure 14.6 and all share common structural elements of an amino-alcohol head group and hydrophobic tails.

KRP-203 (Novartis AG/Kyroin Pharmaceuticals Co. Ltd), an $S1P_{1,4,5}$ agonist ($S1P_1$ $EC_{50} = 0.1$ nM), is currently being evaluated for the treatment of multiple sclerosis and transplantation rejection in Phase I clinical trials. In a skin allograft study in rats, **KRP-203** prolonged graft survival for more than 30 days following 1 mg/kg q.d. dosing, the same efficacy as fingolimod in this model.[52]

CS-0777 is also being evaluated clinically in patients with RRMS. Unlike fingolimod-P, the *in-vivo* active agonist **CS-0777-P** is approximately 320-fold more selective for $S1P_1$ ($EC_{50} = 1.8$ nM) over $S1P_3$ and 19-fold selective over $S1P_5$.[53] Pre-clinically, a 0.1 mg/kg q.d. p.o. dose is sufficient to elicit maximal

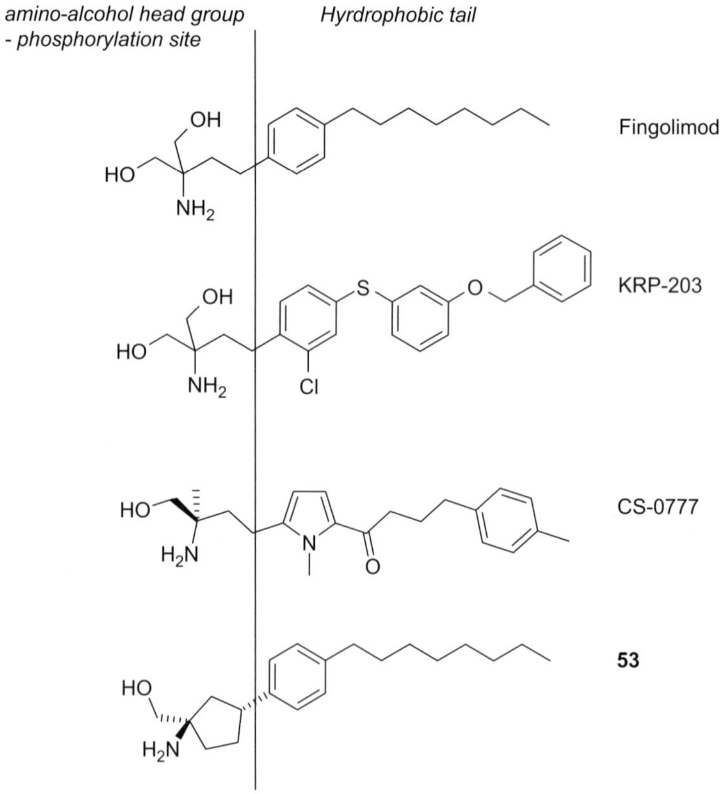

Figure 14.6 Common structural elements of prodrug S1P receptor agonists.

efficacy in a rat EAE model. Following p.o. administration of a 0.1 mg/kg dose in rats, **CS-0777** is slowly converted to **CS-0777-P**, with T_{max} of the phosphate reached at 9.5 hours, in contrast to **CS-0777** with T_{max} achieved at 4.7 hours post-dose (Table 14.12.). The C_{max} and AUC are 10 times higher for the phosphate metabolite than for the parent and the bioequivalence of the metabolite is greater than 80% (calculated as the AUC ratio of **CS-0777-P** after oral and i.v. administration of **CS-0777**).

Evaluation of a 1 mg/kg dose of **CS-0777** in an open-label Phase I study in MS patients resulted in efficacy comparable to fingolimod for the reduction of lymphocyte counts.[54] The ratio of **CS-0777** to its phosphate metabolite in humans is consistent with the observations in the rat, although **CS-0777-P** has a much longer half-life in humans (Table 14.13).

CS-0777 is well tolerated in the clinic with few reports of adverse events. The most common adverse event is transient asymptomatic bradycardia following first dose administration, similar to fingolimod. This effect occurs despite a large selectivity window *in vitro* for $S1P_1$ over $S1P_3$. This mirrors other clinical observations that $S1P_1$-selective agonists still induce bradycardia in humans and that this effect is most likely mechanistically related to $S1P_1$ agonism.

A third amino-alcohol prodrug, **53** (Figure 14.6), was prepared as a constrained analogue of fingolimod.[55] Similar to fingolimod, **53** undergoes phospholrylation *in vivo* by SphK2 to form an active agonist of the $S1P_1$ receptor. Full lymphopenia can be induced in mice following a 1 mg/kg p.o. dose of **53** and the effects are sustained for more than 48 hours. Interestingly, however, phosphorylated **53-P** is an *antagonist* of the $S1P_3$ receptor, a unique selectivity profile.

Table 14.12 Pharmacokinetic parameters of **CS-0777** and **CS-0777-P** in rats.

Compound	Dose (mg/kg)[a]	C_{max} (ng/mL)	T_{max} (h)	$t_{1/2}$ (h)	AUC_{0-inf} (ng·h/mL)
CS-0777	0.1	1.89	4.75	5.4	31.7
	1.0	1.82	3.25	10.4	346
CS-0777-P	0.1	26.4	9.5	10.9	686
	1.0	220	9.0	11.6	6280

[a]Dose of parent **CS-0777**.

Table 14.13 Pharmacokinetic parameters of **CS-0777** and **CS-0777-P** in humans.

Compound	Dose (mg/kg)[a]	C_{max} (ng/mL)	$t_{1/2}$ (h)	AUC_{0-inf} (ng·h/mL)
CS-0777	2.5	0.20	143	30.9
CS-0777-P	2.5	3.3	175	764
	1.0	1.69	211	365

[a]Dose of parent **CS-0777**.

14.4   S1P Receptor Antagonists

The observation that $S1P_1$ receptor agonists cause down-regulation of the receptor following binding has led to the adoption of a "functional antagonist" model to explain the observed pharmacodynamic effects. Based on this model direct antagonists of the target were expected to offer efficacy similar to agonists, with the potential to decouple side-effects driven by the initial agonistic response prior to receptor down-regulation/desensitization. Phosphonate ester **54** (Figure 14.7), a close structural analogue of fingolimod, has been identified as a full and competitive antagonist at the $S1P_1$ receptor ($K_i = 70$ nM as determined by EC_{50} shift of fingoldimod-P in a GTPγS binding assay; $K_i = 18$ nM relative to **SEW2871**) that is neither an agonist nor antagonist at the $S1P_2$, $S1P_3$ or $S1P_5$ receptors.[56] *In-vitro* modulation of signalling targets downstream from the $S1P_1$ receptor, such as ERK and Akt phosphorylation, are also competitively inhibited by **54** following agonism by fingolimod-P and **SEW2871**.

In mice a 10 mg/kg injection of **54** results in a plasma concentration of 268 nM at 5 hours post-dose with a half-life of 73 minutes. Interestingly, i.v. administration of a 20 mg/kg dose of **54** in mice shows no significant effect on either lymphocyte numbers or percentage of lymphocytes within total white blood cells. Lymphopenia induced following the oral administration of **SEW2871** at doses up to 20 mg/kg (**SEW2871** induces maximal lymphopenia at 3 mg/kg) could be abolished following a single 10 mg/kg injection of **54** at the time of oral administration and is consistent with observations reported for other $S1P_1$ antagonists.[57,58]

The conclusion that antagonism of the $S1P_1$ receptor does not recapitulate the lymphostatic response observed with $S1P_1$ agonists has called into question the long-accepted mechanism of action of agonism of the $S1P_1$ receptor. An alternative theory that has been proposed to explain the action of $S1P_1$ agonists is that lymphocyte egress from the secondary lymph is controlled in part by a stromal element functioning as a physical "gate" over the intrinsic lymphostatic mechanism. Two-photon microscopy in lymph node explants indicates that lymphocytes migrate across a stromal "gate" to egress the node. These gates are unaffected by $S1P_1$ antagonists but collapse following $S1P_1$ agonism and are subsequently restored following $S1P_1$ antagonism. This alternate mechanism of action is currently a matter of debate in the S1P field. It is important to note, however, that the possibility that the constitutive activity or priming signal of a small number of $S1P_1$ receptors on lymphocytes is pharmacologically relevant

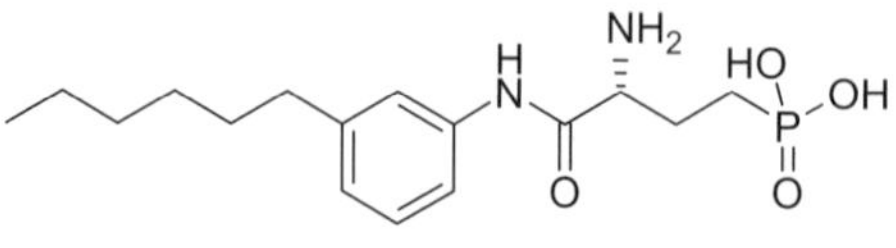

Figure 14.7 $S1P_1$ Selective antagonist **54**.

and responsible for the lack of *in-vivo* efficacy observed for compound **54** and other antagonists.

14.5 Conclusion and Clinical Landscape

In the more than two decades since the discovery of fingolimod, the field of S1P receptor agonists has undergone tremendous expansion. A simple literature search for "S1P receptor" now returns more than 3000 hits ranging from immunological research, oncology, neurosciences and of course drug development and clinical data. This progression is startling considering that the mechanism of action for fingolimod remained clouded during much of its early development and may remain so today.

The clinical efficacy and safety associated with fingolimod is also remarkable, as other immunological disease modifying agents often suffer from serious tolerability and compliance issues. Most importantly though, the efficacy of fingolimod offers patients suffering from RRMS an improved therapy, as compared to the established standard of care, and the increased convenience of daily oral dosing over bi-weekly injections.

Fingolimod is currently being evaluated in clinical trials for the other more insidious forms of primary and secondary progressive MS. These are characterized by un-relapsed phases of neurological impairment and fingolimod may offer some prospect of therapeutic intervention due to its potential CNS effects.

The search for second-generation agonists has almost exclusively focused on differentiating from fingolimod in terms of safety. Elimination of the most severe adverse event reported with fingolimod, transient asymptomatic bradycardia, in a second-generation compound would offer patients increased convenience and perhaps gain a significant market share of MS therapies. The pre-clinical observation that this cardiovascular effect was driven by the $S1P_3$ receptor in rodents became accepted dogma in the industry and led to substantial efforts directed towards the identification of highly selective agonists. This effort generated direct agonists of the target that do not require bioconversion for activity. However, the seminal "triaryl" scaffold presented numerous optimization challenges for drug-discovery scientists, including selectivity, *in-vivo* half-life, solubility and BBB permeability.

In addition, the clinical observation that $S1P_3$-selective agonists recapitulate the transient bradycardia observed with fingolimod[59] has led to the recognition that this effect is mechanistically linked to $S1P_1$, which has hampered the clinical development of numerous selective second-generation agonists. Clinical differentiation from the now marketed fingolimod will be challenging and may limit opportunities for compounds currently in clinical trials (Table 14.14)[60] to demonstrate superiority and enter the marketplace. Novartis appears to be securing its dominance in the field with the sustained development of BAF-312, a dual $S1P_{1/5}$ agonist, which is currently in Phase II clinical trials and poised to be the next compound to the market. The remaining clinical candidates will

Table 14.14 Clinical landscape.[a]

Compound (developing company)	Development status	Compound class	S1P receptor profile	Comment
Fingolimod (Novartis AG)	Launched	Prodrug	$S1P_{1/3/4/5}$	– Clinical trials in primary progressive MS ongoing
BAF-312 (Novartis AG)	Phase II	Direct agonist	$S1P_{1/5}$	– Most advanced second-generation agonist – Transient bradycardia reported in man despite $S1P_3$ selectivity – Structure unknown
Ponesimod (ACT-1288800, Actellion)	Phase II	Direct agonist	$S1P_{1/3/5}$	– Shorter half life in man compared to fingolimod ($\sim 30\,h$)
ONO-4641(Ono Pharmaceutical)	Phase II	Unknown	Unknown	
KRP-203 (Novartis KG/ Kyorin Pharmaceutical)	Phase I	Prodrug	$S1P_{1/4/5}$	– Transient bradycardia reported in man despite $S1P_3$ selectivity – Clinical trials ongoing for transplantation rejection and RRMS
CS-0777(Daiichi Sankyo)	Phase I	Prodrug	$S1P_{1/4/5}$	– Long half-lives reported in man ($\sim 7\,days$)

[a]Multiple additional Phase I candidates have been reported with little information regarding compound class and profile.

face a difficult task in positioning themselves relative to these compounds. Many pharmaceutical companies are therefore repositioning their S1P clinical and development portfolios towards other immunological diseases, such as rheumatoid arthritis, or into different therapeutic areas, such as oncology, in order to define a clear path to the market.[61]

Finally, the remaining strategy to decouple the mechanistic side-effects of fingolimod has focused on identifying direct antagonists of the target. The surprising lack of efficacy of these antagonists, taken together with their ability to abolish agonist induced lymphopenia, has called into question the very fundamental mechanism of action of fingolimod and $S1P_1$. As the Darwinian biologist T. H. Huxley said, the great tragedy of science is "the slaying of a beautiful hypothesis by an ugly fact", which strongly resonates with the S1P field.

References

1. S. Pyne and N. J. Pyne, *Biochem. J.*, 2000, **349**, 385.
2. S. E. Alvarez, S. Milstien and S. Spiegel, *Trends Endocrinol. Metab.*, 2007, **18**, 300.
3. T. Hla and V. Brinkmann, *Neurology*, 2011, **76**, S3.
4. N. Fukushima, I. Ishii, J. J. Contos, J. A. Weiner and J. Chun, *J. Annu. Rev. Pharmacol. Toxicol.*, 2001, **349**, 385.

5. I. Ishii, N. Fukushima, X. Ye and J. Chun, *Annu. Rev. Biochem.*, 2004, **73**, 321.
6. S. Mandala, R. Hajdu, J. Bergstrom, E. Quackenbush, J. Xie, J. Milligan and R. Shei, *Science*, 2002, **296**, 346.
7. G. J. Card, D. Koehane, C. Rosenbach, M. Hale, J. Lynch, C. L. Rupprecht, K. Parsons, W. Rosen and V. Brinkmann, *Pharmacol. Ther.*, 2007, **115**, 84.
8. M. Matoublan, C. G. Lo, G. Cinamon, M. J. Leneski, Y. Xu, V. Brinkmann, M. L. Allende, R. L. Proia and J. G. Cyster, *Nature*, 2004, **427**, 355.
9. H. Chi, *J. Immunol.*, 2005, **174**, 2485.
10. R. Tao, H. E. Hoover, J. Zhang, N. Honbo, C. C. Alano and J. S. Karliner, *J. Cardiovasc. Pharmacol.*, 2009, **53**, 486.
11. B. Soliven, V. Miron and J. Chun, *Neurology*, 2011, **76**, S10.
12. J. Chun and H. P. Hartung, *Clin. Neuropharmacol.*, 2010, **33**, 91.
13. E. M. Frohman, M. K. Racke and C. S. Raine, *N. Engl. J. Med.*, 2006, **354**, 942.
14. V. Brinkmann, A. Billich, T. Baumruker, P. Heining, R. Schmouder, G. Francis, S. Aradhye and P. Burtin, *Nat. Rev. Drug. Discov.*, 2010, **9**, 2010.
15. L. Steinman, *Nature Rev. Drug. Discov.*, 2005, **4**, 510.
16. K. Adachi, T. Kohara, N. Nakao, M. Arita, K. Chiba, T. Mishina, S. Sasaki and T. Fujita, *Bioorg. Med. Chem. Lett.*, 1995, **5**, 853.
17. T. Fujita, M. Yoneta, R. Hirose, S. Sasaki, K. Inoue, M. Kiuchi, S. Hirase, K. Adachi, M. Arita and K. Chiba, *Bioorg. Med. Chem. Lett.*, 1995, **5**, 847.
18. K. Chiba, Y. Hoshino, C. Suzuki, Y. Masubuchi, Y. Yamagaura, M. Ohtsuki, S. Sasaki and T. Fujita, *Transplant Proc.*, 1996, **28**, 1056.
19. V. Brinkmann, M. D. Davis, C. E. Heise, R. Albert, S. Cottens, R. Hof, C. Bruns, E. Prieschl, T. Baumruker, P. Hiestand, C. A. Foster, M. Zollinger and K. R. Lynch, *J. Biol. Chem.*, 2002, **277**, 21453.
20. Q. Deng, J. A. Clemas, G. Chrebet, P. Fisher, J. J. Hale, Z. Li, S. G. Mills, J. Bergstrom, S. Mandala, R. Mosley and S. A. Parent, *Mol. Pharm.*, 2007, **71**, 724.
21. F. Mullershausen, F. Zecri, C. Cetin, A. Bilich, D. Guerini and K. Seuwen, *Nature Chem. Biol.*, 2009, **5**, 428.
22. M. H. Graler and E. J. Goetzl, *FASEB J.*, 2004, **18**, 551.
23. M. L. Allende, J. L. Drier, S. Mandala and R. L. Proi, *J. Biol. Chem.*, 2004, **279**, 15396.
24. V. Brinkmann, *Br. J. Pharmacol.*, 2009, **158**, 1173.
25. B. Metzler, P. Gfeller, G. Wieczorek, J. Li, B. Nuesslein-Hildesheim, A. Katopodis, M. Mueller and V. Brinkmann, *Int. Immunol.*, 2008, **20**, 633.
26. P. Kivisakk, *Ann. Neurol.*, 2004, **55**, 627.
27. C. W. Lee, J. W. Choi and J. Chun, *Arch. Pharm. Res.*, 2010, **33**, 1567.
28. V. E. Miron, C. G. Jung, H. J. Kim, T. E. Kennedy, B. Soliven and J. P. Antel, *Ann. Neurol.*, 2007, **63**, 61.
29. J. W. Choi, S. E. Gardell, D. R. Herr, R. Rivera, C. W. Lee, K. Noguchi, S. T. Teo, Y. C. Yung, M. Lu, G. Kennedy and J. Chun, *Proc. Natl Acad. Sci. USA*, 2011, **108**, 751.

30. K. Mizugishi, T. Yamashita, A. Olivera, G. F. Miller, S. Speigel and R. L. Proia, *Mol. Cell. Biol.*, 2005, **25**, 11113.
31. V. Marion, *Am. J. Pathol.*, 2010, **176**, 2682.
32. H. J. Kim, V. E. Miron, D. Dukala, R. L. Proia, S. K. Ludwin, M. Traka, J. P. Antel and B. Soliven, *FASEB J.*, 2011, **25**, 1509.
33. C. A. Foster, D. Mechtcheriakova, M. K. Storch, B. Balatoni, L. M. Howard, F. Bornancin, A. Wlachos, J. Sobanov, A. Kinnuen and T. Baumruker, *Brain Pathol.*, 2009, **19**, 254.
34. B. Balatoni, M. K. Storch, E. M. Swoboda, V. Schoenborn, A. Koziel, G. N. Lambrou, P. C. Hiestand, R. Wiessert and C. A. Foster, *Brain Res. Bull.*, 2007, **74**, 307.
35. L. Kappos, J. Antel and G. Comi, *N. Engl. J. Med.*, 2006, **355**, 1124.
36. G. Comi, P. O'Connor, X. Montalban, J. Antel, E. W. Radue, G. Karlsson, H. Pohlmann, S. Aradhye and L. Kappos, *Mult. Scler.*, 2010, **16**, 197.
37. P. O'Connor, *Neurology*, 2009, **77**, 73.
38. L. Kappos, G. Comi, P. O'Connor, X. Montalban, J. Antel, E. W. Radue, G. Karlsson, H. Pohlmann and S. Aradhye, *N. Engl. J. Med.*, 2010, **362**, 387.
39. J. A. Cohen, *N. Engl. J. Med.*, 2010, **362**, 402.
40. J. Hale, C. Lynch, W. Neway, S. Mills, R. Hajdu, C. A. Koehane, M. Rosenbach, J. Milligan, G. Shei, S. Parent, G. Chrebet, J. Bergstrom, D. Card, M. Ferrer, P. Hodder, B. Strulovici, H. Rosen and S. Mandala, *J. Med. Chem.*, 2004, **47**, 6662.
41. L. Yan, J. Hale, C. Lynch, R. Budhu, A. Gentry, S. G. Mills, R. Hajdu, C. A. Keohane, M. J. Rosenbach, J. A. Milligan, G. Shei, G. Chrebet, J. Bergstrom, D. Card, H. Rosen and S. M. Mandala, *Bioorg. Med. Chem. Lett.*, 2004, **14**, 4861.
42. Z. Li, W. Chen, J. J. Hale, C. Lynch, S. G. Mills, R. Hajdu, C. A. Keohane, M. J. Rosenbach, J. A. Milligan, G. J. Shei, G. Chrebet, S. A. Parent, J. Bergstrom, D. Card, M. Forrest, E. Quackenbush, L. A. Wickham, H. Vargas, R. M. Evans, H. Rosen and S. Mandala, *J. Med. Chem.*, 2005, **48**, 6169.
43. L. Yan, P. Huo, G. Doherty, L. Toth, J. J. Hale, S. G. Mills, R. Hajdu, C. A. Keohane, M. J. Rosenbach, J. A. Milligan, G. J. Shei, G. Chrebet, J. Bergstrom, D. Card, E. Quackenbush, A. Wickham and S. M. Mandala, *Bioorg. Med. Chem. Lett.*, 2006, **16**, 3679.
44. E. H. Demont, B. I. Andrews, R. A. Bit, C. A. Campbell, J. Cooke, N. Deeks, S. Desai, S. J. Dowell, P. Gaskin, J. Gray, A. Haynes, D. Holmes, U. Kumar, M. A. Morse, G. J. Osborne, T. Panchal, B. Patel, A. Perboni, S. Taylor, R. Watson, J. Witherington and R. Wills, *ACS Med. Chem. Lett.*, 2011, **2**, 444.
45. E. Demont, S. Arpino, R. A. Bit, C. A. Campell, N. Deeks, S. Desai, S. J. Dowell, P. Gaskin, J. Gray, L. A. Harrison, A. Haynes, T. D. Heightman, D. S. Holmes, P. G. Humphreys, U. Kumar, M. A. Moore, G. J. Osborne, T. Panchal, K. L. Philpott, S. Taylor, R. Watson, R. Willis and J. Witherington, *J. Med. Chem.*, 2011, **57**, 6724.
46. S. A. Hitchcock and L. D. Pennington, *J. Med. Chem.*, 2006, **49**, 7579.

47. S. Shacham, Y. Marantz, S. Bar-Haim, O. Kalid, D. Warshaviak, N. Avisar, B. Indal, A. Heifetz, M. Fichman, M. Topf, Z. Noar, S. Noiman and O. M. Becker, *Proteins*, 2004, **57**, 51.

48. A. Saha, X. Yu, J. Lin, M. Lobera, A. Sharadendu, S. Chereku, N. Schutz, D. Segal, Y. Marantz, D. McCauley, S. Middleton, J. Siu, R. Burli, J. Buys, M. Horner, K. Salyers, M. Schrag, H. Vargas, Y. Xu, M. McElvain and H. Xu, *ACS Med. Chem. Lett.*, 2011, **2**, 97.

49. B. Lanman, V. J. Cee, S. R. Cheruka, M. Frohn, J. Golden, J. Lin, M. Lobera, Y. Marantz, K. M. Muller, S. C. Neira, A. J. Pickrell, D. Rivenzon-Segal, N. Schutz, A. Sharadendu, X. Yu, Z. Zhang, J. Buys, M. Fiorino, A. Gore, M. Horner, A. Itano, M. McElvain, S. Middleton, M. Schrag, H. M. Vargas, H. Xu, Y. Xu, X. Zhang, J. Sui and R. W. Burli, *ACS Med. Chem. Lett.*, 2011, **2**, 102.

50. V. J. Cee, M. Frohn, B. Lanman, J. Golden, K. M. Muller, S. C. Neira, A. Pickrell, H. Arnett, J. Buys, M. Fiorino, A. Gore, M. Horner, A. Itano, M. Lee, M. McElvain, S. Middleton, M. Schrag, D. Rivenson-Segal, H. M. Vargas, H. Xu, Y. Xu, X. Zhang, J. Sui, M. Wong and R. W. Burli, *ACS Med. Chem. Lett.*, 2011, **2**, 107.

51. L. D. Pennington, K. Sham, A. J. Pickrell, P. E. Harrington, M. J. Frohn, B. A. Lanman, A. B. Reed, M. D. Croghan, M. R. Lee, H. Xu, M. McElvain, Y. Xu, X. Zhang, M. Fiorino, M. Horner, H. G. Morrison, H. A. Arnett, C. Fotsch, M. Wong and V. J. Cee, *ACS Med. Chem. Lett.*, 2011, **2**, 752.

52. H. Shimizu, M. Takahashi, T. Kaneko, T. Murakami, Y. Hakamata, S. Kudou, T. Kishi, K. Fukuchi, S. Iwanami, K. Kuriyama, T. Yasue, S. Enosawa, K. Matsumoto, I. Takeyoshi, Y. Morishita and E. Kobayashi, *Circulation*, 2005, **111**, 222.

53. T. Nishi, S. Miyazaki, T. Takemoto, K. Suzuki, Y. Iio, K. Nakajima, T. Ohnuki, Y. Kawase, F. Nara, S. Inaba, T. Izumi, H. Yuita, K. Oshima, H. Doi, R. Inoue, W. Tomisato, T. Kagari and T. Shimozato, *ACS Med. Chem. Lett.*, 2011, **2**, 368.

54. J. Moberly, S. Rohatagi, H. Zahir, C. Hsu, R. J. Noveck and K. E. Truitt, *J. Clin. Pharmacol.*, 2011, *in press*, doi:10.1177/0091270011408728.

55. R. Zhu, A. H. Snyder, Y. Kharel, L. Schaffer, Q. Sun, P. C. Kennedy, K. R. Lynch and T. L. Macdonald, *J. Med. Chem.*, 2007, **50**, 6428.

56. M. G. Sanna, S. K. Wang, P. J. Gonzalez, A. Don, D. Marsolais, M. P. Matheu, S. H. Wei, I. Parker, E. Jo, W. Cheng, M. D. Cahalan, C. H. Wong and H. Rosen, *Nat. Chem. Biol.*, 2006, **2**, 434.

57. M. D. Davis, J. J. Clemens, T. L. Macdonald and K. R. Lynch, *J. Biol. Chem.*, 2005, **280**, 9833.

58. F. W. Foss, A. H. Snyder, M. D. Davis, M. Rouse, M. D. Okusa, K. R. Lynch and T. L. Macdonald, *Bioorg. Med. Chem.*, 2007, **15**, 663.

59. P. Gergely, *Mult. Scler.*, 2009, **15**, S125.

60. K. Cusack and R. H. Stoffel, *Curr. Opin. Drug Discovery Dev.*, 2010, **13**, 481.

61. R. Hohlfeld, F. Barkhof and C. Polman, *Neurology*, 2011, S28.

CHAPTER 15

Tipping the Balance of Sphingosine 1-Phosphate Production: Sphingosine Kinases and Sphingosine 1-Phosphate Lyase as Immune Therapeutic Targets

TAMAS ORAVECZ*[a] AND DAVID AUGERI[b,†]

[a] Lexicon Pharmaceuticals, Inc., 8800 Technology Forest Pl., The Woodlands, TX 77389, USA; [b] 350 Carter Rd, Princeton, NJ 08540, USA
*Email: toravecz@lexpharma.com

15.1 Introduction

Sphingolipid metabolism involves the synthesis and degradation of a number of structurally similar molecules that function as building blocks of cell membranes as well as signalling molecules. Sphingosine 1-phosphate (S1P) is the penultimate breakdown product of that metabolism.[1–5] It is present in all mammalian cells and can serve as a second messenger in signal transduction pathways which regulate cell differentiation, proliferation and apoptosis. S1P is

---

†Current address: 107 Carter Rd., Princeton, NJ 08540, USA.

RSC Drug Discovery Series No. 26
Anti-Inflammatory Drug Discovery
Edited by Jeremy I Levin and Stefan Laufer
© The Royal Society of Chemistry 2012
Published by the Royal Society of Chemistry, www.rsc.org

also released into the extra-cellular milieu through transporters present on a variety of cell types, making it one of the most abundant biologically active lysophospholipids in circulation.[2] Extra-cellular S1P is an agonist ligand of five different G-protein coupled receptors, designated $S1PR_1$–$S1PR_5$, which are differentially expressed on various cell types. Autocrine and paracrine interactions between secreted S1P and its receptors are coupled with downstream signal transduction events, which can modulate a wide range of physiological activities and disease states, including cardiovascular development and disease, neuronal cell survival, cancer development and immunity. The autocrine signalling action of secreted S1P through its receptors is termed "inside-out" signalling, which differentiates it from the biochemical events in which S1P takes the role of a secondary messenger.[3]

In the immune system, changes in intra-cellular and extra-cellular S1P concentrations and gradients can modify lymphocyte migration patterns, alter inflammatory cell responses and affect the barrier function of endothelial cells. The main sources of S1P are erythrocytes in the plasma and lymphatic endothelial cells in the lymph. Perivascular cells in the thymus can also produce S1P, which promotes egress of thymocytes across blood vessel endothelium.[5] In addition, platelets release large amounts of S1P during the coagulation process. Under normal circumstances the pattern of S1P production throughout the immune system is such that it generates a concentration gradient between low S1P content in lymphoid tissues and high S1P concentration in the peripheral circulation. This chemical gradient can serve as a chemotactic trail for directional cell migration, and in fact it is a major regulator of lymphocyte egress and crucial for proper recirculation of lymphocytes from the peripheral lymphoid organs to the blood stream. Agonistic engagement of one of the S1P receptors on lymphocytes, $S1PR_1$, is the signal for this chemotaxis. However, S1P is an agent of action and counter-action in this process. In contrast to its egress-promoting action on lymphocytes, engagement of $S1PR_1$ on endothelial cells is thought to restrain cell migration by increasing the barrier between the tissue environment and circulation, thereby closing the gate for trans-endothelial cell migration. Furthermore, ligand binding to $S1PR_1$ not only initiates signal transduction events but can also lead to down-regulation and intra-cellular degradation of the receptor. Synthetic agonists of $S1PR_1$ can induce prolonged down-regulation and degradation of the receptor resulting in deficient receptor function. The latter effect of synthetic agonists is described as "functional antagonism".

Since S1P-mediated cellular traffic is dependent on proper sensing of an S1P gradient and requires open gates through the endothelial cell layer, dysregulation of $S1PR_1$ expression or disruption of the gradient by changes in S1P levels can lead to retention of cells within lymphoid tissues. Absence of functional $S1PR_1$ signalling eventually leads to inhibition of lymphocyte egress from primary and secondary lymphoid tissues, resulting in peripheral lymphopenia and immunosuppression. The modulation of immune function achieved by systemic redistribution of lymphocytes and by S1P's effect on cell activation state offers new opportunities for developing agents to treat autoimmune and inflammatory diseases. The S1P pathway has raised particular interest as an immune therapeutic modality after the phosphorylated metabolite of FTY720 (FTY720-P), an S1P receptor agonist, showed remarkable efficacy in various *in-vivo* models of immune disease. FTY720 (fingolimod, Gilenya®), developed by Novartis, completed successful clinical trials leading to its regulatory approval for the treatment of multiple sclerosis.[6] While agonists and antagonists of the S1P receptor family, as discussed in the previous chapter, have been the subject of recent pharmaceutical discovery focus in the anti-inflammatory arena, enzymes of the S1P metabolic pathway may provide alternative intervention points for therapeutic applications.

15.2   Sphingosine 1-Phosphate Homeostasis

Endogenous levels of S1P can be directly influenced by modulating the activity of the enzymes involved in its synthesis and degradation (Figure 15.1).[7–10] Sphingosine, derived from the breakdown of ceramide, is phosphorylated by two sphingosine kinases (SK), SK1 and SK2, producing S1P. Five phosphatases have been described that are capable of catalyzing the dephosphorylation of S1P. Three of these enzymes belong to a class of lipid phosphate phosphatases (LPP1-3) with broad substrate specificities, while the other two phosphatases, S1P phosphatase (SPP) 1 and 2, are S1P-specific. Irreversible degradation of S1P is carried out by a single enzyme, S1P lyase (S1PL), which cleaves S1P into ethanolamine phosphate and a long-chain aldehyde.[1,2,7–10]

Sphingolipid-metabolizing enzymes are subject to multiple levels of control. Although widely expressed, relative levels of individual proteins vary by tissue, cell type and subcellular localization. Furthermore, their distribution between

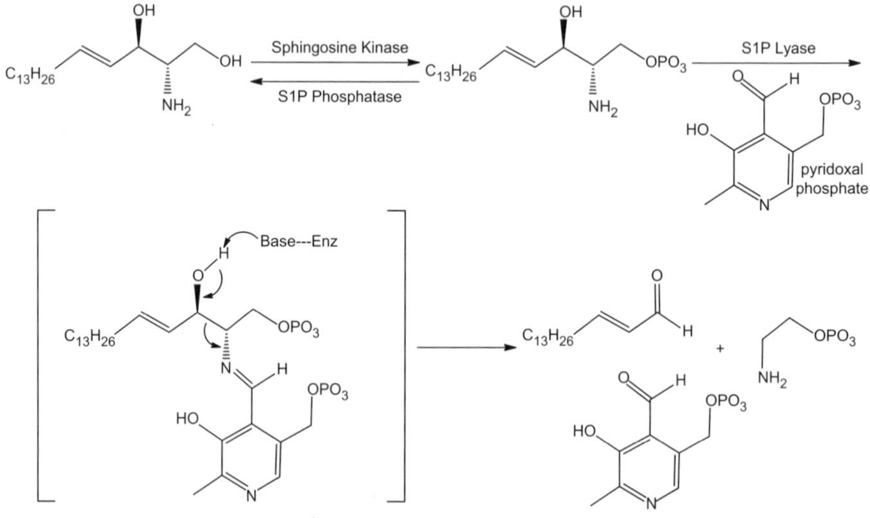

Figure 15.1   S1P metabolism.

the cytosol, endoplasmic reticulum, plasma membranes and the nucleus is regulated by signalling events and cell activation status. For example, SK1 is a cytosolic enzyme but can be translocated to the plasma membrane upon cell activation-induced phosphorylation.[11–13] SK2 is relatively separated from SK1 inside the cell. It resides in the nucleus but can be transported to the cytoplasm after phosporylation.[14,15] On the other hand, the S1P-degrading phosphatases and S1PL all reside in the ER.[16–18] Relative compartmentalization of enzyme activities may generate spatially separated and stoichiometrically independent intra-cellular pools of S1P and its degradation products. Such segregation of S1P metabolism and distribution of S1P inside the cell can be essential in determining the outcome of S1P production and guide the fate of cellular response to extra-cellular stimuli. As an example, S1P can be discharged easily from the cell before being degraded by phosphatases, or S1PL in the ER, if it is produced by SK1 which is translocated to the plasma membrane. It is not surprising, therefore, that the differentially expressed and locally separated SK1 and SK2 enzymes display distinct, and frequently opposite, regulatory functions in a number of processes that govern inflammation and cell apoptosis.[7,8,19]

In addition to variations in the expression pattern of S1P-metabolizing enzymes, different cell types have different capacities to discharge S1P stores into the extra-cellular environment and circulation. More than 90% of S1P found in the circulation is either in HDL and LDL particles or bound to albumin.[20] The HDL-bound S1P has been shown to possess biological activity.[20] It has been proposed that HDL may allow presentation of S1P to receptors adjacent to HDL receptors on the cell surface.[20] Thus, interaction of S1P with plasma carriers can add an additional layer of regulatory mechanism to S1P physiology. Finally, enzymes of the S1P metabolic pathway may also have secreted isoforms, which could contribute to S1P homeostasis in the plasma or the interstitial environment.[21–23]

It is plausible to assume that tight regulation of local and systemic S1P homeostasis by a multitude of factors has evolved to make efficient use of this ubiquitous cellular product, and enable S1P to exert pleiotropic effects and play diverse roles in mammalian physiology. At the same time, such complexity makes it challenging to reconcile *in-vitro* and *ex-vivo* experimental data on the function of S1P with *in-vivo* results that were obtained with animals displaying altered S1P metabolism as a result of genetic engineering or pharmaceutical treatment.

15.3 The Sphingosine 1-Phosphate Network and Immune Regulation

Certainly for immunologists, complexity in a physiological network is not a particular surprise or deterrent. Indeed, the view is emerging that S1P and its metabolizing enzymes act as both inducers as well as inhibitors of inflammation in a context-dependent manner, which places S1P in the company of other secreted factors that display a Yin-Yang role in the inflammatory process.

Fortunately, the use of a plethora of genetic and pharmacological tools, ranging from small inhibitory RNA (siRNA) and genetically engineered animals to pharmacological enzyme inhibitors and S1P mimics, has facilitated research in the field, which had produced remarkable advances in recent years. In the next sections we will provide an overview of the current experimental evidence on how S1P, SK1, SK2 and S1PL steer the inflammatory processes in two opposite directions.

15.3.1 The Pro-inflammatory Function of the Sphingosine 1-Phosphate Pathway

The largest body of evidence for a pro-inflammatory role of the S1P network is centred around a fundamental signal transduction event that is required to mount a successful inflammatory response at the site of bacterial infection: engagement of toll-like receptors (TLR) by bacterial products, including lipopolysaccharide (LPS). TLR-induced innate immunity is mediated by signalling cascades leading to the activation of the mitogen-activated protein kinase (MAPK) family of Ser/Thr protein kinases, including p38 MAPK, which controls cytokine release during innate and adaptive immune responses through the activation of multiple transcription factors. Failure to terminate such inflammatory reactions may lead to detrimental systemic effects, including septic shock and autoimmunity.

In studies of TLR signalling, S1P is required for an optimal inflammatory response under various experimental conditions. This pro-inflammatory S1P is thought to be produced primarily by the action of SK1 on sphingosine. SK1 message and activity is increased during inflammatory challenges, including LPS challenge of macrophages and microglia, and S1P content is also increased after TLR2 and TLR4 signalling in phagocytes.[22,24] Secretion of tumour necrosis factor (TNF)-α is a hallmark of TLR activation and induction of SK1 by TNF-α has also been documented.[25–29] The signal transduction event initiated by TNF-α requires the assembly of a multi-component signalling complex, which incorporates a ubiquitin ligase, TNF receptor-associated factor 2 (TRAF2). The ubiquitin ligase activity of TRAF2 is required for efficient signal transduction and the enzyme uses S1P as a cofactor, which is supplied by SK1.[26,30] In endothelial cells, S1P also mediates TNF-α-induced expression of cell adhesion molecules.[25,31] Accordingly, adhesion of monocytes to endothelial cells is attenuated by N,N-dimethylsphingosine (DMS), a non-specific inhibitor of SK, which reduces the expression of adhesion molecules.[25] Bacterial LPS also induces the pro-inflammatory enzyme cyclooxygenase-2 and its main product prostaglandin E2 (PGE2), which is an SK1/S1P-dependent response as well.[7,32] Finally, SK1 also mediates the activity of another inflammatory mediator, the endothelial nitric oxide synthase.[33]

Extreme systemic elevation of S1P levels, as observed in S1PL KO mice, also leads to neutrophilia, increased pro-inflammatory cytokine levels and hypersensitivity to LPS-challenge.[34] However, these pro-inflammatory effects have been described only after complete knockout of S1PL activity. Reconstitution of as low as 10% of S1PL activity by genetic knockin methods is sufficient to protect mice from exacerbated inflammation, even though S1P levels are still significantly elevated.[35] Furthermore, incomplete inhibition of S1PL activity, both by genetic methods and by pharmacological inhibition, decreases rather than increases inflammatory responses *in vivo*.[36–38]

Additional hallmarks of systemic inflammation are blood coagulation and activation of the complement system. During coagulation, the protease thrombin activates the protease-activated receptor (PAR) on dendritic cells, which results in SK1 activation and production of S1P. In turn, S1P signals through S1PR$_3$ (a good example of inside-out signalling), promoting cell migration and disseminated inflammation in severe sepsis.[39] Thrombin also induces endothelial permeability and secretion of the inflammatory mediator PGE2, and both of these processes are blocked by non-specific SK inhibitors.[40,41] Production of PGE2, as well as MCP-1, IL-6 and IL-8 is also inhibited by siRNA of SK1 or expression of a dominant-negative SK1 mutant in lung epithelial cells.[40] In the complement system, the anaphylatoxin C5a acts as a chemoattractant and modulates cytokine production in various cell types. Antisense knockdown of SK1 in human macrophages inhibits C5a receptor-mediated cytokine production and chemotaxis.[42] The therapeutic relevance of the above *in-vitro* findings on the pro-inflammatory function of S1P, primarily generated by SK1, is reinforced by the results of *in-vivo* studies. For example, deletion of SK1 in mice prevented sepsis after LPS challenge.[39]

The contribution of the S1P pathway to allergic reactions is another intense area of investigation. Asthma is one of the most prevalent allergic diseases, and S1P levels are increased in the lungs of asthma patients compared to control subjects.[43] Crosslinking of the high-affinity Fc receptor for IgE (FcεRI) on mast cells is an initiator of allergic reactions and SK activity is important for mast cell activation. Silencing of SK1 by siRNA indicated a dominant role for SK1 in driving mast cell responses.[44,45] Similarly, SK1 expression was required for optimal asthma response in mice challenged with ovalbumin.[46] However, knockout (KO) mouse studies provided contrasting results. One study concluded that SK2 was required for the FcεRI response[47] while another found normal mast cell responses both in SK1-null and in SK2-null strains.[44] Of note, studies in other areas of investigation also found discrepancies in data obtained with KO animals of the two kinases, which may indicate that these mice developed adaptive functional redundancy during embryonic development. Certainly, the single KO animals of SK1 and SK2 develop normally and show normal S1P content in tissues, while the double KO is an embryonic lethal phenotype due to incomplete neurogenesis and vascular development, and lack of systemic S1P.[48,49]

Immune challenges that are models for autoimmune diseases have provided clear evidence that SK1 and SK2 play distinct roles in immune regulation (Table 15.1). Individual knockdown of the two kinases with siRNA resulted in opposite effects on arthritis development in mice challenged with collagen: blockade of SK1 expression offered resistance while silencing SK2 exacerbated the disease.[50] Although data obtained with KO animals did not match these results initially (both KOs showed normal response to arthritic challenge),[51] the

Table 15.1 Effect of genetic inhibition of SK1 and SK2 on inflammatory responses. ↓, decreased response; ↑, increased response; NR, not reported.

	Inflammatory signal	Inflammatory response	Effect of kinase siRNA or KO	
			SK1	SK2
In-vitro studies	TLR ligands	Induction of COX-2, PGE-2	↓ [7,32]	No effect [26]
	TLR ligands	Septic response	↓ [39]	NR
	Thrombin	S1PR$_3$-mediated dendritic cell migration	↓ [39]	NR
	Thrombin	Production of pro-inflammatory mediators	↓ [40]	NR
	Th1 cell differentiation signals	Cytokine production by Th1 cells	↑ [59]	No effect [59]
	N/A	Histone acetylation	No effect [80]	↓ [80]
In-vivo studies	FcεRI cross-linking	Mast cell degranulation	↓ [44,45]; No effect [44]	↓ [47]; No effect [44,45]
	Ovalbumin	Asthma response	↓ [46]	NR
	Collagen	Arthritis development	↓ [50]; No effect [51]	↑ [50]; No effect [51]
	Transgenic TNF-α	Arthritis development	↓ [52]	NR
	DSS	Susceptibility to IBD	↓ [32]	NR
	Adoptive transfer of T-cells	Susceptibility to IBD	NR	↑ [58]
	Dinitrophenyl-IgE	Anaphylactic response	↑ Poor recovery [61]	↓ Rapid recovery [61]

discrepancy can be the result of functional compensation as noted above, or from differences in assay design. Indeed, SK1 KOs on the transgenic human TNF-α background showed decreased arthritis development in another study.[52] In addition, fibroblast-like synoviocytes obtained from SK1 KO mice produced fewer pro-inflammatory mediators than wild-type littermates, which is consistent with a pro-inflammatory role of SK1 in arthritis.[53] There is also an indication that synoviocytes from patients with rheumatoid arthritis (RA) proliferate and produce cytokines when exposed to S1P,[54] and that S1P and SK1 are elevated in the arthritic synovium.[55,56] However, SK2 protein levels were also elevated in synovial fibroblasts of a group of RA patients.[57] This observation is counter to the potential protective role of SK2 in RA that was indicated by the animal studies.[50] Of note, the ability of TNF-α to induce SK2 activation and S1P formation has been documented for lymphocytes as well as macrophages, neutrophils and endothelial cells, but it is still an important open question whether or not this is a feedback response elicited to provide a restraining order for inflammation *via* a mechanism that is different from the action of SK1-derived S1P.

Studies in mouse models of another autoimmune disorder, inflammatory bowel disease (IBD), suggested a similar functional dichotomy in SK activities (Table 15.1). Ablation of SK1 rendered mice resistant to IBD development[32] while SK2 KO T-cells induced more robust IBD in immunocompromised recipients.[58] The effect of SK modulation on the development of autoimmune diseases, including IBD, may be related partly to the described, although controversial, role of SK1 in regulating $CD4^+$ T helper type-1 (Th1) cells. Silencing of SK1 but not SK2 by siRNA enhanced the cytokine production of Th1 cells, and over-expression of SK1 had the opposite effect *in vitro*, indicating that SK1 is a suppressor of Th1 cell differentiation.[59] However, experiments with SK inhibitors, including the non-specific inhibitor DMS and the relatively SK1-specific SKI-II, did not lead to the same conclusion since they blocked the differentiation of pro-inflammatory Th1 cells and at the same time promoted the development of the anti-inflammatory Foxp3 + regulatory T-cell subset in mice transgenic for the S1P receptor $S1PR_1$.[60]

The bulk of the *in-vitro* and *in-vivo* data summarized in the previous paragraphs have been obtained in studies where the readouts of immune function were tailored to address the role of the S1P pathway in immune cell activation and migration. Modulating the pathway in the context of endothelial barrier function may have a different overall effect on inflammation. This role of S1P has been explored in mouse models of anaphylaxis and acute lung injury (ALI), which are associated with inflammation and severe vascular leakage. Both of these models provided additional examples to the list of animal models where SK1 and SK2 display differential function (Table 15.1). Mice deficient for expression of SK2 showed rapid recovery from anaphylaxis while SK1 KOs had poor recovery.[61] In the ALI model, adenoviral over-expression of SK2 augmented vascular leakage and the degree of LPS-induced lung injury, while delivery of SK1 had the opposite outcome.[62] The effect of SK-produced S1P on the endothelial barrier is thought to be mediated by binding of secreted S1P to receptors on the surface of endothelial cells. One study in the ALI model addressed the concentration-dependence of systemic S1P administration on vascular permeability.[63] Intra-tracheal or intravenous (i.v.) administration of S1P at a dose below 0.3 mg/kg protected mice against LPS-induced lung inflammation and permeability while higher doses (above 0.5 mg/kg) resulted in significant alveolar-capillary barrier disruption. These findings may be explained by S1P interacting with different S1P receptors at different concentrations (*e.g.* affinity-driven interactions), which may regulate the endothelial cell barrier in opposite ways. They may also reflect S1P's contrasting actions since S1P acts as a receptor agonist at low concentrations but can behave as a functional antagonist at high doses due to extensive receptor down-regulation.

In summary, a vast amount of experimental data supports the rationale for the development of SK inhibitors to suppress the inflammation-promoting effect of S1P. Challenges may arise in finding the right ratio of specificity between SK1 and SK2 in order to maximize the compounds' anti-inflammatory effect and to avoid the counteracting activities stemming from the opposing functional roles of the two enzyme isoforms (Table 15.1) and of S1P itself. We present further details of these contrasting activities in the next section where we focus on the role of the S1P pathway in suppressing inflammation.

15.3.2 The Anti-inflammatory Function of the Sphingosine 1-Phosphate Pathway

As discussed in the previous section, the production of S1P is increased in a number of inflammatory conditions, which contributes to the pro-inflammatory response. However, evidence is mounting that S1P can also act as an anti-inflammatory agent to counteract and restrain pro-inflammatory processes. Also, changes in the expression of S1P regulating enzymes don't always predict increased S1P levels during an inflammatory response. For instance, treatment of HUVEC cells with TNF-α increased the message of the S1P-degrading enzymes SPP2 and S1PL and not that of the S1P-producing SK1.[64] In fact, expression of SK2 was somewhat reduced in the same study. Our own unpublished data indicate that LPS challenge decreased the expression of SK2 message in the mouse liver, and ovalbumin challenge also suppressed SK2 expression in the asthmatic lung of mice. In humans, an increased level of S1PL mRNA was detected in atopic dermatitis lesions[65] and in the brain of Alzheimer's patients.[66]

We have indicated that the role of SK2 was found to be the opposite of SK1 in certain challenge models associated with inflammation, and inhibition of one or the other isoforms may aggravate rather than alleviate inflammation in arthritis,[50] inflammatory bowel disease,[32,58] anaphylactic shock[61] and LPS-induced lung injury (Table 15.1).[62] In addition, genetic and pharmaceutical inhibition of S1PL *in vivo* indicated that increased S1P content was associated with significant reduction in the ability of the animals to mount an inflammatory response when challenged.[35–38] Importantly, after dosing of rats with an inhibitor of S1PL activity, LX2931, the increase in S1P levels was not systemic, but localized: lymphoid organs registered the largest increases in S1P content (Figure 15.2).[35,37] For instance, the effect of S1PL inhibition on S1P content is relatively specific to lymphoid tissues, which suggests that S1PL plays a more significant role in controlling S1P homeostasis in immune tissues than in other tissues. Therefore, partial inhibition of S1PL activity may produce potentially therapeutic anti-inflammatory effects without other overt physiological effects.[35–37] Based on promising pre-clinical data, the orally delivered LX2931 became the first clinically studied compound targeting S1P production. This compound has been developed by Lexicon Pharmaceuticals, Inc., initially for the treatment of rheumatoid arthritis.[9,36,37] In Phase I clinical trials LX2931 was well tolerated and produced a dose-dependent and reversible reduction of circulating lymphocytes.[37]

Another disease where inhibition of S1PL and increased S1P levels may be protective is cerebral malaria (CM), the life-threatening complication of

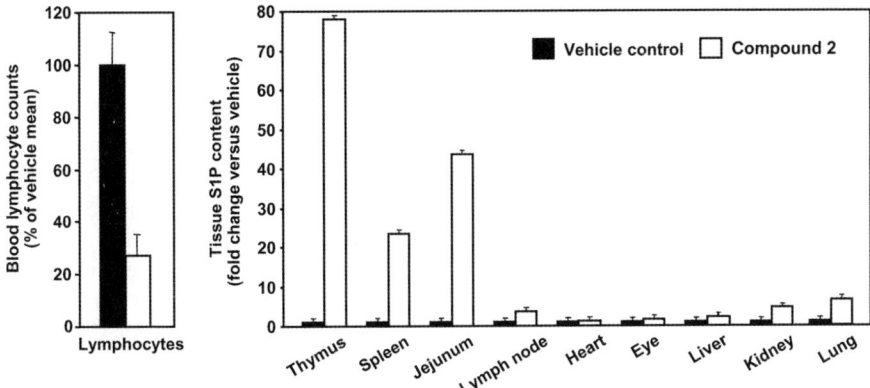

Figure 15.2 The lymphoid system is the primary target of compound **2** activity. Male Sprague-Dawley rats were treated with **2** at 30 mg/kg dose or with vehicle orally q.d. for 3 days (n = 10 each cohort). Blood and tissues were harvested 18 h after final dose.[35,37]

malaria infection. Studies in both humans and animals have demonstrated that dysregulated inflammatory responses to malaria infection and their effects on vascular endothelium play a central role in disease progression and outcome in CM. Mice deficient for S1PL activity or treated with LX2931 showed improved survival in an experimental model of CM compared to wild-type littermates.[38] The protective effect of LX2931 was associated with a significantly decreased concentration of IFNγ in the plasma. Importantly, these results were coupled with human biomarker data, showing that plasma S1P concentration is decreased in children with cerebral malaria compared to those with uncomplicated malaria.[38]

To some extent, the above anti-inflammatory effects of S1P are the consequence of disrupted immune cell migration by neutralization of S1P gradients or functional antagonism of $S1PR_1$ on immune cells, or by closing of the endothelial cell barrier. Indeed, genetic or pharmacologic inhibition of S1PL blocked the deployment of T- and B-cells from immune tissues into the peripheral blood supply.[35–37,67] Local application of S1P by inhalation also blocked the migration of dendritic cells in the lung and suppressed the cardinal features of asthma.[68]

The protective effect of S1P in inflammation *via* promoting vascular integrity is well described. Administration of S1P to mice is protective against LPS-induced lung injury and edema.[69,70] Intra-tracheal instillation of LPS to mice enhanced expression of S1PL and decreased S1P levels in lung tissue. Accordingly, inhibition of S1PL activity in heterozygote S1PL-deficient mice, which increased S1P levels, decreased the amount of protein exudates and interleukin (IL)-6 in the BAL fluid after lung injury, and protected from the challenge.[71] Conversely, deficiency in S1P production, for example by the inhibition of SK activities, induces vascular leakage. Mice engineered to selectively lack S1P in plasma due to mutations in SK enzymes displayed increased vascular leakage and impaired survival after inflammatory challenges, including anaphylaxis.[72] Addition of S1P reversed this effect in SK1 KO mice.[61]

Several studies highlight the functional separation and spatial-temporal regulation of S1P production by the two different SK isoforms, which may differentially affect the inflammatory response and vascular permeability. Thus, SK1 and SK2 expression is temporally separated in the LPS-induced ALI model. SK1 expression was increased in the lung within 6 hours and declined back to baseline levels by 24 hours after intra-tracheal LPS treatment. In comparison, SK2 expression was delayed and gradually elevated during the same 24-hour time period.[62] In the anaphylaxis model, deletion of SK1, but not SK2, alleviated the mast-cell dependent onset of the reaction.[47,61] However, SK2 KO mice recover faster from anaphylaxis than wild-type mice, while SK1 KO mice are more sensitive to the challenge.[61]

The mechanism behind the anti-inflammatory activity of increased S1P content is not limited to tightening of the endothelial barrier and blocking directional cell migration. Elevated S1P levels may also dampen inflammatory responses by inhibiting production of pro-inflammatory mediators, blocking cell adhesion and modulating immune cell differentiation. For instance, S1P reduced LPS-induced secretion of pro-inflammatory cytokines in bone marrow-derived macrophages[73] and blocked TNF-α-mediated adhesion of monocytes to endothelium.[74–76] Moreover, knockdown of S1PL by siRNA increased S1P concentration in human lung endothelial cells and attenuated LPS-mediated phosphorylation of p38 MAPK and IL-6 secretion.[71] The role of S1P in regulating Th1/Th2 cell differentiation that was mentioned in the previous section is further supported by studies on maturing dendritic cells. Extra-cellular S1P inhibited the secretion of TNF-α and IL-12, whereas it enhanced secretion of IL-10. As a consequence, S1P-treated mature dendritic cells skewed T-cell differentiation toward a Th2-type of response.[77]

A number of different signalling events have been suggested to govern the anti-inflammatory effects of S1P. One report indicated that binding of S1P to one of the S1P receptors, S1PR$_4$, can induce immunosuppressive effects in murine T-cell lines by inhibiting cell proliferation and modifying cytokine secretion patterns.[78] This finding was not supported by another report which studied S1PR$_4$ KO T-cells, but rather implicated dendritic cell-expressed S1PR$_4$ as the main receptor responsible for altered immune responses in different disease models.[79] The absence of the S1P-S1PR$_4$ axis in S1PR$_4$ KO mice increased Th2 cytokine dominated disease activities while diminishing the Th1-type response and the differentiation of the regulatory Th17 cell subset.

The most intriguing biochemical mechanism described to date associates the anti-inflammatory activity of S1P with the inhibition of histone deacetylases (HDAC). A plethora of pro-inflammatory genes are induced by HDACs, which regulate chromatin condensation. Histone H3, which is a target of HDAC 1 and 2, was shown to associate with SK2. S1P produced by SK2 can bind to and inhibit HDAC activity, which represses transcription of pro-inflammatory mediators.[80]

Finally, it is important to note that although the majority of studies assumed a role for S1P in the SK-mediated processes, one study suggested that the anti-inflammatory role of SK1 was independent of the production of S1P. That study showed that SK1 stabilizes Jun kinase (JNK) and in turn inhibits its binding to the JNK-interacting protein 3 and abrogates NF-κB activation *via* an S1P-independent pathway.[81]

Thus, review of the current scientific evidence in the inflammation area supports the concept that the physiological outcome of changes in S1P concentrations is spatio-temporally regulated and source- and context-dependent. The rationale exists to either decrease or increase S1P levels for particular therapeutic applications *via* modification of SK and S1PL activities. Compounds that target S1P production have been developed and are at various stages of pre-clinical and clinical development as anti-inflammatory and anti-neoplastic agents. In the subsequent sections, we will give an outline of the medicinal chemistry efforts in this area and review the *in-vivo* results obtained with the compounds in various animal models of inflammatory diseases.

15.4 Medicinal Chemistry Approaches to Inhibit Sphingosine 1-Phosphate Lyase

Human S1PL is an intra-cellular protein composed of 568 amino acids. It bears 91% similarity to its mouse homologue. The N-terminus is anchored in the membrane-rich endoplasmic reticulum (ER). The C-terminus contains the catalytic site that extends into the cytosol and performs an irreversible retro-aldol degradation of S1P. A recent study reported the crystal structure of yeast S1PL and its prokaryotic homologue from *Symbiobacterium thermophilum*.[10] The yeast S1PL structure was useful for modelling human S1PL and further studies with S1PL mutants helped to identify residues involved in substrate binding. The most recent effort to inhibit S1PL has focused on small-molecule compounds that have achieved *in-vivo* efficacy in animal models and has ultimately led to the discovery of LX2931[37] that has progressed to human clinical trials. A discussion of these small-molecule inhibitors will be followed by a discussion of compounds that showed *in-vitro* activity but were inactive *in vivo*.

The immunosuppressant activity of the food colorant 2-acetyl-4(5)-(1(*R*),2(*S*),3(*R*),4-tetrahydroxybutyl)-imidazole (THI, **1**) had been well documented during the late 1980s and early 1990s (Figure 15.3).[67,82,83] In 2005, Schwab and Cyster[67] published results suggesting that the pharmacologic target of **1** was S1PL. Experimental results showed that the administration of **1** in drinking water correlated to lymphopenia concomitant with increased concentrations of S1P in lymphoid tissues such as spleen and thymus. Thymus levels showed increases in S1P as high as 1000 times above normal levels as measured by LC/LC-MS. Pharmacology observed following the administration of **1** closely resembled the phenotype observed after partial deficiency of S1PL gene expression in genetically modified mice, showing lymphopenia with

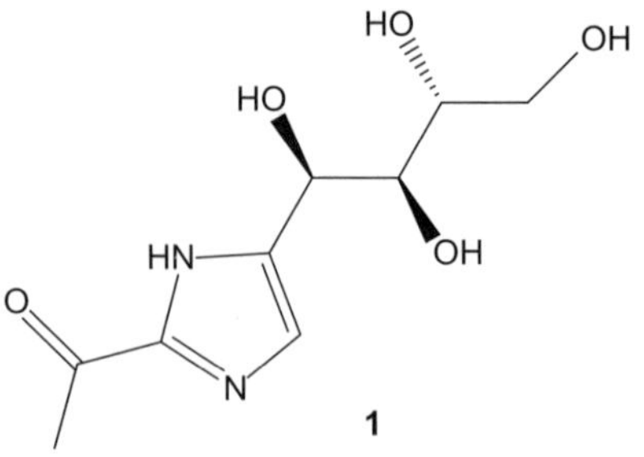

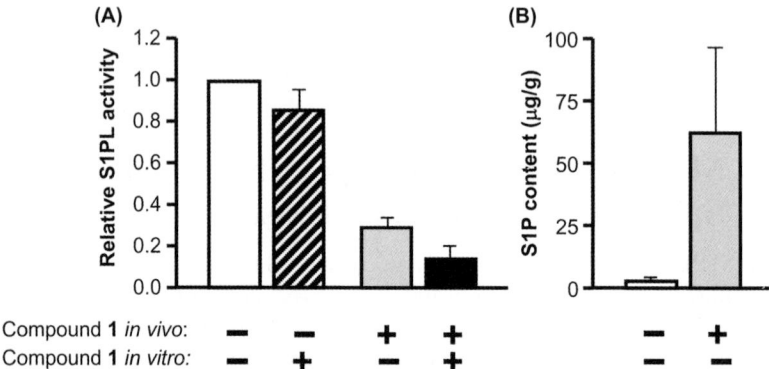

Figure 15.3 Structure of compound 1.

Figure 15.4 Residual S1PL activity and S1P content following oral compound **1** administration. **A.** Mouse spleen lysates were obtained 18 hours after oral administration of 100 mg/kg **1** or vehicle as indicated. Lysates were treated *in vitro* with vehicle or **1** and S1PL activity was determined as described.[36] **B.** S1P content of spleen lysates from the indicated mouse cohorts was measured by radioreceptor-binding assay.[36]

increased concentrations of S1P in the spleen and thymus. However, **1** showed no direct inhibition of S1PL when tested *in vitro* (Figure 15.4A).[36,84]

As a result of the favourable anti-inflammatory phenotype observed with the S1PL-deficient mice, Lexicon Pharmaceuticals, Inc. pursued S1PL as a new target for the treatment of autoimmune disorders. Following the report from Schwab and Cyster,[67] single-dose studies were performed in conjunction with measurement of the PK-PD profile of **1**. To enhance mechanistic understanding of **1**, a biochemical assay measuring S1P turnover was conducted utilizing lysates prepared from mouse spleen tissue following oral administration of **1** (100 mg/kg). Tissues were harvested near peak levels of pharmacology (lymphocyte levels by complete blood cell count (CBC) at 18 hour), and compared directly to vehicle controls. Results showed a significant reduction in lyase activity that was concomitant with a 22-fold increase in S1P levels in spleen tissue[36] (Figure 15.4). Indeed, S1PL showed diminished capacity to perform the retro-aldol catalysis on its substrate S1P. Because the *in-vitro* assay measuring lyase activity (levels of C16 aldehyde and ^{31}P-labeled phosphoethanolamine) showed **1** to be inactive, these combined assay findings suggested that *in-vivo* efficacy was due to a more complex scenario than a simple binary ligand-protein interaction. It was hypothesized that **1** may exert biological activity by interaction with another ligand, peptide or protein that then goes on to inhibit S1PL, but no experimental evidence has been described to date for such interaction.

Further drug discovery efforts to inhibit S1PL *in vivo* focused on **1** as a medicinal chemistry starting point. This drug discovery effort culminated with the development of compounds LX2931 (**2**) and LX2932 (**4**) as investigational inhibitors of S1PL. The design of **4** utilized a ring that would lock the configuration of the oxime in **2** as represented in concept by **3** (Figure 15.5). The isoxazole served as a bioisosteric and isoelectronic functional group of oxime **2** and with the added feature of enhancing hydrolytic stability. A thorough investigation of substituted imidazoles that led to the identification of **2** and **4** benefited from synthetic methods adapted from the Amadori rearrangement[85] and Buchi cyclization methods.[86] Buchi condensation of the aza-sugar with imidates generated from nitriles and sodium methoxide provided hetero-bicyclic tetraols with a variety of C-2 substitution. During the course of synthetic investigations with **1**, it was observed that the formation of aldol side-products resulted from enolization of the ketone, even under mildly acidic conditions. Oxime **2** was found to be stable under conditions that proved problematic for the ketone present in **1**.

The Novartis compound FTY720 has also been reported to inhibit S1PL activity *in vivo* after intra-peritoneal injection of a dose of 1 mg/kg.[87] The effect peaked at 36 hours after FTY720 treatment with 40% inhibition of S1PL activity in the thymus. In various *in-vitro* assays, FTY720 inhibited S1PL

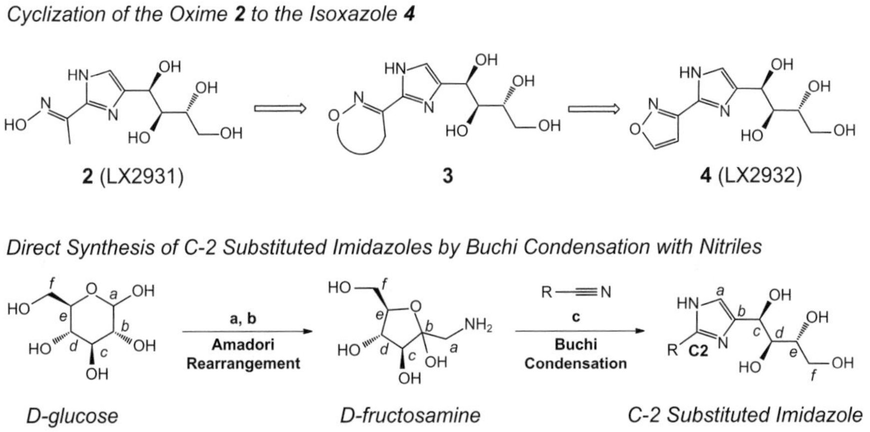

Cyclization of the Oxime 2 to the Isoxazole 4

2 (LX2931)   **3**   **4 (LX2932)**

Direct Synthesis of C-2 Substituted Imidazoles by Buchi Condensation with Nitriles

D-glucose   *D-fructosamine*   *C-2 Substituted Imidazole*

Figure 15.5   Structure of compounds **2** and **4**.

activity at relatively high concentrations with approximate IC_{50} values of 30 to 50 μM.[84,88] In contrast, FTY720-P, which is the active S1P receptor agonist derivative of FTY720, showed only moderate or no inhibition.[87,88]

Several other small-molecule inhibitors of S1PL have appeared in the literature and were reported to inhibit S1PL *in vitro* (Figure 15.6). Early work in the field focused on small chemical changes made to S1P and its reduced congener, dihydrosphingosine 1-phosphate (sphinganine 1-phosphate). Both molecules are natural substrates of S1PL with similar single-digit micromolar affinity. Substitution of a phosphonate group for the terminal phosphate group of S1P transformed the substrates into S1PL inhibitors. This approach led to the synthesis of molecules with much of the substrate recognition elements intact, with only slight modification to the phosphate that would prevent retro-aldol degradation by S1PL. One representative of these compounds is the phosphonate analogue of sphinganine, 1-desoxysphinganine 1-phosphonate, which had K_i of 5 μM against S1PL.[89] A different modification of S1P produced the analogue 2-vinylsphinganine 1-phosphate (**5**), which exhibited an IC_{50} of 2.4 μM against S1PL.[90] An attempt was also made to design an inhibitor that structurally represents the complex between S1PL substrate and cofactor, but the compound (**6**) was inactive *in vitro*.[36] Finally, a high-throughput screen identified compound **7**, which inhibited S1PL with single digit micromolar potency.[36] Unfortunately, none of the compounds mentioned in this paragraph have been shown to induce the lymphopenia that was expected after inhibition of S1PL activity in animals *in vivo*.

Ultimately, the immunological phenotypes observed in genetic models of S1PL deficiency were recapitulated by the oral administration of **2** or **4**.[35–37] Oral dosing of **2** or **4** yielded a dose-dependent decrease in circulating lymphocyte numbers in multiple species and showed a therapeutic effect in rodent

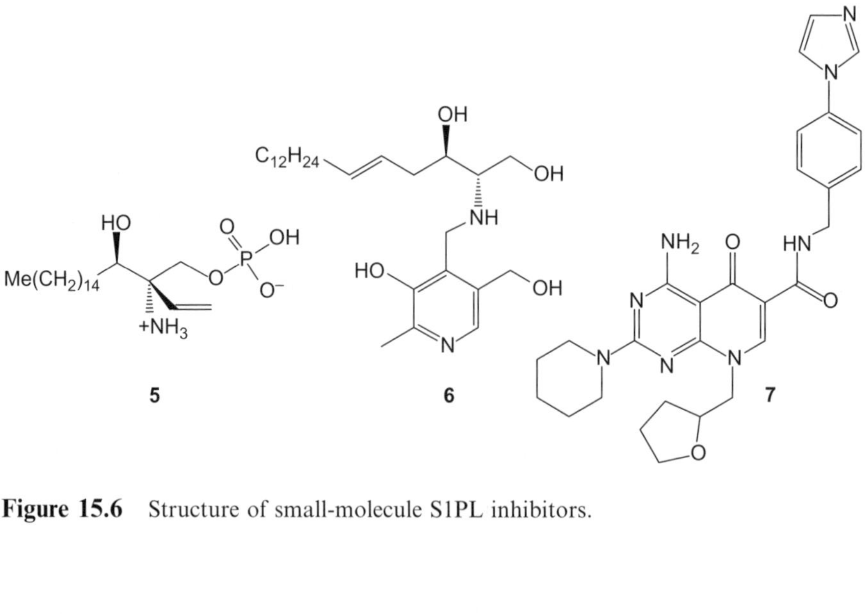

Figure 15.6   Structure of small-molecule S1PL inhibitors.

models of rheumatoid arthritis.[37] Phase I clinical trials indicated that **2**, the first clinically studied inhibitor of S1PL, produced a dose-dependent and reversible reduction of circulating lymphocytes.[37] A description of the medicinal chemistry effort and the study of associated pharmacology in inflammation models follows.

15.4.1 Pharmaceutical Targeting of Sphingosine 1-Phosphate Lyase in Animal Models of Inflammation: PK-PD Relationship, Mechanism and Pharmacology

In-vivo evaluation of **1** revealed that peak lymphopenia occurred between 12 and 24 hours following oral administration to mice.[36] Accordingly, 18 hours after oral administration of a 30 mg/kg dose of **2**, the resulting lymphocyte reduction in peripheral blood was measured by CBC analysis. *E*-Oxime **2** lowered peripheral lymphocyte counts by 60% as compared to vehicle control. The pharmacokinetic profile of **2** shown in Table 15.2 was obtained from a separate PK study.[37] The enhanced pharmacology observed with **2** relative to **1** was observed despite a reduction in overall exposure. This observation indicated that the enhanced lymphopenia observed with **2** is likely a consequence of increased intrinsic potency.

That oxime **2** generated lymphopenia suggested the possibility that other analogues displaying an sp^2-hybridized heteroatom at this position might also provide potent analogues. It was considered that heterocycles containing an appropriately oriented C=N bond within a heterocyclic ring could provide pharmacologically active compounds. To test this hypothesis, isoxazole **4**, the aromatic analogue of **2**, was orally administered to mice at 30 mg/kg and resulted in a 70% reduction in circulating lymphocytes. Pharmacokinetic evaluation of **4** in mice showed a more than two-fold enhancement in exposure (21 µM.hr) over oxime **2** (8.8 µM.hr).[37]

To facilitate advancement of **2** to clinical evaluation, a more detailed pharmacokinetic profile was obtained in multiple species (Table 15.2). Compound **2** was characterized by low clearance and low volume of distribution following pharmacokinetic studies in rodents, dogs, monkeys and, ultimately, humans.[37] Maximum plasma concentrations occurred at approximately 2 hours following oral administration. Oral exposure generally increased with higher species, showing the highest exposure in humans.

In accord with previous studies with **1** (Figure 15.3), compounds **2** and **4** also showed no direct inhibition of S1PL *in vitro*. *In vivo*, the effect of **2** and **4** on disease development and progression was assessed in a collagen (CII)-induced arthritis (CIA) model of RA in mice, using prophylactic and therapeutic dosing regimens. Prophylactic once a day oral dosing of **2** at 100 mg/kg resulted in a significant delay in the onset of arthritis; treated mice had only a minimal increase in clinical scores and joint swelling. As expected from its effect on T-cell trafficking, mice treated with **2** showed an approximately 40–60% decrease in the number of circulating T-cells and an increased percentage of

Table 15.2 Pharmacokinetic parameters of compound 2 in multiple species. T_{max}, time to maximum exposure; C_{max}, maximum exposure; AUC, area under the curve (exposure calculated as $t_0 \rightarrow$ infinity); Vss, volume of distribution; CL, clearance calculated from 1 mg/kg i.v. dose; %F, bioavailability calculated from 1 mg/kg i.v. dose and indicated oral dose; ND, not determined.

Species	Oral dose (mg/kg)	T_{max} (h)	C_{max} (µM)	AUC (µM*h)	Vss (l/kg)	CL (ml/min/kg)	%F
Sprague-Dawley rats	10	2.0	2.7	18.7	0.6	10.2	26.0
Beagle breed dogs	50	2.0	79.5	323.0	0.3	3.5	32.1
Cynomolgous monkeys	30	2.3	16.9	70.1	0.5	4.8	16.4
Healthy human volunteers	~1	2.0	3.0	19.0	ND	ND	ND

CD4 and CD8 single positive cells in the thymus at the conclusion of the experiment. Microscopic analysis of tarsal sections from drug-treated animals showed that **2** significantly reduced synovial hyperplasia, inflammation, erosion and exudate formation. Statistical analysis of histopathology scores confirmed that treatment with **2** effectively reduced joint damage, hyperplasia and immune cell infiltration compared to vehicle control.[37]

The effect of **2** and **4** was further assessed in the CIA model using a therapeutic dosing regime whereby **2**, **4** or vehicle control was administered once 50% of the animals within the group became symptomatic. When administered orally once daily at a 30 mg/kg therapeutic dose, both **2** and **4** gave significant relief from disease as measured by clinical scores and joint swelling, compared to vehicle control. Therapeutic treatment with **2** or **4** at a dose that induces minimal lymphopenia also yielded a trend of reduced disease activity (Figure 15.7A).

The disease-modifying efficacy observed in the presence of relatively mild lymphopenia during both prophylactic and therapeutic treatments indicates that the potent anti-inflammatory effect of S1PL inhibition cannot be explained solely by the effect on the levels of circulating lymphocytes. It is important to note that a 40% decrease in circulating lymphocyte numbers alone is not expected to affect mouse inflammatory challenge models. As discussed in Section 15.3.2, alteration of lymphocyte trafficking likely plays a significant role in immunosuppression *via* S1PL inhibition, but the increased S1P levels may also affect other biological processes in addition to cell migration, contributing to the observed anti-inflammatory effect. These processes may include the inflammatory response of cell types as diverse as endothelial cells, macrophages and NK cells. Importantly, the disease-modifying effect of **2** and **4** was not due to general immunosuppression, since the antibody responses to the injected CII antigen were not blocked by treatment with compound (Figure 15.7B). Taken together, the results suggest that the inhibition of S1PL activity can achieve a significant anti-inflammatory effect without hampering the immune response completely. Of note, the murine CIA results were further supported by an analysis of efficacy of compounds **2** and **4** in the rat adjuvant-induced model of RA (AIA).[37]

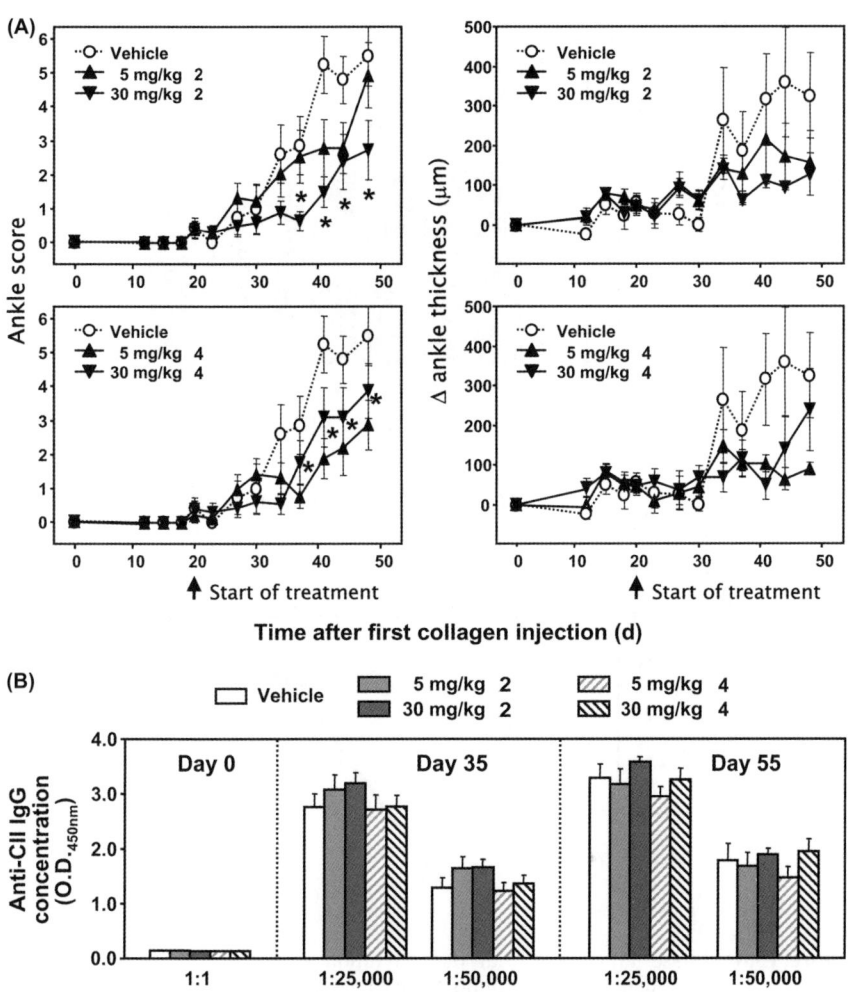

Figure 15.7 Therapeutic dosing of **2** and **4** reduces severity of arthritis in the mouse CIA model. **A.** Average values for joint swelling (left panel) and clinical scores (right panel) taken from the hind limbs for vehicle control (circles) and **2** (triangles) arms are shown with vertical bars representing SEM. *x*-axisrepresents days after collagen immunization. Daily oral dosing was initiated on day 20 (arrow), vehicle control mice were given 10 ml/kg water; **2** was given at the indicated oral doses in 10 ml/kg water. n = 9–10 each cohort; *P value <0.05. **B.** Serum anti-CII Ab titres of mice from the study presented in (**A**) were measured by ELISA at the indicated time points after initial collagen immunization.

Based on the pre-clinical data demonstrating a favourable safety and efficacy profile of **2**, Phase I clinical trials were initiated to determine its safety in human subjects. As a surrogate biomarker of S1PL inhibition, blood lymphocyte

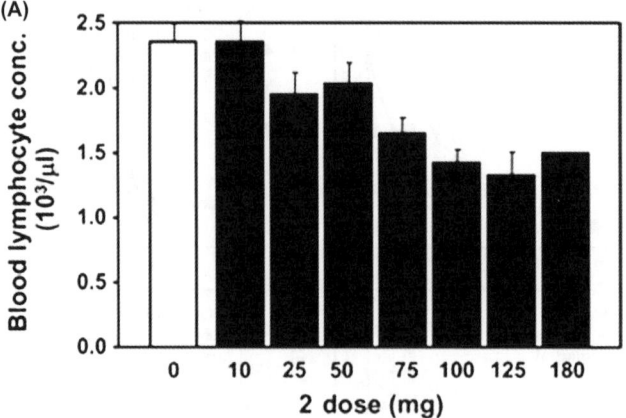

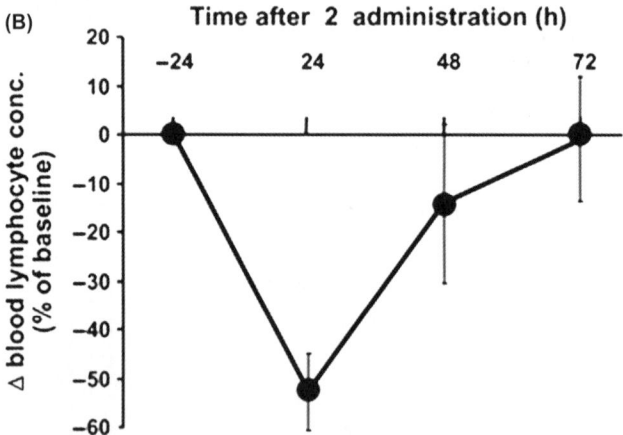

Figure 15.8 Phase Ia clinical trial data: lymphocyte counts are decreased after treatment with compound **2**, with recovery 48 hours after nadir. **A.** Dose-dependent reduction in lymphocyte counts observed after single-dose Phase Ia trial. 159 patients total in single and multiple ascending dose studies; 120 active dosing regimens and 39 placebo regimens, participated in randomized, double-blinded study. **B.** Rapid reduction of blood lymphocytes from a single 125 mg dose (N = 6 patients) followed by a return to baseline in 48 hours after nadir.

populations were determined by CBC analysis. As shown in Figure 15.8A, a single ascending dose study demonstrated a clear dose-responsive relationship, resulting in up to a ∼50% decrease in peripheral lymphocytes at the highest administered dose of 180 mg. The compound was generally well tolerated in all dose groups. Importantly, the induced reduction in circulating blood lymphocytes (N = 6, 125 mg single dose) was reversible (Figure 15.8B) and lymphocyte populations rebounded to pre-dose levels by 48 hours after nadir.[37]

15.5   Medicinal Chemistry Approaches to Inhibit Sphingosine Kinases SK1 and SK2

The two isoforms of SK, SK1 and SK2, share overall homology and produce the common product, S1P, but diverge considerably in their size and cellular localization. Human SK1 is a 384 amino acid protein, which resides in the cytosol and translocates to the plasma membrane following activation.[11–13] In contrast, SK2 is much larger than SK1 (654 amino acids) and can be found in the nucleus and transported to the cytoplasm.[14,15] Both isoforms contain a diacylglycerol kinase catalytic domain and ERK1/2 phosphorylation sites, which mediate their intra-cellular transport and activity.

Pharmaceutical targeting of the sphingosine kinases is a more recent research effort and provides promise for the treatment of hyperproliferative disorders as well as autoimmune and inflammatory diseases, as discussed in Section 15.3. Several different classes of SK1 and SK2 inhibitors have appeared in the literature primarily during the last five or so years and include substrate-based amidines,[91–93] substrate-based dimethylsphingosine,[46,55,94] heterocycle-based SKI-I and SKI-II (also referred to as SKI-2, SK-I, SKI and Ski)[94–100] and adamantyl-based ABC294640.[97,101–104]

The Macdonald research group originally published substrate-based amidines as inhibitors of both SK1 and SK2.[91] Design considerations for this series were based in part on the endogenous substrate S1P and in part on the structure of FTY720. Owing to the notion that FTY720 is phosphorylated to its active form by SK2, FTY720 offered ideas for recognition moieties known to be important for binding to the sphingosine kinases.[105] A morphing of recognition elements led to the design of amidine-based inhibitor **8** (Figure 15.9).[106,107]

While **8** is low micromolar in its *in-vitro* potency and less selective with respect to affinity for SK1 over SK2, the effort was ultimately successful in developing structures that incorporated heterocycles, such as in **9**, which enabled nanomolar potency for SK1 and selectivity for SK1 over SK2 of 180-fold. These compounds significantly lowered endogenous S1P levels in human leukaemia U937 cells at submicromolar concentrations.[92] Evaluation of amidine compounds in mice and rats also helped to probe the *in-vivo* turnover rate of S1P. A rapid decrease in blood S1P levels was observed after compound administration suggesting rapid turnover of circulating S1P.[93]

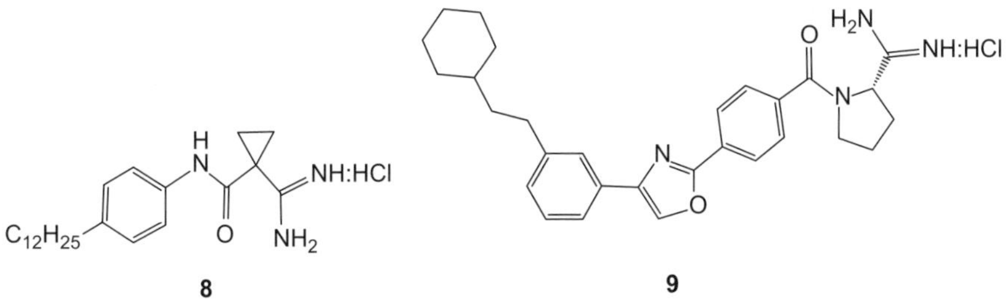

Figure 15.9   Structure of amidine-based SK inhibitors **8** and **9**.

The heterocycle-based SK inhibitors SKI-I (**10**) and SKI-II (**11**) (Figure 15.10) exhibit specificity mainly for inhibition of SK1 over that of SK2. Both **10** and **11** suffered from poor solubility and low bioavailability, which negatively impacted their efficacy as anti-cancer compounds. Medicinal chemistry studies using **10** as a template identified SKI-178 (**12**) (Figure 15.10) with an improved selectivity toward SK1, binding to the enzyme in a non-competitive ATP mode with a K_i of 1.33 µM.[108] Compound **12** was inactive against SK2 up to a concentration of 25 µM, which was near the solubility limit of **12** in the SK assay. An effort to produce a prodrug of **10** and **11** was successful utilizing the phenolic hydroxyl groups of **10** and **11** to prepare aspirinyl salicylate ester prodrugs (**10a** and **11a**) (Figure 15.10) that enhanced their efficacy profile.[95] Both **10a** and **11a** and their parent compounds showed similar cytotoxicity, but **10a** was ~3 times more effective than **10** in lung, breast and pancreatic cell lines. The majority (75%–82%) of the salicylate ester **10a** was hydrolyzed in cell cultures over a 72-hour time period and it was cleaved to afford **10** within 1 hour *in vivo*. The authors suggested that the aspirinyl prodrugs **10a** and **11a** might be effective surrogates for **10** and **11**. Importantly, two independent publications indicate that **11** induces degradation of SK1, which could have significant impact on its efficacy for inhibiting SK1 function.[99,100] *In-vivo* efficacy data are available for **11**, demonstrating inhibitory activity in animal

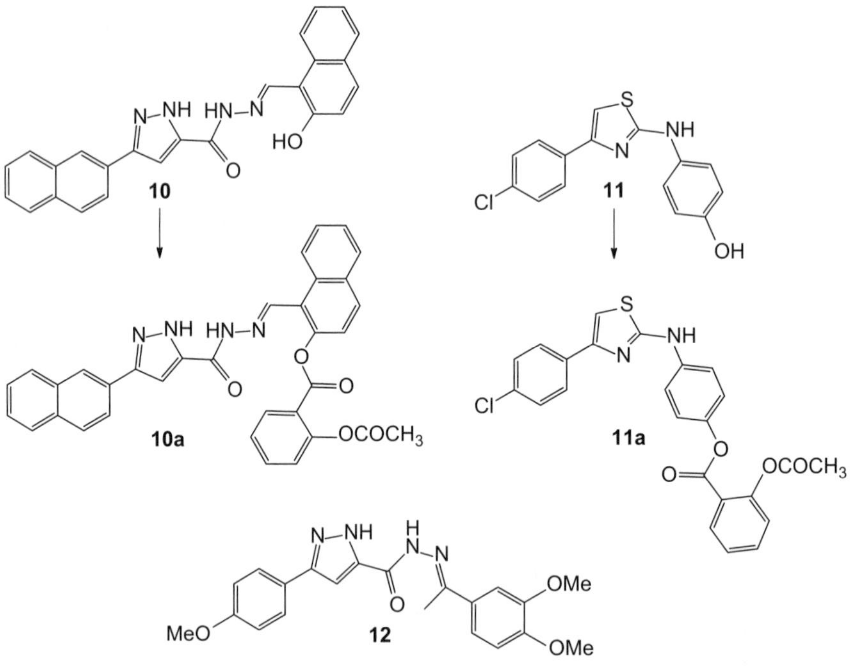

Figure 15.10   Structure of compounds **10** and **11**, prodrugs **10a** and **11a** and compound **12**.

models of asthma in mice[94,96] (at 50 mg/kg dose by i.p. route and by inhalation), and in the haemorrhagic shock-induced model of lung injury in rats (at 30 μmol/l exposure after i.p. administration).[98] All treatments were performed in prophylactic setting.

The adamantyl-based ABC294640 (**13**) (Figure 15.11) induces non-apoptotic cell death and therefore may be useful in anti-cancer drug combinations.[109,110] The compound has been reported to inhibit the activity of recombinant human SK2 with an IC_{50} of approximately 60 μM. In contrast, it showed no effect on the activity of SK1 at concentrations up to at least 100 μM. Apogee Biotechnology Corporation has announced recently an IND filing of **13** for a Phase I clinical trial with plans for treating advanced solid tumours and pancreatic cancer.[111] In addition to pre-clinical cancer models, the *in-vivo* efficacy of **13** has been explored in animal models of IBD,[101,102] arthritis[103] and atherosclerosis.[104] In a dextran sodium sulfate (DSS)-induced mouse model of ulcerative colitis, daily oral administration of 50 mg/kg **13** or another adamantyl-based inhibitor, ABC747080 (**14**) (Figure 15.11), reduced S1P levels in the colon and alleviated disease activity compared to vehicle control.[101] Compound **13** showed similar efficacy in the trinitrobenzene sulfonic acid (TNBS)-induced IBD model, which is a closer approximation of Crohn's disease.[102] Therapeutic daily administration of **13** at 100 mg/kg oral dose also attenuated both CIA in mice and AIA in rats.[103] In addition, intra-peritoneal injection of 25 and 75 mg/kg **13** in rats attenuated the development of diabetic retinopathy, which is partly induced by local inflammation.[97] Counter to the above anti-inflammatory effects, compound **13** failed to affect atherosclerosis in lipoprotein receptor (LDL-R)-deficient mice despite reducing S1P concentration in plasma and decreasing T-cell numbers within atherosclerotic lesions.[104] The lack of anti-atherogenic effect of **13** was thought to be the consequence of increased pro-inflammatory cytokine production and enhanced activation of dendritic cells and T-cells.

Substrate-based sphingolipid 5c (**15**) (Figure 15.12) was discovered to possess sphingosine kinase inhibitory activity.[112] The structure of **15** is consistent with a synthetic chemistry program aimed to yield novel sphingolipids and was prepared *via* oxidation of the corresponding propargylic alcohol. These alcohols, which unfortunately did not inhibit SK1 or SK2, could be prepared with predictably high diastereoselectivity from addition of an acetylenic anion to the Garner aldehyde.[113] It was discovered that acetylenic ketone **15** is a selective inhibitor of SK1 with an IC_{50} of 3.3 μM, and no inhibitory activity against SK2 at 10 μM. Furthermore, the hydroxy-NHBoc carbamate **16** (Figure 15.12),

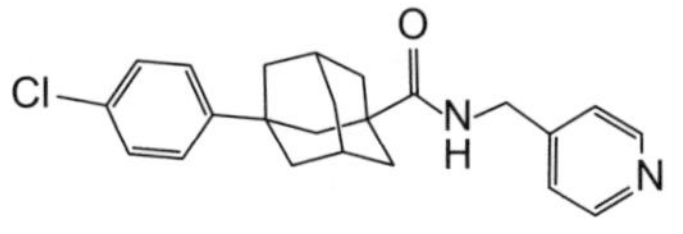

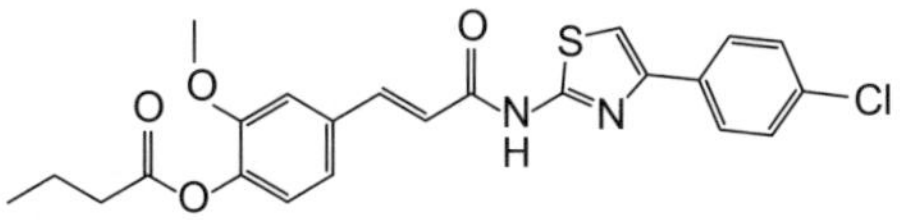

Figure 15.11 Structure of compounds **13** and **14**.

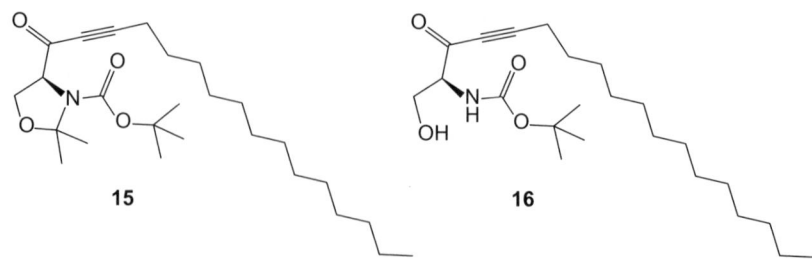

Figure 15.12   Structure of compounds **15** and **16**.

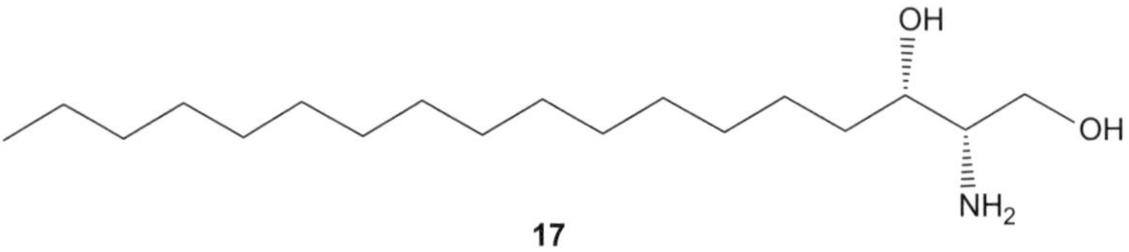

Figure 15.13   Structure of compound **17**.

derived from hydrolysis of **15**, was found to be a more potent inhibitor of SK1 with an IC_{50} of 1.2 µM, though it did inhibit SK2 activity by 54% at 10 µM concentration. Both compounds were more potent than DMS against SK1 ($IC_{50} = 5.7$ µM).

Safingol ((-)-*threo*-dihydrosphingosine, **17**, Figure 15.13) is the L-threo-stereoisomer of endogenous sphinganine. Compound **17** is an inhibitor of protein kinase C (PKC) and SK *in vitro*. Further characterization regarding the specific kinase targeted is lacking. The cell-based activity showing **17** to induce apoptosis and autophagy by increasing conversion of sphingosine to ceramide is consistent with pan sphingosine kinase inhibition. Compound **17** was combined with cisplatin in a Phase I clinical trial in 43 patients with advanced solid tumours.[114] Conclusions drawn from the study showed that cisplatin and **17** can be safely co-administered and that a reversible, dose-dependent hepato-toxicity was observed and found to be consistent with pre-clinical data.[114]

Spiegel and co-workers showed that the substrate-based SK inhibitor **18** (SK1-I, BML-258, Figure 15.14), selectively inhibited SK1 activity by 60% to 70% at 5 µM concentration.[115] SK1 inhibitor **18** combined structural compo-nents of FTY720 with the polar head group of sphingosine. This water-soluble molecule induced apoptosis in various human leukaemia cell lines and primary acute myeloid leukaemia cell cultures while relatively sparing normal peripheral blood mononuclear leukocytes. *In vivo*, **18** inhibited U937 leukaemia growth in a mouse xenograft model. Intra-peritoneal injection of 20 mg/kg **18** decreased tumour size by 50% compared to saline vehicle treatment.[115] Compound **18**

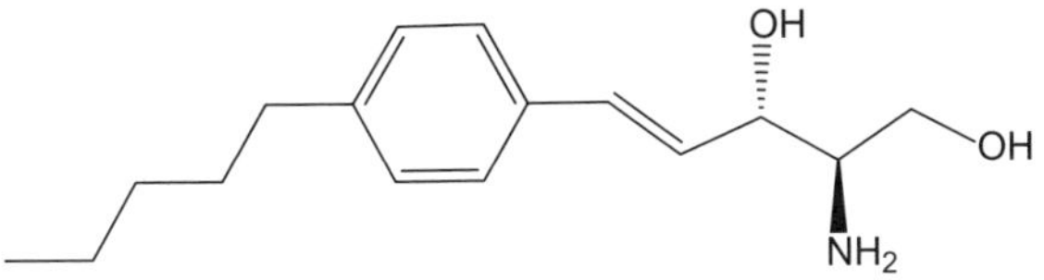

Figure 15.14   Structure of compound **18**.

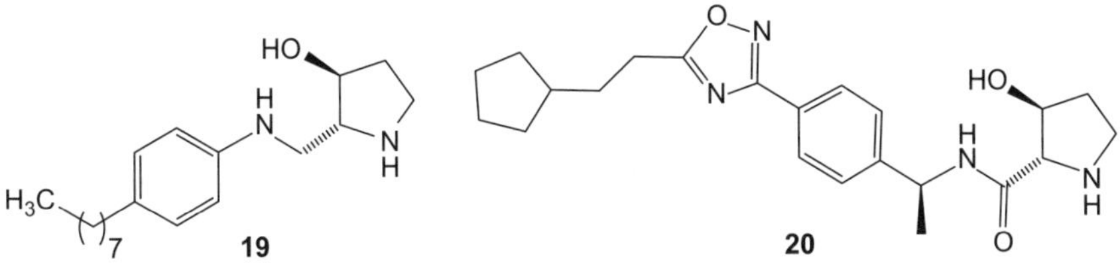

Figure 15.15   Structure of compounds **19** and **20**.

also suppressed the growth of glioblastoma cells *in vitro* and reduced tumour development *in vivo*.[116] Research with **18** showed that it could be considered as an additive agent for both leukaemia and for glioblastoma multi-form.[117]

Additional strides were made in the design of potent substrate-based SK1 inhibitors. Similar to other concurrent approaches to SK inhibitors, the alkylphenyl ring appears to be an interchangeable recognition element with the poly-alkyl chain that characterizes sphingosine. Modification of the sphingosine polar head group to that of a serine or threonine amino acid led Xiang and co-workers to synthesize 3-hydroxypyrrolidine-2-carboxamide-based **19** although **19** suffered from low solubility of <1.5 µg/ml (Figure 15.15).[118] A continued effort to improve solubility characteristics and the ADME profile by heterocyclic substitution of the alkyl chain led to the *N*-(5-alkyl(oxadiazol-3-yl)benzyl)-3-hydroxypyrrolidine-2-carboxamide-based scaffold to afford **20** (Figure 15.15). Studies of **20** revealed an improved solubility profile of 74 µg/ml, an improved cytochrome P450 profile and moderate hepatic clearance.[118,119]

Additional selective inhibitors of SK2 were designed based on recognition elements inherent in the FTY720 structure.[120] A diversity-oriented synthesis provided compounds with varied scaffolds and polar headgroups. Ultimately, **21** and **22** (Figure 15.16) were identified and characterized as potent and selective inhibitors of SK2 containing a quaternary ammonium salt as the polar headgroup. Compounds **21** and **22** showed inhibition of Akt/ERK phosphorylation and may inhibit the SK2-dependent phosphorylation cascade.

Pachastrissamine (**23**) (Figure 15.17), a natural product isolated from the Okinawan marine sponge, is an inhibitor of both SK1 and SK2, as well as the PKC isoforms PKCζ and PKCι. It shows submicromolar cytotoxicity against

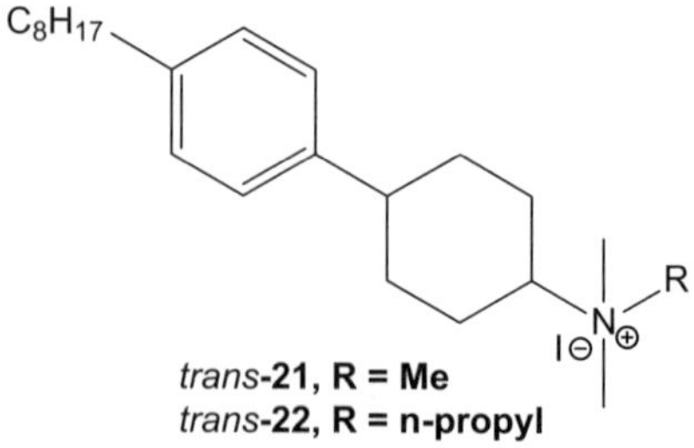

trans-**21, R = Me**
trans-**22, R = n-propyl**

Figure 15.16   Structure of compounds **21** and **22**.

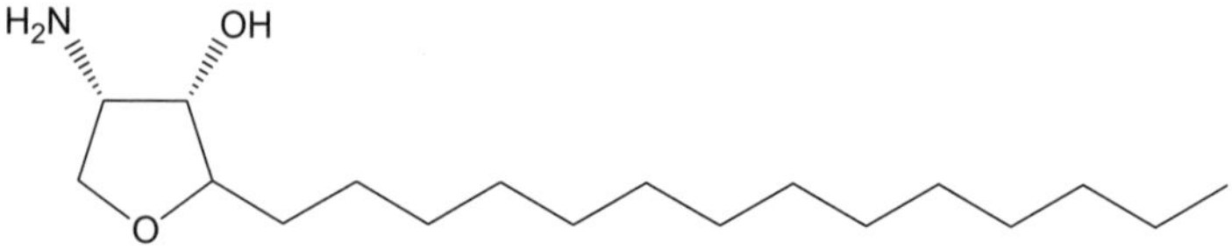

23

Figure 15.17   Structure of compound **23**.

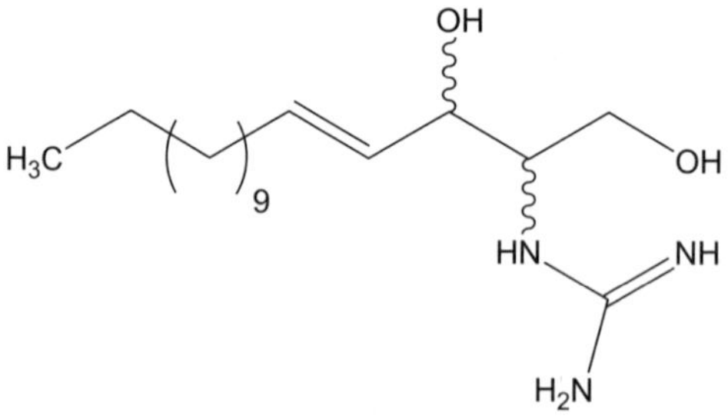

24 (LCL146, (2S, 3R)-Boc-guanidino-Sph, HCl salt)
25 (LCL351, (2R, 3S)-Boc-guanidino-Sph, HCl salt)

Figure 15.18   Structure of compounds **24** and **25**.

several cancer cell lines.[121] Changes in the configuration of stereogenic centres on the tetrahydrofuran ring moderately increased potency. Further optimization of this chemotype is underway.

Finally, a recent patent highlighted two diastereomerically related compounds, **24** (LCL146, (2S,3R)-Boc-guanidino-Sph, HCl salt) and **25** (LCL351, (2R,3S)-Boc-guanidino-Sph, HCl salt) (Figure 15.18).[122] They belong to a series of substrate-based sphingosine kinase inhibitors where the endogenous amino group was modified to an amidine. Compound **25** inhibited SK1 and SK2 *in vitro* with an IC_{50} of 40 nM and 300 nM, respectively. Both molecules inhibited S1P formation in the human breast cancer MCF-7 cell line. The amidine groups improved water solubility of these lipid-based compounds, but bioavailability, potency and animal model efficacy data have yet to be reported.

15.6 Pharmaceutical Targeting of Sphingosine Kinases and Sphingosine 1-Phosphate Lyase in Animal Models of Inflammation

Human, mouse and rat SK and S1PL sequences are highly homologous, therefore animal studies can produce relevant data for the efficacy of compounds targeting these enzymes. All of the published data to date demonstrate suppressive effects of SK and S1PL inhibitors in various models of inflammatory diseases, except in a model of atherosclerosis using the SK2 inhibitor ABC294640. An overview of these data, some of which have already been discussed within the chapter for the specific molecules, is summarized in Table 15.3.

Table 15.3 Reported anti-inflammatory activity of compounds targeting S1P production. All challenges were performed in mice, except for the models of diabetic retinopathy, haemorrhagic shock and AIA, which were done in rats. DNFB, 2,4-dinitrofluorobenzene-induced contact hypersensitivity; NOD, non-obese diabetic mice; OVA, ovalbumin-induced asthma; STZ, streptozotocin-induced diabetes.

Compound target	Disease model	Inflammatory challenge	Compound	References
Sphingosine kinases	Asthma	OVA	DMS	46, 94
		OVA	SKI-II	94, 96
	Atherosclerosis	LDL-R KO genetic background	ABC294640	104
	Diabetic retinopathy	STZ	ABC294640	97
	IBD	DSS	ABC294640, ABC747080	101
		TNBS	ABC294640	102
	Lung injury	Hemorrhagic shock	SKI-II	98
	RA	CIA	DMS	55
		CIA and AIA	ABC294640	103
S1P lyase	Contact hypersensitivity	Oxazolone	THI	123–125
		DNFB	THI	120
	Diabetes	NOD genetic background and cyclophosphamide	THI	126
	Lung injury	LPS	THI	71
	RA	CIA and AIA	LX2931	37

15.7 Concluding Remarks

Enzymes of S1P metabolism such as S1PL, SK1 and SK2, have surfaced as promising new drug targets in the inflammatory and autoimmune disease area, and for treatment of cancer. The potentially opposing effects of S1P on diverse biochemical and cellular pathways justify development of therapeutic modalities that either reduce or amplify the effects of S1P on inflammation and cell apoptosis. Therefore, the rationale exists for the development of small-molecule inhibitors of SK and S1PL activities, which decrease and increase S1P levels, respectively, for different therapeutic applications.

The specific roles of the SK isoenzymes in cancer and inflammation remain to be defined. The majority of studies linking SK activity to disease development have focused on SK1. However, it is still unclear whether selective targeting of SK1 or SK2 will be either required or sufficient for disease modifying effects of SK inhibitors in humans. Small-molecule compounds targeting SK1 and SK2 have already progressed to proof of principle studies using *in-vivo* animal models, which laid the groundwork for advancing SK inhibitors into Phase I human clinical trials. Although the primary focus of drug development for the SK inhibitors is in advanced solid tumours, experience in human trials will certainly aid their exploration in inflammatory and autoimmune diseases.

The enzyme responsible for the irreversible degradation of S1P, S1PL, is in a unique position to regulate the S1P content of tissues. Inhibition of S1PL leads to increased S1P concentration preferentially in lymphoid tissues and significantly blunted immune responses in various *in-vivo* challenge models of inflammation. Lexicon Pharmaceuticals, Inc. has advanced **2**, a S1PL inhibitor, into human trials. In Phase I trials, a single ascending dose study of **2** provided a clear dose-responsive relationship, resulting in up to a ~50% decrease in peripheral blood lymphocytes that was reversible. Phase II clinical trials will focus on the therapeutic efficacy of **2** in RA patients.

In conclusion, the potential therapeutic utility of modifying S1P metabolism is clearly recognized and is being actively explored. It is conceivable to foresee that inhibitors of SK1, SK2 and S1PL will find their own niche in the future for various therapeutic applications ranging from cancer to autoimmune and inflammatory disorders.

Acknowledgements

We are grateful to our colleagues at Lexicon Pharmaceuticals, Inc. for their support and careful review of this work.

References

1. S. Pyne and N. J. Pyne, *Trends Mol. Med.*, 2011, **17**, 463.
2. H. Obinata and T. Hla, *Semin. Immunopathol.*, 2012, **34**, 91.
3. S. Spiegel and S. Milstien, *Nat. Rev. Immunol.*, 2011, **11**, 403.
4. P. Xia and C. Wadham, *Cytokine Growth Factor Rev.*, 2011, **22**, 45.

5. H. Chi, *Trends Pharmacol. Sci.*, 2011, **32**, 16.
6. V. Brinkmann, A. Billich, T. Baumruker, P. Heining, R. Schmouder, G. Francis, S. Aradhye and P. Burtin, *Nat. Rev. Drug Discov.*, 2010, **9**, 883.
7. A. J. Snider, K. A. Orr Gandy and L. M. Obeid, *Biochimie*, 2010, **92**, 707.
8. M. R. Pitman and S. M. Pitson, *Curr. Cancer Drug Targets*, 2010, **10**, 354.
9. M. Serra and J. D. Saba, *Adv. Enzyme Regul.*, 2010, **50**, 348.
10. F. Bourquin, H. Riezman, G. Capitani and M. G. Grütter, *Structure*, 2010, **18**, 1054.
11. K. R. Johnson, K. P. Becker, M. M. Facchinetti, Y. A. Hannun and L. M. Obeid, *J. Biol. Chem.*, 2002, **277**, 35257.
12. S. M. Pitson, P. A. B. Moretti, J. R. Zebol, H. E. Lynn, P. Xia, M. A. Vadas and B. W. Wattenberg, *EMBO J.*, 2003, **22**, 5481.
13. S. M. Pitson, P. Xia, T. M. Leclercq, P. A. B. Moretti, J. R. Zebol, H. E. Lynn, B. W. Wattenberg and M. A. Vadas, *J. Exp. Med.*, 2005, **201**, 48.
14. N. Igarashi, T. Okada, S. Hayashi, T. Fujita, S. Jahangeer and S. Nakamura, *J. Biol. Chem.*, 2003, **278**, 46832.
15. G. Ding, H. Sonoda, H. Yu, T. Kajimoto, S. K. Goparaju, S. Jahangeer, T. Okada and S. Nakamura, *J. Biol. Chem.*, 2007, **282**, 27483.
16. C. Ogawa, A. Kihari, M. Gokoh and Y. Igarashi, *J. Biol. Chem.*, 2003, **278**, 1268.
17. K. R. Johnson, K. Y. Johnson, K. P. Becker, J. Bielawski, C. Mao and L. M. Obeid, *J. Biol. Chem.*, 2003, **278**, 34541.
18. M. Ikeda, A. Kihara and Y. Igarashi, *Biochem. Biophys. Res. Commun.*, 2004, **325**, 338.
19. M. Maceyka, H. Sankala, N. C. Hait, H. Le Stunff, H. Liu, R. Toman, C. Collier, M. Zhang, L. S. Satin, A. H. Merrill Jr, S. Milstien and S. Spiegel, *J. Biol. Chem.*, 2005, **280**, 37118.
20. K. Sattler and B. Levkau, *Cardiovasc. Res.*, 2009, **82**, 201.
21. N. Ancellin, C. Colmont, J. Su, Q. Li, N. Mittereder, S. S. Chae, S. Stefansson, G. Liau and T. Hla, *J. Biol. Chem.*, 2002, **277**, 6667.
22. S. M. Hammad, T. A. Taha, A. Nareika, K. R. Johnson, M. F. Lopes-Virella and L. M. Obeid, *Prostaglandins Other Lipid Mediat.*, 2006, **79**, 126.
23. K. Venkataraman, S. Thangada, J. Michaud, M. L. Oo, Y. Ai, Y. M. Lee, M. Wu, N. S. Parikh, S. M. Hammad, H. G. Crellin, B. X. Wu, J. Melton, V. Anelli and L. M. Obeid, *Prostaglandins Other Lipid Mediat.*, 2008, **85**, 107.
24. D. Nayak, Y. Huo, W. X. Kwang, P. N. Pushparaj, S. D. Kumar, E. A. Ling and S. T. Dheen, *Neuroscience*, 2010, **166**, 132.
25. P. Xia, J. R. Gamble, K. A. Rye, L. Wang, C. S. Hii, P. Cockerill, Y. Khew-Goodall, A. G. Bert, P. J. Barter and M. A. Vadas, *Proc. Natl Acad. Sci. USA*, 1998, **95**, 14196.
26. P. Xia, L. Wang, P. A. Moretti, N. Albanese, F. Chai, S. M. Pitson, R. J. D'Andrea, J. R. Gamble and M. A. Vadas, *J. Biol. Chem.*, 2002, **277**, 7996.

27. A. C. MacKinnon, A. Buckley, E. R. Chilvers, A. G. Rossi, C. Haslett and T. Sethi, *J. Immunol.*, 2002, **169**, 6394.
28. M. Yadav, L. Clark and J. S. Schorey, *J. Immunol.*, 2006, **176**, 5484.
29. L. Zhi, B. P. Leung and A. J. Melendez, *J. Cell Physiol.*, 2006, **208**, 109.
30. S. E. Alvarez, K. B. Harikumar, N. C. Hait, J. Allegood, G. M. Strub, E. Y. Kim, M. Maceyka, H. Jiang, C. Luo, T. Kordula, S. Milstien and S. Spiegel, *Nature*, 2010, **465**, 1084.
31. X. L. Chen, J. Y. Grey, S. Thomas, F. H. Qiu, R. M. Medford, M. A. Wasserman and C. Kunsch, *Am. J. Physiol. Heart Circ. Physiol.*, 2004, **287**, H1452.
32. A. J. Snider, T. Kawamori, S. G. Bradshaw, K. A. Orr, G. S. Gilkeson, Y. A. Hannun and L. M. Obeid, *FASEB J.*, 2009, **23**, 143.
33. C. De Palma, E. Meacci, C. Perrotta, P. Bruni and E. Clementi, *Arterioscler. Thromb. Vasc. Biol.*, 2006, **26**, 99.
34. M. L. Allende, M. Bektas, B. G. Lee, E. Bonifacino, J. Kang, G. Tuymetova, W. Chen, J. D. Saba and R. L. Proia, *J. Biol. Chem.*, 2011, **286**, 7348.
35. P. Vogel, M. S. Donoviel, R. Read, G. M. Hansen, J. Hazlewood, S. J. Anderson, W. Sun, J. Swaffield and T. Oravecz, *PloS ONE*, 2009, **4**, e4112.
36. J. Bagdanoff, M. Donoviel, A. Nouraldeen, J. Tarver, Q. Fu, M. Carlsen, T. C. Jessop, H. Zhang, J. Hazelwood, H. Nguyen, S. Baugh, M. Gardyan, K. Terranova, J. Barbosa, J. Yan, M. Bednarz, S. Layek, L. Courtney, J. Taylor, A. M. Digeorge-Foushee, S. Gopinathan, D. Bruce, T. Smith, L. Moran, E. O'Neill, J. Kramer, Z. Lai, S. D. Kimball, Q. Liu, W. Sun, S. Yu, J. Swaffield, A. Wilson, A. Main, K. G. Carson, T. Oravecz and D. Augeri, *J. Med. Chem.*, 2009, **52**, 3941.
37. J. Bagdanoff, M. Donoviel, A. Nouraldeen, M. Carlsen, T. Jessop, J. Tarver, S. Aleem, L. Dong, H. Zhang, L. Boteju, J. Hazlewood, J. Yan, M. Bednarz, S. Layek, I. B. Owusu, S. Gopinathan, L. Moran, Z. Lai, J. Kramer, S. D. Kimball, P. Yalamanchili, W. Heydorn, K. S. Frazier, B. Brooks, P. Brown, A. Wilson, W. K. Sonnenburg, A. Main, K. G. Carson, T. Oravecz and D. Augeri, *J. Med. Chem.*, 2010, **53**, 8650.
38. C. A. Finney, C. A. Hawkes, D. C. Kain, A. Dhabangi, C. Musoke, C. Cserti-Gazdewich, T. Oravecz, W. C. Liles and K. C. Kain, *Mol. Med.*, 2011, **17**, 717.
39. F. Niessen, F. Schaffner, C. Furlan-Freguia, R. Pawlinski, G. Bhattacharjee, J. Chun, C. K. Derian, P. Andrade-Gordon, H. Rosen and W. Ruf, *Nature*, 2008, **452**, 654.
40. A. Billich, N. Urtz, R. Reuschel and T. Baumruker, *Int. J. Biochem. Cell. Biol.*, 2009, **41**, 1547.
41. K. Itagaki, Q. Zhang and C. J. Hauser, *Shock*, 2010, **33**, 381.
42. A. J. Melendez and F. B. Ibrahim, *J. Immunol.*, 2004, **173**, 1596.
43. A. J. Ammit, A. T. Hastie, L. C. Edsall, R. K. Hoffman, Y. Amrani, V. P. Krymskaya, S. A. Kane, S. P. Peters, R. B. Penn, S. Spiegel and R. A. Panettieri, *FASEB J*, 2001, **15**, 1212.

44. P. N. Pushparaj, J. Manikandan, H. K. Tay, S. C. H'ng, S. D. Kumar, J. Pfeilschifter, A. Huwiler and A. J. Melendez, *J. Immunol.*, 2009, **183**, 221.
45. C. A. Oskeritzian, S. E. Alvarez, N. C. Hait, M. M. Price, S. Milstien and S. Spiegel, *Blood*, 2008, **111**, 4193.
46. W. Q. Lai, H. H. Goh, Z. Bao, W. S. Wong, A. J. Melendez and B. P. Leung, *J. Immunol.*, 2008, **180**, 4323.
47. A. Olivera, K. Mizugishi, A. Tikhonova, L. Ciaccia, S. Odom, R. L. Proia and J. Rivera, *Immunity*, 2007, **26**, 287.
48. K. Mizugishi, T. Yamashita, A. Olivera, G. F. Miller, S. Spiegel and R. L. Proia, *Mol. Cell. Biol.*, 2005, **25**, 11113.
49. B. Zemann, N. Urtz, R. Reuschel, D. Mechtcheriakova, F. Bornancin, R. Badegruber, T. Baumruker and A. Billich, *Immunol. Lett.*, 2007, **109**, 56.
50. W. Q. Lai, A. W. Irwan, H. H. Goh, A. J. Melendez, I. B. McInnes and B. P. Leung, *J. Immunol.*, 2009, **183**, 2097.
51. J. Michaud, M. Kohno, R. L. Proia and T. Hla, *FEBS Lett.*, 2006, **580**, 4607.
52. D. A. Baker, J. Barth, R. Chang, L. M. Obeid and G. S. Gilkeson, *J. Immunol.*, 2010, **185**, 2570.
53. D. A. Baker, L. M. Obeid and G. S. Gilkeson, *Inflamm., Allergy Drug Targets*, 2011, **10**, 464.
54. M. Kitano, T. Hla, M. Sekiguchi, Y. Kawahito, R. Yoshimura, K. Miyazawa, T. Iwasaki, H. Sano, J. D. Saba and Y. Y. Tam, *Arthritis Rheum.*, 2006, **54**, 742.
55. W. Q. Lai, A. W. Irwan, H. H. Goh, H. S. Howe, D. T. Yu, R. Valle-Oñate, I. B. McInnes, A. J. Melendez and B. P. Leung, *J. Immunol.*, 2008, **181**, 8010.
56. X. Pi, S. Y. Tan, M. Hayes, L. Xiao, J. A Shayman, S. Ling and J. Holoshitz, *J. Arthritis Rheum.*, 2006, **54**, 754.
57. K. Kamada, N. Arita, T. Tsubaki, N. Takubo, T. Fujino, Y. Soga, T. Miyazaki, H. Yamamoto and M. Nose, *Pathol. Int.*, 2009, **59**, 382.
58. E. T. Samy, C. A. Meyer, P. Caplazi, C. L. Langrish, J. M. Lora, H. Bluethmann and S. L. Peng, *J. Immunol.*, 2007, **179**, 5644.
59. J. Yang, B. E. Castle, A. Hanidu, L. Stevens, Y. Yu, X. Li, C. Stearns, V. Papov, D. Rajotte and J. Li, *J. Immunol.*, 2005, **175**, 6580.
60. G. Liu, K. Yang, S. Burns, S. Shrestha and H. Chi, *Nat. Immunol.*, 2010, **11**, 1047.
61. A. Olivera, C. Eisner, Y. Kitamura, S. Dillahunt, L. Allende, G. Tuymetova, W. Watford, F. Meylan, S. C. Diesner, L. Li, J. Schnermann, R. L. Proia and J. Rivera, *J. Clin. Invest.*, 2010, **120**, 1429.
62. R. Wadgaonkar, V. Patel, N. Grinkina, C. Romano, J. Liu, Y. Zhao, S. Sammani, J. G. Garcia and V. Natarajan, *Am. J. Physiol. Lung Cell Mol. Physiol.*, 2009, **296**, L603.
63. S. Sammani, L. Moreno-Vinasco, T. Mirzapoiazova, P. A. Singleton, E. T. Chiang, C. L. Evenoski, T. Wang, B. Mathew, A. Husain, J. Moitra, X. Sun, L. Nunez, J. R. Jacobson, S. M. Dudek, V. Natarajan and J. G. Garcia, *Am. J. Respir. Cell. Mol. Biol.*, 2010, **43**, 394.

64. D. Mechtcheriakova, A. Wlachos, J. Sobanov, T. Kopp, R. Reuschel, F. Bornancin, R. Cai, B. Zemann, N. Urtz, G. Stingl, G. Zlabinger, M. Woisetschläger, T. Baumruker and A. Billich, *Cell. Signalling*, 2007, **19**, 748.

65. E. Y. Seo, G. T. Park, K. M. Lee, J. A. Kim, J. H. Lee and J. M. Yang, *J. Invest. Dermatol.*, 2006, **126**, 1187.

66. P. Katsel, C. Li and V. Haroutunian, *Neurochem. Res.*, 2007, **32**, 845.

67. S. R. Schwab, J. P. Pereira, M. Matloubian, Y. Xu, Y. Huang and J. G. Cyster, *Science*, 2005, **309**, 1735.

68. M. Idzko, H. Hammad, M. van Nimwegen, M. Kool, T. Müller, T. Soullié, M. A. Willart, D. Hijdra, H. C. Hoogsteden and B. N. Lambrecht, *J. Clin. Invest.*, 2006, **116**, 2935.

69. B. J. McVerry, X. Peng, P. M. Hassoun, S. Sammani, B. A. Simon and J. G. Garcia, *Am. J. Respir. Crit. Care Med.*, 2004, **170**, 987.

70. X. Peng, P. M. Hassoun, S. Sammani, B. J. McVerry, M. J. Burne, H. Rabb, D. Pearse, R. M. Tuder and J. G. Garcia, *Am. J. Respir. Crit. Care Med.*, 2004, **169**, 1245.

71. Y. Zhao, I. A. Gorshkova, E. Berdyshev, D. He, P. Fu, W. Ma, Y. Su, P. V. Usatyuk, S. Pendyala, B. Oskouian, J. D. Saba, J. G. Garcia and V. Natarajan, *Am. J. Respir. Cell. Mol. Biol.*, 2011, **45**, 426.

72. E. Camerer, J. B. Regard, I. Cornelissen, Y. Srinivasan, D. N. Duong, D. Palmer, T. H. Pham, J. S. Wong, R. Pappu and S. R. Coughlin, *J. Clin. Invest.*, 2009, **119**, 1871.

73. J. E. Hughes, S. Srinivasan, K. R. Lynch, R. L. Proia, P. Ferdek and C. C. Hedrick, *Circ. Res.*, 2008, **102**, 950.

74. D. T. Bolick, S. Srinivasan, K. W. Kim, M. E. Hatley, J. J. Clemens, A. Whetzel, N. Ferger, T. L. Macdonald, M. D. Davis, P. S. Tsao, K. R. Lynch and C. C. Hedrick, *Arterioscler. Thromb. Vasc. Biol.*, 2005, **25**, 976.

75. A. M. Whetzel, D. T. Bolick, S. Srinivasan, T. L. Macdonald, M. A. Morris, K. Ley and C. C. Hedrick, *Circ. Res.*, 2006, **99**, 731.

76. G. Theilmeier, C. Schmidt, J. Herrmann, P. Keul, M. Schafers, I. Herrgott, J. Mersmann, J. Larmann, S. Hermann, J. Stypmann, O. Schober, R. Hildebrand, R. Schulz, G. Heusch, M. Haude, K. von Wnuck Lipinski, C. Herzog, M. Schmitz, R. Erbel, J. Chun and B. Levkau, *Circulation*, 2006, **114**, 1403.

77. M. Idzko, E. Panther, S. Corinti, A. Morelli, D. Ferrari, Y. Herouy, S. Dichmann, M. Mockenhaupt, P. Gebicke-Haerter, F. Di Virgilio, G. Girolomoni and J. Norgauer, *FASEB J.*, 2002, **16**, 625.

78. W. Wang, M. H. Graeler and E. J. Goetzl, *FASEB J.*, 2005, **19**, 1731.

79. T. Schulze, S. Golfier, C. Tabeling, K. Räbel, M. H. Gräler, M. Witzenrath and M. Lipp, *FASEB J.*, 2011, **25**, 4024.

80. N. C. Hait, J. Allegood, M. Maceyka, G. M. Strub, K. B. Harikumar, S. K. Singh, C. Luo, R. Marmorstein, T. Kordula, S. Milstien and S. Spiegel, *Science*, 2009, **325**, 1254.

81. A. Di, T. Kawamura, X. P. Gao, H. Tang, E. Berdyshev, S. M. Vogel, Y. Y. Zhao, T. Sharma, K. Bachmaier, J. Xu and A. B. Malik, *J. Biol. Chem.*, 2010, **285**, 15848.
82. R. Spector and S. Huntoon, *Toxicol. Appl. Pharmacol.*, 1982, **62**, 172.
83. G. F. Houben, M. H. M. Kuijpers, H. Van Loveren, A. H. Penninks, E. J. Sinkeldam and W. Seinen, *Arch. Toxicol.*, 1989, **13**, 183.
84. E. Reina, L. Camacho, J. Casas, P. P. Van Veldhoven and G. Fabrias, *Chem. Phys. Lipids*, 2012, **165**, 225.
85. J. E. Hodge and C. E. Rist, *J. Am. Chem. Soc.*, 1953, **75**, 316.
86. G. Buchi and K. Halweg, *J. Org. Chem.*, 1985, **50**, 1134.
87. P. Bandhuvula, Y. Y. Tam, B. Oskouian and J. D. Saba, *J. Biol. Chem.*, 2005, **280**, 33697.
88. E. V. Berdyshev, J. Goya, I. Gorshkova, G. D. Prestwich, H. S. Byun, R. Bittman and V. Natarajan, *Anal. Biochem.*, 2011, **408**, 12.
89. W. Stoffel and M. Grol, *Chem. Phys. Lipids*, 1974, **13**, 372.
90. A. Boumendjel and S. P. F. Miller, *Tetrahedron Lett.*, 1994, **35**, 819.
91. T. P. Mathews, A. J. Kennedy, Y. Kharel, P. C. Kennedy, O. Nicoara, M. Sunkara, A. J. Morris, B. R. Wamhoff, K. R. Lynch and T. L. Macdonald, *J. Med. Chem.*, 2010, **53**, 2766.
92. A. J. Kennedy, T. P. Mathews, Y. Kharel, S. D. Field, M. L. Moyer, J. E. East, J. D. Houck, K. R. Lynch and T. L. MacDonald, *J. Med. Chem.*, 2011, **54**, 3524.
93. Y. Kharel, T. P. Mathews, A. M. Gellett, J. L. Tomsig, P. C. Kennedy, M. L. Moyer, T. L. Macdonald and K. R. Lynch, *Biochem. J.*, 2011, **440**, 345.
94. T. Nishiuma, Y. Nishimura, T. Okada, E. Kuramoto, Y. Kotani, S. Jahangeer and S. Nakamura, *Am. J. Physiol. Lung Cell. Mol. Physiol.*, 2008, **294**, L1085.
95. A. K. Sharma, U. H. Sk, M. A. Gimbor, J. A. Hengst, X. Wang, J. Yun and S. Amin, *Eur. J. Med. Chem.*, 2010, **45**, 4148.
96. Y. Chiba, H. Takeuchi, H. Sakai and M. Misawa, *J. Pharmacol. Sci.*, 2010, **114**, 304.
97. L. W. Maines, K. J. French, E. B. Wolpert, D. A. Antonetti and C. D. Smith, *Invest. Ophthalmol. Vis. Sci.*, 2006, **47**, 5022.
98. C. Lee, D. A. Zhong, E. Feketeova, K. B. Kannan, J. K. Yun, E. A. Deitch, Z. Fekete, D. H. Livingston and C. J. Hauser, *J. Trauma*, 2004, **57**, 955.
99. S. Ren, C. Xin, J. Pfeilschifter and A. Huwiler, *Cell. Physiol. Biochem.*, 2010, **26**, 97.
100. C. Loveridge, F. Tonelli, T. Leclercq, K. Gat Lim, J. S. Long, E. Berdyshev, R. J. Tate, V. Natarajan, S. M. Pitson, N. J. Pyne and S. Pyne, *J. Biol. Chem.*, 2010, **285**, 38841.
101. L. W. Maines, L. R. Fitzpatrick, K. J. French, Y. Zhuang, Z. Xia, S. N. Keller, J. J. Upson and C. D. Smith, *Dig. Dis. Sci.*, 2008, **53**, 997.
102. L. W. Maines, L. R. Fitzpatrick, C. L. Green, Y. Zhuang and C. D. Smith, *Inflammopharmacology*, 2010, **18**, 73.

103. L. R. Fitzpatrick, C. Green, E. E. Frauenhoffer, K. J. French, Y. Zhuang, L. W. Maines, J. J. Upson, E. Paul, H. Donahue, T. J. Mosher and C. D. Smith, *Inflammopharmacology*, 2011, **19**, 75.

104. F. Poti, M. Bot, S. Costa, V. Bergonzini, L. Maines, G. Varga, H. Freise, H. Robenek, M. Simoni and J. R. Nofer, *Thromb. Haemost.* 2012, **107**, 552.

105. J. A. Hengst, X. Wang, U. H. Sk, A. K. Sharma, S. Amin and J. K. Yun, *Bioorg. Med. Chem. Lett.*, 2010, **20**, 7498.

106. K. R. Lynch, T. L. MacDonald, Y. Kharel, T. P. Mathews and B. R. Wamhoff, University of Virginia, *PCT Pat. Appl.*, WO 2009/146112, 2009.

107. K. R. Lynch, T. L. MacDonald, T. P. Mathews, A. Kennedy and Y. Kharel, University of Virginia, *PCT Pat. Appl.*, WO 2011/020116, 2011.

108. Y. Kharel, S. Lee, A. H. Synder, S. L. Sheasley-O'Neill, M. A. Morris, Y. Setiady, R. Zhu, M. A. Zigler, T. L. Burcin, K. Ley, K. S. Tung, V. H. Engelhard, T. L. Macdonald, S. Pearson-White and K. R. Lynch, *J. Biol. Chem.*, 2005, **280**, 36865.

109. V. Beljanski, C. Knaak and C. D. Smith, *J. Pharmacol. Exp. Ther.*, 2010, **333**, 454.

110. K. J. French, Y. Zhuang, L. W. Maines, P. Gao, W. Wang, V. Beljanski, J. J. Upson, C. L. Green, S. N. Keller and C. D. Smith, *J. Pharmacol. Exp. Ther.*, 2010, **333**, 129.

111. http://apogee-biotech.com/index.php/news/

112. L. Wong, S. S. Tan, Y. Lam and A. J. Melendez, *J. Med. Chem.*, 2009, **52**, 3618.

113. P. Garner and S. Ramakanth, *J. Org. Chem.*, 1986, **51**, 2609.

114. M. A. Dickson, R. D. Carvajal, A. H. Merrill Jr, M. Gonen, L. M. Cane and G. K. Schwartz, *Clin. Cancer Res.*, 2011, **17**, 2484.

115. S. W. Paugh, B. S. Paugh, M. Rahmani, D. Kapitonov, J. A. Almenara, T. Kordula, S. Milstein, J. K. Adams, R. E. Zipkin, S. Grant and S. Spiegel, *Blood*, 2008, **112**, 1382.

116. D. Kapitonov, J. C. Allegood, C. Mitchell, N. C. Hait, J. A. Almenara, J. K. Adams, R. E. Zipkin, P. Dent, T. Kordula, S. Milstein and S. Spiegel, *Cancer Res.*, 2009, **69**, 6915.

117. S. Spiegel, R. E. Zipkin and J. K. Adams, Enzo Biochem, Inc., *US Pat. Appl.*, US 2010/0278741, 2010.

118. Y. Xiang, G. Asmussen, M. Booker, B. Hirth, J. L. Kane Jr., J. Liao, K. D. Noson and C. Lee, *Bioorg. Med. Chem. Lett.*, 2009, **19**, 6119.

119. Y. Xiang, B. Hirth, J. L. Kane Jr., J. Liao, K. D. Noson, C. Lee, G. Asmussen, M. Fitzgerald, C. Klaus and M. Booker, *Bioorg. Med. Chem. Lett.*, 2010, **20**, 4550.

120. M. R. Raje, K. Knott, Y. Kharel, P. Bissel, K. R. Lynch and W. L. Santos, *Bioorg. Med. Chem.*, 2012, **20**, 183.

121. Y. Yoshimitsu, S. Oishi, J. Miyagaki, S. Inuki, H. Ohno and N. Fujii, *Bioorg. Med. Chem.*, 2011, **19**, 5402.

122. A. K. Sharma, *Expert Opin. Ther. Pat.*, 2011, **21**, 807.
123. V. E. Reeve, C. Boehm-Wilcox, M. Bosnic and E. Rozinova, *Int. Arch. Allergy Immunol.*, 1993, **102**, 101.
124. V. E. Reeve, M. Bosnic, C. Boehm-Wilcox and R. B. Cope, *Am. J. Clin. Nutr.*, 1995, **61**, 571.
125. R. Gugasyan, C. Losinno and T. Mandel, *Immunol. Lett.*, 1995, **46**, 221.
126. T. E. Mandel, M. Koulmanda and I. R. Mackay, *Clin. Exp. Immunol.*, 1992, **88**, 414.

Section 5
Steroid Hormone Receptors

CHAPTER 16

Non-steroidal Dissociated Glucocorticoid Receptor Agonists

HOSSEIN RAZAVI[a] AND CHRISTIAN HARCKEN*[b]

[a] Department of Medicinal Chemistry, Boehringer Ingelheim Pharmaceuticals, Inc., 900 Ridgebury Road, Ridgefield, CT 06877, USA; [b] Department of Research Networking & Strategic Planning, Boehringer Ingelheim Pharmaceuticals, Inc., 900 Ridgebury Road, Ridgefield, CT 06877, USA
*Email: christian.harcken@boehringer-ingelheim.com

16.1 Introduction

The powerful effects of adrenal extracts have been known since the first half of the twentieth century. This led to the isolation of glucocorticoids (GCs) in 1936, the synthesis of cortisone (1) in 1946 and the successful administration of cortisone in patients with rheumatoid arthritis in the 1950s (Figure 16.1). The 1950 Nobel Prize in Medicine was awarded for the characterization and isolation of GCs and the discovery of their anti-inflammatory properties. The strong anti-inflammatory effects of the subsequently developed synthetic GCs such as dexamethasone (2) and prednisolone (3) have made these some of the most successful drugs in modern medicine. However, it soon became evident that the long-term and/or high-dose administration of these agents was limited by a number of deleterious side-effects. These unwanted effects include increased susceptibility to infections, muscle wasting, skin thinning, GC-induced osteoporosis, metabolic changes (e.g. weight gain, Cushingoid

RSC Drug Discovery Series No. 26
Anti-Inflammatory Drug Discovery
Edited by Jeremy I Levin and Stefan Laufer
© The Royal Society of Chemistry 2012
Published by the Royal Society of Chemistry, www.rsc.org

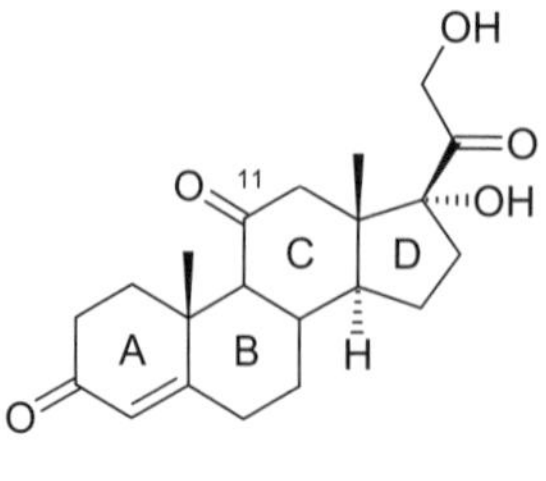

cortisone (1)

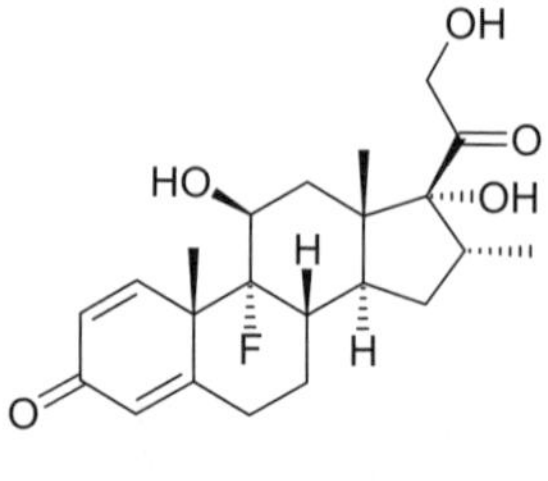

dexamethasone (2)

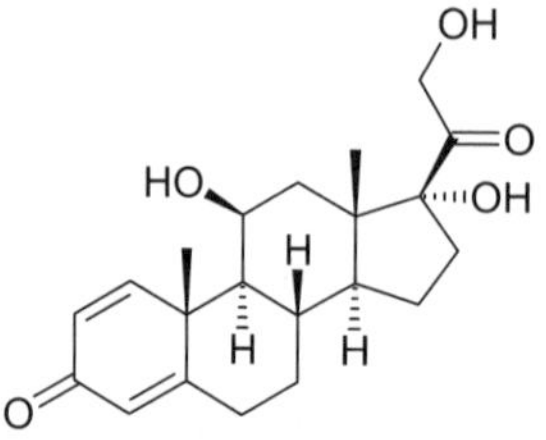

prednisolone (3)

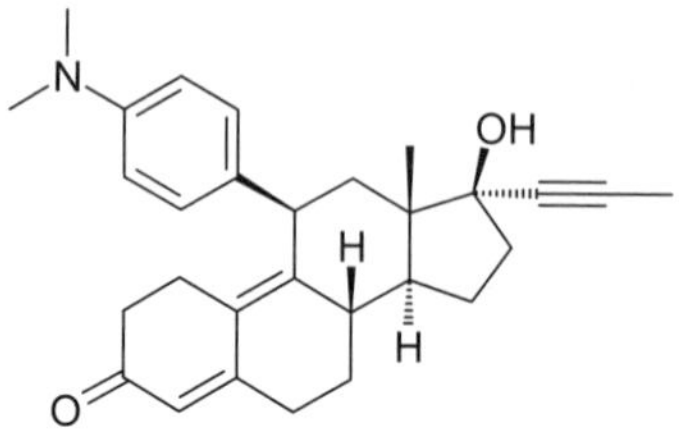

RU486 (4)

Figure 16.1 Steroidal glucocorticoids.

appearance and GC-induced diabetes), ophthalmological effects (*e.g.* cataracts and glaucoma), CNS effects such as hypothalamic-pituitary-adrenal (HPA)-axis suppression and cardiovascular effects.[1] GC-induced diabetes and especially osteoporosis are the most debilitating side-effects that severely limit the use of GCs in chronic diseases such as rheumatoid arthritis (RA).[2] In addition, cross-reactivity of some GCs with the progesterone receptor (PR), androgen receptor (AR) and mineralocorticoid receptor (MR) can elicit off-target pharmacology.[3]

Initially, research efforts were focused on reducing side-effects by optimizing the nuclear hormone receptor (NHR) selectivity of steroidal compounds. In 1994, two distinct functional mechanisms for glucocorticoid receptor (GR)-mediated effects of corticosteroids were postulated.[4] According to this hypothesis, upon cell entry, an agonist binds to the GR to form a receptor-ligand complex. The receptor-ligand complex dimerizes, translocates into the nucleus and acts directly as an endogenous transcription factor by binding to specific DNA sequences, GR responsive elements (GREs) and co-activator proteins, thereby increasing transcription (transactivation, TA) of metabolic and endocrine genes (Figure 16.2). This GR-mediated TA is believed to be the primary contributor to the side-effects of GC therapy.[5–8] Alternatively, the monomeric receptor-ligand complex can translocate into the nucleus and adopt an altered conformation with affinity for transcription factors such as nuclear factor κB (NFκB) and activating protein 1 (AP-1), thus suppressing the gene transcription (transrepression, TR) for pro-inflammatory cytokines including tumour necrosis factor-α (TNF-α) and interleukin 6 (IL-6). Although there is

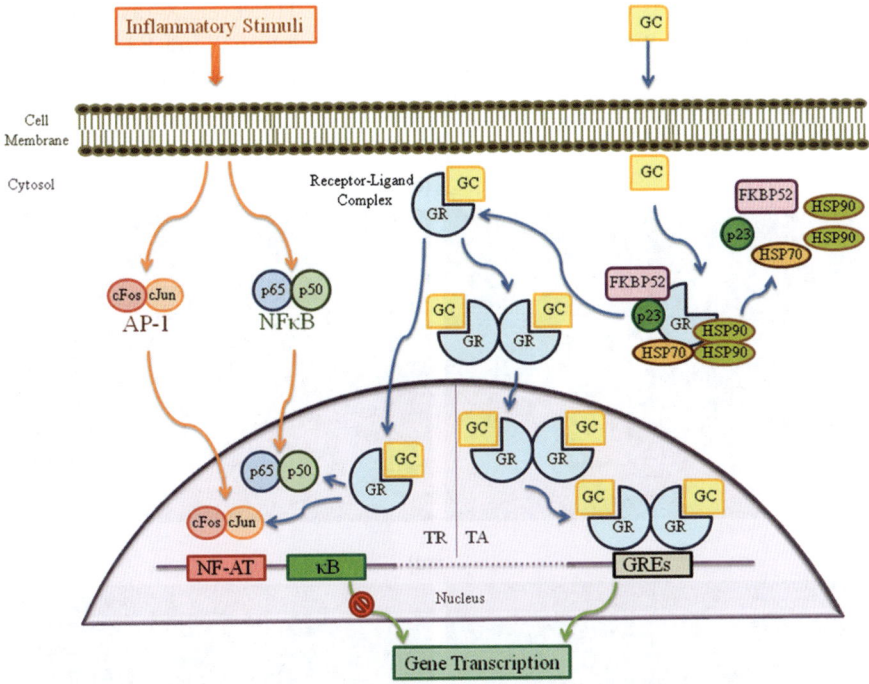

Figure 16.2 Classical mechanism for glucocorticoid signalling.

consensus that the desired anti-inflammatory effects of GCs are mainly mediated *via* TR,[9–12] the underlying molecular mechanisms for GC-induced side-effects are complex and only partially understood. Moreover, certain aspects of the anti-inflammatory effects of GCs are mediated *via* TA pathways.[13] Additional complexities in GR signalling include selective cofactor recruitment,[14,15] post-transcriptional mechanisms[16] and the presence of membrane-bound GR.[17]

The concept of GC's differential molecular regulation has served as the scientific rationale for the discovery of dissociated GR ligands. According to this classical approach, dissociation may be achieved by ligands that preferentially mediate TR while demonstrating little or no TA. A number of publications have described GR partial agonists that displayed slightly lower maximal efficacy in TR, and significantly diminished TA activity. The observed *in-vitro* dissociated profiles have been validated in pre-clinical animal models.[18–20] Although such partial agonists have shown reduced side-effects in clinical trials with topical administration, clinical data on systemically available GR partial agonists have not been published.

Recently, GR ligands with an alternative dissociated profile have been reported.[21–23] These compounds are full or partial TR agonists, and show full antagonism for TA (*vide infra*). Although the published clinical biomarker data for a compound from this class have shown a "dissociation window", it remains to be seen whether it will be maintained in advanced clinical testing.

GR apo-protein is unstable without chaperones. However, X-ray co-crystal structures of GR bound to dexamethasone, a full agonist,[24] and RU-486 (**4**), a full antagonist,[25] have been generated, which have helped to increase the understanding of the mechanism of GR dissociation (Figure 16.3). For instance,

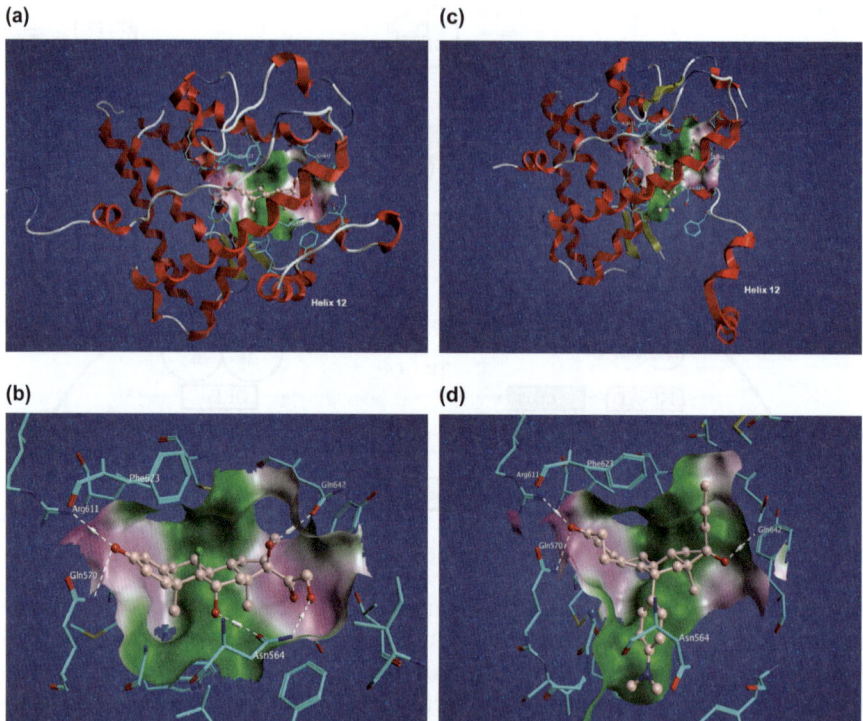

Figure 16.3   a) X-ray structure of GR-LBD in complex with dexamethasone (PDB entry 1M2Z).[24] b) A close-up view of GR-LBD in complex with dexamethasone. Lipophilic regions in the binding pocket are in green, hydrophilic regions are in magenta and potential hydrogen bonds are indicated by white lines. The binding pocket is largely hydrophobic. As depicted in this illustration, the A-ring carbonyl of dexamethasone forms hydrogen bonds with Gln570 and Arg611 residues. The C(11)- and C(21)-hydroxyl groups form hydrogen bonds to Asn564, and the C(17)-hydroxyl forms a hydrogen bond with Gln642. c) X-ray structure of GR-LBD in complex with RU-486 (PDB entry 1NHZ). Binding of RU-486 induces a significant movement of the helix 12, which prevents the binding of the co-activator peptide leading to "active" antagonism. d) A close-up view of GR-LBD in complex with RU-486. The A-ring carbonyl of RU-486 forms hydrogen bonds with Gln570 and Arg611 residues, and the C(17)-hyrdoxyl forms a hydrogen bond with Gln642. Binding of RU-486 to GR leads to the movement of Asn564 and helix 12. All illustrations were generated with the aid of GLIDE, version 5.7 (Schrödinger, LLC, New York, NY, USA, 2011), and rendered with Molecular Operating Environment (MOE), 2010.10 (Chemical Computing Group Inc., 1010 Sherbrooke St. West, Suite #910, Montreal, QC, Canada, H3A 2R7, 2010).

the binding of RU-486 (**4**) to GR has shown to cause a distinct conformational change to the receptor-ligand complex by displacing helix 12; hence, this ligand is referred to as an "active" GR antagonist.[26] The presence of additional induced binding pockets within the GR ligand binding domain (GR-LBD) have been reported.[19,20,27] Recently, the first X-ray structure of GR bound to a non-steroidal full agonist has been disclosed (*vide infra*).[28] Other technologies have also shaped the understanding of ligand-induced conformational changes. For example, the use of hydrogen/deuterium mass spectrometry has illustrated changes in solvent accessibility of different regions of the protein upon ligand binding.[29]

A number of previous reviews have summarized the activities in this field.[30–33] This chapter describes the current status of receptor selective and dissociated non-steroidal GR agonists. Steroidal compounds, and natural products such as Compound A[34] and ginsenosides[35] that have shown GR activity are not included in this article.

16.2 Hydroxy-trifluoromethyl-phenyl-pentanoic Aryl Amides and Related Trifluoromethyl Carbinols

One of the most explored structural classes of non-steroidal GR ligands is the hydroxy-trifluoromethyl-phenyl-pentanoic aryl amide derived compounds (Figure 16.4). The series is assumed to be derived from androgen receptor antagonists such as bicalutamide (**5**).[36] Schering AG (now Bayer-Schering) was the first to discover GR agonists from this series. As an example, **6** (GR $IC_{50} = 3$ nM) demonstrated TR activity in a lipopolysaccharide (LPS)-stimulated IL-8 inhibition assay in THP-1 cells, and *in-vivo* anti-inflammatory efficacy in croton oil-induced ear edema ($ED_{50} = 10$ mg/kg, s.q.).[37,38] The GR activity in this class resided solely with the (*R*)-enantiomer. Optimization in this series led to compounds such as ZK216348 (**7**) that retained GR potency, but lacked the desired NHR selectivity (GR $IC_{50} = 20$ nM, PR $IC_{50} = 20$ nM, MR $IC_{50} = 80$ nM).[39] Compound **7** showed potent *in-vitro* TR activity with moderate efficacy (IL-8 inhibition $IC_{50} = 35$ nM *vs.* 20 nM for prednisolone, 52% maximal efficacy), and a 60-fold separation in TA potency as compared to prednisolone in a tyrosine amino transferase (TAT) induction assay in a rat hepatoma cell line ($EC_{50} = 95$ nM *vs.* 1.5 nM for prednisolone). The compound displayed *in-vivo* prednisolone-like anti-inflammatory efficacy in a croton oil model of ear inflammation upon topical application with an ED_{50} of 0.03 µg/cm^2, and reduced local and systemic side-effects including the undesirable impact on skin thinning, body weight, thymus and spleen weight and blood glucose levels upon chronic dosing.

Further optimization of **7**, such as the replacement of the amide linkage with the corresponding amine and the benzoxazinone A-ring with a quinoline, led to analogues as exemplified by compound **8**, which displayed marked improvement in NHR selectivity (GR $IC_{50} = 1$ nM, PR, MR, AR $IC_{50} > 1000$ nM).[40] From this class, compound **9** (ZK245186, mapracorat) showed the best overall balance of *in-vitro* potency in TR (inhibition of IL-12 secretion $IC_{50} = 7$ nM,

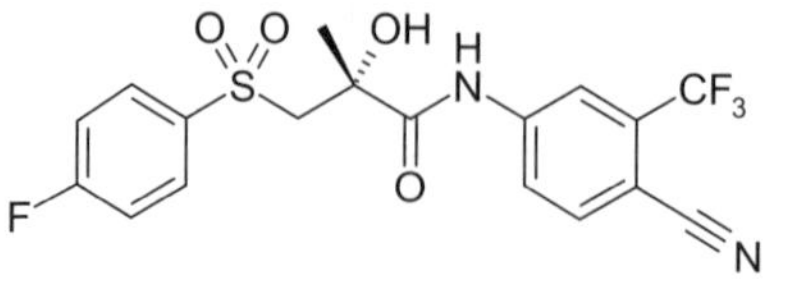

5

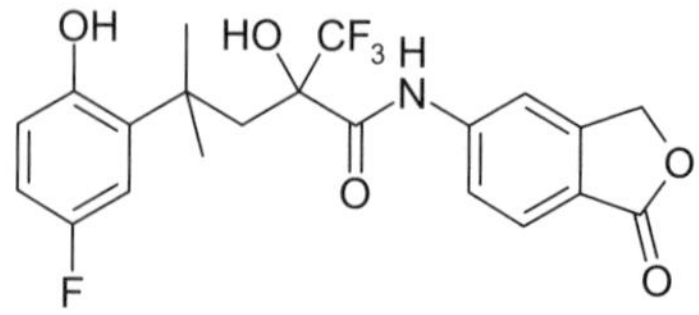

6

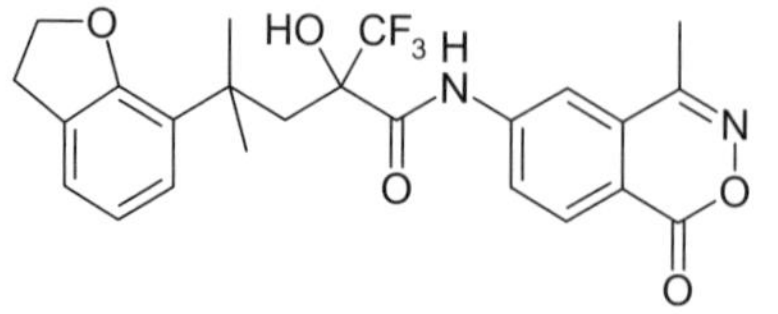

ZK-216348 **(7)**

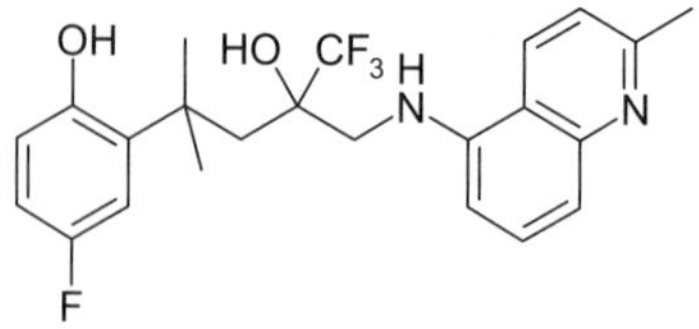

8

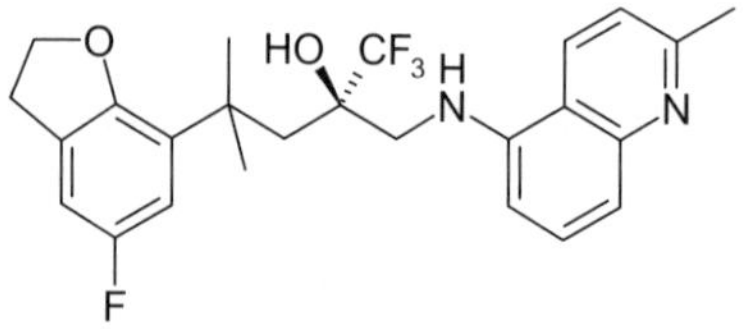

ZK-245186; mapracorat **(9)**

Figure 16.4 Schering's non-steroidal GR ligands.

83% maximal efficacy) and lower efficacy in TA assays measuring activation of the glucocorticoid-responsive mouse mammary tumour virus (MMTV) promoter in HeLa cells and TAT activity (MMTV 63% maximal efficacy, TAT 62% maximal efficacy *vs.* prednisolone).[18] This *in-vitro* profile translated into potent *in-vivo* topical anti-inflammatory activity in a croton oil ear edema model ($ED_{50} = 0.008\%$) with diminished effects on oral glucose tolerance test, and skin thinning as compared to prednisolone. Compound **9** showed rapid systemic clearance with a 2 h half-life upon i.v. dosing and low systemic bioavailability after topical administration.

Mapracorat (**9**) is one of the most advanced dissociated GCs. It was evaluated in Phase II clinical trials for atopic dermatitis by Intendis, a Bayer-Schering subsidiary. An ophthalmological suspension formulation is currently being evaluated by Bausch & Lomb in a Phase III study for post-operative pain and a Phase II trial for inflammation following cataract surgery and dry eye disorder.[41]

An alternative approach for modifying the hydroxy-trifluoromethyl-phenyl-pentanoic aryl amides was taken by Boehringer Ingelheim. Replacing the amide moiety of this chemotype with a saturated aliphatic linker, preferably a methylene group, yielded a new class of potent and dissociated GR partial agonists such as **10** (Figure 16.5).[42] This compound showed *in-vitro* activity in a human foreskin fibroblast (HFF) TR assay measuring IL-1-induced IL-6 production ($IC_{50} = 20\,nM$, 60% maximal efficacy), and reduced efficacy in an aromatase induction TA assay in HFF as determined by the production of β-estradiol in the presence of exogenously added testosterone (10% maximal efficacy). Further optimization in this series afforded compounds such as **11** that showed improved GR potency, NHR selectivity and metabolic stability, as well as potent *in-vivo* anti-inflammatory activity in a mouse LPS-induced TNF-α production model with an ED_{50} of 10 mg/kg, p.o.[43,44] The related aza-indole **12** demonstrated further improvement in potency, which could be rationalized based on hydrogen bonding interactions between the R611 and Q570 residues in the GR-LBD and the pyridine nitrogen of the aza-indole ring, as suggested by docking experiments.[19] Compound **12** showed good *in-vitro* TR activity with reduced TA, acceptable pharmacokinetic properties in rodents (rat i.v. $t_{1/2} = 1.7\,h$, 48% BA), potent *in-vivo* activity in an LPS-challenged acute anti-inflammatory mouse model with an ED_{50} of $<0.2\,mg/kg$, p.o., and prednisolone-like efficacy

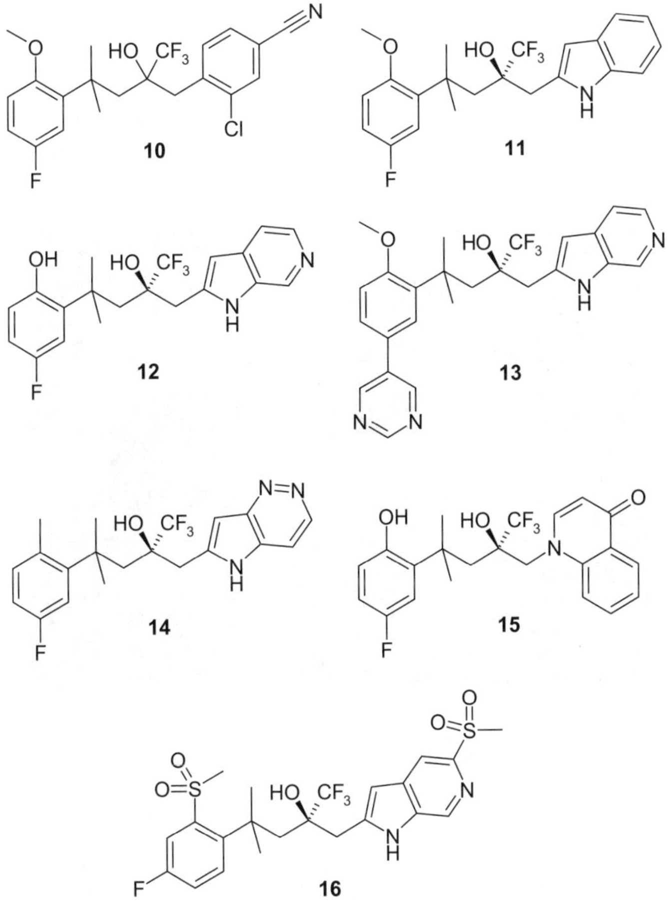

Figure 16.5 Boehringer Ingelheim's glucocorticoid mimetics.

in a therapeutic collagen induced arthritis (CIA) chronic murine model of rheumatoid arthritis ($ED_{50} = 10$ mg/kg, p.o., q.d.). After 5 weeks, significant positive differences in metabolic side-effect markers such as body weight, body fat, triglycerides, free fatty acids and insulin levels were observed in animals treated with **12** as compared to those treated with equi-efficacious doses of prednisolone.

Further improvement in dissociation was observed with compounds that occupy an expanded D-ring pocket in the GR-LBD (Figure 16.6). For instance, **13** showed high TR efficacy in an IL-1-induced IL-6 production assay ($IC_{50} = 50$ nM, 78% maximal efficacy) while displaying minimal efficacy in an aromatase induction TA assay (6% maximal efficacy).

Variation of the A-ring mimetic afforded partial agonists with lower *in-vitro* maximal efficacy, which translated into *in-vivo* dissociation.[45,46] For example, **14** showed an *in-vitro* dissociated profile, as well as efficacy and dissociation of bone relevant side-effect markers in a 5-week mouse therapeutic CIA model (100 mg/kg, p.o., q.d.). Whereas the replacement of the trifluoromethyl group with an alkyl group was previously shown to result in complete loss of agonist activity,[47] such a modification in the aza-indole containing compounds proved to be an effective approach in fine-tuning their partial agonist profile.[48]

Alternative A-ring mimics such as pyridones and quinolones were shown to provide potent and selective GCs.[49] For instance, quinolin-4-one **15**

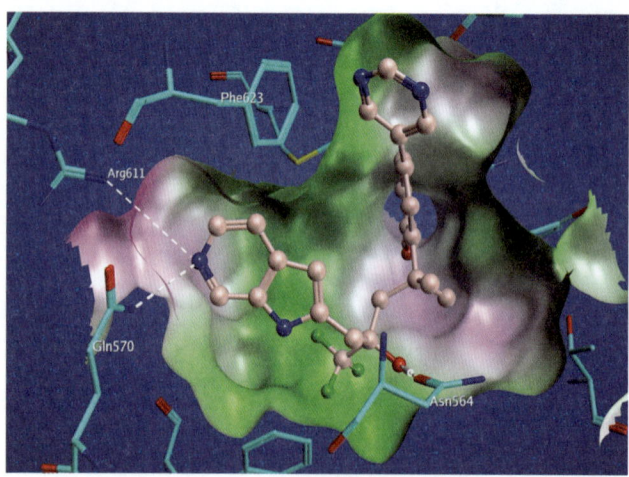

Figure 16.6   Docking pose of compound **13** in the GR-LBD shows that the pyrimidyl moiety of the ligand fills the space occupied by the furanyl moiety of fluticasone furoate. This illustration shows hydrogen bonds between the central hydroxyl group of **13** and Asn564, as well as the N(6) of the aza-indole moiety and the Gln570/Arg611 pair. The docking pose was generated from the fluticasone furoate X-ray structure[27] (PDB entry 3CLD) using GLIDE, and rendered with MOE. Lipophilic regions are in green, hydrophilic regions are in magenta and potential hydrogen bonds are indicated by white lines.

displayed a desirable dissociated *in-vitro* profile as assessed by MMTV-luciferase- (LUC) activation in HeLa cells, and acute *in-vivo* anti-inflammatory activity in an LPS-induced TNF-α production mouse model (55% inhibition at 10 mg/kg, p.o.).

Recent optimization for improved drug development properties such as increased half-life, reduced drug-drug-interaction potential, increased aqueous solubility and diminished hERG inhibition has provided compounds with more elaborate substitution patterns such as **16**.[50–52]

GlaxoSmithKline modelling experiments based on compounds such as **7** (ZK216348) have led to the identification of a novel series whereby a methyl substituent from the *gem*-dimethyl group was formally cyclized onto the D-ring mimetic moiety to yield an indane, tetrahydronaphthalene or benzosuberan (Figure 16.7).[53] For example, suberan **17** showed potent GR binding ($pEC_{50} = 8.57$), moderate TR activity in an NFκB reporter gene assay in an A549 lung epithelial cell line ($pEC_{50} = 6.84$, 60% maximal efficacy), moderate TA antagonism in a MMTV reporter gene antagonist assay in A549 cells ($pIC_{50} = 6.15$), and no MMTV agonism, which suggested a separation between TR and TA. The GR activity of compounds in this series primarily resided in one of the diastereomers of undetermined absolute and relative stereochemistry.

In order to optimize the TR potency and efficacy of this series, substitution on the D-ring mimetic was systematically investigated.[54] Due to synthetic accessibility, these studies were undertaken on analogues bearing the tetrahydronaphthalene moiety. A broad range of substituents was tolerated on the aromatic ring, but improvement in agonism was not detected. Substituents at the C(1)-position of the tetrahydronaphthalene, referred to as "agonist triggers", dramatically increased the agonist potency, ultimately leading to compounds such as **18** that showed potent GR binding ($pEC_{50} = 8.09$) with reduced NHR selectivity (equipotent against PR, 50- to > 100-fold selective over MR, AR and ER). Compound **18** displayed potent and efficacious TR activity in an NFκB reporter gene assay ($pEC_{50} = 8.69$, 92% maximal efficacy), potent TA antagonism in a MMTV antagonist assay with pIC_{50} of 7.27 and minimal TA agonism in an MMTV agonism assay (39% maximal efficacy), which translated into *in-vivo* anti-inflammatory efficacy in a delayed-type hypersensitivity (DTH) mouse model upon topical application with an ED_{50} of 0.05–0.25 µg as compared to 0.27 µg for dexamethasone. The low systemic metabolic stability of these compounds appears to have precluded the evaluation of their oral activity *in vivo*.

To increase systemic exposure, a large parallel screen of A-ring mimetics was undertaken, resulting in the identification of an amide linked *N*-aryl-pyrazole moiety.[55] The preferred compound, **19**, displayed potent GR binding affinity ($pEC_{50} = 8.02$), excellent NHR selectivity (> 100-fold), full TR efficacy in an NFκB reporter gene assay ($pEC_{50} = 8.07$), full TA antagonism in a MMTV reporter gene antagonist assay (93% maximal efficacy) and minimal TA agonism in an MMTV agonist assay (36% maximal efficacy). The first X-ray co-crystal structure of a non-steroidal agonist bound to the GR-LBD was obtained with compound **20** from the *N*-aryl-pyrazole series, which revealed an induced A-ring and an expanded D-ring pockets.[28,56] Analogous to the co-crystal structure of dexamethasone with GR, helix 12 was not displaced upon binding of compound **20** to GR, which was consistent with the full agonist profile of this compound in both TR and TA assays.

Alternatively, previously described A-ring moieties such as indoles, quinolones and quinolines (*vide supra*) were incorporated into the tetrahydronaphthalene series affording hybrids such as compound **21**.[57] This compound showed potent GR binding affinity ($pEC_{50} = 8.30$), subnanomolar potency in an NFκB TR assay ($pIC_{50} = 9.30$), and partial TA agonism and antagonism (MMTV maximal efficacy 48% and 37%, respectively). However, low metabolic stability prevented the evaluation of these compounds *in vivo*.

Further potency gains were attained by employing an *N*-phenylindazole A-ring that has also been reported in other non-steroidal agonist series (*vide infra*).[58] One of the most potent non-steroidal GCs, **22**, belongs to this structural class. It showed subnanomolar TR activity in an NFκB assay ($pEC_{50} = 10.1$, 105% maximal efficacy *vs.* fluticasone $pEC_{50} = 10.4$), but TA

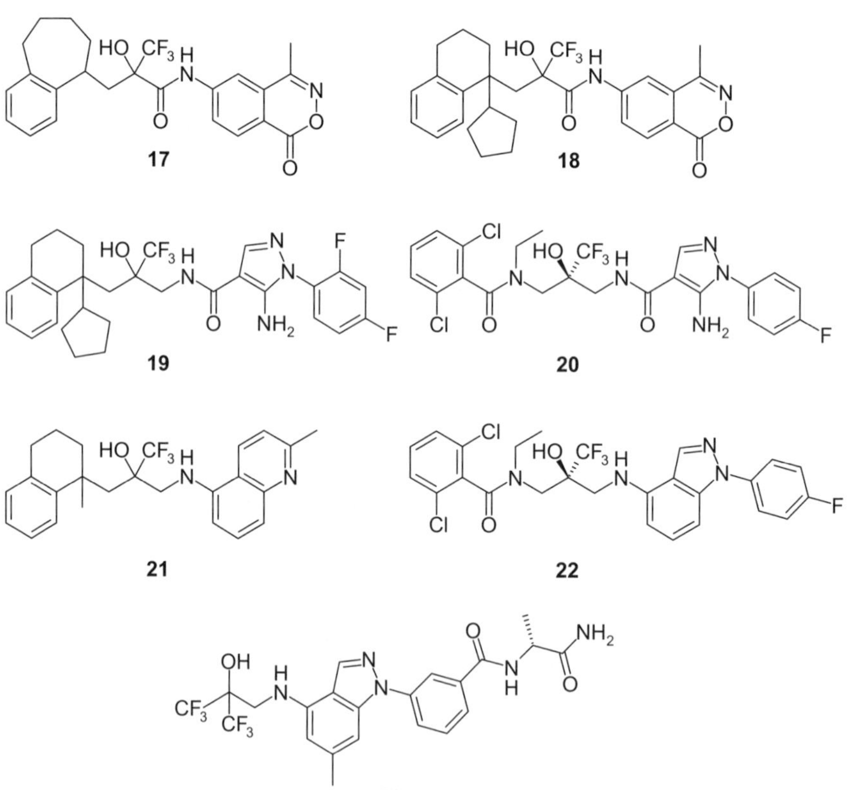

Figure 16.7   GlaxoSmithKline's trifluoromethyl carbinol-containing GR agonists.

data were not reported for this compound. This high-molecular-weight, highly lipophilic compound was not suitable for oral dosing. However, compound **22** could potentially be a candidate for inhaled application.

The recent discovery of a hitherto unoccupied "meta channel" that can accommodate larger polar A-ring substituents led to the identification of analogues with increased aqueous solubility.[20] This approach also resulted in the discovery of compounds with truncated D-ring moieties that only partially occupy the dexamethasone binding pocket. The resultant analogues maintained potent GR binding and agonist activity. As an example, *N*-phenylindazole **23** was potent and efficacious in the NFκB and MMTV assays with pEC_{50}s of 8.9 and 7.3, and maximal efficacies of 101% and 99%, respectively. Generally, the lower molecular weight of these compounds makes them more suitable for oral administration.

16.3 Dihydro-1*H*-[1]benzopyrano[3,4-f]quinolines

At first glance, dihydro-1*H*-[1]benzopyrano[3,4-*f*]quinoline (DBQ), the core of compound **24**, is a polycyclic heteroaromatic scaffold that is comprised of four contiguous fused rings, thus resembling the steroid carbon skeleton. Such structural similarities may contribute to the activity of DBQ-based compounds towards PR.[59-64] This inherent activity was leveraged by Abbott Laboratories and Ligand Pharmaceuticals in their joint venture to discover and develop non-steroidal GR ligands. Lead optimization efforts in the DBQ scaffold led to the discovery of AL-438 (**24**, Figure 16.8), which showed excellent potency and moderate GR selectivity (GR $IC_{50} = 2.5$ nM; PR/GR, AR/GR, ER/GR > 100; MR/GR $= 21$).[65] This compound demonstrated GR-mediated TR as evidenced by TNF/IL-1β-induced E-selectin-LUC activity in HepG2 cells and IL-1β-induced IL-6 production in HSF cells, which translated into anti-inflammatory activity in acute and chronic rat animal models of carrageenan-induced paw edema (CPE at 1, 3, 10 and 30 mg/kg, p.o., $ED_{50} = 11$ mg/kg) and adjuvant-induced arthritis (AIA at 3, 10 and 30 mg/kg, p.o., $ED_{50} = 9$ mg/kg), respectively. AL-438 also displayed an *in-vitro* and *in-vivo* dissociated profile. The compound exhibited a decreased ability to induce aromatase in HSF cells, and to inhibit osteocalcin (OC) and osteoprotegerin in MG63 osteoblast cell lines. Evaluation of the drug candidate and prednisolone in Sprague-Dawley rats at equivalent anti-inflammatory doses (AL-438 at 30 mg/kg, p.o., prednisolone at 10 mg/kg, p.o.) revealed that AL-438 had lesser effects on plasma glucose levels, bone formation and mineral apposition rates.

The results of additional SAR indicated that the C(10)-substituent (see compound **25**) binding pocket could optimally accommodate a moiety comprised of two heavy atoms such as a methoxy group.[66] Such substituents vastly improved the selectivity for GR over PR. Addition of a C(9)-hydroxyl group increased GR potency, but reduced selectivity over PR by approximately 4-fold. A preferred analogue, **25**, displayed potent and efficacious GR agonist activity with acceptable selectivity over PR (GR $IC_{50} = 0.95$ nM;

PR/GR = 157, NF-κB/AP-1 in HEPG2 cells using an E-selectin promoter $EC_{50} = 14$ nM, 96% maximal efficacy). In contrast to the C(9)-unsubstituted analogues, C(9)-hydroxyl derivatives demonstrated cross-species activity by inhibiting concanavalin A promoted T-cell proliferation in rat splenocytes and human whole blood assays. Compound **25** was dose responsive in the *in-vivo* rat CPE anti-inflammatory model with an ED_{50} of 16 mg/kg, p.o. as compared to an ED_{50} of 4 mg/kg, p.o. for prednisolone.

A subsequent disclosure expanded the SAR of the C(5)-allyl moiety of AL-438.[67] Hydrophilic substituents at the C(2′)- and C(3′)-positions of the allyl group resulted in a loss of potency and efficacy in E-selectin TR and MMTV-GRE TA assays, while small lipophilic groups afforded potent GR agonists with improved dissociation profiles, as exemplified by the tri-substituted olefin analogue **26** (GR $IC_{50} = 6$ nM; PR/GR = 267; E-selectin $EC_{50} = 23$ nM, 92% maximal efficacy; MMTV-GRE 16% maximal efficacy). Cyclization of the 2-methyl-2-pentenyl moiety of **26** led to the identification of cyclohexenyl derivatives such as **27**. The preferred absolute configuration at C(5) had previously been established to be (*S*) for AL-438, which corresponded to the (-)-enantiomer.[68] Accordingly, diastereomers of the cyclohexenyl analogue

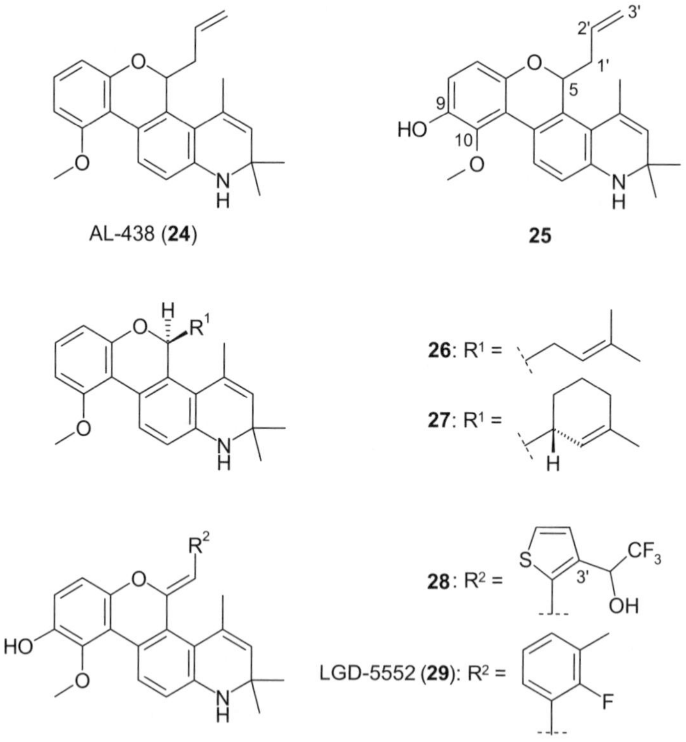

Figure 16.8   Abbott's and Ligand's DBQ-based GR compounds.

with this configuration at the C(5)-position exhibited more potent GR binding activity (((-)-*anti* **27** $IC_{50} = 0.7$ nM). Additionally, **27** showed 1000-fold selectivity for GR over PR, and was potent and efficacious in the E-selectin TR assay ($IC_{50} = 19$ nM; 90% maximal efficacy). As compared to prednisolone, **27** displayed lower levels of aromatase and TAT induction (26% and 49% maximal efficacy, respectively), as well as OC repression (38% maximal efficacy). Consistent with previous observations,[69] incorporation of a C(9)-hydroxyl into **27** resulted in PR agonism in an MMTV-PR-B luciferase reporter gene assay, which surpassed the efficacy of progesterone (190% maximal efficacy relative to progesterone).

The C(5)-position SAR was further extended by Ligand to include analogues with substituted arylidenes that are represented by compounds **28** and **29**.[70] These studies uncovered that substitutions at various positions on the aryl ring have profound effects on the functional activity of the GR ligand. In particular, *ortho*- and *meta*-substitutions furnished analogues with dissociated profiles, as *ortho*-substitution imparted partial agonism/antagonism and *meta*-substituents afforded full antagonist activity in the MMTV-LUC GRE TA assay while efficacy in the E-selectin TR assay was maintained for both substitution patterns. In contrast, *para*-substituted analogues were generally non-dissociated since they typically displayed full efficacy as agonists in the TA and TR assays.

The C(5)-thienylmethylidene derivatives, such as **28**, also displayed antiproliferative activities in RPMI8226 multiple myeloma cells.[71] Studies have indicated that IL-6 is an autocrine growth factor for myeloma cells, thus playing an important role in cell proliferation.[72] Additionally, GCs such as dexamethasone that inhibit IL-6 production are shown to have antiproliferative activity.[73] Analogues from the C(5)-thienylmethylidene series were potent GR binders (0.2 nM $< IC_{50} < 4.4$ nM). Generally, compounds containing polar protic C(3′)-substituents exhibited greater than 100-fold NHR selectivity, while those with C(3′)-hydrophobic groups were less selective. In addition to being selectivity-regulating elements, C(3′)-substituents also acted as function-regulating pharmacophores in this series. Although the majority of analogues showed dexamethasone-like efficacy in the IL-6 TR assay, some were partial or full agonists and others displayed full antagonism in MMTV-GRE activation agonist and antagonist assays, respectively. For example, C(3′)-piperidine amide showed 93% maximal efficacy in the MMTV-GRE agonist assay, while the trifluoromethyl carbinol analogue **28** exhibited 97% maximal efficacy in the MMTV-GRE antagonist assay. Several compounds with desirable *in-vitro* profiles were assessed in a four-week *in-vivo* xenograft cancer model. In this study, nude mice having RPMI8226 tumours were orally treated with various doses of test compounds, and the effects were determined as percent reduction in tumour volume. All tested compounds demonstrated dexamethasone-like activity at their highest dose. For example, thiophene analogue **28** was evaluated at once-daily oral doses of 3, 10 and 30 mg/kg, and displayed significant reduction in tumour burden with the 30 mg/kg dose group showing comparable efficacy to the 2 mg/kg of dexamethasone.

This research culminated in the discovery of the development candidate LGD5552 (**29**), which showed good GR potency ($K_i = 2.4$ nM) and moderate to good selectivity (PR/GR = 360, AR/GR = 379, MR/GR = 62).[23] The candidate was potent and efficacious in the E-selectin TR assay, and displayed dissociation based on low efficacy in the MMTV-GRE agonist assay and activity in the MMTV antagonist assay (E-selectin $EC_{50} = 2$ nM, 100% maximal efficacy, MMTV maximal efficacy < 5%). *In-vitro* gene regulation studies of pyruvate dehydrogenase kinase 4 (PDK4) and phosphoenol pyruvate carboxykinase (PEPCK), biomarkers for glucose homeostasis, in H4IIE liver cells revealed that LGD5552 did not induce PDK4 at any tested concentration, and only reached the prednisolone response level in the transcription of PEPCK at doses of 1 and 10 µM. The observed effects were rationalized in terms of ligand-induced structural changes in the receptor-ligand complex that altered the recruitment of co-activator proteins. To evaluate the *in-vivo* anti-inflammatory efficacy of LGD5552, female DBA/1 mice were treated with once-daily oral doses of 3, 10 and 30 mg/kg of the candidate and prednisolone for 15 days in a therapeutic CIA study. In this study, LGD5552 and prednisolone reduced the severity of disease in a similar dose-responsive manner. To assess the *in-vivo* side-effect profile of LGD5552, male Swiss-Webster mice were dosed orally with the compound or prednisolone for 28 days (3, 10 and 30 mg/kg, q.d.). In contrast to prednisolone-treated mice, LGD5552-dosed animals did not show an increase in body fat at the 10 and 30 mg/kg doses. In addition, histomorphometry analysis of cortical bones from the treated mice revealed a decrease in the rate of bone formation at all doses of prednisolone, and only at the highest dose of LGD5552. In 2007, Ligand reported that the development of this compound was stopped due to the lack of the desired safety profile in GLP preclinical studies.[74]

16.4   Dihydro- and Tetrahydroquinolines

Another starting point for the discovery of new GR scaffolds was the excision of the DBQ's pyranyl ring to unveil a 1,2-dihydroquinoline with a pendant C(6)-methoxyphenyl moiety. This design strategy, in conjunction with translocation of the DBQ's C(5)-pharmacophore to the C(4)-position of the resultant core, was an approach employed at Boehringer Ingelheim (Figure 16.9). Among various substituents examined at C(4), an allyl thioether was the only moiety that imparted affinity for GR. For example, compound **30** had a GR IC_{50} of 360 nM, although it was equipotent against PR.[75] Replacement of the allyl group with phenethyl, and addition of fluorine to the C(6)-methoxyphenyl substituent led to incremental improvement in GR potency and enhanced selectivity over PR as shown in analogues **31** (GR $IC_{50} = 194$ nM; PR $IC_{50} > 2000$) and **32** (GR $IC_{50} = 84$ nM; PR $IC_{50} > 2000$). Installation of a C(4)-allyl ether and substitution at the benzylic ether position furnished dihydroquinoline **33**, which further improved the GR potency and imparted functional activity in the IL-1β-induced IL-6 TR assay (eutomer GR

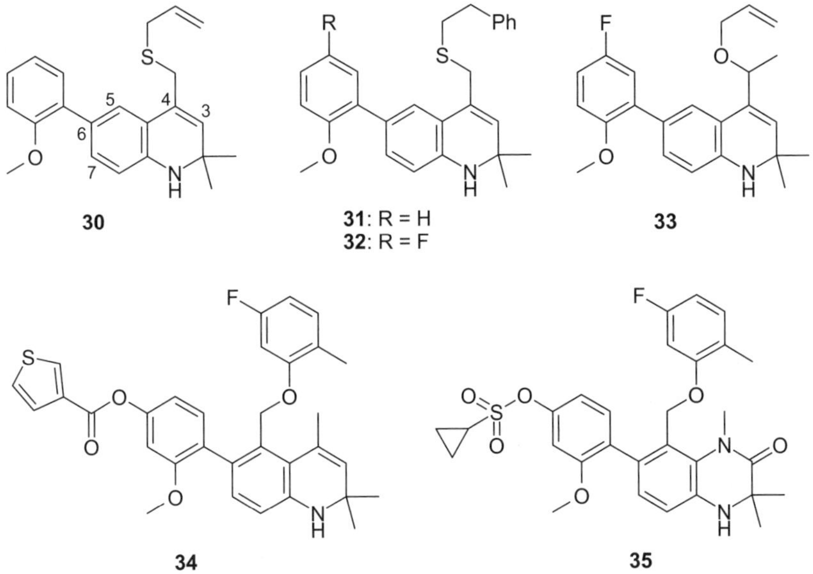

Figure 16.9 Boehringer Ingelheim's and Santen's GR ligands based on the dihydro-quinoline and the tetrahydroquinoxaline cores.

$IC_{50} = 40$ nM; PR $IC_{50} = 450$ nM; IL-6 $EC_{50} = 260$ nM, 46% maximal efficacy.[76]

Recent patent publications from Santen Pharmaceutical have disclosed a series of dihydroquinoline GR agonists useful for the treatment of ocular inflammatory diseases. For example, compound **34** inhibited LPS-induced IL-6 production in human corneal epithelial cells (85% maximal efficacy).[77] Related 1,2,3,4-tetrahydroquinoxaline derivatives have also been reported as GR agonists. For instance, **35** showed dexamethasone-like efficacy in the IL-6 assay (100% maximal efficacy), and *in-vivo* anti-inflammatory activity in a rat CPE model at 10 mg/kg, p.o.[78]

The scope of partially saturated quinolines as GR modulators has been expanded with recent disclosures from Ligand that detail their lead identification and optimization efforts based on a 1,2,3,4-tetrahydroquinoline core (Figure 16.10). Originally a PR antagonist scaffold,[79,80] a C(6)-connected thiophene analogue in this series that had displayed a moderate GR affinity was leveraged for the development of a new GR agonist chemotype.[81] Initially, a C(6)-indole analogue had shown modest GR potency and selectivity, albeit with a lack of cellular activity. An isomeric indole variant **36** displayed potent GR affinity ($K_i = 0.6$ nM), moderate selectivity (PR $K_i = 13$ nM, MR $K_i = 21$ nM, AR $K_i = 57$ nM) and potent TR activity in a TNFα/IL-1β-induced E-Selectin assay ($IC_{50} = 10$ nM, 90% maximal efficacy). In this series, substituents at the C(4')- and C(5')-positions of the indole acted as function-regulating pharmacophores with larger groups such as chloro and methoxy affording analogues that displayed TA antagonism and no appreciable TR activity, while a compound bearing a smaller fluoro moiety showed agonism in both TA and TR assays. However, the C(6′)-methoxy analogue **37** exhibited good GR potency ($K_i = 1.5$ nM), excellent NHR selectivity (PR/GR = 933, MR/GR = 100, AR/GR = 1000), and moderate dissociation (E-Selectin $IC_{50} = 33$ nM, 84% maximal efficacy *vs*. GRE agonist activation $EC_{50} = 48$ nM, 60% maximal efficacy).

Alternatively, NHR cross-reactivity was improved by the addition of a C(3)-hydroxyl group, which also increased TR in the IL-6 assay and enhanced the overall polarity of the molecule. As an example, compound (±)**38** (GR $K_i = 1.7$ nM) showed >470-fold selectivity over PR, MR and AR, while displaying higher potency and efficacy in the IL-6 assay ($IC_{50} = 16$ nM, 96% maximal efficacy) as compared to **36** (IL-6 $IC_{50} = 59$ nM, 67% maximal efficacy).[82] However, **38** suffered from metabolic instability, and did not advance into *in-vivo* models. The lack of microsomal stability was alleviated by the incorporation of fluorines at C(5)- and C(7)-position of the tetrahydroquinoline core.[83] The resultant analogue **39** showed improved microsomal stability (RLM $t_{1/2} = 30$ min., HLM $t_{1/2} = 48$ min.), maintained a desirable *in-vitro* profile (GR $K_i = 2.5$ nM; PR, MR, AR selectivity >500-fold; IL-6 repression $IC_{50} = 11$ nM, 90% maximal efficacy) and displayed anti-inflammatory activity in an acute rat model of CPE ($ED_{50} = 6.6$ mg/kg, p.o.).

Despite its favourable profile, **39** was not a dissociated GR agonist as determined by the MMTV-GRE TA agonist assay. A dissociated profile was achieved based on SAR studies conducted at the C(3)- and C(4)-positions of the tetrahydroquinoline scaffold. In particular, a separation between TR as measured by an IL-6 assay in primary neonatal human dermal fibroblast cells and TA as determined by a PEPCK-LUC assay in H4IIEC3 rat liver cells was

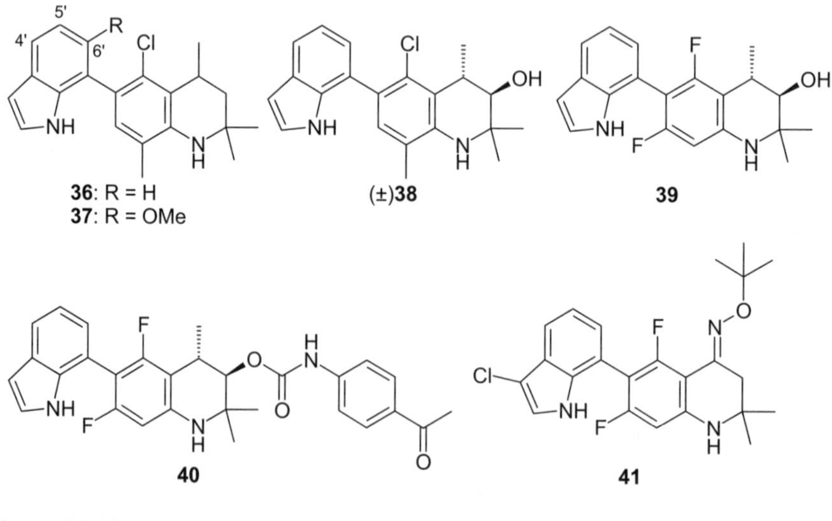

36: R = H
37: R = OMe

(±)**38**

39

40

41

Figure 16.10   Ligand's tetrahydroquinoline-containing GR agonists.

observed for carbamate **40**[84] (IL-6 $IC_{50} = 31$ nM, 95% maximal efficacy; PEPCK $EC_{50} = 182$ nM, 44% maximal efficacy) and oxime **41**[85] (IL-6 $IC_{50} = 19$ nM, 83% maximal efficacy; PEPCK $EC_{50} = 67$ nM, 27% maximal efficacy). Both compounds showed potent GR affinity (**40** $K_i = 3.8$ nM, **41** $K_i = 2$ nM), and moderate to good NHR selectivity (**40** 60 to > 100-fold, **41** 45 to > 100-fold).

16.5 Fluorocortivazol-derived Scaffolds

Modification of existing cores through the incorporation or excision of structural motifs is a strategy that has been successfully employed in lead identification (*vide supra*). Such an expedient was employed in the design of the *N*-aryl pyrazole scaffold starting from known steroidal GR agonists such as fluorocortivazol (**42**) and cortisol. The SAR of this novel GR scaffold has been published by two independent groups.

In 2004, researchers at UCSF published their discovery of the *N*-aryl pyrazole core and exploration of its SAR.[86] The first-generation analogues that showed moderate to low affinity for GR ($IC_{50} > 165$ nM) were derived from the excision of cortisol's C- and D-ring. The addition of a fused *N*-aryl pyrazole group to the A-ring of the first-generation scaffold provided analogues with enhanced GR potency ($IC_{50}s < 20$ nM). In particular, **43** (Figure 16.11) was one of the most potent GR ligands reported with an IC_{50} of 1.8 nM. This analogue displayed agonism in TR assays in osteosarcoma cells (AP-1 $EC_{50} = 1.8$ nM, 71% maximal efficacy; NF-κB $EC_{50} = 22$ nM, 67% maximal efficacy), and dissociation in CV-1 cells measuring TAT induction (44% maximal efficacy). In general, most analogues were not dissociated against MMTV, and some compounds, such as **43** ($EC_{50} = 20$ nM, 112% maximal efficacy), exhibited dexamethasone-like efficacy.

In a separate study, *N*-arylpyrazoles were examined for their ability to modulate the proliferation and differentiation of GC-responsive cells.[87] Proliferation of A549 human lung adenocarcinoma cells due to pro-inflammatory signals is inhibited by steroidal GR agonists such as dexamethasone. However, dexamethasone also induces the differentiation of pre-adipocytes, and inhibits the differentiation of pre-osteoblasts that are indicative of adipose tissue and bone side-effects. In this study, a majority of the tested *N*-arylpyrazoles showed a moderate effect in inhibiting the growth of A549 cells, while a subset did not induce or weakly induced the differentiation of mouse pre-adipocytes. As an example, **43** exerted a weak effect on the proliferation of A549 cells, and the induction of pre-adipocyte differentiation. Interestingly, mouse pre-osteoblast differentiation was not inhibited by any of the tested *N*-arylpyrazoles. Additionally, a differential expression pattern was observed when *N*-arylpyrazoles were tested against a panel of 17 GRE-containing genes in A549 cells. For instance, **43** increased the transcription of all of the dexamethasone-activated genes. However, among the six TNFα-induced genes, **43** only inhibited the expression of IL-6 gene. Also, **43** selectively induced one of the four "glucocorticoid-repressed, but not TNFα-induced" genes. This study highlighted the complexities of GC-mediated signalling in A549 cells. It also bolstered the hypothesis that minor structural changes to a ligand might sufficiently alter the overall topology of its complex with the GR to affect cofactor recruitment, resulting in differential transcription of the GRE containing genes.

Benzothiophene **43** was also independently disclosed by researchers at Merck.[88] A partial agonist in a human and a mouse IL-6 assays (hIL-6 86% maximal efficacy, mIL-6 78% maximal efficacy), **43** was dissociated in a human TAT (21% maximal efficacy) and a mouse glutamine synthetase (GS) assay (33% maximal efficacy). Compound **43** was also efficacious in a mouse LPS model (110% reduction of TNF-α at 30 mg/kg, p.o. as compared to 100% reduction for prednisolone at 3 mg/kg).

Additional analogues from the *N*-aryl pyrazole scaffold that contain a C(1)-secondary alcohol have also been reported by Merck.[89] In general, these compounds displayed a dissociated profile; however, C(1)-tertiary alcohols such as **44** were more potent. Compound **44** showed excellent affinity towards GR ($IC_{50} = 0.8$ nM), TR activity in an IL-6 production assay in A549 epithelial cells ($EC_{50} = 1$ nM, 97% maximal efficacy), and a modest dissociation window in TA assays measuring TAT induction in human HepG2 cells ($EC_{50} = 36$ nM, 69% maximal efficacy) and GS induction in human skeletal muscle cells ($EC_{50} = 14$ nM, 64% maximal efficacy). The potent and efficacious *in-vitro* TR activity of **44** was confirmed *in vivo*, as the compound displayed a potent anti-inflammatory activity in a mouse model of LPS-induced TNF-α production (ED_{50} of 4 mg/kg, p.o., as compared to ED_{50} of 0.5 mg/kg, p.o. for prednisolone).

The related 6,5-bicyclic analogues in this scaffold have also exhibited *in-vitro* dissociated profiles, as exemplified by compound **45** (GR $IC_{50} = 2.1$ nM; hIL-6 $EC_{50} = 5$ nM, 71% maximal efficacy; hTAT 29% maximal efficacy).[90] In line with the SAR observed in the 6,6-bicyclic systems, the C(1) (*S*)-isomer **46** (hIL-6 92% maximal efficacy, hTAT 62% maximal efficacy) was more efficacious and less dissociated in TR assays than the (*R*)-epimer **45**. Both compounds demonstrated prednisolone-like anti-inflammatory activity in a mouse LPS model achieving full inhibition of TNF-α production at 30 mg/kg as compared to prednisolone at 3 mg/kg. However, **45** suffered from the lack of a PK/PD correlation, which was attributed to the *in-vivo* formation of an active metabolite that was presumed to be the benzothiophene dioxide derivative (GR $IC_{50} = 2.8$ nM; hIL-6 $EC_{50} = 2$ nM, 92% maximal efficacy).

An additional publication has detailed the SAR of 6,5- and 6-6-bicyclic *N*-aryl pyrazoles bearing ketal moieties of various sizes. Analogues in this series demonstrated moderate to good GR potency and selectivity. Their typical partial agonist profiles in human and mouse IL-6 assays were sufficient to elicit *in-vivo* anti-inflammatory responses. As an example, **47** exhibited moderate potency and efficacy in a TR assay (mIL-6 $EC_{50} = 61$ nM, 56% maximal efficacy), which translated into *in-vivo* potency in a mouse LPS-induced TNF-α production model ($ED_{50} = 14.1$ mg/kg, p.o.).[91]

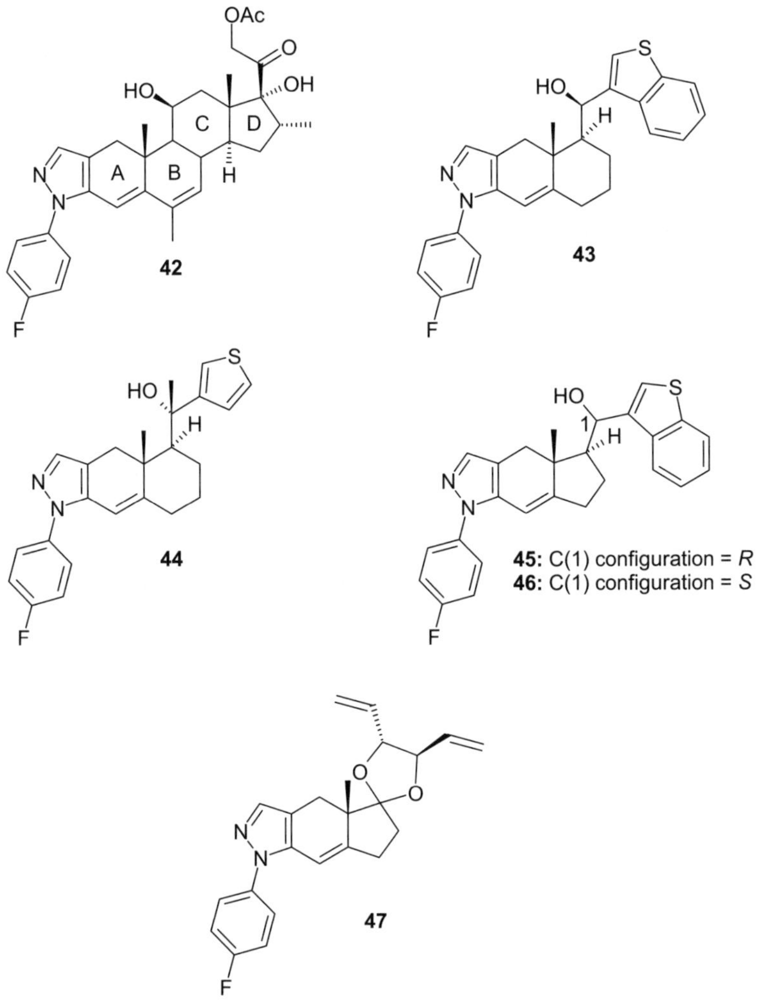

Figure 16.11 *N*-aryl pyrazole GR ligands from UCSF and Merck, and Fluoro-cortivazol (**42**).

Merck has remained active in this field as evident from their recent patent applications, which have disclosed a series of GR partial agonists that contain sulfonamide, amide and urea linkers (Figure 16.12).[92,93] Compounds of the preferred embodiments showed GR potency (inflection points or i.p. < 300 nM), TR efficacy of 40% to 80% in a TNFα-β-lactamase reporter gene assay, and maximal TA efficacy of less than 60% in a MMTV-LUC assay. Examples of the preferred embodiments included **48** (GR i.p. = 1.8 nM; TR i.p. = 187 nM, 78% maximal efficacy, TA i.p. = 247 nM, 38% maximal efficacy) and **49** (GR i.p. = 7.9 nM; TR i.p. = 182 nM, 77% maximal efficacy; TA i.p. = 187 nM, 18% maximal efficacy).

Other Merck patent applications have disclosed hydroxylated derivatives of hexahydrocyclopentyl indazole such as **50** (GR i.p. = 2.96 nM; TR i.p. = 125 nM, 73% maximal efficacy; TA i.p. = 197 nM, 37% maximal efficacy).[94] There has been a significant amount of patent activity around this compound including a process application,[95,96] leading to speculations that **50** was a potential development candidate. A more recent Merck application has claimed dihydroxylated analogues that show unexpected properties such as better "solubility and *in-vivo* selectivity" as compared to the mono-hydroxy derivatives.[97] As an example, **51** (GR i.p. = 6.7 nM; TR i.p. = 95 nM, 71% maximal efficacy; TA i.p. = 116 nM, 65% maximal efficacy) had 148 μg/mL solubility at pH 2, displayed anti-inflammatory activity that was greater or equal to 40% of the maximum response by dexamethasone at 0.1 mg/kg or methylprednisolone at 1.3 mg/kg in a rat oxazolone-challenged contact dermatitis model and did not elevate serum glucose levels as compared to vehicle.

Bristol-Myers Squibb has also been actively seeking a dissociated GR agonist (*vide infra*). A recent publication described their approach from a fluorocortivazol-derived hexahydroimidazo[1,5b]isoquinoline scaffold.[98] In contrast to the *N*-aryl pyrazoles, the C(1)-secondary alcohol analogues in this series

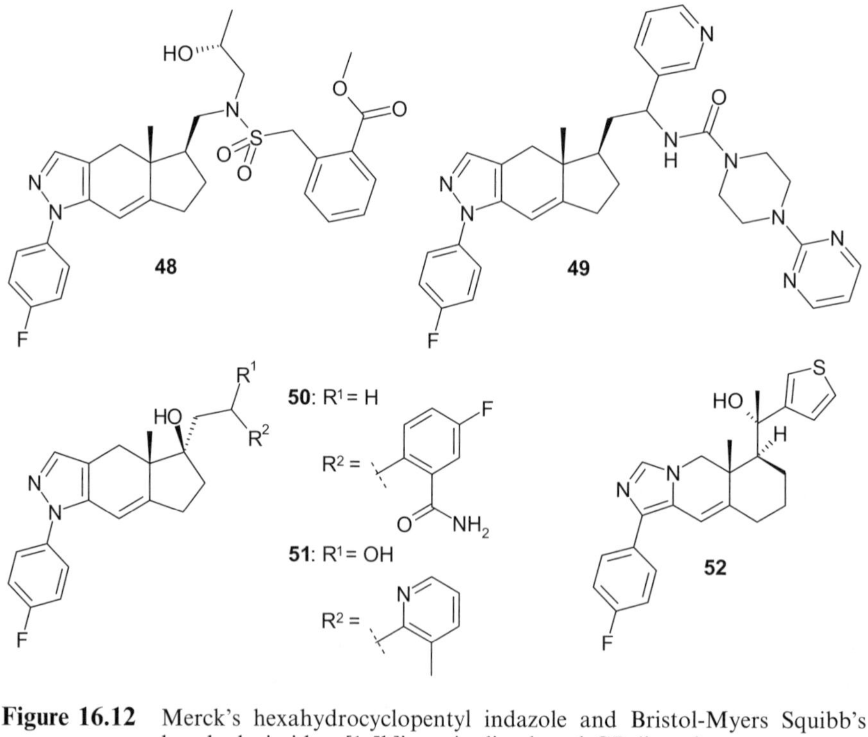

Figure 16.12 Merck's hexahydrocyclopentyl indazole and Bristol-Myers Squibb's hexahydroimidazo[1,5b]isoquinoline-based GR ligands.

showed affinity for GR, but lacked TR activity. Functional activity was rescued by additional substitution at the C(1)-position to form tertiary alcohol derivatives. For instance, compound **52**, a direct comparator of **44**, was a partial agonist in both phorbol myristate acetate (PMA)-induced AP-1-LUC and IL-1β-induced endothelial leukocyte adhesion molecule (ELAM)-LUC TR reporter gene assays in human A549 lung epithelial cells, and dissociated in the agonist mode of a GR activated GAL-4-LUC TA assay in a NP-1 HeLa cell line (GR $K_i = 1$ nM; AP-1 $EC_{50} = 5$ nM, 61% maximal efficacy; ELAM $EC_{50} = 37$ nM, 68% maximal efficacy; NP-1 agonist $EC_{50} > 10$ μM).

16.6 *N*-Aryl Indazole Analogues

In 2006, Boehringer Ingelheim disclosed a series of novel GR agonists that contained an *N*-aryl indazole A-ring mimetic.[99,100] A 2010 report from Albany Molecular Research Institute and Albany Medical College described their SAR studies in this series (Figure 16.13).[101] A preferred compound in this study, the eutomer of **53**, showed good potency for GR ($IC_{50} = 34$ nM), greater than 1000-fold selectivity over other tested NHRs, and potent TR with moderate separation against TA (IL-6 $EC_{50} = 19$ nM, 87% maximal efficacy; MMTV-LUC reporter gene assays $EC_{50} = 350$ nM, 70% maximal efficacy).

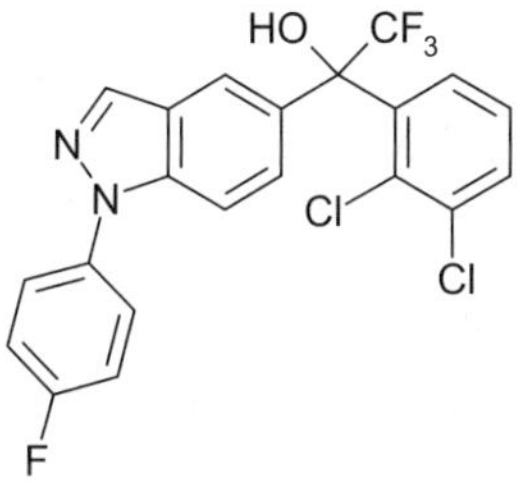

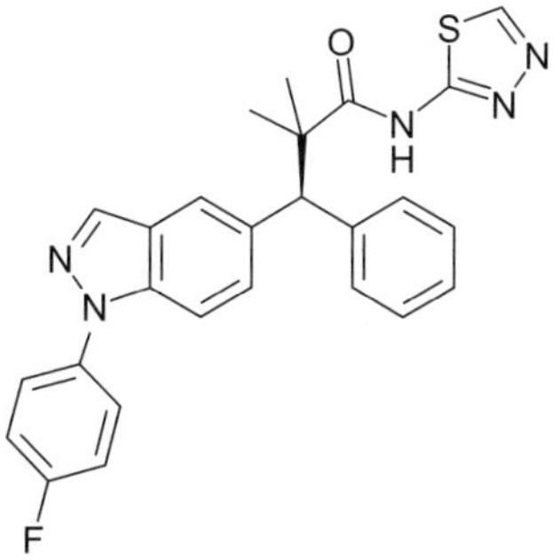

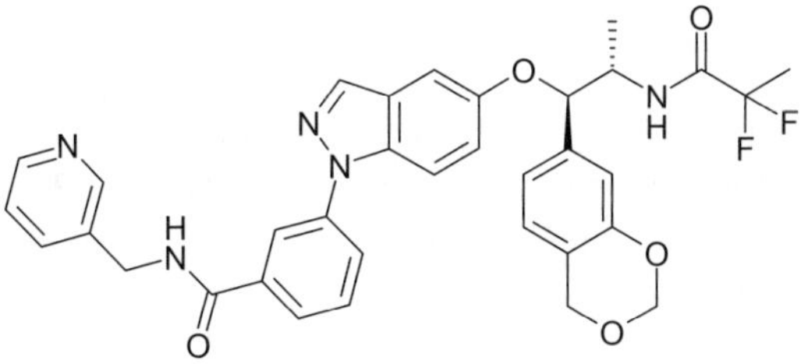

Figure 16.13 *N*-aryl indazole-containing GR ligands.

Moreover, this compound was potent in a mouse acute inflammatory model of LPS-induced TNF-α production (ED$_{50}$ = 6 mg/kg, p.o.).

A phenyl indazole GR scaffold has also been disclosed in patent publications by Bristol-Myers Squibb.[102,103] In contrast to the aforementioned analogue, this series of compounds lacks the central trifluoromethyl carbinol. A representative example, compound **54**, showed good GR binding and TR potency (GR IC$_{50}$ = 5 nM, AP-1 EC$_{50}$ = 2.4 nM). An indazole-containing series has also been reported by AstraZeneca/Bayer-Schering in the patent literature. For example, **55** displayed excellent GR binding and TR potency (GR IC$_{50}$ = 0.8 nM, AP-1 IC$_{50}$ = 0.02 nM).[104–106]

16.7 Additional Fused Cores

Bristol-Myers Squibb identified a series of dihydro-9,10-ethano-anthracene-11-carboxamides as novel GR ligands through virtual screening (Figure 16.14).[107] The initial hit was optimized to provide amidothiazole **56**, which showed potent GR binding (IC$_{50}$ = 1 nM), considerable PR activity (IC$_{50}$ = 31 nM) and partial MR antagonism (49% maximal efficacy). This compound exhibited a partial GR agonist profile in a PMA-induced AP-1-LUC repression assay in human A549 lung epithelial cells (EC$_{50}$ = 5.3 nM, 73% maximal efficacy) and reduced GRE activation in a GAL-4-LUC assay in NP-1 HeLa cells (EC$_{50}$ = 117 nM, 31% maximal efficacy).

In an effort to increase synthetic tractability, further optimization of this scaffold led to truncated analogues such as diphenylpropanamides and tricyclic xanthenes.[108] For example, tricyclic xanthene **57** displayed potent GR binding affinity (IC$_{50}$ = 0.8 nM), moderate PR selectivity (IC$_{50}$ = 53 nM) and a typical partial agonist profile with potent TR (AP-1 repression EC$_{50}$ = 7.8 nM, 82% maximal efficacy) and reduced TA as measured by TAT induction in a cultured hepatoma cell line (EC$_{50}$ = 91 nM, 45% maximal efficacy). This compound also showed good oral *in-vivo* anti-inflammatory efficacy in a CPE model with an ED$_{50}$ of 44 mg/kg as compared to an ED$_{50}$ of 19 mg/kg for prednisolone, as well as reduced side-effects in biomarkers such as TAT and serum glucose concentration six hours after a single oral dose. This series has evolved to include compounds such as **58a**, which displayed higher TR potency and possible improvement in drug-like properties (AP-1 EC$_{50}$ = 3.36 nM).[109]

Clinical development of a glucocorticoid has not been reported by BMS. However, BMS has disclosed the development of amorphous formulations for two compounds, BMS-776532 and BMS-791826, in support of Phase I clinical trials.[110] Recently, BMS-776532 (**58b**, GR K$_i$ = 1.9 nM; AP-1 TR EC$_{50}$ = 33 nM, 70% maximal efficacy; Gal4 TA EC$_{50}$ = 242 nM, 32% maximal efficacy) and BMS-791826 (**58c**, GR K$_i$ = 1.6 nM; AP-1 TR EC$_{50}$ = 18 nM, 79% maximal efficacy; Gal4 TA EC$_{50}$ = 99 nM, 57% maximal efficacy) were reported to be glucocorticoid receptor modulators that displayed anti-inflammatory activity in a 21-day rat prophylactic AIA model with ED$_{50}$s of 17 mg/kg, p.o. and 8 mg/kg, p.o., respectively.[111]

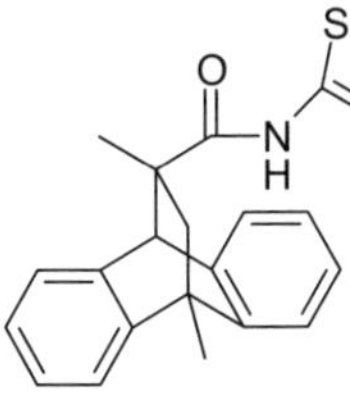

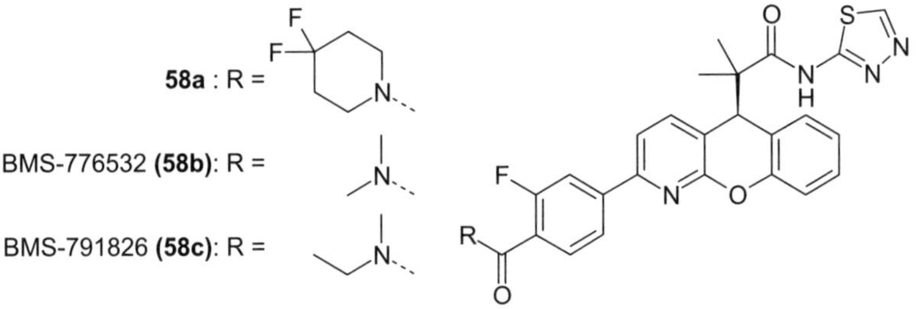

Figure 16.14 Bristol-Myers Squibb's non-steroidal GR agonists.

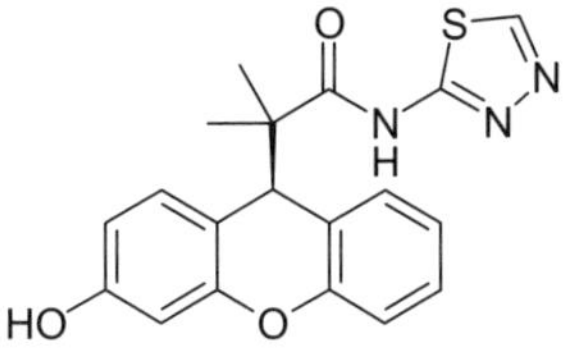

Derived from a GR antagonist scaffold that took inspiration from steroidal structures, Pfizer identified an octahydrophenanthrene series of GR agonists (Figure 16.15). The lead compound, **59**, was a potent GR ligand ($IC_{50} = 6.2$ nM) that exhibited potent TR in an IL-1-stimulated IL-8 production assay in SW1353 human chondrosarcoma cells ($EC_{50} = 34$ nM, 75% maximal efficacy) and reduced TA of the MMTV promoter in the same cell line (9% maximal efficacy).[112] However, **59** suffered from low ER selectivity ($ER\alpha$ $EC_{50} = 93$ nM), and contained a number of metabolically labile and potentially reactive functional groups.

Modification of the phenolic hydroxy group of **59** afforded ethers that showed increased potency and selectivity. Furthermore, the introduction of an additional hydroxyl group at the C(3)-position of the octahydrophenanthrene core sufficiently increased potency to allow for modification of the allyl and propynyl substituents, ultimately leading to **60**.[113] This compound showed potent GR binding affinity ($IC_{50} = 1.5$ nM), excellent nuclear receptor selectivity, partial agonism in an IL-1-induced MMP-13 production assay in human SW1353 human chondrosarcoma cells ($IC_{50} = 38$ nM, 38% maximal efficacy) and very weak TA (MMTV-LUC reporter gene, 6% maximal efficacy).

A compound from this series, possibly Pfizer's early clinical candidate PF-251802 (**61**), has been disclosed in a patent application.[114] Compound **61** displayed very potent TR activity in an IL-1-stimulated IL-6 production assay in A549 lung epithelial cells ($IC_{50} = 0.36$ nM, 86% maximal efficacy), and was reported to be a TA antagonist.[22,115] It showed *in-vivo* potency and efficacy in a mouse therapeutic CIA model (ED_{50} 0.4 mg/kg, p.o., q.d.) while exhibiting reduced bone-related side-effects as evidenced by OC suppression ($ED_{50} = 2.9$ mg/kg). The compound was administered as the parent drug PF-251802 (**61**)

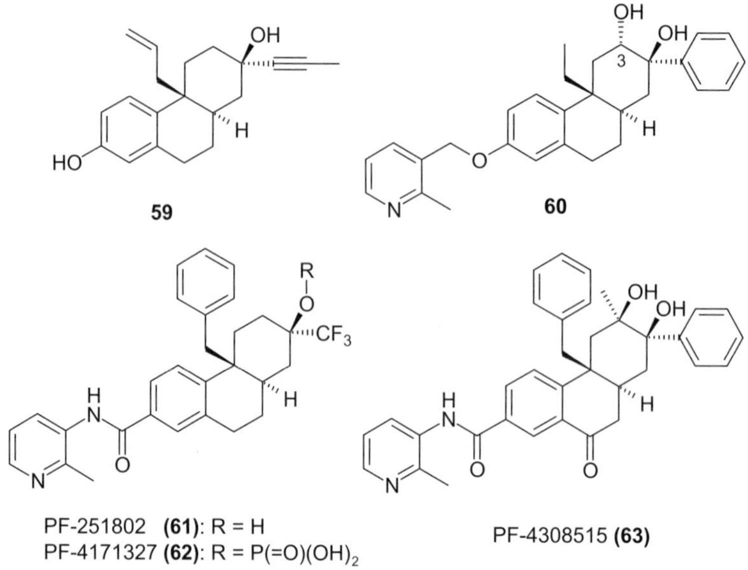

Figure 16.15   Pfizer's octahydrophenanthrene GR agonists.

or its phosphate ester prodrug PF-4171327 (**62**).[116] Compound **62** showed dose-linear pharmacokinetics with a half-life of 23–37 hours, and lack of serious or adverse events in Phase I clinical trials. Based on biomarker data, PK/PD analysis indicated that doses up to 10 mg, q.d. would provide osteocalcin suppression less than or equal to that of prednisolone at 5 mg, q.d., and cortisol suppression and neutrophilia similar to or greater than that of prednisolone at 20 mg, q.d. According to Pfizer's website, compound **62** is currently in Phase II.[117]

A single compound, **63**, from the same series was recently disclosed in a patent application suggesting that it might be a development candidate.[118] Diol **63** was also mentioned in a recent publication as PF-515,[22] possibly referring to a compound that was disclosed in a 2010 poster presentation as PF-4308515.[119] Compound **63** showed partial *in-vitro* TR efficacy in an IFNγ and TNFα production assay in LPS-stimulated human whole blood (73% and 42% at 100 nM, respectively). *In vivo*, the compound showed higher potency than prednisolone in a mouse therapeutic CIA study achieving ED_{50} of 0.3 mg/kg, p.o., q.d., and reduced side-effects in a 28-day mouse side-effect model.[22] The compound was claimed to be dissociated as compared to prednisolone based on a "dissociation index" that calculated fold differences between the equivalent side-effect doses relative to equivalent efficacious anti-inflammatory doses (2.8 for osteocalcin and 44 for insulin). According to the US government's clinical trials website, PF-4308515 (**63**) advanced into Phase I in 2009.[120]

Researchers at Eli Lilly have also discovered a GR agonist series based on a tricyclic core (Figure 16.16). Compound **64** was identified as a singleton screening hit with activity for multiple NHRs (GR IC_{50} = 231 nM, MR

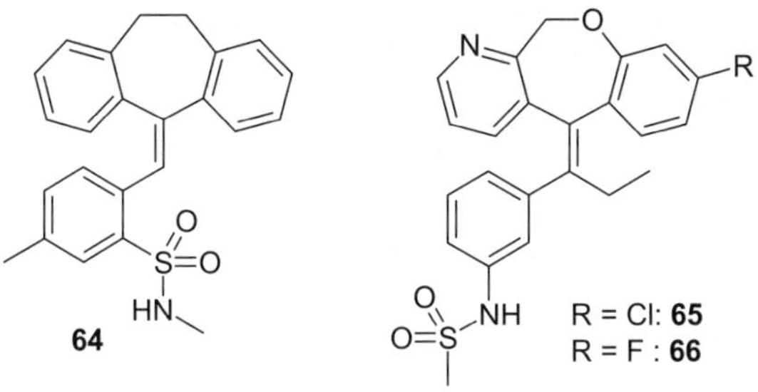

Figure 16.16 Eli Lilly's non-steroidal GR agonists.

$IC_{50} = 6$ nM, PR $IC_{50} = 30$ nM). Optimization of this compound yielded the clinical candidate **65**, which completed Phase Ib trials in 2008.[121] However, further development of this compound has not been reported. Sulfonamide **65** and its multiple salt forms were disclosed in a 2008 patent application. A potent GR binder ($K_i = 0.35$ nM) with modest NHR selectivity (AR $K_i = 16$ nM, PR $K_i = 31$ nM), **65** has shown potent TR activity in an IL-1 stimulated IL-6 inhibition assay in a human fibroblast cell line ($EC_{50} = 4.8$ nM, >90% maximal efficacy). Additionally, this compound has displayed a more potent *in-vivo* anti-inflammatory effect in a CPE model with an ED_{50} of 0.31 mg/kg as compared to ED_{50} of 6.6 mg/kg for prednisolone, and a lower effect on serum OC after acute oral dosing.[122] Compound **66**, the 8-fluoro analogue of **65**, was also extensively characterized but appeared less selective and approximately 10-fold less potent.[123]

16.8 Sulfonamide-linked Scaffolds

Publications from Boehringer Ingelheim have described a series of novel sulfonamide-based GR ligands containing an α-methyltryptamine (Figure 16.17).[124,125] The initial lead identification efforts surrounding this chemotype, which were focused on a hit from ultra-high-throughput screening, improved the inherent GR binding within this scaffold. However, obtaining acceptable cellular activity was elusive. Incorporation of a central trifluoromethyl group as a function-regulating pharmacophore (*vide supra*), and installation of an *ortho*-amino substituent on the phenyl sulfonyl moiety, which was predicted by modelling experiments to be within the hydrogen bonding distance of the N564 residue of the protein, led to an improvement in GR binding and functional activity (Figure 16.18). For example, compound **67** showed potent GR binding ($IC_{50} = 4$ nM) with good selectivity over PR ($IC_{50} > 2$ µM), and a partial agonist profile in an IL-1-induced IL-6 production assay in HFF cells ($IC_{50} = 30$ nM, 80% maximal efficacy).[126,127] Alternatively, enhanced TR activity was attained with the introduction of a cyano substituent on the pendant indole ring that was suggested by docking experiments to interact favourably with the R611 and Q570 residues of the GR-LBD (**68**, GR

$IC_{50} = 5$ nM; PR $IC_{50} = 750$ nM; IL-6 $IC_{50} = 15$ nM, 85% maximal efficacy).[50] However, this series was marred by high *in-vitro* microsomal clearance and potent CYP inhibition.

Researchers at AstraZeneca/Bayer-Schering also disclosed their drug-discovery efforts on this scaffold in a number of patent applications that encompassed various heteroaryl A-ring mimetics, as well as heterocyclic and extended linkers.[128–132] Examples such as **69**, **70** and **72** displayed good affinity for GR with IC_{50}s of 3 nM, 3 nM and 17 nM, respectively. Although functional data for these compounds were not furnished, a recent GlaxoSmithKline publication showed that the 6-methyl indazole variant, **71**, possessed potent and efficacious TR activity (NFκB $pIC_{50} = 8.9$, 101% maximal efficacy).[133] The study showed that various polar *N*-phenyl substituents such as cyano, sulfone and amides were tolerated with some *meta*-substituents offering high efficacy in TR and super-agonist activity (*i.e.* >100% maximal efficacy as compared to dexamethasone) in TA assays. As an example, **73** showed good GR potency and a non-dissociated profile in cellular functional assays (GR $pIC_{50} = 7.5$; TR NFκB $pIC_{50} = 8.6$, 103% maximal efficacy; TA MMTV agonism $pIC_{50} = 7.3$,

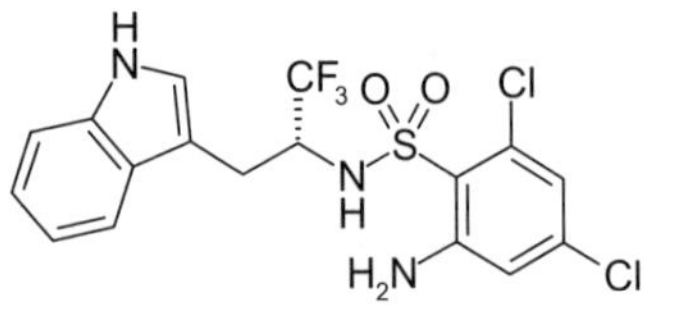

67

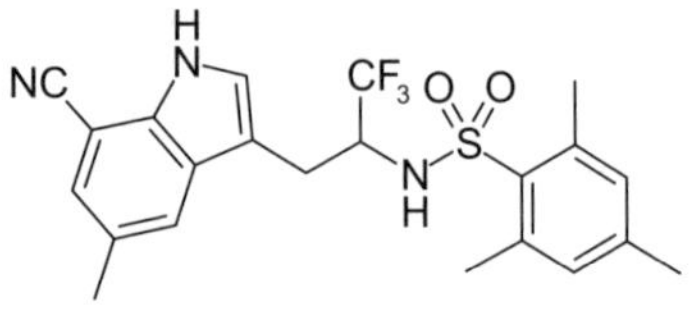

68

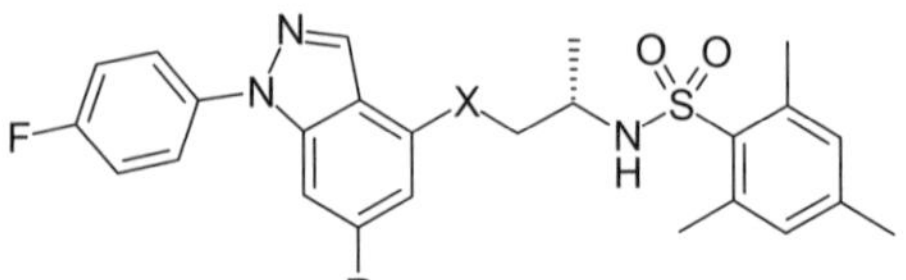

69: X = NH, R = H
70: X = O, R = H
71: X = NH, R = Me

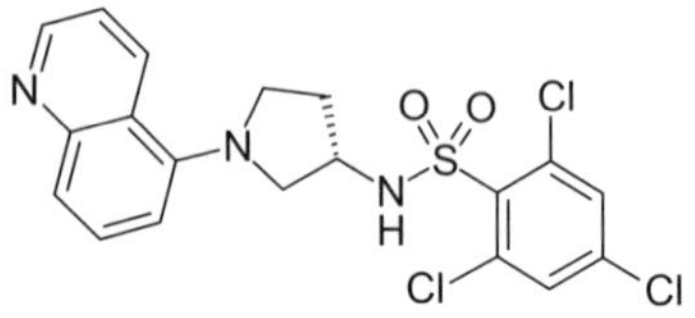

72

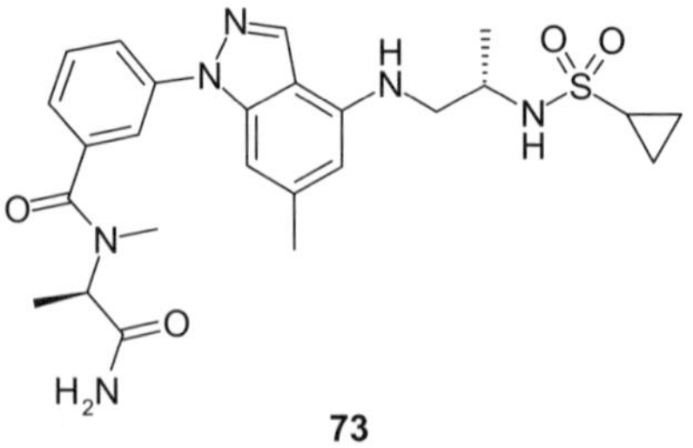

73

Figure 16.17 Sulfonamide-based GR ligands.

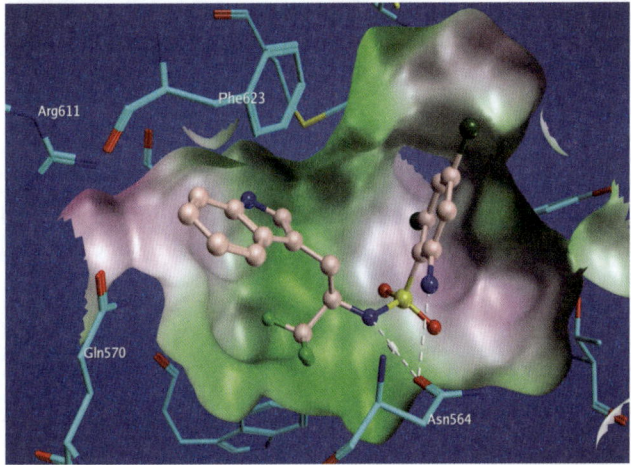

Figure 16.18   Docking pose of compound **67** in the GR-LBD shows that the sulfonamide and the aniline NH groups form hydrogen bonds with Asn564. The docking pose was generated from the fluticasone furoate X-ray structure[27] (PDB entry 3CLD) using GLIDE, and rendered with MOE. Lipophilic regions are in green, hydrophilic regions are in magenta and potential hydrogen bonds are shown by white lines.

111% maximal efficacy). Consistent with the previously observed high *in-vitro* clearance in the sulfonamide-linked scaffold (*vide supra*), compounds reported in this study suffered from poor rat PK profiles as characterized by low exposure and low bioavailability.

16.9   Conclusion

Over the last two decades, many non-steroidal glucocorticoid scaffolds have evolved and diverged from a few initial staring points and screening hits. An impressive structural diversity has led to the identification and pre-clinical evaluation of a multitude of GR agonists. Despite seemingly insurmountable obstacles in fine-tuning the pharmacological profile and achieving the desired nuclear receptor selectivity while maintaining favourable drug-like properties and pharmacokinetic parameters, the immense investment in this field of research has led to a number of discoveries. In particular, the concept of transactivation antagonism has stood out as a promising approach to achieve dissociation. In addition, the X-ray co-crystal structures of non-steroidal ligands bound to GR have led to the discovery of additional binding pockets within the ligand binding domain, and have provided novel opportunities for further optimization. The advancement of the first systemically available non-steroidal GR agonists into clinical trials over the past few years has provided encouraging early-phase biomarker data, albeit with some disappointing set-backs. Seventeen years after proposing that the desired anti-inflammatory effects of GCs could be separated from their unwanted side-effects to achieve functional selectivity, the first clinical proof of this concept may finally be within reach.

Acknowledgements

The authors gratefully acknowledge Dr John Proudfoot for his critique of this work, and Dr Jörg Bentzien for providing the modelling and X-ray illustrations in this manuscript.

References

1. H. Schäcke, W.-D. Döcke and K. Asadullah, *Pharmacol. Ther.*, 2002, **96**, 23.
2. M. Pisu, N. James, S. Sampsel and K. G. Saag, *Rheumatology*, 2005, **44**, 781.
3. A. McMaster and D. W. Ray, *Nat. Clin. Pract. Endocrinol. Metab.*, 2008, **4**, 91.
4. H. M. Reichardt, J. P. Tuckermann, M. Göttlicher, M. Vujic, F. Weih, P. Angel, P. Herrlich and G. Schütz, *EMBO J.*, 2002, **20**, 7168.
5. J. Zhou and J. A. Cidlowski, *Steroids*, 2005, **70**, 407.
6. I. M. Adcock, *Pulm. Pharmacol. Ther.*, 2000, **13**, 115.
7. S. Heck, M. Kullmann, A. Gast, H. Ponta, H. J. Rahmsdorf, P. Herrlich and A. C. B. Cato, *EMBO J.*, 1994, **13**, 4087.
8. J. A. Katzenellenbogen and B. S. Katzenellenbogen, *Chem. Biol.*, 1996, **3**, 529.
9. M. Resche-Rigon and H. Gronemeyer, *Curr. Opin. Chem. Biol.*, 1998, **2**, 501.
10. P. J. Barnes, *Clin. Sci.*, 1998, **94**, 557.
11. L. Buckbinder and R. P. Robinson, *Curr. Drug Targets: Inflammation Allergy*, 2002, **1**, 127.
12. K. L. Burnstein and J. A. Cidlowski, *Mol. Cell. Endocrinol.*, 1992, **83**, C1.
13. M. Perretti, *Blood*, 2007, **109**, 852.
14. S. van der Laan and O. C. Meijer, *Eur. J. Pharmacol.*, 2008, **585**, 483.
15. K. Ronacher, K. Hadley, C. Avenant, E. Stubsrud, S. S. Simons, Jr., A. Louw and J. P. Hapgood, *Mol. Cell. Endocrin.*, 2009, **299**, 219.
16. R. Newton and N. S. Holden, *Mol. Pharmacol.*, 2007, **72**, 799.
17. C. Maier, D. Rünzler, J. Schindelar, G. Grabner, W. Waldhäusl, G. Köhler and A. Luger, *J. Cell Science*, 2005, **118**, 3353.
18. H. Schäcke, T. M. Zollner, W.-D. Döcke, H. Rehwinkel, S. Jaroch, W. Skuballa, R. Neuhaus, E. May, U. Zügel and K. Asadullah, *British J. Pharmacol.*, 2009, **158**, 1088.
19. D. Riether, C. Harcken, H. Razavi, D. Kuzmich, T. Gilmore, J. Bentzien, E. J. Pack, Jr., D. Souza, R. M. Nelson, A. Kukulka, T. N. Fadra, L.

Zuvela-Jelaska, J. Pelletier, R. Dinallo, M. Panzenbeck, C. Torcellini, G. H. Nabozny and D. S. Thomson, *J. Med. Chem.*, 2010, **53**, 6681.

20. K. Biggadike, R. K. Bledsoe, D. M. Coe, T. W. J. Cooper, D. House, M. A. Iannone, S. J. F. Macdonald, K. P. Madauss, I. M. McLay, T. J. Shipley, S. J. Taylor, T. B. Tran, I. J. Uings, V. Weller and S. P. Williams, *Proc. Natl Acad. Sci. USA*, 2009, **106**, 18114.
21. R. A. M. Quax, R. P. Peeters and R. A. Feelders, *Endocrinology*, 2011, **152**, 2927.
22. X. Hu, S. Du, C. Tunca, T. Braden, K. R. Long, J. Lee, E. G. Webb, J. D. Dietz, S. Hummert, S. Rouw, S. G. Hedge, R. K. Webber and M. G. Obukowicz, *Endocrinology*, 2011, **152**, 3123.
23. J. N. Miner, B. Ardecky, K. Benbatoul, K. Griffiths, C. J. Larson, D. E. Mais, K. Marschke, J. Rosen, E. Vajda, L. Zhi and A. Negro-Vilar, *Proc. Natl Acad. Sci. USA*, 2007, **104**, 19244.
24. R. K. Bledsoe, V. G. Montana, T. B. Stanley, C. J. Delves, C. J. Apolito, D. D. McKee, T. G. Consler, D. J. Parks, E. L. Stewart, T. M. Willson, M. H. Lambert, J. T. Moore, K. H. Pearce and H. E. Xu, *Cell*, 2002, **110**, 93.
25. B. Kauppi, C. Jakob, M. Farnegardh, J. Yang, H. Ahola, M. Alarcon, K. Calles, O. Engstrom, J. Harlan and S. Muchmore, *J. Biol. Chem.*, 2003, **278**, 22748.
26. G. A. Schoch, B. D'Arcy, M. Stihle, D. Burger, D. Bär, J. Benz, R. Thoma and A. Ruf, *J. Mol. Biol.*, 2010, **395**, 568.
27. K. Biggadike, R. K. Bledsoe, A. M. Hassell, B. E. Kirk, I. M. McLay, L. M. Shewchuk and E. L. Stewart, *J. Med. Chem.*, 2008, **51**, 3349.
28. K. P. Madauss, R. K. Bledsoe, I. McLay, E. L. Stewart, I. J. Uings, G. Weingarten and S. P. Williams, *Biorg. Med. Chem. Lett.*, 2008, **18**, 6097.
29. L. Frego and W. Davidson, *Protein Sci.*, 2006, **15**, 722.
30. H. Schäcke, M. Berger, T. G. Hansson, D. McKerrecher and H. Rehwinkel, *Expert Opin. Ther. Pat.*, 2008, **18**, 339.
31. M. Berlin, *Expert Opin. Ther. Pat.*, 2010, **20**, 855.
32. J. Regan, H. Razavi and D. Thomson, *Annu. Rep. Med. Chem.*, 2008, **43**, 141.
33. K. De Bosscher, *J. Steroid Biochem. Mol. Biol.*, 2010, **120**, 96.
34. K. De Bosscher, W. Vanden Berghe, I. M. E. Beck, W. Van Molle, N. Hennuyer, J. Hapgood, C. Libert, B. Staels, A. Louw and G. Haegeman, *Proc. Natl Acad. Sci. USA*, 2005, **102**, 15827.
35. J. Du, B. Cheng, X. Zhu and C. Ling, *J. Immunol.*, 2011, **187**, 942.
36. Y. He, D. Yin, M. Perera, L. Kirkovsky, N. Stourman, W. Li, J. T. Dalton and D. D. Miller, *Eur. J. Med. Chem.*, 2002, **37**, 619.
37. H. Schäcke, H. Hennekes, A. Schottelius, S. Jaroch, M. Lehmann, N. Schmees, H. Rehwinkel and K. Asadullah, *Ernst Schering Res. Found. Workshop*, 2002, **40**, 357.
38. M. Lehmann, K. Krolikiewicz, W. Skuballa, P. Strehlke, F. Kalkbrenner, R. Ekerdt and C. Giesen, Schering AG, *PCT Int. Appl.*, WO 2000/032584, 2000.

39. H. Schäcke, A. Schottelius, W.-D. Döcke, P. Strehlke, S. Jaroch, N. Schmees, H. Rehwinkel, H. Hennekes and K. Asadullah, *Proc. Natl Acad. Sci. USA*, 2004, **101**, 227.

40. S. Jaroch, M. Berger, C. Huwe, K. Krolikiewicz, H. Rehwinkel, H. Schäcke, N. Schmees and W. Skuballa, *Bioorg. Med. Chem. Lett.*, 2010, **20**, 5835.

41. M. E. Cavet, K. L. Harrington, K. W. Ward and J.-Z. Zhang, *Mol. Vision*, 2010, **16**, 1791.

42. D. Kuzmich, T. Kirrane, J. Proudfoot, Y. Bekkali, R. Zindell, L. Beck, R. Nelson, C.-K. Shih, A. J. Kukulka, Z. Paw, P. Reilly, R. Deleon, M. Cardozo, G. Nabozny and D. Thomson, *Bioorg. Med. Chem. Lett.*, 2007, **17**, 5025.

43. Manuscript in preparation.

44. Y. Bekkali, R. Betageri, M. J. Emmanuel, A. Hammach, C. Harcken, T. M. Kirrane, D. Kuzmich, T. W. Lee, P. Liu, U. R. Patel, H. Razavi, D. Riether, H. Takahashi, D. S. Thomson, J. Wang, R. Zindell and J. R. Proudfoot, Boehringer Ingelheim, Inc., USA, *PCT Int. Appl.*, WO 2005/030213, 2005.

45. R. Betageri, T. Gilmore, D. Kuzmich, T. M. Kirrane, D. Wiedenmayer, J. Regan, A. J. Kukulka, R. M. Nelson, L. Zuvela-Jelaska, D. Souza, J. Proudfooot, R. Dinallo, M. Panzenbeck, C. Torcellini, H. Lee, E. Pack, C. Harcken, G. Nabozny and D. S. Thomson, *Bioorg. Med. Chem. Lett.*, 2011, **21**, 6842.

46. Manuscript in preparation.

47. R. Betageri, Y. Zhang, R. M. Zindell, D. Kuzmich, T. M. Kirrane, J. Bentzien, M. Cardozo, A. J. Capolino, T. N. Fadra, R. M. Nelson, Z. Paw, D.-T. Shih, C.-K. Shih, L. Zuvela-Jelaska, G. Nabozny and D. S. Thomson, *Biorg. Med. Chem. Lett.*, 2005, **15**, 4761.

48. Manuscript in preparation.

49. J. Regan, T. W. Lee, R. M. Zindell, Y. Bekkali, J. Bentzien, T. Gilmore, A. Hammach, T. M. Kirrane, A. J. Kukulka, D. Kuzmich, R. M. Nelson, J. R. Proudfoot, M. Ralph, J. Pelletier, D. Souza, L. Zuvela-Jelaska, G. Nabozny and D. S. Thomson, *J. Med. Chem.*, 2006, **49**, 7887–7896.

50. D. S. Thomson, Discovery of a Novel Class of Glucocorticoid Receptor Modulators, 239th American Chemical Society National Meeting, San Francisco, CA, March, 2010.

51. P. Norman, *Expert Opin. Ther. Pat.*, 2011, **21**, 1137.

52. R. Betageri, T. Bosanac, M. J. Burke, C. Harcken, S. Kim, D. Kuzmich, T. W. Lee, Z. Li, P. Liu, J. Lord, H. Razavi, J. T. Reeves and D. Thomson, Boehringer Ingelheim, Inc., USA, *PCT Int. Appl.* WO 2009/149139, 2009.

53. M. Barker, M. Clackers, D. A. Demaine, D. Humphreys, M. J. Johnston, H. T. Jones, F. Pacquet, J. M. Pritchard, M. Salter, S. E. Shanahan, P. A. Skone, V. M. Vinader, I. Uings, I. M. McLay and S. J. F. Macdonald, *J. Med. Chem.*, 2005, **48**, 4507.

54. M. Barker, M. Clackers, R. Copley, D. A. Demaine, D. Humphreys, G. G. A. Inglis, M. J. Johnston, H. T. Jones, M. V. Haase, D. House, R. Loiseau, L. Nisbet, F. Pacquet, P. A. Skone, S. E. Shanahan, D. Tape, V. M. Vinader, M. Washington, I. Uings, R. Upton, I. M. McLay and S. J. F. Macdonald, *J. Med. Chem.*, 2006, **49**, 4216.

55. M. Clackers, D. M. Coe, D. A. Demaine, G. W. Hardy, D. Humphreys, G. G. A. Inglis, M. J. Johnston, H. T. Jones, D. House, R. Loiseau, D. J. Minick, P. A. Skone, I. Uings, I. M. McLay and S. J. F. Macdonald, *Bioorg. Med. Chem. Lett.*, 2007, **17**, 4737.

56. H. A. Barnett, D. M. Coe, T. W. J. Cooper, T. I. Jack, H. T. Jones, S. J. F. Macdonald, I. M. McLay, N. Rayner, R. Z. Sasse, T. J. Shipley, P. A. Skone, G. I. Somers, S. Taylor, I. J. Uings, J. M. Woolven and G. C. Weingarten, *Bioorg. Med. Chem. Lett.*, 2009, **19**, 158.

57. K. Biggadike, M. Boudjelal, M. Clackers, D. M. Coe, D. A. Demaine, G. W. Hardy, D. Humphreys, G. G. A. Inglis, M. J. Johnston, H. T. Jones, D. House, R. Loiseau, D. Needham, P. A. Skone, I. Uings, G. Veitch, G. G. Weingarten, I. M. McLay and S. J. F. Macdonald, *J. Med. Chem.*, 2007, **50**, 6519.

58. K. Biggadike, M. Caivano, M. Clackers, D. M. Coe, G. W. Hardy, D. Humphreys, H. T. Jones, D. House, A. Miles-Williams, P. A. Skone, I. Uings, V. Weller, I. M. McLay and S. J. F. Macdonald, *Bioorg. Med. Chem. Lett.*, 2009, **19**, 4846.

59. J. P. Edwards, S. J. West, K. B. Marschke, D. E. Mais, M. Gottardis and T. K. Jones, *J. Med. Chem.*, 1998, **41**, 303.

60. J. P. Edwards, L. Zhi, C. L. F. Pooley, C. M. Tegley, S. J. West, M. W. Wang, M. M. Gottardis, C. Pathirana, W. T. Schader and T. K. Jones, *J. Med. Chem.*, 1998, **41**, 2779.

61. L. Zhi, C. M. Tegley, J. P. Edwards, S. J. West, K. B. Marschke, M. M. Gottardis, D. E. Mais and T. K. Jones, *Bioorg. Med. Chem. Lett.*, 1998, **8**, 3365.

62. L. Zhi, C. M. Tegley, B. Pio, J. P. Edwards, T. K. Jones, K. B. Marschke, D. E. Mais, B. Risek and W. T. Schrader, *Bioorg. Med. Chem. Lett.*, 2003, **13**, 2071.

63. L. Zhi, J. D. Ringgeberg, J. P. Edwards, C. M. Tegley, S. J. West, B. Pio, M. Motamedi, T. K. Jones, K. B. Marschke, D. E. Mais and W. T. Schrader, *Bioorg. Med. Chem. Lett.*, 2003, **13**, 2075.

64. L. Zhi, C. M. Tegley, B. Pio, J. P. Edwards, M. Motamedi, T. K. Jones, K. B. Marschke, D. E. Mais, B. Risek and W. T. Schrader, *J. Med. Chem.*, 2003, **46**, 4104.

65. M. J. Coghlan, P. B. Jacobson, B. Lane, M. Nakane, C. W. Lin, S. W. Elmore, P. R. Kym, J. R. Luly, G. W. Carter, R. Turner, C. M. Tyree, J. Hu, M. Elgort, J. Rosen and J. N. Miner, *Mol. Endocrinol.*, 2003, **17**, 860.

66. P. R. Kym, M. E. Kort, M. J. Coghlan, J. L. Moore, R. Tang, J. D. Ratajczyk, D. P. Larson, S. W. Elmore, J. K. Pratt, M. A. Stashko, H. D. Falls, C. W. Lin, M. Nakane, L. Miller, C. M. Tyree, J. N. Miner, P. B. Jacobson, D. M. Wilcox, P. Nguyen and B. C. Lane, *J. Med. Chem.*, 2003, **46**, 1016.

67. S. W. Elmore, J. K. Pratt, M. J. Coghlan, Y. Mao, B. E. Green, D. D. Anderson, M. A. Stashko, C. W. Lin, D. Falls, M. Nakane, L. Miller, C. M. Tyree, J. N. Miner and B. Lane, *Bioorg. Med. Chem. Lett.*, 2004, **14**, 1721.
68. S. W. Elmore, M. J. Coghlan, D. D. Anderson, J. K. Pratt, B. E. Green, A. W. Wang, M. A. Stashko, C. W. Lin, C. M. Tyree, J. N. Miner, P. B. Jacobson, D. M. Wilcox and B. C. Lane, *J. Med. Chem.*, 2001, **44**, 4481.
69. C. W. Lin, M. Makane, M. Stashko, D. Falls, J. Kuk, L. Miller, R. Huang, C. Tyree, J. N. Miner, J. Rosen, P. R. Kym, M. J. Coghlan, G. Carter and B. C. Lane, *Mol. Pharmacol.*, 2002, **62**, 297.
70. R. J. Ardecky, A. R. Hudson, D. P. Phillips, J. S. Tyhonas, C. Deckhut, T. L. Lau, Y. Li, E. A. Martinborough, S. L. Roach, R. I. Higuchi, F. J. Lopez, K. B. Marschke, J. N. Miner, D. S. Karanewsky, A. Negro-Vilar and L. Zhi, *Bioorg. Med. Chem. Lett.*, 2007, **17**, 4158.
71. A. R. Hudson, S. L. Roach, R. I. Higuchi, D. P. Phillips, R. P. Bissonnette, W. W. Lamph, J. Yen, Y. Li, M. E. Adams, L. J. Valdez, A. Vassar, C. Cuervo, E. A. Kallel, C. J. Gharbaoui, D. E. Mais, J. N. Miner, K. B. Marschke, D. Rungta, A. Negro-Vilar and L. Zhi, *J. Med. Chem.*, 2007, **50**, 4699.
72. M. Kawano, T. Hirano, T. Matsuda, T. Taga, Y. Horii, K. Iwato, H. Asaoku, B. Tang, O. Tanabe, H. Tanaka, A. Kuramoto and T. Kishimoto, *Nature*, 1988, **332**, 83.
73. H. Ishikawa, H. Tanaka, K. Iwato, O. Tanabe, H. Asaoku, M. Nobuyoshi, L. Yamamoto, M. Kawano and A. Kuramoto, *Blood*, 1990, **75**(7), 15.
74. Ligand Pharmaceuticals, Inc., Press Release, March 15, 2007. Available from: http://thomsonpharma.com (accessed October 2011).
75. H. Takahashi, Y. Bekkali, A. J. Capolino, T. Gilmore, S. E. Goldrick, R. M. Nelson, D. Terenzio, J. Wang, L. Zuvela-Jelaska and J. Proudfoot, *Bioorg. Med. Chem. Lett.*, 2006, **16**, 1549.
76. H. Takahashi, Y. Bekkali, A. J. Capolino, T. Gilmore, S. E. Goldrick, P. V. Kaplita, L. Liu, R. M. Nelson, D. Terenzio, J. Wang, L. Zuvela-Jelaska, J. Proudfoot, G. Nabozny and D. Thomson, *Bio. Med. Chem. Lett.*, 2007, **17**, 5091.
77. M. Kato, M. Takai, T. Matsuyama, T. Kurose, Y. Hagiwara, M. Matsuda, T. Mori, K. Imoto and A. Dota, Santen Pharmaceutical Co., Ltd., Japan, *US Pat. Appl.*, US 2011/0118260, 2011.
78. M. Kato, M. Takai, T. Matsuyama, T. Kurose, Y. Hagiwara, K. Oki, M. Matsuda and T. Mori, Santen Pharmaceutical Co., Ltd. Japan, *US Pat. Appl.*, US 2011/0166151, 2011.
79. C. L. F. Pooley, J. P. Edwards, M. E. Goldman, M. W. Wang, K. B. Marschke, D. L. Crombie and T. K. Jones, *J. Med. Chem.*, 1998, **41**, 3461.
80. L. Zhi, C. M. Tegley, B. Pio, S. J. West, K. B. Marschke, D. E. Mais and T. K. Jones, *Bioorg. Med. Chem. Lett.*, 2000, **10**, 415.
81. S. L. Roach, R. I. Higuchi, M. E. Adams, Y. Liu, D. S. Karanewsky, K. B. Marschke, D. E. Mais, J. N. Miner and L. Zhi, *Bioorg. Med. Chem. Lett.*, 2008, **18**, 3504.

82. S. L. Roach, R. I. Higuchi, A. R. Hudson, M. E. Adams, P. M. Syka, D. E. Mais, J. N. Miner, K. B. Marschke and L. Zhi, *Bioorg. Med. Chem. Lett.*, 2011, **21**, 168.

83. A. R. Hudson, R. I. Higuchi, S. L. Roach, M. E. Adams, A. Vassar, P. M. Syka, D. E. Mais, J. N. Miner, K. B. Marschke and L. Zhi, *Bioorg. Med. Chem. Lett.*, 2011, **21**, 1697.

84. S. L. Roach, R. I. Higuchi, A. R. Hudson, A. Vassar, V. H. S. Grant, R. Lamer, C. Hooper, D. Rungta, P. M. Syka, D. E. Mais, K. B. Marschke and L. Zhi, *Bioorg. Med. Chem. Lett.*, 2011, **21**, 1658.

85. A. R. Hudson, R. I. Higuchi, S. L. Roach, L. J. Valdez, M. E. Adams, A. Vassar, D. Rungta, P. M. Syka, D. E. Mais, K. B. Marschke and L. Zhi, *Bioorg. Med. Chem. Lett.*, 2011, **21**, 1654.

86. N. Shah and T. S. Scanlan, *Bioorg. Med. Chem. Lett.*, 2004, **14**, 5199.

87. J.-C. Wang, N. Shah, C. Pantoja, S. H. Meijsing, J. D. Ho, T. S. Scanlan and K. R. Yamamoto, *Genes & Dev.*, 2006, **20**, 689.

88. C. F. Thompson, N. Quraishi, A. Ali, J. R. Tata, M. L. Hammond, J. M. Balkovec, M. Einstein, L. Ge, G. Harris, T. M. Kelly, P. Mazur, S. Pandit, J. Santoro, A. Sitlani, C. Wang, J. Williamson, D. K. Miller, T.-T. D. Yamin, C. M. Thompson, E. A. O'Neill, D. Zaller, M. J. Forrest, E. Carballo-Jane and S. Luell, *Bioorg. Med. Chem. Lett.*, 2005, **15**, 2163.

89. A. Ali, C. F. Thompson, J. M. Balkovec, D. W. Graham, M. L. Hammond, N. Quraishi, J. R. Tata, M. Einstein, L. Ge, G. Harris, T. M. Kelly, P. Mazur, S. Pandit, J. Santoro, A. Sitlani, C. Wang, J. Williamson, D. K. Miller, C. M. Thompson, D. M. Zaller, M. J. Forrest, E. Carballo-Jane and S. Luell, *J. Med. Chem.*, 2004, **47**, 2441.

90. C. F. Thompson, N. Quraishi, A. Ali, R. T. Mosley, J. R. Tata, M. L. Hammond, J. M. Balkovec, M. Einstein, L. Ge, G. Harris, T. M. Kelly, P. Mazur, S. Pandit, J. Santoro, A. Sitlani, C. Wang, J. Williamson, D. K. Miller, T.-T. D. Yamin, C. M. Thompson, E. A. O'Neill, D. Zaller, M. J. Forrest, E. Carballo-Jane and S. Luell, *Bioorg. Med. Chem. Lett.*, 2007, **17**, 3354.

91. C. J. Smith, A. Ali, J. M. Balkovec, D. W. Graham, M. L. Hammond, G. F. Patel, G. P. Rouen, S. K. Smith, J. R. Tata, M. Einstein, L. Ge, G. S. Harris, T. M. Kelly, P. Mazur, C. M. Thompson, C. F. Wang, J. M. Williamson, D. K. Miller, S. Pandit, J. C. Santoro, A. Sitlani, T. D. Yamin, E. A. O'Neill, D. M. Zaller, E. Carballo-Jane, M. J. Forrest and S. Luell, *Bioorg. Med. Chem. Lett.*, 2005, **15**, 2926.

92. W. P. Dankulich, M. L. Kaufman, D. M. Mcmaster, R. S. Meissner and H. J. Mitchell, Merck & Co., Inc., USA, *PCT Int. Appl.*, WO 2009/111214, 2009.

93. H. J. Mitchell, D. M. Hurzy and R. S. Meissner, Merck Sharp & Dohme Corp., USA, *PCT Int. Appl.*, WO 2010/138421, 2010.

94. C. J. Bungard, J. J. Manikowski, J. J. Perkins and R. Meissner, Merck & Co., Inc., USA, *PCT Int. Appl.*, WO 2008/051532, 2008.

95. C. J. Bungard, J. J. Manikowski, J. J. Perkins and R. Meissner, Merck & Co., Inc., USA, *PCT Int. Appl.*, WO 2008/060391, 2008.

96. Q. Chen, S. Fujimori, J. M. Janey, J. Limanto, R. Naccache, A. F. Nolting, Z. J. Song, N. Strotman, L. Tan and M. Weisel, Merck & Co., Inc., USA, *PCT Int. Appl.*, WO 2009/054925, 2009.

97. H. J. Mitchell, W. P. Dankulich, M. L. Kaufman and R. S. Meissner, Merck Sharp & Dohme Corp., USA, *PCT Int. Appl.*, WO 2011/053567, 2011.

98. H.-Y. Xiao, D.-R. Wu, M. F. Malley, J. Z. Gougoutas, S. F. Habte, M. D. Cunningham, J. E. Somerville, J. H. Dodd, J. C. Barrish, S. G. Nadler and T. G. Murali Dhar, *J. Med. Chem.*, 2010, **53**, 1270.

99. I. A. Mugge, M. J. Burke, M. S. Ralph, D. S. Thomson, A. Hammach, J. A. Kowalski and J. M. Bentzien, Boehringer Ingelheim International GmbH, Germany, Boehringer Ingelheim Pharma GmbH & Co. KG, Germany, *PCT Int. Appl.*, WO 2006/135826, 2006.

100. D. Kuzmich, D. Disalvo and H. Razavi, Boehringer Ingelheim International GmbH, Germany, Boehringer Ingelheim Pharma GmbH & Co. KG, Germany, *PCT Int. Appl.*, WO 2008/070507, 2008.

101. M. Bai, G. Carr, R. J. DeOrazio, T. D. Friedrich, S. Dobritsa, K. Fitzpatrick, P. R. Guzzo, D. B. Kitchen, M. A. Lynch, D. Peace, M. Sajad, A. Usyatinsky and M. A. Wolf, *Bioorg. Med. Chem. Lett.*, 2010, **20**, 3017.

102. J. E. Sheppeck, J. L. Gilmore, T. G. M. Dhar and H.-Y. Xiao, Bristol-Myers Squibb Company, USA, *PCT Int. Appl.*, WO 2008/057856, 2008.

103. J. E. Sheppeck, J. L. Gilmore, T. G. M. Dhar, H.-Y. Xiao, J. Wang, B. V. Yang and L. M. Doweyko, Bristol-Myers Squibb Company, USA, *PCT Int. Appl.*, WO 2008/057857, 2008.

104. M. Berger, J. Dahmen, A. Eriksson, B. Gabos, T. Hansson, M. Hemmerling, K. Henriksson, S. Ivanova, M. Lepistö, D. McKerrecher, M. Munck af Rosenschöld, S. Nilsson, H. Rehwinkel and C. Taflin, AstraZeneca AB, Sweden, Bayer Schering Pharma Aktiengesellschaft, Germany, *PCT Int. Appl.*, WO 2008/076048, 2008.

105. M. Berger, J. Dahmen, K. Edman, T. Hansson, M. Hemmerling, N. Hossain, H. Johansson, M. Lepistö, S. Nilsson and H. Rehwinkel, AstraZeneca AB, Sweden, Bayer Schering Pharma AG, Germany, *PCT Int. Appl.*, WO 2009/142569, 2009.

106. M. Berger, J. Dahmen, K. Edman, A. Eriksson, T. Hansson, M. Hemmerling, N. Hossain, T. Klingstedt, M. Lepistö, S. Nilsson and H. Rehwinkel, AstraZeneca AB, Sweden, Bayer Schering Pharma AG, Germany, *PCT Int. Appl.*, WO 2009/142571, 2009.

107. B. V. Yang, W. Vaccaro, A. M. Doweyko, L. M. Doweyko, T. Huynh, D. Tortolani, S. G. Nadler, L. McKay, J. Somerville, D. A. Holloway, S. Habte, D. S. Weinstein and J. C. Barrish, *Bioorg. Med. Chem. Lett.*, 2009, **19**, 2139.

108. B. V. Yang, D. S. Weinstein, L. M. Doweyko, H. Gong, W. Vaccaro, T. Huynh, H.-Y. Xiao, A. M. Doweyko, L. Mckay, D. A. Holloway, J. E. Somerville, S. Habte, M. Cunningham, M. McMahon, R. Townsend, D. Shuster, J. H. Dodd, S. G. Nadler and J. C. Barrish, *J. Med. Chem.*, 2010, **53**, 8241.

109. D. S. Weinstein, P. Chen, T. G. M. Dhar, J. Duan, H. Gong, B. Jiang, B. V. Yang and A. M. Doweyko, Bristol-Myers Squibb, USA, *US Pat. Appl.*, 2009/0075995, 2005.

110. M. L. Adams, J. Huang, A. Patel, L. Gao, V. Guarino, S. Agrawal, R. Latek, S. Yin, D. A. Conlon, U. Thienel and M. Hussain, *The AAPS Journal*, 2010, **12**(S2). Available from http://www.aapsj.org/

111. D. S. Weinstein, H. Gong, A. M. Doweyko, M. Cunningham, S. Habte, J. H. Wang, D. A. Holloway, C. Burke, L. Gao, V. Guarino, J. Carman, J. E. Somerville, D. Shuster, L. Salter-Cid, J. H. Dodd, S. G. Nadler and J. C. Barrish, *J. Med. Chem.*, 2011, **54**, 7318.

112. B. P. Morgan, A. G. Swick, D. M. Hargrove, J. A. LaFlamme, M. S. Moynihan, R. S. Carroll, K. A. Martin, E. Lee, D. Decosta and J. Bordner, *J. Med. Chem.*, 2002, **45**, 2417.

113. R. P. Robinson, L. Buckbinder, A. I. Haugeto, P. A. McNiff, M. L. Millham, M. R. Reese, J. F. Schaefer, Y. A. Abramov, J. Bordner, Y. A. Chantigny, E. F. Kleinman, E. R. Laird, B. P. Morgan, J. C. Murray, E. D. Salter, M. D. Wessel and S. A. Yocum, *J. Med. Chem.*, 2009, **52**, 1731.

114. R. V. Devraj, G. A. De Crescenzo, X. Hu, K. D. Jerome, G. M. Obukowicz, L. Olson, R. V. Rucker and R. K. Webber, Pfizer, Inc., USA, *PCT Int. Appl.*, WO 2008/093227, 2008.

115. X. Hu, A Novel Glucocorticoid Receptor Ligand: Dissociating the Good from the Bad, Keystone Symposia: Nuclear Receptors: Sinaling, Gene Regulation and Cancer (X7) 2010, Keystone, CO.

116. T. Stock, D. Fleishaker, A. Mukherjee, V. Le, J. Xu and B. Zeiher, Evaluation of safety, PK and PD of a SGRM in healthy volunteers, ACR Meeting 2009, Philadelphia.

117. Pfizer Pipeline, November 10, 2011. Available from: http://www.pfizer.com/files/research/pipeline/2011_1110/pipeline_2011_1110.pdf (accessed January 2012).

118. P. V. Rucker, Pfizer, Inc., USA, *PCT Int. Appl.*, WO 2010/013158, 2010.

119. K. R. Long, J. S. Daniels, X. Hu, M. A. Melton, E. G. Webb, R. K. Webber and M. G. Obukowicz, PF-04308515, a Dissociated Agonist of the Glucocorticoid Receptor is Efficacious in Mouse Models of Inflammation with Reduced Side Effects for Biomarkers of Osteoporosis and Insulin Resistance, Keystone Symposia: Nuclear Receptors: Sinaling, Gene Regulation and Cancer (X7) 2010, Keystone, CO.

120. A Phase 1 Single-Dose Escalation Study of PF-04308515 in Healthy Volunteers, October 2, 2009. Available from http://www.clinicaltrials.gov/ (accessed November 2011).

121. M. Coghlan, Preclinical Chracterization of a SGRM, National Medicinal Chemistry Symposium 2008, Pittsburgh.

122. M. W. Carson and M. J. Coghlan, Eli Lilly and Co., USA, *PCT Int. Appl.*, WO 2008/008882, 2008.

123. M. W. Carson and M. J. Coghlan, Eli Lilly and Co., USA, *PCT Int. Appl.*, WO 2009/089312, 2009.

124. D. R. Marshall, Boehringer Ingelheim Pharmaceuticals, Inc., USA, *PCT Int. Appl.*, WO 2004/019935, 2004.
125. D. R. Marshall, G. Rodriguez, D. S. Thomson, R. Nelson and A. Capolina, *Bioorg. Med. Chem. Lett.*, 2007, **17**, 315.
126. R. Betageri, D. Disalvo, D. S. Thomson, D. Kuzmich, J. Regan and J. Kowalski, Boehringer Ingelheim Pharmaceuticals, Inc., USA, *PCT Int. Appl.*, WO 2006/071609, 2006.
127. D. S. Thomson, 233rd American Chemical Society National Meeting, Chicago, IL, March, 2007.
128. H. Bladh, K. Henriksson, V. Hulikal and M. Lepistö, AstraZeneca AB, Sweden, *PCT Int. Appl.*, WO 2006/046914, 2006.
129. H. Bladh, J. Dahmen, T. Hansson, K. Henriksson, M. Lepistö and S. Nilsson, AstraZeneca AB, Sweden, Schering AG, Germany, *PCT Int. Appl.*, WO 2007/046747, 2007.
130. H. Bladh, K. Henriksson, V. Hulikal and M. Lepistö, AstraZeneca AB, Sweden, *PCT Int. Appl.*, WO 2006/046916, 2006.
131. M. Bengtsson, H. Bladh, T. Hansson and E. Kinchin, AstraZeneca AB, Sweden, Schering AG, Germany, *PCT Int. Appl.*, WO 2007/114763, 2007.
132. H. Bladh, M. Lepistö, S. Nilsson and C. Taflin, AstraZeneca AB, Sweden, Bayer Schering Pharma Aktiengesellschaft, Germany, *PCT Int. Appl.*, WO 2008/079073, 2008.
133. H. Diallo, D. C. Angell, H. A. Barnett, K. Biggadike, D. M. Coe, T. W. J. Cooper, A. Craven, J. R. Gray, D. House, T. I. Jack, S. P. Keeling, S. J. F. Macdonald, I. M. McLay, S. Oliver, S. J. Taylor, I. J. Uings and N. Wellaway, *Bioorg. Med. Chem. Lett.*, 2011, **21**, 1126.

Subject Index